Studies in Computational Intelligence

Volume 1014

The series "Studies in Computational Intelligence" (SCI) publishes new developments and advances in the various areas of computational intelligence —quickly and with a high quality. The intent is to cover the theory, applications, and design methods of computational intelligence, as embedded in the fields of engineering, computer science, physics and life sciences, as well as the methodologies behind them. The series contains monographs, lecture notes and edited volumes in computational intelligence spanning the areas of neural networks, connectionist systems, genetic algorithms, evolutionary computation, artificial intelligence, cellular automata, self-organizing systems, soft computing, fuzzy systems, and hybrid intelligent systems. Of particular value to both the contributors and the readership are the short publication timeframe and the world-wide distribution, which enable both wide and rapid dissemination of research output.

Indexed by SCOPUS, DBLP, WTI Frankfurt eG, zbMATH, SCImago.

All books published in the series are submitted for consideration in Web of Science.

More information about this series at https://link.springer.com/bookseries/7092

Boris Kovalerchuk · Kawa Nazemi ·
Răzvan Andonie · Nuno Datia · Ebad Banissi
Editors

Integrating Artificial Intelligence and Visualization for Visual Knowledge Discovery

Springer

Editors
Boris Kovalerchuk
Department of Computer Science
Central Washington University
Ellensburg, WA, USA

Răzvan Andonie
Department of Computer Science
Central Washington University
Ellensburg, WA, USA

Ebad Banissi
Department of Informatics
London South Bank University
London, UK

Kawa Nazemi
Department of Media
Darmstadt University of Applied Sciences
Darmstadt, Hessen, Germany

Nuno Datia
Department of Electronics,
Telecommunications and Computers
Engineering
Lisbon School of Engineering
Lisbon, Portugal

ISSN 1860-949X ISSN 1860-9503 (electronic)
Studies in Computational Intelligence
ISBN 978-3-030-93118-6 ISBN 978-3-030-93119-3 (eBook)
https://doi.org/10.1007/978-3-030-93119-3

This Springer imprint is published by the registered company Springer Nature Switzerland AG
The registered company address is: Gewerbestrasse 11, 6330 Cham, Switzerland

Foreword

I am most pleased to recommend this exceptional book to a wide readership of academicians, graduate students, industry experts, and policymakers. This insightful collection of cutting-edge research in Artificial Intelligence (AI) and Visual Knowledge Discovery (VKD) is edited by the foremost experts. It covers broad ground from theoretical developments to applications to real-world problems.

Human-centered and cognitive AI is a new scientific frontier with expanding applications in every domain from manufacturing to health care, from information processing to social sciences. What differentiates this book from others is its methodical treatment and seamless integration of fundamental principles of Visualization and Visual Analytics with emerging domains of Deep Learning (DL) and Machine Learning (ML) in the context of a rapidly evolving Artificial Intelligence domain. Respectively, the paradigm of Visual Knowledge Discovery emerges from the powerful integration of rich context and the deep understanding of underlying principles.

The current book expands on revolutionary ideas presented in the 2018 Springer monograph "Visual Knowledge Discovery and Machine Learning" by Prof. Kovalerchuk. The 2021 edited collection on the same topic takes those fundamental developments to a new level. The book starts with the vision chapter written by the editors, which makes an intricate connection between the past and the future challenges in the domains of visual analytics and artificial intelligence. It is followed by 25 other chapters written by teams of over 70 scientists from around the world.

The book represents the collection of state-of-the-art methods bridging two areas: Visual Knowledge Discovery (VKD) and Artificial Intelligence (AI) under one roof. It is in this synergy where this book truly shines. The editors did a great job explaining how AI and ML domains need better interpretability/explainability of the models, especially as transparency of models based on deep learning becomes lost. Visual methods play a prominent role in such explainability, which is brilliantly showcased in the book. On the other side, real-world problems demand advanced methods to select appropriate visualization and analytics approaches that are efficient at solving the tasks at hand. AI/ML rise to the challenge and provide a tangible solution to this pressing challenge as well.

One of the fundamental new aspects of the emerging Visual Knowledge Discovery domain is applying Machine Learning to the lossless 2-D/3-D visualization spaces, as demonstrated in chapters two through four. This concept is based on the recent development of General Line Coordinates, elaborated initially in the *Visual Knowledge Discovery and Machine Learning* 2018 Springer monograph. The remaining chapters are devoted to highly insightful topics such as integration of visualization with Natural Language Processing, increased explainability of the language models, multi-dimensional visualization, visual knowledge extraction, case studies of the integrated systems, optimization and evaluation of visualization, and open problems in those domains.

The book is unique in its treatment of visual knowledge discovery by introducing design processes for new machine learning algorithms and learning process refinement techniques. It provides an interesting treatment of a lossy dimensionality reduction, which is essential in the models' interpretability and performance. It also takes on the additional challenges outlined in the most recent surveys of visual analytics techniques for machine learning, for improving data quality, feature engineering, intelligent model refinement, and multimodal data understanding.

In summary, the book is an insightful read, a remarkable source of information, an excellent reference for all readers, regardless of their knowledge of the domain. The book has frequent references to the past discoveries in visualization, artificial intelligence, cognitive sciences, and human-driven knowledge representation. As such, it serves a broader educational purpose for its readers. I hope others will enjoy reading it as much as I did!

November 2021

Marina L. Gavrilova
University of Calgary
Calgary, Canada

Preface

Knowledge Discovery is a process as ancient as history itself. *Visualization* is a concept of forming a mental image, in which it can simulate other senses to acquire knowledge. Today, Knowledge Discovery and Visualization are fields on their own and intertwined that capitalize on the processing capabilities of computers and AI to transform complex data into a visual form not just for communication but also for knowledge discovery that leads to decision- and policymaking.

The visualization field has evolved dramatically over the past four–five decades. In recent years, AI has leapt into the application domain after maturing many machine learning fields such as deep learning techniques. AI is the newest phase in a technological revolution that impacts government policies, societies, and the economy unprecedented and unexpected. There is a race for AI-superiorities between technologically advanced governments and big tech companies, and the prediction is that whoever leads in this field would be leading the world. Inevitably, this is creating an even more significant gap between haves and have-nots. Other issues that predictably make profit-seeking that in turn indents privacy disturb democracy and create unbalance. It is a new phase that requires careful threading with policymaking to ensure AI creates a better future for all, not just a few. An innovative human-machine paradigm for knowledge discovery fosters efficiency and development, enabling to do more with less, provision of prediction against pandemics and environmental disaster, and more initialized services.

The collection of articles in this volume embodies current concepts and research directions in integrating AI with applications for knowledge discovery and visual knowledge discovery. This volume includes three parts: the first focuses on supervised and unsupervised learning and visualization, the core technological and conceptual developments that bring visual knowledge discovery to fruition. The second part on integrated systems exemplifies case studies within the context of application domains. The third and final part provides various views on the important aspect of systems' optimization and evaluation. These three parts of this volume attempt to provide a roadmap to knowledge discovery with visual shine.

The book starts with a vision chapter that highlights some issues for future research issues envisaged for AI-Enabled Visual Knowledge Discovery—underpinning the motto "Knowledge is power". The essence of this motto in this collection is technology, framework, visualization of the learning with exploration, and experimentation on how to take knowledge discovery to the next level, to accelerate the discovery pace with the power of visualization, furthermore, how to bring human-machine interaction to a new level.

Articles concerned with supervised and unsupervised Machine Learning and Visualization present a perspective of both techniques for visual knowledge discovery. These articles provide a conceptual framework for research and studies through Machine Learning to Visual Knowledge Discovery and Visual Analytics. In part, this volume offers insight into the complexities of multidimensionality, an integral aspect of knowledge discovery. This section provides real-time examples of knowledge discovery with real-life case studies on how visualization techniques are essential for presenting knowledge and exploration for data-enabled discovery and prediction.

Nazemi et al. provide a theoretical framework and a roadmap based on past published works that combine machine learning with interactive visualizations as a visual analytics model for decision-making. It is aimed at technology management and foresight based on published research work. The concept incorporates Corporate Foresight and Visual Analytics and proposes a machine learning-based technology.

The chapter "Deep Learning Image Recognition for Non-images" by Kovalerchuk, Kalla, and Agarwal presents a new approach for solving non-image Machine Learning problems using powerful deep learning algorithms by transforming these problems into image recognition problems. The proposed CPC-R algorithm converts non-image data into images by visualizing non-image data. Then deep learning CNN algorithms solve the learning problems on these images. This algorithm preserves all high-dimensional information in 2-D images with two times fewer visual elements than alternative approaches.

Wagle and Kovalerchuk address the interpretability problem of machine learning models within safety-critical domain applications, such as medical diagnosis. The proposed interpretable IVLC algorithm is supported by the Interactive Shifted Paired Coordinates Software System (SPCVis), a lossless multidimensional data visualization system. The approach includes a new Coordinate Order Optimizer (COO) and a Genetic algorithm with interactive features for self-service data classification.

McDonald and Kovalerchuk present a non-linear algorithm for visual knowledge discovery in multidimensional data using new Elliptic Paired Coordinate (EPC) visualizations, which preserve all multidimensional information in 2-D. EPC was successfully tested with multiple datasets with a developed interactive software system, EllipseVis. The EPC concept was generalized to the Dynamic Elliptic Paired Coordinates (DEPC) and the compact machine learning methodology based on EPC/DEPC.

Alves et al.'s chapter presents an interactive visualization technique, DeepRings, to help to ensure that deep learning ML models agree with human ground-truth knowledge, i.e., interpretable. DeepRings uses concentric rings, which represent layers of deep learning models. These rings encode feature maps of each layer and show the process of solving computer vision tasks, like supervised, self-supervised, and reinforcement learning. Preliminary evaluations with domain experts highlight positive points and suggest avenues for future work.

Contretas et al. explore the regression prediction problem complicated by the "negative" examples with missing target values. The major source of the missing values is the measuring instrument, which has an insufficient measuring interval range. While usually, such cases are ignored, this chapter shows that taking these cases into account can help to improve the accuracy of estimates of parameters and accuracy of the resulting predictions of the target attribute in different settings.

Current natural language models need to be able to process or generate text and predict missing words, sentences, or relations depending on the task. Such complex models are difficult to interpret and explain to third parties, and visualization is often the bridge that language model designers use to explain their work. Brașoveanu and Andonie provide a comprehensive survey of visualization and explaining natural language models. They look at the visualization of several types of language models based on deep learning and review the basic methodology used to produce these visual representations.

A novel mathematical approach of probabilistic generalization of formal concepts for describing causal clustering models is proposed by Vityaev and Pak. It is based on cyclical causal relations with fixpoints that form clusters and generate clusters' prototypes. The chapter highlights visual data clustering for visual knowledge discovery and leads to visualizing centers of clusters as prototypes using a prototypical theory of categorization and causal models.

Maçãs et al. apply profiling techniques using a self-organizing map to analyze and detect fraudulent activities in banking transactions. The approach is well-suited for real-time processing applications such as banking that is often a tedious and laborious task. The resulting tool, VaBank, provides analysis time-critical of topology and suspicious behaviors. The technique is fine-tuned to aggregate detection for a given period and scale as well as outliers.

Kohonen's Self-Organizing Maps (SOM) efficiently represent multidimensional data but do not always visualize data in a manner that is accessible to those trying to interpret it. Kilgore et al. introduce hSOM to simplify the visualization of multidimensional data. The hSOM improves upon the classical SOM visualization by allowing for the proportion of each output node's instances of a discrete variable to be visualized. The authors also address the problem of the visual noise that can arise out of dense hSOM visualizations.

Graph theories are used as one of the core concepts in analysis and classification tasks. Robert Grove's proposal of Gragnostics uses such concepts for machine learning classification and visual analysis. It aims to address the problem of interpretable graph features. To address scalability and interpretability, he proposes an approach for fast, interpretable graph comparison methods and a set of fast, interpretable structural graph features.

Hagerman et al. introduce VisIRML, a system for classifying news articles by subject matter experts to use in an ambient visualization system. They state that their approach is generalizable to a wider variety of unstructured data. Subject matter experts define topics by iteratively training a machine learning classifier by labeling sample articles facilitated via information retrieval query expansion. The resulting classifier produces high-quality labels better than comparable semi-supervised learning techniques. While multiple visualization approaches were considered to depict these articles, users strongly preferred a map-based representation.

The Time Series model is adequate for visual analytics with some degree of confidence to use it for input, predicted, intermediate factors, model structure, behavior, sensitivity, and quality in one holistic application. Jonker et al.'s work provides ample examples to demonstrate this is ranging from simple financial ratios to nowcasting and economic forecasting and massive transaction analysis. The important point is that it is scalable to explore large-scale structures with millions of nodes by visually representing many node characteristics, on-demand navigation through sub-graphs; hierarchical clustering of nodes; and aggregation of links and nodes.

The Integrating Systems section concentrates on the nature of data, algorithms, and models where integration is essential for knowledge discovery. One key concept highlighted in the articles in this part is that the system does not exist in isolation and is influenced by contexts that are complex to model. Furthermore, these contextual influences vary spatially, temporally, culturally as well as socially. Techniques may be characterized by their simplicity, efficiency, applicability, and generality. However, without contextual knowledge and proper integration, systems will be far from satisfactory and often will have a detrimental effect within the system's context.

Datia et al. report on integrating contextual attributes that contribute to knowledge discovery of air-quality metrics. With this example, knowledge of air quality and contributing factors has become necessary to understand better respiratory diseases that have become a public health concern during the current Covid-19 pandemic. The contextual integration is not just what factors affect the air quality but also its complex spatial and temporal influences. Knowledge of how these spatial and temporal influences give regions authority to decide is paramount. This context is key for managing global warming that has caused significant mayhem globally with disastrous consequences in recent years.

Kaupp et al.'s chapter focuses on integrating context analytics as a concept of integration within manufacturing with a specific case of intelligent factory setting; the context plays an essential role beyond the particular task of manufacturing. The context integration is often complex and often with unrelated, multivariate, and multidimensional datasets. Such integration is one of the prime sources of knowledge discovery for contextual influences that are often overlooked.

Buono and Balducci present in their chapter a pipeline for identifying exciting patterns and malware behaviors based on Android log files. They found out that it is possible to reveal malware families by observing graph topology patterns. In that context, they present the first approach for visualizing the revealed anomalies, whereas they state that visualizing such patterns should consider more the task, the user, and the context. They evaluated their system, which illustrated that average skilled people could understand the number of weaknesses and their severity.

A key aspect of knowledge discovery is exploration and experimentation with what-if scenarios. Berger et al.'s chapter provides this idea in a game-style structure like a soccer game. The concept applies to any situation multivariate graph analysis.

The study aims to edit the graph data visually in the analysis process and identify the influencing variables in the analysis process. Their chapter presents an integrated approach for exploring relations between structure and attributes of graphs, editing the graph data, and investigating changes in the characteristics of their relationships.

Knowledge about air pollution has become one of the critical issues in the climate and environmental crisis that need specific public policy to address these health and sustainability issues. Bachechi et al. present a tool tailored both for citizens and municipalities' policymakers. Named TRAFAIR Air Quality dashboard, it accounts for spatial and temporal and pollution-related dimensions. Spatio-temporal visualization of pollution data is helpful, and with past, historical data analysis, prediction of the pollution into near-future time is projected. It is also a research tool to determine the causal relationship between traffic and traffic, manufacturing, or farming activities.

In their chapter, Nguyen et al. explore the concept of a hybrid model to support multidimensional data visualization. They present a comprehensive review of multidimensional visualization methods and introduce a hybrid model for multidimensional visualization. Notably, the proposed method integrates star plots with scatterplots, showing the selected attributes on each item for better comparison among and within individual items while using scatterplots to show the correlation among the data items.

Essays concerned with the optimization and evaluation aspects present user perspectives of techniques with the aims to provide a conceptual and empirical framework to research and study from the user perspective.

Bouali et al. propose a dimension reduction method Gen-POIViz that visualizes multidimensional data in 2-D with decreased loss of information about the multidimensional dataset. Gen-POIViz uses a radial approach with a circle-based representation of n-D data in 2-D. In Gen-POIViz, anchors are data, unlike RadViz, in which anchors are dimensions. The genetic algorithm searches for a set of Points of Interest (POIs) that minimizes a cost function. The chapter reports the results of experiments with this method for various datasets of increasing size.

Brath et al.'s chapter provides a Knowledge presentation with examples and user study cases that dual-axis charts have a role in visual knowledge presentation and human cognitive ability to process. Furthermore, it presents an argument that dual-axis charts are more effective than single-axis charts for identifying relationships and trends. The authors suggest an automation approach for the creation of dual-axis charts.

Macquisten et al.'s chapter theme is combining interaction with hierarchical visualization techniques to analyze high-dimensional data. It demonstrates the ocean tic microbiome study with case studies and abstraction for large and complex investigation datasets. From a user's perspective, it is easier to identify a suitable dataset and a representation of hierarchical visualization. It further illustrates the visual cues such as size and color to draw user attention to the underlying features of the dataset under investigation. This chapter is a good resource for hierarchical visualization and its effectiveness with dimensionality and scalability of data.

Servin, Kosheleva, and Kreinovich expand the modeling of adversarial learning to teaching cybersecurity topics. The chapter discusses ways to arrange the competition best and take student feelings into account. The proposed mathematical model explains why adversarial teaching works. It shows that, in some practical sense, adversarial teaching is a close-to-optimal teaching strategy. The provided mathematical statements are accompanied by their visualizations.

Nakabayashi and Itoh provide further insight for the visualization analysis of multidimensional data by adopting scatterplot matrix and parallel coordinate plots. The solution explored in this chapter selects important scatterplots from all scatterplots generated, enabling the identification of "outliers" and "regions enclosing non-outlier plots". The technique is useful for users to determine whether to delete outliers from the datasets.

Finally, Lou et al. provide a review of case studies of visualizations in the supply chain and logistics domain that reflect domain-specific research, development for knowledge discovery using visualization. The paper suggests a major application of visualization in the current supply chain literature in supporting decision-making. The directions include visualization of supply chain data using different presentation graphics and tools, data flow presentations through visualization of business processes in the supply chain, contextualized operational visualization of daily activities, risk management and real-time monitoring, together with interactive visualization.

In all, 26 chapters of this book report on the work of over 70 researchers encompassing the listed topics.

Target Audience

The intended audience for this collection includes researchers from industry and academia whose backgrounds reflect a diverse range of ideas, applications, and insights from the knowledge discovery visualization communities. This volume provides them with unique examples of applied AI-related techniques for visual knowledge discovery visualization in areas of scholarship traditionally less associated with visualization, such as history, management, and the humanities. Finally, it reveals the evolving features of visualization, with examples of early visualization that enable us to understand cultural aspects of information and knowledge visualization.

Ellensburg, USA — Boris Kovalerchuk
Darmstadt, Germany — Kawa Nazemi
Ellensburg, USA — Răzvan Andonie
Lisbon, Portugal — Nuno Datia
London, UK — Ebad Banissi

Contents

Visual Knowledge Discovery with Artificial Intelligence: Challenges and Future Directions

Boris Kovalerchuk, Răzvan Andonie, Nuno Datia, Kawa Nazemi, and Ebad Banissi

Abstract Integrating artificial intelligence (AI) and machine learning (ML) methods with interactive visualization is a research area that has evolved for years. With the rise of AI applications, the combination of AI/ML and interactive visualization is elevated to new levels of sophistication and has become more widespread in many domains. Such application drive has led to a growing trend to bridge the gap between AI/ML and visualizations. This chapter summarizes the current research trend and provides foresight to future research direction in integrating AI/ML and visualization. It investigates different areas of integrating the named disciplines, starting with visualization in ML, visual analytics, visual-enabled machine learning, natural language processing, and multidimensional visualization and AI to illustrate the research trend towards visual knowledge discovery. Each section of this chapter presents the current research state along with problem statements or future directions that allow a deeper investigation of seamless integration of novel AI methods in interactive visualizations.

All authors contributed equally.

B. Kovalerchuk (✉) · R. Andonie
Central Washington University, 400 E University Way, Ellensburg, WA 98926, USA
e-mail: boris.kovalerchuk@cwu.edu

R. Andonie
Transilvania University, Bulevardul Eroilor 29, Braşov 500036, Romania
e-mail: razvan.andonie@cwu.edu

N. Datia
ISEL-Instituto Superior de Engenharia de Lisboa and NOVALINCS, Lisbon, Portugal
e-mail: datia@isel.ipl.pt

K. Nazemi
Darmstadt University of Applied Sciences, Haardtring 100, 64295 Darmstadt, Germany
e-mail: kawa.nazemi@h-da.de

E. Banissi
London South Bank University, London SE10AA, UK
e-mail: banisse@lsbu.ac.uk

B. Kovalerchuk et al. (eds.), *Integrating Artificial Intelligence and Visualization for Visual Knowledge Discovery*, Studies in Computational Intelligence 1014,
https://doi.org/10.1007/978-3-030-93119-3_1

1 Introduction

The human processes images 60,000 times faster than text [70], and 90% of information transmitted to the brain is visual [55]. Moreover, Kelts reports that 35% of our brain is devoted to vision [28]. A team of neuroscientists from MIT has found that the human brain can process entire images that the eye sees for as little as 13 ms.[1]

The definition given by the Merriam-Webster Dictionary for visualization is "the formation of mental visual images, the act or process of interpreting visual terms, or of putting them into visible form". Today, the actual content behind visualization and visual analytics is a collection of methods from multiple domains, including Artificial Intelligence/Machine Learning (AI/ML), rather than distinct methods. Within a broad definition, visualization is not a new phenomenon, being used in maps, scientific drawings, and data plots for hundreds of years [20]. Many visualization concepts are imported in computer visualization [67].

Data visualization is the process of translating raw data into images that allow us to gain insight into them. Common general types of data visualization are many ranging from simple charts to Timelines. A complete list and examples of interactive data visualizations can be found at.[2] While this is true, the available tools are very limited in discovering multidimensional knowledge that is the core of ML. Therefore such definitions of information visualization inflate expectations beyond the actual capabilities. They "oversell" its current capabilities but rightly describe its future.

The above definition of visualization is uncertain. It does not convey the significance of "interpreting visual terms" and identifying "putting to the visual form". An alternative definition tells us that visualization is (1) the act or an instance of visualizing or (2) (Psychology) a technique involving focusing on positive mental images to achieve a particular goal.[3] It is a bit more specific. The third definition is given in Wikipedia specifically for graphics: "visualisation or visualization is any technique for creating images, diagrams, or animations to communicate a message" really expresses the meaning that is common in computer science for it. It is clear from this definition that we *visualize the message that already exists*.

In AI/ML, the message of the main interest is a *pattern discovered* in the data and the *prediction* for a new case based on this pattern. In other words, *visualization is a visual representation of already existing data, information, and knowledge, but not a process of discovering new knowledge*. In contrast, the goal of Visual Analytics (VA) and Visual Knowledge Discovery (VKD) is broader; it includes *discovering new messages—knowledge using visual means beyond visualizing given input data, ML algorithm and already discovered patterns*.

Thus, in the context of ML, we are interested in both Visualization of all existing AI/ML messages, and Visual Discovery of new AI/ML messages/knowledge, where visual means enhance AI/ML and AI/ML enhances visual means. In other words,

[1] Anne Trafton, MIT News Office, January 16, 2014.

[2] https://www.tableau.com/learn/articles/interactive-map-and-data-visualization-examples.

[3] https://www.thefreedictionary.com/visualization.

we feel that it is time to expand the focus of visualization from visual message communication to message discovery with visualization means. It is hard to predict if the visualization term will cover both communication and discovery aspects or communication only. The terms of visual analytics and visual knowledge discovery emerged to accommodate the discovery aspects.

Tukey, who introduced Cooley the Fast Fourier Transform, suggested [68] that the idea of visualization helps us see what we have not noticed before. That is especially true when trying to identify relationships and find meaning in vast amounts of collected data: "The greatest value of a picture is when it forces us to notice what we never expected to see."

As we see, Tukey did not try to make a definition; he just noted that visualization allows us to see what we did not notice before in data. For example, we see numbers in the table and do not see any pattern, but visualization immediately shows a linear trend. Unfortunately, such impressive examples do not scale well to multidimensional data critical in AI/ML, as we already mentioned before. Thus, we should not exaggerate the abilities of the current tools and develop new tools which will be efficient for multidimensional pattern discovery in AI/ML.

The term *scientific visualization* refers to the process of representing scientific data. It provides an external aid to improve the interpretation of complex datasets, and to gain insights that may be overlooked by other methods (e.g., statistical methods).

Scientific visualization has evolved as a subset of computer graphics. The emphasis on visualization in computer graphics started in 1987 with the special issue of Computer Graphics, on Visualization in Scientific Computing. We quote from there: "Scientists need an alternative to numbers. A technical reality today and a cognitive imperative tomorrow is the use of images. The ability of scientists to visualize complex computations and simulations is essential to ensure the integrity of analyses, to provoke insights and to communicate those insights with others." Very interesting, at the level of 1987, visualization was already considered as embracing both image understanding and image synthesis: Visualization is a method of computing. It transforms the symbolic into the geometric, enabling researchers to observe their simulations and computations. Visualization offers a method for seeing the unseen. It enriches the process of scientific discovery and fosters profound and unexpected insights." This is probably the most synthetic characteristic of scientific visualization: *to see the unseen*.

Since 1987, there have been several IEEE and ACM SIGGRAPH visualization conferences and workshops. Recent conferences on this topic include the International Conference on Information Visualization.[4] The Visualization Handbook [21] is a textbook that serves as a survey of the field of scientific visualization and computer graphics.

This chapter presents a vision of the future of VKD, including visual analytics. VKD can help humans understand how AI/ML algorithms learn and provide new knowledge discovery avenues. The chapter outlines the future of more traditional

[4] https://iv.csites.fct.unl.pt/au/ International Conference Information Visualization.

visual methods for developing and understanding multiple ML models. We summarize open problems and current research frontiers in visualization relevant to AI/ML.

2 Visualization in ML

Visual analytics for ML has recently evolved as one of the latest areas in the field of visualization. A comprehensive review of progress and developments in visual analytics techniques for ML is available in [75]. The authors classified these techniques into three groups based on the ML pipeline: before, during, and after model building.

According to Yaakov Bressler, Data Scientist at Open Broadway Data, visualizing ML workflows most often only takes place at the final stage, but there are some situations where data visualization would take precedence during each step[5]:

- *In formulating a hypothesis*, exploratory data analysis helps contextualize problems.
- *Invalidating data integrity* by ensuring data is in the correct shape, we can create visualizations that demonstrate continuity, balance, or correct class.
- *Model selection*. Some ML scenarios will have widely different results depending on models' hyper-parameter selection/model architecture. Comparing these models' outputs (the loss functions) with visualization is a prevalent practice.
- *Testing model performance*. Once a model with good performance is identified, ML practitioners generally test this model to see how resistant it is to overfitting. Comparing performance is often done with data visualization.

Today, visualization in ML appears in the following refined stages:

1. New ML algorithm design stage, where visualization supports the *design process of new ML algorithms* (e.g., designing a new ensemble algorithm from the existing algorithms using interactive visual programming with drag and drop).
2. New model discovery by an existing ML algorithm, where visualization supports the use of an algorithm to discover a model for the *given data*:
   - Visualization of *input data* (this is similar to data visualization).
   - Visualization of the *learning process* (e.g., how a decision tree algorithm learns the model by animation or other means).
   - Visualization of the *results* (e.g., a learned model such as decision tree, SVM, CNN and saliency map).
   - Visualization of *learning process refinement* (e.g., pruning a decision tree model).

Stages 1 and 2 assume the *current paradigm* that the ML algorithm operates in the n-dimensional data space, not in the 2-D or 3-D visualization space. It is apparent

[5] https://www.quora.com/Why-is-data-visualization-essential-in-every-step-of-machine-learning.

why 2D or 3D visualization space was not used in Stages 1 and 2. Traditionally visualization space is lossy, and it does not preserve all n-D information [30].

The situation has changed with building a 2-D/3-D visualization space that preserves all n-D information [30]. This creates a *new paradigm* of visual knowledge discovery in this 2-D/3-D visualization space that we call visual space for short. This paradigm is discussed in Sects. 3.4 and 4 below, and in [36].

There are very few monographs on the visualization of ML algorithms [63]. The recent monographs focus primarily on applications using different languages: Python & Mathematica [3], R [72], Julia [58], and Scala [40]. Dedicated software tools are also available.[6] For instance, MLDemos[7] creates an open-source visualization tool for machine learning algorithms in order to understand how several algorithms function; how their parameters affect and modify the results in problems of classification, regression, clustering, dimensionality reduction, dynamical systems and reward maximization.

In,[8] it was pointed out that:

> One of the main limitations of plotting decision boundaries is that they can only be easily visualized in two or three dimensions. Due to these limitations, it might be necessary *to reduce the dimensionality* of our input features (using some form of feature extraction techniques) before plotting the decision boundary.

We want to emphasize the part that we put in italic above on reducing the data dimensionality. It represents typical mainstream practice, not an emerging opportunity. This work did not mention the negative effect of Dimension Reduction, which is generally lossy, as Johnson-Lindenstrauss Lemma [30] shows. Therefore, 2-D decision boundaries likely will distort the n-D boundary that was built by the ML methods. The above citation is from a paper published in October 2020, but demonstrated before in [30, 34], that a lossy Dimension Reduction is not necessary to visualize the n-D decision boundary fully. The concept is further expanded in this volume, in *Non-linear Visual Knowledge Discovery with Elliptic Paired Coordinates* by McDonald and Kovalerchuk, and in *Self-service Data Classification Using Interactive Visualization and Interpretable Machine Learning* by Wagle and Kovalerchuk.

These works had shown that n-D decision boundaries could be represented in 2-D/3-D fully without any information loss of n-D. Therefore, one of the goals of this volume is to attract attention to these new methods that enhance ML capabilities for applications.

The lossy situations only reflect the mainstream practice, not an emerging opportunity to use visual methods as a core of the ML model development that we call VKD.

Visualization of knowledge extraction is a topic with a long history in scientific visualization (see [6]). For instance, a representative paper from that volume extracts the isosurfaces in 3-D data.

[6] https://neptune.ai/blog/the-best-tools-for-machine-learning-model-visualization.

[7] https://basilio.dev/.

[8] https://towardsdatascience.com/machine-learning-visualization-fcc39a1e376a.

However, the topic of visualization of ML outcomes is new, "hot", and growing. A search on IEEE Xplore for "visualization" + "machine learning" in the title results in 43 entries. Of these, over 85% are from the last five years. In contrast, "visualization" + "data" has 2,200 entries. Of these, less than 40% are from the last five years. "Artificial intelligence" and "visualization" have only nine entries.

Recently, a new journal was introduced: Machine Learning and Knowledge Extraction[9] (vol 1 in 2019). This journal fosters an integrated machine learning approach, supporting the whole ML and knowledge extraction and discovery pipeline from data pre-processing to visualizing the results. "Visual Informatics", another new journal published by Elsevier, is dedicated to the visual data acquisition, analysis, synthesis, perception, enhancement and applications.[10]

Looking more closely at the actual visualizations used with ML, we observe that many of them are limited to heatmaps (for saliency maps) and bar charts (for feature importance). Therefore, we can say that the synthesis of ML and visualization is still in its infancy as an interdisciplinary domain, with an expectation of broader integration with a range of real-world applications in the future.

The importance of visual methods in ML has grown recently due to their perceptual advantages over other alternatives for model discovery, development, verification, and interpretation. In ML-model development and verification, visual methods are beneficial to avoid both overgeneralization and overfitting. In ML model interpretation, the human component is a significant part of this process. It is fundamentally human activity, not a mechanical activity. Visuals can naturally support efficient ML explanation. The other process in this area requires overcoming several limitations, such as human understanding of complex multidimensional data and models without "downgrading" it to human perceptual and cognitive limits. Existing methods often lead to the loss of interpretable information, occlusion, clutter and result in quasi-explanations [33].

A list of six challenges and potential research directions in ML visualization was suggested in [75]: improving data quality for weakly supervised learning and explainable feature engineering before model building, online training diagnosis and intelligent model refinement during model building, and multimodal data understanding and concept drift analysis after model building.

3 Visual Analytics, Visual Knowledge Discovery, and AI/ML

Artificial intelligence, machine learning, and visualization are the technological drive for visual knowledge discovery within the application domain. This section presents different aspects of the current and future of this mutual enrichment.

[9] https://www.mdpi.com/journal/make.

[10] https://www.journals.elsevier.com/visual-informatics.

3.1 What is Visual Analytics?

Visual Analytics systems combine AI/ML automated analysis techniques with interactive data visualization to promote analytical reasoning. In this case, the focus is on interactive data analytics using machine learning rather than visualizing machine learning.

Thomas and Cook proposed an early definition of Visual Analytics as "the science of analytical reasoning facilitated by interactive visual interfaces" [66, p. 4]. They emphasized the "overwhelming amounts of disparate, conflicting, and dynamic information" [66, p. 2] particularly for security-related analysis tasks. Visual Analytics focused thereby mainly on *"detecting the expected and discovering the unexpected"* [66, p. 4] from massive and ambiguous data. Thomas and Cook outlined that the main areas of the interdisciplinary field of Visual Analytics are (1) analytical reasoning, (2) visual representation and interaction techniques, (3) data representation and transformation, and (4) production, presentation, and dissemination [66]. Thereby analytical reasoning should enable users to gain *insights* that support *"assessment, planning, and decision making"*. The visual representation and subsequently the interaction techniques should enable users to "see, explore, and understand large amounts of information at once". The data transformation process converts dynamic and heterogeneous data in a way supported by visualizations, production, presentation, and dissemination. It allows to communicate the results of the analysis process to a broad audience [66]. Their definition and the related process should be outlined to focus on information rather than on raw and unstructured data. This indicates that the raw and unstructured data from heterogeneous resources are already synthesized and processed to get information. In this context, information is synthesized and processed data that can easily be visualized. The transformation process synthesizes data from different sources and different types into a unified "data representation" that can be interpreted as information [66, p.11]. Visual Analytics gained throughout the years a series of revised and more specific definitions. Keim et al. commented that defining such an interdisciplinary field is not easy [27] and proposed a different definition:

> Visual analytics combines automated analysis techniques with *interactive visualizations* for an effective *understanding, reasoning and decision making* on the basis of very large and complex data-sets. [27, p. 7]

They suggested the combined use of *automated analysis methods* and interactive visualization, particularly for understanding, reasoning, and decision making. The *automated analysis* in this context relied on *data mining approaches* [27, pp. 41] based on the work of Bertini and Lalanne [5]. They differentiate between the data mining and information visualization process. The data mining process incorporates the steps from data to computational model by transforming data into a computational model, which allows the interpretation and verification of data and generating hypotheses that lead to knowledge. This process has no feedback loops. In contrast, information visualization incorporates the steps of mapping data to a visual model, which allows pattern extraction for generating hypotheses that leads to knowledge and has a feedback loop to all previous steps. This process is based on the initial work of Card et al., who proposed the information visualization reference model [8] with the steps of data transformation, visual mappings, and view transformation, including a feedback loop to all steps.

Keim et al. proposed a Visual Analytics model, based on the introduced two processes of data mining and information visualization [27]. This combined model starts with data and spreads out on two paths: the path to visualization uses mapping, and the path to models uses data mining [27, p. 10]. The main difference is that they included a direct combining of visualizations and models. The path can first go through the data mining path to the computational model and be visualized. Vice versa, a visual mapping can be used for model building [27]. The entire Visual Analytics process provides an interactive process to make use of both the interactive visual representations and data modeling approaches for acquiring knowledge and insights [27]. The role of humans and the possibilities to interact in the stages of the Visual Analytics process remains as they are proposed in the reference model for visualization [8]. The main difference is the interactively combined techniques for visualizing and mining data. In the past years, the limitation of Visual Analytics was lifted by different approaches, particularly machine learning and artificial intelligence. Recent works integrate decision trees and rule-based classifiers [64], time series analysis [42], topic modeling and word embedding [16], statistical methods, machine learning [11, 50], and neural networks [54].

The emerging coupling of artificial intelligence and machine learning approaches in Visual Analytics leads to discover new patterns, new knowledge and enables humans to facilitate in-depth analysis. With such a direct combining of artificial intelligence, machine learning, and interactive visualization methods, Visual Analytics is more than just the "science of analytics reasoning facilitated by interactive visual interfaces", as proposed by Thomas and Cook [66]. It has *changed its problem-solving context* dramatically since the processing of vast amounts of raw data is an integral part of Visual Analytics. Considering that interactive visual interfaces can not lead to analytical reasoning, we define Visual Analytics as follow:

> Visual Analytics is the science of analytical reasoning facilitated by the direct coupling of learning models and information visualization.

Thereby "learning models" consist of any learning method, e.g., unsupervised, semi-supervised, and supervised learning. Information Visualization is, per definition, interactive, as already proposed by Card et al. [8]. Visual Analytics should be seen as a discipline that incorporates both humans and computers, whereas the advanced automatic processing of any data is prominent.

3.2 Human Interaction

Figure 1 illustrates the computer's and human's roles based on our previous work [29]. Below we discuss the change of the human roles from Information design to visual knowledge discovery.

Information Design (ID) or typically non-interactive "*Info-graphics*" are static, non-interactive visual representations of mostly abstract data. The computer has no active role here since the static graphics represent information graphically without interaction capabilities and without pre-processing of data. Info-graphics may also occur in non-computational media. Information Visualization is the interactive visualization of abstract data. The computer pre-processes the data, and the interactivity of the visual interfaces is also enabled through the computer. However, *Information Visualization commonly does not integrate any learning methods*. Visual Analytics commonly incorporates different pre-processing and learning methods. It couples machine learning methods directly with interactive visualizations and enables them to choose and parametrize the learning methods in the best cases. VKD goes one *step beyond* by *deeper coupling advanced learning methods* as described in Sect. 3.4.

Commonly humans' interaction with an interactive system refers to Human-Computer Interaction (HCI). The interaction modalities of HCI vary enormously from simple interactive systems like text-editors to interactions with advanced automated computer systems like ML-based systems. While many types of interactions are common for both, fundamental differences exist. Advanced visual learning systems based on machine learning/artificial intelligence methods require much more human involvement and more complex interaction modalities and goes far beyond the influential "information-seeking mantra" proposed by Shneiderman [61]: Overview First, then Zoom and Filter, and followed by Details on Demand.

Below we discuss the types of human interaction with automated systems from both human and automated system perspectives in Information Design (ID), Information Visualization (IV), Visual Analytics (VA) and Visual Knowledge Discovery

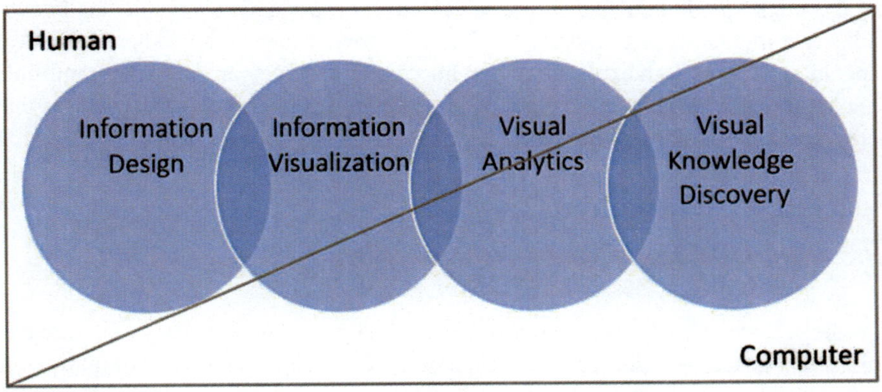

Fig. 1 The humans's and computer's roles in visualization and knowledge discovery disciplines

(VKD). For the human, the automated system is an *assistant*, while for *the automated system*, the human is one of the *sources* of information. The *human perspective* of types of interactions with an automated visual system can be categorized as follow:

(1) In Information Design, a human does not interact with the visual system. The human takes the role of the information *consumer*.
(2) In *Information Visualization*, a human decides which visualization to use and interacts with the visualizations, e.g., through panning, brushing, zooming, and deriving conclusions. The human takes the role of information *consumer*, while simple interactions do not change the information behavior.
(3) In *Visual Analytics (VA)*, a human first defines the problem to the level sufficient to use learning methods and other analytical tools, likely in several iterations. Next, the human plans the work of such automated systems, delegates some tasks to them, and analyzes the results. This is the mixed role of a direct *creator*, a *planner*, and an *analyzer*.
(4) In *Visual Knowledge Discovery (VKD)*, as a field of VA and AI, the role of the human is the same as a mixture of a *creator*, a *planner*, and an *analyzer*, but with important specifics of guiding specialized advanced automated *Visual Knowledge Discovery* ML-tools. See examples in Sects. 4 and 5, where a human interactively discovers and analyzes patterns in the lossless visual space provided by the General Line Coordinates.

In (3) and (4), more time is devoted to *interactive planning* and *managing* the work of the advanced automated ML systems. In making such interactions efficient, *new methodologies, approaches, interactive protocols, and tools* are needed. From *AI/ML and systems perspective*, the types of interactions with humans are defined by the human's need as a source of data, information, and knowledge for such systems. Typically, the primary ML data (training, validation, and testing data) are generated in advance without entering them interactively. However, the ML system needs other information from humans. *Explaining* the machine learning model and its prediction in terms of *domain knowledge* nowadays needs a domain expert as a source of this knowledge, especially if such knowledge is implicit/tacit. Improving *the accuracy* of the model prediction also still needs a domain expert as a source of this knowledge that includes tips to modify attributes, remove irrelevant attributes, search for specific patterns. Both explanation and improving accuracy are tasks with multiple uncertainties. Thus, human interaction is needed when we have *uncertain* tasks that are not ready for automation.

3.3 Future Challenges of Visual Analytics

As the science of analytical reasoning, Visual Analytics combines two main research areas that lead to a variety of future challenges: On the one hand, the computational learning models, and on the other hand, interactive visualization that involves humans in the entire transformation process. The following challenges need to be addressed:

Adaptive visual analytics: Considering that humans perform the analytical reasoning process through complex interactive visual representations, it is necessary to consider the cognitive, mental, and perceptive capabilities of humans. Besides considering humans' abilities, their interests, tasks to be solved, and the data (content) are important to consider. Adaptive Visual Analytics uses machine learning methods to adapt the visual interface, the data, and the interaction concepts to the demands of a specific user or user group [48]. It reduces the complexity of analytical reasoning tasks and leads to more efficient and effective problem-solving. Although several approaches and even implemented systems exist [49], the increasingly complex and visual representations require further research to meet the demands of a specific user or a user group.

Visual parametrization and model adjustment: The direct and deep coupling of learning methods and interactive visualizations enables visual parametrization and learning model adjustment to improve the learning results and to reduce overfitting. Visual Analytics should integrate visual interaction on model-level to allow refinements and model adjustments of the integrated learning method. The adjustment and refinement of the underlying learning method would lead to better results and a better understanding of the integrated methods.

Multi-model integration: Visual Analytics allows integrating a variety of interactive visualization in a juxtaposed and superimposed way. Besides integrating interactive visualizations, Visual Analytics allows integrating more than one learning method, e.g., for data processing, forecasting, or clustering. While single-model Visual Analytics systems provide commonly appropriate analytical reasoning, the problem context and task-solving are strictly limited. Multi-model Visual Analytics should integrate a variety of learning models that allow solving analytical tasks with different models. Ideally, the choice of the underlying model can be performed by users through the interactive visual interface.

Assisted visual analytics: Analytical reasoning tasks may be difficult to solve even with appropriate interactive visualizations and learning methods. Visual Analytics systems could support users through guidance and assistance. Although assisted Visual Analytics seems to be similar to Adaptive Visual Analytics, the main difference is the fact that users get guidance but no automatic adaptation. The complexity of Visual Analytics increases with the number of visual layouts and learning models. Users should get recommendations based on tasks to be solved, data, integrated learning methods, and particularly the capabilities of the integrated visual layouts and learning methods.

Progressive visual analytics: Coupling learning methods and interactive visualizations support the analytical reasoning process in Visual Analytics. However, the reasoning is performed by users. With progressive Visual Analytics, users get intermediate results of the underlying learning methods or any integrated models [2]. The intermediate results are particularly of great interest if supervised learning methods are integrated into a Visual Analytics system. The computational processes are getting more transparent through execution feedback and control [47]. Furthermore, the intermediate results lead particularly in multi-model Visual Analytics to early decisions [19], parameter refinements, and model optimization through interactive visualizations. Progressive Visual Analytics should be applied in multi-model visual analytics systems that allow refinements and parametrization of the underlying learning methods.

Explainable AI through visual analytics: Explainable AI has been gained a lot of popularity in research. It provides information about how and why the inputs led to certain outputs [25] and increase transparency, result tracing, and model improvement [52]. Through the direct combining of learning methods with interactive visualizations, Visual Analytics is predestined for explainable artificial intelligence [33]. Particularly multi-model Visual Analytics systems and those that use supervised methods can use the interactive visualizations of Visual Analytics to explain the underlying learning methods through visual interfaces.

3.4 Visual Knowledge Discovery as an Integral Part of Visual Analytics and Machine Learning

Visual Knowledge Discovery (VKD) is a part of a wider field of Visual Analytics (VA). Respectively the VA includes more tasks and goals that are in the realm of VKD. VA includes any task where we use a combination of analytical and visual tools to solve it. VA tasks can be decision making, prediction, preliminary data analysis, post prediction analysis, post decision making, more exact problem formulation for very uncertain situations are needed. It is a vast field, which is a source of VA's strengths and weaknesses. It is challenging to develop such a diverse field deeply enough in all its parts simultaneously because VA borrows methods from very diverse fields that are also not equally well developed and often with competing methodologies. While VKD also can be given the same far-reaching meaning as it is done for VA, we prefer to keep a scope of VKD narrower, keeping its meaning closely associated with the concept of the knowledge discovery in databases (KDD) used in the machine learning/data mining (ML/DM) community for predictive supervised and unsupervised ML/DM modeling. Therefore, we view VKD as an intersection of VA and ML/DM domains, where we view ML/DM as a part of the broader AI domain. If we do not give VKD this narrower meaning, we can view it much broader, practically equal to VA. For instance, discovering or formulating the decision strategy can also be considered a part of the knowledge discovery. At this time, we prefer a narrower meaning of VKD as a field of visual predictive modeling integrated with supervised and unsupervised ML/DM methodologies. To clarify the differences between the domains, consider an example of stock market trading. In this task, we need to make an investment decision by picking up a set of stocks.

First, we need to collect the data, predict the behavior of different stocks on the market, and finally make an actual investment decision. Different investment strategies lead to different investment decisions. Those investment strategies are partially based on the prediction and its quality. The low quality of the prediction leads to more conservative strategies than a high-quality prediction. In detail, such an example is considered in [73] for USD and EURO trading using VKD as a part of the whole process. Hence, we focus on visual knowledge discovery with visual predictive modeling because the models are not accurate enough and not well-explained. The user cannot trust them and hardly will have solid, well-grounded further steps of VA such as decision making. Traditional visualization methods, which convert n-D data to 2-D data, are lossy, not preserving all multidimensional information [30, 33]. In contrast, the representation of n-D data using General Line Coordinates (GLC) is lossless [30]. This visual representation of n-D data opened the opportunity for full multidimensional machine learning in two dimensions without loss of information (Full 2-D ML).

In simple situations, it allows visually discovering the pattern by observing these n-D data visualized in GLC. In more complex situations, which are common in ML, it allows discovering patterns in 2-D representations using new 2-D ML methods. We envision that it will grow into a whole new field of full 2-D machine learning. It requires developing a whole new class of 2-D ML methods. The firsts set of such methods have been developed and are presented in this volume with different types of General Line Coordinates (Parallel Coordinates, Radial Coordinates, Shifted Paired Coordinates, Elliptic Paired Coordinates, In-Line Coordinates, CPC-R and GLC-L) [13, 30, 32, 35, 36, 41, 71]. The studies in this realm can be traced to Parallel Coordinates [23, 24].

Often 2-D studies in ML cover only simple 2-D examples to illustrate ML algorithms visually. Next, Visual Analytics studies have been very active in exploring parallel coordinates for tasks related to clustering [22], but much less for supervised learning, which needs to be developed further. The studies in this area include [18, 65, 74]. We believe that it is time to consolidate all such studies within a general concept of a full 2D ML methodology. Traditionally 2D studies in machine learning were considered only an auxiliary exploratory data/model visualization with loss of n-D information mostly afterward or before the actual machine learning. It was assumed that in 2-D, we are losing n-D information, and we need complete n-dimensional analysis in n-D space to construct ML models. The full 2-D ML methodology shows that it is not necessary. It expands visual discovery by human-aided ML methods to the full scope of machine learning methods to visualize full patterns analysis and with 2D interaction.

4 Full 2D Machine Learning with Visual Means

As of today, clustering tasks dominate in visual analytics: "Clustering is one of the most popular algorithms that have been integrated within visual analytics applications" [17]. In the IEEE published 2019 survey of clustering and classification for time series in visual analytics [1], the summary table shows 79 papers on clustering and only seven papers on classification (decision trees, SVM, neural networks and others) from 2007 to 2018, i.e., over 10 times dominance of clustering. The earlier review (2017) on state of the art in integrating ML into visual analytics [17] shows 26 papers on clustering, 13 on classification, and nine on regression in the summary table. Together only 21 papers are on supervised learning vs. 26 papers on clustering. All 21 papers on supervised learning (classification and regression) focus on modifying parameters, computation domain, and analytical expectation, i.e. pre and post modeling, not actual model construction tools, which are traditional *existing* ML algorithms in these papers.

Why is it important to change the focus in *visualization* for ML from unsupervised learning (clustering, unsupervised dimension reduction) to supervised learning (classification, regression, supervised dimension reduction)? It is not a move from one type of ML task to another equally important task, but it is the change of the research goal. Major impressive current achievements of ML are in supervised learning not in clustering. The fundamental difference between supervised and unsupervised learning is that, in supervised learning, we have a basis for judging the quality of our solution (i.e., how good or bad the solution is). In contrast, unsupervised learning is considered ill-posed [17]. Therefore, with supervised learning we can progress more efficiently and solve important predictive problems, like medical diagnostics. In clustering, we rarely, if ever, solve such "final" problems.

On the other side, the role of clustering is growing in ML. Many funding agencies now support extensive activities to create large databases of labeled cases, especially in medicine and health care. However, it is a long and expensive process. Clustering is considered as a promising and less expensive approach to assist in solving this problem. For supervised learning, clustering is an important auxiliary supporting task. It is not a predictive task per se, while it has multiple benefits for improving supervised learning. In between, we have semi-supervised learning tasks with some cases unlabeled.

In summary, clustering is not the primary "final" task; it does not predict classes. It only helps to predict, while supervised learning predicts. Therefore, supervised learning is the core of machine learning and focusing on it is well justified for visual analytics and visual knowledge discovery. While clustering plays only a supporting role in supervised learning predictive tasks, it is beneficial in many other tasks like pointing to the outliers that may or may not be related to predictive tasks.

The third review from 2018 on visual analytics in Deep Learning (DL) [22] outlines the next frontiers for deep supervised learning in using visualization during and after training DL models, again assuming existing DL models. A similar idea is presented in the 2020 review [15] with visualization for existing ML models. The most recent review from 2021 on visual methods in ML model interpretation [33] pointed out that the visualization methods used in DL are very limited. Typically they are heatmaps for activations and bar charts for feature importance.

The goal of full 2D machine learning is different. It is developing *new ML models* based on the lossless 2D visual representation (visualization) of n-D data. In this methodology, visuals move *from supporting tools to core knowledge discovery tools*.

Below we present a representative but not a complete list of open problems of full 2D machine learning that we envision as a new research frontier.

Problem 1 Developing *new lossless visualizations* of multidimensional data. Several types of General Line Coordinates [30] are developed, explored, and applied at different levels (parallel coordinates, radial coordinates, shifted paired coordinates, collocated pairs coordinates, in-line coordinates, GLC-L) [13, 30, 32, 33, 35, 36, 41, 71]. Many other GLCs are just defined in [30]. This includes n-gon coordinates and other not paired GLCs, which recently have been implemented as npGLC-Vis software library in [39].

Developing new lossless visualizations of multidimensional data should not be limited by GLCs. Other lossless visual representations of multidimensional data are possible, and some already exist [32]. While GLCs are line (graph) based [30], others are pixel-based lossless visualizations [32] which include GLC-R presented in [18], and (in this volume) in *Deep Learning Image Recognition for Non-images* by Kovalerchuk, Kalla, and Agarwal. Also, it will be interesting to see if other lossless visualizations of n-D data can be developed based on entirely different principles.

Problem 2 How to *select a lossless visualization* of multidimensional data among multiple ones for visual knowledge discovery? There is no silver bullet, and specific types of lossless visualizations will always serve specific data types.

Problem 3 *Discovering patterns* in lossless 2-D representations of n-D data. In essence, this is to develop specific machine learning algorithms for such 2-D representations. These 2-D representations are very specific relative to traditional n-D representations-they rather like discovering patterns in pattern recognition on images, including comparing and matching with templates. In general, 2-D spatial representations play an important role in human reasoning, while spatial reasoning is a special case.

While traditional pattern recognition on images deal with the raster images, GLCs form vector images, therefore specific methods are needed for vector images. The area of matching vector images is known as the map conflation area [37], where a vector road network map from one source is matched with a vector road network map from another source. A similar task is when a vector road network map is matched with the overhead imagery of the area [37].

Thus, multiple existing and new techniques in pattern recognition on vector and raster images can be applied to develop methods specific for total 2-D ML. It has already started, as the publications listed above show.

Problem 4 Simplification of visible results, predictive models and presenting them to the users. If the model is very complex, its visual representation is often quite multifaceted and can exceed human perception abilities.

Problem 5 How to evaluate the accuracy of the predictive model produced in the full 2D ML process? Will we use traditional k-fold cross validation, say with k = 10, or will we develop and use new alternative methods specific to full 2D ML? Lossless visualization opens an opportunity to introduce new visual evaluation for ML methods [31].

Problem 6 Developing full 2D ML methods for *unsupervised* learning. There are multiple specifics in unsupervised learning in this area. We can borrow multiple approaches already developed in visual analytics where a significant portion of work was done for clustering in parallel coordinates.

Problem 7 Development of full 2D ML for the data with specific characteristics: imbalanced, missed values, very high resolution, and others. The problems with the imbalanced data and data with missed values are well known in machine learning; dealing with them in the 2-D space ML has its own challenges. Dealing with high-resolution data is a particular problem for visual knowledge discovery in full 2-D ML [36]. Consider a numeric attribute with 5 digits in every two values, like 34567 and 34568, which differ only in the last digit. Visual separation of these values can be beyond the resolution of visualization and visual knowledge discovery. Moreover, it creates significant computational challenges for full 2-D ML [36], which must be addressed.

Problem 8 How to optimize the full 2D ML process for efficient human perception and interaction? We need to make the process interactive for the domain experts as a self-service. It will allow domain experts to trust the ML predictive model from the very beginning because they built the model themselves.

Problem 9 How to interpret Deep Learning (DL) models with full 2D ML methodology? The problem of interpreting deep learning models today involves visualization at the latest stage where heatmaps are used to show the salient elements of the model or to show the importance of the attributes with simple bar charts. A much deeper GLC visualization options are available with a full 2D machine learning approach, as we describe below. The main idea of today's interpretation is first building a black box model and then trying to interpret it in domain terms with the help of domain experts. The *fundamental problem* of this approach is that the deep learning models are commonly expressed in very different terms than used in domain knowledge. This domain knowledge is rarely represented with many "raw" attributes like pixels or individual words in image and text classification tasks. Often this mismatch makes it practically *impossible* to interpret the deep learning models in terms of existing domain knowledge. Therefore, a fundamentally new approach is needed to resolve challenges. To generate such fundamentally new solutions, we first need to uncover the *cause of this mismatch*. A significant source of the DL success in getting high accuracy is using so-called "raw" features to eliminate feature engineering by domain experts. It dramatically simplifies, speeds up and "industrialises" model development. This significant advantage of DL is also a major obstacle for the interpretation of DL models because only features engineered by domain experts bring domain knowledge to the model. How can we reconcile this *deep contradiction* of deep model methodology? Dropping the use of raw features will return us to traditional machine learning and will nullify DL. Can we continue using only raw features and avoid or decrease conceptual mismatch that prevents interpretation? The fundamentally new methodology is needed to incorporate *the domain knowledge from the very beginning along with raw features*. The advantage of this methodology is in the opportunity to get benefits from both.

This leads us to the formulation of **Problems 10–11**.

Problem 10 How to incorporate the *domain knowledge* from the *beginning* along with basic features to discover DL models? Can total 2D ML help with this?

Problem 11 How to (1) discover new domain knowledge (interpretable features, rules, models) with total 2D ML on raw data and then (2) to use them to *guide* discovering deep learning models on raw data?

Part (1) is **Problem 3** above with specifics that discovered knowledge must be applied to solve (2). One of the possibilities is starting not from randomly generated weights in discovering DL models but with the higher values of the weights in the areas of raw data where interpretable features are discovered in (1). This will increase the chances that the deep learning models will be interpretable. The goal is to exceed the LIME approach [56] that builds local linear classifiers in raw data assuming that any linear model is interpretable, which is not the case in general. It is typically applicable for homogeneous data, but data are heterogeneous in machine learning, especially in medical applications (temperature, blood pressure, pulse, weight, height).

5 Visualization in NLP

A review of visualization techniques for text analysis can be found in [38]. According to this study, 263 text visualization papers and 4,346 text mining papers were published between 1992 and 2017. The authors derived around 300 concepts (visualization techniques, mining techniques, and analysis tasks) and built taxonomies.

The visualization concepts that have the most significant number of related papers are typographic visualizations (text highlighting, word cloud), chart visualizations (bar chart, scatterplot, and graph visualizations (node-link, tree, matrix).

More recently, there has been an increasing interest in visualizing NLP ML models. Using visualization methods in image processing ML tasks is quite intuitive, and saliency maps are a popular tool for gaining insight into the deep learning of images. For instance, Grad-Cam [60] is a standard technique for saliency map visualizations in deep learning. NLP ML visualization is not intuitive when it is applied to text.

NLP models are currently visualized by heatmaps similar to Grad-CAM, looking at the connections between tokens in models that utilize attention mechanisms (see *Visualizing and Explaining Language Models* by Braşoveanu and Andonie, in this volume). These solutions typically look at the importance of individual tokens (words) on the model output. The goal of NLP MLP visualizations is usually to highlight the most significant tokens that have the most important impact on the model output. For instance, what tokens contribute most to the decision that a text is classified as "fake news" [7].

The landscape of NLP has recently changed with the introduction of Transformers (see *Visualizing and Explaining Language Models* by Braşoveanu and Andonie, in this volume). Transformer models can extract complex features from the input data and effectively solve NLP problems. For instance, BERT is a state-of-the-art NLP model developed by Google successfully in NLP tasks such as text classification and sentence prediction.

Explaining the information processing flow and results in a Transformer is difficult because of its complexity. A convenient and very actual approach is visualization. The first survey on visualization techniques for Transformers is *Visualizing and Explaining Language Models* by Braşoveanu and Andonie (in this book). None of the current visualization systems is capable to examine all the facets of the Transformers, since this area is relatively new and there is no consensus on what needs to be visualized. The visualization of NLP neural models is still under development.

Without being exhaustive, we state the following three open problems in NLP ML visualization:

Problem 1 Visualization is often the bridge that language model designers use to explain their work. For instance, coloring of the salient words or sentences, and clustering of neuron activations can be used to quickly understand a neural model. The approach in *Visualizing and Explaining Language Models* by Braşoveanu and Andonie (in this volume) attempts to visualize Transformer representations of n-tuples, equivalent to context-sensitive text structures. It also showcases the techniques used in some of the most popular deep learning techniques for NLP visualizations, with a special focus on interpretability and explainability. Going beyond current visualizations that are model-agnostic, future frameworks will have to provide visualization components that focus on the important Transformer components like corpora, embeddings, attention heads or additional neural network layers that might be problem-specific. For instance, Yun *et al.* applied dictionary learning techniques to provide detailed visualizations of Transformer representations and insights into the semantic structures [76]. We consider visualization of context-sensitive syntactic information and semantic structures as one of the hottest applications of ML in NLP.

Problem 2 There is a subtle interplay between syntactic and semantic information, as outlined in semiotics. In *semiotics*, a *sign* is anything that communicates a meaning, that is not the sign itself to the interpreter of the sign. In-depth definitions can be found in [4, 9, 14, 45, 59]. The triadic model of semiosis, as stated by Charles Sanders Peirce defined semiosis as an irreducible triadic relation between Sign-Object-Interpretant [53]. The recent interest in self-explaining ML models can be regarded as exposure of the self-interpretation and semiotic awareness mechanism [46]. The concept of sign and semiotics offers a promising and tempting conceptual basis to ML and visualization.

Problem 3 We have to make a difference between attention and explanation. These terms are frequently used in ML visualization, for instance, when using saliency maps in deep learning visualization. Such saliency maps generally visualize our attention but do not "explain" the deep learning model [26]. A challenging and open problem is to use visualization as an explanation tool for ML models. In other words, visualising the combination of words according to which the text is classified as "fake news" highlights our attention or even may explain the classification decision.

6 Multidimensional Visualizations and Artificial Intelligence

Visualization plays an essential role in humans' cognitive process [51]. As a tool, visualization helps us mitigate our limitations regarding information overload [57]; by properly transforming data into information, we will ultimately turn it into new knowledge. However, the increase of the available data, generated at different velocities, and presented in many formats, turns a visual representation of most of the current datasets into a difficult task. At the turn of the century, multidimensional datasets were stored into a special kind of database, known as data warehouses, where data were classified into two buckets: facts and dimensions. Datawarehouse was built to scale and present the users' data in an intelligible way. Using a simple yet powerful visualization solution—the pivot table and later the pivot chart—helped many decision-makers get insights into data through Online Analytical Processing (OLAP) [10]. The data are aggregated before being displayed to the analysts. This mechanism allows the pivot table to scale to any kind of dimension, and together with the drill-down (to get more detail) and drill-up (to get less detail) it helps decision-makers to explore data. We can see the relation with Schneiderman' visualization mantra.

Datawarehouse also adopts metadata everywhere used by the visualization interfaces to help users hide data in certain operations and suggest which data goes into the specific part of the pivot table. This usage of contextual (meta) data lowers some manual labor of the users and improves analytical results. Their strength was to have a small concepts model, a single visualization scheme, and a set of restricted interactions. Although datawarehouses and OLAP are still viable tools for some multidimensional data analysis, they assume that everything can be put into a tabular form. Each dimension has a proper semantic and is intelligible to the analysts.

This is not the case with many of the datasets we deal with:

1. Many datasets result from non-human data harvesting, with high requirements on processing velocity [44].
2. Data comes in a variety of formats. They are often unstructured, many of which are non-tabular. It is not easy to aggregate such data without knowing the specific goal.
3. Many variables of the datasets result from feature generation algorithms (e.g. medical imaging), whose semantics are not apparent to the human analyst.
4. Multidimensional datasets, especially the high dimensional ones, often are highly sparse.

We need to improve artificial intelligence (AI) solutions to help users with better and fast data analysis towards visual knowledge exploration (see Fig. 2). AI is about learning, reasoning, and evolving.

From the visualization viewpoint, learning (the Machine Learning part of the AI, and often wrongly used as a synonym) is about learning to visualize better, helping the analysts perform the visual exploration. In contrast, from the machine learning viewpoint, visualization is about visualizing machine learning better. It is important to establish a formal definition of visualization paradigms, namely GLC, to ensure that AI and ML can leverage the data and the formal representation to automatically (or semi-automatically) choose the proper processing steps towards the "proper" visualization.

The reasoning is the capability to use the known knowledge to derive, in context, solutions or new knowledge. Furthermore, Evolving can increase its awareness of the world by updating the knowledge base, removing non-relevant facts and including new ones. Figure 3 illustrates the relation between AI, Machine Learning(ML), and Visualization in the Visual Knowledge Exploration carried on by an analyst.

The context, identified by the dashed line, must be known both to the analyst and the AI algorithms. Thus, contextual data must be stored using an AI tractable representation, such as descriptive logic assertions [12]. That way, every step towards a better visualization can be AI-assisted by using reasoning and optimization, among others. On the one hand, using an interactive approach, humans can still derive new knowledge by interacting in a mixed reality environment, using contextual information to make decisions, assisted by machine learning models. On the other hand, visu-

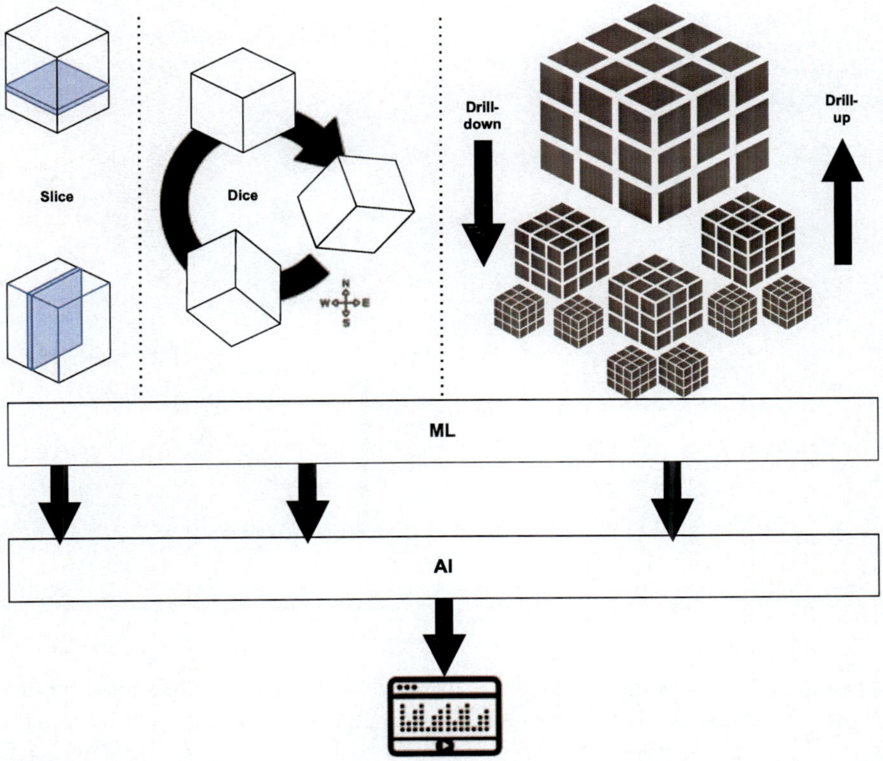

Fig. 2 ML and AI role improving the selection of valid data and visualization for multidimensional datasets. While slice, dice and drill up and down are taken by users, ML/AI should do this automatically to settle the center point of the dataset and allow further exploration by the users from there

alization is key to bridge the gap between computer-generated knowledge/models and human knowledge. The computer-generated knowledge can be transferred to humans through our cognitive processes if one can visualize how it is constructed. It is not a question of how good a machine learning model is but instead pursuing the capabilities to say why the model presents such an answer.

Problem 1 Level of detail—Multidimensional datasets present features in multiple levels of detail (LoD). However, discovering new hidden patterns can be difficult depending on the level of detail presented to the user [62]. To analysts, visualization tools need to incorporate AI and ML algorithms that automatically choose, in the background, the proper LoD to start with the interactive analysis. However, to decide which LoD is the best, a goal function must be minimized/maximized. How can we define them, depending on the task and the visualization scheme?

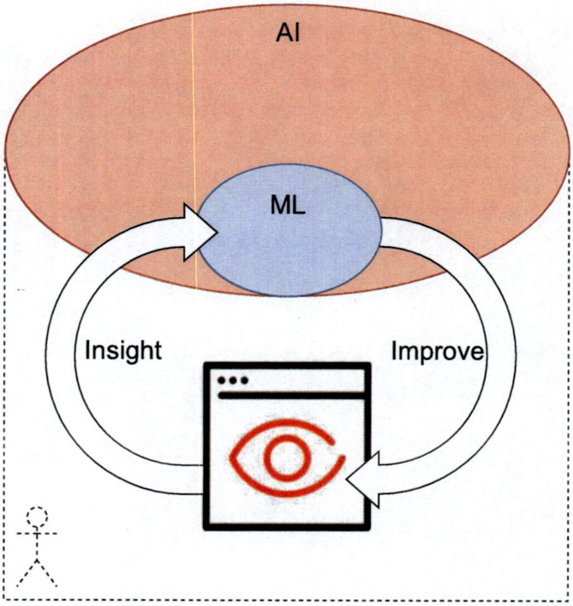

Fig. 3 Relation between AI, ML and visualization towards the visual knowledge extraction

Problem 2 Spatiotemporal patterns—Datasets representing human activity often contain features related to space and time. Thus, the patterns are spatiotemporal. However, to explain why those patterns occur, we need data about many aspects that do not have a proper and standard representation in time and space. Besides, answers may result from applying an ML model to the data, which sometimes transforms the feature space into another multidimensional space, making representing the results difficult.

Problem 3 Semantics for visualization—Most AI reasoning is only applicable if there are proper semantics on the terms used. A knowledge base (KB) is responsible for keeping the information and knowledge used to describe a specific domain's context. It consists of axioms describing conceptual entities and their relations and a set of asserted facts specific to a domain. OLAP has proper semantics and a limited visualization scheme. It can support many analyses, but the lack of visual expressiveness is evident. In order to use complex visualization schemes, there must be a formal definition of the process, a formal definition of the actions, a formal definition of the visual scheme, to name a few. No standard formalization of the visualization aspects exists, making it difficult to have a general solution for visual knowledge exploration. Thus, the task is how to introduce semantics to be able to optimize visualization.

Problem 4 Transfer of visualization knowledge between domains with AI—To have AI-supported Visualization in an interactive environment, reducing the scope is key. A set of formal definitions must be defined for a specific domain, together with a set of key visualizations schemes. However, it is not easy to share knowledge between domains when this is done, even if they share many terms and semantics. Thus, a minimum model of interoperability is needed.

Problem 5 Interpretability of discovered patterns—In a Visual Knowledge Discovery environment, even if the analysts discover some patterns of insight, they must be turned into a set of facts that support those findings. For any domain, including social sciences [43] and medicine [69], the AI-generated result must be interpreted by a human. The interpretability should be done visually to use all our cognitive strengths.

7 Conclusions

This chapter summarizes the current trend and a view of future directions of infusion of AI/ML and visualization. First-time mutual enhancements of AI and Machine Learning (AI/ML) with visualization were codified under *Visual Analytics* concept over 20 years ago, where AI/ML is considered among other analytical methods. In this volume, we use *Visual Knowledge Discovery* to represent the current fusion stage of AI/ML with visualization, where knowledge discovery with supervised ML plays a critical role. From our viewpoint, the term Visual Knowledge Discovery emphasizes the desired results—*knowledge discovered*, while Visual Analytics emphasizes the analytical process.

Recently the prominence of visualization in AI/ML became very evident with the progress in deep learning and machine learning in general, where visualization is a key player in explaining black-box learning models. The following important emerging area of Visual Knowledge Discovery is total ML in the lossless 2D/3D visualization space. It allows building ML algorithms and models, which work on n-D data but in the 2D visualization space without loss of multidimensional information. This emerging capability dramatically expands the opportunities for end-users to build ML models themselves as a self-service bringing deep domain knowledge to the process of explainable ML model discovery.

References

1. Ali, M., Alqahtani, A., Jones, M.W., Xie, X.: Clustering and classification for time series data in visual analytics: a survey. IEEE Access **7**, 181314–181338 (2019)
2. Angelini, M., Santucci, G., Schumann, H., Schulz, H.J.: A review and characterization of progressive visual analytics. Informatics **5**(3) (2018). https://www.mdpi.com/2227-9709/5/3/31. https://doi.org/10.3390/informatics5030031
3. Awange, J., Paláncz, B., Völgyesi, L.: Hybrid Imaging and Visualization. Springer (2020)

4. Bense, M.: Semiotische Prozesse und Systeme in Wissenschaftstheorie und Design. Ästhetik und Mathematik. Agis-Verlag, Baden-Baden (1975)
5. Bertini, E., Lalanne, D.: Investigating and reflecting on the integration of automatic data analysis and visualization in knowledge discovery. SIGKDD Explor. Newsl. **11**(2), 9–18 (2010). https://doi.org/10.1145/1809400.1809404
6. Bonneau, G.P., Ertl, T., Nielson, G.M.: Scientific Visualization: The Visual Extraction of Knowledge from Data, vol. 1. Springer (2006)
7. Braşoveanu, A.M., Andonie, R.: Integrating machine learning techniques in semantic fake news detection. Neural Process. Lett. 1–18 (2020)
8. Card, S.K., Mackinlay, J.D., Shneiderman, B.: Readings in Information Visualization: Using Vision to Think, 1st edn. Morgan Kaufmann (1999)
9. Chandler, D.: Semiotics: The basics. Taylor & Francis (2017)
10. Chaudhuri, S., Dayal, U.: An overview of data warehousing and olap technology. ACM Sigmod Rec. **26**(1), 65–74 (1997). https://doi.org/10.1145/248603.248616
11. Cook, A., Wu, P., Mengersen, K.: Machine learning and visual analytics for consulting business decision support. In: 2015 Big Data Visual Analytics (BDVA), pp. 1–2 (2015). https://doi.org/10.1109/BDVA.2015.7314299
12. Datia, N., Pires, J.M., Correia, N.: Time and space for segmenting personal photo sets. Multimed. Tools Appl. **76**(5), 7141–7173 (2017). https://doi.org/10.1007/s11042-016-3341-2
13. Dovhalets, D., Kovalerchuk, B., Vajda, S., Andonie, R.: Deep learning of 2-d images representing n-d data in general line coordinates. Int. Symp. Affect. Sci. Eng. ISASE **2018**, 1–6 (2018). https://doi.org/10.5057/isase.2018-c000025
14. Eco, U.: A Theory of Semiotics. Indiana University Press (1976)
15. Eisler, S., Meyer, J.: Visual Analytics and Human Involvement in Machine Learning (2020)
16. El-Assady, M., Kehlbeck, R., Collins, C., Keim, D., Deussen, O.: Semantic concept spaces: guided topic model refinement using word-embedding projections. IEEE Trans. Vis. Comput. Graph. **26**(1), 1001–1011 (2020). https://doi.org/10.1109/TVCG.2019.2934654
17. Endert, A., Ribarsky, W., Turkay, C., Wong, B.W., Nabney, I., Blanco, I.D., Rossi, F.: The state of the art in integrating machine learning into visual analytics. Comput. Graph. Forum **36**(8), 458–486 (2017)
18. Estivill-Castro, V., Gilmore, E., Hexel, R.: Constructing interpretable decision trees using parallel coordinates. In: Rutkowski, L., Scherer, R., Korytkowski, M., Pedrycz, W., Tadeusiewicz, R., Zurada, J.M. (eds.) Artificial Intelligence and Soft Computing, pp. 152–164. Springer International Publishing, Cham (2020)
19. Fisher, D., Popov, I., Drucker, S., Schraefel, M.: Trust Me, i'm Partially Right: Incremental Visualization Lets Analysts Explore Large Datasets Faster, pp. 1673–1682. Association for Computing Machinery, New York, NY, USA (2012). https://doi.org/10.1145/2207676.2208294
20. Friendly, M.: A Brief History of Data Visualization, pp. 15–56. Springer Berlin Heidelberg, Berlin, Heidelberg (2008). https://doi.org/10.1007/978-3-540-33037-0_2
21. Hansen, C.D., Johnson, C.R.: Visualization Handbook. Elsevier (2011)
22. Hohman, F., Kahng, M., Pienta, R., Chau, D.H.: Visual analytics in deep learning: an interrogative survey for the next frontiers. IEEE Trans. Vis. Comput. Graph. **25**(8), 2674–2693 (2019). https://doi.org/10.1109/TVCG.2018.2843369
23. Inselberg, A.: Visual data mining with parallel coordinates. Comput. Stat. **13**(1), (1998)
24. Inselberg, A.: Parallel Coordinates Visual Multidimensional Geometry and Its Applications. Springer New York (2009). https://doi.org/10.1007/978-0-387-68628-8
25. Jain, A., Keller, J., Popescu, M.: Explainable ai for dataset comparison. In: 2019 IEEE International Conference on Fuzzy Systems (FUZZ-IEEE), pp. 1–7 (2019). https://doi.org/10.1109/FUZZ-IEEE.2019.8858911
26. Jain, S., Wallace, B.C.: Attention is not explanation. CoRR arXiv:abs/1902.10186 (2019)
27. Keim, D., Kohlhammer, J., Ellis, G., Mansmann, F.: Mastering the Information Age Solving Problems with Visual Analytics. Eurographics Association (2010)
28. Kelts, E.A.: The basic anatomy of the optic nerve and visual system (or, why thoreau was wrong). NeuroRehabilitation **27**, 217–22 (2010)

29. Kohlhammer, J., Nazemi, K., Ruppert, T., Burkhardt, D.: Toward visualization in policy modeling. IEEE Comput. Graph. Appl. **32**(5), 84–89 (2012). https://doi.org/10.1109/MCG.2012.107
30. Kovalerchuk, B.: Visual Knowledge Discovery and Machine Learning. Springer International Publishing (2018). https://doi.org/10.1007/978-3-319-73040-0
31. Kovalerchuk, B.: Enhancement of cross validation using hybrid visual and analytical means with shannon function. In: Beyond Traditional Probabilistic Data Processing Techniques: Interval, Fuzzy etc. Methods and Their Applications, pp. 517–543. Springer (2020)
32. Kovalerchuk, B., Agarwal, B., Kalla, D.C.: Solving non-image learning problems by mapping to images. In: 2020 24th International Conference Information Visualisation (IV), pp. 264–269 (2020). https://doi.org/10.1109/IV51561.2020.00050
33. Kovalerchuk, B., Ahmad, M.A., Teredesai, A.: Survey of Explainable Machine Learning with Visual and Granular Methods Beyond Quasi-Explanations, pp. 217–267. Springer International Publishing, Cham (2021). https://doi.org/10.1007/978-3-030-64949-4_8
34. Kovalerchuk, B., Delizy, F., Riggs, L., Vityaev, E.: Visual Data Mining and Discovery with Binarized Vectors, pp. 135–156. Springer, Berlin (2012). https://doi.org/10.1007/978-3-642-23241-1_7
35. Kovalerchuk, B., Gharawi, A.: Decreasing occlusion and increasing explanation in interactive visual knowledge discovery. In: Yamamoto, S., Mori, H. (eds.) Human Interface and the Management of Information. Interaction, Visualization, and Analytics, pp. 505–526. Springer International Publishing, Cham (2018)
36. Kovalerchuk, B., Phan, H.: Full interpretable machine learning. In: 2021 25th International Conference Information Visualisation (IV) pp. 189–196. IEEE (2021) CoRR arXiv:abs/2106.07568
37. Kovalerchuk, B., Schwing, J.: Visual and Spatial Analysis. Springer (2004)
38. Liu, S., Wang, X., Collins, C., Dou, W., Ouyang, F., El-Assady, M., Jiang, L., Keim, D.A.: Bridging text visualization and mining: a task-driven survey. IEEE Trans. Vis. Comput. Graph. **25**(7), 2482–2504 (2018)
39. Luque, L.E., Ganuza, M.L., Antonini, A.S., Castro, S.M.: npGLC-Vis library for multidimensional data visualization. In: Conference on Cloud Computing, Big Data & Emerging Topics, pp. 188–202. Springer (2021)
40. Manivannan, A.: Scala Data Analysis Cookbook. Packt Publishing (2015)
41. McDonald, R., Kovalerchuk, B.: Lossless visual knowledge discovery in high dimensional data with elliptic paired coordinates. In: 2020 24th International Conference Information Visualisation (IV), pp. 286–291 (2020). https://doi.org/10.1109/IV51561.2020.00053
42. Meschenmoser, P., Buchmüller, J.F., Seebacher, D., Wikelski, M., Keim, D.A.: Multisegva: using visual analytics to segment biologging time series on multiple scales. IEEE Trans. Vis. Comput. Graph. **27**(2), 1623–1633 (2021). https://doi.org/10.1109/TVCG.2020.3030386
43. Miller, T.: Explanation in artificial intelligence: insights from the social sciences. Artif. Intell. **267**, 1–38 (2019). https://doi.org/10.1016/j.artint.2018.07.007
44. Mohammadi, M., Al-Fuqaha, A., Sorour, S., Guizani, M.: Deep learning for IoT big data and streaming analytics: a survey. IEEE Commun. Surv. Tutor. **20**(4), 2923–2960 (2018). https://doi.org/10.1109/COMST.2018.2844341
45. Morris, C., Charles William, M.: Writings on the General Theory of Signs. Mouton, Approaches to semiotics (1972)
46. Muşat, B., Andonie, R.: Semiotic aggregation in deep learning. Entropy **22**(12) (2020). https://doi.org/10.3390/e22121365
47. Mühlbacher, T., Piringer, H., Gratzl, S., Sedlmair, M., Streit, M.: Opening the black box: strategies for increased user involvement in existing algorithm implementations. IEEE Trans. Vis. Comput. Graph. **20**(12), 1643–1652 (2014). https://doi.org/10.1109/TVCG.2014.2346578
48. Nazemi, K.: Adaptive semantics visualization. In: Studies in Computational Intelligence, p. 646. Springer International Publishing (2016). http://www.springer.com/de/book/9783319308159. https://doi.org/10.1007/978-3-319-30816-6
49. Nazemi, K.: Intelligent visual analytics—a human-adaptive approach for complex and analytical tasks. In: Karwowski, W., Ahram, T. (eds.) Intelligent Human Systems Integration, pp. 180–190. Springer International Publishing, Cham (2018)

50. Nazemi, K., Burkhardt, D.: Visual analytics for analyzing technological trends from text. In: 2019 23rd International Conference Information Visualisation (IV), pp. 191–200 (2019). https://doi.org/10.1109/IV.2019.00041
51. Parsons, P., Sedig, K.: Common visualizations: their cognitive utility. In: Handbook of human centric visualization, pp. 671–691. Springer (2014). https://doi.org/10.1007/978-1-4614-7485-2_27
52. Pawar, U., O'Shea, D., Rea, S., O'Reilly, R.: Explainable ai in healthcare. In: 2020 International Conference on Cyber Situational Awareness, Data Analytics and Assessment (CyberSA), pp. 1–2 (2020). https://doi.org/10.1109/CyberSA49311.2020.9139655
53. Peirce, C.S.: Collected papers of charles sanders peirce, vol. 2. Harvard University Press (1960)
54. Pezzotti, N., Höllt, T., Van Gemert, J., Lelieveldt, B.P., Eisemann, E., Vilanova, A.: Deepeyes: progressive visual analytics for designing deep neural networks. IEEE Trans. Vis. Comput. Graph. **24**(1), 98–108 (2018). https://doi.org/10.1109/TVCG.2017.2744358
55. Potter, M.C., Wyble, B., Hagmann, C.E., McCourt, E.S.: Detecting meaning in rsvp at 13 ms per picture. Atten. Percept. Psychophys. **76**(2), 270–279 (2014)
56. Ribeiro, M.T., Singh, S., Guestrin, C.: Model-agnostic interpretability of machine learning. arXiv:1606.05386 (2016)
57. Roetzel, P.G.: Information overload in the information age: a review of the literature from business administration, business psychology, and related disciplines with a bibliometric approach and framework development. Bus. Res. **12**(2), 479–522 (2019). https://doi.org/10.1007/s40685-018-0069-z
58. Salceanu, A.: Julia Programming Projects: Learn Julia 1.x by Building Apps for Data Analysis, Visualization, Machine Learning, and the Web. Packt Publishing (2019)
59. Sebeok, T.: Signs: An Introduction to Semiotics. Toronto Studies in Semiotics. University of Toronto Press (1994)
60. Selvaraju, R.R., Cogswell, M., Das, A., Vedantam, R., Parikh, D., Batra, D.: Grad-cam: visual explanations from deep networks via gradient-based localization. In: 2017 IEEE International Conference on Computer Vision (ICCV), pp. 618–626 (2017). https://doi.org/10.1109/ICCV.2017.74
61. Shneiderman, B.: The eyes have it: a task by data type taxonomy for information visualizations. In: VL, pp. 336–343 (1996)
62. Silva, R.A., Pires, J.M., Datia, N., Santos, M.Y., Martins, B., Birra, F.: Visual analytics for spatiotemporal events. Multimed. Tools Appl. **78**(23), 32805–32847 (2019). https://doi.org/10.1007/s11042-019-08012-2
63. Simoff, S.J., Böhlen, M.H., Mazeika, A. (eds.): Visual Data Mining: Theory. Techniques and Tools for Visual Analytics. Springer-Verlag, Berlin (2008)
64. Streeb, D., Metz, Y., Schlegel, U., Schneider, B., El-Assady, M., Neth, H., Chen, M., Keim, D.: Task-based visual interactive modeling: decision trees and rule-based classifiers. In: IEEE Transactions on Visualization and Computer Graphics, p. 1 (2021). https://doi.org/10.1109/TVCG.2020.3045560
65. Tam, G.K.L., Kothari, V., Chen, M.: An analysis of machine- and human-analytics in classification. IEEE Trans. Vis. Comput. Graph. **23**(1), 71–80 (2017). https://doi.org/10.1109/TVCG.2016.2598829
66. Thomas, J.J., Cook, K.A.: Illuminating the Path: The Research and Development Agenda for Visual Analytics. National Visualization and Analytics Ctr (2005). http://www.worldcat.org/isbn/0769523234
67. Tufte, E.: The Visual Display of Quantitative Informations, 2nd edn. Graphics Press, Cheshire, Conn (2001)
68. Tukey, J.W., et al.: Exploratory Data Analysis, vol. 2. Reading, Mass (1977)
69. Vellido, A.: The importance of interpretability and visualization in machine learning for applications in medicine and health care. Neural Comput. Appl. **32**(24), 18069–18083 (2020). https://doi.org/10.1007/s00521-019-04051-w
70. Vogel, D., Dickson, G., Lehman, J.: Persuasion and the role of visual presentation support: the UM/3M study. Working Papers Series. Management Information Systems Research Center, School of Management, University of Minnesota (1986)

71. Wagle, S.N., Kovalerchuk, B.: Interactive visual self-service data classification approach to democratize machine learning. In: 24th International Conference on Information Visualisation, IV 2020, Melbourne, Australia, September 7-11, 2020, pp. 280–285. IEEE (2020). https://doi.org/10.1109/IV51561.2020.00052
72. Wiley, M., Wiley, J.F.: Advanced R Statistical Programming and Data Models. Springer (2019)
73. Wilinski, A., Kovalerchuk, B.: Visual knowledge discovery and machine learning for investment strategy. Cogn. Syst. Res. **44**, 100–114 (2017). https://doi.org/10.1016/j.cogsys.2017.04.004
74. Xu, Y., Hong, W., Chen, N., Li, X., Liu, W., Zhang, T.: Parallel filter: a visual classifier based on parallel coordinates and multivariate data analysis. In: Huang, D.S., Heutte, L., Loog, M. (eds.) Advanced Intelligent Computing Theories and Applications. With Aspects of Artificial Intelligence, pp. 1172–1183. Springer, Berlin (2007)
75. Yuan, J., Chen, C., Yang, W., Liu, M., Xia, J., Liu, S.: A survey of visual analytics techniques for machine learning. Comput. Vis. Media **7**(1), 3–36 (2021)
76. Yun, Z., Chen, Y., Olshausen, B.A., LeCun, Y.: Transformer visualization via dictionary learning: contextualized embedding as a linear superposition of transformer factors. arXiv:2103.15949 (2021)

Machine Learning and Visualization

Visual Analytics for Strategic Decision Making in Technology Management

Kawa Nazemi, Tim Feiter, Lennart B. Sina, Dirk Burkhardt, and Alexander Kock

Abstract Strategic foresight, corporate foresight, and technology management enable firms to detect discontinuous changes early and develop future courses for a more sophisticated market positioning. The enhancements in machine learning and artificial intelligence allow more automatic detection of early trends to create future courses and make strategic decisions. Visual Analytics combines methods of automated data analysis through machine learning methods and interactive visualizations. It enables a far better way to gather insights from a vast amount of data to make a strategic decision. While Visual Analytics got various models and approaches to enable strategic decision-making, the analysis of trends is still a matter of research. The forecasting approaches and involvement of humans in the visual trend analysis process require further investigation that will lead to sophisticated analytical methods. We introduce in this paper a novel model of Visual Analytics for decision-making, particularly for technology management, through early trends from scientific publications. We combine Corporate Foresight and Visual Analytics and propose a machine learning-based Technology Roadmapping based on our previous work.

K. Nazemi (✉) · L. B. Sina · D. Burkhardt
Darmstadt University of Applied Sciences, 64295 Darmstadt, Germany
e-mail: kawa.nazemi@h-da.de

L. B. Sina
e-mail: lennart.b.sina@h-da.de

D. Burkhardt
e-mail: dirk.burkhardt@h-da.de

T. Feiter · A. Kock
Techniche Universität Darmstadt, 64295 Darmstadt, Germany
e-mail: tim.feiter@tu-darmstadt.de

A. Kock
e-mail: kock@tim.tu-darmstadt.de

B. Kovalerchuk et al. (eds.), *Integrating Artificial Intelligence and Visualization for Visual Knowledge Discovery*, Studies in Computational Intelligence 1014,
https://doi.org/10.1007/978-3-030-93119-3_2

1 Introduction

Visual Analytics and information visualization enable, through the combination of automated analysis methods and interactive visualizations, the decision-making process in various domains [29]. Visual Trend Analytics incorporates, in particular, the temporal dimension of data and enables identifying, detecting, and predicting technological trends to support strengthening the competitiveness of firms. Technological developments have an important impact on strategic decision-making. The early awareness of possible upcoming or emerging technological trends could strengthen enterprises' competitiveness and market positioning. The anticipation of trends into the cooperate strategy is called corporate foresight [20], and the application of corporate foresight is positively related to firms' long-term performance [50]. Qualitative methods like technology roadmaps are commonly applied but show severe weaknesses compared to qualitative approaches that are rarely applied. Visual Trend Analytics includes prediction algorithms that facilitate corporate foresight with accessible, emerging trends. This enables profound strategic decision-making and a higher acceptance rate beneath managers. Therefore, we further investigate the conceptual study of Lee et al. [30] and develop a general three-step managing process that investigates technology trends.

There exist a variety of methods for analyzing emerging or decreasing trends and predict possible future scenarios. Existing technologies and approaches commonly focus, in particular, on measuring and computing some values for identifying emerging trends [41] or providing better prediction results [37], the process of strategic decision-making is often not considered in such analytical systems. However, the process of managing information, technologies, innovations, and emerging trends are crucial for decision-making. Another important aspect for identifying emerging technology trends is data. Social media, news, company reports, and blogs refer commonly to those technologies that already reached their climax or are already available at the market. Early technology trends are often propagated first in research and scientific publications. Therefore, these data should be considered early signals and trends [43]. Although scientific publications and their value for identifying early trends are apparent, accurate analysis and identification of emerging trends out of textual scientific publications are rarely proposed. The gathering and analysis of this continuously increasing knowledge pool is a very tedious and time-consuming task and borders on the limits of manual feasibility.

We propose in this paper a new approach for Visual Analytics for decision making by incorporating some main ideas from innovation management. We first give a literature review of existing approaches and systems that are mining and visualizing trends. The literature review reveals the missing inclusion of management and decision-making approaches in such analytical systems. Our general model tries to fill this gap by a first attempt and combines an appropriate model to enable strategic decision-making. The outcome is a model with three main steps for integrating innovations in firms. This model is enhanced by a more technical approach that illustrates the process of Visual Analytics and illustrates the main steps of our approach.

This chapter enhances the three-fold contribution of our previous work [41], consisted of (1) a model for gathering trends from text to visual interactive analysis representations, (2) the identification of upcoming or emerging trends based on text, and (3) an approach for visual interaction through different data models and related interactive visual representations to explore the potentials of technologies and detect new insights, with an advanced Visual Analytics approach and system that enables decision making for technology and innovation management with advanced methods in terms of analysis and integrates the ideas of technology management in a Visual Analytics system.

2 Related Work

The literature review in this paper follows the procedure recommended by Webster and Watson [57]. We performed a "concept centric" review with the main concepts of technology forecasting methodologies, machine learning techniques, and methods for visualizations. Therefore, our literature review is subdivided into two main sections, a review of visualization techniques and a review of visual forecasting methods. The review is complemented with approaches from technology and innovation management to bridge the gap between the disciplines.

2.1 *Trend and Text Visualization*

Current trend mining methods provide useful indications for discovering trends [1, 14, 15, 19, 32] Nevertheless, the interpretation and conclusion for serious decision making still require a human's knowledge acquisition abilities. Therefore, the representation of trends is one of the most important aspects of analyzing trends. Common approaches often include basic visualization techniques. Depending on the concrete results, line graphs, bar charts, word clouds, frequency tables, sparklines, or histograms convey different aspects of trends. *ThemeRiver* represents thematic variations over time in a stacked graph visualization with a temporal horizontal axis [22]. The variation of the stream width indicates the strength of a specific topic over time. *Tiara* uses a similar approach, with the difference that it includes additional features such as magic lenses and an integrated graph visualization [35]. *ParallelTopics* includes a stacked graph for visualizing topic distribution over time [12]. Although the system was not designed for discovering trends but rather for analyzing large text corpora, it allows users to interactively inspect topics and their strength over time and thus allows the exploration of important trend indicators in the underlying text collection. *Parallel Tag Clouds* (PTC) is based on multiple word clouds that represent the contents of different facets in the document collection [10]. Temporal facets can be used to identify certain keywords' differences over time and infer the dynamics of themes in a text collection. Another extension of word clouds is *SparkClouds* that includes a sparkline for each word [31]. These sparklines indicate each term's temporal distribution and allow conclusions about the topic trends. A user study reveals that participants are more effective with *SparkClouds* compared to three other visualization techniques in tasks related to trend discovery [31]. A similar approach [36] also includes co-occurrence highlighting. In contrast to *SparkClouds*, this technique includes a histogram for representing the temporal relevance of each tag. Additional overlays in the histograms show the co-occurrences over time for a selected word to enable a more comprehensive analysis of trend indicators. Han et al. introduce with *PatStream* a visual trend analysis system for technology management [21]. Their system measures similarity between pairs of patents using the cosine metrics and extends the work of Heimerl et al. [23] in particular in regards to visualization. The evolution and structure of topics that indicate the trends is visualized through a *Streamgraph*, which was already proposed in the previous works of Heimerl et al. [23]. In contrast to this previous work, *Patstream* breaks down the streams into vertical time slices, which represent periods of years. These time slices are based on their introduced concept that uses the term score, the ratio between the radiative frequency of a term in the given patent collection, and its relative frequency in a general discourse reference corpus [21]. Although their concept makes use of term frequencies, title score, and claims score [21], the most useful approach seems to be the term score. Thus it relies on a relative score and investigates the entire document or patent corpus. The topic stream visualization is similar to a stacked graph with the included term (topics) in the area-based visual representation. As they hierarchically cluster patents according to their textual similarities, users are able to zoom in into a cluster through a level-slider. Besides the main visual representation, the stream visualization, they provide four other visual representations, such as a "scatterplot" with brushing and linking [21].

2.2 Technology Forecast Analytics

The literature review for "Technology Forecast" was performed by keyword searches to identify literature and then backward and forward reviews of the identified literature according to Webster and Watson [57]. The initial search was performed by combining keywords from different categories. The search was structured as follows: $S_1 =$ (Technology) AND (Trends OR Forecast OR Foresight OR Intelligence) AND (Methods OR Visualization OR Analysis OR Model OR Discovery). After the first search and analysis, the results were used to search for more individual keywords, which often represented identified elements of the categories. $S_2 =$ Visual Analytics, Text Mining, Trend Mining, NLP, Patent Analysis, Tech Mining, Data Mining, Technology Roadmapping, Patent Network, Technology Radar, Trend Analysis, Bibliometrics, TRIZ. The search terms in S_2 were combined with those in S_1. The search terms were applied to six different databases: IEEE Xplore, ACM Digital Library, SpringerLink, ScienceDirect, CiteSeerX, and GoogleScholar. Overall, 100 publications were chosen for further analysis. A second selection procedure was done by reading the papers and investigating the appropriateness of our research. Out of the 100 publications, 42 were chosen for a deeper investigation. In the third step, we further reduced our review according to the following criteria:

- only papers were investigated that were published after 2016
- papers that used existing frameworks with new data were removed if the used framework was older than five years
- If authors had multiple publications on the same topic, the one with the most citations were used

An et al. proposed a novel approach to derive "technology intelligence" from patents, which can be used to forecast technology [2]. The advancement of text mining techniques has enabled the analysis of the descriptive parts of patent documents and therefore extended the scope of patent analysis [9]. Conventional patent analysis approaches are based on keyword analysis and use specific algorithms to specify keyword sets. The relationship of these keyword sets is analyzed with text mining techniques to describe the technological content of the patent. Previous patent analysis used a subject-action-object (SAO) model to analyze the semantic structure of the keywords. This approach is limited in that it increases the complexity too much and ignores non-functional relationships [2]. As an improvement, the authors propose a preposition-based semantic analysis to develop a technology-relation-technology (TRT) network [2]. This approach uses the fact that the number of prepositions in the English language is limited and can be used to identify the functional and non-functional relationship between technological terms. A network is used to structure the data systematically and to easier visualize and quantify the relationship between different analysis units [46]. A keyword-based network analysis relies on unstructured data and can be used to identify vacant technology and map technology evaluation by analyzing the degree and type of relationship between technological keywords [2, 60]. In the first step of creating a TRT network, the relevant patent data, structured and unstructured, needs to be extracted. Afterward, the data is preprocessed, and the noun-preposition phrases in the abstracts are extracted [2]. With the help of NLP libraries used for text parsing, the target technological keywords are used to filter the noun-preposition phrases. In the third step, using the noun-preposition phrases are used to identify "technological keyword—preposition—technological keyword" structures, where the first keyword is extracted from the noun phrase while the second keyword is extracted from the preposition phrase [2]. These structures are clustered into several groups based on the used preposition, which should define the relationship of the keywords. To verify the relationship, the structures are transformed into TRT structures. In the final step, the TRT network is developed. Domain experts choose the target keywords for the network out of all the keywords from the TRT structures [2]. A relationship matrix for each relationship type is constructed, where the frequency of the TRT based on that type is transformed into binary values. The value 1 indicates a relationship between the keywords, 0 indicates no relationship [2]. This information is used as the input for the network. Trend analysis can be conducted on this network by using the year of application and observing the evolution of technological structures based on the change in technological keywords, and their relationship [2]. This allows forecasting changes in technologies. The network can also be used to forecast possibly new technology by comparing TRT networks with similar structures in different areas and comparing the keyword sets. The framework is tested on electronic vehicles in the paper [2].

Li et al. proposed an approach that makes use of Twitter data additionally to doing patent analysis for identifying technology trends [34]. It uses data mining to gather Twitter data for sentiment analysis to detect and classify opinions, and emotional reactions to different technology [34]. The main component is the author-topic over time (ATOT) model, a combination of the topic over time (TOT) and author-topic (AT) model, which are based on the Latent Dirichlet Allocation (LDA) [6, 34]. For the ATOT model, the topics, authors, and time information of the documents are combined to form a three-dimensional analysis model to study the evolution of authors' research interests [34]. The approach consists of three processes, patent analysis, Twitter data mining, and the combination of both to identify technology trends. For the first process, a patent database is chosen as a data source. The patents are then converted to raw text format [34]. These are clustered via the "Lingo algorithm" using semantic hierarchical clustering, a vector space model, and singular value decomposition [34, 54]. In the second process, the Twitter data is mined by using the Twitter Search API [34]. This data is also preprocessed, removing duplicates and cleaning redundant information. The tweets are classified according to time. In the third step of the second process, a word document distribution is created [34]. The preprocessed Twitter data is analyzed with the ATOT model, and a topic-feature words probability distribution is obtained. Nouns, adjectives, and adverbs are extracted for keywords for each year to obtain awareness keywords for the sentiment analysis. The profiles of the Twitter users are analyzed, and the ATOT model and experts identify their profession [34]. It is then analyzed which type of professions focused on which topics over time to identify the changing pattern in their topic of interest. In the third process, the results of both processes, the evolution map and trend analysis of the patent analysis and the sentiment and pattern analysis of the Twitter data mining, are combined in a differences analysis [34]. This is done by first doing a comparative analysis of the results, comparing the gaps, and combining the results of changing patterns in the topic of interest to identify developing trends of emerging technologies. The final comparison analysis is also mapped out over time in a combined evolution map, where the results of the patent analysis are in one half, and the results of the Twitter data are one the other half [34].

Li et al. also proposed an approach that makes use of a patent network to forecast through analogy and social network analysis visualization [33]. Forecasting by analogy is transferring previously identified patterns of change in similar technology onto the technology to be forecasted [3]. It is a form of trend analysis that uses historical data in related fields to project the development. The forecasting in the framework is done in five steps [33]. First, a bibliometric analysis is performed on many sources, including scientific publications, patents, and news from the internet. The analysis aims to identify the development trend of the target technology, including a pattern of change. In the second step, time and trends are normalized based on the maximum number of patterns in the technology life cycle and divided into five-year periods.

This is done to facilitate the comparison of technologies developed in different time periods and to identify the characteristics of each technology development period [33]. Afterward, collaboration networks of patents are created, with the companies as nodes and the co-parenting behavior as edges. Sub-networks are created for each period and compared via two centrality metrics, average degree and density. The analogy forecasting is done by analyzing these metrics and using the visualization of the networks to project the future development of the target technology by extrapolating the historical data of the identified similar technology [33].

Yang et al. proposed an approach for semantic analysis of keywords with the "Subject-Action-Object model" [59]. The forecasting is performed by extrapolating technology information in trend analysis and visualizing the output on a technology roadmap [59]. The technology roadmap is a structured and graphical method to explore and communicate the relationships between evolving and developing markets, products, and technologies over time [49]. They use the Web of Science as the data source. The framework is divided into two parts. In the first part, the semantic information is extracted. In the second part, the technology roadmap is constructed. To extract the semantic information, firstly, the principles of semantic information extraction have to be defined. This includes specifying the relevant technology field. After that, a preprocessing of the data is performed. These are annotated, and the SAO structure of the data is extracted algorithm. To create the semantic-based technology roadmap, the technological factors have to be first defined [59]. They use fuzzy matching functions or using the SAO frequency statistics. After the identification of the technological factors, the relationship of the SAO structures is identified. The relationship can be temporal or correlative. A cross-correlation map is used to show the temporal relation or each year, which indicates the order in which the specific technology occurs and can be used for the trend analysis [59]. A factor map is created with a principal component analysis (PCA) to visualize the correlative relations based on SAO co-occurrence. With the help of experts, the technology roadmap is created and used to identify the trends of technology development [59].

Hurtado et al. proposed an approach that makes use of association analysis for discovering topics from a text and time series analysis to forecast their evolving trend [25]. The approach consists of six steps. In the first step, the text corpus is preprocessed, by removing stop words, verbs, using stemming and lemmatization [25]. The sentences are represented as a vector where each dimension corresponds to a keyword, and its value is a binary indicator relating to keyword occurrence. The second step consists of frequent pattern mining by using association analysis on the matrix. The rules discover a set of antecedent items. Because of redundant words, not each association rule (pattern) is treated as a topic [25]. A refinement process is added so that the final topics are all unique. In the third step, a topic incidence matrix is built by collecting the data from each year using the topics and binary incidence matrix. The generated vectors record the number of times a topic appears in each year's paper. A temporal topic correlation analysis is performed on the new matrix. This generates a data set of correlation (correlation co-efficient) among all topics and measures the strength of correlation between two random variables [25]. The correlation coefficient is used to build a network, where each node denotes a topic, and the edge denotes the degree of correlation. To find strongly correlated topics, a threshold value is set to remove edges with weak correlation. A clique percolation method algorithm is used to find groups of topics with strong correlations connected to form a node group. To forecast topic trends, a time series analysis is used [25]. A forecasting model will use historical time series data of a selected field to predict future evolution. It is possible to select multiple fields, which improves forecasting accuracy by having more historical data and making use of the temporal correlation between the different topics [25].

Zhang et al. used text mining for data extraction and processing, machine learning for topic analysis and expert knowledge, and a technology roadmap for forecasting and visualization of the analyzed topics [62]. It is therefore separated into three sectors. In the first sector, the data gathered and preprocessed. An updated Term Clumping process is used to retrieve core terms, and a Term-Record-Matrix is created [61, 62]. A Term Frequency Inverse Document Frequency (TFIDF) analysis is done on these core terms to help the oncoming clustering process [51, 62]. The second section is the topic analysis. The topic analysis is done in four steps, setting up a training set of labeled data for machine learning to use a data-driven K-Means-based clustering method. At first, a cluster validation model is established, which focuses on Total Precision as the target value. In the second step, the features are selected, and a weighting model is created. This is done manually on a multitude of sets. The weighting is only applied to half of the sets in order to calculate similarity [62]. Afterward, a K-Local Optimum algorithm is used as a K-means model for the clustering. It identifies the record with the highest similarity value and uses it as the Centroid of the cluster [62]. The results of this clustering are topics. The topics are weighed using the results of TFIDF analysis, which is also used as the Y-axis of topics in the forthcoming technology roadmap. The semantic relationship between topics is calculated via similarity measure in the K-Means Optimum model. This is used to indicate an evolution in specific fields [62]. In the third section, expert knowledge is used to help with the forecasting of the quantitative results and used to create the technology roadmap [62]. This is done in multiple round interviews, workshops, or seminars. The technology roadmap uses historical data, so both the historical and the possible future evolution of trends are visible. The framework is then applied to research concerning big data technology [62].

Nguyen et al. proposed an approach that makes use of "Term Frequency and Proportional Document Frequency" (TF * PDF) analysis for detecting hot topics and trends from patents with the help of the "Internation Patent Classification" (IPC) ranking [45, 58]. Another key component is the usage of the Aging Theory to calculate the variation of trends over time [7, 45]. The approach can be divided into four phases [45]. The first phase is the preparation and data collection, where the IPC is defined and the patents are downloaded. In the second phase, the patents are transformed into structured data. Their keywords are extracted with the help of stop word removal, and an NLP Part-of-Speech Tagger [45, 55]. In the third phase, the frequency of the term and the term variation over time are considered the main characteristics. To measure the frequency, the TF * PDF method is used [45]. To measure the term variation over time, a term life cycle model is used. The birth, growth, decay, and death of each term are measured with the help of three mathematical functions. To get weight, the TF * PDF and term variation over time are combined [45]. The terms in a candidate list will then get ranked with the combined weight. The top-ranked terms are identified as "hot terms" that reflect the hot topics in the corpus. In the last step, a patent timeline is created and divided into yearly time slots [45]. For each year, a trend is represented by the normalized weight of occurrence from a term in N documents. These results are visualized by line graphs, which illustrate the change over time for each trend and compare it to other trends. It also can be seen whether a trend is growing or is decreasing. The framework was applied on patents from 1976 to 2005 to identify the ten hottest very stable trends [45].

Cho and Daim integrated a frequency analysis in a technology diffusion model [8]. The Fisher-pry model [17], and growth curves were used for visualization, and forecasting calculation [8]. The idea is to use historical data for technology trajectory utilizing a mathematical model [5]. The authors propose a six-step approach to forecast technology based on growth patterns. The first step identifies technology trends in the market. This is done manually by creating literature reviews, taxonomies, and market reports of the target technology and market [8]. The next step relies heavily on expert knowledge [8]: an expert panel discussion is used to analyze the market structure and provide potential keywords. Data Mining is used to gather patent data from a multitude of sources. This data is discussed by experts again. The third step is a general gathering of data by using bibliometric data. The keywords used to get data from "Web of Science" were applied for research and patents. In the fourth step, the frequency of publications is calculated [8]. This includes a preprocessing of the data, as unrelated publications are removed. The frequency seems to be calculated without weight, though the authors do not mention the specific calculation method. The frequency is calculated for each used database [8]. To calculate growth curves and identify growth patterns, a mathematical model is needed. To simulate market penetration, technology diffusion models are used [18]. The Fisher-Pry model simulates technology advances as competitive substitutions of one method to satisfy the need for another and is similar to biological system growth [8, 17]. The growth of technology is slow in the beginning, then rapid until an upper limit is reached. After that, the growth slows down again. The upper limit is estimated by using historical analogies [8, 17]. It forecasts the growth rate of the substitution technology based on the technological advantage over the old technology. The data for the mathematical model was gathered in the steps before, and the diffusion rate is calculated. In the last step, the growth patterns of each database are consolidated by identifying the time lag. Growth curves are created to forecast and visualize the technology trend [8].

Considering the Visual Analytics approaches, the most advanced approach is the work of Heimerl and colleagues [23]. It provides more than one view, uses relative scores and co-occurrences, and visualizes the temporal spread of the topics with the related categories. Furthermore, it provides a kind of process of functionalities to support trend analysis and technology management, particularly for patents. They propose a five-step approach derived from the works of Ernst [13], and Joho et al. [26] that starts with (1) obtaining an overview of different technology topics in a given field, (2) identifying relevant trends according to individual information needs, (3) evaluating the importance of technological trends, (4) observing the behavior and productivity of different players relevant to a specific trend, and (5) spotting new technologies related to a trend. Although this work was the only one, by our best of knowledge, that considers human tasks and is propagating a kind of procedure, the system itself does not really support the users in the proposed way. The approach and the according system "*PatStream*" just provides a dashboard of four static visualizations. The interaction capabilities are limited, a real overview is not given, and changing visual representations to gather aspect-oriented visualizations is not provided.

As the literature review revealed, several algorithms exist for gathering trends from text, forecasting technologies, topics and trends, and various approaches to visualize the extracted terms, trends, and forecasting results. From the visualization point of view, the systems are commonly designed to illustrate the trend or frequency of terms (even without the temporal dimension). An analytical visualization system that enables humans to analyze the innovation and technology management process through different data models and selectable and appropriate visual structures could not be found.

3 General Method

Corporate foresight is a dynamic capability describing an organization's ability to anticipate and proactively adapt to future developments [16, 20] and thus an organization's innovative capabilities. A company's proficiency in corporate foresight significantly affects its success [50]. Corporate foresight comprises many activities, but the most common approach is applying methods such as scenario planning, technology roadmapping, or the Delphi approach, creating an understanding of the future for specific objectives. In recent years, foresight methods are commonly adopted and integrated into the organizational strategy formulation. Usually, these methods aim at detecting discontinuities or future projections [16]. A common problem with these approaches is the dependency on expert knowledge and opinions. Consequently, several studies try to apply big data and modern technology to anticipate future trends. Recent research discusses how to support the identification of technological trends, a foundation of corporate foresight, by text mining methods [27, 38].

Only a few studies rely on specific corporate foresight methods that are known in an organizational context. The improvement of existing corporate foresight methods based on automated and data-driven approaches enables easy integration into existing organizational processes and reduces managers' adoption barriers. This can increase the acceptance of surprising results in an organizational context. This opportunity is rarely used, and further research is necessary [27]. Subsequently, we propose a method for technological trend prediction based on an established corporate foresight method—technology roadmapping.

Technology roadmapping is a method for technology planning that aims to align an organization to technology developments strategically. The roadmaps, which have diverse forms and visualizations, can be used to explore and discuss upcoming relational aspects between technologies [48]. The most common visualizations show the technological maturity (e.g., technology, product, and market) on the vertical axis and the time-dependent progress on the horizontal axis [30]. The different technologies can be linked based on improvements or similarities. The visualization can change based on the specific purpose [30, 48]. Consequently, technology trends show a significant impact on these visualizations' characteristics. Internal emergent technologies will impact future products, and external research achievements may threaten existing solutions. Consequently, organizations should foster awareness about these topics and create mechanisms to efficiently map these trends and include them in their technology roadmaps.

In the field of technology road-mapping using visual analytics, only two approaches can be identified. Pepin et al. [47] develop a dynamic topic extraction and create a visualization based on a Sankey diagram. Based on a Twitter data set, they divide the tweets into explicit phases and extract topics for each period. The relational processes between the technologies are calculated based on topic similarity and a specific threshold. In contrast, Kayser et al. [28] calculate several visualizations supporting individual steps in the roadmap creation and miss the objective of a comprehensive method.

In our study, we develop a roadmapping process based on the customization model of Lee, and Parker [30] who describe different types and use cases of roadmaps. They define three phases—classification, standardization, and modularization—to structure existing roadmapping approaches. We classify our visualization methodology for trend identification based on their framework to enhance the adoption willingness in an organizational context. This helps to choose targeted methods in an organizational context for corporate foresight activities. The classification phase defines the roadmap's functional purpose. They differentiate forecasting, planning, and administration as managerial use cases of technology roadmapping. Forecasting is thereby the most plausible activity because the assessment of technologies' future-readiness as roadmapping's main objective relies mainly on predicting technological trends. Second, a technology roadmap is a foundation for strategic planning activities. It reduces uncertainties and creates clear objectives. Third, better communication based on visualization and a feeling for the overall vision support administrative processes. Lee and Parker propose several roadmap types in the standardization phase based on the differentiation of products and technologies. The final modularization phase aims to match the objectives of the initial phase to the standardized visualization methods.

Considering the technological capabilities of automation and possible data sources, we reduce Lee and Parkers' [30] framework and ensure the compatibility of the automation with the managerial strategy alignment through a defined objective and standardized output. The application of automated methods is primarily beneficial if individuals cannot perceive and process a high amount of information, for example, a data set of many text documents. Up-to-date language processes are especially useful for latent topics that are not explicitly prevalent in the data set. Especially product and strategically relevant topics are not ubiquitous in such organizational data sets.

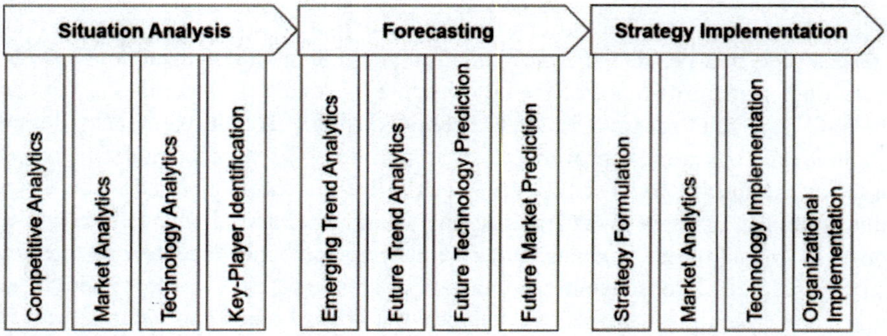

Fig. 1 The general model of innovation management through Visual Analytics

The knowledge about strategic objectives and products is usually defined in specific documents, and there is no need for further specifications based on language processing. Hence, the administrative objective is not adequately addressed with automated processes. Additionally, product knowledge is difficult to separate from incorporated technologies, and product-specific content cannot be visualized. The main objectives for automated technology roadmapping are forecasting and planning, focusing on technology rather than concrete products. Lee and Parker [30] propose a single possible customized roadmap considering the mentioned requirements: The technology trend roadmap visualizes technology trends over time in different technology areas. In this visualization, each technology is assessed for its importance and technological coverage in the organization. The technology trend map can be created fully automated, while the technology portfolio map improves in quality considering the objective data from trend analysis.

To create a technology trend map and use the information for the technology portfolio, we follow the following steps: Situation Analysis, Forecasting, and Strategy Implementation (see Fig. 1).

In the first step, the current situation is analyzed. This includes competitive analytics, market analytics, technology analytics, and key-player analytics. These steps require automated methods that allow the analysis of the market, competitors, key players, and technologies.

In the second step, forecasting of possible future scenarios should be performed. This step includes emerging trend analytics, future trend analytics, future technology prediction, and future market prediction. For this, enhanced predictive analytics methods are used based on historical data to gather a probability of future scenarios. The probabilities allow at least to validate the hypotheses from step one or even gathering information about probable future scenarios.

The last step of the model focuses on the "strategy implementation". The implementation process is based on the gathered information through the first two steps and is performed in a more "organizational" way, which commonly leads to "strategy formulation". Based on the formulation, a comprehensive market analysis can be performed, technology implementation through technology development, the organizational implementation that leads to an ideation process that can be supported through the exploratory character of the Visual Analytics system. The tasks can be combined and are primarily performed by humans. The ideation process as part of "strategy formulation" can be supported through Visual Analytics.

4 Visual Analytics Approach

We introduced in Sect. 3 the general method and will investigate in this section the according to Visual Analytics approach. Our approach includes six main steps as illustrated in an abstract way in Fig. 2 [41]. We describe these steps assuming that scientific data are used for early trend identification, and these have to be crawled through the Internet. Thus, we focus our attention primarily upon enabling users to interactively gather an overall topic trend evolution and different perspectives (e.g., geographical or semantic) on data to inspect and analyze potential technological trends.

In this section, we explain the processing exemplary based on the DBLP database. We chose the DPLB indexing database since it does not provide any abstracts or full-texts and makes the data gathering process more difficult so that the process

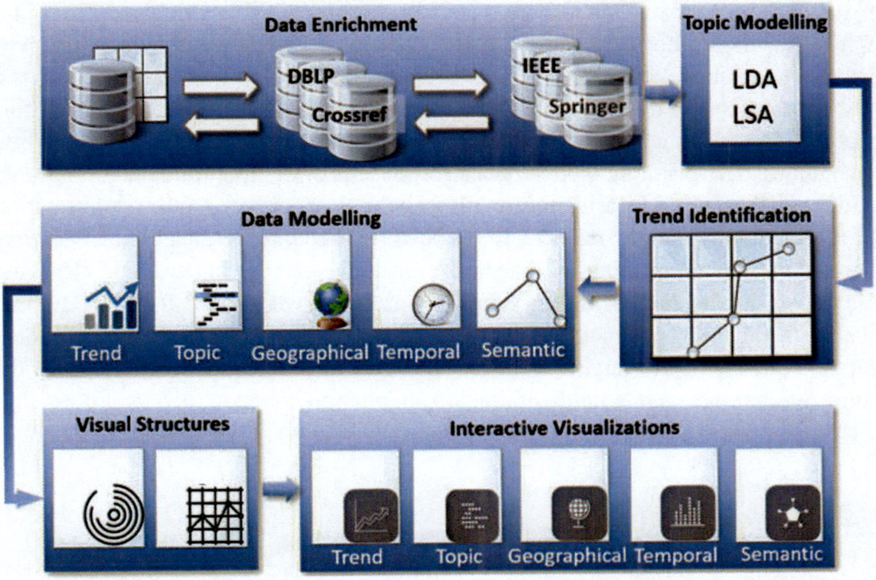

Fig. 2 Our transformation process from raw data to interactive visual representations consisting of six main steps (adapted from [41])

of data gathering can be illustrated too. The DBLP is a research paper index for computer science-related publications.

4.1 Data Enrichment

In the first step of "data enrichment," we use web-based resources as a baseline to gather initial information. Our Visual Analytics system gathers data from different resources, e.g. "DBLP", "Crossref", "Springer", "ACM" or "IEEE". The most complex procedure is gathering data from "DBLP", since it is an indexing database and does not contain any abstracts, full-texts, or further meta-data, e.g., geographical information of the authors. The "DBLP" data are stored first in a database and provides at least "title", "year", and authors' names of all indexed publications. In more and more cases, an "Document Object Identifier" (DOI) can be gathered too that allows a unique identification of certain publications.

For a proper analysis of the given data, enhancements of data quality are necessary. To enhance the quality of data, we gather additional data from the Web. The data collection used as a basis is a combination of multiple different data sets. The individual data sets offer data of varying quality and content. We, therefore, balance out the limitation of the original data basis of "DBLP" by augmenting the available data with additional information for each publication. For this purpose, the system has to figure out where data resources are located on the Web or which online digital library has more information about a particular publication. We integrated the publisher named above. The primary data collection contains a link to the publisher's resource and is used to identify the digital library and the location of additional information. This information can be gathered either through a web service or crawling techniques. The resulting web services response is well structured and commonly contains all required information, while crawling techniques require confirmation of robot policies, and the results have to be normalized. Nevertheless, the data may contain duplicates, missing, or inaccurate data. Therefore, standard data cleansing techniques are applied. With this step, we enrich the data of DBLP with additional metadata, including abstracts and text directly from the publishers, and include some citation information through "CrossRef" that should enable the identification of the most relevant papers in a field with regards to citation count.

4.2 Topic Modelling

In the previous step, we gathered at least abstracts for a major part of the "DBLP" entries and some open access full text for some entries from a general public source like CEUR-WS or the *Springer* database. Based on these enriched data, we are able to perform information extraction from text to generate topics. We conducted a preliminary study with 2.670 full-text articles and their corresponding abstracts and used the Latent Dirichlet Allocation and Latent Semantic Analysis (LSA) [11]. Both algorithms were used with and without lemmatization for full-text articles and abstracts. The best results could be gathered through LDA without lemmatization even though topic extraction from abstracts of the publications [42]. We applied the Latent Dirichlet Allocation for topic extraction according to Blei et al. [6]. The topic extraction through LDA was performed with single words and n-grams that consisted of two or more terms (n-grams). Overall, 500 topics were generated automatically in a data-set of more than seven million documents. For each document, we set a number of 20 for words and 20 for generating n-grams. The according labels of the topic-based trends were generated through the highest scored n-gram, word, or word-combination as the label for a particular topic. A word combination consists of two words, where the score value of each of the words is significantly higher than the scores of the first three n-grams together. The score itself is the value for the distribution of a topic in a finite set of documents. This could be the result-set of a search query or even the entire data set. This kind of labeling has the advantage that reliable and sense-full topic-based trends can be generated with statistical methods. It is not language-dependent (like LDA), and the generation of the labels is fast and easy. This could be enhanced with semantic approaches and linguistic corpora and would provide far better accuracy and sense in terms of semantics, at least for the most prominent languages. A topic-based trend (element) has one more and crucial dimension: time. Temporal information of the topics and their distribution is extracted through the publication date of a document or the dates of a set of documents. With this procedure, we generate labeled and time-stamped topics that can be used to identify trends. Figure 3 illustrates on the left side the measured score of the n-gram "big data". Thereby, the score of the n-gram is higher than the sum of the scores of the second and third words. In such a case, the according n-gram is used as a label since only the first word has a significantly higher score but not the second word. The document set consists here of the search results of the term "Visual Analytics". On

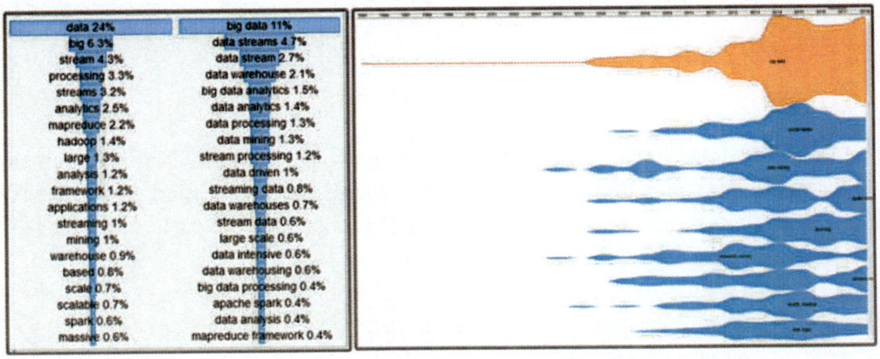

Fig. 3 Topic labeling and topic-based trends: The labeling of topics is performed through the distribution score of words and n-grams in a finite set of documents (left) and the topic-based trends can be gathered through the distribution of the topics over the years in finite set of documents (right)

the right side of Fig. 3 the temporal spread of topics related to the same document-set is visualized according to temporal values and temporal spread. Thereby "big data" is prevalent and has the highest score of all topics related to the finite document set.

4.3 Trend Identification and Forecasting

We extracted topics through LDA, labeled them through a statistical method, and enriched those topics with temporal data. If we tried to identify trends based on the frequency of the topics over the years, we would not get any appropriate trends. Nearly the number of all topics will increase through the years. This is because the number of publications increased in the last years dramatically. Table 1 illustrates the actual number of publications at the time of writing this paper in the DBLP database for every five years.

We worked out in our previous work [41] a five-year periodic model that measures through the slopes of the regression probabilities for the subsequent five periods (25 years) and validated the results with historical data. We started with the normalization of the topic frequencies. We, therefore, calculate for each year the normalized number of documents containing a topic. The normalized topic $\tilde{t}_y$ for a particular year y was calculated by dividing the occurrence of a topic t_y by the number of documents d_y in that year [41]. After having the normalized frequency of documents containing the topic, the entire years with documents with a specific $\tilde{t}$ are split into periods of a fixed length $x > 1$, limiting the length of the period to the time of the first occurrence of the topic, if necessary. So at the current year y_c, each period p_k covers the previous years $[y_c - x \cdot (k + 1), y_c - x \cdot k]$ [41]. For each period, we calculate the regression of the normalized topic frequencies and take the gradient (slope) as an indicator for the trend. Equation (1) calculates the slope for a topic t in a period p_k, based on the normalized topic frequencies $\tilde{t}_y$, where $\bar{t}$ is the mean of the normalized topic frequencies and $\bar{y}$ is the mean of years in the time period.

$$b_{\tilde{t},k} = \frac{\sum_{y \in p_k} (y - \bar{y}) \cdot (\tilde{t}_y - \bar{t})}{\sum_{y \in p_k} (y - \bar{y})^2} \tag{1}$$

Each calculated slope $b_{\tilde{t},k}$ is weighted through two parameters. The first parameter is the coefficient of determination R_k^2 of the regression. The second parameter is a weight ω_k that is determined with a function that decreases for earlier periods [41].

Table 1 Number of publications in DBLP every five year

	Documents in DBLP					
Years	1995	2000	2005	2010	2015	2020
Documents	13,775	38,908	111,022	201,245	365,426	507,375

This parameter was calculated through a linear function $\omega_k = max\left(0, 1 - \frac{k}{4}\right)$ and through an exponential function $\omega_k = \frac{1}{2^k}$ [41], whereas we found out that the linear function provides more reliable emerging trends.

The final weighting for a topic t is then computed from the slopes $b_{\tilde{i},k}$, the coefficients of determination $R_{\tilde{k}}^2$, and the weights ω_k of each of the K periods as follows:

$$\omega = \frac{1}{K} \cdot \sum_{i=1}^{K} b_{\tilde{i},k} w_k R_{\tilde{k}}^2 \tag{2}$$

Besides identifying emerging trends, we tested different methods to forecast topic-based trends based on historic trend evolution. The historical evolution was tested through various statistical methods, e.g., regression or ARIMA, and various machine learning methods to get the best mean prediction probability (MPP). We integrated "Dense" models, regression models, and a variety of neural networks (NN), including "Graph Neural Networks" [52]. The most appropriate results could be determined through LSTM-based [24] Recurrent Neural Networks (RNN) to forecast long periods. Thereby the "labeled topic" was used as an input variable to determine the forecasting quality. The forecasting is still work in progress. Through different parametrization, the quality could be significantly improved.

4.4 Data Modeling

The analysis process according to our general method requires the identification of various factors, e.g. key-player, competitors, or geographic spread. To meet these requirements, we integrated "aspect-oriented data models" that focus on certain aspects, e.g. temporal or geographical that are given in the data or are gathered through the step of "data enrichment". We generated five data models, "*Semantics Model*", "*Temporal Model*", "*Geographical Model*", "*Topic Model,*" and "*Trend Model*" [41]. Thereby, the "semantic data model" [39] serve as the primary data model for storing all information. It adds structure and relation between data elements to generate graphs from the data and visualize relations in the data set. This model is used in visual layouts of semantics structures, textual list presentation, and facet generation. A graph representation of data is generated containing all publications with their attributes and relations to accomplish this.

Multiple temporal visualizations use the temporal data model. Here, multiple aspects of the information in the data collection need to be accessible based on the time property. For the overview of the whole result set in a temporal spread, the temporal model needs to map publication years to the number of publications in a particular year. This temporal analysis is not only necessary for the entirety of the available result set, but it is also vital to analyze specialized parts of faceted aspects. Based on these faceted attributes, detailed temporal spreads need to be part of the temporal model for all attributes of each facet type. The temporal spread analysis needs to be available for each facet in the underlying data. With this information, temporal visualizations can be built more easily. These are then able to show a ranking over time or demonstrate comparisons of popularity over time. The temporal model allows us to measure trends and forecast possible future evolutions of trends through time-series analysis and machine learning approaches. The geographical data model contains geographical information of the available data. The complexity of this model is lower than that of the temporal model, as the geographical visualization only needs quantity information at the country level. This data model provides information about the origin country of the authors of publications. Although the data are enriched with information from different databases as described, there are many data entities without the country information. To face this problem, we introduced two approaches, if no country information could be gathered: (1) we take the affiliation of the authors for gathering the country, and (2) we take publications from the same author from the same discipline based on the extracted topics and the same year plus and minus one to estimate the country. The year of publication is important since many researchers change the affiliation and with the affiliation the country. The topic model contains detailed information about the generated topics as described above. The semantic model contains publications with all assigned properties and relations, including topics, and the topic model supplements this data by offering insights into the assigned topics. Like already mentioned in Sect. 4.2, the information about each topic contains the top 20 most used words and phrases with the assigned probability of usage for each word (see Fig. 3). The inclusion of the most used phrases can help the user immensely in the reformulation of the search query to find additional information on topics of interest. Nevertheless, the primary purpose of the topic model is to gather relevant information about technological developments and the used approaches within a development. The topic model is commonly correlated to the temporal model and also provides the temporal spread of topics. Figure 3-right illustrates the temporal spread of topics related to the search term "Visual Analytics". The trend model is generated through the trend identification process described in Sect. 4.3 in combination with the temporal model. It illustrates the main trends either as an overview of "top trends" identified through the described weight calculation or after a performed query. In the second case, the same procedure is applied with the difference that the document corpus is not the entire database but only the results referring to the queried term.

4.5 Visual Structure

The visual structure enables a fully automatic selection of visual representations based on the underlying data model. We applied the procedure of visual adaptation according to our previews work [44] with the three steps of *semantics*, *visual layout*, and *visual variable*. As proposed in [44], we start the visual transformation for generating a visual structure with the semantics layer. Thus our system is not yet adaptive. We investigate the data characteristics for choosing an appropriate visual layout. Based on the chosen visual layout, we identified visual variables according to Bertin [4] that are applied to a certain visual layout. This procedure allows us to enhance the system with adaptive behavior and reduces the complexity of integrating new visualizations.

5 Visual Analytics for Decision Making

We described in the previous section the overall procedure of gathering data, extracting topics and trends, and modeling data to enable an interactive visual approach for decision making. We will introduce the user interface of our system in this section, including some visualizations that enable the process of decision-making and the interaction design of our visual analytics system.

5.1 User Interface Design

The user interface (UI) of our visual analytics system consists of four areas (see Fig. 4). All areas are dynamic, particularly in terms of data and data entity selection. The top area (1) provides search functionalities, including an advanced search for dedicated search in certain fields and "assisted search". The "assisted search" enhances users' query based on the resulted top five phrases and words of the top-ranked topic [43]. This allows extending the search with similar and related topics to the formulated query. The top area provides, besides search functionalities, the choice of all databases and color schemes for the entire user interface and all visual-

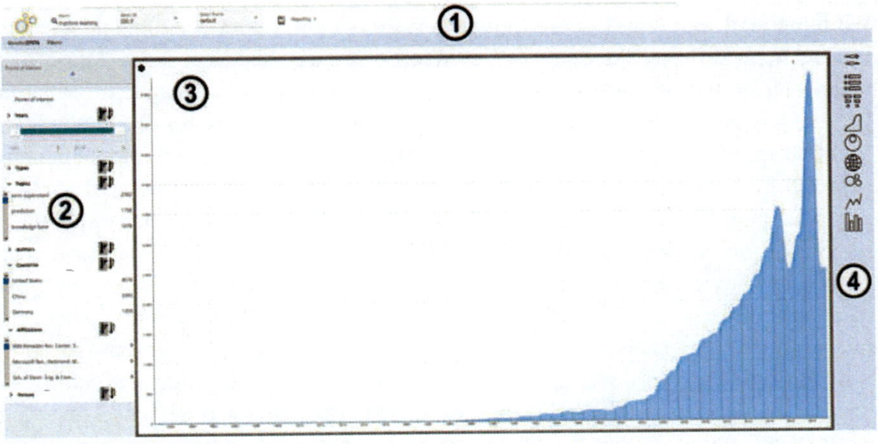

Fig. 4 The UI of our system with its four areas

izations. This is enhanced with a "reporting functionality" to enable the generation of reports for decision-makers who do not interact with the system.

At the left (2), the facets of the underlying data are generated and illustrated automatically. This area also includes the number of results that are automatically adapted to the selected facets, a logical facet selection, and the "graphical search" functionality (see Fig. 12). The logical facet selection allows users to reduce the amount of the results to get the most appropriate documents for a specific task and provide beside a visual interaction an overview of data entities for reducing or enhancing the visualized results.

In the center area (3), the main visualization(s) are placed that are either automatically selected by the type of data, the search query (see Sect. 4.4), or by the user himself in the right area (4), where a dynamic set of visualization are available based on the data and their structure. In Fig. 4 the temporal overview of the entire data is visualized. This area allows the placement of more than one visualization to generate visual dashboards.

The right area (4) provides the functionality to choose either one visualization as illustrated in Fig. 4 or create through drag and drop an arrangement of several visualizations. This area shows icons of visualizations that are supported through the underlying data. For example, if geographic data are provided through country names or longitudes and latitudes, an icon for geographical visualization appears. The according visualizations are related to the data that should be visualized. In the case of a query, these data are the results of the query. If the result-set does not provide any geographical or semantic information, the according visual layout disappears from that area.

5.2 *Visual Representations and Visual Interaction*

We integrated into our general approach various data models that enable users to interact with different aspects of the underlying data (see Sect. 4.4,). These data models allow us to provide several interactive visual layouts that enable information gathering from different perspectives and support decision-making. We applied two different approaches for interacting with visualizations. Thereby, two complementary approaches were integrated to support users' information acquisition and analysis process. First, the *information-seeking mantra* by Shneiderman [53] is applied to provide an overview followed by zoom and filter and then details on demand. This procedure allows seeing either emerging trends or the most recent search terms typed by other users as a "word cloud" to gather first information of the underlying data and interact until they reached their intended goals. Second, the approach proposed by van Ham and Perer [56] was applied that enables searching the database, getting the context, and get more detailed information. This approach was designed for graph exploration. We think that the entire process of interacting with visualizations can profit from this approach due to its complementary interacting process compared to the overview-first approach. Figure 5 illustrates the start screen of our system

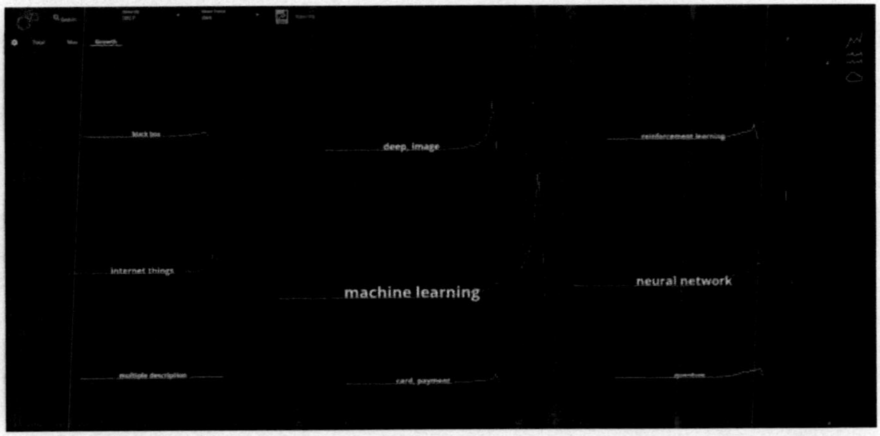

Fig. 5 The initial starting screen of our system applying "SparkClouds" for emerging trends in the entire data-base. Each item is selectable and leads to a search

with an overview on emerging topics as "SparkClouds". Thereby a different color scheme was chosen. The starting screen after a search or choice of the initial screen is illustrated in Fig. 4.

A simple temporal visual layout for overview purposes that visualizes the number of documents over the years of the entire search results is illustrated in Fig. 4. The visualization can be enhanced with several statistical values to allow a more detailed view of the data. These include linear and Loess regression (locally estimated scatter-plot smoothing), neighborhood-weighting through color and values, and minimum, average, and maximum values. Figure 6 illustrates such a temporal visualization. Thereby the term "machine learning" was chosen as a search term through the initial start screen, and the mouse hovers the year 2017.

Temporal visual layouts that illustrate the temporal topic distribution of the according documents over the years instead of the numbers of documents use the two data models, the topic model, and the temporal model. Our temporal topic river is such a visualization and separates all the topics and trends for a more comprehensible view. Instead of layering (stacking) the items on top of each other with no space between them, we represent each facet item with a "river". Each river has a center-line and a uniform expansion to each side based on frequency distribution over time. Additionally, placing multiple rivers next to each other makes spotting differences in temporal data sets straightforward. Tasks like comparing the impact of various authors, topics, or trends on a search term become easier. Figure 7 illustrates our topic river for two different data sets of the same database. We arranged the visual layouts on top of each other. The above topic river illustrates the temporal spread of the topics, and the river on the bottom illustrates the temporal spread of publications of those countries that published the most works in the area of machine learning.

For analyzing trends, it is crucial to gather the knowledge of the underlying topics, technologies, etc. emerged during the time or lost their relevance. To enable a fast

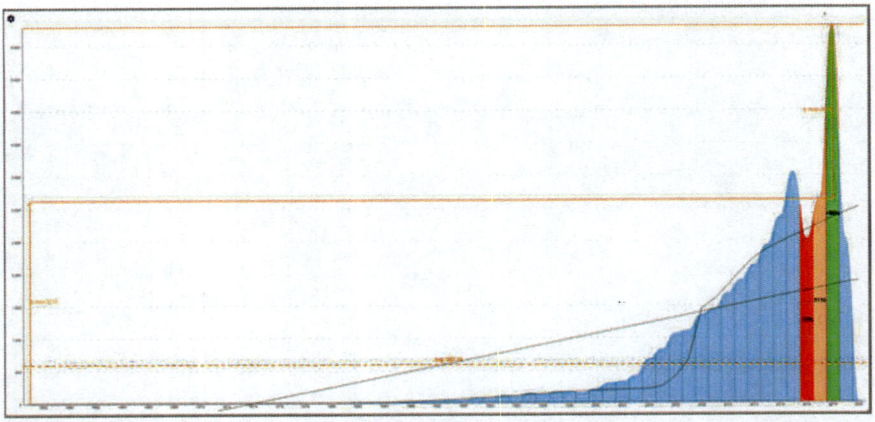

Fig. 6 Statistical values in simple visual layouts to gain more insights. A chosen year is illustrated in orange, increased numbers of publications in green and decreased in red. The illustration contains further statistical values, e.g., the LOESS-Regression, the linear regression etc.

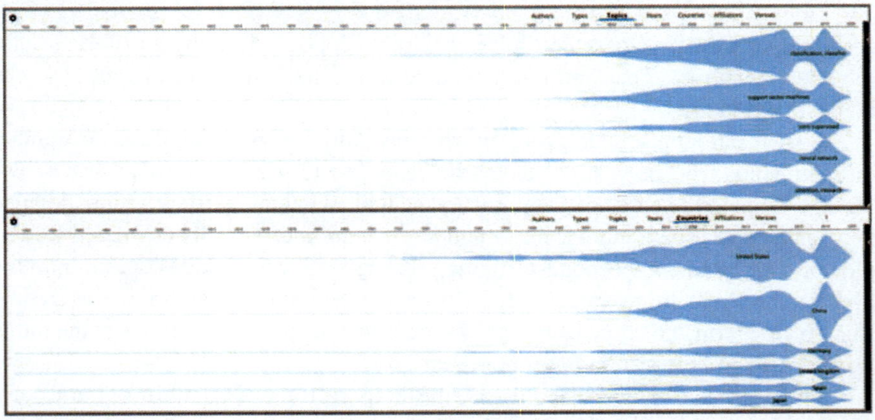

Fig. 7 Stacked river: on top, a topic river illustrates the temporal topic spread and the temporal spread of publications per country on the bottom. Thereby, the number of topics published in a particular year or from a specific country is taken for measuring the spread

and comprehensible analysis, we integrated a temporal ranking (see Fig. 8). Besides the introduced configuration areas, this visual layout offers the ability to specify the number of rows to be visualized. The visual layout is divided horizontally into columns for each year of the analyzed time period. The arrangement is based on the number of publications having a facet item as a property of the selected facet type, sorted in descending order from top to bottom.

The order only represents the ranking. The width of each rectangle represents additional, more concrete information about the relative amount. With these position and form indicators, the user can quickly determine facet items with high influences

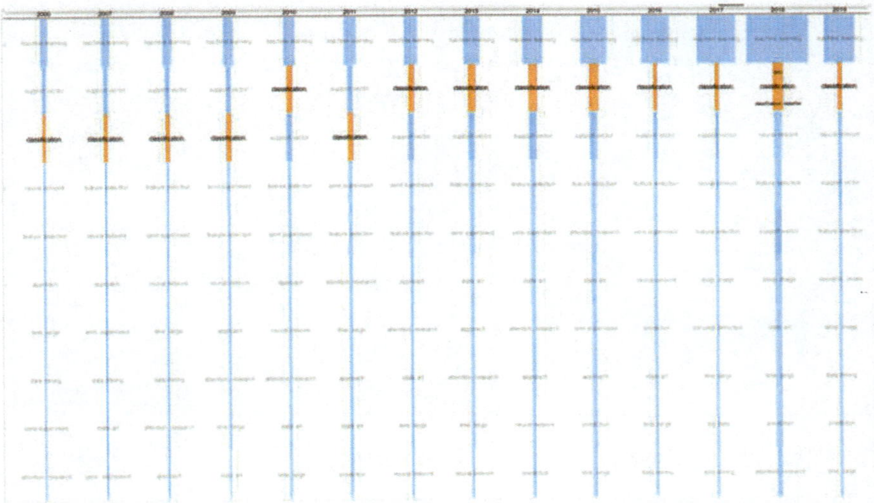

Fig. 8 Temporal spread of topics and the related words and phrases as columns

each year. In Fig. 8 the temporal data model, the topic model, and the semantic model are merged to get the visualized information. Thereby, the top related topics of documents from the result set of the query "machine learning" are visualized, and by selecting one topic, the temporal ranking for each year is highlighted.

For a non-temporal visualization of topic distribution, we have integrated a set of various visualizations. This set should allow to figure out which topics are prevalent in a particular database or a data-set as a result of a search query. The main focus lies on the general occurrence of topics instead of the temporal spread. We reduce the temporal information to enable a more focused view of the topics and their distribution. Figure 9 illustrates a dashboard of two identical topic distributions, left as a pie-chart and right as a bar chart. The simple views allow gathering non-temporal information faster if this is necessary for a certain analytical task.

Our system provides a semantic visual layout that offers the ability to understand relations between authors (co-author information) and between topics (which topic relations) and the semantic correlation between the information entries. Commonly semantic relations are visualized with node-link graphs, leading to complex visualizations and reducing the analysis capability. We integrated beside such node-link visualizations a circle layout that arranges the entities as a spiral starting from the center of the screen. The size of each element indicates the amount of publication per author, whereas the degree option indicates the number of distinct relation targets within the facet type.

The semantic data model provides detailed relational information about individual facet items, which can be accessed through user interaction. After selecting a circle, all relational information within the same facet type is highlighted. This leads to a real-time loading of all co-authors in the result set. Further, users are able to get

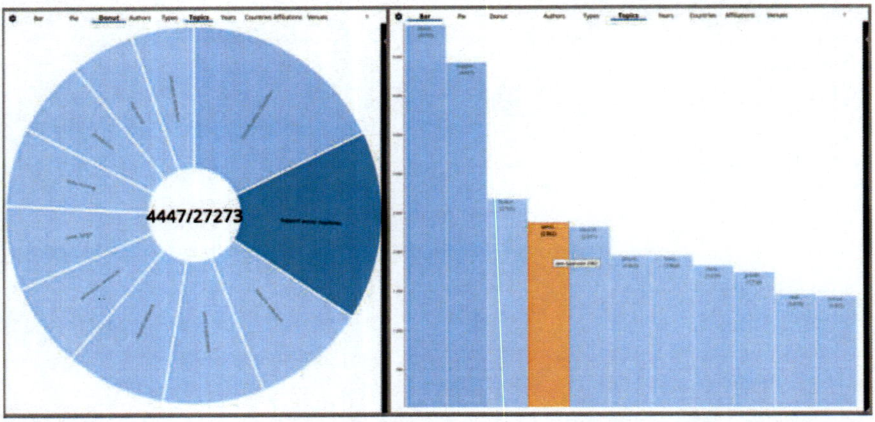

Fig. 9 Examples for non-temporal topic-distribution visualizations

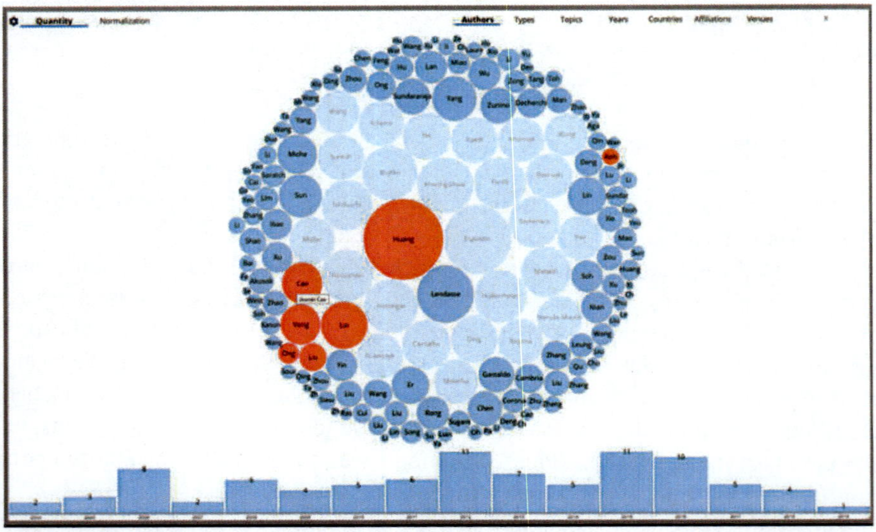

Fig. 10 Co-authorship with relations temporal information

an insight about correlations within the semantic relations through mouse-over that refines the in the figured cased the co-authorship of certain authors through the color. We integrated the disambiguation of authors' names since authors from the same discipline could have the same surname. To distinguish the authors, we compare the first names, the affiliations (with relation to the year), the country, and the co-authorship. Figure 10 illustrates the co-authorship relations based on the search term "machine learning". At first glance, the users can see the most publishing visual layout that visualizes topics related to search terms based on the frequency of their appearance.

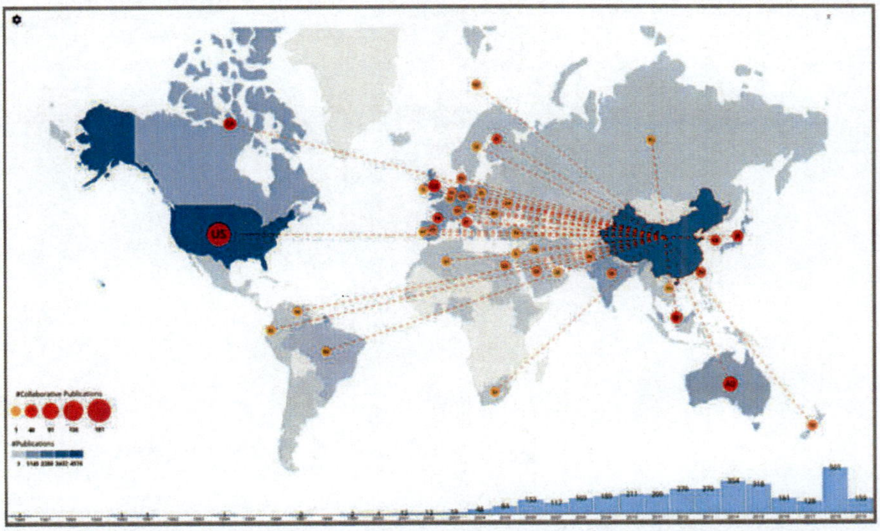

Fig. 11 Geographical search based on the geographical, semantic and temporal data models

Besides temporal, semantic, and topic visualizations, we integrated an advanced geographic visualization that illustrates the number of publications on certain search-term on the country level through saturation. Besides this, the visual interface makes use of the semantic and temporal data models. Figure 11 illustrates the geographical result-set of the search-term "machine learning". Thereby the user clicked on China to see the temporal spread of publications from China and the relations to other countries. This relation is measured through the co-authors of the publications of Chinese authors and the origin countries of the co-authors. Figure 11 illustrates on the bottom the temporal spread of publications of which at least one author's origin country is China. The legend of the visualization shows that the United States published the highest number, and China is primarily collaborating with the United States on "machine learning".

5.3 *Visual Search*

We integrated a "visual search" or graphical search to enable a more advanced search and analysis process. The visual search functionality allows users to formulate terms that are relevant for them, create so-called visual *points-of-interests* (POI) and see at a glance the number of documents that contain the created points of interest. In Fig. 12 the user searched for "machine learning" and created several visual POIs. The defined POIs are visualized on the right side and can be included in the main search term "machine learning" per drag and drop. The color is the indicator for a certain POI and allows users to see how many publications are in the database with

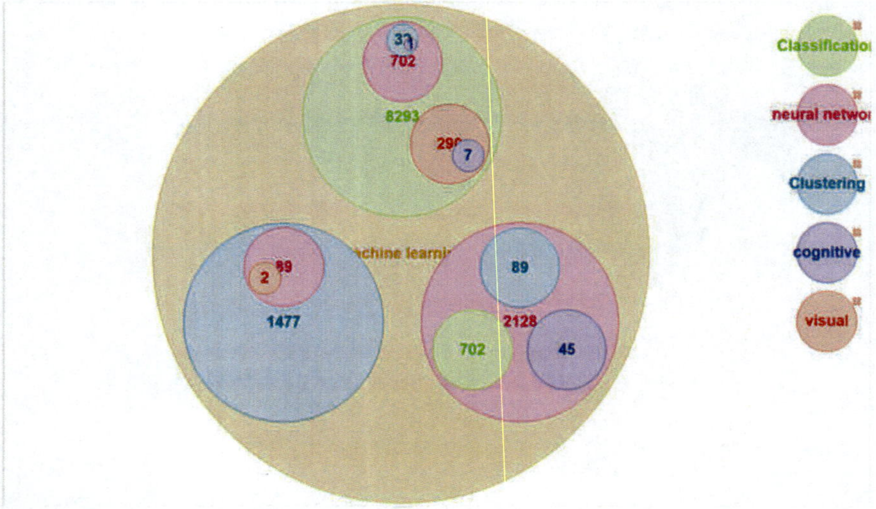

Fig. 12 The visual search for enabling a search within the result set

the created POIs. The number represents the search result quantity within the search result set, so the user is able to define and redefine such POIs for his purposes. In Fig. 12 the main search term is "machine learning". The user is able to see at a glance the results for machine learning documents containing the "classification", "neural network", "clustering", "cognitive" or "visual". With a nested method, they can see that 8293 publications contain the phrase "classification" 702 of these publications include neural networks. Within this search result, 32 publications contain the term "clustering", and one of these publications is related to visual. The users are able just to double click within the circles and get the list of the documents. In the case of machine learning combined with classification, neural networks, clustering, and visual, there is just one publication, so that the list will provide just one publication. This way of interaction allows not only reduces the search result set, but it also gives an overview of the number of related topics, and the number of those relations [40].

5.4 *Reporting*

We integrated into our system a reporting functionality that enables users to generate reports for non-analysts. This can be performed in two different ways. Each visualization can be saved with the parameters defined by users in our integrated reporting tool. These visualizations are stored permanently to enable the generation of a report. Users might want to integrate the entire dashboard for a certain report. For that reason, they are able to capture the entire dashboard for reporting issues. The reporting functionality aims at creating reports for presentations or reports for

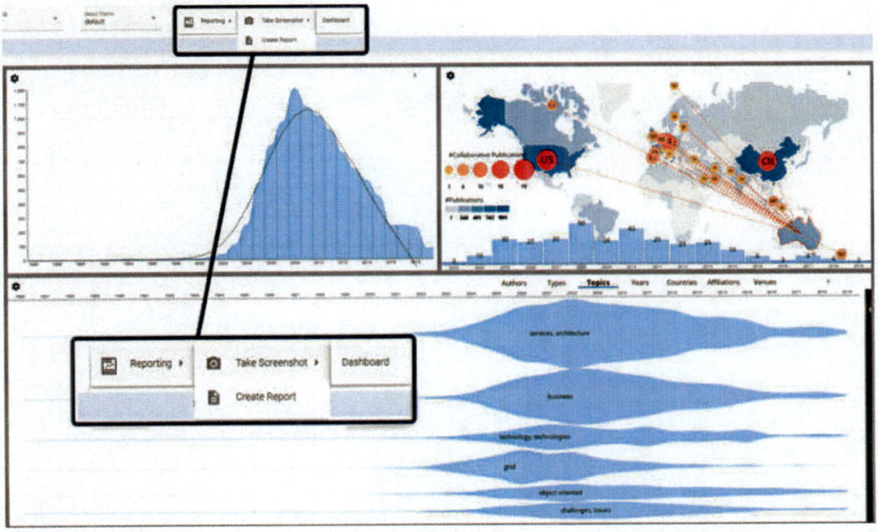

Fig. 13 Capturing the entire dashboard for reporting

decision-makers on the management level. There is no need for decision-makers to interact with the system. Besides this, the reporting functionality creates a "snapshot" at a specific time. Since the data are changing, the reports can be used to validate hypotheses.

For creating a report, first, the figure is stored either through the user interface for the entire screen, regardless if there are one or N visualizations placed on the screen, or through the reporting button on each visualization. Figure 13 illustrates the creation of such a dashboard-capture. The user searched for the term "service oriented architecture" and placed three visualizations on the dashboard. Through the capturing functionality, this dashboard is saved as an image.

After capturing all required visualizations, the users are able to generate a report through the reporting tool. Thereby a web-based HTML editor is provided with predefined text snippets, e.g., for timestamps, search queries, or parameterizations of the visualization. All saved images and text snippets can be found in the reporting tool of our system.

Figure 14 illustrates the reporting tool of our system. On the right side, the HTML editor allows creating the reports and write comments. On the left side, there are the captured figures and text snippets. With just one click, the users are able to generate a report. The dashboard in Fig. 14 illustrates clearly that there is a decreasing trend of using such kinds of web service architectures. Particularly grids were not used in last years. We currently provide exports as PDF, Word documents, and images. We will integrate the raw data used in the reports as JSON and CSV to enable a simple exchange of the data for several systems.

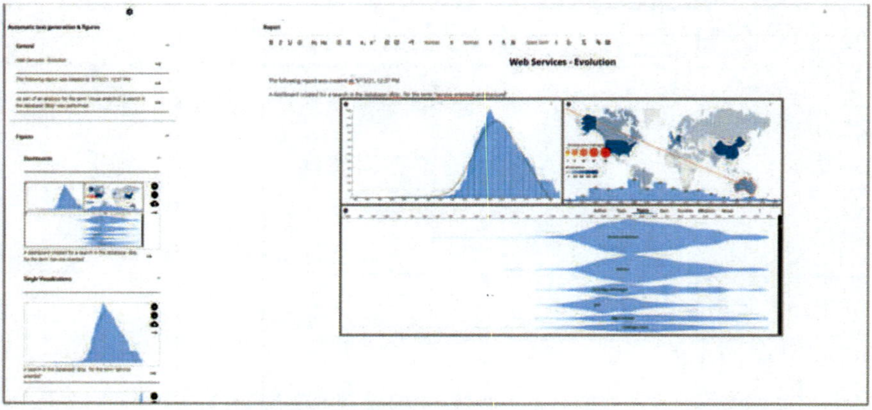

Fig. 14 Generating a report

6 Conclusions

We proposed in this paper a new approach for Visual Analytics for decision making by incorporating some main ideas from innovation management. We first gave an extensive literature review of existing approaches and systems that are mining and visualizing trends. The literature review revealed the missing inclusion of management and decision-making approaches in such analytical systems, particularly for technology management. Our general model filled this gap by a first attempt and combined an appropriate model to enable strategic decision-making. The outcome is a model with three main steps for integrating innovations in firms. This model is enhanced by a more technical approach that illustrated the process of Visual Analytics and the main steps of our approach. Our main contribution is an advanced Visual Analytics approach and system based on our previous work [41] that enables decision making for technology and innovation management with advanced methods in terms of analysis and integrates the ideas of technology management in a Visual Analytics system.

Acknowledgements This work was conducted within the research group on Human-Computer Interaction and Visual Analytics (https://vis.h-da.de).

References

1. Agrawal, R., Srikant, R.: Mining sequential patterns. In: Proceedings of the Eleventh International Conference on Data Engineering (1997)
2. An, J., Kim, K., Mortara, L., Lee, S.: Deriving technology intelligence from patents: preposition-based semantic analysis. J. Inf. **12**(1), 217–236 (2018). https://doi.org/10.1016/j.joi.2018.01.001

3. Armstrong, J.S.: Forecasting by extrapolation: conclusions from 25 years of research. Interfaces **14**(6), 52–66 (1984). https://doi.org/10.1287/inte.14.6.52
4. Bertin, J.: Semiology of graphics. University of Wisconsin Press (1983)
5. Blackman, A.W.: A mathematical model for trend forecasts. Technol. Forecast. Soc. Change **3**, 441–452 (1971). https://doi.org/10.1016/s0040-1625(71)80031-8
6. Blei, D.M., Ng, A.Y., Jordan, M.I.: Latent dirichlet allocation. J. Mach. Learn. Res. **3** (2003). http://www.jmlr.org/papers/v3/blei03a.html
7. Chen, C.C., Chen, Y.T., Sun, Y., Chen, M.C.: Life cycle modeling of news events using aging theory. In: Machine Learning ECML 2003, pp. 47–59. Springer, Berlin (2003)
8. Cho, Y., Daim, T.: OLED TV technology forecasting using technology mining and the fisher-pry diffusion model. Foresight **18**(2), 117–137 (2016). https://doi.org/10.1108/fs-08-2015-0043
9. Choi, S., Kim, H., Yoon, J., Kim, K., Lee, J.Y.: An SAO-based text-mining approach for technology roadmapping using patent information. R&D Manag. **43**(1), 52–74 (2012). https://doi.org/10.1111/j.1467-9310.2012.00702.x
10. Collins, C., Viegas, F., Wattenberg, M.: Parallel tag clouds to explore and analyze faceted text corpora. In: VAST 2009 (2009). https://doi.org/10.1109/VAST.2009.5333443
11. Deerwester, S., Dumais, S.T., Furnas, G.W., Landauer, T.K., Harshman, R.: Indexing by latent semantic analysis. J. Am. Soc. Inf. Sc. **41**(6), 391–407 (1990)
12. Dou, W., Wang, X., Chang, R., Ribarsky, W.: Paralleltopics: a probabilistic approach to exploring document collections. In: VAST 2011 (2011). https://doi.org/10.1109/VAST.2011.6102461
13. Ernst, H.: Patent information for strategic technology management. World Pat. Inf. **25**(3), 233–242 (2003). https://doi.org/10.1016/S0172-2190(03)00077-2
14. Feldman, R., Dagan, I.: Knowledge discovery in textual databases (KDT). In: Proceedings of the First International Conference on Knowledge Discovery and Data Mining (1995)
15. Feldman, R., Aumann, Y., Zilberstein, A., Ben-Yehuda, Y.: Trend graphs: visualizing the evolution of concept relationships in large document collections. In: Zytkow, J.M., Quafafou, M. (eds.) Principles of Data Mining and Knowledge Discovery, pp. 38–46. Springer, Berlin (1998)
16. Fergnani, A.: Corporate foresight: A new frontier for strategy and management. Acad. Manag. Perspect. **0**(0) (2020). https://doi.org/10.5465/amp.2018.0178
17. Fisher, J., Pry, R.: A simple substitution model of technological change. Technol. Forecast. Soc. Change **3**, 75–88 (1971). https://doi.org/10.1016/s0040-1625(71)80005-7
18. Fourt, L.A., Woodlock, J.W.: Early prediction of market success for new grocery products. J. Mark. **25**(2), 31 (1960). https://doi.org/10.2307/1248608
19. Glance, N.S., Hurst, M., Tomokiyo, T.: Blogpulse: automated trend discovery for weblogs. In: In WWW 2004 WS on Weblogging. ACM (2004)
20. Gordon, A.V., Ramic, M., Rohrbeck, R., Spaniol, M.J.: 50 years of corporate and organizational foresight: looking back and going forward. Technol. Forecast. Soc. Change **154**, (2020). https://doi.org/10.1016/j.techfore.2020.119966
21. Han, Q., Heimerl, F., Codina-Filba, J., Lohmann, S., Wanner, L., Ertl, T.: Visual patent trend analysis for informed decision making in technology management. World Pat. Inf. **49**, 34–42 (2017). http://www.sciencedirect.com/science/article/pii/S0172219017300455. https://doi.org/10.1016/j.wpi.2017.04.003
22. Havre, S., et al.: Themeriver: visualizing thematic changes in large document collections. IEEE TVCG **8**(1), 9–20 (2002)
23. Heimerl, F., Han, Q., Koch, S., Ertl, T.: Citerivers: visual analytics of citation patterns. IEEE Trans. Vis. Comput. Graph. **22**(1), 190–199 (2016). https://doi.org/10.1109/TVCG.2015.2467621
24. Hochreiter, S., Schmidhuber, J.: Long short-term memory. Neural Comput. **9**(8), 1735–1780 (1997). https://doi.org/10.1162/neco.1997.9.8.1735
25. Hurtado, J.L., Agarwal, A., Zhu, X.: Topic discovery and future trend forecasting for texts. J. Big Data **3**(1), 7 (2016)

26. Joho, H., Azzopardi, L.A., Vanderbauwhede, W.: A survey of patent users: an analysis of tasks, behavior, search functionality and system requirements. In: Proceedings of the Third Symposium on Information Interaction in Context, IIiX '10, pp. 13–24. ACM, New York, USA (2010). https://doi.org/10.1145/1840784.1840789
27. Kayser, V., Blind, K.: Extending the knowledge base of foresight: the contribution of text mining. Technol. Forecast. Soc. Change **116**, 208–215 (2017). https://doi.org/10.1016/j.techfore.2016.10.017
28. Kayser, V., Goluchowicz, K., Bierwisch, A.: Text mining for technology roadmapping—the strategic value of information. Int. J. Innov. Manag. **18**(03), 1440004 (2014). https://doi.org/10.1142/S1363919614400040
29. Keim, D., Kohlhammer J., Ellis G., Mansmann, F. (eds.): Mastering the Information Age: Solving Problems with Visual Analytics. Goslar, Eurographics Association (2010)
30. Lee, S., Park, Y.: Customization of technology roadmaps according to roadmapping purposes: overall process and detailed modules. Technol. Forecast. Soc. Change **72**(5), 567–583 (2005)
31. Lee, B., Riche, N.H., Karlson, A.K., Carpendale, S.: Sparkclouds: visualizing trends in tag clouds. IEEE TVCG **16** (2010). https://doi.org/10.1109/TVCG.2010.194
32. Lent, B., Agrawal, R., Srikant, R.: Discovering trends in text databases. In: Proceedings of KDD '97 (1997)
33. Li, S., Garces, E., Daim, T.: Technology forecasting by analogy-based on social network analysis: the case of autonomous vehicles. Technol. Forecast. Soc. Change **148**, (2019). https://doi.org/10.1016/j.techfore.2019.119731
34. Li, X., Xie, Q., Jiang, J., Zhou, Y., Huang, L.: Identifying and monitoring the development trends of emerging technologies using patent analysis and twitter data mining: the case of perovskite solar cell technology. Technol. Forecast. Soc. Change **146**, 687–705 (2019). https://doi.org/10.1016/j.techfore.2018.06.004
35. Liu, S., et al.: Tiara: interactive, topic-based visual text summarization and analysis. ACM Trans. Intell. Syst. Technol. **3**(2), 1–28 (2012). https://doi.org/10.1145/2089094.2089101
36. Lohmann, S., Burch, M., Schmauder, H., Weiskopf, D.: Visual analysis of microblog content using time-varying co-occurrence highlighting in tag clouds. In: Proceedings of the International Working Conference on Advanced Visual Interfaces, AVI '12, pp. 753–756. ACM, New York, USA (2012). https://doi.org/10.1145/2254556.2254701
37. Muhlroth, C., Grottke, M.: Artificial intelligence in innovation: how to spot emerging trends and technologies. IEEE Trans. Eng. Manag. 1–18 (2020). https://doi.org/10.1109/tem.2020.2989214
38. Mühlroth, C., Grottke, M.: A systematic literature review of mining weak signals and trends for corporate foresight. J. Bus. Econ. **88**(5), 643–687 (2018). https://doi.org/10.1007/s11573-018-0898-4
39. Nazemi, K., Burkhardt, D., Retz, R., Kuijper, A., Kohlhammer, J.: Adaptive visualization of linked-data. In: Advances in Visual Computing, pp. 872–883. Springer (2014)
40. Nazemi, K., Burkhardt, D.: A visual analytics approach for analyzing technological trends in technology and innovation management. In: Advances in Visual Computing, pp. 283–294. Springer International Publishing (2019)
41. Nazemi, K., Burkhardt, D.: Visual analytics for analyzing technological trends from text. In: 2019 23rd International Conference Information Visualisation (IV). IEEE (2019). https://doi.org/10.1109/iv.2019.00041
42. Nazemi, K., Klepsch, M., Burkhardt, D., Kaupp, L.: Comparison of full-text articles and their-corresponding abstracts for visual trend analytics. In: Proceedings of the 24rd International Conference Information Visualisation. IEEE (2020). (To appear)
43. Nazemi, K., Retz, R., Burkhardt, D., Kuijper, A., Kohlhammer, J., Fellner, D.W.: Visual trend analysis with digital libraries. In: Proceedings of the 15th International Conference on Knowledge Technologies and Data-Driven Business—i-KNOW'15. ACM Press (2015). https://doi.org/10.1145/2809563.2809569
44. Nazemi, K.: Adaptive semantics visualization. In: Studies in Computational Intelligence, vol. 646. Springer International Publishing (2016). http://www.springer.com/de/book/9783319308159. https://doi.org/10.1007/978-3-319-30816-6

45. Nguyen, K.: Hot topic detection and technology trend tracking for patents utilizing term frequency and proportional document frequency and semantic information. In: 2016 International Conference on Big Data and Smart Computing (BigComp), pp. 223–230. IEEE (2016). https://doi.org/10.1109/BIGCOMP.2016.7425917
46. Otte, E., Rousseau, R.: Social network analysis: a powerful strategy, also for the information sciences. J. Inf. Sci. **28**(6), 441–453 (2002). https://doi.org/10.1177/016555150202800601
47. Pépin, L., Kuntz, P., Blanchard, J., Guillet, F., Suignard, P.: Visual analytics for exploring topic long-term evolution and detecting weak signals in company targeted tweets. Comput. Ind. Eng. **112**, 450–458 (2017)
48. Phaal, R., Farrukh, C.J., Probert, D.R.: Technology roadmapping–a planning framework for evolution and revolution. Technol. Forecast. Soc. Change **71**(1), 5–26 (2004)
49. Phaal, R., Farrukh, C.J., Probert, D.R.: Technology roadmapping–a planning framework for evolution and revolution. Technol. Forecast. Soc. Change **71**(1–2), 5–26 (2004). https://doi.org/10.1016/s0040-1625(03)00072-6
50. Rohrbeck, R., Kum, M.E.: Corporate foresight and its impact on firm performance: a longitudinal analysis. Technol. Forecast. Soc. Change **129**, 105–116 (2018). https://doi.org/10.1016/j.techfore.2017.12.013
51. Salton, G., Buckley, C.: Term-weighting approaches in automatic text retrieval. Inf. Process. Manag. **24**(5), 513–523 (1988). https://doi.org/10.1016/0306-4573(88)90021-0
52. Scarselli, F., Gori, M., Tsoi, A.C., Hagenbuchner, M., Monfardini, G.: Computational capabilities of graph neural networks. IEEE Trans. Neural Netw. **20**(1), 81–102 (2009). https://doi.org/10.1109/TNN.2008.2005141
53. Shneiderman, B.: The eyes have it: a task by data type taxonomy for information visualizations. In: VL, pp. 336–343 (1996)
54. Stefanowski, J., Weiss, D.: Carrot2 and language properties in web search results clustering. In: Advances in Web Intelligence, pp. 240–249. Springer, Berlin (2003)
55. Toutanova, K., Manning, C.D.: Enriching the knowledge sources used in a maximum entropy part-of-speech tagger. In: EMNLP '00: Proceedings of the 2000 Joint SIGDAT Conference on Empirical Methods in Natural Language Processing and Very Large Corpora: Held in Conjunction with the 38th Annual Meeting of the Association for Computational Linguistics, EMNLP '00, p. 63-70. Association for Computational Linguistics, USA (2000). https://doi.org/10.3115/1117794.1117802
56. Van Ham, F., Perer, A.: Search, show context, expand on demand: supporting large graph exploration with degree-of-interest. IEEE Trans. Vis. Comput. Graph. **15**, 953–960 (2009)
57. Webster, Watson: Analyzing the past to prepare for the future: writing a literature review. MIS Q. **26**(2), 13–24 (2002)
58. World Intellectual Property Organization: Guide to the International Patent Classification. Electronic Print (2019). (Version 2019)
59. Yang, C., Zhu, D., Zhang, G.: Semantic-based technology trend analysis. In: 2015 10th International Conference on Intelligent Systems and Knowledge Engineering (ISKE). IEEE (2015). https://doi.org/10.1109/iske.2015.43
60. Yoon, B., Park, Y.: A text-mining-based patent network: analytical tool for high-technology trend. J. High Technol. Manag. Res. **15**(1), 37–50 (2004). https://doi.org/10.1016/j.hitech.2003.09.003
61. Zhang, Y., Porter, A.L., Hu, Z., Guo, Y., Newman, N.C.: Term clumping for technical intelligence: a case study on dye-sensitized solar cells. Technol. Forecast. Soc. Change **85**, 26–39 (2014). https://doi.org/10.1016/j.techfore.2013.12.019
62. Zhang, Y., Zhang, G., Chen, H., Porter, A.L., Zhu, D., Lu, J.: Topic analysis and forecasting for science, technology and innovation: methodology with a case study focusing on big data research. Technol. Forecast. Soc. Change **105**, 179–191 (2016). https://doi.org/10.1016/j.techfore.2016.01.015

Deep Learning Image Recognition for Non-images

Boris Kovalerchuk, Divya Chandrika Kalla, and Bedant Agarwal

Abstract Powerful deep learning algorithms open an opportunity for solving non-image Machine Learning (ML) problems by transforming these problems into the image recognition problems. The CPC-R algorithm presented in this chapter converts non-image data into images by visualizing non-image data. Then deep learning CNN algorithms solve the learning problems on these images. The design of the CPC-R algorithm allows preserving all high-dimensional information in 2-D images. The use of pair values mapping instead of single value mapping used in the alternative approaches allows encoding each n-D point with 2 times fewer visual elements. The attributes of an n-D point are divided into pairs of its values and each pair is visualized as 2-D points in the same 2-D Cartesian coordinates. Next, grey scale or color intensity values are assigned to each pair to encode the order of pairs. This is resulted in the heatmap image. The computational experiments with CPC-R are conducted for different CNN architectures, and methods to optimize the CPC-R images showing that the combined CPC-R and deep learning CNN algorithms are able to solve non-image ML problems reaching high accuracy on the benchmark datasets. This chapter expands our prior work by adding more experiments to test accuracy of classification, exploring saliency and informativeness of discovered features to test their interpretability, and generalizing the approach.

Keywords Compute vision · Convolutional neutral networks · Deep learning · Machine learning · Raster images · Visualization · Non-image data · Data conversion

B. Kovalerchuk (✉) · D. C. Kalla
Department of Computer Science, Central Washington University, Ellensburg, WA, USA
e-mail: borisk@cwu.edu

D. C. Kalla
e-mail: DivyaChandrika.Kalla@cwu.edu

B. Agarwal
Indian Institute of Technology Kharagpur, Kharagpur, West Bengal, India

B. Kovalerchuk et al. (eds.), *Integrating Artificial Intelligence and Visualization for Visual Knowledge Discovery*, Studies in Computational Intelligence 1014,
https://doi.org/10.1007/978-3-030-93119-3_3

1 Introduction

The success in solving image recognition problems by Deep Learning (DL) algorithms is very evident. Moreover, DL architectures designed for some types of images have been efficient for other types of images. This chapter expands such knowledge transfer opportunity by converting non-image data to images by using visualization of non-image data. It can solve a wide variety of Machine Learning problems [1, 2] by converting a non-image classification task into the image recognition task and solving it by efficient DL algorithms. This chapter expands [3] as follows: (1) adding more experiments with more benchmark datasets to test accuracy of classification, (2) exploring saliency and informativeness of discovered features to test their interpretability, and (3) generalizing the approach.

This chapter starts from the concepts of single value and pair value mappings. Sections 2, 3 and 4 present result of experiments with Wisconsin breast cancer data (WBC), Swiss roll data, Ionosphere, Glass, and Car data. Section 5 presents results with Saliency maps and Section. 6 presents results with Informative cells feature importance models. Section 7 explores advantages of Spiral filling of colliding pairs. Section 8 presents results for frequencies of informative cells. Section 9 presents comparisons with other studies. Generalization of CPC-R methodology, future work and conclusion are presented in Sects. 10 and 11.

1.1 Single Value Mapping

In a *single value mapping* [3], each *single* value a_i of the n-D point $\mathbf{a} = (a_1, a_2, \ldots, a_n)$ is mapped by function F to a cell/pixel p_i with coordinates (p_{i1}, p_{i2}), $F(a_i) = (p_{i1}, p_{i2})$. The coordinate value a_i is assigned as the intensity of pixel p_i. Before the value a_i can be normalized to be an intensity value, $I(p_i) = a_i$. Next, we define the *mapping density* relative to the image size. For $n = 3$, consider the image of size 2 × 2 pixels with mapping a_1 to pixel (1, 1), a_2 to pixel (1, 2) and a_3 to pixel (2, 1) with pixel (2, 2) left empty (see Fig. 1a, b). Such sequential mapping is a dense mapping, e.g., it needs only 10 × 10 pixels for 100-D point $\mathbf{a}$. The use of a larger 3 × 3 image with $F(a_1) = (1, 1)$, $F(a_2) = (2, 2)$, and $F(a_3) = (3, 3)$ leads to density 3/9, where these 3 pixels contain relevant information. See Fig. 1c, d.

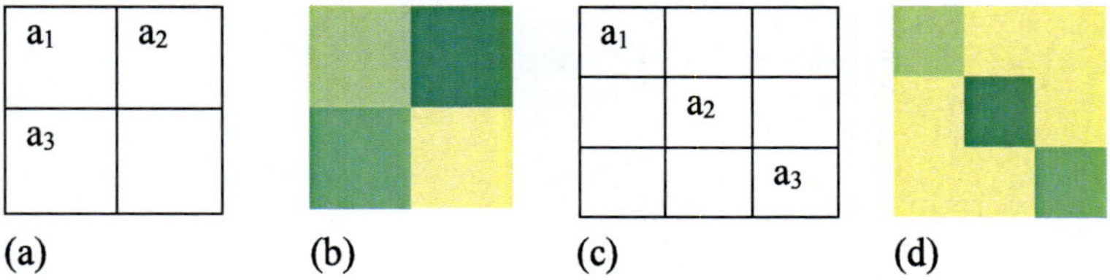

Fig. 1 Sequential (**a,b**) and diagonal (**c,d**) mappings

These examples illustrate only a few options to map all attributes of **a** to pixels in the $k \times k$ pixels image, while multiple other mappings are possible. It leads to the *optimization task* of optimal mapping attributes of **a** to pixels that will maximize the *classification accuracy* by a given machine learning algorithm. The formal mathematical exploration of this task still is a future task. The practical approach taken in [2] is discovering relations between attributes a_i and putting related attributes to adjacent pixels avoiding exhaustive search of the optimal mapping. Its novelty is that PCA, K-PCA or t-SNE mapping is used not as a visualization itself, but only to identify the *location* of the pixels on the 2-D plot. While, in general, it is productive approach, a specific relation discovered by such *unsupervised* methods can be irrelevant to the supervised classification task. Next, the distances that PCA, and t-SNE discover to locate pixels nearby can corrupt n-D distances [4].

Also, recently, a single value mapping has been successfully applied to timeseries [5], where different univariate time series for the same sample are displayed on the same plot and run by a single CNN. The points can be connected to form a graph. Alternatively, the univariate time series are displayed in separate plots and run by several "Siamese" CNNs with the same convolutional and pooling layers. Then before flattening, the outputs of these CNNs are joined and run by the same classifier section to produce a single output.

1.2 Graph Mappings

The algorithm GLC-L in [1] represents an n-D point **a** as a 2-D *graph*, which is a sequence of connected straight lines under different angles where line L_i has the length of a_i of n-D point **a**. The disadvantage of this approach is the *size* of its raster image because drawing graph edges requires many pixels. Methods in [1, 2, 5] use images that they produce as inputs to CNN algorithms for *image recognition* and all reported success on different datasets. Their disadvantage is a single value mapping requiring a cell (pixel) for each a_i of n-D point **a**. In general, single value mappings can produce a *sparce mapping* where only a small fraction of pixels represent each n-D point **a**. It is illustrated in Figs. 2 and 3, which show other single-value mapping options. Figure 2a shows an option from [6], which encodes each value a_i as bars of (max-a_i) going from the top that we call a *reversed bar.* Figure 2b shows an equivalent single-value mapping using *parallel coordinates* for the same max-a_i values. In general, it is possible to map each a_i itself without making max-a_i. Figure 3 presents 5-D point $\mathbf{a} = (2, 4, 5, 1, 6)$ in lossless GLC-L and Collocated Paired Coordinates (CPC) [7].

Figure 4 shows *partial pair values mappings* [6], where Fig. 4a shows a heatmap of differences x_i-x_j and Fig. 4b uses this heatmap as a background for the reversed bar chart shown in Fig. 2a.

The proposed **algorithm CPC-R** uses a *pair values mapping with preserving* all n-D information. It has advantages over single value mappings such as proposed

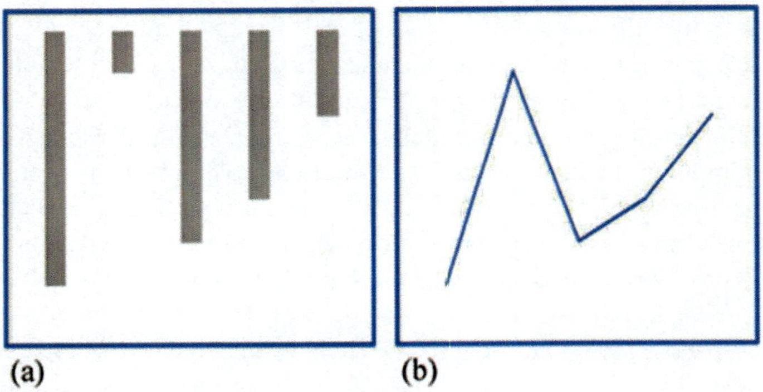

Fig. 2 5-D point $\mathbf{x} = (x_1, x_2, x_3, x_4, x_5) = (6, 1, 5, 4, 2)$ in reversed bar graph (**a**), i.e., showing bars of (max-a_i) going from the top, and **b** in parallel coordinates

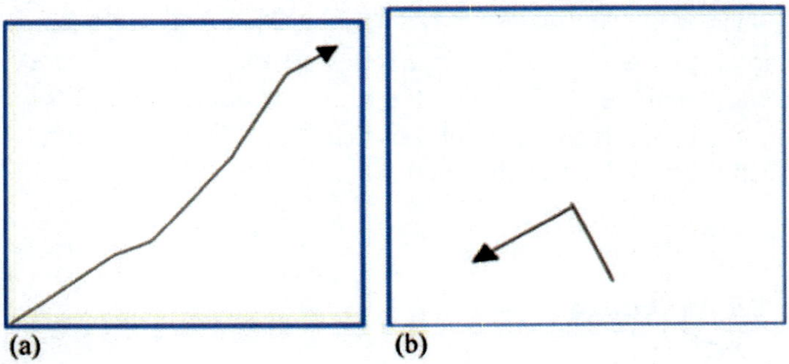

Fig. 3 5-D point **a** = (2, 4, 5, 1, 6) in GLC-L (**a**) and Collocated Paired Coordinates (**b**)

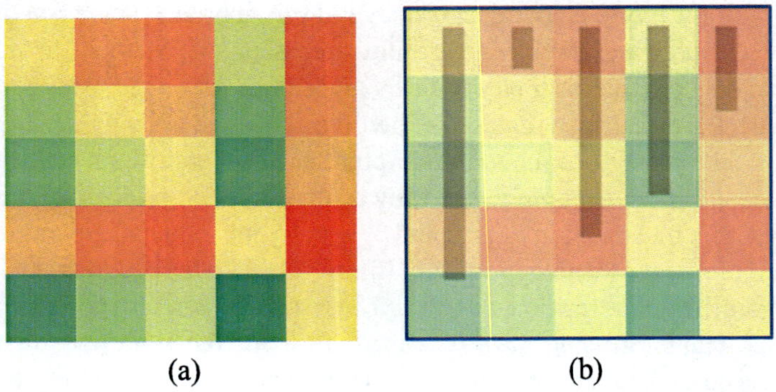

Fig. 4 **a** Heatmap of differences x_i-x_j used as a background for **b** that is a reversed bar chart

in [1]: (1) *two times less* cells (pixels), (2) smaller images, and (3) less occlusion because CPC-R does not need the graph edges.

1.3 Pair Values Mapping

A *pair values mapping H* maps a pair of values (a_i, a_j) to a single pixel p, $H(a_i, a_j) = (p_{1ij}, p_{2ij})$. Thus, the values a_i and a_j are encoded by the *location* of the pixel (p_{i1}, p_{i2}). In a simple version of H mapping, the coordinates of the pixels are values a_i and a_j, $(p_{1ij}, p_{2ij}) = (a_i, a_j)$ and $(p_{1i,i+1}, p_{2i,i+1}) = (a_i, a_{i+1})$. It puts similar pairs next to each other. This requires converting a_i and a_j to integers by *rounding* them. When n is odd we form the last pair (a_n, a_n) by repeating a_n. The *grey scale intensities* of the pixels are used to specify the order of the pairs, where the first pair is black; the second pair is dark grey and so on. The last pair is very light grey. A color sequence is also used as an alternative producing a *heatmap*. The values of intensities and colors can be optimized at the learning stage. We denote this mapping algorithm as **CPC-R algorithm** and images that it produces as **CPC-R images**, because it is inspired by visualization of n-D points by *Collocated Paired Coordinates* (CPC) in vector graphics [7]. The letter R indicates that CPC-R are raster non vector images.

The benefit of the CPC-R pair values mapping is in representing an n-D point *without loss of information* by using only $\lceil n/2 \rceil$ pixels, which is two times less pixels than in a single pixel mapping. It makes images *simpler.* The CPC-R algorithm allows to optimize the location of the pixels by using alternative pairing (a_i, a_j) not only (a_i, a_{i+1}) to increase classification accuracy. A simplest version of CPC-R algorithm tests a fixed number of alternative pairings randomly generated with a given classification algorithm.

1.4 Pair Values Mapping Algorithm CPC-R

A large part of this chapter presents the experiments to classify the real world and simulated data with the CPC-R and CNN algorithms. We tested the efficiency of CPC-R+CNN algorithm successfully by evaluating the accuracy of the classification on images produced from benchmark datasets that have been discretized to map to the CPC-R images.

The steps of the **base CPC-R algorithm version 1.0** are as follow:

S1. *Split attributes* of an n-D point $\mathbf{x} = (x_1, x_2, \ldots, x_n)$ to consecutive pairs (x_1, x_2), (x_3, x_4), …, (x_{n-1}, x_n). If n is odd repeat the last attribute of **x** to make n even.
S2. *Set up a cell size* to locate pairs in the image. Each cell can consist of a single pixel or dozens of pixels.
S3. *Locate pairs* (x_i, x_{i+1}) in the image at the cell coordinate pair (x_i, x_{i+1}) in the image.

S4. *Assign the grey scale intensity* to cells from black for (x_1, x_2) and very light grey for (x_{n-1}, x_n). Alternatively, *assign heatmap colors.*
S5. *Optional: Combine* produced images with context information in the form of average images of classes.
S6. *Call/Run ML/DL* algorithm to discover a predictive model.
S7. *Optional: Optimize* intensities in S4 and pairing coordinates beyond the sequential pairs (x_i, x_{i+1}) used in S1 by the methods ranged from random generating a fixed number of alternatives to genetic algorithms with testing ML prediction accuracy for these alternatives.

Full restoration of the order of the pairs and their values from a CPC-R image is possible due to the order of intensities if all pairs differ. The equal pairs will *collide* in the image location. The base algorithm keeps only the intensity of the *last pair* from a set of colliding pairs. Other versions preserve more information about the colliding pairs.

According to this algorithm a small image with 10 × 10 pixels can represent a 10-D point, when each attribute has 10 different values and each cell uses only a single pixel in step 2. It will locate five grey scale pixels in this image.

The generation of raster images in CPC-R for the Wisconsin Breast Cancer (WBC) data [8] is illustrated in Fig. 5. We consider a 10-D point **x** = (8, 10, 10, 8, 7, 10, 9, 7, 1, 1) constructed from the 9-D point by copying x_9 =1 to a new value x_{10} = 1 to make 5 pairs. The process of generation of CPC-R image for this **x** is as follows:

- Forming consecutive pairs of values from **x** = (8, 10, 10, 8, 7, 10, 9, 7, 1, 1): (8, 10), (10, 8), (7, 10), (9, 7), (1, 1).
- Filling cell (8, 10) for pair (x_1, x_2) = (8, 10) according to the grey scale value for the 1st pair (black).
- Filling cell (10, 8) for pair (x_3, x_4) = (10, 8) according to the grey scale value for the 2nd pair (very dark grey).

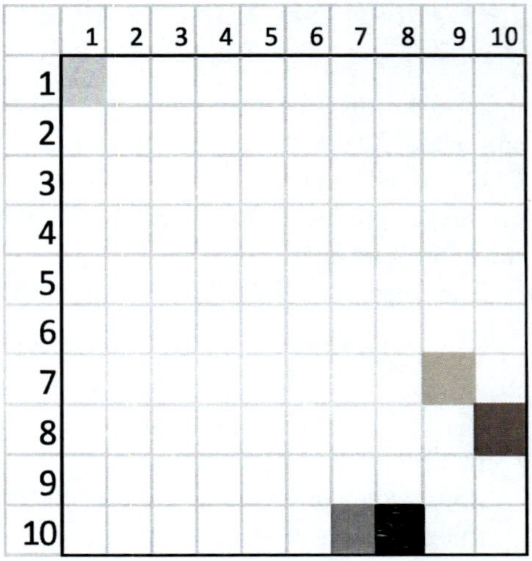

Fig. 5 10-D point (8, 10, 10, 8, 7, 10, 9, 7, 1, 1) in CPC-R

- Filling cell (7, 10) for pair $(x_5, x_6) = (7, 10)$ according to the grey scale value for the 3rd pair (grey).
- Filling cell (9, 7) for pair $(x_7, x_8) = (9, 7)$ according to the grey scale value for the 4th pair (light grey).
- Filling cell (1, 1) for pair $(x_9, x_{10}) = (1, 1)$ according to the grey scale value for the 5th pair (very light grey).

It is a lossless visualization if values of all pairs (x_i, x_{i+1}) are not identical. The next versions of the CPC-R algorithm deal with colliding pairs and other steps in more elaborated ways. The version 2.1 for elaborated steps 3 is below.

CPC-R version 2.1 with using **adjacent** cells in step 3:

3.1. *Set image coordinate system* (origin at upper left, or low left).
3.2. *Locate non-colliding pairs* (x_i, x_{i+1}) in the image at cell coordinate values (x_i, x_{i+1}).
3.3. *Set the starting adjacent cell* (right, left, top or bottom) for collided pairs.
3.4. *Set order of filling of adjacent cells* (clockwise or counterclockwise).
3.5. *Locate colliding pairs* (x_i, x_{i+1}) in the cell adjacent to (x_i, x_{i+1}) according to 3.3 and 3.4.

This version uses four adjacent cells (right, left, top, and, bottom) and is called a ***cross filling***. Section 7 presents another option, which is called a *serpent filling*. It puts collided pairs to all 8 adjacent cells not only four cells that we use here.

CPC-R version 2.2 with **splitting** cells in steps 3 and 4:

Step 3:

3.1. *Set image coordinate system* (origin upper left, or low left).
3.2. *Locate non-colliding pairs* (x_i, x_{i+1}) in the image at cell coordinate values (x_i, x_{i+1}).
3.3. *Split* colliding cells to vertical strips.

Step 4:

4.1. *Assign* the grey scale intensity from black for (x_1, x_2) and very light grey for (x_{n-1}, x_n) for *non-colliding pairs.* Alternatively, *assign heatmap colors* for non-colliding pairs.
4.2. Assign the grey level intensity of the respective pair of values (x_i, x_{i+1}) to the strip for *colliding pairs.*

For example, two strips 4×8 will be assigned to a cell 8×8 for two collided pairs. Three strips 3×8, 3×8, and 2×8 are assigned for each three collided pairs. Similarly, 4–5 strips are generated for 4–5 collided pairs, respectively.

CPC-R version 2.3 with the **darkest** intensity of colliding pairs is used in step S4:

4.1. For all *non-colliding pairs assign* the grey scale intensity starting from black for (x_1, x_2) and very light grey for (x_{n-1}, x_n). Alternatively, *assign heatmap colors* for non-colliding pairs.

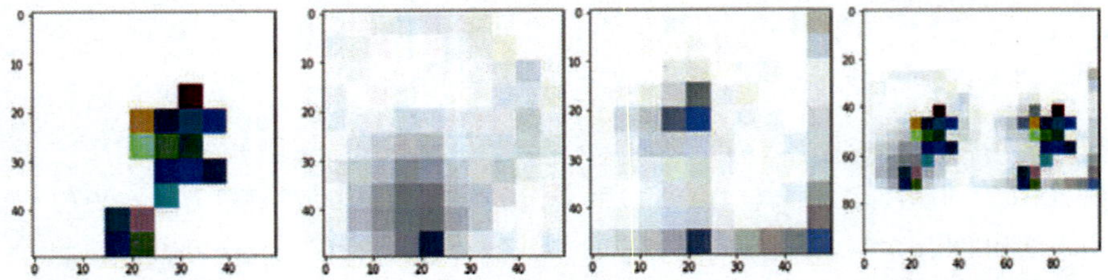

(a) Image formed after marking adjacent cells to handle collisions.

(b) G mean image – mean of all images in training set of some fold with G label.

(c) B mean image – mean of all images in training set of a given fold with B label.

(d) Final double image with image (a) on the top of mean images (b) and (c).

Fig. 6 Building double color image with context of two means for 34-D ionosphere data

4.2. For *colliding pairs assign* the darkest grey level intensity of all pairs with equal values (x_i, x_{i+1}).

CPC-R version 3.0 with **context** incorporated in step 5:

S5. *Put* an image of each n-D point on the top of both G mean (mean image of the training cases of class 1), and B mean (mean image of the training cases of class 2). Put both images side-by-side to form a double image.

The image size after forming a double image is (w) × (2·w). Then (2·w) × (2·w) image is formed by padding the double image. Figure 6 shows steps of constructing a double image superimposed with mean images for each class from training data with padding. This is a way to add *context* to the images for analysis by humans and deep learning algorithms. The difference between two means in its full 34-D representation is shown in Fig. 6d *without degrading it to a lossy single distance number* to the mean n-D points.

CPC-R version 4.0 with optimized intensities and pairing of coordinates (x_i, x_j) added in step 7:

S7. *Optimize* intensities and pairing coordinates beyond sequential pairs (x_i, x_{i+1}) by algorithms ranged from random generating a fixed number of alternatives and testing ML prediction accuracy on training data for these alternatives to genetic algorithms. In all experiments below with data from UCI ML repository. We use tenfold Cross Validation (CV) with such optimization.

2 Experiments with WBC Data

Several experiments reported below have been conducted with WBC data [8] to test feasibility of the CPC-R approach for CNN architectures with different number of pixels per cell, which represents each pair (x_i, x_{i+1}). We first assign the grey scale intensities to the pixels as follows: $I(x_1, x_2) = 0$ (darkest), $I(x_3, x_4) = 51$, $I(x_5, x_6) =$

102, $I(x_7, x_8) = 153, I(x_9, x_{10}) = 204$, then we optimize these intensities and pairing coordinates (x_i, x_j) for accuracy in some experiments.

2.1 Initial Experiments

Experiments **E1** and **E2** have been conducted with CPC-R **1.0** and **grey scale** values defined above. We used the CNN architecture also used in [1] that we denote as CNN64 for short and summarize below, to be able to compare results with [1]:

- Convolutional layer with 64 output channels, a kernel shape of 2 × 2, stride of 2 × 2 and RELU activation.
- Convolutional layer with 64 output channels, a kernel shape of 2 × 2, stride of 2 × 2 and RELU activation.
- Pooling layer with pooling size of 2 × 2.
- Drop out layer with 0.4 fraction of input units to drop.
- Convolutional layer with 128 output channels, a kernel shape of 2 × 2, stride of 2 × 2 and RELU activation.
- RELU Convolutional layer with 128 output channels, a kernel shape of 2 × 2, stride of 2 × 2 and relu activation.
- Pooling layer with pooling size of 2 × 2.
- Drop out layer with 0.4 fraction of input units to drop.
- Fully connected layer with 256 output nodes and RELU activation
- Drop out layer with 0.4 fraction of input units to drop.
- Fully connected layer with number of output nodes equal to the number of classes, with SoftMax activation.

Experiments **E2** involve the InceptionResNetV2 CNN architecture [9], which we denote as ResNetV2, with the following properties: Flatten → Dense (256, relu) → Dense(1, sigmoid). The impact of image sizes and origins of coordinates: upper left corner (ULC) or low left corner (LLC) is explored in E1 and E2 (see Table 1 with results).

Table 1 Achieved accuracies of experiments E1 and E2 in tenfold cross validation

Exp.	Model	Image	Origin	Accuracy
E1	CNN64	100 × 100	ULC	95.61
E2	ResNetV2	80 × 80	LLC	94.45

Table 2 Achieved accuracies in experiments C1 and C2 with putting colliding values to adjacent cells in tenfold CV

Exp.	Model	Image	Accuracy
C1	ResNetV2	80 × 80	95.61
C2	ResNetV2	320 × 320	96.20
C2	LeNet5	50 × 50	94.88

2.2 *Experiments Dealing with Colliding Values*

The impact of colliding values along with image sizes on accuracy using ResNetV2 and LeNet5 architecture [10] is explored in experiments C1 and C2. See Table 2.

Experiments **C1** are conducted with CPC-R **2.1**, grey scale images, different orders of filling **the adjacent cells**, and two CNN architectures. *Slight increase in accuracy* was observed over E1 and E2, showing that this way to resolve collision is beneficial. We also observed that accuracy of 80 × 80 images is about the same as of larger 160 × 160 images.

The impact of the **splitting of colliding cells** was explored in Experiments C2 with vertical strips. C2 results are slightly better than E1 and E2, which do not address collision. The LeNet5 architecture provided 94.88% accuracy for smaller images, 50 × 50, than other CNNs.

2.3 *Experiments with Double Images and Padding*

The impact of adding **context** via double images on the accuracy was explored in **Experiments D1–D8** with CPC-R **3.0** for different architectures, image sizes, and other parameters listed below for each experiment setting D1–D8. We tested the hypothesis that adding the context can increase accuracy.

While we extensively varied the parameters of experiments D1–D8, to save space, Table 3 contains only results for the parameters with the best accuracies. The settings of experiments are as follows:

D1 with superimposed colliding values (v1.0),
D2 with filling adjacent cells (v2.1),
D3 with splits of cells (v2.2),
D4 with darkest cells for colliding pairs (v2.3),
D5 with red levels and grey means,
D6 with red levels and colored means,
D7 with grey cases and colored means,
D8 with all colored images.

The accuracy in these D experiments with double images is improved over experiments E and C with single images. The accuracy of D2 is not only slightly less than D1, but it also with much larger double images. The accuracy of D3 is less than for both D1 and D2, likely due to the smaller image size. Splitting cells necessitates

Table 3 Achieved accuracies of tenfold cross validation experiments D1–D8 with double images

Exp.	Model	Double image	Accuracy
D1	LeNet5	60 × 60	97.22
D2	LeNet5	200 × 200	96.77
D3	LeNet5	60 × 60	95.90
D3	CNN64	200 × 200	95.90
D4	CNN64	160 × 160	96.19
D5	LeNet5	60 × 60	97.66
D6	LeNet5	60 × 60	97.80
D6	CNN64	100 × 100	97.36
D7	LeNet5	60 × 60	97.51
D8	LeNet5	60 × 60	97.37

larger images. The advantages of D1 and D2, which are without splitting cells, are also visible relative to D4 that has lower accuracy.

Exploring the impact of the color is the goal of the next experiments. We choose randomly intensity of each color point (3 intensities for 3 channels of a point). Thus, 15 random values for 5 cells are generated. We used images with **red** levels, **grey means** and filling **adjacent** cells in experiments **D5**. A *slight increase* of accuracy relative to the previous experiments D1–D4 was observed (see Table 3). Images with **red** levels for n-D points and **colored means** were used in experiments **D6**. Images with **grey** cases and **colored means** were used in experiments **D7. Color** cases and **colored means** were used in the experiments **D8** (see Fig. 6). In the experiments **D9**, we generated the images, with **red** levels, and **colored means** and **optimized intensities and pairing** of coordinates (x_i, x_j).

The *optimization of pairing was most beneficial* to improve the accuracy, while the accuracy can vary due to the DL random elements as Table 3 shows. We used a simple optimization method with random generating a fixed number of perturbed alternatives and testing ML prediction accuracy on training data for these alternatives. The use of a more sophisticated methods will likely further improve accuracy due to abilities to explore more alternatives.

2.4 MLP Experiments

The importance of using DL models versus simpler MLP models was explored with the following Multilayer Perceptron (MLP) architecture that is simplification of CNN64:

- Layer with 64 output channels and RELU activation. This hidden layer with 64 nodes and a rectified linear activation function.
- Layer with 64 output channels and RELU activation. This hidden layer with 64 nodes and a rectified linear activation function.

- Dropout layer with fraction of input units to drop is 0.4.
- Layer with 128 output channels and RELU activation.
- Drop out layer with fraction of input units to drop is 0.4.
- Fully connected layer with number of output nodes equal to the number of classes, with SoftMax activation.

The **8 experiments** with **MLP** using single and double images allowed to achieve accuracy from 70.31 to 76.51% in tenfold cross validation using the same setting as we used for DL models above. At first glance it clearly shows the advantages of using CNN models that gave from **94.45** to **97.8%** accuracy versus much lower accuracy of MLP models. However, experiments with other data in Sect. 4 had shown that this conclusion is not so certain for those data. In some experiments MLP was a winner.

3 Experiments with Swiss Rolls Data

The goal of this experiment is to test abilities of CPC-R algorithm to classify data from two Swiss roll datasets (2-D and 3-D) [11]. These data are commonly used to test abilities of the algorithms to discover patterns of the data located on a low-dimensional manifold being isometric to the Euclidean space. Each 2-D Swiss roll data point (x_1, x_2) is represented in CPC-R images a single point (see Fig. 7a). To enrich the image, we generated an image with a "plus" centered at the points. See Fig. 7b.

Each Swiss roll 3-D point is visualized, in CPC-R, as two 2-D points: (x_1, x_2) and (x_3, x_3). See Fig. 8a. It also can be visualized as three 2-D points: (x_1, x_2), (x_2, x_3), and (x_3, x_1). See Fig. 8b. Respectively, we generated CPC-R images, for 3-D Swiss roll, with "pluses" at three 2-D points: (x_1, x_2), (x_2, x_3), and (x_3, x_1) (Fig. 4c). The max accuracies achieved are 97.87 and 97.56% for 2D and 3D Swiss rolls, respectively, in our experiments (see Table 4).

(a) image with a single point (b) image with a "plus" centered at the point

Fig. 7 CPC-R images for 2-D Swiss roll point (x_1, x_2)

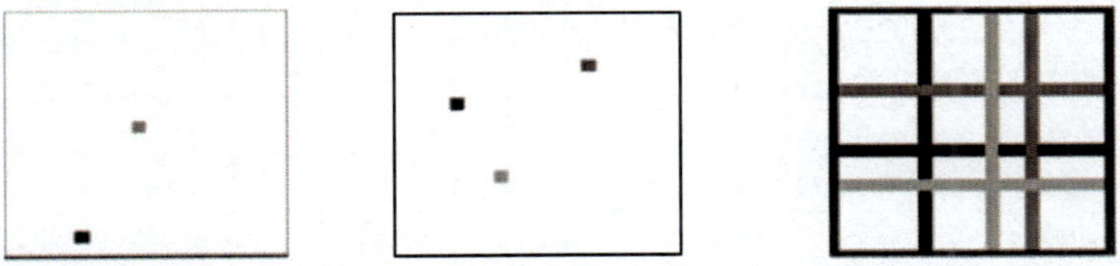

(a) 3-D point as two points (x_1, x_2), (x_3, x_3).

(b) 3-D point as three points: (x_1, x_2), (x_2, x_3), and (x_3, x_1)

(c) 3-D point with "pluses" at (x_1, x_2), (x_2, x_3), and (x_3, x_1)

Fig. 8 CPC-R images for 3-D Swiss roll point (x_1, x_2, x_3)

Table 4 Results for Swiss roll 2D and 3D

Model	Data	Image	Max accuracy by varying intensities
LeNet5	2-D roll	Point	96.56
LeNet5	2-D roll	"plus" at point	97.88
LeNet5	3-D roll	2 points	97.56
LeNet5	3-D roll	3 points	94.69
LeNet5	3-D roll	3 "pluses" at point	96.31

4 Experiments with Ionosphere, Glass and Car Data

The setting of these experiments is the same as for WBC data experiments above with results summarized in Tables 5 and 6 for tenfold cross validation for accuracy on the validation data. These tables report **achieved** accuracies for Ionosphere [12], Glass [13] and Car data [14], which are the best results obtained in a series of experiment of a given type. For instance, the experiments of type E2 were run for ResNetV2 for 60 and 100 epochs, images of sizes 30×30, 50×50, and 100×100, different 10-fold cross validation settings for CNN64, ResNetV2 and MLP, while Table 5 reports a single best result that is 89.98 and 50×50 images.

In Table 5, the best accuracy is 95.89% for the MLP classifier for Ionosphere data in E3 and the best accuracy is 96.86% for ResNetV2 classifier for Glass data in E2. This table shows that MLP was a winner only in E3 experiment for Ionosphere data being competitive in other experiments.

The car data set [14] includes 1728 instances of 4 classes with 6 attributes. The data have been normalized in [0, 10] interval. We conducted multiple experiments to explore the impact of combinations of properties of CPC-R images on accuracy of classification. The Experiments E1-E7 with the Car data used MLP and CNN classifiers. Table 6 summarizes the best accuracy results with the placement of the collided pairs using *cross* filling process defined in Sect. 1.4. These experiments involve optimization of coordinate pairing and intensities of pairs. The achieved accuracies are between 79.68 and 96.8% for the Car data. The experiments with the MLP classifier provided lower accuracies than other classifications.

Table 5 Achieved accuracies of tenfold cross validation experiments with single and double images for ionosphere and glass data

Ionosphere data				Glass data			
Exp.	Model	Image size	Accuracy	Exp.	Model	Image size	Accuracy
E1	ResNetV2	100 × 100	86.98	E1	CNN64	100 × 100	96.45
E1	MLP	50 × 50	74.58	E1	MLP	100 × 100	89.8
E2	ResNetV2	50 × 50	89.98	E2	ResNetV2	100 × 100	**96.86**
E3	MLP	100 × 100	**95.89**	E2	MLP	50 × 50	94.6
E3	CNN64	50 × 50	94.13	E4	MLP	50 × 50	94.7
E4	MLP	100 × 100	90.91	E4	CNN64	50 × 50	95.9
E5	ResNetV2	50 × 50	91.80	E5	ResNetV2	50 × 50	94.89
E5	MLP	100 × 100	85.75	E5	MLP	100 × 100	92.54
C1	CNN64	100 × 100	89.04	C1	ResNetV2	100 × 100	96.01
C1	MLP	50 × 50	87.43	C1	MLP	50 × 50	94.6
D5	ResNetV2	100 × 100	92.03	D5	CNN64	50 × 50	94.13
D6	CNN64	100 × 100	93.46	D6	ResNetV2	100 × 100	95.06
D7	CNN64	50 × 50	91.73	D7	CNN64	100 × 100	95.9
D7	MLP	100 × 100	90.91	D7	MLP	100 × 100	94.7
D8	CNN64	100 × 100	92.46	D8	ResNetV2	100 × 100	96.8
D8	MLP	50 × 50	90.29	D8	MLP	200 ×200	92.78

5 Saliency Maps

The goal of this experiment is exploring abilities to discover most informative attributes in CPC- R images by using the saliency maps. Specifically, we explore the approach based on the gradient $\frac{\delta output}{\delta input}$ of the output category with respect to an input image that allows observing how the output value changes with a minor change in the input image pixels. This approach expects that visualizing these gradients with the same shape as the image will produce some intuition of attention [15]. These gradients highlight the input regions that can cause a major change in the output and highlight the salient image regions that contribute towards the output [16].

5.1 Image-Specific Class Saliency Visualization

The mechanism of this saliency approach is as follows [17]. Pixels are ranked with a given image I_0, a class c, and a classification ConvNet with the class score function $Sc(I)$, influencing the score $\mathrm{Sc}(I_0)$, e. g., a linear score model for the class c

$$S_c(I) \approx \mathrm{w}_c^T I + b_c$$

Table 6 Results of E1-E7 experiments for Car data with cross filling of collided pairs in 0-fold cross validation

Exp.	Image	Epochs	Model	Accuracy	Exp.	Image	Epochs	Model	Accuracy
E1	50 × 50	100	CNN64	94.79	E4	100 × 100	100	MLP	88.97
E1	100 × 100	100	MLP	83.68	E5	100 × 100	150	CNN64	**96.8**
E2	100 × 100	100	LeNet5	91.01	E5	50 × 50	50	LeNet5	92.03
E2	50 × 50	100	CNN64	90.60	E5	100 × 100	100	MLP	83.87
E2	100 × 100	100	MLP	80.01	E6	50 × 50	50	LeNet5	92.04
E3	50 × 50	150	LeNet5	96.75	E6	50 × 50	50	CNN64	94.08
E3	100 × 100	100	CNN64	93.99	E6	100 × 100	100	MLP	85.87
E3	100 × 100	100	MLP	87.09	E7	50 × 50	100	CNN64	94.07
E4	100 × 100	100	CNN64	96.28	E7	50 × 50	50	LeNet5	96.28
E4	50 × 50	100	LeNet5	94.63	E7	100 × 100	100	MLP	87.91
E4	100 × 100	100	MLP	88.97	–	–	–	–	

where I is an image in a one-dimensional form (vectorized), w_c is the weight vector, and b_c is the bias of the model. Within this model each w values "defines the importance of the corresponding pixels of an image I for the class c" [17]. Then this statement is softened in [17] to accommodate the fact that in deep convolution networks, the class score Sc(I) is a non-linear function of I. Thus, now Sc(I) is considered as a linear function only locally that is computed as the first-order Taylor expansion, where w is the derivative of S_c with respect to the image I at the point (image) I_0:

$$\mathrm{w} = \frac{\delta Sc}{\delta I}$$

Next, the saliency maps are visualized with the highest-class score (top class prediction) on a test image from the data set. We conducted such saliency experiments with CNN on the randomly selected ionosphere CPC-R images to check the importance of the corresponding pixels in these images.

5.2 *Saliency Visualization with Ionosphere Data Set (Experiment E8)*

In this experiment computing saliency maps includes two steps:

1. Generating an image that *maximizes the class's score*, which visualizes the CNN [17, 18].
2. Computing the class saliency map *specific to a given image* and class.

Both steps are based on computing the gradient of a class score to the input image. The experiment has been conducted using Keras-vis with its components Visualize_saliency and Visual_cam implemented with backpropagation modifiers [19].

Several deep learning architectures have been trained on Ionosphere data [12] to produce saliency maps. This dataset includes 351 instances of 2 classes with 34 attributes each used in tenfold cross validation.

The list of those networks and accuracies obtained in this process of producing saliency maps are presented in Table 7. The analysis of resulting saliency maps shows that maps for all of them are extremely similar for each image as Fig. 9 illustrates.

Figure 9 shows the representative saliency maps result for CNN64 and MLP architectures defined in Sect. 2 for one of the CPC-R images from the Ionosphere data: (a) original image (b) Guided backpropagation, (c) Grad-CAM, which localizes class discriminative regions, and (d) Guided Grad-CAM that combines (b) and (c). The gray scale intensities are normalized from 0 to 1. We can easily locate the dark pixels in Fig. 9c (Grad-CAM).

In the Gradient Weighted Class Activation Mapping (Grad-CAM) [20, 21] the class-specific gradient information is flowing into the final convolutional layer of CNN to generate the localization map of important regions in the image. We also explored the combination of the Grad-CAM with Guided backpropagation to create a high-resolution class discriminative visualization known as Guided Grad-CAM [20].

Grad-CAM localizes the relevant image regions, but does not show finely grained importance, like pixel-space gradient visualization methods such as Guided backpropagation do. For example, Grad-CAM can easily localize the darkest pixel in the

Table 7 The results of Saliency experiment E8 with Ionosphere data

Model	Image size	Epochs	Cross validation	Accuracy
CNN64	50 × 50	50	Tenfold	94.4
CNN64	50 × 50	100	Stratified tenfold	94.58
CNN64	100 × 100	50	Tenfold	91.6
CNN64	100 × 100	100	Stratified tenfold	91.46
ResNetV2	100 × 100	100	Stratified tenfold	93.46
ResNetV2	50 × 50	50	Tenfold	95.01
MLP	50 × 50	100	Stratified tenfold	88.07
MLP	100 × 100	50	Tenfold	90.75

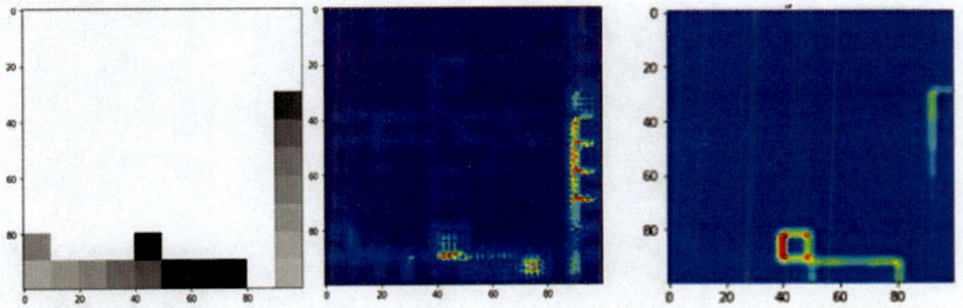

(a) Base greyscale image

(b) Guided Backpropagation, CNN64

(c) Grad-CAM, CNN 64

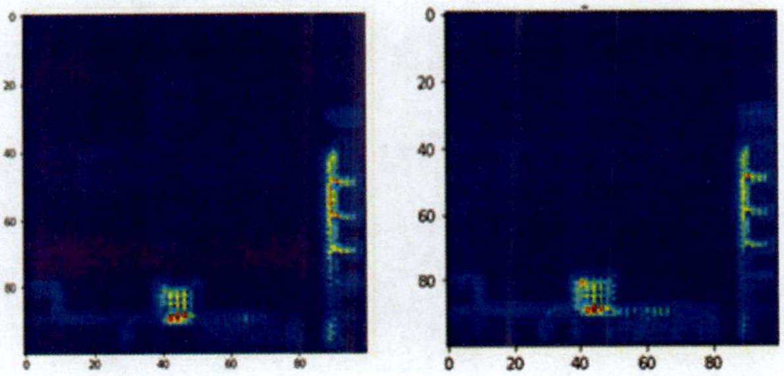

(a) Guided Grad-CAM, CNN64

(b) Guided Grad-CAM, MLP

Fig. 9 Saliency maps for CNN64 and MPL

input region but not others. To combine the best aspects of both, Guided Backpropagation and Grad-CAM visualizations are fused via point-wise multiplication [20]. Guided Backpropagation was chosen over deconvolution because backpropagation visualizations being generally noise-free [22].

Figure 10 shows saliency results for more CPC-R images from the Ionosphere data, where Fig. 10a presents CPC-R images produced using the color coding. The CPC-R images in the first and last rows are from class B and CPC-R images in rows 2–4 are from class G. It is visible that these randomly selected CPC-R images from the validation set are very different without obvious visual patterns that would allow to classify them to class B or G, but trained CNN64 was able to classify them correctly.

It is visible that the most salient pixels correspond to the darkest pixels in CPC-R images. However, the saliency of these pixels is simply an **artifact** of the CPC-R coding schema where the first pair (x_1, x_2) is the darkest one and the last pair (x_{n-1}, x_n) is the lightest one. Thus, this saliency provides a *distorted importance* of pixels focusing only on darker start pairs like (x_1, x_2), (x_3, x_4) and others next to them. The fact that all considered CNN architectures provided quite high accuracy in tenfold cross validation tells that those dark pairs significantly contributed to the output, but it does not justify that they are more important and relevant than others.

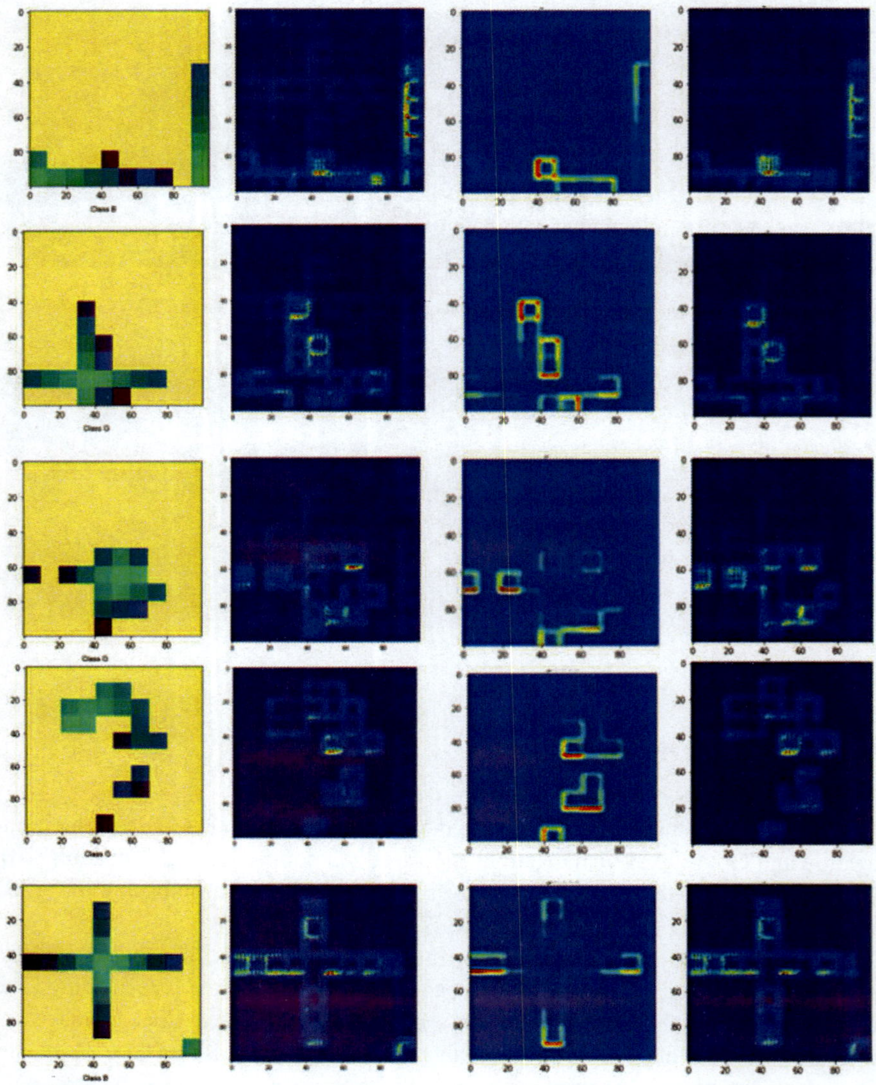

(a) Color image (b) Guided Backpropagation (c) Grad-CAM (d) Guided Grad-CAM

Fig. 10 Saliency experiments with Ionosphere data

This is illustrated in the next section where we employ an alternative method to find importance of the pixels.

6 Informative Cells: Feature Importance

6.1 Concept of Informative CPC-R Cells

Experiments in the previous section shows that saliency maps did not produced the justified interpretable feature importance order on CPC-R images focusing only on dark pixels. Another common reason of saliency maps failure is their local approach with small patches without large context, and sensitivity to different contrasts [23]. Thus, we need another method that will be able to deal with these issues better. The idea of the method presented in this section is defining the importance of features by estimating the change of prediction accuracy due to the exclusion of the individual features. The large cells (super-pixels) are used as such features that can capture a larger context. This approach is in line with Super CNN approach that is a hierarchical super pixel CNN for salient object detection [23].

The experiment below presents the accuracy of classification with covering respective cells of CPC-R images of ionosphere data that are made white. This approach views a cell as the most informative if covering it leads to largest decrease in classification accuracy. These cells are called **covered cells**. Below this method is called the **Informative Cell Covering (ICC)** algorithm. A 5 $\times$ 5 grid with 25 cells was created for each CPC-R image. Figure 11 and the list below show this configuration. Then 25 images were created from each CPC-R image where a respective cell was made white.

1	Cell 21	Cell 21	Cell 22	Cell 22	Cell 23	Cell 23	Cell 24	Cell 24	Cell 25	Cell 25
2	Cell 21	Cell 21	Cell 22	Cell 22	Cell 23	Cell 23	Cell 24	Cell 24	Cell 25	Cell 25
3	Cell 16	Cell 16	Cell 17	Cell 17	Cell 18	Cell 18	Cell 19	Cell 19	Cell 20	Cell 20
4	Cell 16	Cell 16	Cell 17	Cell 17	Cell 18	Cell 18	Cell 19	Cell 19	Cell 20	Cell 20
5	Cell 11	Cell 11	Cell 12	Cell 12	Cell 13	Cell 13	Cell 14	Cell 14	Cell 15	Cell 15
6	Cell 11	Cell 11	Cell 12	Cell 12	Cell 13	Cell 13	Cell 14	Cell 14	Cell 15	Cell 15
7	Cell 6	Cell 6	Cell 7	Cell 7	Cell 8	Cell 8	Cell 9	Cell 9	Cell 10	Cell 10
8	Cell 6	Cell 6	Cell 7	Cell 7	Cell 8	Cell 8	Cell 9	Cell 9	Cell 10	Cell 10
9	Cell 1	Cell 1	Cell 2	Cell 2	Cell 3	Cell 3	Cell 4	Cell 4	Cell 5	Cell 5
10	Cell 1	Cell 1	Cell 2	Cell 2	Cell 3	Cell 3	Cell 4	Cell 4	Cell 5	Cell 5
0	**1**	**2**	**3**	**4**	**5**	**6**	**7**	**8**	**9**	**10**

Fig. 11 Formation of informative cells

$(x_1, y_{10}), (x_1, y_9), (x_2, y_{10}), (x_2, y_9)$ = Cell 1;
$(x_3, y_{10}), (x_3, y_9), (x_4, y_{10}), (x_4, y_9)$ = Cell 2;
$(x_5, y_{10}), (x_5, y_9), (x_6, y_{10}), (x_6, y_9)$ = Cell 3;
$(x_7, y_{10}), (x_7, y_9), (x_8, y_{10}), (x_8, y_9)$ = Cell 4;
$(x_9, y_{10}), (x_9, y_9), (x_{10}, y_{10}), (x_{10}, y_9)$ = Cell 5;
$(x_1, y_8), (x_1, y_7), (x_2, y_8), (x_2, y_7)$ = Cell 6;
$(x_3, y_8), (x_3, y_7), (x_4, y_8), (x_4, y_7)$ = Cell 7;
$(x_5, y_8), (x_5, y_7), (x_6, y_8), (x_6, y_7)$ = Cell 8;
$(x_7, y_8), (x_7, y_7), (x_8, y_8), (x_8, y_7)$ = Cell 9;
$(x_9, y_8), (x_9, y_7), (x_{10}, y_8), (x_{10}, y_7)$ = Cell 10;
$(x_1, y_6), (x_1, y_5), (x_2, y_6), (x_2, y_5)$ = Cell 11;
$(x_3, y_6), (x_3, y_5), (x_4, y_6), (x_4, y_5)$ = Cell 12;
$(x_5, y_6), (x_5, y_5), (x_6, y_6), (x_6, y_5)$ = Cell 13;
$(x_7, y_6), (x_7, y_5), (x_8, y_6), (x_8, y_5)$ = Cell14;
$(x_9, y_6), (x_9, y_5), (x_{10}, y_6), (x_{10}, y_5)$ = Cell 15;
$(x_1, y_4), (x_1, y_3), (x_2, y_4), (x_2, y_3)$ = Cell 16;
$(x_3, y_4), (x_3, y_3), (x_4, y_4), (x_4, y_3)$ = Cell 17;
$(x_5, y_4), (x_5, y_3), (x_6, y_4), (x_6, y_3)$ = Cell 18;
$(x_7, y_4), (x_7, y_3), (x_8, y_4), (x_8, y_3)$ = Cell 19;
$(x_9, y_4), (x_9, y_3), (x_{10}, y_4), (x_{10}, y_3)$ = Cell 20;
$(x_1, y_2), (x_1, y_1), (x_2, y_2), (x_2, y_1)$ = Cell 21;
$(x_3, y_2), (x_3, y_1), (x_4, y_2), (x_4, y_1)$ = Cell 22;
$(x_5, y_2), (x_5, y_1), (x_6, y_2), (x_6, y_1)$ = Cell 23;
$(x_7, y_2), (x_7, y_1), (x_8, y_2), (x_8, y_1)$ = Cell 24;
$(x_9, y_2), (x_9, y_1), (x_{10}, y_2), (x_{10}, y_1)$ = Cell 25.

6.2 *Comparison of Guided Back Propagation Salient Pixels with ICC Informative Cells*

The covered cells, which led to the most significant drop of the accuracy, are considered as the *most informative* in the ICC approach. Table 8 shows the accuracy of classification of Ionosphere data with covered cells in the ascending order of the accuracy.

In Table 8, the lowest accuracy is 80.35% (cell 13), which is considered as the most informative in this approach. Figure 12 shows 4 types of images for cell 13: (a) original CPC-R images from the Ionosphere data, (b) CPC-R images superimposed with heatmap when cell 13 was fully covered, (c) saliency maps for the same CPC-R image without making cell 13 white, and (d) saliency maps for the same CPC-R image with making cell 13 white.

The second lowest accuracy for the ICC method is 81.78% (cell 23), which is considered as the second most informative cell. Figure 13 contains the same four types of images as Fig. 13, but for cell 23. The highest accuracy for ICC method is 86.62% which is cell 17, which supposed to be the *least informative* cell. Figure 14 presents four types of images for this cell.

Table 8 The classification accuracy of Ionosphere data with covered cells (images 100 × 100, epochs 30, tenfold cross validation)

Cells	Accuracy	Cells	Accuracy	Cells	Accuracy	Cells	Accuracy
13	80.35	12	83.77	22	84.34	19	85.50
23	81.78	3	83.85	10	84.61	1	85.77
25	82.07	21	84.05	18	84.61	24	85.77
20	82.34	4	84.06	2	84.62	17	86.62
16	82.35	9	84.06	7	84.50	–	–
11	83.50	15	84.06	8	84.64	–	–
6	83.77	5	84.34	14	85.50	–	–

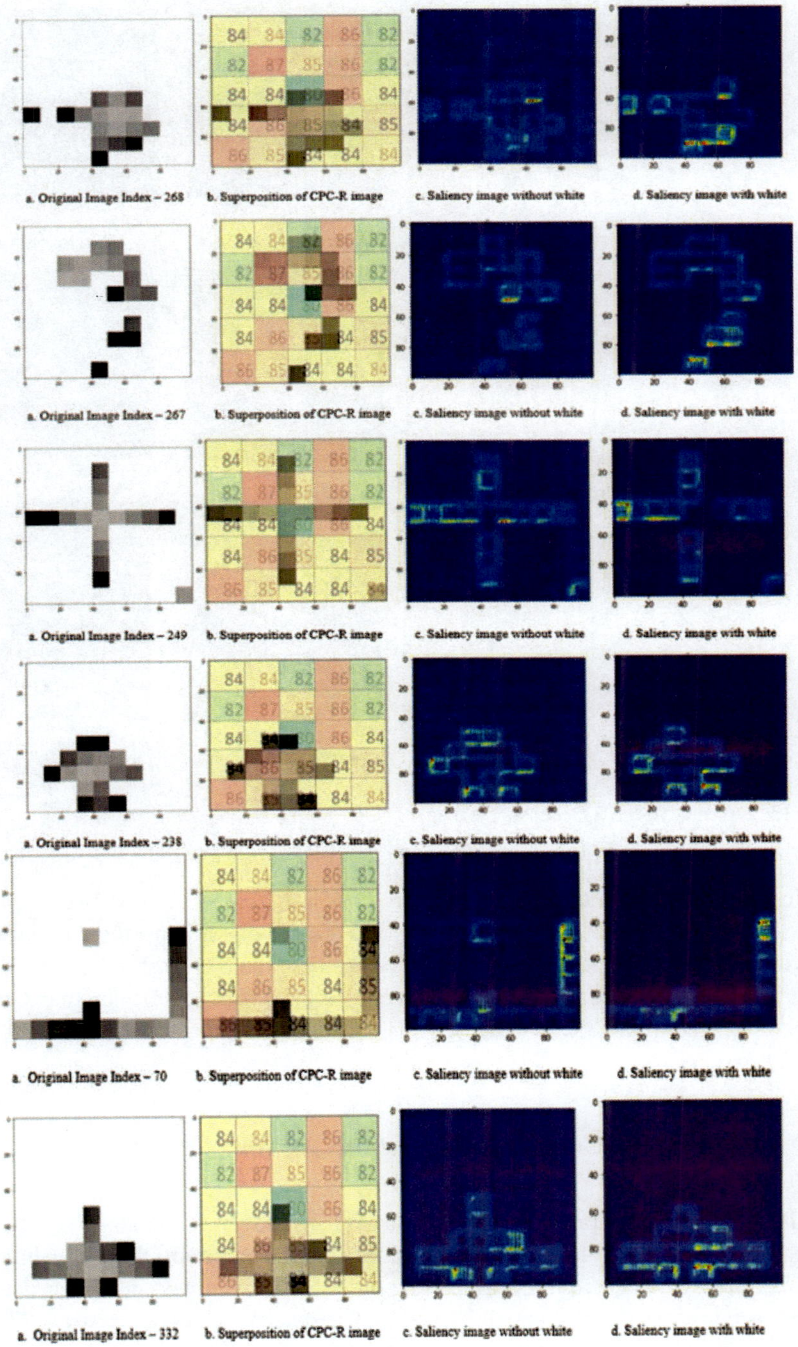

Fig. 12 Informative cell 13 (Ionosphere data)

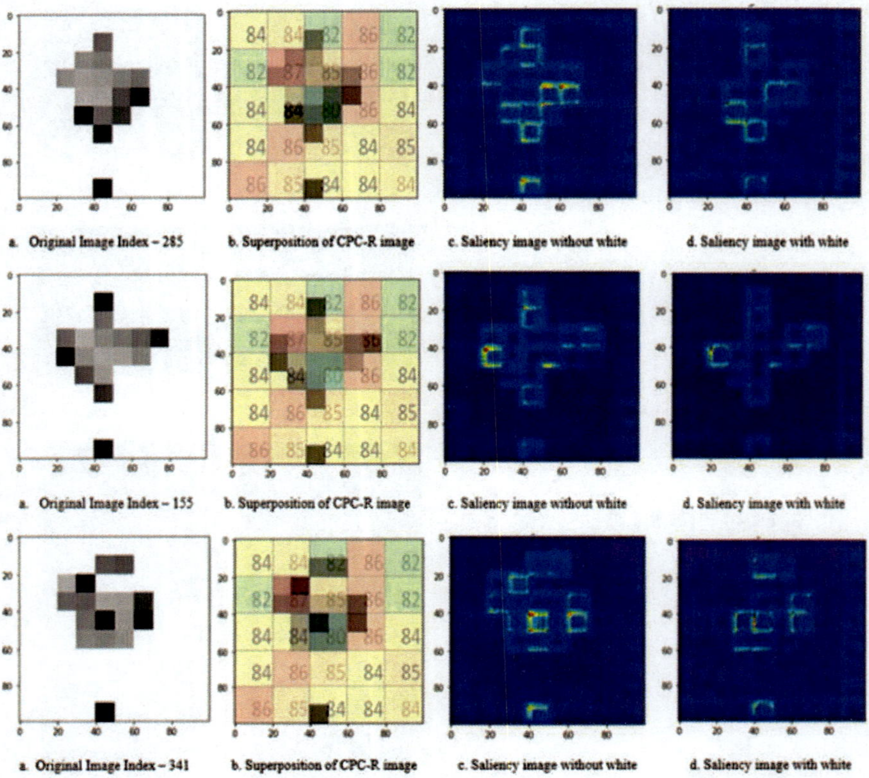

Fig. 13 Informative cell 23 (Ionosphere data)

In contrast with salient darkest cells in Sect. 5, Figs. 12, 13, and 14b show that cell 13, 23, and 17 contain pairs from CPC-R images that are *not the darkest ones*, i.e., are not most salient in these images according to the Guided Back Propagation. This illustrates that ICC algorithm discovers cells that are more relevant than Guided Back Propagation in CPC-R images.

6.3 Informative Cell Covering with Glass and Car Data

Table 9 shows the accuracy of classification of Glass [13] and Car data [14] with their CPC-R images 100 $\times$ 100, covered cells, 60 epochs, and tenfold cross validation. These cells are presented in the ascending order of the accuracy. For Glass data the lowest accuracy is 86.35% for cell 18, and for Car data the lowest accuracy is 87.85% for cell 13, which are the most informative cells.

Figure 15 illustrates informative cells for glass data and Fig. 16 summarizes informative cells for all three datasets (Ionosphere, glass, and car datasets).

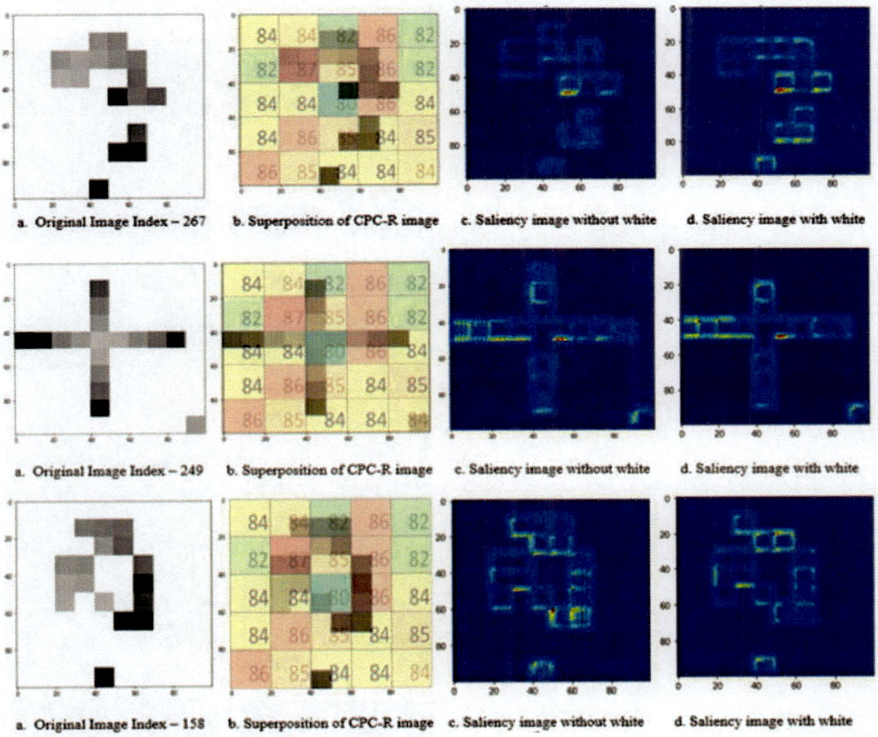

Fig. 14 Informative cell 17 (Ionosphere data)

Table 9 The classification accuracy of glass data with covered cells

Glass data				Car data			
Cells	Accuracy	Cells	Accuracy	Cells	Accuracy	Cells	Accuracy
18	86.35	22	89.54	13	87.85	16	90.58
13	86.37	10	89.68	17	88.07	10	90.98
23	87.01	2	89.99	18	88.17	12	90.93
17	87.34	8	90.06	23	88.58	8	91.65
16	87.55	7	90.19	14	88.99	9	91.66
21	87.69	14	90.35	21	88.99	25	91.88
25	88.07	1	90.58	4	89.02	19	91.91
6	88.16	24	90.89	6	89.22	20	92.11
3	88.29	11	90.89	3	89.52	1	92.11
5	88.56	12	91.09	11	89.53	5	92.11
9	89.01	20	91.19	22	89.99	2	92.11
4	89.18	19	91.27	24	90.01	15	93.68
15	89.54	–	–	7	90.04	–	–

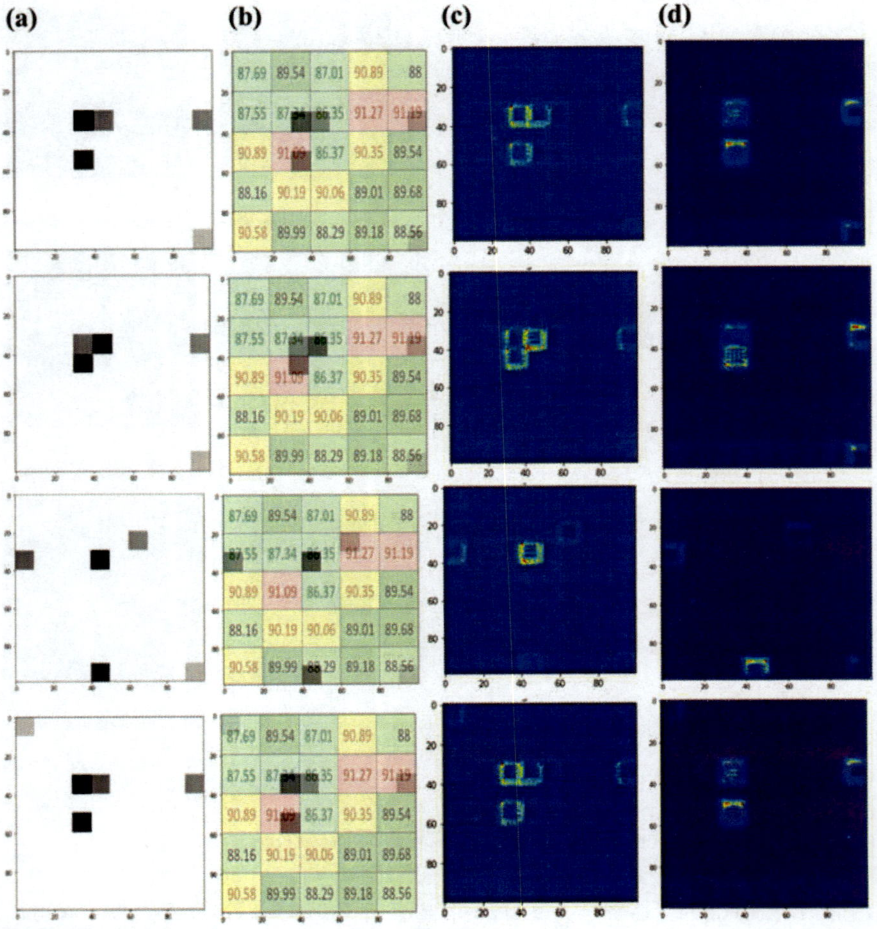

Fig. 15 Glass data informative cells: **a** Original image index, **b** Superposition of CPC-R image, **c** Saliency image without white, **d** Saliency image with white

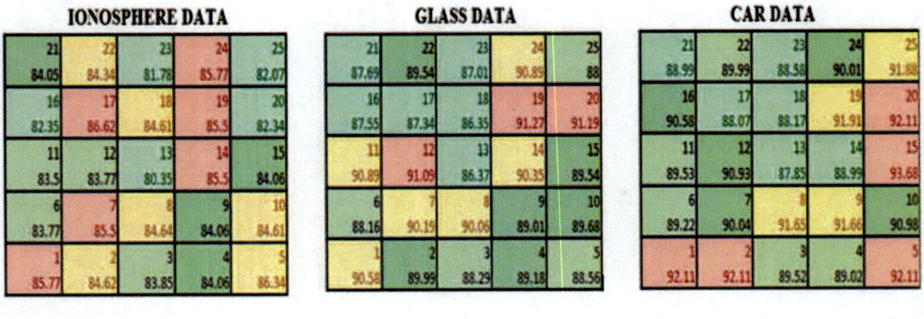

Fig. 16 Informative cells

7 Spiral Filling of Colliding Pairs

This experiment with Ionosphere data uses the adjacent cells method to represent colliding pairs with the order of filling the adjacent cells that differs from the previous experiment **E2** with the same data reported in Sect. 4. A new *spiral filling order* used in this experiment is:

right → down → left → up → lower right → lower left → upper right → upper left

Figure 17a illustrates the method used in E2, and Fig. 17b illustrates a new spiral filling. Table 10 shows the accuracy results of the new method.

The spiral filling provided a better accuracy of 95.47% in comparison with the best results of the previous "cross" adjacent cells experiment (89.98%) in experiment E2 (see Table 5) with Ionosphere data. Table 10 presents the results of cell accuracy for each covered cell that are made white and new adjacent order. There is not much difference between the accuracies for the cells in this new adjacent order. The accuracy ranges from 88.21 to 91.15% with still cell 13 is most informative and cell 17 is among the least informative. The cells in the ascending order by accuracy values are shown in Table 11.

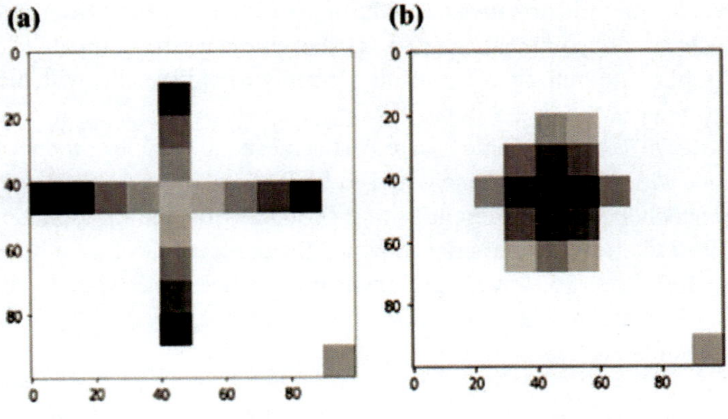

Fig. 17 **a** Adjacent cells with ***cross*** filling, **b** Adjacent cells with new *spiral* filling

Table 10 Ionosphere data experiment (E2 type) with spiral filling of adjacent cells

Image size	Epochs	Model	Cross validation	Accuracy
50 × 50	30	ResNetV2	Tenfold	95.47
50 × 50	50	CNN64	Tenfold	94.03
50 × 50	100	ResNetV2	Stratified Tenfold	93.46
100 × 100	30	CNN64	Tenfold	92.61
100 × 100	50	MLP	Tenfold	91.91

Table 11 Ionosphere data informative cells results with spiral filling of adjacent cells in tenfold cross validation (image size 50 × 50, 30 epochs) for CNN64

Cells	Accuracy	Cells	Accuracy	Cells	Accuracy	Cells	Accuracy	Cells	Accuracy
13	88.21	21	88.75	25	89.28	8	89.61	10	89.8
18	88.32	22	89.06	3	89.32	15	89.64	17	89.8
2	88.43	23	89.16	4	89.61	19	89.75	5	89.91
11	88.46	12	89.18	6	89.61	20	89.75	24	90.08
14	88.57	16	89.19	7	89.61	9	89.79	1	91.15

Table 12 Results of experiments of types E3-E8 for Ionosphere data with spiral filling of collided cells, 50 × 50 images and 50 run epochs

Exp.	Model	Accuracy	Exp.	Model	Accuracy	Exp.	Model	Accuracy
E3	LeNet5	93.85	E5	LeNet5	95.85	E8	LeNet5	95.98
E3	CNN64	93.15	E6	LeNet5	94.54	E8	CNN64	94.56
E4	CNN64	93.98	E6	CNN64	93.78	–	–	–
E5	CNN64	**96.02**	E7	LeNet5	92.98	–	–	–

As was pointed out above the lowest accuracy for this method is 88.21% for cell 13, which is the most informative cell. Multiple experiments have been conducted to explore the impact of combinations of properties of images on accuracy of classification. The Table 12 summarizes the results. It contains only results with the different order that produced the best accuracies.

The above 10 results in Table 12 are best achieved accuracies produced with the spiral filling with the highest one equal to 96.02% while the lowest results in all experiments conducted in different setting is 89.98% for the ionosphere data. The experiments with the MLP classifier provided lower accuracies than other classifications (see Table 10). The above experiments involve both optimizations of the order of pairs (x_i, x_j) and their intensities.

8 Frequency for Informative Cells

In Sect. 6 we identified the most informative cells for several datasets using the ICC method. The goal of this section is identifying which pairs of attributes (x_i, x_j) are most frequent in the most informative cells. This most frequent pairs of attributes are interpreted as **most informative pairs of attributes**.

This process of consists of the following steps:

Table 13 Ionosphere data frequency for cell 13 with and spiral order of filling adjacent cells

Pairs	Values					Pairs	Values				
	(5, 5)	(5, 6)	(6, 5)	(6, 6)	Total		(5, 5)	(5, 6)	(6, 5)	(6, 6)	Total
x_5, x_6	18	8	19	3	48	x_{29}, x_{30}	5	3	8	7	23
x_7, x_8	2	15	19	4	40	x_{31}, x_{32}	4	5	8	6	23
x_1, x_2	38	0	0	0	38	x_{17}, x_{18}	3	4	5	9	21
x_3, x_4	20	0	15	3	38	x_{23}, x_{24}	2	2	6	7	17
x_9, x_{10}	5	12	15	3	35	x_{21}, x_{22}	1	2	4	7	14
x_{11}, x_{12}	2	12	8	6	28	x_{15}, x_{16}	4	0	6	2	12
x_{13}, x_{14}	2	10	6	8	26	x_{19}, x_{20}	0	3	2	6	11
x_{33}, x_{34}	5	4	7	10	26	x_{27}, x_{28}	2	1	3	4	10
x_{25}, x_{26}	4	3	7	11	25	–	–	–	–	–	–

1. Identifying pairs of values (v_i, v_j) of pairs of attributes (x_i, x_j) that belongs to the most informative cell.
2. Computing frequency of each pair of values (v_i, v_j) for each pair of attributes (x_i, x_j) that are in that cell.
3. Order frequencies in descending order for each pair of attributes (x_i, x_j)
4. Find most frequent pairs (x_i, x_j)
5. Identify pair of values (v_i, v_j) for the most frequent pairs (x_i, x_j).

We start showing this process for cell 13 which is the most informative for ionosphere data (see Table 11). By ICC method design each cell includes values of four pairs of attributes (see Fig. 11). For cell 13 these values are (5, 5), (5, 6), (6, 5) and (6, 6) in ionosphere data. The frequencies of these values are summarized in Table 13. It shows that most frequent and informative is pair (x_5, x_6) that appears 48 times with the most frequent pairs of values $(x_5, x_6) = (6, 5)$ and $(x_5, x_6) = (5, 5)$ that have respective frequencies 19 and 18. The next most informative pair is (x_7, x_8) that appears 40 times and the pairs (x_1, x_2) and (x_3, x_4) follow them with frequency 38.

Now we can compare the results for cell 13 with saliency results in Sect. 5 for the same ionosphere data where the darkest pairs (x_1, x_2) and (x_3, x_4) are the most salient ones with pairs (x_5, x_6) and (x_7, x_8) follow them. Thus, in fact, the saliency map distorted the ICC importance order of attributes. It puts (x_1, x_2) and (x_3, x_4) ahead of (x_5, x_6) and (x_7, x_8) due to the fact that the CNN models were trained using "darkest first" coding schema, where (x_1, x_2) is the darkest pair. Training the models using "lightest first" coding schema, where (x_1, x_2) is the lightest pair would reverse the importance of pairs (x_1, x_2) and (x_3, x_4) putting them among least informative pairs. Thus, high salience of (x_1, x_2) and (x_3, x_4) in Sect. 5 is rather accidental.

Note that, not only pairs (x_5, x_6) and (x_7, x_8) are most informative, but their values 5 and 6 are most informative. Thus, the Informative Cell Covering method on CPC-R images allows discovering *informative pairs* of attributes and their *specific values* not *only individual* attributes that traditional attribute covering methods do in n-D data.

Table 14 CPC-R ionosphere data representation

CPC-R value	Actual attribute values	CPC-R value	Actual attribute values
3	[−0.4, −0.2)	6	[0.2, 0.4)
4	[−0.2, 0)	7	[0.4, 0.6)
5	[0, 0.2)	8	[0.6, 0.8)

Table 15 Frequency Results for Cells 14, 18, and 2

Pairs	Cell 14	Cell 18	Cell 2	Pairs	Cell 14	Cell 18	Cell 2
x_1, x_2	0	0	0	x_{19}, x_{20}	38	14	4
x_3, x_4	34	6	0	x_{21}, x_{22}	39	20	7
x_5, x_6	24	3	2	x_{23}, x_{24}	24	19	7
x_7, x_8	42	2	1	x_{25}, x_{26}	40	20	5
x_9, x_{10}	39	8	18	x_{27}, x_{28}	34	12	3
x_{11}, x_{12}	34	10	17	x_{29}, x_{30}	24	26	3
x_{13}, x_{14}	29	11	7	x_{31}, x_{32}	20	27	2
x_{15}, x_{16}	55	7	7	x_{33}, x_{34}	23	43	4
x_{17}, x_{18}	27	15	8	–	–	–	–

Next the ICC schema and CPC-R images allow using a *less detailed data* representation than the actual measurement of the attributes as Table 14 shows. These super-pixels produced high accuracy CNN models with CNN on ionosphere data. Table 15 contains the total frequency results for the next most informative cells 14, 18 and 2.

The *cell 14* consists of boxes with values (7, 5), (5, 6), (8, 8) and (8, 6). It's most frequent/ informative pair (x_{15}, x_{16}) that appears 55 times. The next most informative pair is (x_7, x_8) that appears 42 times. The *cell 18* consists of boxes (5, 3), (5, 4), (6, 3), (6, 4) with the most informative pair (x_{33}, x_{34}) that appears 43 times. The next most informative pair is (x_{31}, x_{32}) that appears 27 times. The *cell 2* consists of boxes (3, 9), (3, 10), (4, 9), (4, 10) with the most informative pair (x_9, x_{10}) that appears 18 times. The most informative pair after it is (x_{11}, x_{12}) that appears 17 times.

In summary the results of this analysis show that most informative pairs of attributes are (x_{15}, x_{16}) with frequency 55 and values 7 or 8, pair (x_5, x_6) with frequency 48 and values 5 or 6, and pair (x_{33}, x_{34}) with frequency 43 times and values in 3 or 4. Thus, this algorithm shows *informative pairs* of attributes *not informative individual attributes*. It highlights *mutual dependence* of the attributes and their *joint impact* on classification accuracy.

9 Comparisons with Other Studies and CPC-R Domain

9.1 Comparisons with Other Studies

The diversity of the ways to verify the results often makes the direct comparison of different methods practically impossible. Some accuracies are reported in publications without identifying its method. Other publications report tenfold cross validation, but typically without providing actual 10 folds of the data. Different random splits of data into these 10 folds, different time limitations on model optimization lead to different accuracies. Table 18 shows comparisons of the accuracies of models presented in this chapter with other classification models.

Other variations include using the ROC curve, F measure, precision, and recall. For these reasons, we compare our results only with published tenfold Cross Validation (CV) accuracies. Tables 16 and 17 summarize the comparison, while a more detailed comparison is presented below.

WBC data. The experiments on WBC data suggest that to get best results with CPC-R approach one needs to optimize: (1) pairing of coordinates, (2) ordering of pairs, (3) values of intensity of cells that encode pairs and (4) using the LeNet5 architecture on CPC-R images. It is a simplest among the explored network architectures that allowed the smallest size of the CPC-R images of 30 $\times$ 30 pixels. The obtained accuracy of 95.9–97.8% for WBC data for tenfold CV in D1-D9 experiments suggest that D1-D9 are the best CPC-R settings to be used. There accuracies are **in the range** of the current published results: such as 94.74, 94.36 and 95.27%, for interpretable C4.5, J4 and fuzzy decision trees, respectively, on tenfold CV [24]. Higher accuracy from 97.97 to 99.51% on the tenfold CV also summarized in [24] for less interpretable methods such as SVM and Neural Networks.

Swiss roll. Table 16 shows that CPC-R accuracies are slightly higher than obtained in [1] for tenfold CV of CNN models on the original numeric data without transforming them to images, and on images constructed [1].

The reported max accuracies for 2-D Swiss roll [25] vary dramatically from 51% to 71.24% and 96.6% reported in [25] for Autoencoder, PCA and Isomap, respectively. Our accuracies are at the same level. Beyond Swiss rolls, similar spirals have been explored using DL models in [26] with reported very high accuracy reached.

Our results show CPC-R image with DL methods can *model manifolds such Swiss roll in the CPC-R visual form as images*. Thus, we expanded the methods to model and discover manifolds.

Table 16 Comparison of accuracies with results from [1]

	Numeric data [1]	Images [1]	CPC-R
Swiss roll 2-D	72.50	97.43	97.87
Swiss roll 3-D	96.18	97.55	97.56
WBC	96.92	97.22	97.66

Table 17 Comparison of different classification models

Classification algorithm accuracy	Accuracy
Breast cancer data	
CPC-R with cross filling	**95.58**
CPC-R with spiral filling	**97.89**
GLC-R 5[3]	95.61
Deep learning in mammography [31]	94
DWS-MKL [32]	96.9
LMDT algorithm [33]	95.74
Ionosphere data	
CPC-R with cross filling	**95.01**
CPC-R with spiral filling	**96.02**
DWS-MKL [32]	92.3
ITI algorithm [33]	93.65
Deep extreme learning machine and its application in EEG classification[34]	94.74
Glass data	
CPC-R with cross filling	**96.80**
C4.5 algorithm [33]	70.23
Glass classification using artificial neural network/ANN model [35]	96.7
Comparative analysis of classification algorithms using WEKA/MLP [36]	67.75
Car data	
CPC-R with cross filling	**96.8**
Performance comparison of data mining algorithms [37]	93.22
A large-scale car dataset for fine-grained categorization and verification [38]	83.22

Ionosphere data. The range of reported accuracies for ionosphere data is from 93% to 100% on training and validation data [24] for 70% to training and 30% to validation data and tenfold CV from different sources: 93% for MLP, 94.87% for C4.5, 94.59% for Rule Induction, 97.33% for SVM without converting n-D data to images. We obtained 94.13% by using CPC-R that is in the range of the published results.

Therefore, the CPC-R methodology is competitive with advantages of using CPC-R images uniformly with any DL and MLP algorithm along with lossless n-D data visualization as CPC-R image.

Glass data. The reported results range from 68.2% for C4.5 to 98.13% for random forest [27–29] but without presenting the way of getting these numbers, or by using the area under the ROC curve and F [30] with tenfold CV. The indirect comparison shows that our achieved 96.01% accuracy in tenfold cross validation is also competitive and in the range of results reported in the literature.

The max accuracies for ionosphere and glass data are different. The experiments with CPC-R images resulted in accuracies between 69.8 and 95.47% for the ionosphere data and from 84 to 96.8% for glass data for different architectures. The achieved accuracy result for the saliency experiment is 94.4%.

Experiments E1 and E2 reported above had shown better accuracy with Inception ResNetV2 architecture for both datasets. The achieved accuracy for the ionosphere data is 95% in E10 and glass data is 97% in E2, which means the CPC-R can be considered as competitive with other algorithms in accuracy. The accuracy is low with small images 30×30 in experiments E2 and E3, where each pair occupied a single pixel. MLP architecture provided lower accuracies than other classifications. The best of the accuracies involves both optimizations of pairs in the order of the coordinates and their intensities. In summary, the proposed CPC-R algorithm is a competing alternative to other Machine Learning algorithms as conducted experiments had shown.

9.2 CPC-R Application Domain

In the previous section we compared accuracies obtained in this work with published in the literature showing that they are competitive. Can we expect that this competitiveness will sustain on other data? This is still the open questions for classical methods, which exist for many years. The experimental comparison on datasets is always limited in the number of experiments, size and variability of data used. Therefore, it is desirable to explore theoretical options to justify the method.

Below we explore this option based on two theorems which state that:

- Neural Networks are universal approximators of continuous functions (Universal approximation theorem), and
- Every multivariate continuous function can be represented as a superposition of continuous functions of one variable (Kolmogorov-Arnold superposition theorem [39]).

By combining functions of one variable we can have continuous functions of two variables. The CPC-R pairs (x_i, x_j) represent this situation. Both theorems are existence theorems not constructive ones. How can it be done constructively? One way is expanding experiments with CPC-R and CNN to new data. Another way is analyzing relevant experiments already conducted with positive results. One of them is another pairs-based method [40] known as the extended Standard Generalized Additive Models called GA2M-models. It consists of univariate terms and a small number of pairwise interaction terms. These authors announced: "the surprising result that GA2M-models have almost the same performance as the best full-complexity models on a number of real datasets".

Is it surprising is that CNN can find patterns in CPC-R images? CNN was associated with discovering *local features* due to the layer design that generalizes nearby pixels. CPC-R images also localize similar pairs by design and increase the neighborhoods and CNN efficiency by (1) locating collided cells in the adjacent cells, and (2) using multiple pixels to represent each pair/cell (x_i, x_j) by increasing the size of each cell of the CPC-R image. Next, often individual pairs are enough [40] and local neighborhoods not necessary. We use perturbation (see experiments D9) to ensure that important pairs are not missed. Also, weights of features assigned by CNN capture importance of features that are far away from each other. All these aspects point to the sources of CNN's success on CPC-R images.

CPC-R domain. *Where is the domain of application of the CPC-R methodology and current versions of the CPC-R algorithm?* It is classification tasks where the *sets of pair relations* in the data are sufficient for discovering efficient classification models by respective ML algorithms. Table 18 elaborates this statement. The base version of CPC-R 1.0 algorithm can be efficient for data where pairs (x_i, x_j) for each n-D point **x** are quite unique. i.e., without equal pairs $(x_i, x_j) = (x_k, x_m)$ when $i \neq k$ and $j \neq m$, or with rare and unimportant such equal pairs. For the data where the adjacent cells for each pair (x_i, x_j) in the CPC-R image are free or almost free from other pairs (x_t, x_s) from n-D point **x** the version of the CPC-R algorithm 2.0 is used. For the data where the adjacent cells are occupied by other pairs the split cells version of CPC-R algorithm 2.0 is applicable. For the data where multiple pairs collide the version of CPC-R 2.0 that selects the most important one can be used with its intensity assigned to the cell.

If above presented versions do not produce high accuracy, then versions of CPC-R 3.0 algorithm that adds context need to be used. It adds the background average images of different classes if they are distinct. And finally, version 4.0 is applied to optimize intensities and attribute pairs. The theoretical basis to ensure that such datasets and classification models exist provide mentioned Kolmogorov-Arnold and universal approximator theorems. The diversity of datasets used in our experiments shows that CPC-R methodology can handle variety of datasets.

Table 18 CPC-R use scenarios

CPC-R version	Use scenario
1.0 Base version	Data without pair collision or unimportant collision
2.0 Collision treatment	2.1 Adjacent cells are almost free–use adjacent cells 2.2 Adjacent cells are occupied–split cells 2.3 First pair is most important–use darkest intensity
3.0 Adding context	Average images of different classes are distinct
4.0 Optimization	Data with non-optimal intensities and attribute pairs

10 Generalization of CPC-R Methodology and Future Work

10.1 Generalizations

From CPC-R images to GLC images. General Line Coordinates (GLC) lossless visualizations use polylines (directed graphs) in 2-D to represent each n-D point [7]. In contrast each CPC-R image encodes an n-D point using intensities of pixels or their colors without connecting nodes by edges. The same CPC-R intensity-based methodology is applicable to any GLC representation of n-D such as Shifted Paired Coordinates (SPC) [7] to produce images that we denote as **GLC-R** and **SPC-R** images, where R stands for *raster*. Respectively, this methodology contains two major steps: (1) producing GLC-R images for each n-D point and then (2) feeding these GLC-R images to CNN or other algorithms to be classified. Another aspect of generalization of CPC-R images is considering it as a part of the full 2-D Machine Learning methodology proposed in [41].

Context generalization. In experiments D1-D8 we used the *mean* CPC-R images of classes as a background for CPC-R images of each n-D point to add context. Let's denote this image B_{mean}. In this approach some background cells are occluded by the cells of the CPC-R images of the n-D point, loosing this background information. We can decrease such loss of context by computing the *differences* between CPC-R image intensities and the mean image intensities and producing the image of these differences. Let's denote this image as B_{dif}. It can be put side-by-side with the image B_{mean} forming a new image that provide more context. Another option to add context is using *frequencies* of each pair within each class. The high frequency of some pairs of values in many images of a given class can indicate patterns specific for this class. The *frequency* of each pair within each class can be computed and the higher frequency is encoded by the higher intensity producing a version of a heatmap image. Let's denote this image as B_{freq}. It can be combined with B_{mean} and B_{dif} by putting it side-by-side with them in a joint image to be used by CNN.

Concentrating cases of a class in local 2-D area. When we have only two attributes (X, Y), each case of each class is a 2-D point in a standard Cartesian coordinates. The ML expectation is that points of one class will concentrate in one area and points of another class will concentrate in another area and, respectively, a linear or nonlinear discrimination function can be discovered such that the points of one class are in one subspace, say, divided by a hyperplane, from the points of another class. These points can be spread in the subspace. Respectively, the salient points can be far from each other.

Similarly, in the multidimensional situation, we expect a pattern such that all points of class 1 will be in one subspace and points of class 2 will be in another subspace. However, the Collocated Paired Coordinates (CPC) space defined in [7] may not provide this property, because the graph x* of n-D point **x** can spread on the CPC space and may not reveal this closeness between n-D points in the same way as in the n-D space.

Therefore, the algorithm is needed that will concentrate cases of the same class in a local area in 2-D visualization space. Such algorithm was proposed for Shifted Paired Coordinates (SPC) in [7] where the center of the class **c** is represented losslessly as a single point in 2-D. all n-D cases that are in the n-D hypercube centered in **c**, will be in a square in SPC. This property is also true for SPC-R images.

The next question is how to assign the intensity to the center of the class, which integrates all pairs (nodes) that have different intensities. The same issue exists for other n-D points that combine two or more pairs (nodes) to a single 2-D point. For instance, consider a graph x* of the n-D point **x** with three 2-D points where one 2-D point $\mathbf{p}_{1,2}$ represents two pairs $\mathbf{p}_1$ and $\mathbf{p}_2$ (two graph nodes). One option is assigning *max* or *mean* of the intensities of these nodes,

$$\mathrm{I}(\mathbf{p}_{1,2}) = \max(\mathrm{I}(\mathbf{p}_1), \mathrm{I}(\mathbf{p}_2)), \mathrm{I}(\mathbf{p}_{1,2}) = (\mathrm{I}(\mathbf{p}_1) + \mathrm{I}(\mathbf{p}_2))/2,$$

which is a lossy approach. Other options are: (i) *splitting* the cell assigned to $\mathbf{p}_{1,2}$ to strips of different intensities or (ii) using *adjacent cells* to put the values. Both have been done for colliding cells and the later worked better (see Sects. 2 and 3).

Saliency analysis with SPC-R images. The CNN saliency analysis in SPC-R can be more meaningful than in CPC-R because CNN focuses on local features and SPC-R has more opportunity to localize parts of the graphs than CPC-R.

Incomplete data with GLC-R. Incomplete datasets are common in machine learning. One of the ways to deal with them is introducing an artificial value that can encode empty spots in the data. For instance, the attribute values can be in the range [0, 10] then −1 value can be assigned to empty values in this attribute, which can be visualized.

10.2 Toward CPC-R Specific Algorithms

Can simpler or more specialized models substitute CNN for CPC-R images? So far, our experiments had shown that MLP provided lower accuracy with one exclusion. In experiment E2 for Glass data and experiment E3 for Ionosphere data, the MLP provided the better accuracy than CNNs) reported in Sect. 4 in Table 5. while further studies can find simpler MLP or other models.

The traditional deep learning CNN algorithms do well on *rich natural scenes* where convolutional layers exploit the local spatial coherence in the image. These learning algorithms search for hierarchy of informative *local groups of pixels.* In contrast, MLP ignores the spatial information by using *flattened vectors* as inputs allowing to capture non-local features. The CNN algorithms capture the *non-local relations* between features *only at the later flattened layers* of CNN.

The *CPC-R artificial images are much simpler* and with less variability than natural images. Each CPC-R image consists of a set of squares (cells) with fixed intensity of all pixels inside of each square. Thus, discovering local features in CPC-R images is simpler and the full power of deep learning algorithms seems *redundant for discovering such simpler local features.* On the other side, an algorithm for CPC-R images needs to *discover complex non-local relations* between squares in CPC-R images.

CNN on CPC-R images aggregates artificial intensities of pixels to produce aggregated features such as vertical and horizontal lines by using respective masks. Thus, these masks *rediscover* edges of squares that are already known. See Figures in Sect. 5 for the saliency maps. So, this feature discovery step can be removed or modified to make them more useful. Modified feature discovery steps can include CPC-R specialized masks. The CPC-R *specialized masks* should aggregate pairs of values (x_i, x_j) and their intensities, which are artificial values in contrast with natural images, where the intensity is actual amount of light that is coming to a physical image cell.

In general, the future work is developing a specialized image recognition algorithm to discover patterns in CPC-R images. In accordance with the analysis above, the design of these algorithms should focus on *modifying* and *simplifying convolutional layers* or even removing them, but with *more complex layers that capture not-local relations.* Therefore, it was suggested to use a MLP neural network without deep learning convolutional layers at all. However, so far, it did produce a better accuracy with one exclusion as we reported above. This indicated that full removal of convolutional layers is likely should be avoided, but CPC-R specific layers need to be developed and explored. Another possible avenue of the future work is defining an optimal size of cells in CPC-R images. A preliminary hypothesis is that MLP will work better on smaller cells than CNN.

Future CPC-R specific algorithms need to aggregate the values of coordinates meaningfully by developing CPC-R relevant masks. For example, in a coordinate pair $(x_i, x_j) = (3, 7)$ with intensity $I(3, 7) = 0.3$, both 3 and 7 have meaning as values of x_i and x_j, but 0.3 is just a coding that preserves ordering of the pairs of attributes. We can use another value that also preserves the order of pairs.

Commonly CNN aggregate intensities of 2×2 adjacent pixels by computing their max. For CPC-R images it is a form of generalization of artificial intensities of four adjacent pairs. Consider, an example of adjacent pairs $(x_3, x_4) = (3, 7)$, $(x_5, x_6) = (4, 7)$, $(x_9, x_{10}) = (3, 6)$ and $(x_7, x_8) = (4, 6)$ with respective intensities 0.9, 0.5, 0.4, and 0.3. The max of these values is 0.9, which means that we ignore all pairs, but $(x_3, x_4) = (3.6)$. It has a meaning in natural scenes focusing on most distinct pixels, but in CPC-R images its distinction is only a result of the coding schema where the increased values of intensities are assigned to preserve the order of pairs (x_i, x_j). This max pair $(x_3, x_4) = (3, 7)$ can be as informative as ignored pairs or less informative than they are. We could use the opposite order that would give priority to $(x_9, x_{10}) = (3.6)$. Thus, for CPC-R, max aggregation does not show the most distinct pair, but a pair that is closest to the beginning of the n-D point $\mathbf{x}$. In natural scenes, a pixel with higher intensity is the most prominent pixel in the vicinity and CNN amplifies it.

Despite of this counter-intuitive masks CNN produces a high accuracy on CPC-R images that seems like a unexplained magic. This leads us to the question of explainability of CNN and Deep Learning in general. We hope that answering this question for CPC-R in the future work will be helpful to understand the general reason of CNN and DL efficiency.

10.3 CPC-R Anonymous ML Methodology

The CPC-R images strip detailed information of values of attributes, not showing numeric values of the attributes. To restore attribute values from CPC-R images one needs to know (1) whether the coding schema is increasing or decreasing, (2) the location of the origin of the coordinates, (3) pixel conversion schema (float and integer numbers are converted to pixel coordinates). This opened the opportunity for data anonymization for machine learning. Moreover, additional coding elements can be added specifically for data anonymization.

In the *CPC-R anonymous ML methodology*, the CPC-R algorithm converts numeric n-D data into anonymized CPC-R images and CNN builds a model on them without explicitly using numeric values of the attributes of the cases. Such images can be transferred to the SaS (Software as Service) centers for processing by powerful algorithms on the high-performance platforms.

11 Conclusion

This chapter had shown that the combination of CNN and MLP algorithms with CPC-R images, produced from numeric n-D data by CPC-R algorithm, is a *feasible* and beneficial ML methodology. Its advantages include *lossless visualization* of n-D data, and the abilities to add *context* to the visualization by overlaying the images of n-D points with the mean images of competing classes. The CNN was most successful in these context-enhanced CPC-R images in our experiments accompanied by optimization of pairing of attributes and pixel intensities. The CPC-R methodology allows *visual explanation* of the discovered models by tracing back to the informative pairs of attributes (x_i, x_j) and visualizing them using a heatmap. Data anonymization also can be accomplished by using CPC-R images that represent numeric n-D data as anonymized images, which is important for many machine learning tasks. Enhancing CPC-R methodology in multiple ways outlined in Sect. 10 is the goal of the future research. The Python code at GitHub can be made available upon request.

References

1. Dovhalets, D., Kovalerchuk, B., Vajda, S., Andonie, R.: Deep learning of 2-D images representing n-D data in general line coordinates. In: Intern. Symp. on Affective Science and Engineering, pp. 1–6 (2018). https://www.jstage.jst.go.jp/article/isase/ISASE2018/0/ISASE2018_1_18/_pdf
2. Sharma, A., Vans, E., Shigemizu, D., Boroevich, K.A., Tsunoda, T.: Deep insight: a methodology to transform a non-image data to an image for convolution neural network architecture. Nat. Sci. Rep. **9**(1), 1–7 (2019)
3. Kovalerchuk, B., Agrawal, B., Kalla, D.: Solving non-image learning problems by mapping to images, 24th International Conference Information Visualisation, Melbourne, Victoria, Australia (2020), pp. 264–269, IEEE. https://doi.org/10.1109/IV51561.2020.00050
4. van der Maaten, L.: Dos and don'ts of using t-SNE to understand vision models, CVPR 2018, Tutorial
5. Rodrigues, N.M., Batista, J.E., Trujillo, L., Duarte, B., Giacobini, M., Vanneschi, L., Silva, S.: Plotting time: on the usage of CNNs for time series classification (2021). arXiv:2102.04179
6. Sharma, A., Kumar, D.: Non-image data classification with convolutional neural networks (2020). arXiv:2007.03218
7. Kovalerchuk, B.: Visual knowledge discovery and machine learning. Springer (2018)
8. Wolberg, W., Mangasarian, O.: UCI ML repository: Breast Cancer Wisconsin Data Set (1991). https://archive.ics.uci.edu/ml/datasets/breast+cancer+wisconsin+%28original%29
9. Szegedy, C., Ioffe, S., Vanhoucke, V., Alemi, A.A.: Inception-v4, inception-resnet and the impact of residual connections on learning. In: 31st AAAI Conference on Artificial Intelligence (2017), https://www.aaai.org/ocs/index.php/AAAI/AAAI17/paper/viewPDFInterstitial/14806/14311
10. LeCun, Y., Bottou, L., Bengio, Y.: Gradient-based learning applied to document recognition. IEEE Proc. **86**(11), 2278–2324 (1998)
11. Balasubramanian, M., Schwartz, E.L.: The isomap algorithm and topological stability. Science **295**(5552), 7–7 (2002)
12. Asuncion, A., Newman, D.: Ionosphere data set (2007). https://archive.ics.uci.edu/ml/datasets/Ionosphere
13. Spiehler, V.: Glass identification data set (1987). https://archive.ics.uci.edu/ml/datasets/Glass+Identification
14. Bohanec, M., Zupan, B.: UCI machine learning repository: car evaluation data set (1997). https://archive.ics.uci.edu/ml/datasets/car+evaluation
15. Ernst, N.: Saliency map. Scholarpedia **2**(8), 2675 (2007)
16. Radhakrishna, A., Sabine, S.: Saliency detection for content-aware image resizing. In: 2009 16th IEEE International Conference on Image Processing (ICIP), pp. 1005–1008. IEEE (2009)
17. Simonyan, K., Vedaldi, A., Zisserman, A.: Deep inside convolutional networks: visualising image classification models and saliency maps (2013). arXiv:1312.6034
18. Seunghoon, H., Tackgeun, Y., Suha, K., Bohyung, H.: Online tracking by learning discriminative saliency map with convolutional neural network. In: International on Conference on Machine Learning, pp. 597–606. PMLR (2015)
19. Kotikalapudi, Raghavendra and contributors, Keras-vis (2017). https://github.com/raghakot/keras-vis
20. Selvaraju Ramprasaath, R., Michael, C., Abhishek, D., Ramakrishna, V., Devi, P., Dhruv, B.: Grad-cam: visual explanations from deep networks via gradient-based localization. In: Proceedings of the IEEE International Conference on Computer Vision, pp. 618–626 (2017)
21. Schreiber, A.: Saliency maps for deep learning, Part 1: Vanilla Gradient (2019). https://andrewschrbr.medium.com/saliency-maps-for-deep-learning-part-1-vanilla-gradient-1d0665de3284
22. Kim, B., Seo, J., Jeon, S., Koo, J., Choe, J., Jeon, T.: Why are saliency maps noisy? Cause of and solution to noisy saliency maps. In: IEEE CVF International Conference on Computer Vision Workshop, pp. 4149–4157 (2019)

23. He, S., Lau, R.W., Liu, W., Huang, Z., Yang, Q.: SuperCNN: a superpixelwise convolutional neural network for salient object detection. Int. J. Comput. Vision **115**(3), 330–344 (2015)
24. Kovalerchuk, B., Gharawi, A.: Decreasing occlusion and increasing explanation in interactive visual knowledge discovery. In: International Conference on Human Interface and the Management of Information, pp. 505–526. Springer (2018)
25. Van Der Maaten, L., Postma, E., Van den Herik, J.: Dimensionality reduction: a comparative. J Mach Learn Res. **10**(66–71), 13 (2009)
26. Smilkov, D., Carter, S., Sculley, D., Viégas, F.B., Wattenberg, M.: Direct-manipulation visualization of deep networks (2017). arXiv:1708.03788
27. Aldayel, M.S.: K-Nearest Neighbor classification for glass identification problem. In: 2012 International Conference on Computer Systems and Industrial Informatics, pp. 1–5. IEEE (2012)
28. Khan, M.M., Arif, R.B., Siddique, M.A., Oishe, M.R.: Study and observation of the variation of accuracies of KNN, SVM, LMNN, ENN algorithms on eleven different datasets from UCI machine learning repository. In: 2018 4th International Conference on iCEEiCT, pp. 124–129. IEEE (2018)
29. Mohit, R.R., Katoch, S., Vanjare, A., Omkar, S.N.: Classification of complex UCI datasets using machine learning algorithms using hadoop. In: IJCSSE, vol. 4, pp. 190–198 (2015)
30. Prachuabsupakij, W., Soonthornphisaj, N.: Clustering and combined sampling approaches for multi-class imbalanced data classification. In: Advances in IT and Industry Applications, pp. 717–724. Springer (2012)
31. Becker, S., Marcon, M., Ghafoor, S., Wurnig, C., Frauenfelder, T., Boss, A.: Deep learning in mammography: diagnostic accuracy of a multipurpose image analysis software in the detection of breast cancer. Invest. Radiol. **52**(7), 434–440 (2017)
32. Junbao, L., Tingting, W., Huayou, S.: Dws-mkl: Depth-width-scaling multiple kernel learning for data classification. Neurocomputing **411**, 455–467 (2020)
33. Eklund, P., Hoang, A.: A performance survey of public domain supervised machine learning algorithms. Austr. J. Intell. Inform. Syst. **9**(1), 1–47 (2002)
34. Ding, S., Zhang, N., Xu, X., Guo, L., Zhang, J.: Deep extreme learning machine and its application in EEG classification. Math. Probl. Eng. (2015)
35. El-Khatib, M.J., Abu-Nasser, B.S., Abu-Naser, S.S.: Glass Classification using Artificial Neural Network (2019). http://dstore.alazhar.edu.ps/xmlui/bitstream/handle/123456789/144/ELKGCUv1.pdf?sequence=1&isAllowed=y
36. Arora, R.: Comparative analysis of classification algorithms on different datasets using Weka. Int. J. Comp. Appl. **54**(13) (2012)
37. Awwalu, J., Ghazvini, A., Bakar, A.A.: Performance comparison of data mining algorithms: a case study on car evaluation dataset. Int. J. Comput. Trends Technol. **13**(2) (2014)
38. Yang, L., Luo, P., Change Loy, C., Tang, X.: A large-scale car dataset for fine-grained categorization and verification. In: Proceedings of the IEEE CVPR Conference, pp. 3973–3981 (2015)
39. Braun J.: On Kolmogorov's Superposition Theorem and its Applications, p. 192. SVH Verlag (2010)
40. Lou, Y., Caruana, R., Gehrke, J., Hooker, G.: Accurate intelligible models with pairwise interactions. In: 19th SIGKDD, pp. 623–631. ACM (2013)
41. Kovalerchuk, B., Phan, H.: Full interpretable machine learning in 2D with inline coordinates. In: 25th International Conference Information Visualisation, Australia (2021) Vol. 1, pp. 189-196, IEEE, https://doi.org/10.1109/IV53921.2021.00038

Self-service Data Classification Using Interactive Visualization and Interpretable Machine Learning

Sridevi Narayana Wagle and Boris Kovalerchuk

Abstract Machine learning algorithms often produce models considered as complex black-box models by both end users and developers. Such algorithms fail to explain the model in terms of the domain they are designed for. The proposed *Iterative Visual Logical Classifier* (IVLC) is an interpretable machine learning algorithm that allows end users to design a model with more confidence and without compromising the accuracy. Such technique is especially helpful for tasks like cancer diagnostics with high cost of errors. With the proposed interactive and lossless multidimensional visualization, end users can identify the pattern and make explainable decisions, which is not possible in black box machine learning methodologies. The interpretable IVLC algorithm is supported by the Interactive Shifted Paired Coordinates Software System (SPCVis), which is a lossless multidimensional data visualization system. The interactivity provides flexibility to the end user to perform data classification as self-service without a machine learning expert. Interactive pattern discovery is challenging for data with hundreds of dimensions/features. To overcome this problem, this chapter proposes an automated classification approach combined with new Coordinate Order Optimizer (COO) algorithm and a Genetic algorithm. The COO algorithm automatically generates the coordinate pair sequences that best represent the data separation and the genetic algorithm optimizes the IVLC algorithm by automatically generating the areas for data classification. The feasibility of the approach is shown by experiments on benchmark datasets covering both interactive and automated processes.

Keywords Shifted paired coordinates · Interactive data visualization · Democratized machine learning · Iterative visual logical classifier · Coordinate order optimizer · Genetic algorithm

---

S. N. Wagle · B. Kovalerchuk (✉)
Department of Computer Science, Central Washington University, Ellensburg, WA, USA
e-mail: borisk@cwu.edu

S. N. Wagle
e-mail: sridevi.wagle@cwu.edu

B. Kovalerchuk et al. (eds.), *Integrating Artificial Intelligence and Visualization for Visual Knowledge Discovery*, Studies in Computational Intelligence 1014,
https://doi.org/10.1007/978-3-030-93119-3_4

1 Introduction

The exponential growth of Machine Learning (ML) has resulted in smart applications making critical decisions without human intervention. However, people with no technical background find it often difficult to rely on ML models, since many of them are *black-boxed* [1]. *Interpretability* plays a crucial role in deciphering behind-the-scenes actions in a ML algorithm, giving more clarity to the end user with more confidence compared to black-box ML models [2].

Interpretable ML techniques combined with *visualization of high-dimensional data* unfold numerous ways to discover deep hidden data patterns. However, the traditional visualization of multidimensional data is limited for this being *lossy* and *irreversible* not allowing to see multidimensional data fully without loss of n-D information. With the help of new lossless n-D data visualization, it is now possible to maintain structural integrity of the data. The major advantage is that these lossless techniques are completely reversible, representing n-D data in 2-D without loss of n-D information. Thus, the end user can see those data fully and discover patterns in them without loss of n-D information. It is not possible with methods like Principal Component Analysis when data are visualized with in the first two principal components [3, 4].

In this chapter, we focus on interpretable data classification techniques using a combination of interactive data visualization and analytical rules. The Shifted Paired Coordinates (SPC) system [4] is a lossless and a more compact way to visualize multi-dimensional data when compared to other lossless data visualization like Parallel Coordinates. SPC allows discovering patterns more efficiently since the number of lines required to display the data is reduced by half [4].

However, as the size of the data increases, pattern discovery becomes challenging due to occlusion. To reduce occlusion and expose hidden patterns, data representation needs to be reorganized. This can be achieved by applying interactive techniques such as changing the order of coordinates, swapping within the coordinate pair etc. These interactive capabilities allow the end-users to intervene and optimize the classification model generated by the machine thereby improving the overall model performance. To further leverage the classification technique, areas within the coordinate pairs are discovered either interactively or automatically, unravelling the deep hidden patterns in the data. Analytical rules are then built on the areas discovered to classify the data.

The data classification approach using analytical rules is implemented using IVLC algorithm [5] wherein the areas and analytical rules are generated in iterations until all the data are covered. This implementation works for smaller datasets, the areas and analytical rules can be generated interactively by the end users. This chapter expands this work to large datasets by adding automation and more visual operations.

For the classification of larger data with hundreds of dimensions, interactive methods alone do not suffice. Due to high occlusion for such datasets, discovery of patterns purely on interactive methods becomes challenging where deep hidden patterns cannot be detected by humans. In this chapter, automation for the classification approach is implemented in two stages:

First Stage: Order of coordinates are optimized using COO algorithm. It is used to find the best coordinate orders for the Shifted Paired Coordinates System where the data separation can be visually identified along the vertical axis of each coordinate pair.

Second Stage: This stage involves generating the areas within the coordinate pairs with high purity or fitness i.e., areas with high density of data belonging to same class are generated. The genetic algorithm is used for the automatic generation of areas with high purity. It involves generation of random areas and that are mutated and altered over generations to obtain the areas with high purity or fitness [6].

Several experiments are conducted on benchmark datasets like Wisconsin Breast Cancer (WBC), Iris, Seeds and Air Pressure System (APS) in Scania Trucks. The experiments are performed using both interactive and automated approach using tenfold cross validation with worst case heuristics [7] where the initial validation set contains the data from one class overlapping with another class and vice versa. The results obtained with both interactive and automated techniques were on par with the published results in other studies.

This chapter is organized as follows. Section 2 presents the interactive shifted paired coordinates visualization system, Sect. 3 presents the IVLC algorithm and the worst-case k-fold cross validation approach. Section 4 reports experiments with interactive data classification approach. Section 5 describes automation of classification using Coordinate Order Optimizer and Genetic Algorithms. Section 6 reports experiments with automated data classification approach. Section 7 summarizes experimental results in comparison with published results and Sect. 8 concludes the chapter with its summary and future directions.

2 Interactive Shifted Paired Coordinates System

The proposed system is based on the Shifted Paired Coordinates (SPC) [4] that are described in this section.

The Shifted Paired Coordinates visualize data losslessly using coordinate pairs where each pair of data is represented on a two-dimensional plane. Figure 1 represents an 8-D data (8, 1, 3, 9, 8, 3, 2, 5) using SPC, where pair (8, 1) is visualized in (X_1, X_2), the next pair (3, 9) is visualized in (X_2, X_4) and so on. Then these points are connected to form a graph.

Same data can be displayed in multiple ways using different combinations of pairs of coordinates. Figure 2a, b represent the same data with (X_8, X_1), (X_3, X_7), (X_5, X_2) and (X_6, X_4) sequence of coordinates and (X_2, X_6), (X_3, X_7), (X_8, X_1) and (X_5, X_4) sequence of coordinates respectively.

Figure 3 represents a real-world dataset from UCI Machine Learning Repository [8] consisting of 683 cases of breast cancer data. Coordinate X_2 is duplicated to get even number of coordinates. Although Shifted Paired Coordinates system provides lossless data visualization, discovering patterns in the data becomes challenging due

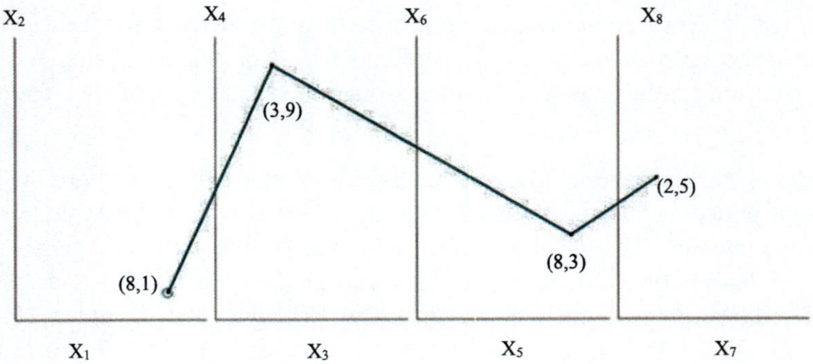

Fig. 1 Representation of 8-D data (8, 1, 3, 9, 8, 3, 2, 5) in SPC

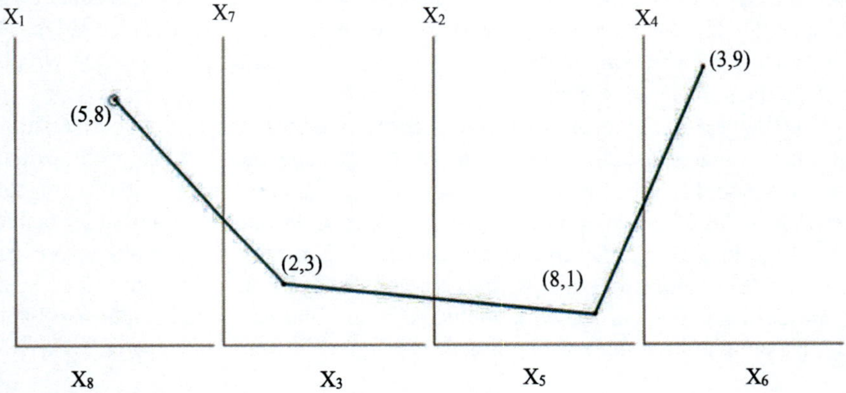

(a). 8-D data with (X_8, X_1), (X_3, X_7), (X_5, X_2) and (X_6, X_4) coordinate pair sequence.

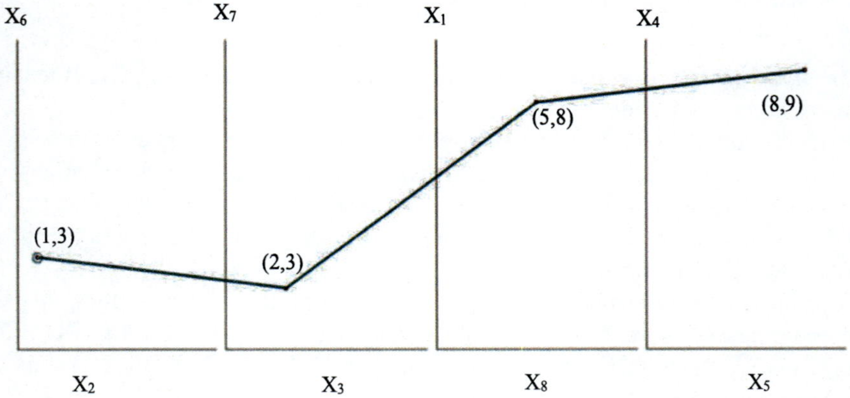

(b). 8-D data with (X_2, X_6), (X_3, X_7), (X_8, X_1) and (X_5, X_4) coordinate pair sequence.

Fig. 2 Representation of 8-D data (8, 1, 3, 9, 8, 3, 2, 5) in SPC with different coordinate pair sequences

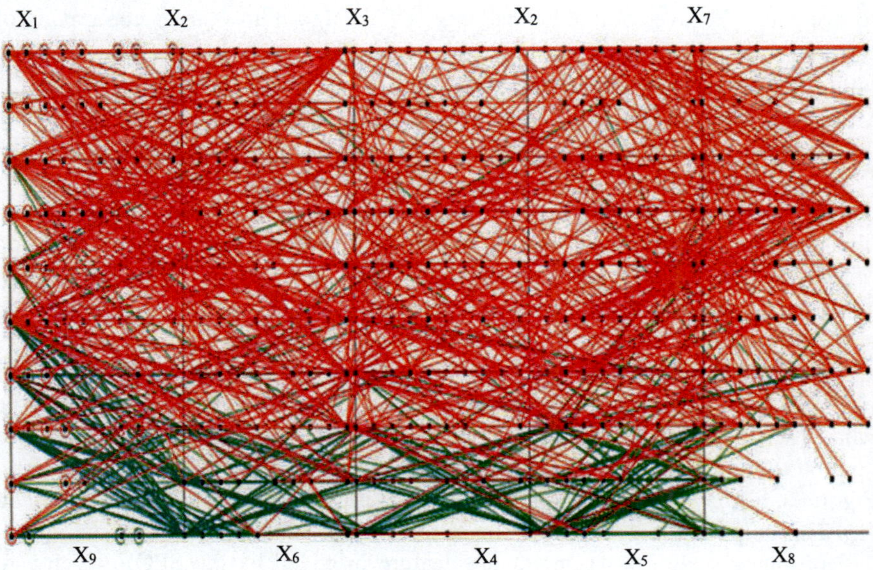

(a). WBC data with red class on top.

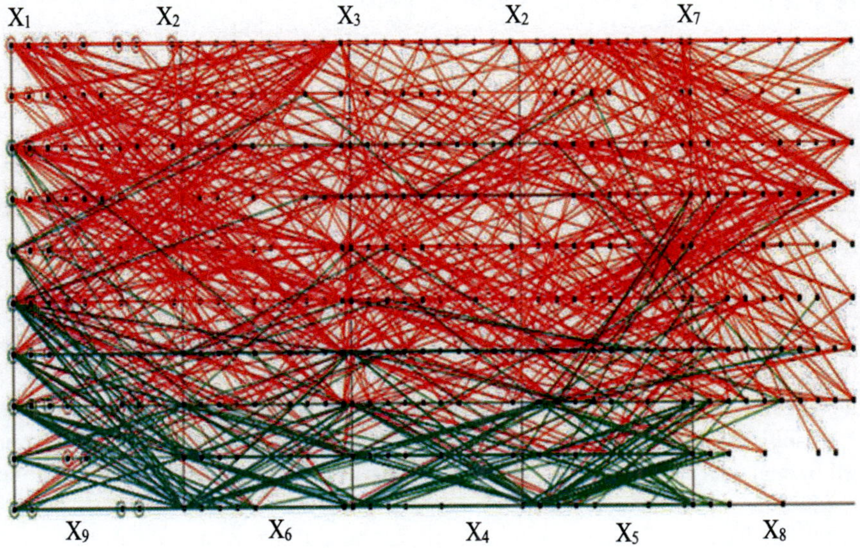

(b). WBC data with green class on top.

Fig. 3 Wisconsin breast cancer (WBC) 9-D dataset visualized in SPCVis to occlusion. To overcome this challenge, IVLC algorithm is used comprising of interactive controls to reorient the data to make the pattern discovery easier along with analytical rules to classify the data.

The lossless n-D visualization is achieved by representing data in *Interactive Shifted Paired Coordinates System* (SPCVis). Reordering the coordinates is one of the interactive features provided by the SPCVis software system. Using this feature, the coordinates are reordered in such a way that the class separation is prominent along the vertical coordinates. The discovering of coordinates to get good separation of classes is performed interactively by the user.

There are several interactive features provided to the end user like reversing data, non-linear scaling etc. For instance, if $\mathbf{x}$ is an n-D point where $\mathbf{x} = (x_1, x_2, x_3, \ldots\ldots x_n)$, the reverse of x_1 would display the data as $(1 - x_1, x_2, x_3, \ldots\ldots x_n)$ when x_1 is in [0, 1]. Also, SPCVis software system provides the user ability to click and drag the whole (X_i, X_j) plot to desired location until occlusion is reduced. Another interactive control allows a user to display data of user selected class on top of another class. Figure 3a displays WBC data with red class on top and Fig. 3b with green class on top. This helps the user to observe the pattern of individual classes more clearly.

Non-linear scaling is an interactive feature provided by the SPCVis Software where only a part of the user selected coordinate is scaled differently. The generalized formula for an n-D point x_j is given in Eq. (1),

$$x_j' = \begin{cases} x_j, & \text{if } x_j < k \\ x_j + r \times graph\,Width, & \text{if } k \le x_j < 1 \end{cases} \tag{1}$$

here k is a constant and $0 < k < 1$, r is the resolution of the data i.e., the shortest distance between the data points. The value of k is set by the user. The data used for SPC are normalized to [0, 1].

Figure 4 displays the breast cancer data set after applying the non-linear scaling at $r = 0.1$, $k = 0.6$ on X_1 and $k = 0.3$ on X_2, X_5, and X_7 coordinates.

Another interactive feature provided by SPCVis software is Non-orthogonal coordinate system that has a coordinate inclined at an angle other than 90° with respect to the other coordinate. Figure 5 displays a simple 2-D graph with Y coordinate inclined at an angle of 30° with respect to its previous Y coordinate. Figure 6 displays non-orthogonal coordinate representation with horizontal coordinates X_8 and X_5 inclined at −30°.

Interactive controls like non-linear scaling, non-orthogonal coordinates etc. allow improving visual discrimination of classes. However, interactive visualization alone does not completely perform the data separation. It only provides a base for it.

This chapter uses the IVLC algorithm that generates analytical rules to perform further class separation after reordering the coordinates. This algorithm generates these rules mainly using the threshold values generated from non-linear scaling. The rules belong to the class of rules proposed in [9].

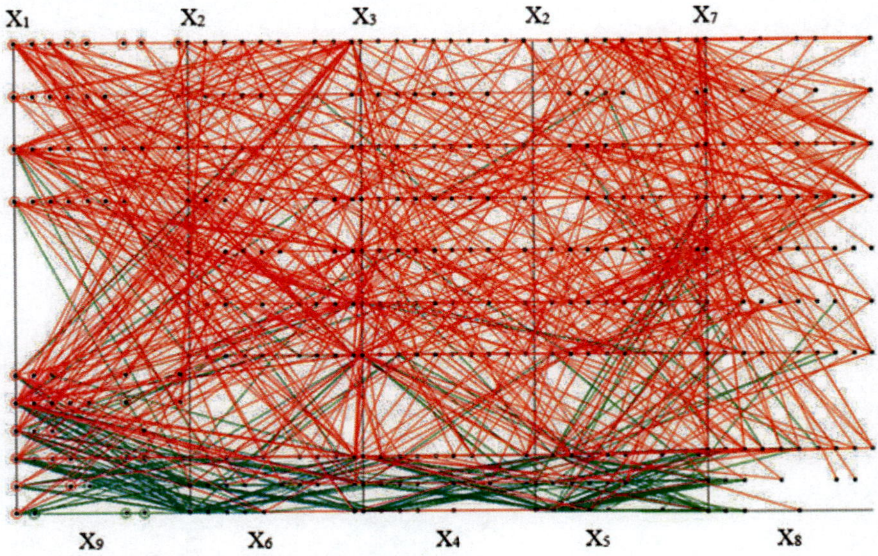

Fig. 4 WBC 9-D data after non-linear scaling on all the vertical coordinates

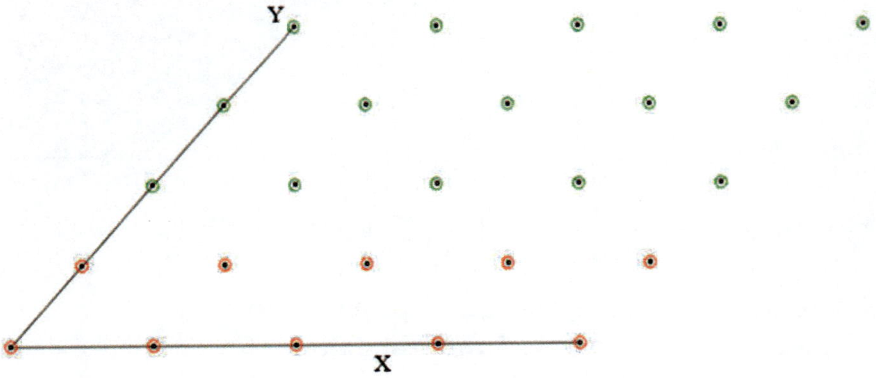

Fig. 5 Non-orthogonal display of 2-D data (Y = 30°)

Visualizing data with larger dimensions become challenging in SPC. To display data of larger dimensions, a modified version of the SPC called as **Serpent Coordinate System** (**SCS**) is proposed. It is visualized in a grid like structure to accommodate all the dimensions on a single screen. Air Pressure System (APS) for Scania Trucks [8] consists of 2 classes and 170 dimensions wherein 4 dimensions were removed since all the data points under those columns were 0 and not informative. The data with 166 dimensions are displayed in Fig. 7 and the coordinate labels corresponding each coordinate pair are displayed in Table 1.

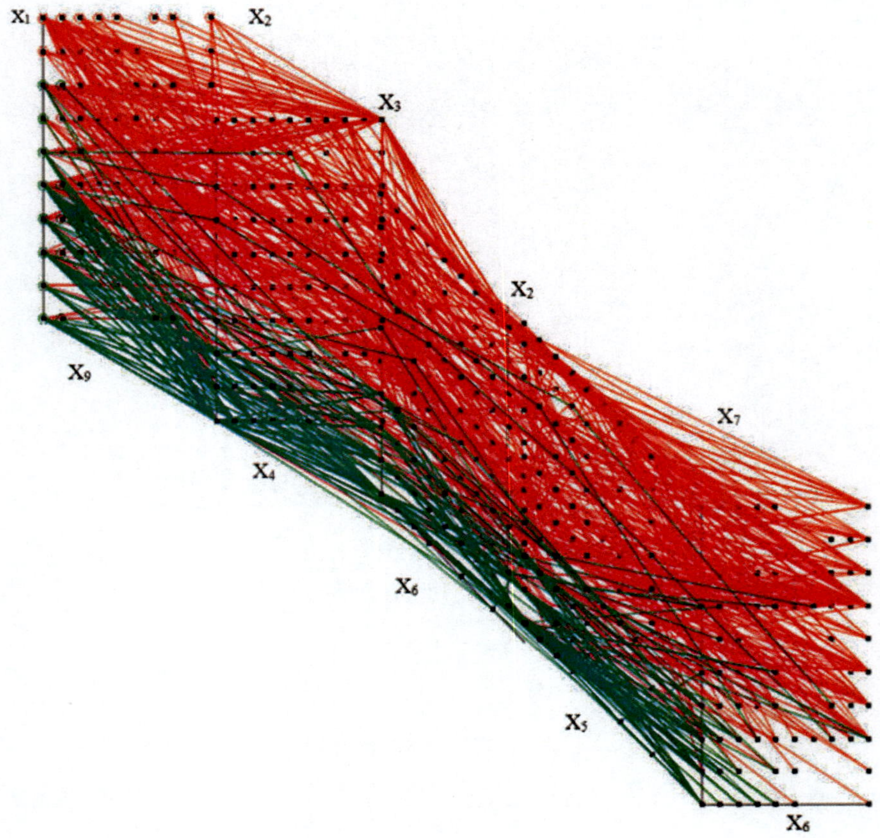

Fig. 6 Non-orthogonal display of WBC 9-D data (X_6 and X_5 inclined at −30°)

3 Iterative Visual Logical Classifier Algorithm and Worst-Case *k*-Fold Cross Validation

The approach proposed in this chapter contains interactive and automatic parts. This section presents the interactive part, where the Iterative Visual Logical Classifier algorithm is described in Sect. 3.1 and the worst-case k-fold cross validation, which allows evaluating the classifier in an innovative way is described in Sect. 3.2.

3.1 Iterative Visual Logical Classifier Algorithm

Below we present the *Iterative Visual Logical Classifier* algorithm that classifies data in iterations. As discussed in the previous section, once we reorder the coordinates

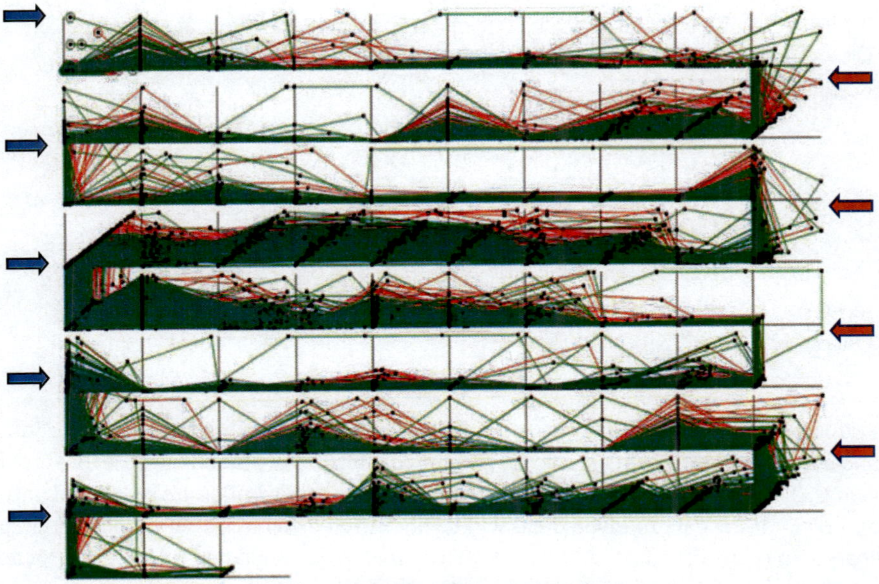

(a). APS data with green class on top.

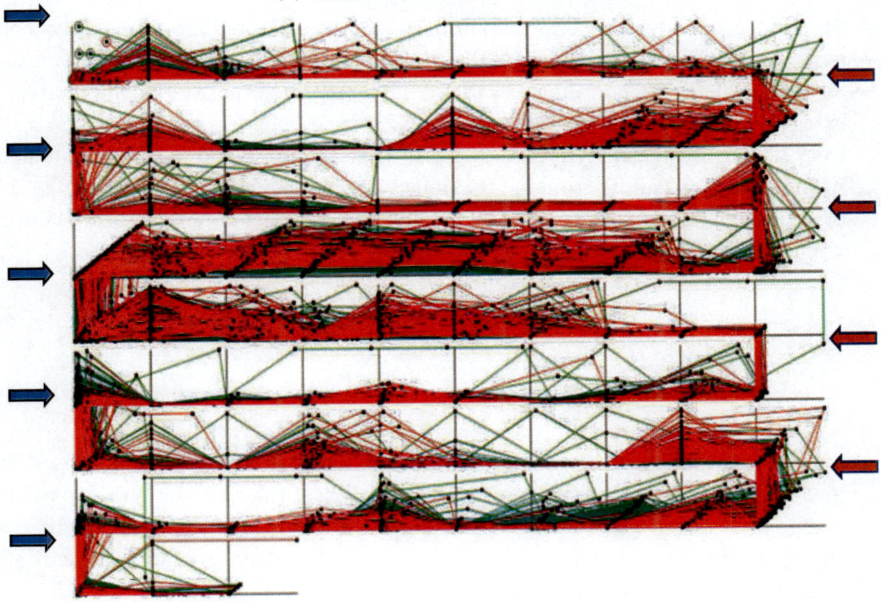

(b). APS data with red class on top.

Fig. 7 APS failure at scania trucks (166 dimensions) visualized in serpent coordinate system

Table 1 Coordinate labels for serpent coordinate system (SCS) for Fig. 7a, b

(X_1, X_2)	(X_3, X_4)	(X_5, X_6)	…	(X_{15}, X_{16})	(X_{17}, X_{18})	(X_{19}, X_{20})
(X_{21}, X_{22})	(X_{23}, X_{24})	(X_{25}, X_{26})	…	(X_{35}, X_{36})	(X_{37}, X_{38})	(X_{39}, X_{40})
(X_{41}, X_{42})	(X_{43}, X_{44})	(X_{45}, X_{46})	…	(X_{55}, X_{56})	(X_{57}, X_{58})	(X_{59}, X_{60})
(X_{61}, X_{62})	(X_{63}, X_{64})	(X_{65}, X_{66})	…	(X_{75}, X_{76})	(X_{77}, X_{78})	(X_{79}, X_{80})
(X_{81}, X_{82})	(X_{83}, X_{84})	(X_{85}, X_{86})	…	(X_{95}, X_{96})	(X_{97}, X_{98})	(X_{99}, X_{100})
(X_{101}, X_{102})	(X_{103}, X_{104})	(X_{105}, X_{106})	…	(X_{115}, X_{116})	(X_{117}, X_{118})	(X_{119}, X_{120})
(X_{121}, X_{122})	(X_{123}, X_{124})	(X_{125}, X_{126})	…	(X_{135}, X_{136})	(X_{137}, X_{138})	(X_{139}, X_{140})
(X_{141}, X_{142})	(X_{143}, X_{144})	(X_{145}, X_{146})	…	(X_{155}, X_{156})	(X_{157}, X_{158})	(X_{149}, X_{160})
(X_{161}, X_{162})	(X_{163}, X_{164})	(X_{165}, X_{166})	–	–	–	–

to find good vertical separation and obtain the vertical threshold values from non-linear scaling, the analytical rules are generated interactively based on these threshold values. This process of reordering and generating analytical rules are continued until we cover all the data in given dataset. The algorithm generates a set of interpretable analytical rules for data classification. All steps can be conducted by the end user as a self-service. The steps performed for the classifier are:

Step 1: Reorder the coordinates to find a good vertical separation for classes and perform non-linear scaling to get the threshold values along the vertical coordinates. Reordering of coordinates and non-linear scaling is performed interactively using SPCVis software system.
Step 2: Generate the analytical rules mainly based on the threshold values obtained from non-linear scaling from the previous step. For example, if we denote the set of areas generated as R_{class1} and R_{class2}, then the classification rules for n-D point $\mathbf{x} = (x_1, x_2, x_3, \ldots, x_n)$ are:

$$\text{If } x_i \in R_{class1}, \text{ then } \mathbf{x} \in \text{class } 1 \tag{2}$$

$$\text{If } x_j \in R_{class2}, \text{ then } \mathbf{x} \in \text{class } 2 \tag{3}$$

Step 3: The data that do not follow the rules generated in step 2 are used as input for next step. Also, in this step, the analytical rules can be tuned to avoid overgeneralization [10]. Figure 8b displays the generation of R_1 area where a large part of the area is empty. The area can be reduced by generating the area R_1' of smaller dimension to avoid overgeneralization (see Fig. 8c).
Step 4: Repeat the steps above until all the data are covered.

The output of Iterative Visual Logical Classifier algorithm results in series of rectangular areas. The outputs of IVLC algorithm are displayed in Fig. 8.

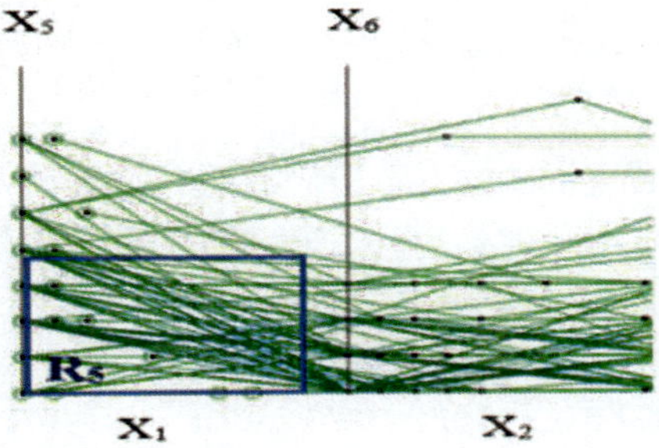

(a). Example of area (R_5) generated by IVLC for Wisconsin Breast Cancer (WBC) 9-D dataset.

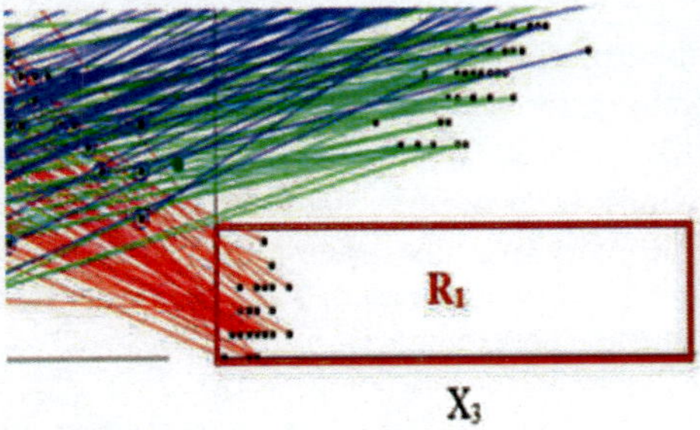

(b). Overgeneralized Area R1 generation (larger part of the area is empty) for 4-D iris dataset.

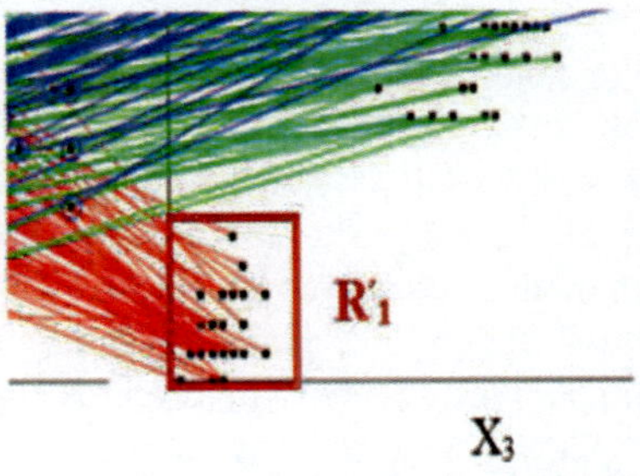

(c). Optimized Area R•1 generation for 4-D iris dataset

Fig. 8 Outputs of iterative visual logical classifier algorithm

3.2 *Model Evaluation with Worst-Case* k*-Fold Cross Validation Approach*

Although Cross Validation is a common technique used for model evaluation, it comes with its own challenges. Due to the random split of training and validation data, we might observe a bias in the estimated average error rate. Also, if we consider all the possible splits, it becomes computationally challenging to find all the combinations since the number of splits grows exponentially with the number of given data points [7].

In order to overcome this challenge, we use a worst case heuristics technique to split the data into training and validation sets in *k*-fold cross validation. The worst case fold contains the data of one class that are similar to cases of the opposing class [7] making classification more challenging with higher number of misclassification than in the traditional random *k*-fold cross validation. If the algorithm produces high accuracy in the worst-case fold, then the average case accuracy produced by the traditional random *k*-fold cross validation is expected to be greater.

In this chapter, the worst-case fold is extracted using visual representation of the data in Shifted Paired Coordinates System. As already mentioned, the data are displayed in such a way that they tend to be separated along the vertical axes in the SPC. For instance, if the dataset contains class *A* and class *B* with class *B* at the bottom and class *A* at the top in SPC visualization, then the worst-case validation split contains the cases with class *A* displayed on the bottom along with class *B* and vice versa. Since tenfold cross validation is used in our classification model, first validation fold contains the top 10% of the worst-case data n-D points, next validation fold contains the next 10% of the worst-case data and so on.

4 Experiments with Interactive Data Classification Approach

The goal of this section is presenting the experiment conducted on the benchmark datasets with the iterative data classification approach described in Sect. 3 above.

The first data set is the Iris data [8]. It has 4 dimensions (sepal length, petal length, sepal width and petal width) with a total of 150 cases. The data consists of three classes namely setosa, versicolor and virginica, each class consisting 50 cases. Figure 9 displays the data in SPCVis software system. The four dimensions are denoted as X_1, X_2, X_3, and X_4 coordinates.

Class 1 separation is defined by the rule in (4):

$$\text{If } (x_4, x_3) \in \mathrm{R}_1', \text{ then } \mathbf{x} \in \text{class } 1. \tag{4}$$

The optimized coordinate order for separation of classes 1 and 3 are (X_1, X_3) and (X_2, X_4) with X_3 and X_4 as vertical coordinates. Separation criteria for class 2 and class 3 is given in (5):

$$\text{If } (x_1, x_2, x_3, x_4) \in \mathrm{R}_2 \text{ then } \mathbf{x} \in \text{class } 2, \text{ else } \mathbf{x} \in \text{class } 3. \tag{5}$$

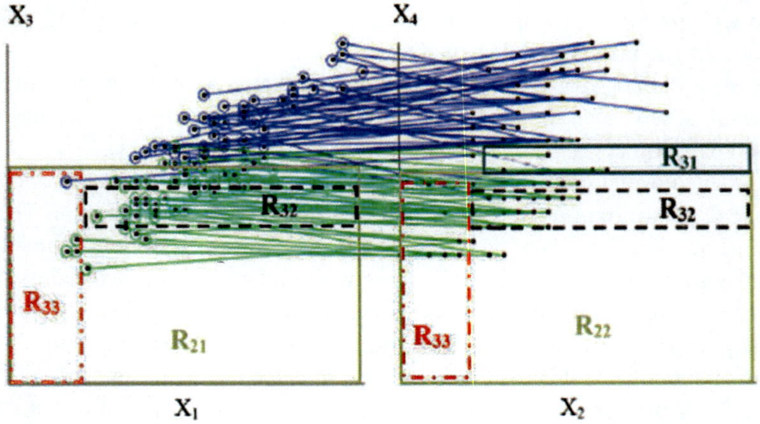

Fig. 9 Visualization of rule for R_1' on Iris dataset (4-D) for class 1 separation

We can further refine rule defined by $R_2' = R_{21}$ & R_{22} by adding another area R_3 to form a new optimized rule in (6):

$$\text{If } (x_1, x_2, x_3, x_4) \in R_2' \text{ or } R_3, \text{ then } \mathbf{x} \in \text{class 2, else } \mathbf{x} \in \text{class 3.} \tag{6}$$

Figure 10 visualizes the above rule for separation classes 2 and 3. The accuracy obtained in tenfold cross validation with worst case split is **100%**.

The second dataset is Wisconsin Breast Cancer (WBC) dataset [8] contains 699 cases of data with 9 features. In this dataset, 16 cases were incomplete and hence were removed. Remaining data with 683 cases consists of 444 benign cases and 239 malignant instances. Figure 3 displays WBC data after loading in SPCVis software system. Figure 11 visualizes analytical rules for R_5 and R_6 generated for benign class classification.

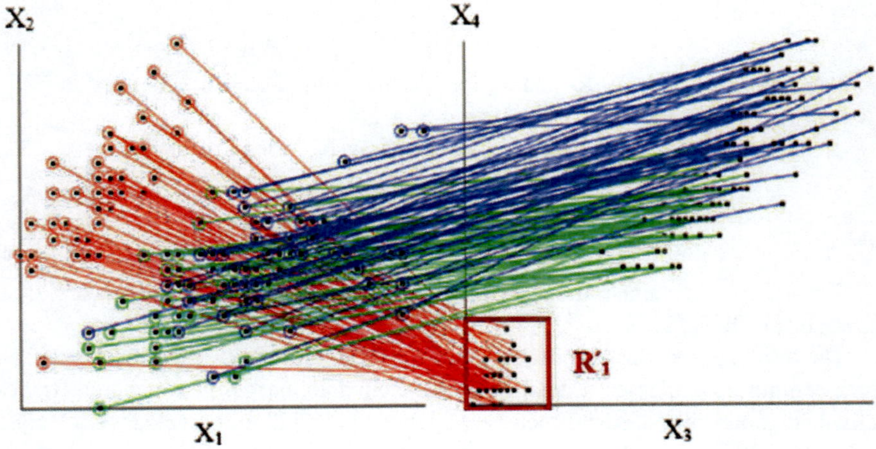

Fig. 10 Visualization of rule for R_2 and R_3 on Iris dataset (4-D) for classes 2 and 3 separation

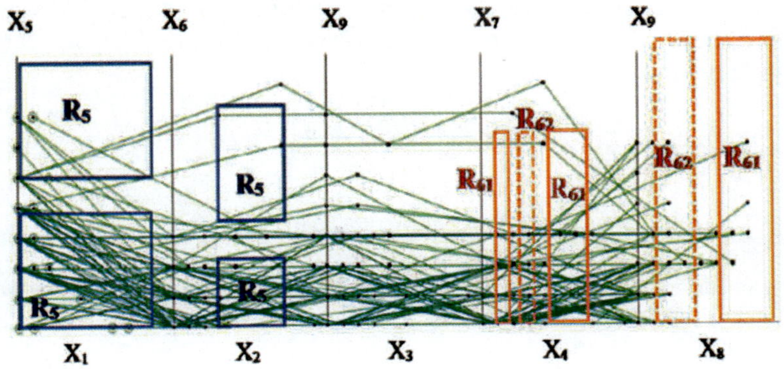

Fig. 11 Visualization of rules for R_5 and R_6 on WBC dataset (9-D)

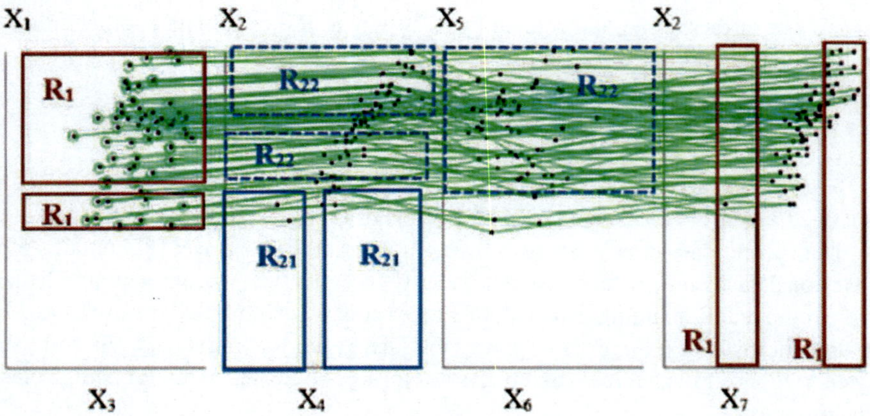

Fig. 12 Visualization of rules for R_1 and R_2 on Seeds dataset (7-D) for class 1 separation with all cases from class 2

The analytical rules for R_5 and R_6 for classification of class 1 is defined in (7).

$$\text{If } (x_1, x_5, x_2, x_6, x_4, x_7, x_8, x_9) \in R_5 \ \& \ R_6, \ \text{ then x} \in \text{class 1} \tag{7}$$

$$R_6 = R_{61} \text{ or } R_{62}. \tag{8}$$

The accuracy obtained after tenfold cross validation technique with worst case heuristics is **99.56%**.

The third dataset consists of seeds data with 7 dimensions and 210 instances. The data contain three classes: Kama, Rosa and Canadian, based on the characteristics of wheat kernels like seed perimeter area, length and width of kernel etc. Each class consists of 70 instances [8]. The data are loaded, and the coordinates are reordered to find the prominent class separation along vertical coordinates. The analytical rules are generated based on the vertical separation. Figure 12 displays seeds data with areas for analytical rules R_1 and R_2 for class 2 (green) separation. Due to odd number of coordinates X_2 coordinate is duplicated as the 8th coordinate. The rule for class 2 classification is defined in (9).

$$\mathbf{r_2} : \text{If } (x_1, x_7, x_4, x_2, x_6) \in (R_1 \text{ or } R_2), \ \text{ then } \mathbf{x} \in \text{class 2} \tag{9}$$

$$R_2 = R_{21} \text{ or } R_{22} \tag{10}$$

The accuracy obtained after tenfold cross validation technique with worst case heuristics is **100%**.

5 Data Classification Using Automation

This section presents the second part of the proposed approach i.e., the automated classification approach. It augments the interactive part making the classification process more complete and efficient.

In most of the scenarios, interactive approach to data classification works well for small dataset. When handling large data, this approach becomes tedious and time consuming. Also, due to the display of large number of data cases within limited space would lead to occlusion and it becomes challenging for the end users to find the patterns. To overcome this problem, automation is used with patterns discovered automatically with minimal human intervention. Automation is implemented using: (1) Coordinate Order Optimizer (COO) algorithm, and (2) Genetic algorithm (GA).

These algorithms are used in combination with non-linear scaling, to enhance the data interpretability. Figure 13 summarizes the automated data classification approach.

Coordinate Order Optimizer Algorithm. It optimizes the order of coordinates by primarily using Coefficient of Variation (CV) [11] parameter, also called as relative standard deviation (RSD) defined as the ratio of the standard deviation σ to the mean μ of the given data sample,

$$C_v = \frac{\sigma}{\mu} \tag{11}$$

CV is a standardized measure of dispersion of data distribution. It is computed individually per coordinate for each class. Next, the mean of all CVs of classes of each coordinate is calculated. Lesser the mean of CV, lesser the data dispersion along the coordinate. The coordinate with the least mean of CV is considered the *best coordinate*. Hence, the coordinates are arranged in the descending order of the mean CV values. Figure 14 displays WBC data before and after order of coordinates is optimized, respectively. In the optimized order of coordinates (Fig. 14b), the green class is settled at the bottom and red class on the top, while in the Fig. 14a, more green lines are at the top along with red class and more red lines in the bottom along with green lines.

Fig. 13 Overview of automation for data classification in SPCVis

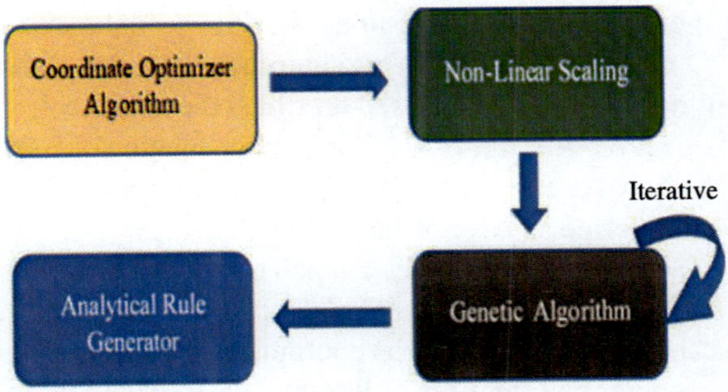

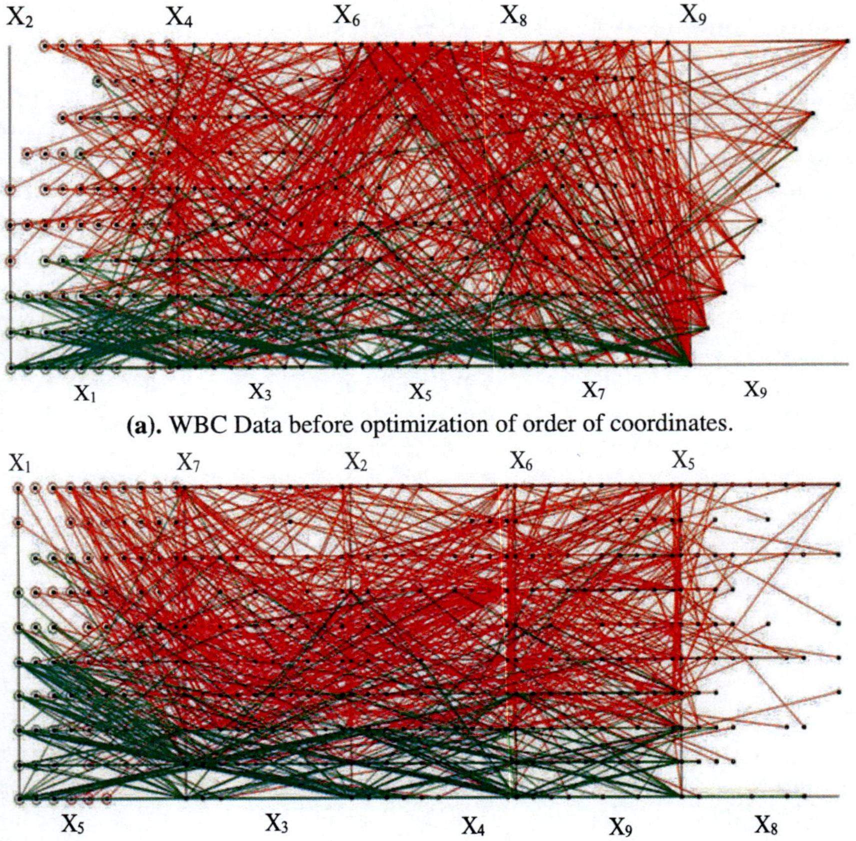

(a). WBC Data before optimization of order of coordinates.

(b). WBC Data after optimization of order of coordinates (lesser green cases on top).

Fig. 14 Visualization of WBC data before and after applying COO algorithm

Non-Linear Scaling: The threshold for all the vertical coordinates is calculated from the average of the bottom class taken from vertical coordinates. Non-linear scaling is performed using Eq. (1) for all the vertical coordinates. This improves data interpretability and provides better visualization of separation of classes. Figure 4 shows the output of non-linear scaling that enhances the visual separation of classes.

Genetic Algorithm: This algorithm is used to generate the optimized areas of high fitness or purity [12] based on which the analytical rules are created for further classification. An overview of the implementation of Genetic Algorithm in our approach is shown in Fig. 15 for discovering the areas for classification. In this context, areas are referred as Area of Interest (AOI).

Initial Population Selection: The initial population contains randomly generated rectangles (see Fig. 16) defined by the fixed ratio *r* of each coordinate. For instance, *r* can be 0.1 of length of the coordinate. The data are normalized to [0, 1] interval. The generation of the Area of Interest (AOI) in the SPCVis using genetic algorithm is an

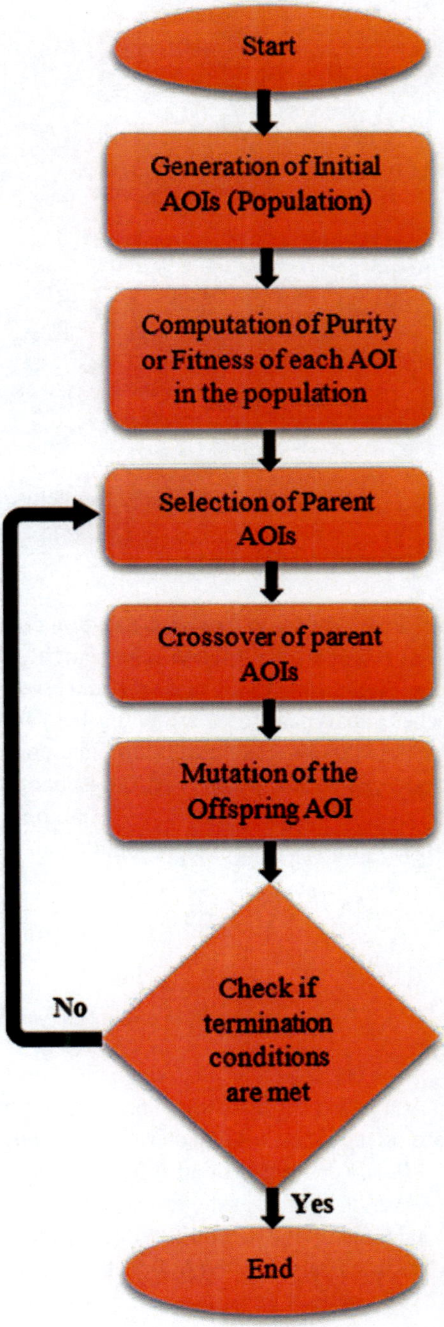

Fig. 15 Genetic algorithm flow chart used in SPCVis data classification

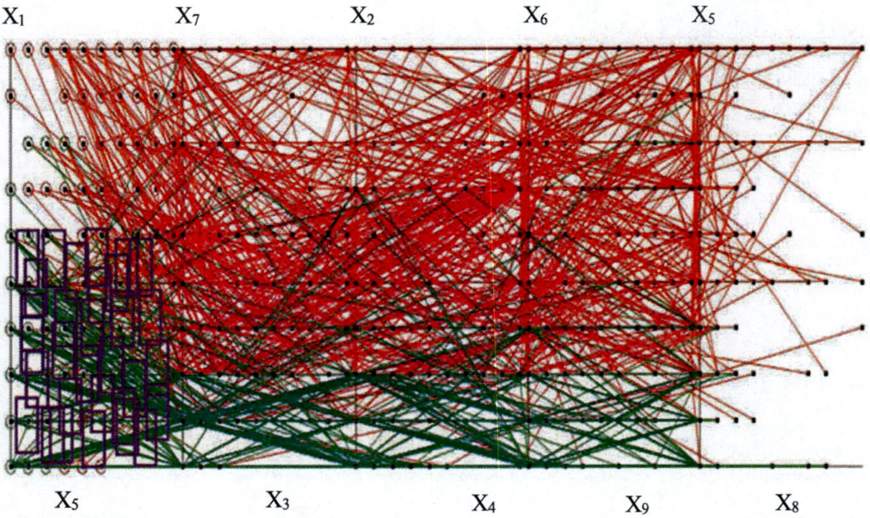

Fig. 16 Random generation of areas in WBC data

iterative process where each iteration creates a generation [6] of a new set of AOIs. Before generating the AOIs, a search space [13] is defined containing maximum cases belonging to the class on which we build analytical rules. Consider a situation where analytical rules are being built to classify class k. Let $x_{\text{imax}}(k)$ and $x_{\text{imin}}(k)$ be the maximum and minimum data points respectively belonging to coordinate X_i and $x_{\text{jmax}}(k)$ and $x_{\text{jmin}}(k)$ be the maximum and minimum data points respectively belonging to coordinate X_j. The number of Areas of Interests N_{AOI} generated to classify class k within a coordinate pair (X_i, X_j) is given in (12) and (13),

$$N_{\text{AOI}} = \frac{S(k)}{r^2} \tag{12}$$

$$S(k) = (x_{\text{imax}}(k) - x_{\text{imin}}(k))\left(x_{\text{jmax}}(k) - x_{\text{imin}}(k)\right) \tag{13}$$

Parents Selection: This stage involves selection of the AOIs or parents to generate a AOI called the offspring, i.e., combing two areas to form a bigger area. The parents are selected based on two criteria (1) *purity* or fitness, and (2) *proximity*.

Purity or Fitness of an AOI with respect to class k for a given pair of coordinates (X_i, X_j) is defined as the ratio of the number of data points belonging to class k to the total number of data points within the AOI. Let AOI_t be an AOI in the coordinate pair (X_i, X_j). The purity $P_k(\text{AOI}_t)$ of a single AOI_t with respect to class k is as follows,

$$P_k(\text{AOI}_t) = \frac{N_k(\text{AOI}_t)}{N(\text{AOI}_t)} \tag{14}$$

where, $N_k(AOI_t)$ is the number of points (x_i, x_j) in AOI_t in (X_i, X_j) that belong to lines from class C_k,

$$N_k(AOI_t) = \left\| \left\{ (x_i, x_j) : (x_i, x_j) \in AOI_t \ \& \ C_k \right\} \right\| \tag{15}$$

and $N(AOI_t)$ is the total number of points (x_i, x_j) within a given AOI_t in (X_i, X_j),

$$N(AOI_t) = \left\| \left\{ (x_i, x_j) : (x_i, x_j) \in AOI_t \right\} \right\| \tag{16}$$

After computing the purity of the AOIs, the parents with closest *Proximity* (nearest parents) are selected for the next stage, i.e., the crossover. While proximity can be defined in multiple ways, in our experiments we applied the Euclidian distance between the mid-point of two areas within the given coordinate pair commonly used in Genetic algorithms for similar tasks [13]. For instance, given the mid-point of two AOIs $\left\{ (x_{im1}, x_{jm1}), (x_{im2}, x_{jm2}) \right\}$ within coordinate pair (X_i, X_j), the proximity of two AOIs is given in (17):

$$Proximity = \sqrt{(x_{im1} - x_{im2})^2 + (x_{jm1} - x_{jm2})^2} \tag{17}$$

Crossover: Once the parents with highest *Purity* and closest *Proximity* (nearness) are selected, the parent AOIs are combined to form a new AOI (offspring). A single parent AOI_{Pg} is represented below:

$$AOI_{Pg} = P_g(x_1, x_2, y_1, y_2) \tag{18}$$

Here x_1, x_2, y_1, y_2 are left, right, bottom ant top coordinates of the rectangular AOI_{Pg} and is represented in the Fig. 17.

A crossover of two parents AOI_{P1g} and AOI_{P2g} from generation g to produce an offspring AOI_{O1g} within a given coordinate pair (X_i, X_j) is represented as:

$$AOI_{O1g} = F\left(AOI_{P1g}, AOI_{P2g}\right) \tag{19}$$

The two parents are defined as:

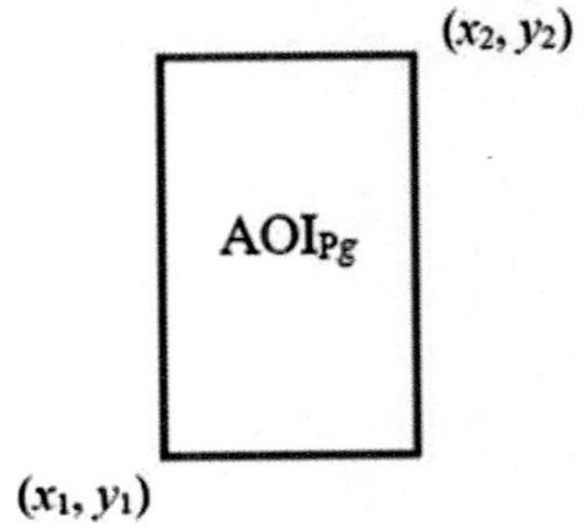

Fig. 17 Representation of single parent AOI_{Pg} from generation g

$$\mathrm{AOI}_{\mathrm{P1}g} = \mathrm{P}_{1g}(x_{11}, x_{12}, y_{11}, y_{12}) \tag{20}$$

$$\mathrm{AOI}_{\mathrm{P2}g} = \mathrm{P}_{2g}(x_{21}, x_{22}, y_{21}, y_{22}) \tag{21}$$

The function $F(\mathrm{AOI}_{\mathrm{P1}g}, \mathrm{AOI}_{\mathrm{P2}g})$ is defined in (22) as an envelope around these two AOIs.

$$\mathrm{F}\big(\mathrm{AOI}_{\mathrm{P1}g}, \mathrm{AOI}_{\mathrm{P2}g}\big) = \{\min(x_{11}, x_{21}), \max(x_{12}, x_{22}), \min(y_{11}, y_{21}), \max(y_{12}, y_{22})\} \tag{22}$$

Figure 18 displays different types of crossovers of the parent AOIs (with and without overlapping, or diagonally overlapping).

Mutation: In genetic algorithm, certain characteristics of the offspring generated from the previous generation are modified (mutated) in order to speed up the process of reaching an optimized solution. This includes modifying the characteristics of the offspring either by flipping, swapping or shuffling the properties that represent the offspring. For instance, if the offspring is represented by bits, then the mutation by flipping would include switching some of the bits from 0 to 1 or vice versa [14].

Since the objective of mutation is to generate an offspring with better characteristics than its parents, in our proposed technique, we generate the mutated offspring by interactively *generating a new parent AOI* with high *Purity* and close *Proximity* with the automatically generated parent AOI. This results in an offspring with better characteristics compared to its parents in terms of size and purity. Figure 19a represents the automatically generated parent AOI in purple (straight line) and interactively generated parent AOI (dotted lines). The resulting mutated offspring AOI has superior characteristics compared to its parents with high purity and larger area compared to its previous generation as displayed in Fig. 19b.

Termination: Since genetic algorithm process is iterative, there are several conditions based on which the process can be terminated. In our proposed method, we use two techniques as the termination criteria: (1) areas with highest fitness or purity (100%) are generated, or (2) manual inspection termination in SPCVis. If either of the two criteria is met, the process is terminated.

Analytical Rule Generator: This is similar to the second step performed in IVLC algorithm as discussed in Sect. 3 of this chapter. The only change that the areas used here are generated from genetic algorithm whereas in Sect. 3, the areas used are interactively generated.

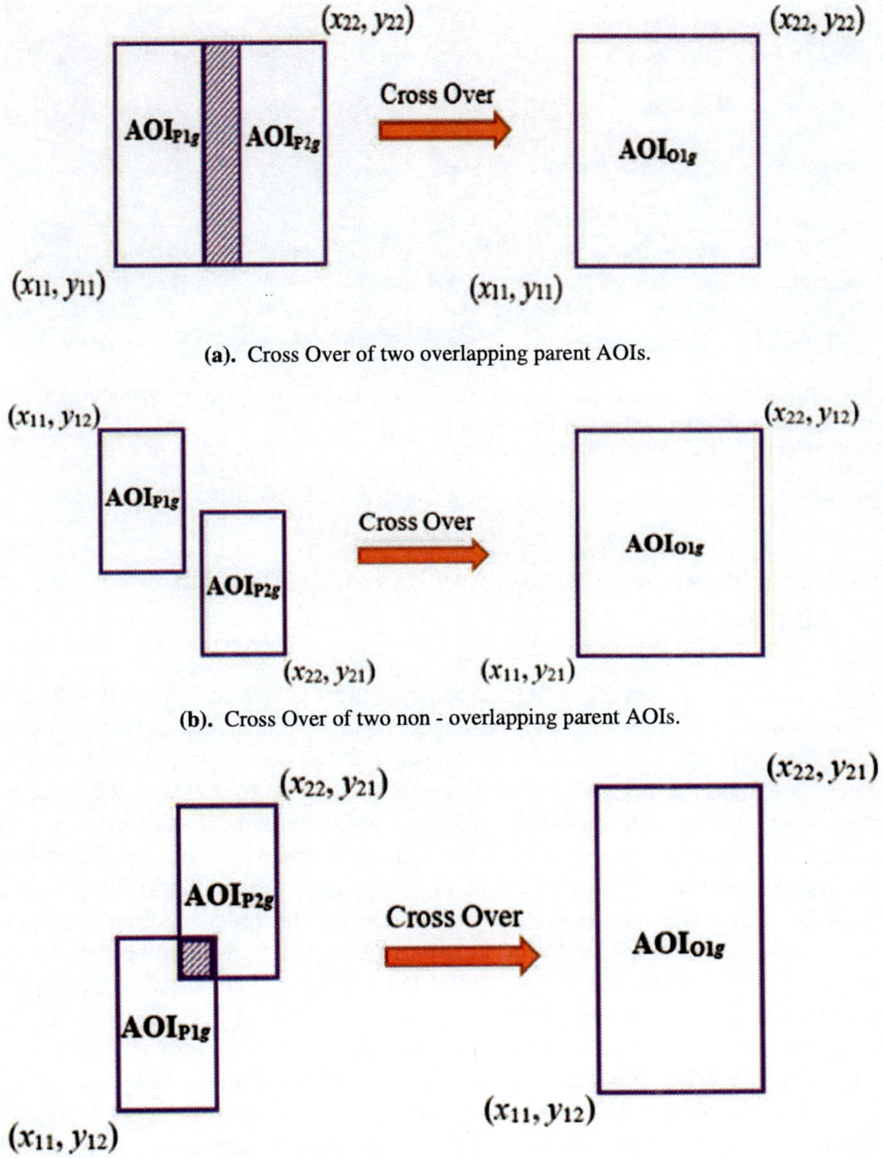

(c). Cross Over of two diagonally overlapping parent AOIs.

Fig. 18 Different types of cross overs of two parent AOIs to generate offspring AOI

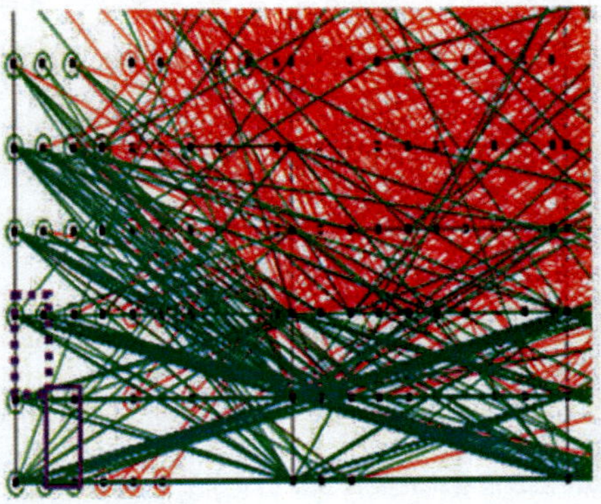

(a). Parent AOI (dotted lines) generated automatically (straight lines) and interactively (dotted lines) with high purity in generation g.

(b). Mutated Offspring in generation $g + 1$.

Fig. 19 Visualizations of consecutive generations of AOIs in WBC data

6 Experiments with Automated Data Classification Approach

The goal of this section is presenting the experiment conducted on the benchmark datasets with the automated data classification approach described in Sect. 5 above.

The experimental framework with implementation details are as follows: The automated approach begins with running COO algorithm to optimize the order of coordinates that separate the data along the vertical coordinate of a given coordinate pair. Next, nonlinear scaling is performed on the vertical coordinates to enhance the visual separation. Then the genetic algorithm is run to automatically generate the AOIs. The parameters considered for genetic algorithm are population size N_{AOI}, number of generations (i.e., the number of times GA ran after successful generation of offspring AOIs), fitness function $P_k(AOI_t)$ and genetic operator $F(AOI_{P1g}, AOI_{P2g})$. These parameters are discussed in detail in Sect. 5.

Experiments are conducted with same data sets used in interactive classification approach, i.e., WBC, Iris and Seeds datasets. In addition to these data sets, experiments are also conducted on Air pressure system failure at Scania trucks. This dataset consists of 60,000 cases with 170 features. Compared to interactive approach, the automated techniques provided better results with smaller number of areas and less iterations.

The experiment is conducted on Iris dataset, as discussed in Sect. 5. Class 1 is classified after running the COO algorithm and genetic algorithm. The optimized order of the coordinates for class 1 classification is (X_4, X_3) and (X_1, X_2). It is displayed in Fig. 20. The rule for class 1 (green) separation is:

$$\mathbf{r}_1 : \text{If } (x_4, x_3) \in R_{11}, \text{ then } \mathbf{x} \in \text{class } 1 \tag{23}$$

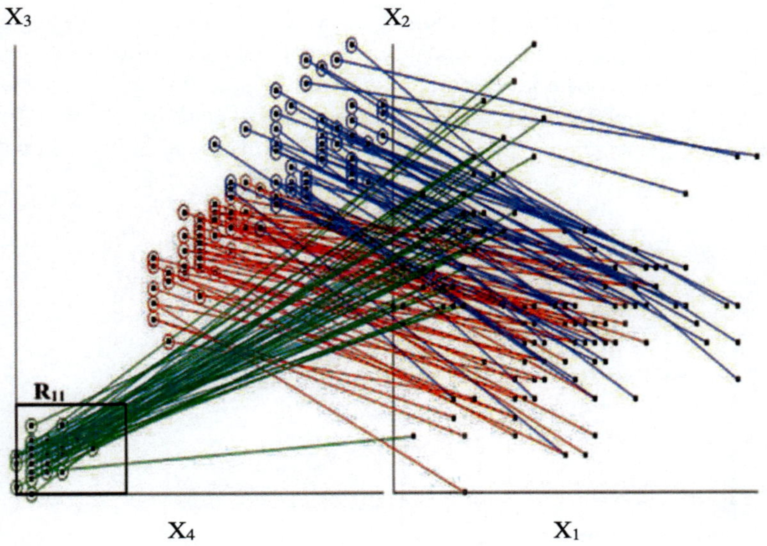

Fig. 20 Visualization of Iris data with class 1 separation rule

After class 1 separation, Coordinate Order Optimizer is run again for classifying the remaining classes. The optimized order of coordinates for class 2 and class 3 separation are (X_1, X_3) and (X_2, X_4). The visualization of classes 2 and 3 after reordering the coordinates is shown in Fig. 21.

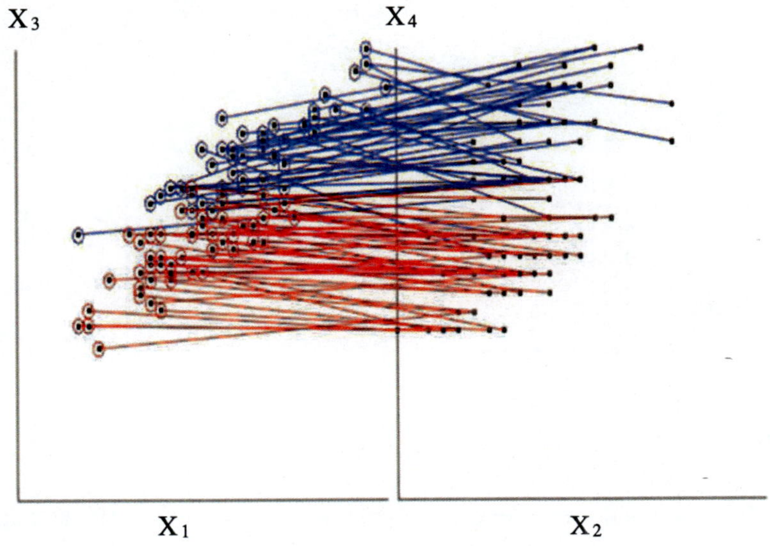

Fig. 21 Visualization of Iris data with classes 2 and 3 after reordering the coordinates

Figure 21 clearly displays separation of classes 2 and 3 along the vertical coordinates. This visualization is further enhanced by applying the non-linear scaling with following thresholds on coordinates: 0.7 on X_3 and 0.71 on X_4.

Genetic algorithm is run on class 2 and class 3 data to generate the areas. Visualization of non-linear scaling along with the areas are displayed in Fig. 22a with 10

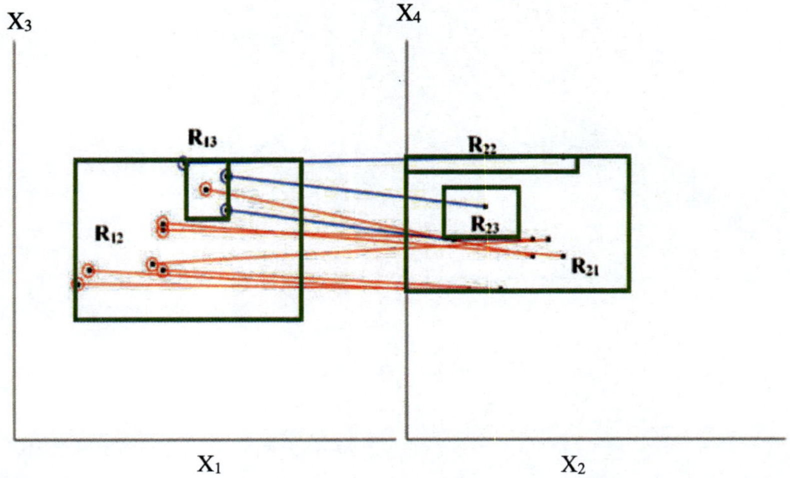

(a). Visualization of rule $\mathbf{r}_2$ on Iris dataset for classes 2 and 3 separation with 10 cases.

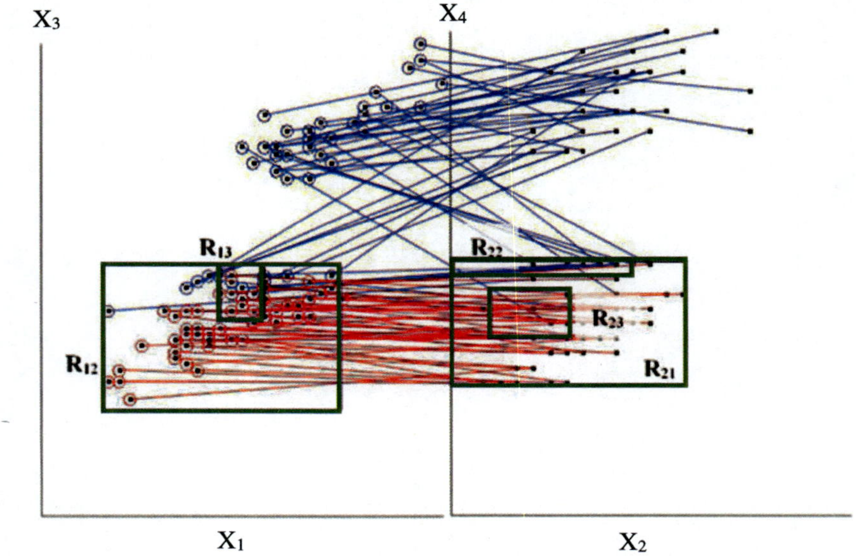

(b). Visualization of rule $\mathbf{r}_2$ on Iris dataset for classes 2 and 3 separation with all the cases.

Fig. 22 Visualization of rule $\mathbf{r}_2$ on Iris dataset for classes 2 and 3 separation

Table 2 Parameters of the areas generated for Iris data classification

Rectangle	Left	Right	Bottom	Top	Coordinate pair
R_{11}	0.0	0.3	0.0	0.2	(X_1, X_3)
R_{12}	0.16	0.75	0.3	0.7	(X_1, X_3)
R_{13}	0.45	0.56	0.55	0.7	(X_1, X_3)
R_{21}	0.0	0.59	0.37	0.71	(X_2, X_4)
R_{22}	0.0	0.45	0.67	0.71	(X_2, X_4)
R_{23}	0.1	0.3	0.5	0.63	(X_2, X_4)

instances. Figure 22b displays the same visualization with all the cases from class 2 and class 3.

The area R_2 for classes 2 and 3 classification is:

$$R_2 = R_{12} \ \& \ R_{21} \ \& \ \neg R_{13} \ \& \ \neg R_{23} \ \& \ \neg R_{22} \tag{24}$$

The rule $\mathbf{r_2}$ for class 2 classification for Iris data is defined below:

$$\mathbf{r_2} : \text{If } (x_1, x_2, x_3, x_4) \in R_2, \text{ then } \mathbf{x} \in \text{class 2} \tag{25}$$

The rule $\mathbf{r_3}$ for class 3 classification for Iris data is defined below:

$$\mathbf{r_3} : \text{If } (x_1, x_2, x_3, x_4) \in R_3, \text{ then } \mathbf{x} \in \text{class 3} \tag{26}$$

where

$$R_3 = (\neg R_{11} \ \& \ \neg R_2) \tag{27}$$

The area parameters generated for Iris data classification is listed in Table 2. The accuracy obtained for Iris data classification with tenfold cross validation using worst-case heuristics approach is **100%**.

The second dataset is WBC dataset, as discussed in Sect. 5. Running the COO algorithm produced the following order of coordinates: (X_5, X_1), (X_3, X_7), (X_4, X_2), (X_9, X_6) and (X_8, X_5). Figure 23a displays WBC data visualized in SPCVis with 12 cases of class 1 data along with the areas.

The rectangle R_{km} is mth rectangle in the kth pair of coordinates. For instance, in Fig. 23a, rectangle R_{24} is a 4th rectangle in the second pair of coordinates that is (X_3, X_7). Figure 23b displays all the cases from both classes of WBC data along with the non-linear scaling with following thresholds on coordinates: 0.6 on X_1, 0.25 on X_7 and X_2 and 0.3 on X_6.

The areas R_1–R_3 are defined as follows:

$$R_1 = R_{11} \ \& \ \neg R_{14} \ \& \ R_{41} \ \& \ \neg R_{42} \ \& \ (\neg R_{31} \text{ or } \neg R_{23} \text{ or } \neg R_{32}) \tag{28}$$

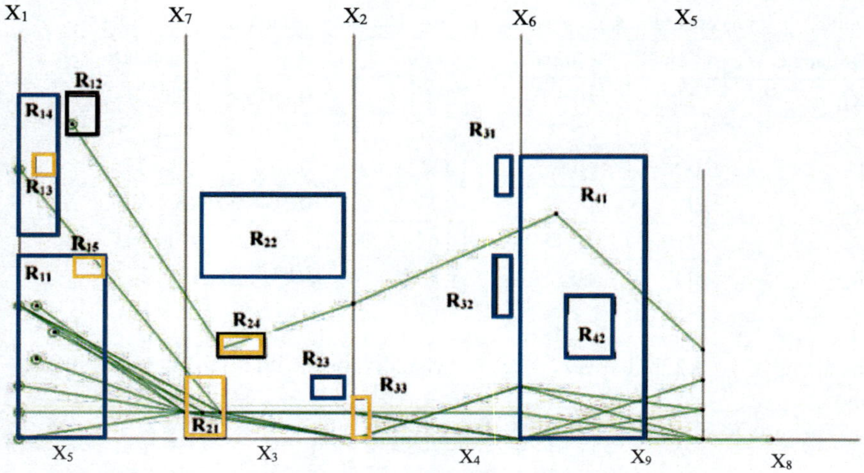

(a). Visualization of rule **r** on WBC dataset with 12 cases.

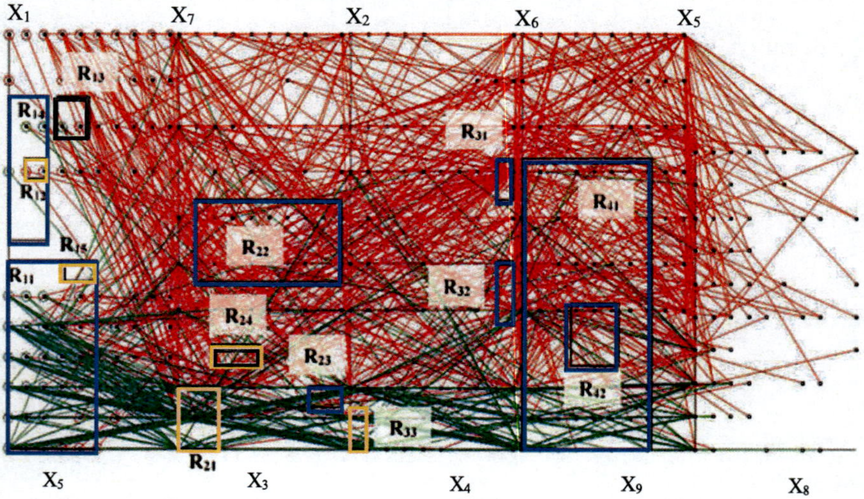

(b). Visualization of rule **r** on WBC dataset with all the cases.

Fig. 23 Visualization of rule **r** on WBC dataset for class 1 separation

$$R_2 = R_{12} \ \& \ \neg R_{15} \ \& \ (R_{21} \text{ or } R_{24} \text{ or } R_{33}) \tag{29}$$

$$R_3 = R_{13} \ \& \ R_{24} \tag{30}$$

The rule **r** is defined below:

$$\mathbf{r} = \text{If } (x_1, x_2, x_3, x_4, x_5, x_6, x_7, x_9) \in R_1 \text{ or } R_2 \text{ or } R_3 \text{ then } \mathbf{x} \in \text{Class } 1, \text{ else } \mathbf{x} \in \text{Class } 2 \tag{31}$$

The area coordinates are given in the Table 3. The accuracy obtained after tenfold cross validation technique with worst case heuristics is **99.71%**.

Seeds data, as discussed in Sect. 5 consists of 7 dimensions. Due to odd number of dimensions, X_7 is duplicated to display the data in SPCVis (see Fig. 24).

Table 3 Parameters of the rectangles generated for WBC data classification

Rectangle	Left	Right	Bottom	Top	Coordinate pair
R_{11}	0.0	0.55	0.0	0.45	(X_5, X_1)
R_{12}	0.3	0.5	0.75	0.85	(X_5, X_1)
R_{13}	0.1	0.25	0.65	0.7	(X_5, X_1)
R_{14}	0.0	0.25	0.5	0.85	(X_5, X_1)
R_{15}	0.40	0.55	0.40	0.45	(X_5, X_1)
R_{21}	0.0	0.25	0.0	0.15	(X_3, X_7)
R_{22}	0.1	1.0	0.4	0.6	(X_3, X_7)
R_{23}	0.8	1.0	0.1	0.15	(X_3, X_7)
R_{24}	0.2	0.5	0.2	0.25	(X_3, X_7)
R_{31}	0.9	1.0	0.6	0.7	(X_4, X_2)
R_{32}	0.9	1.0	0.3	0.45	(X_4, X_2)
R_{33}	0.0	0.1	0.0	0.12	(X_4, X_2)
R_{41}	0.0	0.8	0.0	0.7	(X_9, X_6)
R_{42}	0.3	0.6	0.2	0.35	(X_9, X_6)

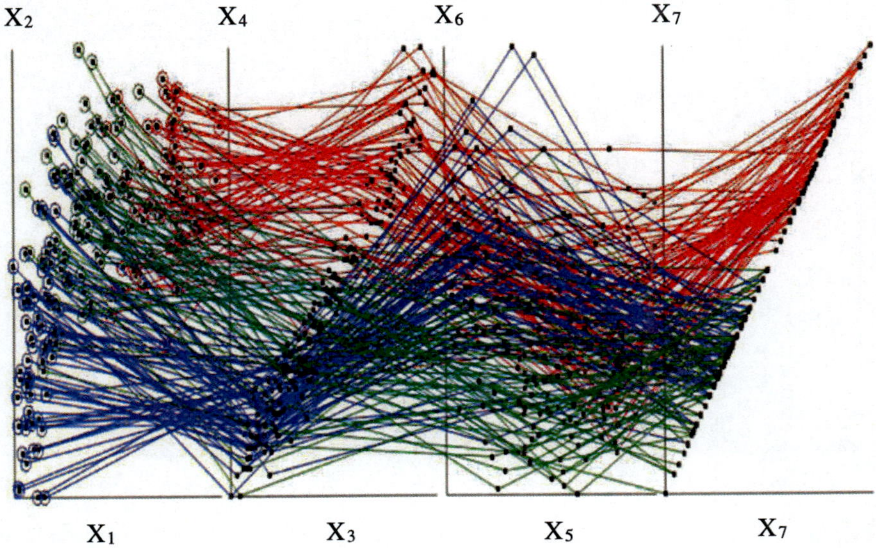

Fig. 24 Visualization of Seeds dataset (7-D) with all the three classes in SPCVis

Analytical rules for class 3 (blue) are generated after running the COO algorithm and genetic algorithm. The optimized order of the coordinates for class 1 classification is (X_3, X_1), (X_7, X_5), (X_6, X_2), and (X_3, X_4). Figure 25 represents the optimized order of coordinates.

Non-linear scaling is then performed with following thresholds on coordinates: 0.4 on X_1, 0.45 on X_5, 0.4 on X_2 and 0.4 on X_4. The visualization of classes 2 and 3 after reordering the coordinates and non-linear scaling is shown in Fig. 26.

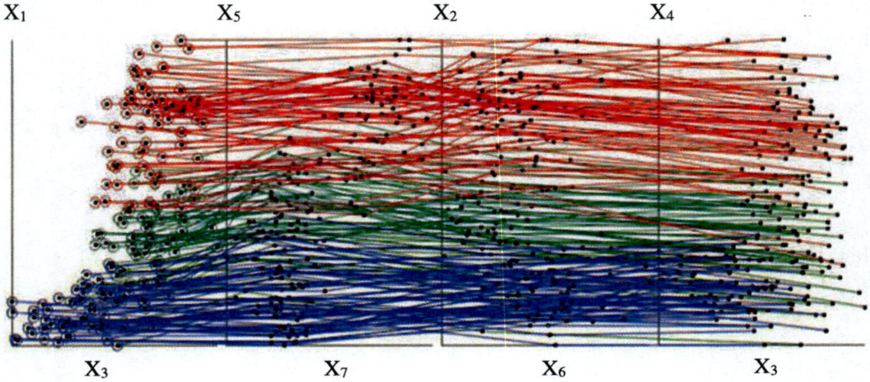

Fig. 25 Visualization of Seeds dataset (7-D) with all the three classes in SPCVis after coordinate order optimization

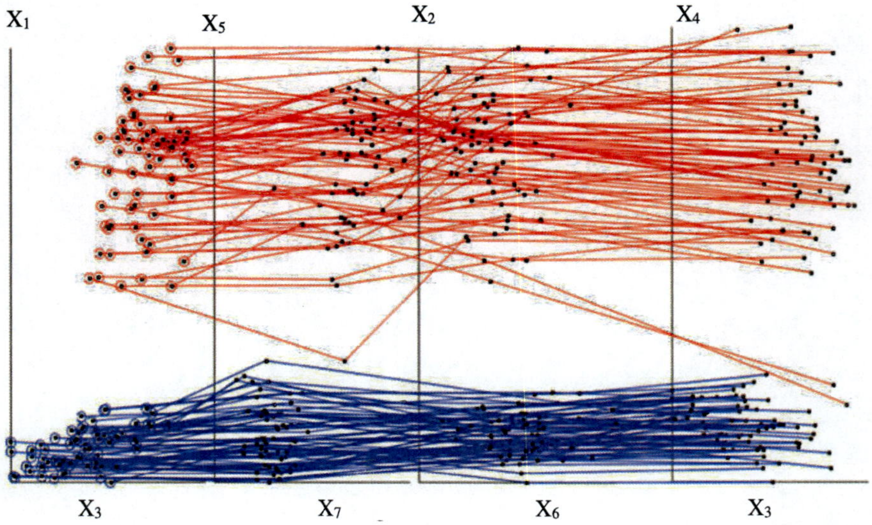

Fig. 26 Visualization of Seeds dataset with classes 2 (red) and 3 (blue) after performing non-linear scaling on optimized order of coordinates

Areas are generated by running genetic algorithm and analytical rules are built using these generated areas. Figure 27a visualizes all the three classes with non-linear scaling and the areas for class 3 (blue) separation. The area is defined below:

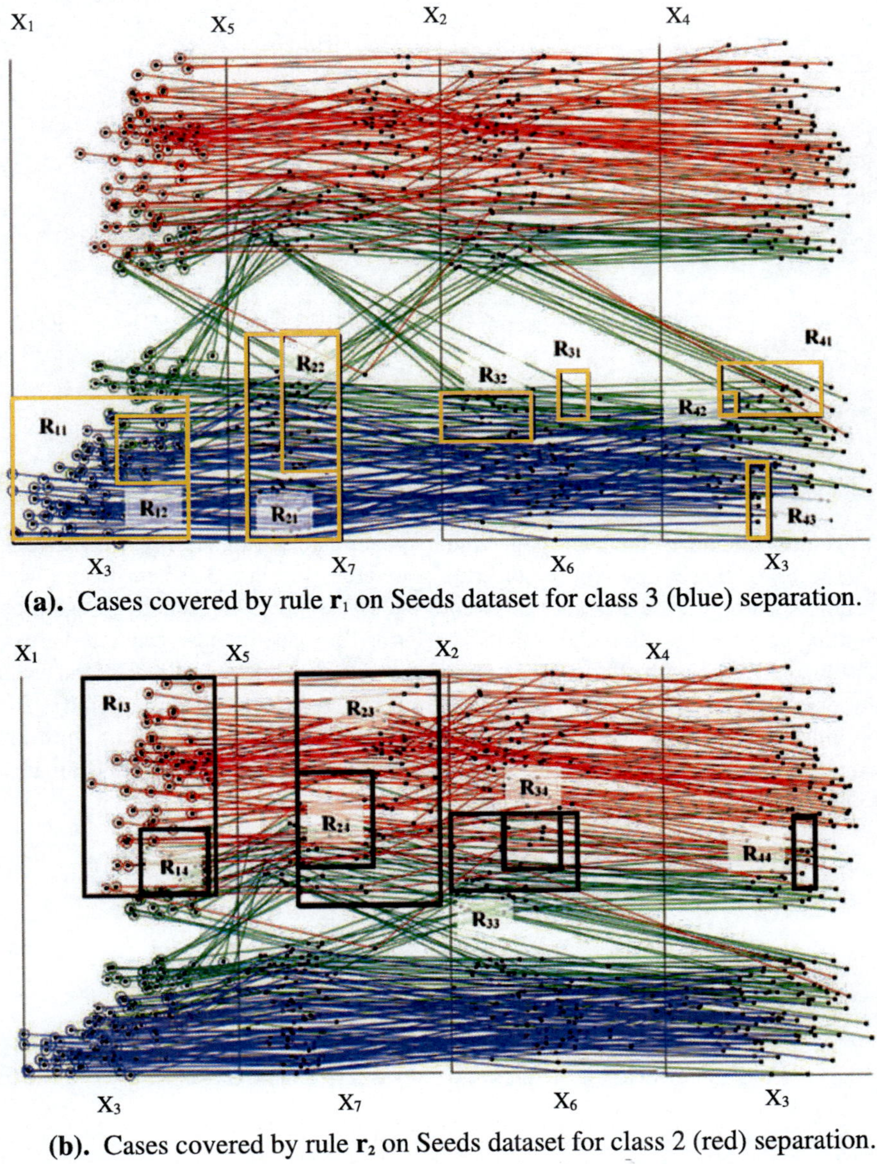

Fig. 27 Visualization of rules $\mathbf{r}_1$ and $\mathbf{r}_2$ on Seeds dataset for classes 2 and 3 separation with all the cases

Table 4 Parameters of the areas generated for classification in Seeds data in 1st iteration

Rectangle	Left	Right	Bottom	Top	Coordinate pair
R_{11}	0.0	0.85	0.0	0.3	(X_3, X_1)
R_{12}	0.5	0.85	0.12	0.26	(X_3, X_1)
R_{21}	0.1	0.55	0.0	0.43	(X_7, X_5)
R_{22}	0.27	0.55	0.2	0.43	(X_7, X_5)
R_{31}	0.58	0.72	0.25	0.35	(X_6, X_2)
R_{32}	0.0	0.45	0.21	0.3	(X_6, X_2)
R_{41}	0.32	0.8	0.25	0.36	(X_3, X_4)
R_{42}	0.32	0.4	0.25	0.3	(X_3, X_4)
R_{43}	0.45	0.55	0.0	0.15	(X_3, X_4)

$$R_1 = R_{11} \ \& \ R_{21} \ \& \ R_{41} \ \& \ (\neg R_{42} \text{ or } \neg R_{31}) \ \& \ (\neg R_{12} \ \& \ \neg R_{22} \ \& \ (\neg R_{32} \text{ or } \neg R_{43})) \tag{32}$$

The rule $\mathbf{r_1}$ for class 3 classification is given as:

$$\mathbf{r_1} : \text{If } (x_1, x_2, x_3, x_4, x_5, x_6, x_7) \in R_1, \text{ then } \mathbf{x} \in \text{class } 3 \tag{33}$$

The parameters of the areas generated by genetic algorithm are listed in Table 4.

The cases that do not follow the rule generated for class 3 is sent to the next iteration where rules are generated for other classes. Areas are generated again by running genetic algorithm and analytical rules are built using these generated areas. The visualization with all the three cases along with non-linear scaling and the areas are displayed in Fig. 27b. In this case the rule is generated for class 2 (red). The optimized order of the coordinates remains the same. Non-linear scaling with the same threshold as in the previous iteration is performed on the vertical coordinates. The rule for class 2 (red) separation is:

$$R_2 = R_{13} \ \& \ R_{23} \ \& \ ((\neg R_{14} \ \& \ \neg R_{24} \ \& \ \neg R_{33}) \ \& \ (\neg R_{34} \ \& \ \neg R_{44})) \tag{34}$$

$$R_3 = (\neg R_1 \ \& \ \neg R_2) \tag{35}$$

The rule $\mathbf{r_2}$ for class 2 separation is given below:

$$\mathbf{r_2} : \text{If } (x_1, x_2, x_3, x_4, x_5, x_6, x_7) \in R_1, \text{ then } \mathbf{x} \in \text{class } 2 \tag{36}$$

The rule $\mathbf{r_3}$ for class 1 separation is given below:

$$\mathbf{r_3} : \text{If } (x_1, x_2, x_3, x_4, x_5, x_6, x_7) \in R_3, \text{ then } \mathbf{x} \in \text{class } 1 \tag{37}$$

Table 5 Parameters of the areas generated for classification in Seeds data in 2nd iteration

Rectangle	Left	Right	Bottom	Top	Coordinate pair
R_{13}	0.3	0.94	0.45	1.0	(X_3, X_1)
R_{14}	0.58	0.91	0.45	0.62	(X_3, X_1)
R_{23}	0.3	1.0	0.42	1.0	(X_7, X_5)
R_{24}	0.3	0.67	0.52	0.76	(X_7, X_5)
R_{33}	0.0	0.63	0.46	0.65	(X_6, X_2)
R_{34}	0.27	0.55	0.51	0.65	(X_6, X_2)
R_{44}	0.63	0.74	0.46	0.63	(X_3, X_4)

The accuracy obtained for Seeds data classification with this approach and applying tenfold cross validation using worst-case heuristics validation split is **100%.** The parameters of the areas generated by genetic algorithm are listed in Table 5.

APS (Air Pressure System) failure for Scania Trucks from UCI Repository [8] consists of two classes. Class 1 corresponds to the failure in Scania trucks that is not due to the air pressure system and class 2 corresponds to the failure in Scania trucks that is due to the air pressure system. This data consists of 60,000 cases and 170 dimensions.

However, the data set contains a large number of missing values. These missing values are replaced by calculating 10% of the maximum value of the corresponding column and multiplying the result by -1. Also, there are 4 columns of data with all 0 values. After imputation and removing the columns without any information, the final data contains 60,000 cases with 166 dimensions. As discussed in Sect. 2, visualizing 166 dimensions becomes very challenging. Hence, we use Serpent Coordinate System (SCS) to visualize high dimension data as shown in Fig. 7.

Data classification using interactive approach becomes very tedious due large number of dimensions and instances. Hence, for this dataset, classification is performed using automation technique. After running the COO algorithm, the result contains 166 coordinates arranged from the most to least optimized coordinates. Here, we start with extracting top 4 coordinates and perform analysis on the data with 2 pairs of coordinates displayed in SPCVis.

The coordinates are gradually increased by pairs until we get the desired results. In this case, 12 coordinates were selected for further analysis. They are X_7, X_{140}, X_{70}, X_{75}, X_{145}, X_{74}, X_{141}, X_{71}, X_{143}, X_{76}, X_{122}, and X_{69}. APS dataset with top 12 coordinates is displayed with green class on top is displayed in Fig. 28a and red class on top is displayed in Fig. 28b.

The visualizations in Fig. 28 display high degree of occlusion even with the best order coordinates. Although there is fair amount of separation observed in the first pair of coordinates. Data in the remaining five pairs are highly occluded. Since there is no clear vertical separation between red class and green class, non-linear scaling becomes insignificant and hence not performed on this data set. However, genetic algorithm is still run on this data to generate areas with high purity. The resulting visualization after running genetic algorithm is displayed in Fig. 29.

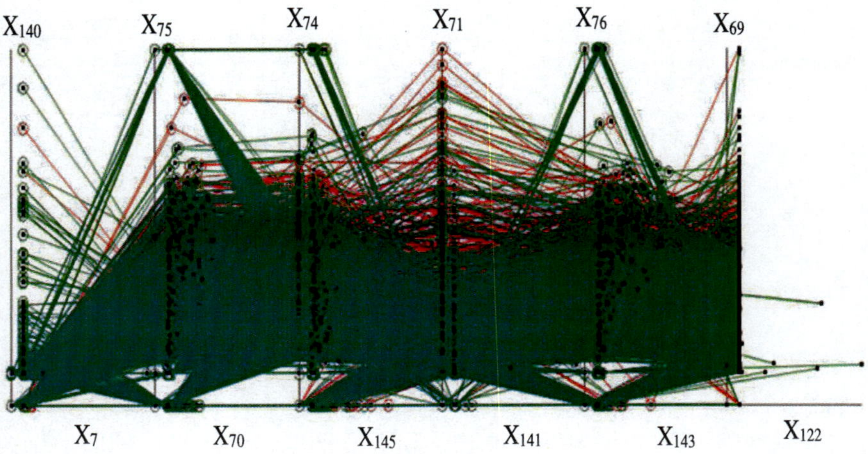

(a). APS data with green class on top.

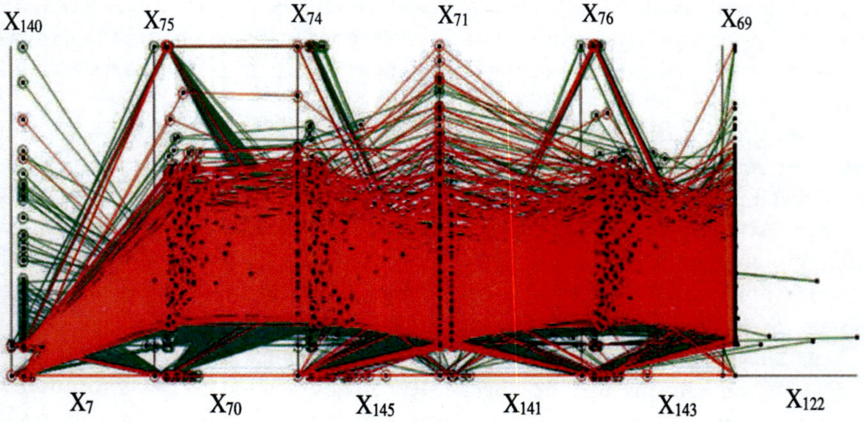

(b). APS data with red class on top.

Fig. 28 Visualization of 12 best coordinates of APS data in SPCVis

Due to two main reasons, the data pattern cannot be interpreted by the end users in this situation: (1) high density of data within the areas generated and (2) small size of the areas generated by genetic algorithm.

To address these issues, we use zooming and averaging. Zoom interactive features wherein the small areas data can be zoomed to view the data more clearly. Figure 30 displays the zoomed image of are R_{31}.

The zoomed visualization in Fig. 30a solves the problem partially. Although, the data are distinctly visible, the pattern is still hidden. To view the overall distribution of red and green class data, the average of individual class within the area is performed. Figures 30b and 31 display the averaged red and green class with R_{31} area.

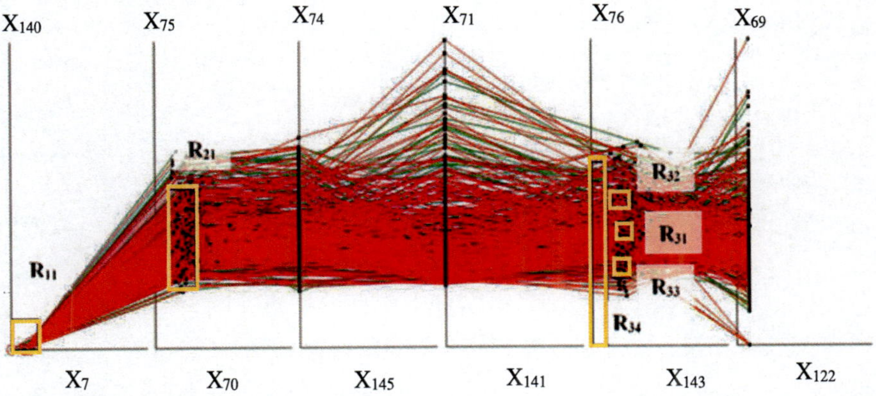

Fig. 29 Visualization of APS data with areas generated by Genetic Algorithm for red class classification

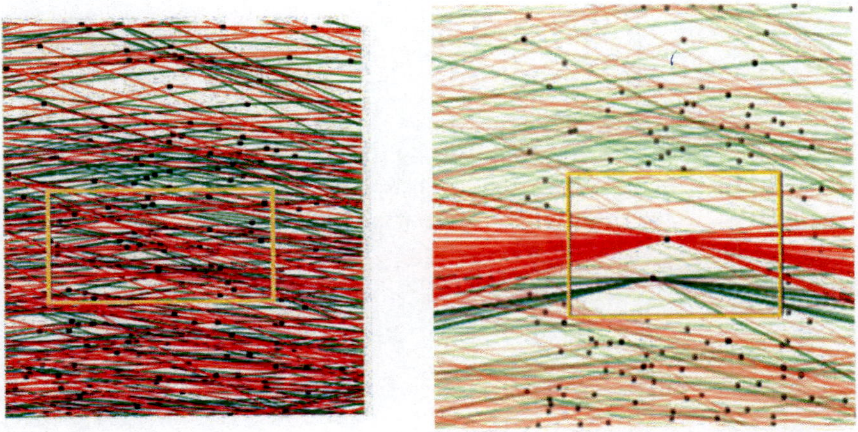

(a). Visualization of zoomed R_{31} area in the APS failure data without averaging.

(b). Visualization of zoomed R_{31} area in the APS failure data after averaging.

Fig. 30 Visualization of R_{31} area in the APS failure data with zooming and averaging

Averaging is performed on all the areas generated by the algorithm. Since the areas are generated in the first, second and fifth pair of coordinates, we can disregard the coordinate pairs in between second and fifth, resulting in only four pairs of coordinates. Figure 32 shows the overall visualization of red class classification.

The area R_1 generated for red class 2 (red) classification is defined below:

$$R_1 = R_{11} \;\&\; R_{21} \;\&\; (R_{51} \text{ or } R_{52} \text{ or } R_{53}) \;\&\; \neg R_{54} \tag{38}$$

The rule $\mathbf{r_1}$ for class 2 (red) classification is defined below.

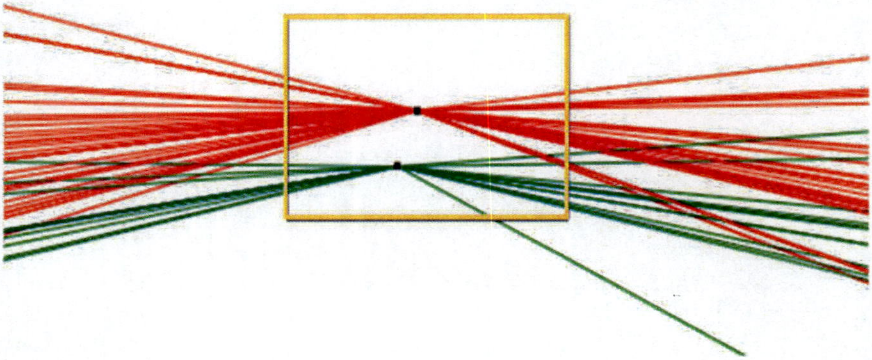

Fig. 31 Visualization of zoomed R_{31} area in the APS failure data with averaged classes with the area (without the surrounding data)

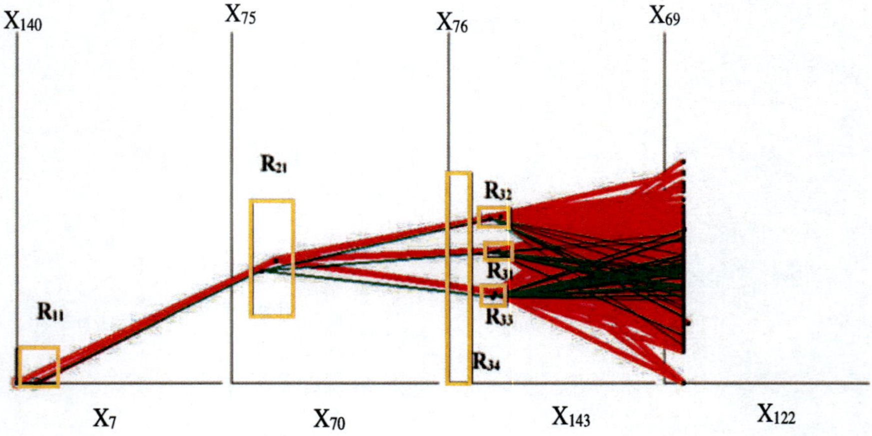

Fig. 32 Visualization of rule $\mathbf{r_1}$ for red class classification in the APS truck data set

$$\mathbf{r_1} : \text{If } (x_7, x_{140}, x_{70}, x_{75}, x_{143}, x_{76}) \in R_1, \text{ then } \mathbf{x} \in \text{class 2} \tag{39}$$

Data that do not follow $\mathbf{r_1}$ rule are sent to the next iteration. The Coordinate Order Optimizer is run on 166 coordinates again to get their optimized order for the remaining data. The resulting order of coordinates is X_{61}, X_{164}, X_{103}, X_{97}, X_9, X_{39}, X_1 and X_{26}. The data are visualized in Fig. 33a.

In the second iteration, the green class tends to be clustered at the bottom and red towards the top. Since the separation along vertical coordinates is clearly visible, we performed the non-linear scaling to get better data interpretation with thresholds 0.15 on X_{164}, 0.2 on X_{97}, 0.25 on X_{39}, and 0.25 on X_{26}. Then the data are visualized with non-linear scaling and analytical rules discovered (see Fig. 33b).

The area R_2 for class 2 (red) classification is defined below:

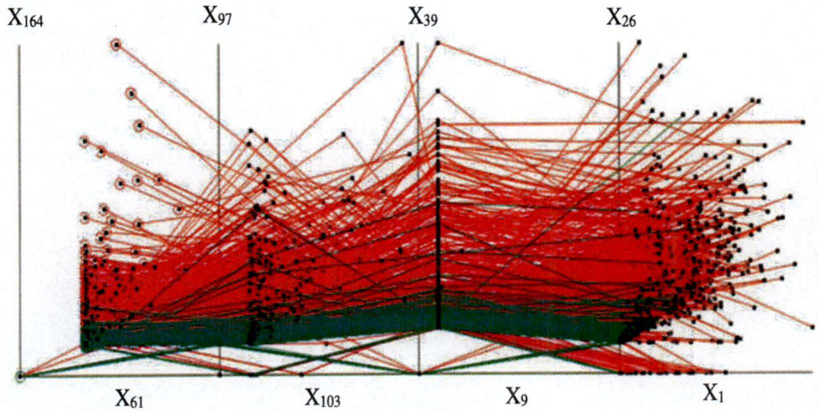

(a). Visualization of APS truck data set with top 8 coordinates.

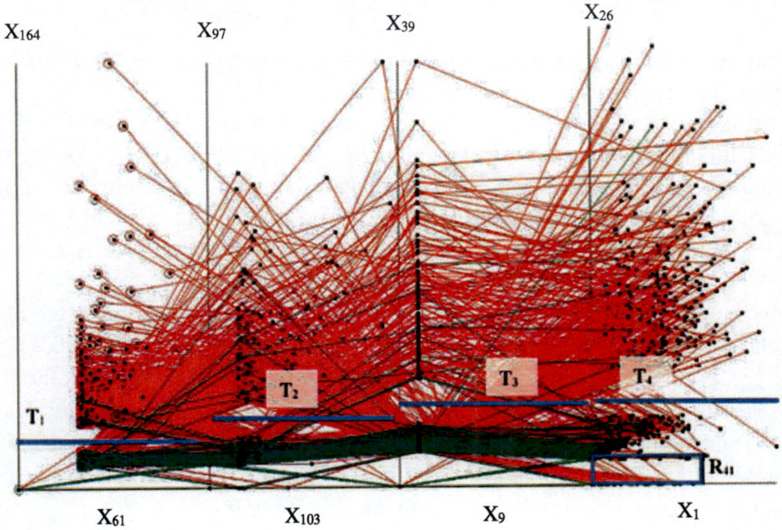

(b). Visualization of $\mathbf{r}_2$ in APS truck data set with top 8 coordinates with non-linear scaling.

Fig. 33 Visualization of APS truck data set in in the second iteration

$$R_2 = T_1 \;\&\; T_2 \;\&\; T_3 \;\&\; (T_4 \text{ or } R_{41}) \tag{40}$$

The rule $\mathbf{r_2}$ for APS data classification is given below:

$$\mathbf{r_2} : \text{If } (x_{61}, x_{164}, x_{108}, x_{97}, x_9, x_{39}, x_1, x_{26}) \in R_2, \text{ then } \mathbf{x} \in \text{class 2 else class 1.} \tag{41}$$

(T_1, T_2, T_3 and T_4 are the threshold values of non-linear scaling)
Table 6 lists the area parameters used in both iterations for APS data classification.

Table 6 Parameters of the rectangles generated for classification in APS data

Rectangle	Left	Right	Bottom	Top	Coordinate pair
R_{11}	0.0	0.2	0.0	0.1	(X_7, X_{140})
R_{21}	0.1	0.3	0.19	0.52	(X_{70}, X_{75})
R_{31}	0.18	0.3	0.36	0.4	(X_{143}, X_{76})
R_{32}	0.15	0.3	0.45	0.5	(X_{143}, X_{76})
R_{33}	0.16	0.28	0.22	0.28	(X_{143}, X_{76})
R_{34}	0.0	0.11	0.0	0.6	(X_{143}, X_{76})
R_{41}	0.0	0.6	0.0	0.1	(X_1, X_{26})

7 Experimental Results and Comparison with Published Results

The results obtained are compared with the published results that use both black-box and interpretable techniques (see Table 7).

From Table 7, we can see that the classification accuracy obtained with the proposed method is on par with the published results and in some cases, have performed better than the published results. Since the interactive technique is more challenging for classifying data of larger size, we used only automated classification for such dataset (APS truck data). The results produced in this chapter are listed in bold.

From the results in Table 7, we can clearly see that the accuracies obtained from our proposed method is better than black box machine learning models [19, 20, 22] and on par with interpretable models [4]. However, the accuracy for APS failure at Scania Trucks is slightly lesser compared to the accuracy in [22] using Deep Neural Network, which is a black box model. Despite of lesser accuracy, our proposed model is favorable due to its transparency, the ability to use the model as self-service and the ability to interpret the model by non-technical end users.

8 Conclusion

In this chapter, we demonstrated the power of lossless data visualization in our proposed interpretable data classification techniques that are implemented both interactively and automatically. We observed that the interactive data classification technique works well for data with lesser cases and dimensions but fail to perform well for data with higher number of cases like the APS failure truck dataset. High degree of occlusion was observed and was challenging to discover pattern interactively. This issue was successfully addressed by our newly proposed automated interpretable technique using *Coordinate Order Optimizer* (COO) Algorithm and

Table 7 Comparison of Different Classification Models

Classification algorithms	Accuracy %
Breast Cancer data (9-D)	
Iterative visual logical classifier (Automated)	**99.71**
Iterative visual logical classifier (Interactive)	**99.56**
SVM [15]	96.995
DCP/RPPR [16]	99.3
SVM/C4.5/kNN/Bayesian [17]	97.28
Iris Data (4-D)	
Iterative visual logical classifier (Automated)	**100**
Iterative visual logical classifier (Interactive)	**100**
Multilayer visual knowledge discovery (Kovalerchuk)	100
k-Means + J48 classifier [18]	98.67
Neural Network [19]	96.66
Seeds Data (7-D)	
Iterative visual logical classifier (Automated)	**100**
Iterative visual logical classifier (Interactive)	**100**
Deep neural network [20]	100
K- nearest neighbor [21]	95.71
APS Failure at Scania Trucks (170-D)	
Iterative visual logical classifier (Automated)	**99.36**
Deep neural network (DNN) [22]	99.50
Random forest [23]	99.025
Support vector machine (SVM) [23]	98.26

Genetic Algorithm (GA) where the areas were generated automatically rather than interactively.

We also demonstrated the power of interactive features that improved the visualization due to which discovering patterns in the data became much easier. With non-linear scaling, zooming and averaging, the visualization was improved to increase data interpretability. The SPCVis software successfully visualized the larger dataset using modified Shifted Paired Coordinates System called as Serpent Coordinate System (SCS).

Our proposed techniques can be further leveraged by incorporating more interactive features like non-orthogonal coordinates, data reversing etc. Shifted Paired

Coordinates (SPC) and Serpent Coordinate System (SCS) visualization helps us to discover only specific patterns in the data and our future goal is to incorporate more General Line Coordinate Visualizations.

References

1. Feurer, M., Klein, A., Eggensperger, K., Springenberg, J.T., Blum, M., Hutter, F.: Auto-sklearn: efficient and robust automated machine learning. In: Automated Machine Learning, pp. 113–134. Springer, Cham (2019)
2. Hutter, F., Kotthoff, L., Vanschoren, J.: Automated Machine Learning: Methods, Systems, Challenges. Springer Nature (2019)
3. Kovalerchuk, B., Ahmad, M.A., Teredesai, A.: Survey of explainable machine learning with visual and granular methods beyond quasi-explanations, In: Pedrycz, W., Chen, S.M. (eds.) Interpretable Artificial Intelligence: A Perspective of Granular Computing, pp. 217–267, Springer (2021)
4. Kovalerchuk, B.: Visual Knowledge Discovery and Machine Learning. Springer (2018)
5. Wagle, S., Kovalerchuk, B.: Interactive visual self-service data classification approach to democratize machine learning. In: 24th International Conference IV Information Visualisation, pp. 280–285. IEEE (2020). https://doi.org/10.1109/IV51561.2020.00052
6. Banzhaf, W., Nordin, P., Keller, R.E., Francone, F.D.: Genetic Programming: An Introduction. Morgan Kaufmann Publishers, San Francisco (1998)
7. Kovalerchuk, B.: Enhancement of cross validation using hybrid visual and analytical means with Shannon function. In: Beyond Traditional Probabilistic Data Processing Techniques: Interval, Fuzzy etc. Methods and Their Applications, pp. 517–543. Springer (2020)
8. Dua, D., Graff, C.: UCI Machine Learning Repository, http://archive.ics.uci.edu/ml. University of California, School of Information and Computer Science, Irvine, CA (2019)
9. Kovalerchuk, B., Gharawi, A.: Decreasing occlusion and increasing explanation in interactive visual knowledge discovery. In: Human Interface and the Management of Information. Interaction, Visualization, and Analytics. Lecture Notes in Computer Science Series, vol. 10904, pp. 505–526. Springer (2018)
10. Kovalerchuk, B., Grishin, V.: Reversible data visualization to support machine learning. In: Human Interface and the Management of Information. Interaction, Visualization, and Analytics. LNCS, vol. 10904, pp. 45–59. Springer (2018)
11. Everitt B.: The Cambridge Dictionary of Statistics. Cambridge University Press (1998)
12. Cowgill, M.C., Harvey, R.J., Watson, L.T.: A genetic algorithm approach to cluster analysis. Comput. Math. Appl. **37**(7), 99–108 (1999)
13. Bouali, F., Serres, B., Guinot, C., Venturini, G.: Optimizing a radial visualization with a genetic algorithm. In: 2020 24th International Conference Information Visualisation (IV), pp. 409–414. IEEE (2020 Sep. 7)
14. Rifki, O., Ono, H.: A survey of computational approaches to portfolio optimization by genetic algorithms. In: 18th International Conference on Computing in Economics and Finance 2012 (2012)
15. Christobel, A., Sivaprakasam, Y.: An empirical comparison of data mining classification methods. Int. J. Comput. Inf. Syst. **3**(2), 24–28 (2011)
16. Neuhaus, N., Kovalerchuk, B., Interpretable machine learning with boosting by Boolean algorithm. In: Joint 2019 8th International Conferences on Informatics, Electronics & Vision (ICIEV) & 3rd International Conferences on Imaging, Vision & Pattern Recognition (IVPR), pp. 307–311 (2019)
17. Salama, G.I., Abdelhalim, M., Zeid, M.A.: Breast cancer diagnosis on three different datasets using multi-classifiers. Int. J. Comput. Inf. Technol. **01**(01) (2012)

18. Kumar, V., Rathee, N.: Knowledge discovery from database using an integration of clustering and classification. Int. J. Adv. Comput. Sci. Appl. **2**(3), 29–33 (2011)
19. Swain, M., Dash, S.K., Dash, S., Mohapatra, A.: An approach for iris plant classification using neural network. Int. J. Soft Comput. **3**(1), 79 (2012)
20. Eldem, A.: An application of deep neural network for classification of wheat seeds. Eur. J. Sci. Technol. **19**, 213–220 (2020)
21. Sabanc, K., Akkaya, M.: Classification of different wheat varieties by using data mining algorithms. Int. J. Intell. Syst. Appl. Eng. **4**(2), 40–44 (2016)
22. Zhou, F., Yang, S., Fujita, H., Chen, D., Wen, C.: Deep learning fault diagnosis method based on global optimization GAN for unbalanced data. Knowl. Based Syst. **187**, 104837 (2020)
23. Rafsunjani, S., Safa, R.S., Al Imran, A., Rahim, M.S., Nandi, D.: An empirical comparison of missing value imputation techniques on APS failure prediction. IJ Inf. Technol. Comput. Sci. **2**, 21–29 (2019)

Non-linear Visual Knowledge Discovery with Elliptic Paired Coordinates

Rose McDonald and Boris Kovalerchuk

Abstract It is challenging for humans to enable visual knowledge discovery in data with more than 2–3 dimensions with a naked eye. This chapter explores the efficiency of discovering predictive machine learning models interactively using new Elliptic Paired coordinates (EPC) visualizations. It is shown that EPC are capable to visualize multidimensional data and support visual machine learning with preservation of multidimensional information in 2-D. Relative to parallel and radial coordinates, EPC visualization requires only a half of the visual elements for each n–D point. An interactive software system EllipseVis, which is developed in this work, processes high-dimensional datasets, creates EPC visualizations, and produces predictive classification models by discovering dominance rules in EPC. By using interactive and automatic processes it discovers zones in EPC with a high dominance of a single class. The EPC methodology has been successful in discovering non-linear predictive models with high coverage and precision in the computational experiments. This can benefit multiple domains by producing visually appealing dominance rules. This chapter presents results of successful testing the EPC non-linear methodology in experiments using real and simulated data, EPC generalized to the Dynamic Elliptic Paired Coordinates (DEPC), incorporation of the weights of coordinates to optimize the visual discovery, introduction of an alternative EPC design and introduction of the concept of incompact machine learning methodology based on EPC/DEPC.

Keywords Machine learning · Data visualization · Knowledge discovery · Elliptic paired coordinates

R. McDonald · B. Kovalerchuk (✉)
Department of Computer Science, Central Washington University, Ellensburg, WA, USA
e-mail: Borisk@cwu.edu

R. McDonald
e-mail: Rose.McDonald@cwu.edu

B. Kovalerchuk et al. (eds.), *Integrating Artificial Intelligence and Visualization for Visual Knowledge Discovery*, Studies in Computational Intelligence 1014,
https://doi.org/10.1007/978-3-030-93119-3_5

1 Introduction

The efficient use of visualization in Machine Learning (ML) requires preservation of multidimensional information and **Elliptic Paired Coordinates (EPCs)** is one of the visualization methods that preserves this n–D information [1, 2].

Commonly *n–D points* are mapped *to 2-D points* for visualization, which *can approximate n–D information partially preserving it* in the form of *similarities* between n–D points by using MDS, SOM, t-SNE, PCA and other *unsupervised* methods [3–6]. Moreover, both n–D similarities injected by these methods and their 2-D projections for visualization can be irrelevant to a given learning task [1]. In [5] t-SNE is used to visualize not only the input data but also the activation (output) of neurons of a selected hidden layer of the MLP and CNN trained models. This combines unsupervised t-SNE method with the information learned at this layer. However, projection of the multidimensional activation information of the layer to 2-D for visualization is lossy. It only partially preserves the information of that layer. Moreover, the layer itself only partially preserves n–D input information. Thus, some information that is important for classification can be missed.

An alternative methodology is *mapping n–D points to 2-D graphs* that preserves all n–D information [1, 7–9]. Elliptic Paired Coordinates belong to the later. In both methodologies the respective visual representations of n–D data are used to solve predictive classification tasks [10].

The advantages of EPC include: (1) *preserving* all n–D information, (2) capturing *non-linear dependencies* in the data, and (3) requiring fewer visual elements than methods such as parallel and radial coordinates.

This chapter expands our prior work [2] on EPC in: (1) successful *testing* the EPC approach and methodology in additional experiments with both real and simulated data, (2) generalizing EPC to the *Dynamic* Elliptic Paired Coordinates (DEPC), (3) generalizing EPC and DEPC by incorporating the *weights* of coordinates to optimize the visual discovery in EPC/DEPC, alternating side ellipses and introducing the concept of *incompact* machine learning methodology based on EPC/DEPC.

The chapter is organized as follows. Section 2 describes the concept of elliptic paired coordinates. Section 3 presents a visual knowledge discovery system based on the elliptic paired coordinates. Section 4 describes the results of experiments with multiple real data sets. Section 5 presents experiments with synthetic data. Section 6 provides generalization options for elliptic paired coordinates and Sect. 7 concludes the paper with its summary and future work.

2 Elliptic Paired Coordinates

2.1 Concept

In EPC coordinate axes are located on ellipses (see Fig. 1). In [1] the EPC shows in Fig. 1 is called EPC-H, in this chapter we omit H. In Fig. 1, a short green arrow losslessly represents a 4-D point P = (0.3,0.5,0.5,0.2) in EPC, i.e., this 4-D point can be restored from it. The number of nodes in EPC is two times less than in parallel and radial coordinates with less occlusion of lines. For comparison see point P in Fig. 2 in radial and parallel coordinates with 4 nodes instead of 2 nodes in EPC. The dark blue ellipse C_E in Fig. 1 contain four coordinate curves X_1–X_4. It is called the **central ellipse.**

Figure 1 shows only one of the options how we can split an ellipse or a circle to the segments. Here each coordinate starts at the horizontal red marks on the right or on the left edges of the central ellipse, where X_1, X_4 go up and X_2, X_3 go down from respective points. An alternative *sequential way* of directing coordinates is that X_1 starts from the top, ends in the right middle point, where X_2 starts and go down to the bottom point, where X_3 starts as so on. This is a simpler way when we have more than 4 attributes to be mapped to the ellipse. In general, the direction of each coordinate can be completely independent of the directions of the other coordinates. While all these possibilities exist for EPC, we use either one which is shown in Fig. 1 or a

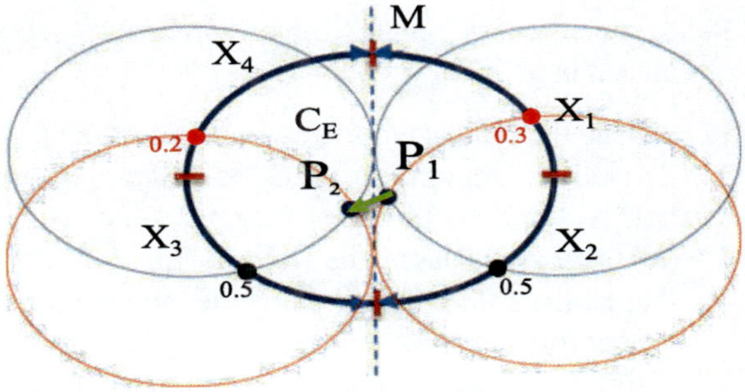

Fig. 1 4-D point P = (0.3,0.5,0.5,0.2) in 4-D EPC as green arrow $P_1 \rightarrow P_2$. Red marks separate coordinates in the blue coordinate ellipse

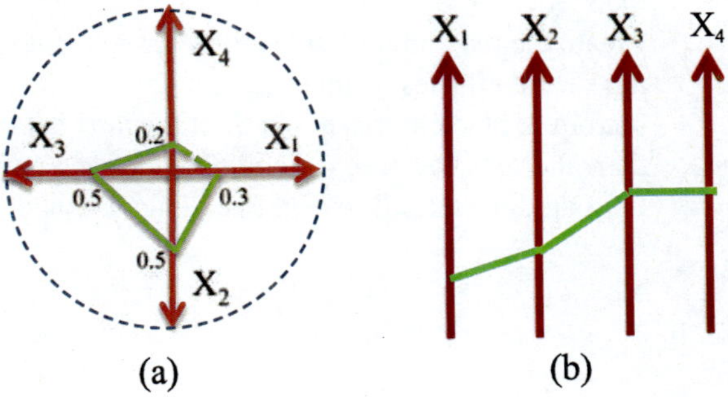

Fig. 2 4-D point P = (0.3,0.5,0.5,0.2) in radial **a** and parallel coordinates **b**

sequential order from the top of the central ellipse. Another option called Dynamic EPC is present later in section 6.1.

Four **side ellipses** of the size of the central blue ellipse are used to build the green arrow from point P_1 to point P_2, $P_1 \rightarrow P_2$. The middle vertical line M is guiding line for side ellipses. They touch line M. Moving these side ellipses along line M produces different 4-D points. The thin red side ellipse on the right goes through $x_1 = 0.3$, touch line M and has the size of the central ellipse C_E.

The thin blue side ellipse on the right is built in the same way for $x_2 = 0.5$. The point P_1 is the **crossing point** where these red and blue side ellipses cross each other in C_E. The point P_1 represents pair (x_1, x_2)= (0.3,0.5). The point P_2 that represents pair $(x_3, x_4) = (0.5,0.2)$ is constructed in the same way for $x_3 = 0.5$ and $x_4 = 0.2$ by generating respective red and blue side ellipses on the left. Next, the arrow from P_1 to P_2 is made (see a short green arrow in Fig. 1), which losslessly visualize 4-D point P = (0.3,0.5,0.5,0.2) (Fig. 2).

It means that we can reverse these steps to restore the 4-D point P from ($P_1 \rightarrow P_2$) arrow. Visually we slide the red right-side ellipse to cross point P_1. Its crossing point with the central blue ellipse is $x_1 = 0.3$. Similarly, we get $x_2 = 0.5$ with the right-side blue ellipse. The same process allows restoring P_2, with red and blue left-side ellipses.

2.2 Elliptic Paired Coordinates Algorithm

Below we present steps of EPC algorithm for n–D data referencing Fig. 1 and notation introduced in it:

1. Create the *central ellipse*, C_E, with a vertical line M bisecting it.
2. Divide the circumference of C_E into *n equal sectors* by the number of dimensions $X_1, X_2, \ldots, X_n$ to be represented.
3. *Normalize* all data on the scale [0,1].
4. Graph the *data points* x_i along the circumference of the C_E in the appropriate sector for each n–D point $\mathbf{x} = (x_1, x_2, \ldots, x_n)$.
5. Pair data points by dimension, starting at (x_1, x_2). If there is an odd number of pairs of coordinates, use the horizontal bisector N for the middle pair instead of the vertical bisector M.
6. Create the red and blue side ellipses, such that they touch the line M. Have the top of red ellipses to intersect the data point x_i on the central ellipse and the bottom of blue ellipses to intersect the next data point on the central ellipse and then alternate between red and blue for each coordinate in sequence.
7. Find the Intersects Between Each Pair, Connect Them.

2.3 EPC Mathematical Formulation for n–D Data

Steps 6 of EPC algorithm requires constructing thin ellipses. Their width W and heigh H are equal to them for the central ellipse C_E, but it requires computing their centers. Let point (A,B) be the center of such ellipse. It can be found by soling for (A,B) the general equation for an ellipse,

$$\frac{(x-A)^2}{W^2} - \frac{(y-B)^2}{H^2} = 1$$

with given the W, H, x- and y-coordinates of the point on the ellipse. Then we can calculate the intercepts of the ellipses to form the points P_1 and P_2. Let the center of the central ellipse be the point (cx, cy) then formulas for A and B for side ellipses are as follows.

Right and Left Ellipses. The +/– assignment is dependent on the ellipse being to the right or left of the line M, respectively,

$$A = cx \pm W/2$$

The +/– assignment (red/blue respectively) depends on the ellipse intersecting at the top or bottom of the thin ellipse.

$$B = y \pm \sqrt{H^2\left(1 - \frac{(x-A)^2}{W^2}\right)}$$

Top and Bottom Ellipses. If the number of coordinates n is even, but the number of pairs of coordinates $n/2$ is odd (e.g., 5 pairs from 10 coordinates), a horizontal line N across the center ellipse is used similarly to M for placing two ellipses for the selected pair (x_i, x_j). These ellipses touch line N (being above or below line N) and the intersect point of these ellipses is computed. If the selected pair is $(x_{n/2}, x_{(n/2)+1})$, e.g., (x_5, x_6) for $n = 10$ then the bottom ellipse is constructed with

$$B = cy - H/2$$

If the selected pair is (x_1, x_n), e.g. (x_1, x_{10}) for $n = 10$ then then the top ellipse is constructed with

$$B = cy + H/2$$

For the value A the $\pm$ assignment depends on the ellipse being to the right or left of the line M, respectively.

$$A = x \pm \sqrt{W^2\left(1 - \frac{(y-B)^2}{H}\right)}$$

2.4 Elliptic Paired Coordinates Pseudo-Code

The pseudo-code is as follows.

```
CreateEPC(W, H, cx, cy):
   DrawEllipse(W,H, cx,cy,)
   // Draw ellipse CE using (x−A)²/W² − (y−B)²/H² = 1 with (A, B) = (cx, cy).
   DrawDottedVerticalLineM(cx,cy) //vertical line M at point (cx,cy)
   DrawDottedHorizontallLineN(cx.cy) //horizontal line N at point (cx,cy)
   ComputeEllipseSectorLength(D) = (CE circumference) / (number of coordinates, n)
   If n is odd then n++ for all x(n), x(n) = x(n-1)
      // Duplicate the last coordinate to make n even.
   Normalize(x(i)):  x(i) = x(i) – min(i) / (max(i) – min(i)) // normalize each x(i) of
     // Coordinate Xi, where min(i) and max(i) are min and max for coordinate Xi in data.

For i= 1:n
    DrawDataPointOnEllipse(x(i))
      // Draw data point value x(i) along CE's axis on sector(i).
            If n is even then
                 If (i < n/2) DrawRightEllipse(i) Else DrawLeftEllipse(i)
                      Else if (i = 0 to n/2 – 1) DrawRightEllipse(i)
                       Else if (i = n/2 – 1 to n/2 + 1) DrawBottomEllipse(i)
                            Else DrawLeftEllipse(i)
    DrawEPC_Point(i,i+1)) // For each pair (x1,x2), (x3,x4),…,(xi,xi+1), (xn-1,xn)
      // Compute a crossing point Pi,i+1 of two ellipses.
       If n is odd then
          If (i < n/2) DrawTopEllipse(i) Else DrawBottomEllipse(i)
             // For each data point dimension pair Pi, draw Pi.1 to Pi.2.

DrawGraph() // Draw arrows to connect points P1,2, P3,4,…, Pi,i+1,…, Pn-1,n.
```

3 EPC Visual Knowledge Discovery System

3.1 Dominance Rectangular Rules (DR2) Algorithm in EPC

Section 2 described how n–D data points are visualized in EPC losslessly. This section presents the algorithm that discovers Dominance Rectangular Rules (DR2) based on interactive or automatic finding the rectangular areas where a single class dominates other classes. In EPC an n–D point $\mathbf{x} = (x_1, x_2, \ldots, x_n)$ is represented as a graph x*, where each node encodes a pair of values (x_i, x_{i+1}).

Consider a set of n–D point visualized in EPC, a class Q and a rectangle R in the EPC, then the forms of the rule **r** that we explore in this chapter are:

Point Rule r: If a node of graph x* of n–D point **x** is *in* R then **x** is in class Q.

Intersect Rule r: If graph x* of n–D point **x** *intersects* R then **x** is in class Q.

A point rule captures a non-linear relation of *2 attributes* that are encoded in the graph node in R. An intersect rule captures a non-linear relation of *4 attributes* that form a line that crosses R. This line is the edge that connects two nodes of the graph x*.

We discover both types of rules using the DDR algorithm. The steps of the **DRR algorithm** are:

1. Visualize data in EPC;
2. Set up parameters of the dominant rectangles: coverage, dominance precision thresholds and type (intersect or point);
3. Search for dominant rectangles:
   - 3.1 Automatic search: setting up the size of the rectangle, shift rectangle with a given step over the EPC display, compute coverage and precision for each rectangle, record rectangles that satisfy thresholds;
   - 3.2 Interactive search: draw a rectangle of any size and at any location in the EPC display, collect coverage and precision for this rectangle;
4. Remove/hide cases that are in the accepted rules and continue search for new rules if desired.

The automatic search involves the *dominance search parameters*: coverage (recall in the class) and precision thresholds. A user gets only rules that satisfy these thresholds. We typically used at least 10% coverage and 90% precision.

3.2 EllipseVis: Interactive Software System

In **EllipseVis** users can interactively conduct visual analysis of datasets in EPC and discover classification rules. EllipseVis is available at GitHub [11]. A user can more easily detect patterns to classify data by moving the camera, and hiding various elements of the visualization.

In the automatic mode EllipseVis can do calculations to find classification patterns. EllipseVis checks if most data points within the rectangle belongs to a single class when dividing the EPC space into smaller rectangular sections. We call such rectangles *class dominant rectangles* or shorter as dominant rectangles. Respectively rules that correspond to dominance rectangles discovered by EllipseVis are called dominance rules. The quality of rules is defined by the total number of points in the rectangle (rule coverage) and the percentage of dominant class compared to the others (precision of classification). In the interactive mode a user creates the dominance rules in EllipseVis. After a rule is created, the data points that satisfy the rule, i.e., inside the rectangle, are removed from the rest of the data and further rules are discovered without them.

Several interactive capabilities are implemented in EllipseVis such as:

- Camera move (pan and zoom)
- Setting of a dominance rectangle: minimum coverage and precision; rectangle dimensions;
- Toggle whether the rectangles are calculating data on their points or intersect lines; Automatically find dominance rectangle rules, Clear rectangles.
- Hide or show elements: intersect lines, dominance rectangles, cycle between showing all lines, lines not within rules, and line within rules, Side, thin red and thin blue ellipses.

4 Experiments With Real Data

The goal of experimentation is exploring efficiency of EPC on several benchmark datasets from UCI ML repository [12].

4.1 Experiment With Iris Data

All 150 cases of Iris 4-D data of 3 classes are shown in Fig. 3 losslessly in EPC with a small insert showing them inside of the 4-D elliptic coordinates. Discovered classification rules $\mathbf{r_1}$–$\mathbf{r_3}$ are represented by rectangles R_1–R_3 in a zoomed part in Fig. 3. Two misclassified red cases are presented zoomed in Fig. 4. The comparison with Iris data in parallel coordinates shown in Fig. 5 demonstrates the advantage of EPC for Iris data where the data of three classes are more distinct that helps in search for rectangular rules. This advantage is a result of non-linear transformations in EPC in comparison with parallel coordinates.

The rule $\mathbf{r}_1$ covers 52 cases (50 correct 2 misclassified cases), the rule $\mathbf{r}_2$ covers 48 cases (all 48 correct cases) and the rule $\mathbf{r}_3$ covers 50 cases (all 50 correct cases). These rules correctly classified 148 out of 150, i.e., 98.67%. Often each individual rule discovered in EPC cover only a fraction of all cases of the dominant class. For such situations we use the **weighted precision** formula to compute the total precision of k rules

$$\sum_{i=1}^{k}(p_i c_i) / \sum_{i=1}^{k} c_i \tag{1}$$

where p_i is precision of rule $\mathbf{r}_i$ and c_i is the number of cases covered by $\mathbf{r}_i$.

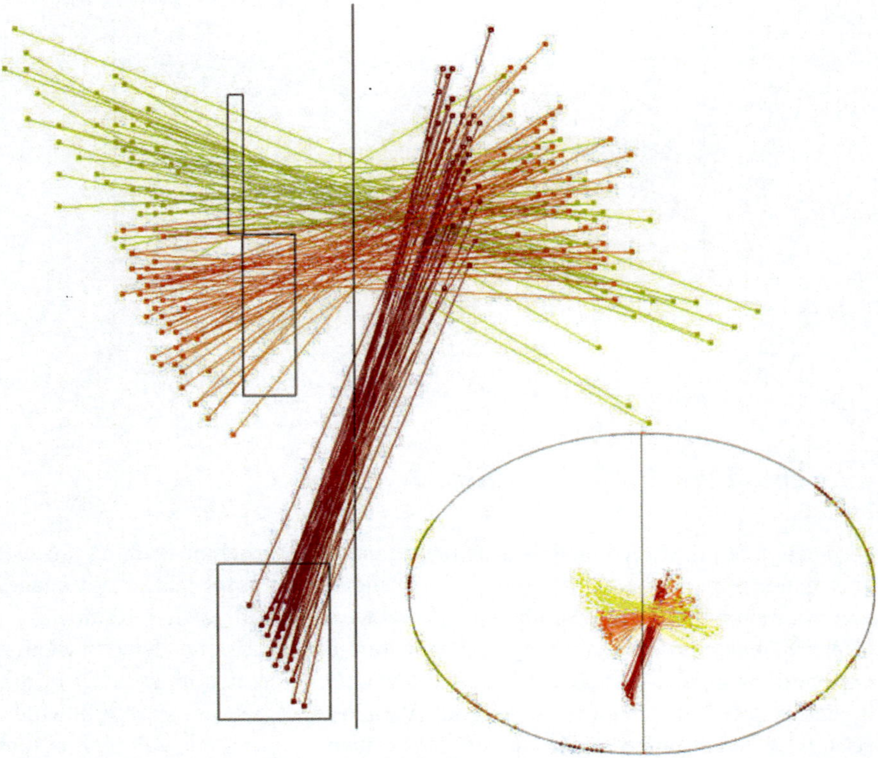

Fig. 3 Three Iris data classes classified by interactively created dominance rules (intersect-based)

Fig. 4 Zoomed overlapped area from Fig. 3

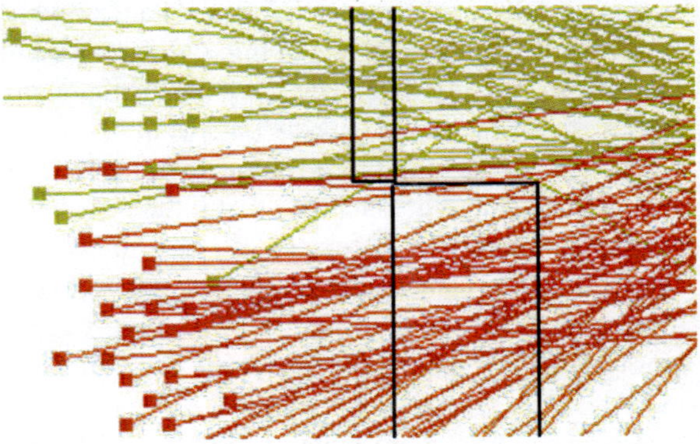

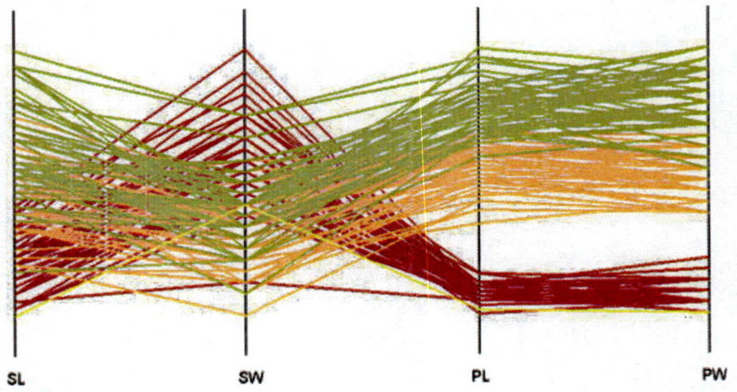

Fig. 5 Iris data in parallel coordinates

4.2 Visual Verification of Rules

Verification of discovered models is an important part of machine learning process. This section presents a simplified visual rule verification approach that has advantages over traditional k-fold cross validation (CV) as we show with the Iris example.

A *random* split data to training and validation data to 10 folds is conducted in a common tenfold CV approach to test the quality of prediction accuracy on the validation data. This random split does not guarantee that the worst-case split will be discovered and evaluated that is important for many applications with high cost or errors. In other words, cross validation can provide of the predictive model. EllipseVis allows to simplify or even eliminate cross validation.

Consider Iris data in Fig. 3 and randomly select 90% of lines to training and remaining 10% of them to validation data. Let these 90% of lines (training cases) cover the overlap area shown in Fig. 4, then we build rules $\mathbf{r}_1$-$\mathbf{r}_3$ as shows in Fig. 3 using these training data. The accuracy of these rules is 100% on validation data because these validation data are outside of the overlap area, confirming these rules.

In the opposite situation, when the training data do not include lines in the overlap area, but validation data include them, the discovered dominance rules (rectangles) can differ from $\mathbf{r}_1$–$\mathbf{r}_3$. Rule $\mathbf{r}_1$ can be shifted lower or $\mathbf{r}_2$ can be shifted higher and misclassify some validation cases. These cases are misclassified because the training data are not representative for these validation data. This is a **worst-case** cross validation split of data to training and validation when the training data are not representative for the validation data.

The visual EPC representation allows to find and see the worst split, and use this worst split, say, for selecting 15 Iris cases (10%) in the overlap area to a worst fold that include all overlap cases. A trivial **lower bound** of worst-case classification accuracy of this fold is 0 when all cases that are in this fold are misclassified. In contrast 9 other folds of 10-fold CV will likely be recognized with 100% accuracy by a well-designed and trained rules/models when considered as validation folds, because these folds are outside of the overlap area. Thus, the average 10-fold CV

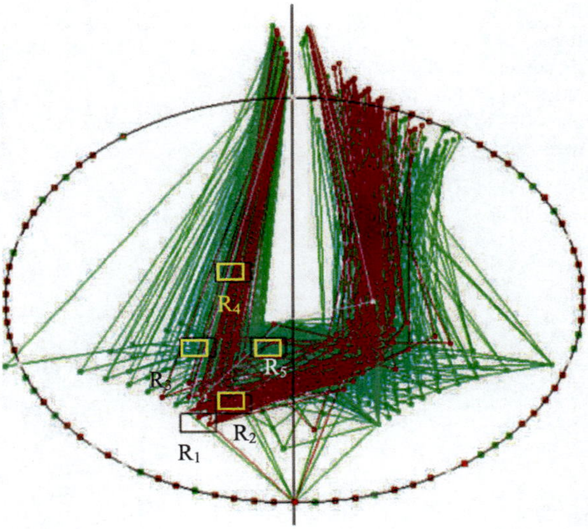

Fig. 6 WBC cases covered by automatically discovered rules $\mathbf{r}_1$–$\mathbf{r}_5$ with rectangles R_1–R_5: 658 out of 683 cases (96.34%) with 95.13% precsion

accuracy will be 90%. So, finding accurate rules such as $\mathbf{r}_1$–$\mathbf{r}_3$ on the full dataset likely indicates that 10-fold CV result will be similar. Thus, instead of CV we can focus on finding visually the worst split and evaluate its accuracy on validation data [1]. It will likely be greater than 0 in general as it is the case for the iris data. Therefore, the 10-fold average will be greater than 90%. A more general and detailed treatment of the worst-case visual cross validation based on the Shannon function can be found in [13].

4.3 *Experiment With Wisconsin Breast Cancer Data*

The goal of this experiment is testing EPC capabilities for the Wisconsin Breast Cancer (WBC) dataset [12] that consists of 683 full 9-D cases: 444 benign (red) and 239 malignant (green) cases as shown in Figs. 6, 7, 8, and 9 in EPC. To get an even number of coordinates we doubled coordinate X_9 creating X_{10}. Table 1 and Figs. 6, 7, 8 show the *automatically discovered* rules in the EllipseVis system. Together five simple rectangular rules cover **96.34%** of cases with weighted precision **95.13%**. Figure 9 shows an example where interactive rule discovery is less efficient than automatic requiring two times more rules and larger rectangles.

4.4 *Experiment with Multi-class Glass Data*

The 10-D multi-class Glass Identification dataset consists of 214 cases of 6 imbalanced classes of glass [12] used in criminal investigations by the U.S. Forensic

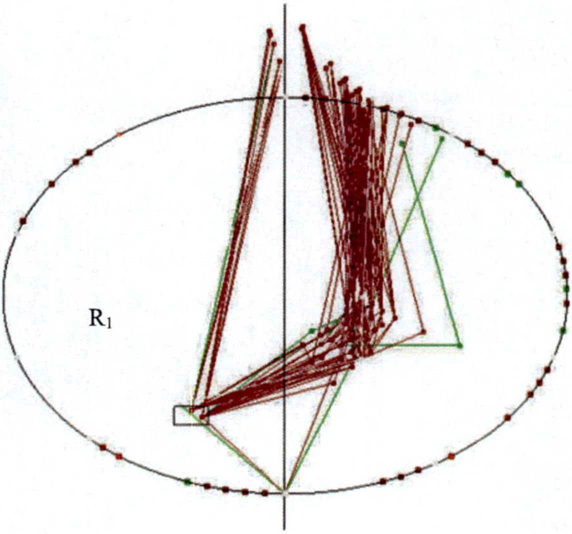

Fig. 7 Cases that satisfy rule $\mathbf{r}_1$ based on the rectangle R_1 that covers 285 cases out of 444 cases of this class (64.18%) with 98.59% precision

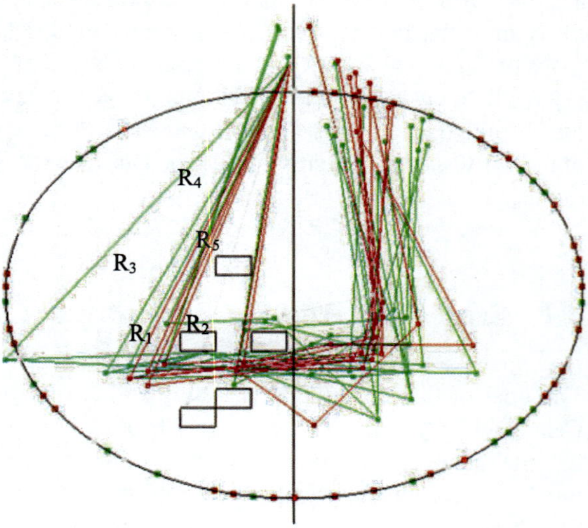

Fig. 8 Automatic rule discovery: remaining cases that do not satisfy rules $\mathbf{r}_1$–$\mathbf{r}_5$ based on the rectangles R_1–R_5 with 96.34% of total coverage/recall

Science Service. Figure 10 visualizes all Glass classes in EPC. Below we show results of one class vs. all other classes.

All other classes versus class 5. Table 2 shows all three rules discovered. They cover **87.06%** of cases of all other classes with weighted precision **98.29%**.

All other classes versus class 6. Figure 11 shows the result of rule discovery to separate cases of all other classes from class 6. The EllipseVis system found three rectangles, R_1–R_3. and respective rules $\mathbf{r}_1$–$\mathbf{r}_3$. Total all three rules **cover 99.51%** of cases with weighted **precision 95.59%** (see Table 3).

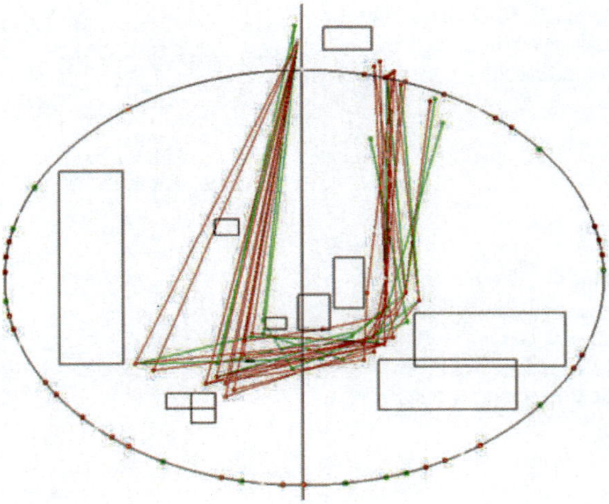

Fig. 9 Interactive "manual" rule discovery: remaining cases that do not satisfy discovered rules based on the shown rectangles with 96.63% of total coverage/recall with two times more rules

Table 1 Results of automatic WBC rules discovery in EPC (intersect-based)

Rule	Class	Coverage/recall in class, %	Precision, %
r_1	B	64.18	98.59
r_2	B	33.10	92.51
r_1 or r_2	**B**	**97.28**	
r_3	M	42.67	92.17
r_4	M	37.23	92.13
r_5	M	14.64	97.14
r_3 or r_4 or r_5	**M**	**94.54**	
All rules	**B, M**	**96.34**	**95.13**

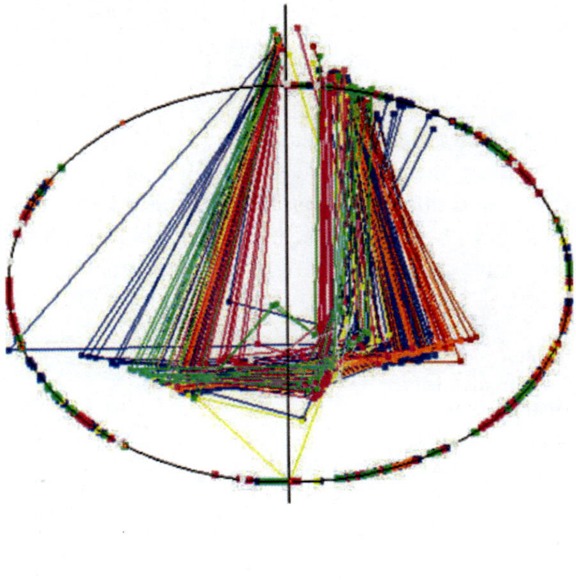

Fig. 10 All glass data of six classes in EPC

Table 2 10-D Glass rules for class 5 versus all others (point-based)

Rule	Class	Coverage/recall in class, %	Precision, %
$\mathbf{r}_1$	All but 5	37.31	100
$\mathbf{r}_2$	All but 5	38.81	98.72
$\mathbf{r}_3$	All but 5	10.95	90.91
All rules	All but 5	**87.06**	**98.29**

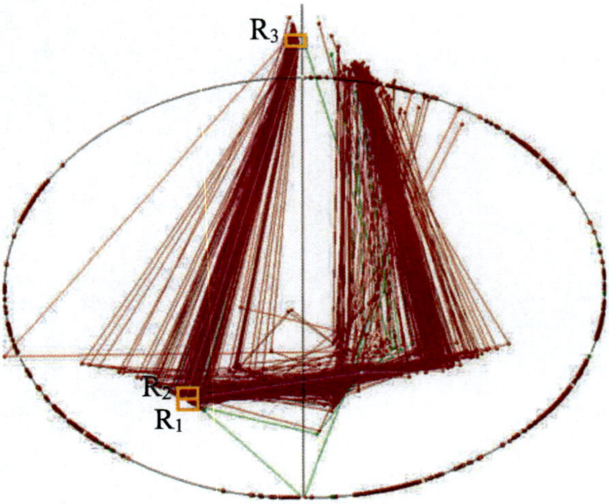

Fig. 11 Extracting rule for glass: all other classes versus class 6. Cases that satisfy extracted 3 rules $\mathbf{r}_1$–$\mathbf{r}_3$ based on the rectangles R_1–R_3

Table 3 10-D Glass rules for class 6 versus all others (point-based)

Rule	Class	Coverage/recall in class, %	Precision, %
$\mathbf{r}_1$	All but 6	50.24	99.03
$\mathbf{r}_2$	All but 6	18.05	94.59
$\mathbf{r}_3$	All but 6	31.22	90.63
All rules	All but 6	**99.51**	**95.59**

All other classes versus class 7. Total all three rules **cover 87.57%** of cases with total **precision 97.31%** weighted by coverage (see Table 4). In addition, EllipseVis discovered rule $\mathbf{r}_1$ of **class 7 versus all others** that covered 79.31% of all cases of class 7 with 91.30% precision (see Table 5). These rules are intercept-based and respectively, capture the non-linear relations between 4 attributes, which form a

Table 4 Glass rules for class 7 versus all others (point-based)

Rule	Class	Coverage/recall in class, %	Precision, %
$\mathbf{r}_1$	All but 7	14.05	96.15
$\mathbf{r}_2$	All but 7	54.59	100.00
$\mathbf{r}_3$	All but 7	18.92	91.43
All rules	All but 7	**87.57**	**97.53**

Table 5 Glass rules for class 7 versus all others (intersect-based)

Rule	Class	Coverage/recall in class, %	Precision, %
r_1	**7**	**79.31%**	**91.30**

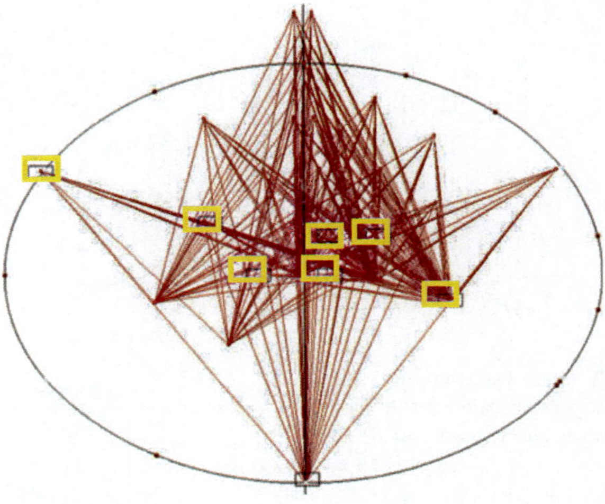

Fig. 12 Extracting rules for car data: cases of class unacc (red) that satisfy extracted 8 rules $\mathbf{r}_1$–$\mathbf{r}_8$ based on the rectangles R_1–R_8. (yellow)

line that intersects the dominance rectangle), while point-based rules capture them between 2 attributes.

4.5 Experiment with Car Data

Figures 12 and 13 and Table 6 show the results for Car data with total **coverage of 91.24 and 100% precision.**

4.6 Experiment with Ionosphere Data

Tables 7 and 8 and Figs. 14 and 15, show results with Ionosphere data [12] for both intercept- and point-based rules. These rules cover both classes with similar coverage (**78.63, 71.51**) and precision (**91.37, 94.02**).

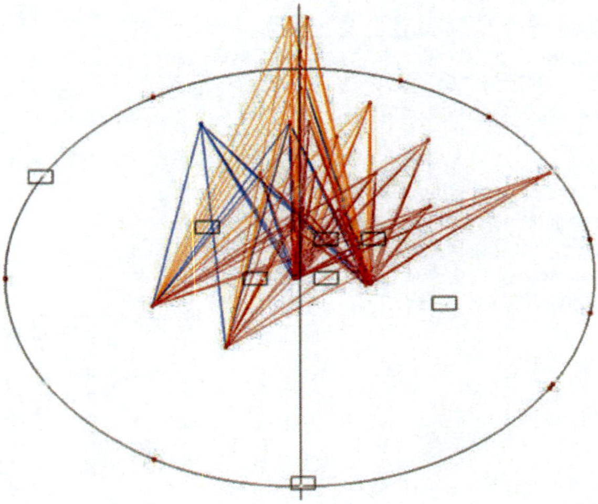

Fig. 13 Remaining car cases of three other classes that do not satisfy rules $\mathbf{r}_1$–$\mathbf{r}_8$ based on the rectangles R_1–R_8

Table 6 Car data rule experimentation results for class unacc (point-based)

Rule	Class	Coverage/recall in class, %	Precision, %
$\mathbf{r}_1$	UNACC	15.87	100
$\mathbf{r}_2$	UNACC	15.87	100
$\mathbf{r}_3$	UNACC	15.87	100
$\mathbf{r}_4$	UNACC	23.80	100
$\mathbf{r}_5$	UNACC	7.93	100
$\mathbf{r}_6$	UNACC	3.97	100
$\mathbf{r}_7$	UNACC	3.97	100
$\mathbf{r}_8$	UNACC	3.97	100
All rules	UNACC	91.24	100

Table 7 34-D Ionosphere data dominance rule experimentation results (intersect-based)

Rule	Class	Coverage/recall in class, %	Precision, %
$\mathbf{r}_1$	b	30.95	97.87
$\mathbf{r}_2$	b	34.13	90.70
$\mathbf{r}_1$ or $\mathbf{r}_2$	b	**65.08**	
r_3	g	18.22	90.24
$\mathbf{r}_4$	g	27.11	90.16
$\mathbf{r}_5$	g	40.89	90.22
$\mathbf{r}_3$ or $\mathbf{r}_4$ or $\mathbf{r}_5$	g	**86.22**	
All rules	b,g	**78.63**	**91.37**

Table 8 34-D Ionosphere data dominance rule experimentation results (point-based)

Rule	class	Coverage/recall in class, %	Precision, %
$\mathbf{r}_1$	b	29.37	100.00
$\mathbf{r}_2$	b	30.16	97.37
$\mathbf{r}_1$ or $\mathbf{r}_2$	b	**59.53**	
$\mathbf{r}_3$	g	54.67	90.24
$\mathbf{r}_4$	g	23.56	96.23
$\mathbf{r}_3$ or $\mathbf{r}_4$	g	**78.23**	
All rules	b,g	**71.51**	**94.02**

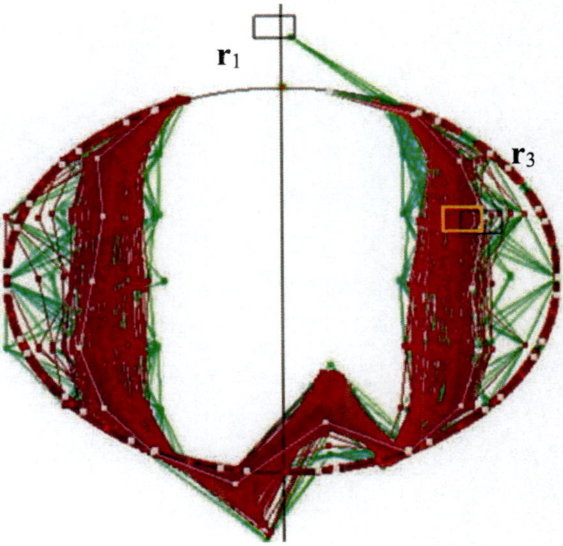

Fig. 14 Ionosphere data with rules $\mathbf{r}_1$ (green) and $\mathbf{r}_3$ (red)

4.7 Experiment with Abalone Data

Figure 15 and Table 9 present the results for Abalone data [12] on full data and Table 10 shows in split 70–30% to training and validation data for two classes (green and red) (Fig. 16).

4.8 Experiment With Skin Segmentation Data

This experiment is to explore abilities to build EPC rectangular rules on a large skin segmentation dataset [12], which contain 245,000 cases of three dimensions and two classes. Below we present discovered rules for these data based on two methods: (1)

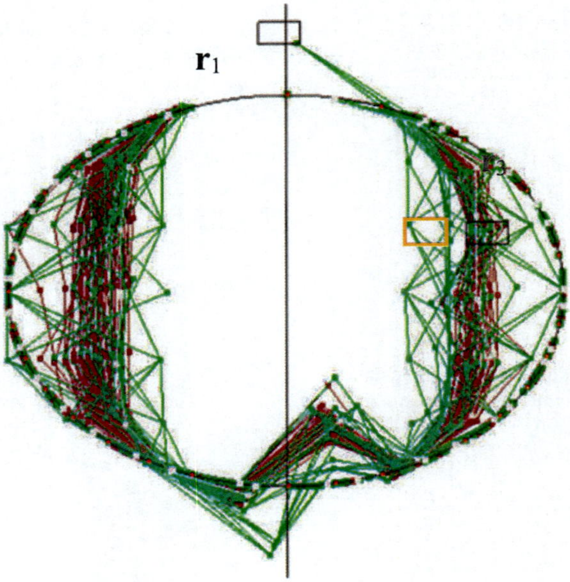

Fig. 15 Data that satisfy Rule $\mathbf{r}_1$ (green class) for Ionosphere with data with cases from red class that satisfy Rule $\mathbf{r}_3$ from red class on the background

Table 9 8-D Abalone data dominance rule experimentation results (point-based)

Rule	class	Coverage/recall in class, %	Precision, %
$\mathbf{r}_1$	1	45.12	92.83

Table 10 8-D Abalone data 70:30% split (point-based)

Rule	Class	Training		Validation	
		Coverage, %	Precision, %	Coverage, %	Validation, %
$\mathbf{r}_1$	1	12.12	91.18	12.59	94.12
$\mathbf{r}_2$	1	11.28	90.08	12.59	88.24

the number of points within the rectangle and (2) the number of lines that intersects the rectangle.

We use the following notation: skin class (class 1, red) and non-skin class (class 2, green). In Figs. 17, 18, 19, and 20, the darker green color shows the cases of the green class which have already been included in previous rules. The current version of EPC requires an even number of attributes. Therefore, a new attribute x_4 is generated to be equal to 1.0 for all cases on the normalized [0,1] scale. The rendering 245,000 cases in EPC is relatively slow in the current implementation: approximately 20 s, compared to less than one second for much smaller WBC data with less than 700 cases. Approximately 50,000 cases are skin class.

The discovered **point-based rules and intersect line-based rules** are presented, respectively, in Table 11 and Figs. 17 and 18, and Table 12 and Figs. 19 and 20.

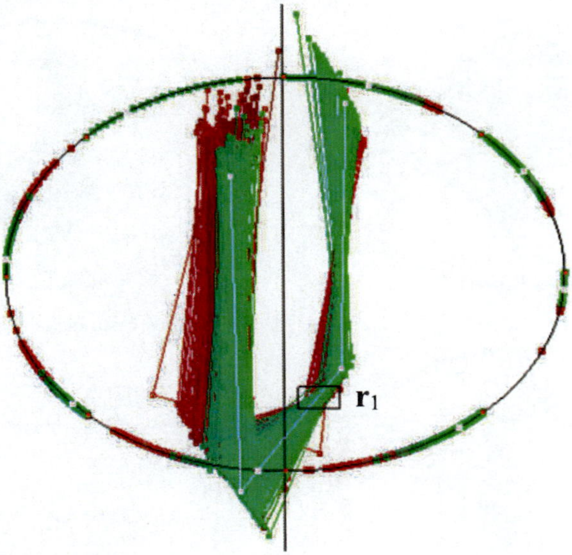

Fig. 16 Abalone data with rule $\mathbf{r}_1$

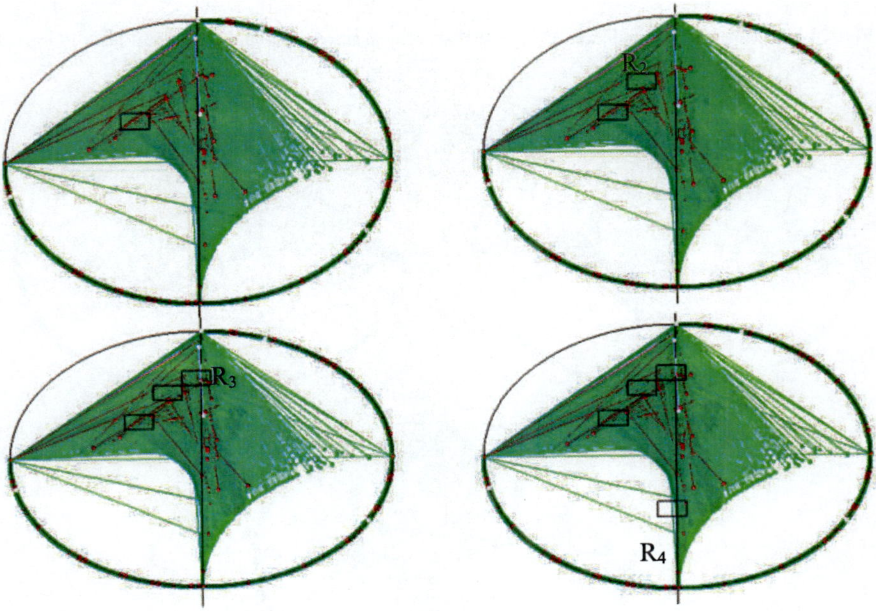

Fig. 17 Skin segmentation data in EPC with point-based rules r_1–r_4 and green cases in front of red cases

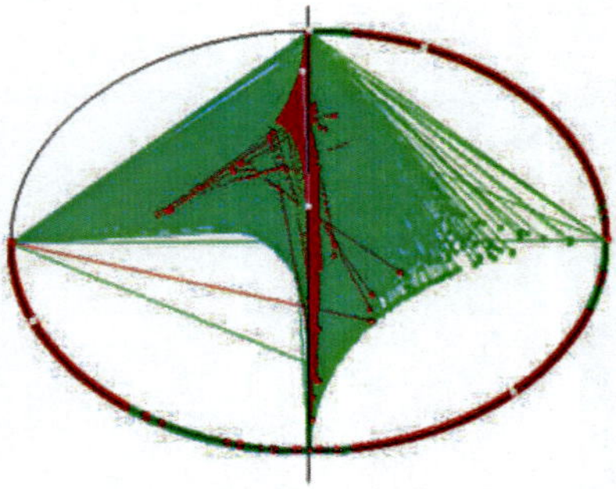

(a) All cases

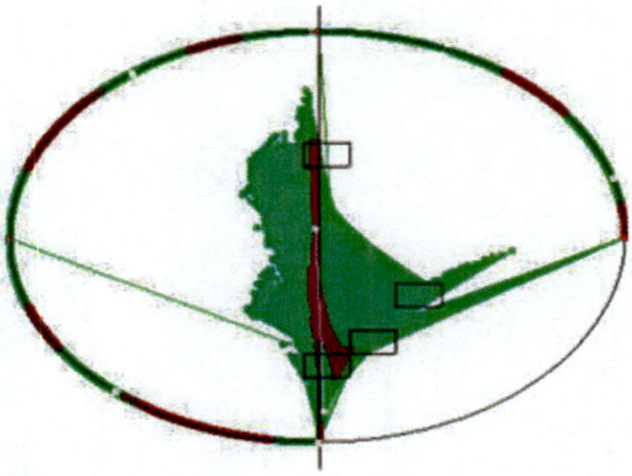

(b) Only cases included in rules r_1-r_4.

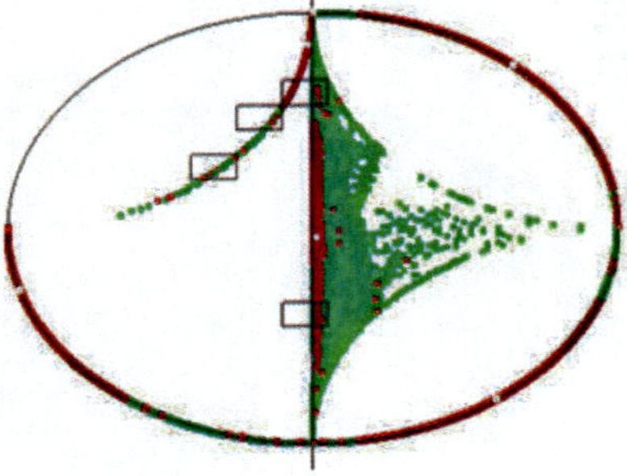

(c) Only points displayed with lines omitted.

Fig. 18 Skin segmentation data in EPC with point-based rules r_1–r_4 and red cases in front of green cases

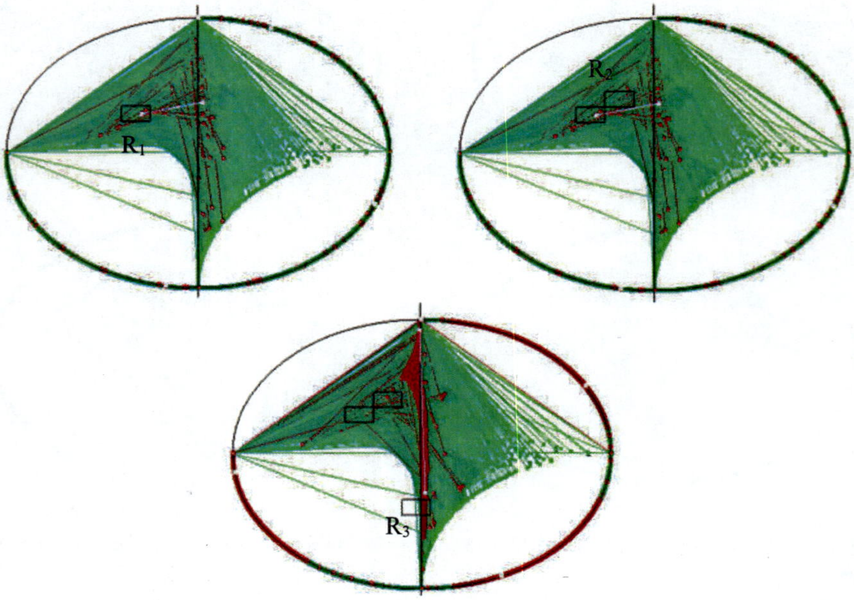

Fig. 19 Skin segmentation data in EPC with intersect-based rules r_1–r_3 and green cases in front of red cases

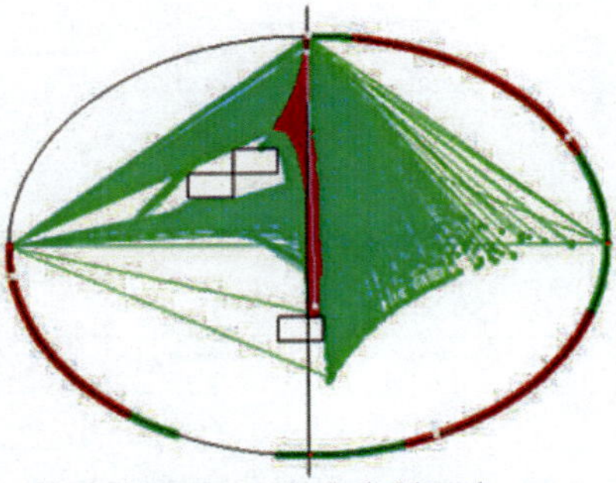

(a) Only cases not included in rules r_1-r_3.

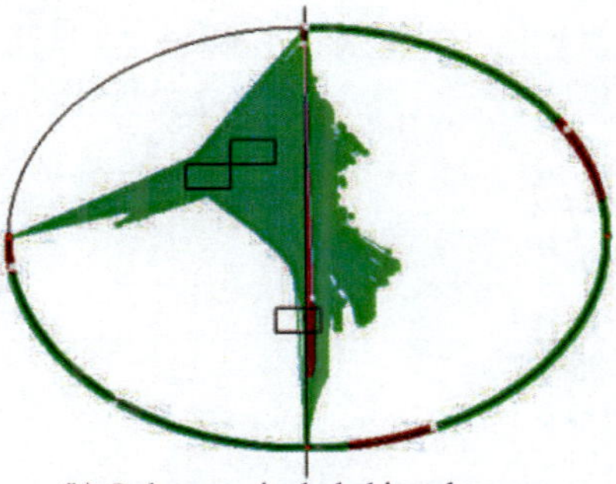

(b) Only cases included in rules r_1-r_3.

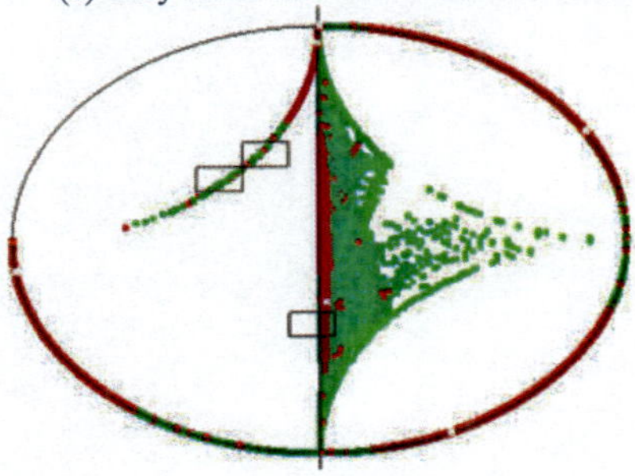

(c) Only points displayed, lines omitted.

Fig. 20 Skin segmentation data in EPC with intersect-based rules r_1–r_3 and red cases in front of green cases

Table 11 Skin Segmentation data dominance rule experimentation results (point-based)

Rule	Class	Coverage/recall, %	Coverage/recall, #	Precision %
r_1	2	10.56	25,878	100
r_2	2	10.83	26,545	99.94
r_3	2	32.32	79,199	90.9
r_4	2	7.69	18,838	95.34
All Rules	2	61.4	150,460	**94.62***

* Weighted precision

Table 12 Skin Segmentation data dominance rule experimentation results (intersect-based)

Rule	Class	Coverage/recall, %	Coverage/recall, #	Precision %
r_1	2	17.64	43,230	100
r_2	2	11.27	27,622	100
r_3	2	13.09	32,093	92.87
All Rules	2	42.00	102,945	**97.78***

* Weighted precision

Table 13 Dominance rule experimentation results

Experiment	n–D	Classes	Rules	Recall, %	Precision, %
Iris	4	3	3	100	98.66
Cancer	9	2	5	96.33	95.13
Glass 1	10	2*	3	87.06	98.29
Glass 2	10	2*	3	99.51	95.59
Glass 3	10	2*	3	87.57	97.53
Glass 4	10	2*	1	79.31	91.30
Car	6	4	8	91.24	100.00
Ionosphere 1	34	2	5	78.63	91.37
Ionosphere 2	34	2	4	71.51	94.02
Abalone	8	2	1	45.12	92.83
Skin 1	4	2	4	61.40	94.62
Skin 2	4	2	3	42.00	97.78

*One class versus all other classes

Figures 17 and 19 shows non-skin cases (green) in front of the skin cases (red) with automatically and sequentially discovered rectangles for green class rules. Figures 18 and 20 show the opposite order of the green and red cases.

While the precision or rules for Skin segmentation data shown in Tables 11 and 12 is quite high (94.62, 97.78%), the coverage is relatively low (61.4, 42%). It is likely related to the *large size* of this dataset (245,000 cases) where the data are hardly can be *homogeneous* to be covered by few rectangles, which cover over 10% of the data each.

The *higher dimensions* (Ionosphere data) is another likely reason why that data are non-homogenous requiring more rectangles. *Multiple imbalanced classes* with some extreme classes which contain a single case (Abalone data) is another possible reason why that data are non-homogenous requiring more rectangles too. These issues suggest areas of further development and improvement for EPC. We explore generalizations of EPC in Sect. 6 to address these issues in the future work.

5 Experiment With Synthetic Data

The goal of this experiment is to explore the abilities of EPC to represent data as simple easy recognizable shapes like straight horizontal lines, rectangles, and others by using synthetic data.

Experiment S1: complex dependence. This experiment was conducted with a **set A** of 9 synthetic 8-D data points $\mathbf{x}_1 = (0.9, 0.1, 0.9, 0.1,\ldots)$, $\mathbf{x}_2 = (0.8, 0.2, 0.8, 0.2,\ldots)$, …., $\mathbf{x}_9 = (0.1, 0.9, 0.1,0.9,\ldots)$, which have complex dependencies within each 8-D points and between points. The non-linear dependence within each 8-D point is:

$$\text{if } j \text{ is odd then } x_{i+1,j} = k, \text{ else } x_{i+1,j} = 1 - k$$

and non-linear dependence between consecutive 8-points $\mathbf{x}_i$ and $\mathbf{x}_{i+1}$ is:

$$\text{if } j \text{isodd then } x_{i+1,\,j} = x_{i,\,j} - 0.1, \text{ else } x_{i+1,\,j} = x_{i,\,j} + 0.1$$

These data are shown in EPC and parallel coordinates Fig. 21 where it is visible that their pattern is more complex in parallel coordinates than in EPC.

Experiment S2: linear dependence. For this experiment, we generated a **set B** of nine 4-D points $\mathbf{x_1} = (0.1, 0.1, 0.1, 0.1)$, $\mathbf{x_2} = (0.2, 0.2, 0.2, 0.2)$, …, $\mathbf{x_9} = (0.9, 0.9, 0.9, 0.9)$ with equal values within each 4-D point. Points $\mathbf{x_i} = (x_{i1}, x_{i2}, x_{i3}, x_{i4})$ have simple linear dependences: $x_{ij} = x_{ik}$ and $x_{i+1,\,j} = x_{i,\,j} + 0.1$.

Figure 22 shows these 4-D points as lines of different colors in EPC and parallel coordinates. In parallel coordinates, 9 parallel lines show these 4-D points. These lines do not overlap and respectively simpler for visual analysis than EPC. This is expected because each of these 4-D points satisfy simple linear dependences, while EPC are designed for non-linear dependencies.

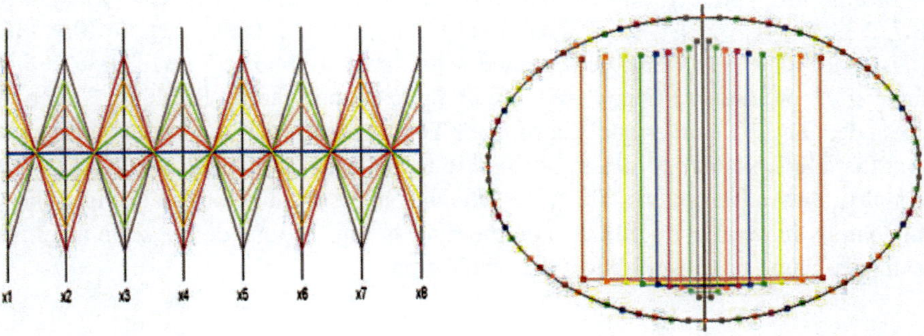

Fig. 21 8-D synthetic dataset A in parallel coordinates and EPC

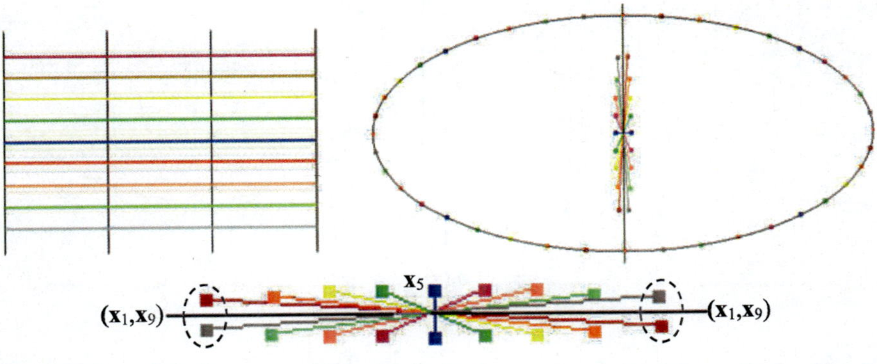

Fig. 22 Experiment S2 with data in parallel coordinates, EPC and zoomed and rotated EPC

In EPC, the dark blue 4-D point $\mathbf{x}_5$ in the middle is the shortest line perpendicular to M axis and $\mathbf{x}_4$ and $\mathbf{x}_6$ are next to it and $\mathbf{x}_1$ and $\mathbf{x}_9$ are far away from them. Similarly, lines for these points are located close or far away in parallel coordinates. Thus, both EPC and parallel coordinates capture similarities and differences by putting similar 4-D points next to each other and different far away in the visualization.

There is also a difference between EPC and parallel coordinates that is especially visible when $\mathbf{x}_1$ and $\mathbf{x}_9$ belong to the same class and we would like to have a visualization where they are close to each other in this visualization. Parallel coordinates do not do this. In Fig. 22, they are far away in parallel coordinates, while in EPC $\mathbf{x}_1$ and $\mathbf{x}_9$ are next to each other. Thus, this example shows that EPC allows visualizing cases of **non-compact classes** next to each other in contrast with parallel coordinates.

Since the graph is dependent on the central ellipse, changing the horizontal and/or vertical dimensions of this ellipse results in a stretched/shrunk graph. This helps to make the data points and lines more distinct in the respective axis being stretched, i.e., increased, and conversely compresses data for a decreased axis. Increased data point distinction within the ellipse is currently done by zooming in using the camera controls; effectively increasing both axes' dimensions.

Experiment S3: complex dependence. For this experiment, we generated a **set C** of nine 4-D points $\mathbf{x}_1$ = (0.1, 0.9, 0,1, 0.9), $\mathbf{x}_2$ = (0.2, 0.8, 0.2, 0.8), $\mathbf{x}_3$ = (0.3, 0.7, 0.3, 0.7), $\mathbf{x}_4$ = (0.4, 0.6, 0.4, 0.6), $\mathbf{x}_5$ = (0.5, 0.5, 0,5, 0.5), …, $\mathbf{x}_9$ = (0.9, 0.1, 0,9, 0.1). This is an inverse dependence within each 4-D point. The relation within and between these 4-D points are more complex than in the set B. Figure 23 shows these data in EPC and parallel coordinates The importance of this example is in the fact that all these 4-D points are located in the straight horizontal line producing a simple preattantive pattern. This is beneficial if all of them belong to a single class. In contrast in parallel coordinates on the right in Fig. 23 thue do not form a simple compact patter, but cover almost the whole area.

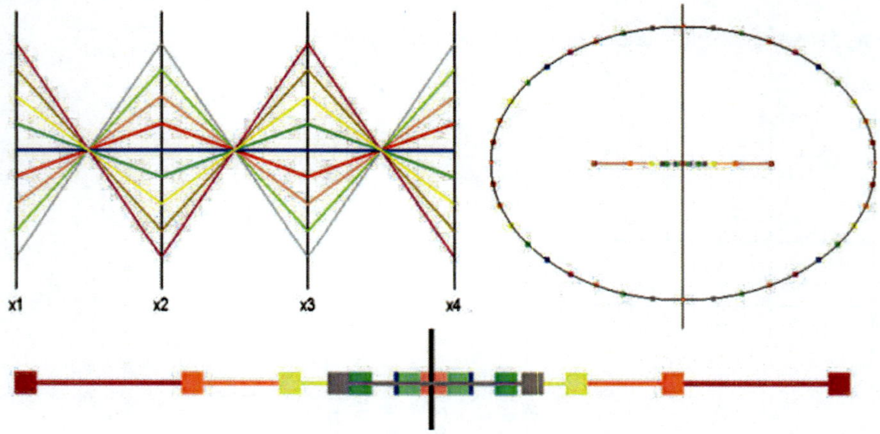

Fig. 23 Experiment S3 with data in parallel coordinates, EPC and zoomed EPC

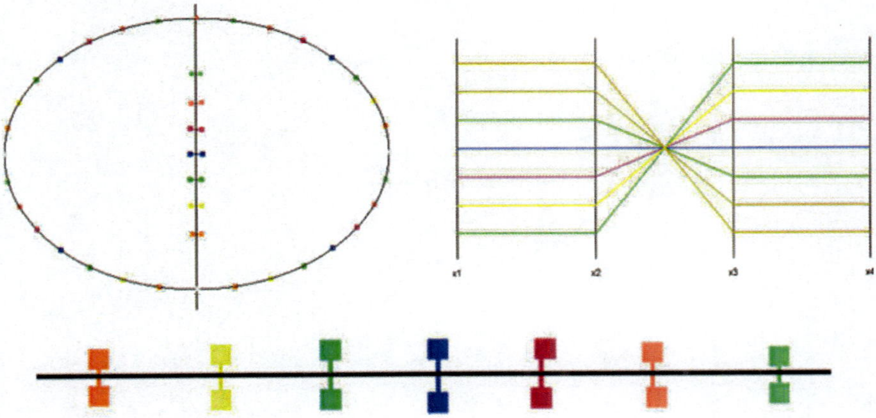

Fig. 24 Experiment S4 with data in parallel coordinates, EPC and zoomed and rotated EPC

Experiment S4: complex dependence. In this experiment we visualized seven 4-D points with the following property: $(x_1, x_2, x_3, x_4) = (x_1, x_2, 1\text{-}x_1, 1\text{-}x_2)$. These points shown in Fig. 24 are (1/8, 1/8, 7/8, 7/8) (light green), (2/8, 2/8, 6/8, 6/8) (brown), (3/8, 3/8, 5/8, 5/8) (magenta), (4/8, 4/8, 4/8, 4/8) (blue),(5/8, 5/8, 3/8, 3/8) (darker green), (6/8, 6/8, 2/8, 2/8) (yellow) and (7/8, 7/8, 1/8, 1/8) (red).

In contrast with the Fig. 23 all these 4-D points are located vertically in EPC. If these points represent one class and points in Fig. 23 represent another class, then we easily see the difference between classes—one is horizontal another one is vertical overlap that can be observed quickly pre-attentively.

Figure 6 also shows this data in parallel coordinates. The patterns in in parallel coordinates are more complex. Each 4-D point is represented as a polyline that consists of 4 points and three segments that connect them while in EPC each 4-D point requires two points and one segment to connect them. Next, the polylines in parallel coordinates in this figure are not straight lines and therefore cannot be observed pre-attentively.

In addition, these polylines cross each other cluttering the display. In contrast these data in EPC in Fig. 24 are simple straight horizontal lines that do not overlap that can be observed pre-attentively.

Comparing these data in parallel coordinates in Figs. 23 and 24 allows to see the difference clearly when these two images are put side-by-side. However, when all data of both classes are in the single parallel coordinates the multiple lines cross each other and the difference between classes is muted. In contrast in EPC the lines of two classes overlap only in single the middle point.

Algorithm Example above show us that simple visual patterns are possible in EPC for the data that are quite complex in other visualizations. The next question is how to find n–D points that have a given simple visual pattern in EPC like a short straight line, a rectangle, or others.

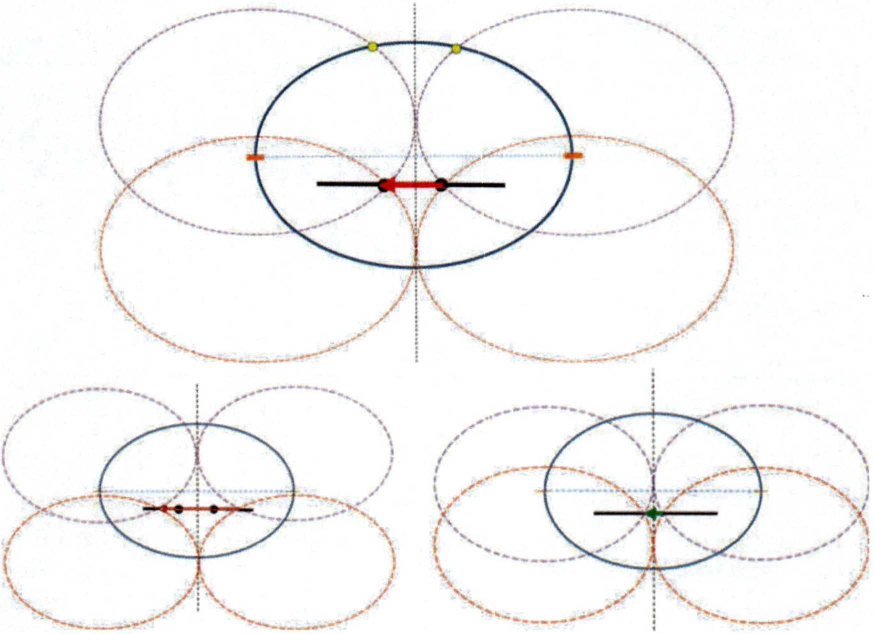

Fig. 25 Three 4-D points mapped to the back line as arrows in EPC

The steps of the algorithm are as follows for a horizontal line (see Fig. 25 for illustration):

(1) Select a horizontal black line in EPC;
(2) Pick up a point P_1 on the horizontal black line;
(3) Shift the red right side ellipse to reach this point P_1;
(4) Shift the blue right ellipse to the same point P_1;
(5) Shift the blue left ellipse to touch the right blue ellipse;
(6) Shift the red left ellipse to touch the right red ellipse;
(7) Find the point where left red and blue ellipses cross each other and mark this point as P_2 (now both P_1 and P_2 are on the black line);
(8) Draw an arrow from P_1 to P_2;
(9) Find the points where blue and red ellipses cross the main ellipse
(10) Mark these points by x_1, x_2, x_3, x_4 to indicate that they represent values of x_1, x_2, x_3, x_4 of the 4-D point $\mathbf{x}$ that is on the black line.
(11) Repeat 1–10 for any other point on the back line.

Finding values in (9) requires taking the equation of the main ellipse and the equation of respective side ellipse and solving the system of these equations to find a crossing point that belongs to both. A general property of all 4-D points that are on the horizontal black line is that that $x_1 = x_3$, and $x_2 = x_4$. In other words, let $\mathbf{a}$ be a 4-D point on this line then all other points can be expresses as follows: $\mathbf{y} = (a_1 + e_1, a_2 + e_2, a_3 + e_1, a_4 + e_2)$, where e_1 and e_2 are function of the shifts of $\mathbf{y}$ relative to $\mathbf{a}$.

6 Generalization of EPC

This section presents two generalizations of elliptic paired coordinates: the **dynamic elliptic paired coordinates (DEPC)** and **weights of coordinates**. The first one allows to represent more complex relations and the second one allows making selected attributes more prominent that can reflect their importance.

6.1 Dynamic Elliptic Paired Coordinates

In DEPC the location of each next point x_{i+1} **depends** on the location of the previous point x_i. DEPC differs from EPC that is *static*, i.e., the location of points x_i does not depend on the location of the other points. Figure 1 shows a 4-D point $\mathbf{x} = (x_1, x_2, x_3, x_4)$ = (0.3, 0.4, 0.2, 0.6) in the dynamic elliptic paired coordinates in Fig. 26 the value of x_1 is located at 0.3 on the blue central ellipse. The value of x_2 is located at 0.3 + 0.4 = 0.7 on the same blue ellipse starting from the origin, the value of x_3 is located at 0.7 + 0.2 = 0.9, and the value of x_4 is located at 0.9 + 0.6 = 1.5, i.e., each next coordinate value starts at the location of the previous point. Figure 27 shows the same 4-D points in the static EPC.

The difference from the static EPC is that we do not need to create separate sections for each attribute on the EPC central ellipse, but start from the *common origin* at the top of the central ellipse (the grey dot in Fig. 26) and add value for the next attribute to the location of the previous one. Then the process of construction

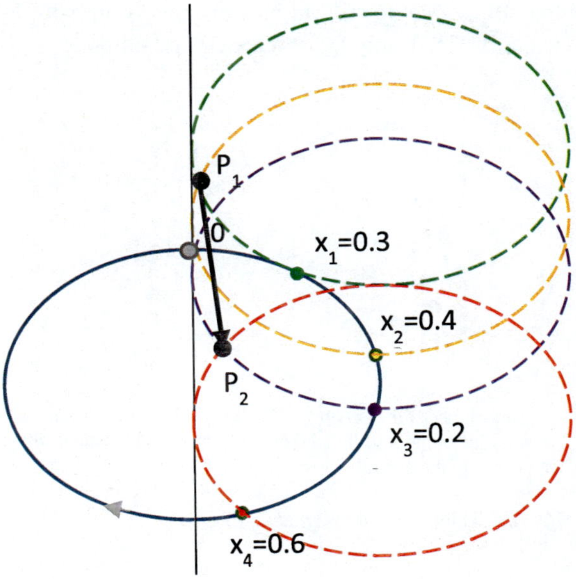

Fig. 26 4-D point $\mathbf{x}$ = (0.3, 0.4, 0.2, 0.6) in dynamic EPC

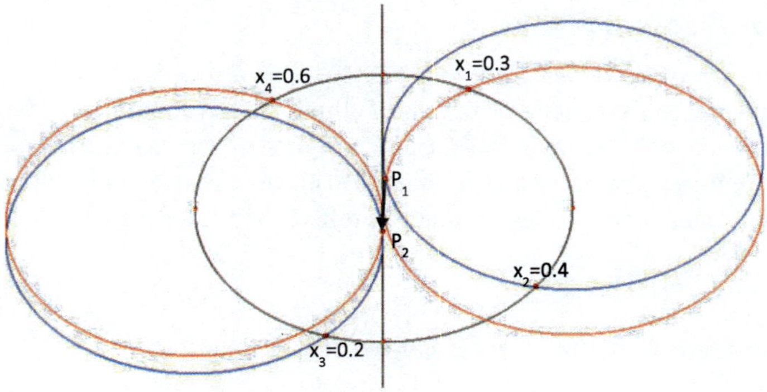

Fig. 27 4-D point $\mathbf{x} = (0.3, 0.4, 0.2, 0.6)$ in static EPC

of points P_1 and P_2 is the same as in the static EPC. The advantages of DEPC is that it allows discovering more complex non-linear relations than EPC.

There are several options to locate side ellipses relative to the point x_i on the central ellipse: (1) the center of the side ellipse is above the point x_i (ellipse goes up), or (2) the center of the side ellipse is below the point A (ellipse goes down). Next, (1) and (2) can be in different combination as shows in Fig. 28. Figure 28a shows the case when the red ellipses go down for values x_2 and x_4, but the blue ellipses go up for values x_1 and x_3. Figure 3b shows the case when all ellipses go down and Fig. 28c shows the case when all ellipses go up. Respectively points P_1 and P_2 are differently located relative to the central ellipse as shows in these figures. Both points P_1 and P_2 are in the central ellipse in Fig. 3a, P2 is outside in Fig. 3b and P1 is outside in Fig. 3c. In Fig. 1 all side ellipses go up (centers above the x_i point), and in Fig. 2 red ellipses go down, and blue ellipses go up with P_1, P_2 within the central ellipse, while in Fig. 1 only P_2 is in the central ellipse.

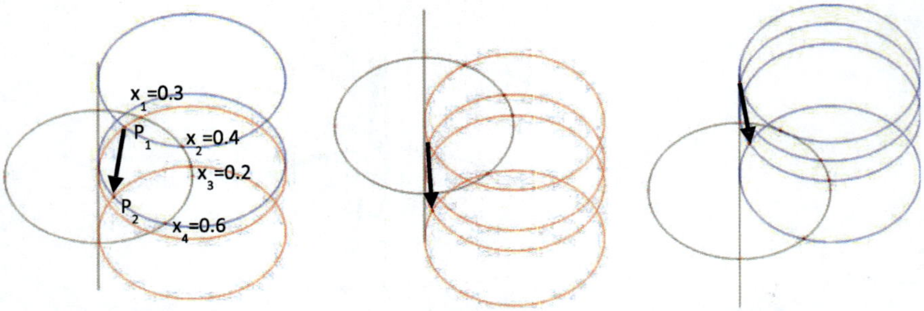

(a) Red ellipses for x_1,x_3 go down, blue ellipses for x_2 , x_4 go up, P_1, P_2 in the central ellipse.

(b) All ellipses go down, P_2 outside of the central ellipse.

(c) All ellipses go up, P_1 outside of the central ellipse.

Fig. 28 Three version of dynamic EPC

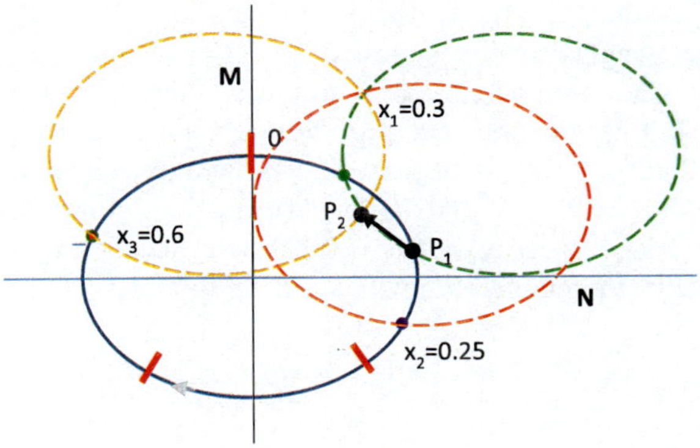

Fig. 29 EPC with mixture of right and top side ellipses

6.2 EPC with Odd Number of Coordinates and Alternative Side Ellipses

EPC coordinates have been defined in Sect. 2. Below we present alternative definitions. Such alternative definitions expand the abilities to represent n–D data in EPC differently opening an opportunity to find representation that will be most appropriate for particular n–D data and specific machine learning task on this data.

The EPC defined in Sect. 2 require an even number of coordinates. For the odd number of coordinates, we artificially add another coordinate either by copying one of coordinates or set up an additional coordinate be equal to a constant for all cases.

However, there is a way to visualize in EPC the odd number of coordinates as Fig. 29 illustrates for the 3-D point $\mathbf{x} = (x_1, x_2, x_3) = (0.3, 0.25, 0.6)$. Here ellipses for x_1 and x_3 are build relative to the horizontal line N (right (yellow) and left (green) side ellipses), but ellipse for x_2 is built relative to vertical line M (right (red) ellipse).

This idea of using a mixture of top, bottom, left and right ellipses is expandable to the situations with any number of coordinates odd or even. Some of these ellipses can be used for some coordinates while others for other coordinates in multiple possible combinations. The number of these combinations is quite large resulting in a wide variety of visualizations with an opportunity to optimize the visualization for both human perception and machine learning.

6.3 EPC With Weighs of Coordinates

The introduction of weights of the attributes x_i allows to build EPC visual representations where some attributes will be more prominent than others. A *base option* is multiplying every value of x_i by its weigh w_i with using $w_i x_i$ in EPC instead of original x_i to build a visual representation of n–D point **x**. This option can fail when the value of $w_i x_i$, will be out of the range of the ellipse segment assigned to the coordinate x_i. It can also exceed the whole ellipse. This can happen when the values of weights are not controlled.

In a **controlled static option**, *the length of segments* of the ellipse associated with each coordinate x_i are *proportional* to its weight w_i. For example, consider, four segments for attributes x_1–x_4 with respective weigh w_i as 4, 2, 6, and 5. Then 4/17, 2/17, 6/17, and 5/17 will be fractions of the ellipse circumference assigned to x_1–x_4, respectively. If $x_1 = 0.3$ then it's location x_{iw} will be 0.3*4/17 fraction of the ellipse circumference with a general formula:

$$x_{iw} = w_i x_i / \left(\sum_{i=1}^{n} w_i \right) \tag{2}$$

A **controlled dynamic option** is using the values x_{iw} from (1) instead of x_i in DEPC without creating separate segments for each coordinate.

Assigning and optimizing weights. Weights can be assigned by a user interactively or can be optimized by the program in the following way by first making all weights $w_i = 1$ and then adjusting them with steps $\delta(w_i)$. After each adjustment of the set of weights, the program computes *accuracy* of classification and other quality indicators on the training data in search for the best set of weights that maximize the selected quality indicators. The adjustments can be conducted randomly, adaptively, by using genetic algorithms and others.

The *compactness* of location of cases each class and far away from case of other classes is one of such quality indicators.

6.4 Incompact Machine Learning Tasks

The two generalizations presented above open the opportunity for solving **incompact machine learning** tasks—tasks with cases of the same class that are far away from each other in the feature space. A traditional assumption in machine learning is the **compactness hypothesis** that points of one class are located of next each other in the feature space, in other words, similar real-world objects have to be close in the feature space [14]. This is a classical assumption in *k*-nearest neighbor algorithm, Fisher Linear Discrimination Function (LDF) and other ML methods.

Experiment S2 in Sect. 5 had shown an example of n–D points $\mathbf{x}_1$ and $\mathbf{x}_9$ that are far away from each other in the feature space are close to each other in EPC space. Thus, EPC has capabilities needed for incompact ML. The use of *weights* for attributes can enhance these capabilities. Thus, we want to find area A_1 in the EPC where the cases of a one class will be compactly located in spite likely being far away in the original feature space and cases of another class will concentrate in another area A_2, similarly, as we had shown in Figs. 5, 6 and 18 in Sect. 5.

While we have these positive examples, it is desirable to prove mathematically that EPC and DEPC have enough power for learning such distinct areas A_1 and A_2 with optimization of weights for the data with a wide range of properties. For instance, this can be data where each class consists of the cases that belong to multidimensional normal distributions located far away from each other.

The rectangles we discovered for real data in Sect. 4 partially satisfy this goal. It is not a single rectangle for each class but several rectangles. Next, these rectangles contain only a part of each n–D point. In contrast the compactness hypothesis expects that full n–D points will be in some local area of the feature space. In other words, the compactness hypothesis assumes a space where each n–D point is a single point in this n–D space. In contracts in EPC each n–D point is graph in 2-D EPC space. Thus the compactness hypothesis in EPC space must differ from the traditional formulation requiring only a part of the graphs of cases of a class localized in the rectangle. Next the minimization of the number of rectangles is another important task, while reaching a single rectangle per class can be impossible. Also, these rectangles may not cover all cases of the class as we have seen in Sect. 4. All these rectangles are not in the predefined locations, because they discovered with fixed weights without optimization of weights. Optimization of weighs can make the compactness hypothesis richer allowing rectangles to be in the predefined locations by adjusting weighs combined with dynamic EPC.

7 Summary and Conclusion

The results of all experiments with EPC using the EllipseVis are summarized in Table 13 showing precision of rules from 91 to 100% with recall from 42 to 100% and 1–8 rules per experiment. They produced a small number of simple, visual rules which show cases clearly where a given class dominates. The end users and domain experts who are not machine learning experts can discover these rules themselves using EllipseVis doing end-user self-service. The visual process does not require mathematical knowledge of ML algorithms from users to produce these visual rules in the EllipseVis system (Table 13).

In addtion, the accuracy of the presented results are competitive with reported in [1, 8] and exceed some of the other published resutls, while the goal is not competing with them in accuracy, but showing opprtunity for the the end users and domain experts to discover understandable rules themselves as *self-service* without programming and studying mathematical intricacies of ML algorithms. Another brenefit for the end-users is ablities to find the visually the worst split of the data for training and validation and evaluating the accuracy of classification on this split getting the worst case estimate for algorithm accuracis it is demonstrated with the Iris data. Several options of the further development are outlined in Sect. 6 that include generalization of elliptic paired coordinates to the dynamic elliptic paired coordinates, incorporating and optimizing weights of these coordinates and intorducing incompact machine learning tasks. As development and evaluation of EllipseVis and EPC continues, we will be able to gain a better understanding of EPC's capabilities to construct a data visualization for visual knowledge discovery.

References

1. Kovalerchuk, B.: Visual knowledge discovery and machine learning. Springer (2018)
2. McDonald, R., Kovalerchuk, B.: lossless visual knowledge discovery in high dimensional data with elliptic paired coordinates. In: 2020 24th International Conference Information Visualisation (IV), pp. 286–291. IEEE (2020). https://doi.org/10.1109/IV51561.2020.00053
3. van der Maaten, L.J.P., Hinton, G.E.: Visualizing data using t-SNE. J. Mach. Learn. Res. **9**, 2579–2605 (2008)
4. Liu, S., Wang, X., Liu, M., Zhu, J.: Towards better analysis of machine learning models: a visual analytics perspective. Vis. Inform. **1**(1), 48–56 (2017)
5. Rauber, P.E., Fadel, S.G., Falcao, A.X., Telea, A.C.: Visualizing the hidden activity of artificial neural networks. IEEE Trans. Visual. Comput. Graph. **23**(1), 101–110 (2016)
6. Yuan, J., Chen, C., Yang, W., Liu, M., Xia, J., Liu, S.: A survey of visual analytics techniques for machine learning. Comput. Vis. Media **25**, 1–34 (2020)
7. Inselberg, A.: Parallel coordinates: visual multidimensional geometry and its applications. Springer Science & Business Media (2009)
8. Kovalerchuk, B., Gharawi, A.: Decreasing occlusion and increasing explanation in interactive visual knowledge discovery. In: HIMI 2018, LNCS 10904, pp. 505–526. Springer, (2018)
9. Kovalerchuk, B., Ahmad, M.A., Teredesai, A.: Survey of explainable machine learning with visual and granular methods beyond quasi-explanations. In: Pedrycz, W., Chen, S.M. (eds) Interpretable Artificial Intelligence: A Perspective of Granular Computing, pp. 217–267. Springer (2021) https://arxiv.org/abs/2009.10221
10. Ming, Y., Qu, H., Bertini, E.: Rulematrix: visualizing and understanding classifiers with rules. IEEE Trans. Visual Comput. Graph. **25**(1), 342–352 (2018)
11. McDonald, R.: Elliptic paired coordinates for data visualization and machine learning https://github.com/McDonaldRo/sure-epc
12. Dua, D., Graff, C.: UCI machine learning repository Irvine. University of California, CA (2019). https://archive.ics.uci.edu/ml/index.php
13. Kovalerchuk, B.: Enhancement of cross validation using hybrid visual and analytical means with Shannon function, In: Beyond Traditional Probabilistic Data Processing Techniques: Interval, Fuzzy etc. Methods and Their Applications, pp. 517–543. Springer, (2020). https://doi.org/10.1007/978-3-030-31041-7
14. Arkad′ev, A.G., Braverman, ĖM.: Computers and pattern recognition. Thompson Book Company, Washington, D.C. (1967)

Convolutional Neural Networks Analysis Using Concentric-Rings Interactive Visualization

João Alves, Tiago Araújo, Bianchi Serique Meiguins, and Beatriz Sousa Santos

Abstract The goal of this paper is to present the interactive web visualization technique DeepRings. The technique has a radial design, using concentric rings to represent the layers of a deep learning model, where each circular ring encodes the feature maps of that layer. The proposed technique allows to perform analysis of tasks over time regarding a single model or a comparison between two distinct models, thus contributing to a better understanding of the behavior of such models. The design supports several training methods designed to solve Computer Vision tasks, like supervised learning and self-supervised learning, as well as reinforcement learning. Additional charts highlight similarity metrics, and interaction techniques such as filtering help reduce the analysis data. Finally, preliminary evaluations were conducted with domain experts highlighting positive points and aspects that can be improved, suggesting avenues for future work.

Keywords Interpretability · Deep learning · Visualization · Interaction · Evaluation

1 Introduction

A significant amount of progress in machine learning (ML) has been made in the last decade. These techniques have impacted areas like personal assistants, logistics, surveillance systems, high-frequency trading, health care, and scientific research. Transferring decision processes to an Artificial Intelligence (AI) based system might lead to faster and more consistent decisions, freeing human resources for more creative tasks [1].

J. Alves (✉) · T. Araújo · B. S. Santos
University of Aveiro, DETI-IEETA, Aveiro, Portugal
e-mail: jbga@ua.pt

T. Araújo · B. S. Meiguins
Federal University of Pará, PPGCC, Belém, Brazil

B. Kovalerchuk et al. (eds.), *Integrating Artificial Intelligence and Visualization for Visual Knowledge Discovery*, Studies in Computational Intelligence 1014,
https://doi.org/10.1007/978-3-030-93119-3_6

While many AI systems have already been deployed, what remains a truly limiting factor for a broader adoption of AI technology is the inherent and undeniable risks that come with giving up human control and oversight to 'intelligent' machines [2]. Clearly, for sensitive tasks involving critical infrastructures and affecting human well-being or health, it is crucial to limit the possibility of wrong, non-robust, and unsafe decisions and actions [3]. As such, it is of uttermost importance to validate the behavior of an AI system before deploying it. Hence, it is possible to establish guarantees that it will continue to perform as expected when deployed in a real-world environment. With this objective in mind, several ways for humans to verify the agreement between the AI decision structure and their ground-truth knowledge have been explored [1, 4–6].

Simple models such as shallow decision trees or response curves are readily interpretable, but their predicting capability is limited [7]. More recent Deep Learning-based Neural Networks (DNNs) provide far superior predictive power but at the price of behaving like a 'black-box' where the underlying reasoning is much more challenging to extract. Moreover, deep learning is increasingly used in decision-making tasks due to its high performance on previously-thought complex problems and a low barrier to entry for building, training, and deploying neural networks [8]. The case for transparency in decisions produced by a DNN has been made in many settings, including government policy, business, charity, and algorithms [9]. This topic is also given a keen interest in laws such as the General Data Protection Regulation[1] (introduced in the EU in 2018) which seeks to provide users with meaningful information about algorithmic decisions.

Explainable AI (XAI) has developed as a subfield of AI, focused on exposing complex AI models to humans systematically and in an interpretable manner. Interpretability as a way to explain these AI models does not have a clear definition centring around human understanding, varying according to the aspect of the model to be understood: its internal workings [10], operations [11], mapping of data [12], or representation [13]. Some XAI techniques have already proven helpful by revealing to the user unsuspected flaws or strategies in commonly used ML models [8, 14]. However, many questions remain on whether these explanations are robust, reliable, and sufficiently comprehensive to fully assess the AI system's quality.

Deep learning and interpretability have heavily influenced image classification tasks, where Convolutional Neural Networks (CNN) are a key component. A type of recent approach in this domain attempts to identify the parts of a given image which are most salient, i.e. those parts which, in a sense, were most responsible for leading to the system's prediction [13, 15]. CNN are composed of layers of neurons, and while they process an input image, each layer extracts a set of features grouped in feature maps [16].

Aiming to provide a representation that considers this natural structure of a CNN, a visualization able to present feature maps from several convolutional layers at once and convey their hierarchical structure is proposed and evaluated in this work. In previous work, we proposed a visual idiom composed of a set of concentric rings

[1] GDPR Legal Text—https://eur-lex.europa.eu/eli/reg/2016/679/oj.

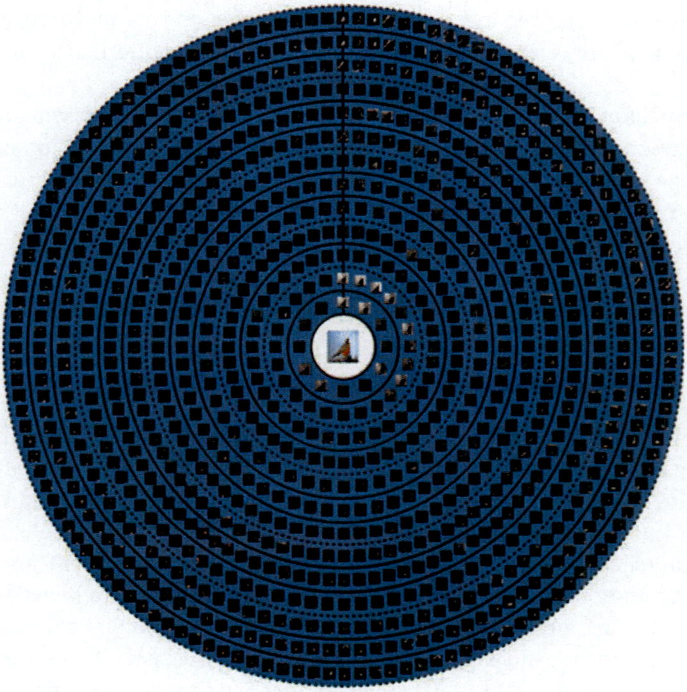

Fig. 1 DeepRings. The main unit of the visual idiom depicts a set of concentric rings, each ring has feature maps referent to a layer [17]. In the center it features an input image, and each consequent ring inward to outward shows layers beginning to end. Each square is a feature map referent to a layer, sorted by a metric. This example presents a VGG16 model [18]

encoding the feature maps that highlight the overview of the network's feature maps and allows interaction for them [17]. Our method does not focus on a single task, being generic enough allowing to visualize model representations learned with any type of training procedure.

Figure 1 shows the main characteristics of the idiom proposed for the DeepRings visualization. Besides the work presented before, we introduce new features based on this visualization. The novel features are:

- Feature maps visualization of different models with the same architecture
- Feature maps visualization of the same model with different sorting criteria
- Visualization of models in different training stages
- Support to visualize models trained from scratch with self-supervised and reinforcement learning techniques.

We propose some case studies and usage scenarios with different interaction techniques in a two-stage evaluation method. We use the think-aloud protocol with domain experts in a free form interview in the first evaluation method. In the second one, we interviewed domain experts using a task-based method to guide the interviews. The results showed that the layout presents the network overview efficiently, the interactions are relevant for the use cases, and the feature maps can be easily inspected.

The remaining of this paper is structured as follows: Sect. 2 presents the related works. Section 3 describes the architecture containing the visualization design and the machine learning system. Section 4 presents the evaluation methods, case studies and scenarios used. In Sect. 5, we discuss the impact of our approach. Finally, in Sect. 6, we draw conclusions and present ideas for future work.

2 Related Works

Using vision as a medium, we can help the operationalization of data analysis by presenting them through charts, guiding a decision-making process. Visual representations allow the user to make discoveries and decisions; build presentations about patterns (trends, intervals, outliers), groups, and individual items.

The use of computational support to visualize and interact with abstract data amplifies or reinforces human cognition, enabling the user to gain knowledge about the data and its relationships [19], facilitating tasks of research, analysis, communication, comparison, and exploration to extend the discovery of patterns, outliers, and trends [20]. Many stages of ML algorithms can use visual analytics as a way to facilitate communication during different stages of processing [21].

The works of [22, 23] show that common visualization solutions in literature do not comprise the features of a DNN representation. A complete state of the art report [24] on multilayer networks presents a graph-based approach with implicit hierarchies, and even on this work there is no direction for the DNNs visualization. A new representation based on the structure of DNN graphs may be needed as none is found in the literature.

A type of approach to interpret neural network predictions for images is via feature visualization. This technique studies what each neuron codes for, or what information its firing represents. The intuition behind these approaches is that inspecting the preferred stimuli of a unit can shed light on what the neuron is doing [6, 25]. They often focus on explaining predictions showing feature maps from one single convolutional layer at a time [6, 26]. As large-scale model predictions are often computed from a consecutive number of layers that learn hierarchical representations [6], the limitation of not presenting all feature maps at once can lead the user to miss the bigger picture.

One of the first works to explore visualization of complete deep networks is Yosinski's work [25]. It uses tabs to keep track of activation of each layer. It is possible to observe gradient ascent on input, and the images on the dataset that are most activated by the selected channels. Recent works [8] use feature maps from layer to layer to build semantic graphs.

Visual analytics and interpretability techniques designed to explain image classification have been leveraged to provide insight into representations learned in Reinforcement Learning (RL) tasks with raw image pixels as input [27–31]. These techniques were extended to explain generative models like Variational Auto Encoders (VAEs) [32] and to design a platform that instructs users in deep learning-related subjects like CNNs [33]. Liu et al. [32] method explains Variational Auto Encoders (VAEs) visually employing gradient-based attention. Taking advantage of this attention mechanism, this method can localize anomalies in images and improve latent space disentanglement.

DQNViz [27] is designed to help domain experts understand the experiences of a Deep Q-Network (DQN) agent [34], allowing them to identify action and reward patterns, potentially helpful in understanding the behavior of the agent, evaluating the model quality, and improving the training performance. Hilton et al. [28] used dimensionality reduction and attribution techniques to perceive which objects are detected by the model, and how they influence the value function and policy. They used them to understand why the agent failed to achieve a maximum reward and why the value function was sometimes inaccurate. Rupprecht et al. [30] presented a method for synthesizing visual inputs of interest, which could represent situations in which specific actions are necessary. Their method consisted of learning a generative model over the state space and using it to optimize a target function for the states of interest, showing that it can generate insights for various environments where RL methods were used.

The DeepEyes system [35] is a Progressive Visual Analytics system that supports the design of neural networks during training. The system facilitates the identification of problems, such as superfluous filters or layers, and information that is not being captured by the network. Temporal analysis is also a theme in Deep Learning, and the work of [36] presents a visualization of classification during the iterative development pipeline of a DNN model.

Some works focus on the method of visualization, using metrics for models and model layers, to understand how the model behaves in different situations. Grad-CAM [37] is one of these methods, that focus on visual explanations for CNN models. It uses the gradients of a target concept to highlight important regions in an image. The IFeaLiD tool [38] provides a visualization that focus on CNN layers, and encodes the similarity between the feature vectors of individual pixels of an input image in a heat map display.

Many works show a user-based approach for evaluation of visualizations and tools. The CNNVis [39] presents a hybrid visualization to show multiple facets of each neuron and the interactions between them. This work presents two detailed case studies based on common situations of domain experts. ActiVis [40] is a system that relies on multiple coordinated views for interpreting large-scale deep learning models and results. These views are mainly a computation graph overview of the model architecture, and a neuron activation view for pattern discovery and comparison. Two case studies are used in the evaluation.

NeuralVis [41] is an instance-based visualization tool for DNN. It has a diverse range of functionalities, as it allows to visualize the structure of DNN models and their data transformation process. This work uses a task-based user study to guide exploration through the tool. These works use domain expert interviews as a form of system and visualization evaluation, with one to three case studies each. The TensorFlow Graph Visualizer [42] presents a system to visualize DNN graphs and shows user feedback based on real-world usage of its visualizations. The graph visualizations are even used on official tutorials of the Tensorflow library.

These works all present different aspects of DNN representations. For images, CNNs are inherently hierarchical, and many works present this structure using a graph. While adequate for this context, a graph representation needs screen space for nodes and edges. Representations using space filling could be used to represent the natural flow of feature maps through the hierarchy using this implicit hierarchy [43, 44].

Aiming at providing a representation that considers this structure, a visual idiom able to present feature maps from several convolutional layers at once and able to convey their hierarchical structure is proposed in prior work [17]. In this work, besides the central rings featuring the layers, this visual idiom supports interaction in many scenarios: sorting methods for feature maps, selection of layers for further inspection and comparison between different models. These new contributions are essential to enhance humans ability to perceive commonalities and differences between the representations learned by different models.

3 Design and Prototype

In this section, we propose and describe a web-based visualization platform and the architecture linking it to a feature map generation engine. The visualization is based on a concentric-ring design, where the number of rings depends on the number of layers of a specific CNN architecture, and the feature maps are encoded accordingly.

3.1 Design

Our design, depicted in Fig. 2, is a concentric-ring design and each ring has several image placeholders embedded near its outer border. The hierarchical radial design emphasizes the implicit hierarchy of CNNs, where feature maps are forwarded in the network. Each ring uses its internal space to order feature maps clockwise, following a sorting algorithm. The usage of radial design for hierarchies is well established in visualization literature [43, 44]. A feature map, in this context, refers to a two-dimensional output matrix activation regarding a specific filter in the convolutional part of the network.

After computing the feature maps, these placeholders are replaced by them using the following criteria: Each ring contains the feature map of one convolutional layer - inner rings contain the feature maps from the first layers and as we move away from the center the feature maps correspond to those computed in deeper layers. The number of placeholders is static for each layer, but in each transition it increases as

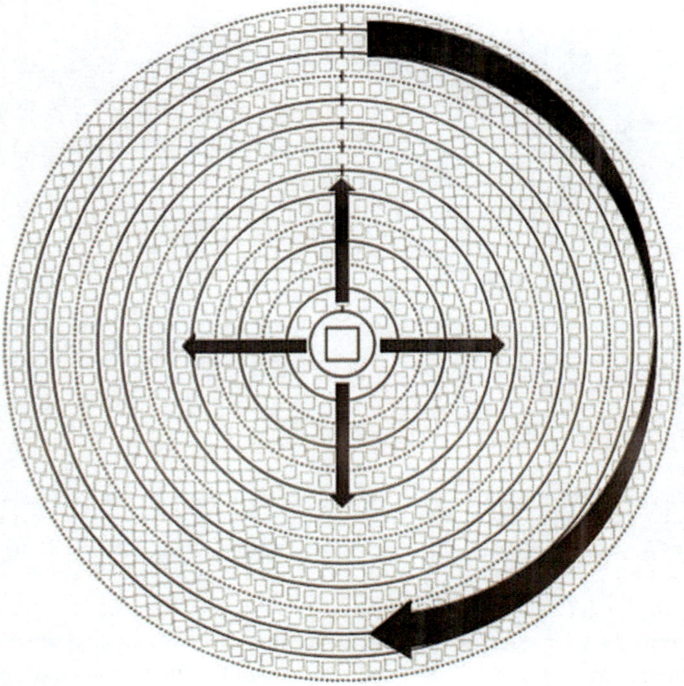

Fig. 2 DeepRings main design, presenting reading order. Layers start at the center, and feature maps follow a clockwise rotation using some sorting method. Each square encodes a feature map. In the center it features an input image, and each consequent ring inward to outward shows layers beginning to end

we move towards the final layers. This decision was made because CNNs have more feature maps as we get close to the final layers.

In this visualization, the user can define how many and which layers to visualize and specify what activation metric is considered more relevant. It is possible to choose between two metrics defined by the number of neurons activated or the intensity value of the activated neurons, respectively. The most relevant filter per layer based on the user-defined metric is shown directly above the visualization center, and as we rotate clockwise we find with a decreasing importance degree the remaining filters. To create the visualization shown in Fig. 1 the metric used was the intensity value of the activated neurons and all the layers from the VGG16 architecture [18] are displayed.

The DeepRings can be used in a set, following a certain context to ease comparisons, like along time or changing sorting methods. Right now, the design supports the same model in different settings, like how the training is done (supervised, self-supervised), the sorting method chosen and the training along the time. Figure 3 presents some of these scenarios.

In the end, a layer band is presented for both cases of comparison, side by side and along the time. A layer band is another representation for a single layer in a horizontal

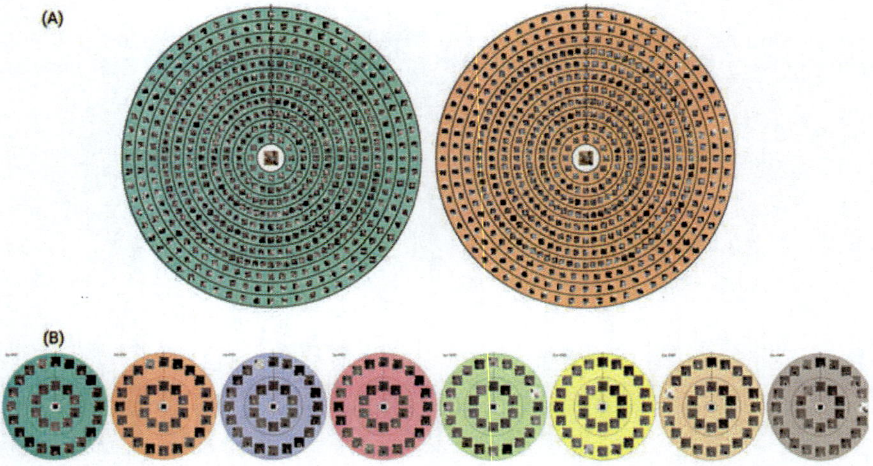

Fig. 3 Methods of overview comparison between models. Each layer of a model is coordinated with the layers of the others, to expand in a layer band. **A** DeepRings of the same model with different sorting methods. **B** DeepRings of the training progress along time of a RL model

fashion. Once the user selects a layer (ring) of one of the models, it turns red and it updates the feature maps of the respective layer bands. The side-by-side model also presents a bar chart, which encodes the euclidean distance between feature maps in the same order position. Figure 4 presents the layer bands of selected layers of two scenarios.

A ML engine is required to obtain the feature maps from a specific image uses a prebuilt or a user-defined model. The ML engine receives an image from the client and using a preloaded model to compute the feature maps. After this operation, the server sends back this information to be displayed using the proposed visualization.

3.2 *Prototype*

Our goal with the proposed prototype is to have an interactive visual tool to help machine learning practitioners to obtain a better understanding of what type of representations were learned by the CNN when a specific input image is given. Based on previous deep learning practitioners' feedback, we decided to develop a tool that enables the users to compare representations learned by the same model architecture trained with different losses. The prototype presented also allows to compare model representations learned in different training stages.

In this section, we will address the two main components of our prototype: The visualization component and the engine used to compute the feature maps produced by the convolutional layers that are shown by the former.

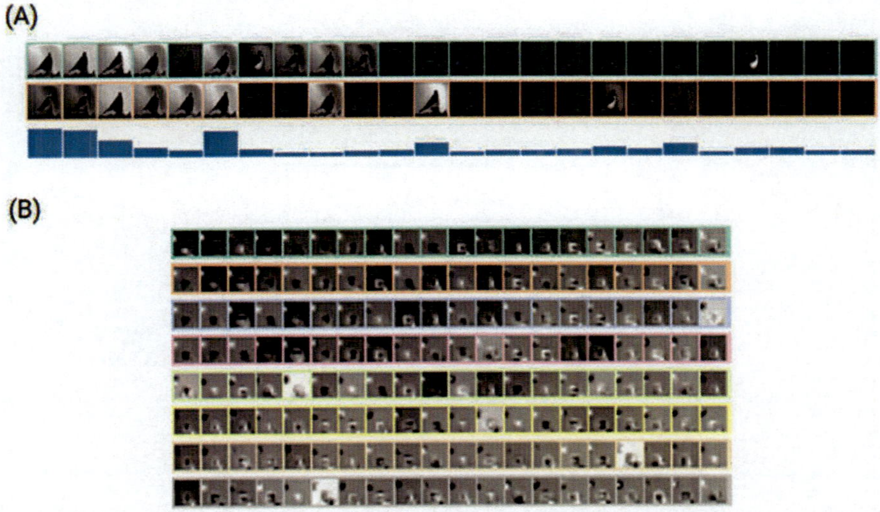

Fig. 4 Available layer bands for different scenarios. **A** Layer band of the same model with different sorting methods. It also presents a bar chart to encode the euclidean distance between feature maps of the same column. **B** Layer band of the training progress along time of a RL model. Time is seen from top to bottom

We used D3.js[2] to develop the visualization as it is a flexible library, supported by a robust and well-established framework. As a machine learning platform, we opted for using TensorFlow[3] as the first version of this library had been already used in our previous work. To establish the communication between these two technologies we used a Flask web server.[4] The system architecture together with an example of information exchange between client and server is depicted in Fig. 5. Using these technologies, loading a complete VGG16 model, including the feature maps generations takes between 1.3 and 1.5 s, depending on the video card, with layer selection times between 0.1 and 0.3 s, depending of the number of feature maps to copy.

3.2.1 Visualizations

The layered hierarchy of DNNs mapped to the visualization layers with its respective feature maps allows the observation of patterns between them. In Fig. 6 we can see parts of the bird body and its contours activated in several layers of the network, indicating these features are used throughout the network. The visualization shown presents only eight out of the thirteen convolutional layers with the collapsed ones represented by thin red semi-rings.

[2] D3 Website—https://d3js.org/ (Accessed: 28 December 2021).

[3] TensorFlow Website—https://www.tensorflow.org/ (Accessed: 28 December 2021).

[4] Flask Website—https://flask.palletsprojects.com (Accessed: 28 December 2021).

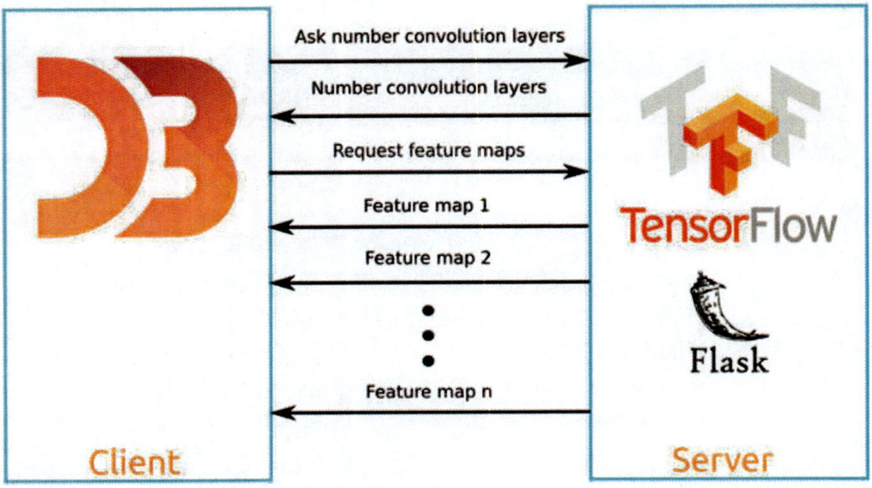

Fig. 5 System Architecture and Information Flow. Front-end visualization uses D3.js to create the visualization. The information required to be displayed is requested to a Flask server which computes the feature maps using TensorFlow

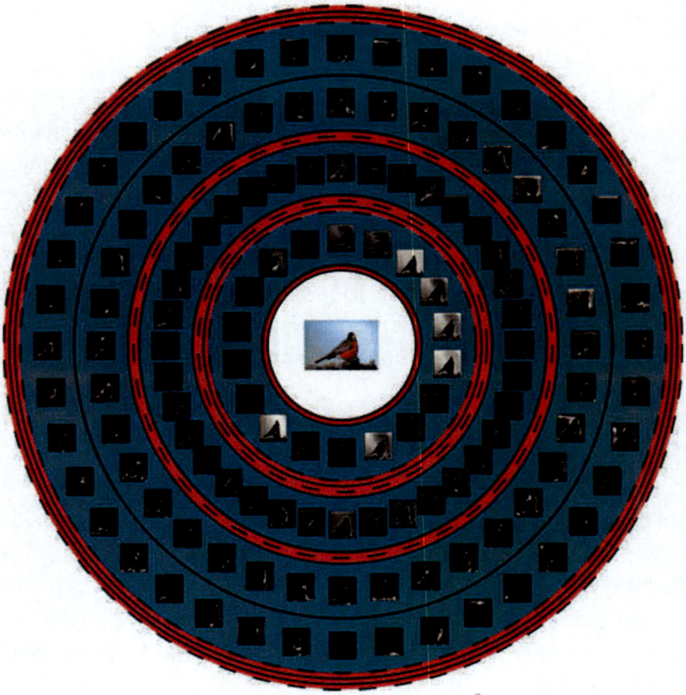

Fig. 6 DeepRings displaying only the feature maps of the layers selected by the user. Red rings are collapsed layers. This example shows eight user selected layers of 13 of a VGG16 model

The proposed visualization design does not scale to very deep networks. If it is necessary to display all single layers, the area of each ring tends to become smaller when the number of layers increases, as illustrated in Fig. 1. However, this issue may be alleviated since the user can select which layers and the visualization options of a specific feature map will be displayed. This can be used to analyze specific layers further within the same representation

A problem with deploying CNNs in critical domains that require low latency is the system response time. In Fig. 6, we can notice that the activation of several feature maps in the last layer is very similar between them. This suggests the existence of redundant filters in the CNN, which leads to unnecessary computation. Removing these filters would potentially alleviate the computational burden leading to a reduction of the system response time.

The visualization layout is also able to show for the user that the learned features are hierarchical. The outer rings (final layers) are composed of activation patterns from previous layers. The behavior of learning hierarchical features representations is present in most modern CNN architectures, and it is easily visible using this layout.

3.2.2 Feature Map Computation Engine

To display the visualization shown previously, we decided to offload the feature map production to an external server, not only due to the computational requirements, but also to produce a modular architecture. We think this is an important design choice, because it allows other visualizations to leverage the information produced by this server.

This component is responsible for generating the required information needed to answer the requests produced by the visualization component. It can provide useful information regarding different model architectures trained with different learning techniques in different datasets. For example, in Fig. 1, we sourced the feature maps produced using a VGG16 [18] architecture with weights obtained after training this architecture in the ImageNet [45] dataset. Although in this particular example we have used a pre-trained Keras model, we also trained models from scratch using standard supervised learning, self-supervised contrastive learning and even RL training procedures. Using supervised learning, the objective was to correctly classify the label class by minimizing a cross-entropy loss. When training with self-supervised techniques, we aimed at the same objective using a two-phase training procedure, where in the first one an encoder is pre-trained to optimize the supervised contrastive loss, described in [46]. To train these models, we used the Keras API from the Tensorflow2 library.

Besides image classification, we also trained a model with Soft-Actor-Critic (SAC) [47], a RL algorithm, which aims to solve a relative simple grid world scenario where the agent should be capable of reaching a green object with a plus shape, while avoiding a red cross. Figure 7 shows one successful episode regarding

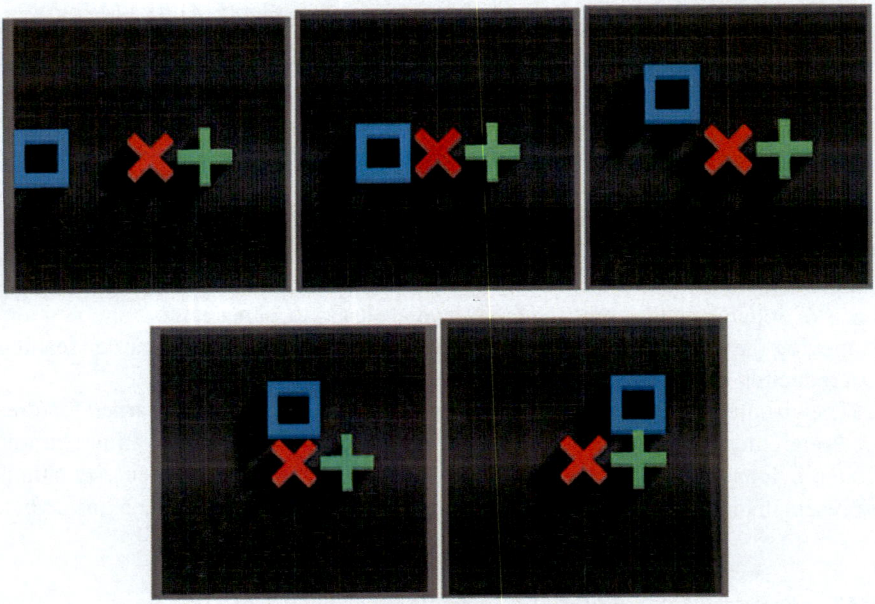

Fig. 7 Virtual environment used to train the Reinforcement Learning agent. The agent objective in this task is to reach the green object with a plus shape, while avoiding the red cross. The first image (upper left) shows the start of the episode and the remaining ones show the agent state in each timestep before reaching its objective

this task. For training this specific model, we used Unity[5] to design the task scenario and the ML-agents toolkit [48] to train the agent.

One of the services provided by this component allows the user to see the evolution of learned representation throughout the training process in an offline way. This service can be requested for either the supervised trained model as well as the model trained using RL. For this effect, throughout the training process, the models are saved to the storage device in predefined time intervals, to be subsequently used in the feature map computation. For the supervised learning model, the intermediate models are saved in every five epochs, while the RL model is saved in every 20000 steps.

4 Evaluation Methods

The evaluation of the DeepRings design was based on two case-studies, each one with a distinct evaluation method. The first evaluation method is an adapted thinking aloud protocol, with domain experts sharing opinions and feedback about interaction

[5] Unity Website—https://unity.com/.

with layers and sorting methods. The second use case is a guided user interview, using tasks to lead the interview, focused on comparison of models and layers. The visualization and models used in each step are described next, followed by each evaluation step.

4.1 Models and Visualizations

The first case-study, besides displaying the concentric-ring based visualization, (Fig. 8) is also interactive, allowing the user to point over the feature maps presented in the rings to visualize (on the left) a detailed version of them. Furthermore, the user can select the convolutional layers to visualize by selecting and deselecting each one of the check boxes associated with each layer.

In this case-study, we used the VGG16 model architecture with pre-trained ImageNet weights. In Fig. 8 the initial feature maps activate almost in the whole image, but it gets more specific as we move forward. This shows that general purpose filters are used at the beginning and get more refined as the image advances. The widgets displayed together with the DeepRings in Fig. 8 are only used in the Thinking Aloud use case, as a way to allow the experts to interact in a simple way with the model. They are not part of the DeepRings visual idiom core.

In the second case-study, we present how the prototype can be used to perform comparisons between feature maps. Figure 9 presents the first scenario, a comparison between the same architecture using different training modes, supervised in the left and self-supervised in the right. This visualization also features a layer band, that

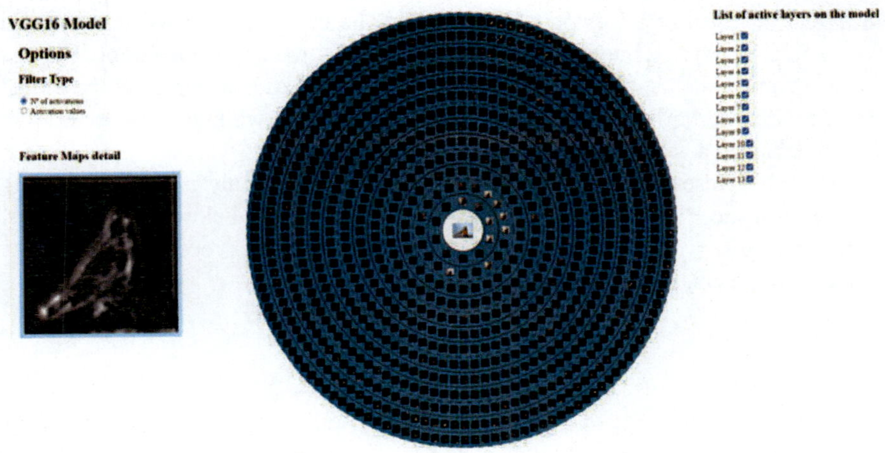

Fig. 8 Visualization platform—concentric-ring based visualization generated using feature maps ordered by a specific metric is displayed in the center; on the left-details of the feature map selected and the possible sorting metrics; on the right-selection of the convolutional layers to visualize. Input image is Robin, by Chris Heald

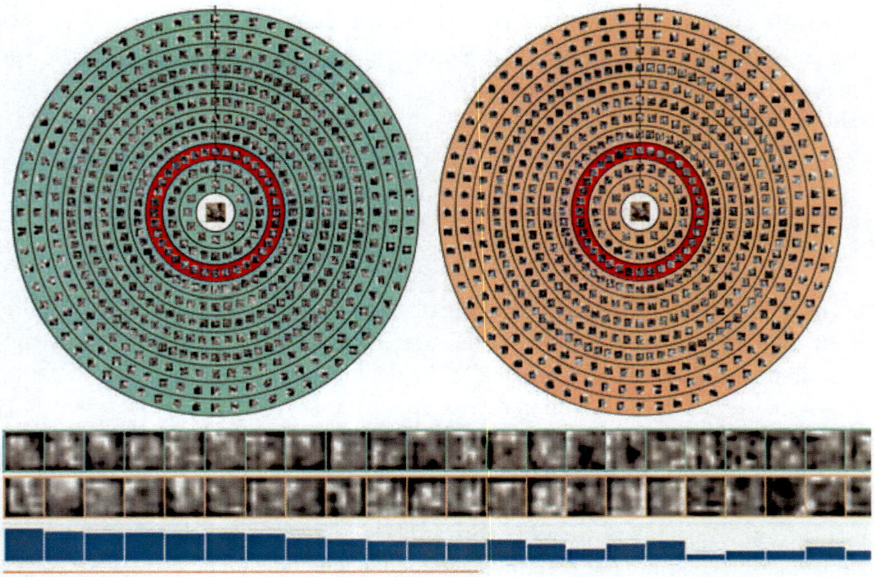

Fig. 9 First scenario from the second case study. Comparison between feature maps drawn from models with the same architecture trained using supervised (left) and self-supervised (right) training procedures. Below the Deep Rings visualization, with two bands of layers highlighting the feature maps in the user selected layer, and a bar chart encoding the euclidean distance, as a visual difference metric, among feature maps from the same layer

highlights feature maps following user interaction, and a bar chart encoding the euclidean distance between the feature maps on the same layer. This is the visual difference metric used to compare feature maps. The architecture used in this scenario was a ResNet-like architecture as described in [49] with 53 convolutional layers trained from scratch in the CIFAR-10 dataset. The validation accuracy of the model trained with standard supervised techniques was 73.94%, while the self-supervised model achieved a validation accuracy of 77.78%.

Figure 10 presents the second scenario, a temporal analysis of the learning evolution of the RL model with the layer band of all steps below the DeepRings. In this scenario, we used the Soft Actor-Critic [47] training procedure to learn the model parameters.

$$T(p) = \begin{cases} 1 & A(p) > 0 \\ 0 & A(p) \leq 0 \end{cases}$$

$$c_1 = \sum_{p \in P_{ij}} T(p) \tag{1}$$

$$c_2 = \sum_{p \in P_{ij}} A(p) \tag{2}$$

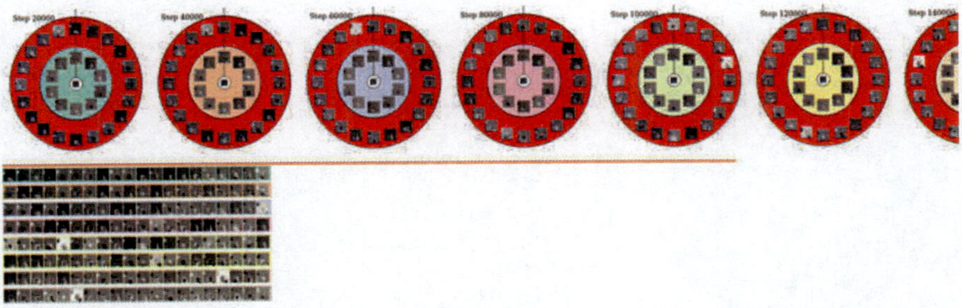

Fig. 10 Second scenario from the second case study. Temporal analysis of the learning evolution of the RL model with the layer band of all steps below the DeepRings visualization

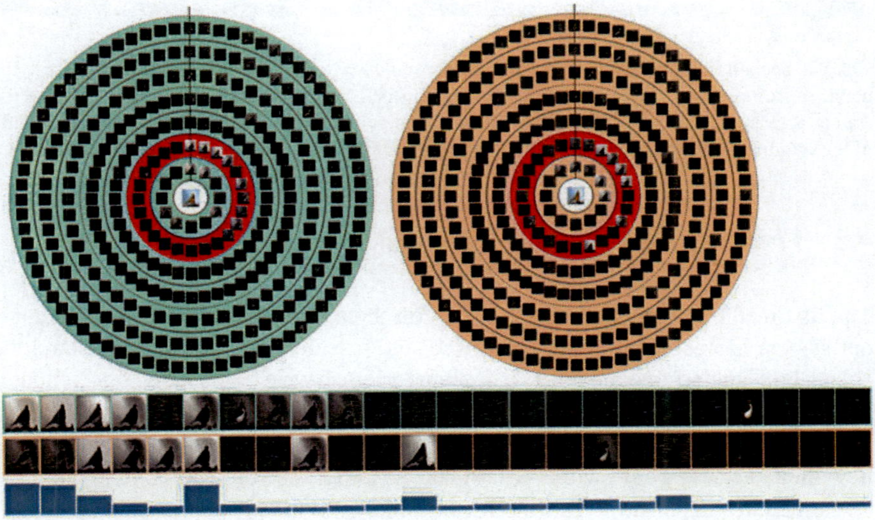

Fig. 11 First exploratory analysis from the second case study. Display of VGG16 feature maps obtained from training on ImageNet using both sort methods: number of activations (left) and the sum of activations values (right) with the bar chart for comparison

Figure 11 shows the first exploratory analysis, displaying layers from a VGG16 model trained on ImageNet using both sort methods, number of activations (left) and sum of activations (right) with the bar chart for comparison. These metrics are defined respectively in Eqs. 1 and 2. P is the set of all pixel coordinates present in all activation maps. i and j are respectively the layer number and the index of each feature map. In this context, A is the function mapping the pixel coordinates to its value. Figure 12 presents the second exploratory analysis, similar to the first one where we used an image taken from the scenario where the RL agent operates. We instructed the network to classify this image with the purpose of obtaining the feature maps produced in this case, as it clearly represents and out-of-distribution training example.

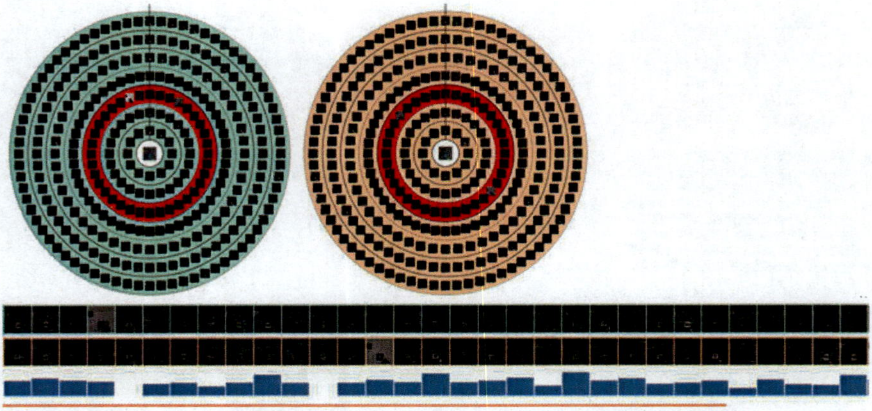

Fig. 12 Second exploratory analysis from the second case study. Display of VGG16 feature maps obtained from training on ImageNet using both sort methods: number of activations (left) and the sum of activations values (right) with the bar chart for comparison. The input image is now outside of the training distribution, as it represents one image taken from the RL scenario

4.2 *Thinking Aloud*

The first method is an exploratory study with three domain experts to understand if our system has potential to help end-users and researchers to better understand the "black box" underlying model. The domain experts have a background on Computer Vision and have been using Deep Learning on their works for more than two years. Before starting the evaluation, we presented the DeepRings, with its interactions and how the interview would work. After that, we asked the participants to explore the visualization and performed a simplified Thinking Aloud [50] observation protocol, with no direct tasks to perform, followed by some questions. The following questions were asked:

- What do you observe as positive and negative aspects?
- What do you think of the interactions?
- What features do the visualization lack?
- What can be discovered using the visualization?
- What can be improved?
- Are there any conceptual weaknesses?

4.3 *User Interview*

The guided user interview was used in this step of evaluation, with two domain experts, but they feature the same expertise as the ones of the first method. These domain experts were not the same of the first method. Before starting the evaluation, we presented the DeepRings, with its interactions and how the interview would work. A video of the RL task used in the second scenario training was also shown. The domain experts have a background on Computer Vision and have used Deep Learning in their works for more than two years, and knowledge about Reinforcement Learning. We use two scenarios, one for comparison with two models and the other showing temporal changes. The last question of every scenario and exploratory analysis is an open comment question, leaving free time to make any comment.

Scenario 1—Two models with different training and the same architecture

1. Compare activation maps from one model to another and highlight differences in activations. Do you see differences or similarities?
2. Use the bar chart to describe the differences between activation maps across architectures. Is there a pattern?
3. Highlight the layers in the overview where the features are most similar.
4. Highlight features in the activation maps that are maintained throughout the network. What kind of feature or in what position is it kept?

Scenario 2—Training of a model along the time

1. Compare activation maps from one time to another and highlight differences in activations. Highlight the differences or similarities.
2. Highlight the layers in the view of the activation maps where the features are most similar. Is there a pattern?
3. Highlight features in the activation maps that are maintained throughout the training. What type of feature or in what position is it kept?

Exploratory analysis 1—Two models with the same architecture but different feature map ordering criteria

1. Which of the sorting metrics do you think best explains the network behavior?
2. Does the bar chart reflect well the differences and similarities of the feature maps?

Exploratory analysis 2—RL image classification with the model trained on the ImageNet dataset (VGG16)

1. Would you trust this model to create a latent space with features suitable for producing actions?
2. Do you think this network is capable of capturing relevant feature maps to solve the task shown in the video?

5 Results and Discussion

The results of each evaluation method are detailed below, including the insights obtained based on the domain experts analysis and feedback.

5.1 Thinking Aloud

While all participants perceived minor bugs and display errors, they also highlighted some important aspects of lack of information. The sessions took between one and two hours each. The class label and certainty are missing aspects of the image input. The visualization also needed an indicator of where the order of feature maps starts on the rings. It was not clear that only a specific number of feature maps per layer were presented leading the domain experts to assume that they had already visualized all of them.

Even without the information of missing feature maps, the domain experts highlighted that filtering by the best on a specific criterion helps the user find patterns, not overwhelming the user when it shows only the best ones. The domain experts praised the overview presentation of the network in a circular shape, as it also shows the hierarchical structure. Quoting one of the domain experts: *With the visualization, a user starts from the input and observes the abstractions, observing that it goes from the shape to class abstraction, being also able to perform background removal.* They also suggested new features to be included, and the ones aligned with the application roadmap are highlighted in the next section on the future works. Another remark done by one of the domain experts was the fact that this representation type allows to spot errors during training as it allows the user to quickly perceive possible erroneous feature maps while the network is learning. With a pre-trained network, the expert also pointed out that the visualization suggests that pruning some network layers could be helpful to reduce inference time as the final layers seem to have less relevant information.

5.2 User Interview

The participants highlighted some important aspects of design and prototype layout. They also suggested important changes due the lack of information in some tasks. The sessions took between one and two hours. The main aspects of the experts feedback are shown below.

5.2.1 Scenario 1

Both domain experts noted that from the third layer onwards it gets very hard to identify features, as feature maps become imperceptible because the input image is very small. As the feature maps decrease in size, the visual difference metric turns out to be more important to the analysis. Even with small input images and without perceptible feature maps the network was able to lean a meaningful representation having achieved a mean validation score of 75%.

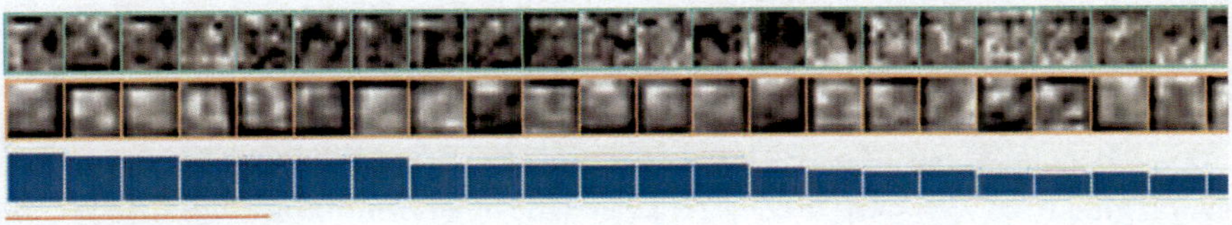

Fig. 13 Layer band presenting feature maps of two different models, but within the same layer (15) of a ResNet, focusing on different features, but nonetheless each model focus in only one feature each

Although the feature maps were not clearly visible, the user agreed that the representations learned are different. Many feature maps are similar, and the visual difference metric bar chart helps searching for different ones, and even helps to identify that the network learns the same features in different places.

The image difference visual metric using the euclidean distance does not match reality very well. The domain experts noted many cases where the images are alike, but the metric fails to present them efficiently. In many cases, the visual difference metric starts making sense only at the tail of the end layers, where feature maps alike are closer, while different ones are more distant in euclidean space. So right now the scenario lacks an efficient visual difference metric for comparison or a way to exchange between different metrics.

One domain expert noted that the visual difference metric revealed to be useful at the outer (end) layers, and it should show more feature maps in the starting layers. This last comment was already mentioned on the thinking aloud protocol. One domain expert analysis was heavily based on the presented bar chart, and was able to identify a singular pattern in the bar charts on many layers.

One domain expert pointed to an interesting phenomenon in the layer 15 of both models. In the supervised model, almost all feature maps recovered by the sort method focus on the bird structure, while in the self-supervised model it highlights the background. It is quite curious even with different training procedures, that both models of the same architecture highlight different features, but each one chooses a single feature, loosely remembering studies on convergence [51]. The layer band with these features is presented in Fig. 13.

The main suggestions of the experts refer to the layer band of feature maps below the Deep Rings. By clicking in one of them the same feature map should be highlighted in the other row if it exists, and the layer band lacks a mechanism for feature map identification to facilitate communication.

5.2.2 Scenario 2

A domain expert noticed that the first layer was similar across the training process while the second layer representations evolved over time. The participant pointed out that in the first timesteps there is more black feature maps and more sharpness in the feature maps while the final ones are more blurred. In the beginning, the model is not paying much attention to the good object and the agent itself while the last model pays attention first to the positive and then to the negative. The activation maps of the last layer from the final model tends to highlight the relevant objects.

One domain expert was not a RL expert, but easily spotted two temporal groups, one ranging from 20 k steps till 80 k steps, and the other from 100 k steps till the end. This intuition matches the rewards of these steps, where from the 20 k steps till 80 k steps (minimum reward) the reward is decreasing, then from the step 100 k onwards the rewards start increasing.

Layout-wise, the temporal representation of layer bands should be column by line, not line by column. The bands should be oriented vertically instead of its current horizontal disposition. The prototype should show the steps rewards. Once more, it was already noted that the end result of the network should be visible. For detailed analysis, zoom should be provided in this scenario, like in the scenario of thinking aloud protocol.

5.2.3 Exploratory Analysis 1

Both domain experts agree that the euclidean distance is good to enhance the perception that two feature maps are equal and unlike the first scenario, the metric now proved to be more helpful. This opinion may be related to the increased input image size (224×224) compared with the first scenario (32×32), so the domain experts can see clearly the features compared to the ResNet architectures from Scenario 1. However, the euclidean distance still falls short when two feature maps are visually similar not being always able to reflect well the human perceived distance.

One domain expert pointed out that the metric that sorts the feature maps by the sum of activation values can show more relevant features without losing the details than other metrics that sort them by the number of activations. The natural sparsity of the end layers hinders visual partition of features.

5.2.4 Exploratory Analysis 2

The RL expert said that he would not trust the model to create a reliable latent space based on the produced feature maps. The other domain expert expressed a biased view towards the size of the VGG16 architecture, as he had already seen the RL scenario previously with a CNN composed of just two layers. However, the former noticed that the model weights could be used as initialization parameters to learn how to perform the task in a reduced amount of time. He also mentioned that the visualization could be used to prune the network as the final two consecutive layers seemed to have redundant information. An important commentary made by expert was that it is possible to have a notion of which layers should be trained to fine-tune the network performance and which layers should be frozen. This would allow a training method with a human in the loop, intervening in the training method with contextual feedback.

6 Conclusion

Deep learning as an AI technique is increasingly used in decision-making tasks, and for this reason it is important to understand how neural networks learn their internal representations. In this work, we presented a visual idiom, DeepRings, that provides an overall perspective over the feature maps of a CNN in a single image, showing which features a deep learning model has considered making predictions. While it shows a complete overview of the layers, it also allows the user to inspect layers and feature maps through interaction. The visual idiom also supports the comparison of different models. This representation crystallizes the knowledge regarding the learning of hierarchical features, while revealing the existence of redundant filters in CNN models.

In both case studies, a total of five domain experts evaluated the design, showing that the layout and interaction are efficient to solve a diverse number of tasks, being also helpful in a free form exploration.

The presented DeepRings version still has some weaknesses in some components, mainly in the way of comparison of feature maps. The user needs a way to define how the comparison of layers and features maps should be done, as the euclidean distance is a simple metric that does not cover all comparison cases.

As future work, we plan to integrate the possibility to change dynamically the model architecture as well as the number of feature maps to be visualized per layer. A better temporal analysis is necessary, to show more information and present training information in a better way. In addition, it is key to let the user define their own metrics for the feature maps, to compare and sort, allowing them to obtain new insights about the model.

At a technical perspective, DeepRings is a standalone visualization, but it can easily be integrated within a Computer Vision pipeline to help the inspection of models during training, allowing network pruning, layer freezing and fine-tuning. The DeepRings visual idiom and design can be combined with other XAI techniques to greatly enhance the human comprehension of convolution neural networks.

Acknowledgements We thank everyone involved in discussion groups and case studies for their time and expertise. This research was developed in the scope of the Ph.D. grant [2020.05789.BD], funded by FCT—Foundation for Science and Technology. It was also supported by IEETA—Institute of Electronics and Informatics Engineering of Aveiro, funded by National Funds through FCT, in the context of the project [UID/CEC/00127/2019]. This study was also supported by PPGCC—UFPA—Computer Science Graduate Program of Federal University of Para, funded by National Funds through the CAPES Edital no 47/2017.

References

1. Samek, W., Müller, K.-R.: Towards Explainable Artificial Intelligence, pp. 5–22. Springer International Publishing, Cham (2019)
2. Amodei, D., Olah, C., Steinhardt, J., Christiano, P.F., Schulman, J., Mané, .: Concrete problems in AI safety. CoRR. arXiv:abs/1606.06565 (2016)
3. Lapuschkin, Sebastian, Wäldchen, Stephan, Binder, Alexander, Montavon, Grégoire., Samek, Wojciech, Müller, Klaus-Robert.: Unmasking clever hans predictors and assessing what machines really learn. Nat. Commun. **10**(1), 1096 (2019)
4. Bach, Sebastian, Binder, Alexander, Montavon, Grégoire., Klauschen, Frederick, Müller, Klaus-Robert., Samek, Wojciech: On pixel-wise explanations for non-linear classifier decisions by layer-wise relevance propagation. PLoS One **10**(7), 1–46 (2015)
5. Fong, R.C., Vedaldi, A.: Interpretable explanations of black boxes by meaningful perturbation. In: 2017 IEEE International Conference on Computer Vision (ICCV), (2017)
6. Zeiler, M.D., Fergus, R.: Visualizing and understanding convolutional networks. In: Fleet, D., Pajdla, T., Schiele, B., Tuytelaars, T. (eds.) Computer Vision—ECCV 2014, pp. 818–833. Springer International Publishing, Cham (2014)
7. Hehn, T.M., Kooij, J.F.P., Hamprecht, F.A.: End-to-end learning of decision trees and forests. Int. J. Comput. Vis. **128**(4), 997–1011 (2020)
8. Hohman, F., Park, H., Robinson, H., Chau, D.H.P.: Summit: scaling deep learning interpretability by visualizing activation and attribution summarizations. IEEE Trans. Vis. Comput. Graph. **26**(1), 1096–1106 (2020)
9. Mortier, R., Haddadi, H., Henderson, T., McAuley, D., Crowcroft, J.: Human-data interaction: the human face of the data-driven society. SSRN Electron. J. (2014)
10. Gilpin, L.H., Bau, D., Yuan, B.Z., Bajwa, A., Specter, M., Kagal, L.: Explaining explanations: an overview of interpretability of machine learning. In: 2018 IEEE 5th International Conference on Data Science and Advanced Analytics (DSAA), pp. 80–89 (2018)
11. Biran, O., Cotton, C.: Explanation and justification in machine learning: a survey. In: IJCAI-17 Workshop on Explainable AI (XAI), vol. 8 (2017)
12. Montavon, Grégoire., Samek, Wojciech, Müller, Klaus-Robert.: Methods for interpreting and understanding deep neural networks. Digit. Signal Process. **73**, 1–15 (2018)
13. Ribeiro, M.T., Singh, S., Guestrin, C.: "why should i trust you?" Explaining the predictions of any classifier. In: Proceedings of the 22nd ACM SIGKDD International Conference on Knowledge Discovery and Data Mining, KDD '16, pp. 1135–1144. Association for Computing Machinery, New York, USA (2016)
14. Selvaraju, R.R., Cogswell, M., Das, A., Vedantam, R., Parikh, D., Batra, D.: Grad-CAM: visual explanations from deep networks via gradient-based localization. Int. J. Comput. Vis. **128**(2), 336–359 (2020)
15. Zintgraf, L.M., Cohen, T.S., Adel, T., Welling, M.: Visualizing deep neural network decisions: prediction difference analysis. In: 5th International Conference on Learning Representations 2017, (2017)
16. Yann, L., Yoshua, B., Geoffrey, H.: Deep learning. Nature **521**(7553), 436–444 (2015)
17. Alves, J., Araújo, T., Marques, B., Dias, P., Santos, B.S.: Deeprings: a concentric-ring based visualization to understand deep learning models. In: 2020 24th International Conference Information Visualisation (IV), pp. 292–295 (2020)
18. Simonyan, K., Zisserman, A.: Very deep convolutional networks for large-scale image recognition. arXiv:1409.1556 (2014)
19. Card, S., Mackinlay, J.D., Shneiderman, B.: Information visualization. Hum.-Comput. Interact. Des. Issues, Solut. Appl. **181**, (2009)
20. Munzner, T.: Visualization Analysis and Design. CRC Press (2014)
21. Hohman, F., Kahng, M., Pienta, R., Chau, D.H.: Visual analytics in deep learning: an interrogative survey for the next frontiers. IEEE Trans. Vis. Comput. Graph. **25**(8), 2674–2693 (2019)

22. Komarek, A., Pavlik, J., Sobeslav, V.: Network visualization survey. In: Núñez, M., Nguyen, N.T., Camacho, D., Trawiński, B. (eds.) Computational Collective Intelligence, pp. 275–284. Springer International Publishing, Cham (2015)
23. Shaobo, Y., Lingda, W.: A key technology survey and summary of dynamic network visualization. In: 2017 8th IEEE International Conference on Software Engineering and Service Science (ICSESS), pp. 474–478. IEEE (2017)
24. McGee, F., Ghoniem, M., Melançon, G., Otjacques, B., Pinaud, B.: The state of the art in multilayer network visualization. Comput. Graph. Forum **38**(6), 125–149 (2019)
25. Yosinski, J., Clune, J., Nguyen, A., Fuchs, T., Lipson, H.: Understanding neural networks through deep visualization. In: Deep Learning Workshop, International Conference on Machine Learning (ICML), (2015)
26. Olah, C., Mordvintsev, A., Schubert, L.: Feature visualization. Distill, (2017)
27. Wang, J., Gou, L., Shen, H.W., Yang, H.: DQNViz: a visual analytics approach to understand deep Q-networks. IEEE Trans. Vis. Comput. Graph. **25**(1), 288–298 (2019)
28. Hilton, J., Cammarata, N., Carter, S., Goh, G., Olah, C.: Understanding RL vision. Distill, (2020). https://distill.pub/2020/understanding-rl-vision
29. Such, F.P., Madhavan, V., Liu, R., Wang, R., Castro, P.S., Li, Y., Zhi, J., Schubert, L., Bellemare, M.G., Clune, J., et al.: An atari model zoo for analyzing, visualizing, and comparing deep reinforcement learning agents. In: Proceedings of IJCAI 2019, (2019)
30. Rupprecht, C., Ibrahim, C., Pal, C.J.: Finding and visualizing weaknesses of deep reinforcement learning agents. In: International Conference on Learning Representations (ICLR), (2020)
31. Gupta, P., Puri, N., Verma, S., Kayastha, D., Deshmukh, S., Krishnamurthy, B., Singh, S.: Explain your move: understanding agent actions using specific and relevant feature attribution. In: International Conference on Learning Representations (ICLR), (2020)
32. Liu, W., Li, R., Zheng, M., Karanam, S., Wu, Z., Bhanu, B., Radke, R.J., Camps, O.: Towards visually explaining variational autoencoders. In: IEEE/CVF Conference on Computer Vision and Pattern Recognition (CVPR), (2020)
33. Wang, Z.J., Turko, R., Shaikh, O., Park, H., Das, N., Hohman, F., Kahng, M., Chau, D.H.P.: CNN explainer: learning convolutional neural networks with interactive visualization. IEEE Trans. Vis. Comput. Graph. **27**(2), 1396–1406 (2021)
34. Mnih, V., Kavukcuoglu, K., Silver, D., Rusu, A.A., Veness, J., Bellemare, M.G., Graves, A., Riedmiller, M., Fidjeland, A.K., Ostrovski, G., Petersen, S., Beattie, C., Sadik, A., Antonoglou, I., King, H., Kumaran, D., Wierstra, D., Legg, S., Hassabis, D.: Human-level control through deep reinforcement learning. Nature **518**(7540), 529–533 (2015)
35. Pezzotti, N., Höllt, T., Van Gemert, J., Lelieveldt, B.P.F., Eisemann, E., Vilanova, A.: Deepeyes: progressive visual analytics for designing deep neural networks. IEEE Trans. Vis. Comput. Graph. **24**(1), 98–108 (2017)
36. Chae, J., Gao, S., Ramanathan, A., Steed, C.A., Tourassi, G.: Visualization for classification in deep neural networks. Technical Report, Oak Ridge National Lab.(ORNL), Oak Ridge, TN (United States), (2017)
37. Selvaraju, R.R., Cogswell, M., Das, A., Vedantam, R., Parikh, D., Batra, D.: Grad-cam: visual explanations from deep networks via gradient-based localization. In: Proceedings of the IEEE international conference on computer vision, pp. 618–626. IEEE (2017)
38. Zurowietz, M., Nattkemper, T.W.: An interactive visualization for feature localization in deep neural networks. Front. Artif. Intell. **3**(49), (2020)
39. Liu, Mengchen, Shi, Jiaxin, Li, Zhen, Li, Chongxuan, Zhu, Jun, Liu, Shixia: Towards better analysis of deep convolutional neural networks. IEEE Trans. Vis. Comput. Graph. **23**(1), 91–100 (2016)
40. Kahng, M., Andrews, P.Y., Kalro, A., Chau, D.H.: Activis: visual exploration of industry-scale deep neural network models. IEEE Trans. Vis. Comput. Graph. **24**(1), 88–97 (2017)
41. Zhang, X., Yin, Z., Feng, Y., Shi, Q., Liu, J., Chen, Z.: Neuralvis: visualizing and interpreting deep learning models. In: 2019 34th IEEE/ACM International Conference on Automated Software Engineering (ASE), pp. 1106–1109. IEEE (2019)

42. Wongsuphasawat, K., Smilkov, D., Wexler, J., Wilson, J., Mane, D., Fritz, D., Krishnan, D., Viégas, F.B., Wattenberg, M.: Visualizing dataflow graphs of deep learning models in tensorflow. IEEE Trans. Vis. Comput. Graph. **24**(1), 1–12 (2017)
43. Woodburn, L., Yang, Y., Marriott, K.: Interactive visualisation of hierarchical quantitative data: an evaluation. In: 2019 IEEE Visualization Conference (VIS), pp. 96–100. IEEE (2019)
44. Schulz, Hans-Jorg., Hadlak, Steffen, Schumann, Heidrun: The design space of implicit hierarchy visualization: a survey. IEEE Trans. Vis. Comput. Graph. **17**(4), 393–411 (2010)
45. Deng, J., Dong, W., Socher, R., Li, L.-J., Li, K., Fei-Fei, L.: ImageNet: a large-scale hierarchical image database. In: CVPR'09, (2009)
46. Khosla, P., Teterwak, P., Wang, C., Sarna, A., Tian, Y., Isola, P., Maschinot, A., Liu, C., Krishnan, D.: Supervised contrastive learning. In: Larochelle, H., Ranzato, M., Hadsell, R., Balcan, M.F., Lin, H. (eds.) Advances in Neural Information Processing Systems, vol. 33, pp. 18661–18673. Curran Associates, Inc. (2020)
47. Haarnoja, T., Zhou, A., Abbeel, P., Levine, S.: Soft actor-critic: off-policy maximum entropy deep reinforcement learning with a stochastic actor. In: Dy, J., Krause, A. (eds.) Proceedings of the 35th International Conference on Machine Learning, vol. 80, pp. 1861–1870. PMLR (2018)
48. Juliani, A., Berges, V.-P., Teng, E., Cohen, A., Harper, J., Elion, C., Goy, C., Gao, Y., Henry, H., Mattar, M., et al.: Unity: a general platform for intelligent agents. arXiv:1809.02627 (2018)
49. He, K., Zhang, X., Ren, S., Sun, J.: Identity mappings in deep residual networks. In: Leibe, B., Matas, J., Sebe, N., Welling, M. (eds.) Computer Vision—ECCV 2016, pp. 630–645. Springer International Publishing, Cham (2016)
50. Riitta, J.: Think-aloud protocol. Handb. Transl. Stud. **1**, 371–374 (2010)
51. Li, Y., Yosinski, J., Clune, J., Lipson, H., Hopcroft, J.E.: Convergent learning: do different neural networks learn the same representations? In: FE@NIPS, pp. 196–212, (2015)

"Negative" Results—When the Measured Quantity Is Outside the Sensor's Range—Can Help Data Processing

Jonatan Contreras, Francisco Zapata, Olga Kosheleva, Vladik Kreinovich, and Martine Ceberio

Abstract In many real-life situations, we know the general form of the dependence $y = f(x, c_1, \ldots, c_m)$ between physical quantities, but the values c_i need to be determined experimentally, based on the results of measuring x and y. In some cases, we do not get any result of measuring y since the actual value is outside the range of the measuring instrument. Usually, such cases are ignored. In this paper, we show that taking these cases into account can help data processing—by improving the accuracy of our estimates of c_i and thus, by improving the accuracy of the resulting predictions of y.

1 A Brief Introduction

In many real-life situations, we know, from experience, that the value of a quantity y is determined by the values of related quantities $x = (x_1, \ldots, x_v)$. For example, we often know that the future value y of a physical quantity—e.g., of the temperature at a given location—is determined by the current values of this and related quantities: temperature, wind speed and direction, humidity, etc.—at this and nearby locations. In many such situations, we know the general type of this dependence, i.e., we know

J. Contreras (✉) · F. Zapata · O. Kosheleva · V. Kreinovich · M. Ceberio
University of Texas at El Paso, 500 W. University, El Paso, TX 79968, USA
e-mail: jmcontretas2@utep.edu

F. Zapata
e-mail: fcozpt@outlook.com

O. Kosheleva
e-mail: olgak@utep.edu

V. Kreinovich
e-mail: vladik@utep.edu

M. Ceberio
e-mail: mceberio@utep.edu

B. Kovalerchuk et al. (eds.), *Integrating Artificial Intelligence and Visualization for Visual Knowledge Discovery*, Studies in Computational Intelligence 1014,
https://doi.org/10.1007/978-3-030-93119-3_7

that this dependence takes the form $y = f(x, c_1, \ldots, c_n)$, where the parameters c_i need to be determined from experiments:

- we measure the values of x and y in different situations, and
- we find the value of the parameters c_i which are consistent with the results of these measurements.

Measurements are never 100% accurate, so we can only find the values c_i with some accuracy—and this accuracy also needs to be determined based on the same measurement results. The more measurements we process, the more accurate the results. However, as we process a large number of measurements, we will eventually encounter situations when we could not measure y—because the corresponding value was outside the sensor's range. Usually, in data processing, such situations are simply dismissed—or, at best, used as an additional check of the resulting model. In this paper, we show that such "negative" data can help in all aspects of data processing—not only to select an appropriate model, but also to estimate the model's accuracy.

Preliminary results of this research first appeared in our conference paper [2]. In the current paper, we extend the 1-D case analyzed in [2] into a general multi-D case. We have also added an explicit explanation of where such "negative" data points come from.

2 Formulation of the Problem: A More Detailed Description

What are the main objectives of science and engineering. Crudely speaking, the main objective of science is to *predict* future events. This is the ultimate goal by which we gauge the quality of each new scientific theory: if this theory can make a new prediction, and this prediction is experimentally confirmed, this confirms the theory. This is how General Relativity was confirmed—by measuring the gravitation-related deviation of the light from the straight line. This is how all theories are checked and confirmed.

The main objective of engineering is to come up with designs and/or control strategies that will *improve* our future. To understand which gadgets, which strategies will lead to the desired improvement, we need to predict how exactly the state of the world will change under each such strategy.

For both objectives, we must be able:

- *given* the initial conditions x—the initial state of the world plus, in the engineering case, the description of the changes that we plan to make,
- to *predict* the value of each quantity y that characterizes the future state of the world (or at least the future state of the system in which we are interested).

Comment. In general, the values of the quantity y can be all possible real values—or all the values from some finite (or semi-infinite) interval. Such quantities are known as *continuous*.

In some case, possible values of y are limited to a discrete set. For example, electric charge can only take values which are proportional to the elementary charge. Such quantities are known as *discrete*.

In this paper, we concentrate on continuous quantities. However, our ideas and formulas can be easily extended to the discrete cases as well. For example, for discrete quantities, in the case of probabilistic uncertainty:

- instead of the probability density function (pdf), we can use its discrete analogue—probabilities of different values, and
- instead of the requirement that the integral of the pdf is equal to 1, we have the requirement that the sum of the probabilities is equal to 1.

In most practical situations, we do not know the dependence of y on x. In some cases we know the equations (or even explicit formulas) that relate the available information x and the desired quantity y. In such cases, in principle, we have an algorithm for predicting y. A typical situation of this type is celestial mechanics, where we can predict, e.g., solar eclipses and re-appearance of comets hundreds of years ahead.

In some cases, we may know the dependence, but there is no feasible algorithm for making the corresponding prediction. For example, it is possible to predict—reasonably reliably—in what direction a tornado will turn in the next 15 min. However, such computations require several hours of computations on a high-performance computer. This is a good confirmation of the current tornado models, but from the practical viewpoint, this result is so far useless: what is the purpose of these hour-long computations if in 15 min we will already know where the tornado moved. The hope is that as computers get faster and faster, we will eventually be able to make the corresponding computations practical.

In many other cases, we do not know how exactly y depends on x. In such cases, we need to determine this dependence based on the known observations $\left(x^{(k)}, y^{(k)}\right)$, in which we measured (or otherwise estimated) both x and y. In many such situations, we know that the dependence of y on x belongs to a known family of functions, i.e., that $y = f(x, c_1, \ldots, c_n)$ for some values of the parameters c_i, but these values need to be determined from experiment.

In other situations, we do not know the family, so we use appropriate machine learning tools to come up with the desired dependence $y = f(x)$. In mathematical terms, the results of machine learning—e.g., the results of deep learning—can also be described as $y = f(c_1, \ldots, c_n)$, where c_i are the parameters that change during training, and the expression $f(x, c_1, \ldots, c_n)$ describes the result of applying, to the input x, the machine learning algorithm with parameters c_i.

Need to take uncertainty into account. Our knowledge of x and y usually comes from measurements, and measurements are never absolutely accurate: the measurement result $\widetilde{y}$ is, in general, different from the actual value y. In some cases, we know the probabilities of measurement errors $\Delta y \stackrel{\text{def}}{=} \widetilde{y} - y$. In such cases, we have a probabilistic information about y; see, e.g., [16].

In other cases, we do not know these probabilities, all we know is the upper bound Δ on the absolute value of the measurement error $|\Delta y| \le \Delta$. In such cases, once we know the measurement result $\widetilde{y}$, the only thing that we can conclude about the actual value y is that this value is located in the interval $[\widetilde{y} - \Delta, \widetilde{y} + \Delta]$.

In this paper, we will consider both cases—of probabilistic and of interval uncertainty.

Important comment about the accuracies of measuring x and y. Usually:

- we set up the values x, and
- measure the corresponding values y.

The values x we set up ourselves, we know exactly the range of the corresponding values, so we can select the most appropriate measuring instrument, and get reasonably accurate values.

In contrast, we only have a crude idea about the future values y; based on this idea, we can guess the range of y, but this guess may be wrong, so the selected measuring instrument may be not the most accurate one for this range. As a result, in general, the accuracy with which we know y is much worse that the accuracy with which we know x.

So, it makes sense—at least in the first reasonable approximation—to ignore the measurement errors corresponding to measuring x—since they are much smaller than the errors of measuring y—and assume that the values x are known exactly.

Positive and negative examples: what are they and why it is important. In most practical situations, we have measurement results describing both x and y. We will call the corresponding examples $\left(x^{(k)}, y^{(j)}\right)$ *positive examples.*

However, sometimes :

- we know the values x, but
- we could not measure y.

Indeed, each measuring instrument has a range of values for which it has been designed:

- a ruler cannot directly measure lengths which are too long;
- scales cannot measure weights which are too heavy—they will simply crush the scales—or too light: they will not be noticed by the scale at all.

In each experiment, we try to select a measuring instrument that will be able to measure the corresponding value—but remember that we are dealing with situations in which we do not know the actual dependence, so we cannot exactly predict the value y. Because of this, sometimes, we have wrong expectations, we select a wrong instrument. Thus, as we have mentioned earlier, the actual signal may not be in the area of the sensor's not-the-best accuracy.

Moreover, the signal can be outside the interval on which we selected and/or trained the measuring instrument: the actual value could be smaller than this interval's lower endpoint or it could be larger that the interval's upper endpoint.

In this case, for the only information about the measured quantity y is that this quantity is *not* located in the (known) interval range of the measuring instrument. We will call such situations *negative* examples.

Usually, negative examples are ignored. Usually, practitioners simply ignore such negative examples—mostly because there are no methods for using them.

At best, such examples are used as an additional conformation of the resulting values c_i: by checking that for each such example, the corresponding value $y^{(k)} = f\left(x^{(k)}, c_1, \ldots, c_n\right)$ is indeed outside the range of the corresponding measuring instrument.

What we do in this paper. Since the measurement results are approximate, the resulting estimates for the parameters c_i are also approximate. From the practical viewpoint, it is important to know this uncertainty—since it affects the accuracy of the resulting future predictions $y = f(x, c_1, \ldots, c_n)$.

In this paper, we show that negative example can help to gauge this uncertainty—and thus, help us make more accurate predictions.

3 Case of Interval Uncertainty

Data processing under interval uncertainty, case of positive examples: a brief reminder. Following our assumption about the accuracies of measuring x and y, we assume that:

- we know the exact values $x^{(k)}$, while
- the values $y^{(k)}$ are only known with interval uncertainty—i.e., for each k, we know the interval $\left[\underline{y}^{(k)}, \overline{y}^{(k)}\right]$ that contains the actual (unknown) value $y^{(k)}$.

In this case, our objective is to find the values $c = (c_1, \ldots, c_n)$ for which the following inequality is satisfied for all k:

$$\underline{y}^{(k)} \leq f\left(x^{(k)}, c_1, \ldots, c_n\right) \leq \overline{y}^{(k)}, 1 \leq k \leq K. \tag{1}$$

Data processing under interval uncertainty, case of positive examples: algorithms. For each i, we want to find the range $\left[\underline{c}_i, \overline{c}_i\right]$ of possible values of c_i. To find this range, we need to solve the following two constraint optimization problems:

- to find $\underline{c}_i$, we minimize c_i under the constraints (1); and
- to find $\overline{c}_i$, we maximize c_i under the constraints (1).

In the general non-linear case, this problem is NP-hard (even finding one single combination c that satisfies all the constraints (1) is, in general, NP-hard); see, e.g., [7]. In such cases, constraint solving algorithms (see, e.g., [4]) can lead to approximate ranges: e.g., to enclosures $\left[\underline{c}'_i, \overline{c}'_i\right] \supseteq \left[\underline{c}_i, \overline{c}_i\right]$ for the actual range.

The problem of computing the ranges $[\underline{c}_i, \overline{c}_i]$ becomes feasible if we consider families that linearly depend on the parameters c_i, i.e., families of the type

$$f(x, c_1, \ldots, c_n) = f_0(x) + c_1 \cdot f_1(x) + \ldots + c_n \cdot f_n(x). \quad (2)$$

In this case, inequalities (1) become linear inequalities in terms of the unknowns c_i:

$$\underline{y}^{(k)} \le f_0\left(x^{(k)}\right) + c_1 \cdot f_1\left(x^{(k)}\right) + \ldots + c_n \cdot f_n\left(x^{(k)}\right) \le \overline{y}^{(k)}, \quad 1 \le k \le K \quad (3)$$

In this case, e.g., the range $[\underline{c}_i, \overline{c}_i]$ of possible values of c_i can be obtained by solving the following two *linear programming* problems – i.e., problems of optimizing a linear function under linear constraints:

- to find $\underline{c}_i$, we need to minimize c_i under the linear constraints (3); and
- to find $\overline{c}_i$, we need to maximize c_i under the linear constraints (3).

There are efficient feasible algorithms for solving linear programming problems; see, e.g., [3, 8]. So, the corresponding regression problem can indeed be efficiently solved.

How can we take "negative" intervals into account. As we have mentioned earlier:

- in addition to "positive" intervals—i.e., intervals that contain the y-values $y^{(k)}$, $k = 1, \ldots, K$,
- we can also have "negative" intervals $\left(\underline{y}^{(\ell)}, \overline{y}^{(\ell)}\right)$, $\ell = K + 1, \ldots, L$—i.e., intervals that are known *not* to contain the corresponding values $y^{(\ell)}$.

In this case, in addition to the condition (1), we also have an additional condition that must be satisfied for each ℓ from $K + 1$ to L:

$$f\left(x^{(\ell)}, c_1, \ldots, c_n\right) \le \underline{y}^{(\ell)} \text{ or } \overline{y}^{(\ell)} \le f\left(x^{(\ell)}, c_1, \ldots, c_n\right). \quad (4)$$

In this case, we need to find the values $c = (c_1, \ldots, c_n)$ that satisfy *both* constraints (1) and (4).

An example showing that negative intervals can help. Let us first show that the use of negative example can indeed improve the accuracy of the resulting estimates for c_i.

Let us consider a very linear model $y = c_1 \cdot x$, and the simplest case when we have only two observations: for two values $x = -1$ and for $x = 1$, we have the same interval of possible values of y: $y \in [-1, 1]$. One can easily see that in this case, the set of possible values of c_1 is the interval $[-1, 1]$.

In particular, for $x = 2$, the only information that we can conclude based on this estimate for c_1, is that $y \in [-2, 2]$.

Suppose now that, in addition to the above two positive examples, we also have a negative example: namely, we know that for $x = 2$, the value y cannot be in the interval $(-3, 2)$—the range of the corresponding sensor. In this case, for this x, the set of possible values of y narrows down to a single value $y = 2$. Correspondingly, the set of possible values of c_1 narrows down from the original interval $[-1, 1]$ to a single value $c_1 = 1$.

In this case, the accuracy of the resulting predictions $y = c_1 \cdot x$ drastically increases, from very inaccurate predictions $y \in [-x, x]$ to a very accurate prediction $y = x$.

Towards algorithms: the problem is not easy. Now that we know that negative intervals can drastically improve the prediction accuracy, it is desirable to look for algorithms that will take such intervals into account.

It would have been nice to have a general feasible algorithm. However, unfortunately, one can prove that if we take into account possible negative intervals, then even for linear constraints, the problem of finding the bounds $\underline{c}_i$ and $\overline{c}_i$ becomes NP-hard.

The proof of this NP-hardness is rather straightforward. Indeed, it is known that the following problem is NP-hard (see, e.g., [7, 15]):

- *given* natural numbers $s_1, \ldots, s_n$ and s,
- *find* a subset of the values s_i that adds up to s.

In other words, we need to find the values $c_i \in \{0, 1\}$ (describing whether to take the i-th value s_i or not) for which $\sum\limits_{i=1}^{n} c_i \cdot s_i = s$.

This problem can be easily reformulated as an interval problem with positive and negative examples. For this purpose, we take a linear model

$$y = c_1 \cdot x_1 + \ldots + c_n \cdot x_n$$

and the following examples:

- a positive example in which $x_i = s_i$ for all i and $y \in [s, s]$; consistency with this positive example means that
$$s = \sum_{i=1}^{n} c_i \cdot s_i;$$
- n additional positive examples; in the i-th example, $x_i = 1$, $x_j = 0$ for all $j \neq i$, and $y \in [0, 1]$; consistency with each such example means that $c_i \in [0, 1]$; and
- n negative examples; in the i-th example, $x_i = 1$, $x_j = 0$ for all $j \neq i$, and $y \notin (0, 1)$; consistency with each such example means that $c_i \notin (0, 1)$.

Together with the previous consistency, this means exactly that $c_i \in \{0, 1\}$.

Since we cannot compute the exact range, let us compute the enclosure for the range. NP-hard implies that, unless P = NP (which most computer scientists believe to be impossible), no feasible algorithm is possible that would always compute the exact ranges for c_i—or even check whether the data is consistent with the model.

Since we cannot compute the exact range $\left[\underline{c}_i, \overline{c}_i\right]$, a natural idea is to compute an *enclosure* $\left[\underline{C}_i, \overline{C}_i\right] \supseteq \left[\underline{c}_i, \overline{c}_i\right]$. Let us show how we can do it.

Computing enclosure: first algorithm. Each negative interval $\left(\underline{y}^{(\ell)}, \overline{y}^{(\ell)}\right)$ means that the actual value of $y^{(\ell)}$ is either in the interval $\left(-\infty, \underline{y}^{(\ell)}\right]$ or in the interval $\left[\overline{y}^{(\ell)}, \infty\right)$. So, the following is a natural algorithm:

- we can add, to K positive intervals, the first of these two semi-infinite intervals, solve the corresponding linear programming problem, and get ranges $\left[\underline{C}_i^{(\ell),-}, \overline{C}_i^{(\ell),-}\right]$ for the coefficients c_i;
- we can also add, to K positive intervals, the second of these two semi-infinite intervals, solve the corresponding linear programming problem, and get ranges $\left[\underline{C}_i^{(\ell),+}, \overline{C}_i^{(\ell),+}\right]$ for the coefficients c_i.

We know that the actual value $y^{(\ell)}$ is either in the first *or* in the second of the semi-infinite intervals. Therefore, the actual range $\left[\underline{c}_i, \overline{c}_i\right]$ of possible values of each c_i is contained in the *union* of the two intervals:

$$\left[\underline{C}_i^{(\ell)}, \overline{C}_i^{(\ell)}\right] = \left[\underline{C}_i^{(\ell),-}, \overline{C}_i^{(\ell),-}\right] \bigcup \left[\underline{C}_i^{(\ell),+}, \overline{C}_i^{(\ell),+}\right]. \tag{5}$$

In order words, we get the enclosure $\left[\underline{C}_i^{(\ell)}, \overline{C}_i^{(\ell)}\right]$, where:

$$\underline{C}_i^{(\ell)} = \min\left(\underline{c}_i^{(\ell),-}, \underline{c}_i^{(\ell),+}\right) \text{ and } \overline{C}_i^{(\ell)} = \max\left(\overline{c}_i^{(\ell),-}, \overline{c}_i^{(\ell),+}\right). \tag{6}$$

The actual value c_i belongs to *all* these intervals, so we can conclude that it belongs to the intersection $\left[\underline{c}_i, \overline{c}_i\right]$ of all these intervals:

$$[\underline{C}_i, \overline{C}_i] = \bigcap_{\ell=K+1}^{L} \left[\underline{C}_i^{(\ell)}, \overline{C}_i^{(\ell)}\right]. \tag{7}$$

For this new enclosure, we have:

$$\underline{C}_i = \max_{\ell} \underline{C}_i^{(\ell)} \text{ and } \overline{C}_i = \min_{\ell} \overline{C}_i^{(\ell)}. \tag{8}$$

If this intersection is empty, this means that the model is inconsistent with observations.

Computing enclosure: second algorithm. In the above algorithm, at each step, we only take into account *one* negative example. Instead, we can take into account *two* negative examples. Then, for each pair (ℓ, ℓ') of negative examples, we have four possible cases:

- we can have the case $a = --$ when $y^{\ell} \in \left(-\infty, \underline{y}^{(\ell)}\right]$ and $y^{\ell'} \in \left(-\infty, \underline{y}^{(\ell')}\right]$;
- we can have the case $a = -+$ when $y^{\ell} \in \left(-\infty, \underline{y}^{(\ell)}\right]$ and $y^{\ell'} \in \left[\overline{y}^{(\ell')}, \infty\right)$;

- we can have the case $a = +-$ when $y^\ell \in \left[\overline{y}^{(\ell)}, \infty\right)$ and $y^{\ell'} \in \left(-\infty, \underline{y}^{(\ell')}\right]$; and
- we can have the case $a = ++$ when $y^\ell \in \left[\overline{y}^{(\ell)}, \infty\right)$ and $y^{\ell'} \in \left[\overline{y}^{(\ell')}, \infty\right)$.

For each of these four cases $a = --, -+, +-, ++$, we can add the corresponding two semi-infinite intervals to K positive intervals, and find the enclosures $\left[\underline{C}_i^{(\ell,\ell'),a}, \overline{C}_i^{(\ell,\ell'),a}\right]$ for the actual range $\left[\underline{c}_i, \overline{c}_i\right]$ of each coefficient c_i. Then, we can conclude that the actual value of c_i belongs to the union of these four intervals:

$$\left[\underline{C}_i^{(\ell,\ell')}, \overline{C}_i^{(\ell,\ell')}\right] = \bigcup_a \left[\underline{C}_i^{(\ell,\ell'),a}, \overline{C}_i^{(\ell,\ell'),a}\right], \tag{9}$$

i.e., we take

$$\underline{C}_i^{(\ell,\ell')} = \min_a \underline{C}_i^{(\ell,\ell'),a} \text{ and } \overline{C}_i^{(\ell,\ell')} = \max_a \overline{C}_i^{(\ell,\ell'),a}. \tag{10}$$

The actual value c_i belongs to *all* these enclosures. So, we can conclude that the actual value c_i belongs to the intersection $\left[\underline{C}_i, \overline{C}_i\right]$ of all these enclosures:

$$\left[\underline{C}_i, \overline{C}_i\right] = \bigcap_{K+1 \le \ell, \ell' \le L} \left[\underline{C}_i^{(\ell,\ell')}, \overline{C}_i^{(\ell,\ell')}\right], \tag{11}$$

i.e., we take

$$\underline{C}_i = \max_{\ell,\ell'} \underline{C}_i^{(\ell,\ell')} \text{ and } \overline{C}_i = \min_{\ell,\ell'} \overline{C}_i^{(\ell,\ell')}. \tag{12}$$

By applying this algorithm:

- on the one hand, we get, in general, a better range—with smaller excess width;
- however now, instead of considering $O(L - K)$ cases as in the first algorithms, we need to consider $O\left((L - K)^2\right)$ cases.

Possible other algorithms. We can get even more accurate estimates for the range if we consider all possible triples, 4-tuples, etc., of negative intervals. However, in this case, we will need to consider $O\left((L - K)^3\right)$, $O\left((L - K)^4\right)$, etc. cases—i.e., we get a much longer computation time.

4 What If We Are Interested in Several Quantities

Negative information that we analyzed so far: a reminder. In the previous text, we considered negative examples corresponding to the case when a value of a quantity y cannot be detected by a sensor tuned for values from some interval $(\underline{y}, \overline{y})$.

In this case, we can conclude that the actual value y is outside this interval.

There are other types of negative information. A similar—but somewhat more complicated—situation occurs if we train, e.g., a camera to a certain area or a microphone to a certain area of spatial directions. In such cases, we are not simply limiting the range of possible values of a single quantity y. Instead, we simultaneously limit the value of two or more quantities $y_1, \ldots, y_m$ to corresponding intervals:

$$y_1 \in \left(\underline{y}_1, \overline{y}_1\right), \ldots, y_m \in \left(\underline{y}_m, \overline{y}_m\right). \tag{13}$$

In this case, in contrast to the previously analyzed case, if we do not detect anything, we cannot conclude that, e.g., the value y_1 is necessarily not in the corresponding interval $\left[\underline{y}_1, \overline{y}_1\right]$—this value may well be within this interval, but one of the other quantities is outside its interval. All we know is that the tuple $y = (y_1, \ldots, y_m)$ is *not* located inside the corresponding box

$$y \notin \left(\underline{y}_1, \overline{y}_1\right) \times \ldots \times \left(\underline{y}_m, \overline{y}_m\right). \tag{14}$$

Because of this inter-relation between different variables, to deal with such situations, we can no longer concentrate on one of the quantities y_i—there are no restrictions on each value y_i per se—we need to simultaneously consider *all* related quantities $y_1, \ldots, y_m$.

Let us describe the resulting problem in precise terms: general case. We know that several quantities $y_1, \ldots, y_m$ depend on the quantities $x = (x_1, \ldots, x_v)$. We assume that this dependence is described by functions from a certain family of functions, characterized by parameters $c_1, \ldots, c_n$:

$$y_1 = f_1(x, c_1, \ldots, c_n), \ldots, y_m = f_m(x, c_1, \ldots, c_n). \tag{15}$$

We have several measurements in which the vector $y = (y_1, \ldots, y_m)$ was inside the corresponding box, i.e., when we had, for all k from 1 to K:

$$\begin{gathered} \underline{y}_1^{(k)} \le f_1\left(x^{(k)}, c_1, \ldots, c_m\right) \le \overline{y}_1^{(k)}, \\ \ldots \\ \underline{y}_m^{(k)} \le f_m\left(x^{(k)}, c_1, \ldots, c_m\right) \le \overline{y}_m^{(k)}. \end{gathered} \tag{16}$$

We also have negative examples, for which, for each ℓ from $K + 1$ to L, the following condition must be satisfied:

$$\begin{gathered} f_1\left(x^{(\ell)}, c_1, \ldots, c_n\right) \le \underline{y}_1^{(\ell)}, \text{ or } \overline{y}_1^{(\ell)} \le f_1\left(x^{(\ell)}, c_1, \ldots, c_n\right), \text{ or} \\ \ldots, \text{ or} \\ f_m\left(x^{(\ell)}, c_1, \ldots, c_n\right) \le \underline{y}_m^{(\ell)}, \text{ or } \overline{y}_m^{(\ell)} \le f_m\left(x^{(\ell)}, c_1, \ldots, c_n\right). \end{gathered} \tag{17}$$

The question is to find the values $c = (c_1, \ldots, c_n)$ that satisfy all the constraints (16) and (17).

In particular, for each of the parameters c_i, we need to find the range $\left[\underline{c}_i, \overline{c}_i\right]$ of possible values of this parameter:

- Each value $\underline{c}_i$ can be obtained by minimizing c_i under the constraints (16) and (17).
- Similarly, each value $\overline{c}_i$ can be obtained by maximizing c_i under the constraints (16) and (17).

Important case when the dependence on parameters is linear. As we have mentioned, in the general case, the corresponding problems are NP-hard even when we do not have negative examples. In this no-negative-examples case, however, the problem becomes feasible if we consider the common situations in which the dependence on the parameters c_i is linear, i.e., in which

$$\begin{aligned} y_1 &= f_{1,0}(x) + c_1 \cdot f_{1,1}(x) + \ldots + c_n \cdot f_{1,n}(x), \\ &\ldots \\ y_m &= f_{m,0}(x) + c_1 \cdot f_{m,1}(x) + \ldots + c_n \cdot f_{m,n}(x). \end{aligned} \tag{18}$$

In this case, the condition (16) corresponding to each measurement k takes the form:

$$\begin{aligned} \underline{y}_1^{(k)} \le f_{1,0}\left(x^{(k)}\right) + c_1 \cdot f_{1,1}\left(x^{(k)}\right) + \ldots + c_n \cdot f_{1,n}\left(x^{(k)}\right) \le \overline{y}_1^{(k)}, \\ \ldots \\ \underline{y}_m^{(k)} \le f_{m,0}\left(x^{(k)}\right) + c_1 \cdot f_{m,1}\left(x^{(k)}\right) + \ldots + c_n \cdot f_{m,n}\left(x^{(k)}\right) \le \overline{y}_m^{(k)}. \end{aligned} \tag{19}$$

Similarly, the condition (17) corresponding to each measurement ℓ takes the form

$$\begin{aligned} f_{1,0}\left(x^{(\ell)}\right) + c_1 \cdot f_{1,1}\left(x^{(\ell)}\right) + \ldots + c_n \cdot f_{1,n}\left(x^{(\ell)}\right) \le \underline{y}_1^{(\ell)}, \text{ or} \\ \overline{y}_1^{(\ell)} \le f_{1,0}\left(x^{(\ell)}\right) + c_1 \cdot f_{1,1}\left(x^{(\ell)}\right) + \ldots + c_n \cdot f_{1,n}\left(x^{(\ell)}\right), \text{ or} \\ \ldots, \text{ or} \\ f_{m,0}\left(x^{(\ell)}\right) + c_1 \cdot f_{m,1}\left(x^{(\ell)}\right) + \ldots + c_n \cdot f_{n,n}\left(x^{(\ell)}\right) \le \underline{y}_m{}^{(\ell)}, \text{ or} \\ \overline{y}_m{}^{(\ell)} \le f_{m,0}\left(x^{(\ell)}\right) + c_1 \cdot f_{m,1}\left(x^{(\ell)}\right) + \ldots + c_n \cdot f_{m,n}\left(x^{(\ell)}\right). \end{aligned} \tag{20}$$

For each ℓ, we know that one of $2m$ possible inequalities (20) is satisfied.

How can we solve this problem? As we have mentioned, in the presence of negative examples, even for this case—when the dependence on the parameters is linear—the exact computations of the bounds $\underline{c}_i$ and $\overline{c}_i$ is an NP-hard problem already for $m = 1$.

However, we can use the same ideas as in the previous section and come up with a feasible algorithm for computing an *enclosure* for the desired range $[\underline{c}_i, \overline{c}_i]$, i.e., for computing the interval that *contains* the desired range. To be more precise, similarly to the above case $m = 1$, we have a family of feasible algorithms that can bring us closer and closer to the desired range.

First algorithm. The first algorithm in this sequence is when on each step, we take into account only one negative example ℓ. For each ℓ:

- we add one of the $2m$ inequalities (20) to the system (19); thus, we get $2m$ problems of minimizing c_i and $2m$ problems of maximizing c_i; for each of $2m$ pairs of linear programming problems, we thus find an interval of possible values of c_i;
- since *one* of these inequalities is satisfied, we can conclude that the desired range is contained in the *union* of the resulting $2m$ intervals; we can compute this union by computing the smallest of $2m$ lower endpoints and the largest of $2m$ upper endpoints.

For each ℓ, we know that the actual range is contained in the corresponding union—thus, it is contained in the *intersection* of these unions. To compute such an intersection:

- we compute the largest of the lower endpoints corresponding to different ℓ, and
- we compute the smallest of the upper endpoints corresponding to different ℓ.

More accurate—but more time-consuming—algorithms. To get a more accurate estimate of the desired range $[\underline{c}_i, \overline{c}_i]$, instead of taking only *one* negative example ℓ into account in each linear programming problem, we take *two* such negative examples ℓ and ℓ' into account. We have $2m$ possible inequalities for ℓ and we have $2m$ possible inequalities for ℓ', so we have $(2m)^2$ pairs of possible inequalities.

For each pair (ℓ, ℓ'):

- we add one of $2m$ inequalities (20) corresponding to ℓ and one of $2m$ inequalities corresponding to ℓ' to the system (19); thus, we get $(2m)^2$ problems of minimizing c_i and $(2m)^2$ problems of maximizing c_i; for each of $(2m)^2$ pairs of linear programming problems, we thus find an interval of possible values of c_i;
- since *one* of these pairs of inequalities is satisfied, we can conclude that the desired range is contained in the *union* of the resulting $(2m)^2$ intervals; we can compute this union by computing the smallest of $(2m)^2$ lower endpoints and the largest of $(2m)^2$ upper endpoints.

For each pair (ℓ, ℓ'), we know that the actual range is contained in the corresponding union—thus, it is contained in the *intersection* of these unions. To compute such an intersection:

- we compute the largest of the lower endpoints corresponding to different pairs (ℓ, ℓ'), and
- we compute the smallest of the upper endpoints corresponding to different pairs (ℓ, ℓ').

Instead of pairs, we can consider triples, quadruples, etc. Every time we consider tuples with one more element, the computation time increases—but we get more accurate enclosures.

5 Case of Probabilistic Uncertainty

Probabilistic uncertainty means that for each measurement k, we know the probabilities of different possible values of the measurement error

$$\Delta y^{(k)} = y^{(k)} - y = y^{(k)} - f\left(x^{(k)}, c_1, \ldots, c_n\right),$$

i.e., we know, e.g., the probability density function $\rho_k\left(y^{(k)} - y\right)$ describing these probabilities.

In this case, the probability that a model $y = f(x, c_1, \ldots, c_n)$ is consistent with the k-th observation is proportional to $\rho_k\left(y^{(k)} - f\left(x^{(k)}, c_1, \ldots, c_n\right)\right)$. It is usually assumed that different measurements are independent. Thus, the probability that a model is consistent with all K observations is equal to the product of the corresponding probabilities

$$\prod_{k=1}^{K} \rho_k\left(y^{(k)} - f\left(x^{(k)}, c_1, \ldots, c_n\right)\right). \tag{21}$$

A natural idea is to select the values $c_1, \ldots, c_n$ for which this probability is the largest possible. This is known as the Maximum Likelihood method.

If we want to find the range of possible values of c, then we must look for all the values c for which the expression (21) is larger than or equal to a certain threshold value ρ_0—where this threshold value can be determined, e.g., from the condition that the probability to be outside this region is equal to some pre-selected value $\varepsilon > 0$.

What if we have negative examples? In this case, instead of considering all possible combinations c, we need to consider only combinations which are consistent with all $L - K$ negative examples, i.e., combinations that satisfy the property (4) for all $\ell = K + 1, \ldots, L$.

6 Case of Fuzzy Uncertainty

What is fuzzy uncertainty: a brief reminder. In some cases, we do not know the probabilities of different values of the measurement errors, but we have expert estimates of which values are possible. These expert estimates are usually formulated by using imprecise ("fuzzy") words from natural language. For example, in the situation when the guaranteed upper bound 0.1 on the absolute value of the measurement error, an expert may say that this absolute value is not much larger than 0.05. To formalize such imprecise ("fuzzy") knowledge, Lotfi Zadeh invented special techniques—that he called *fuzzy*; see, e.g., [1, 5, 6, 10, 12–14, 17].

Degree of certainty and a membership function. In fuzzy techniques:

- for each imprecise expert statement about a quantity,
- we ask an expert to estimate, on a scale from 0 to 1, his/her degree of confidence that the expert's statement holds for this value (e.g., that 0.06 is not much larger than 0.05).

A function that assigns this degree to each possible value is called a *membership function*.

"And"- and "or"-operations. The degrees of confidence $a, b, \ldots$ in individual statements $A, B, \ldots$ enable us also to estimate degrees of confidence in composite statements such as $A \,\&\, B$, $A \vee B$, etc.

The algorithms $f_\&(a, b)$ and $f_\vee(a, b)$ for such estimates are called *"and"- and "or"-operations*, or, for historical reasons, *t-norms* and *t-conorms*.

The most widely used "and"-operations are $\min(a, b)$ and $a \cdot b$.

Data processing under fuzzy uncertainty: a brief reminder. In line with our general assumption, let us assume that:

- we know the values $x^{(k)}$ exactly, and
- we know the corresponding y-valued $y^{(k)}$ with fuzzy uncertainty—i.e., that for each example k and for each possible value y of this quantity, we know our degree of confidence $\mu_k\left(y^{(k)} - y\right)$ that the corresponding value $y^{(k)} - y$ of the measurement error is possible.

In this case, the degree to which a model $y = f(x, c_1, \ldots, c_n)$ is consistent with the k-th observation is equal to $\mu_k\left(y^{(k)} - f\left(x^{(k)}, c_1, \ldots, c_n\right)\right)$. By applying an appropriate "and"-operation, we can conclude that the degree to which a model is consistent with all K observations is equal to

$$f_\&\left(\mu_1\left(y^{(1)} - f\left(x^{(1)}, c\right)\right), \ldots, \mu_K\left(y^{(K)} - f\left(x^{(K)}, c\right)\right)\right). \tag{22}$$

A natural idea is to select the values $c = (c_1, \ldots, c_n)$ for which this degree is the largest possible—and if we want to find the range of possible values of c, to select the range of all the values for which the degree (22) is larger than or equal to a certain threshold μ_0.

What if we have negative examples? In this case, instead of considering all possible combinations c, we need to consider only combinations which are consistent with all $L - K$ negative examples, i.e., combinations that satisfy the property (4) for all $\ell = K + 1, \ldots, L$.

7 Conclusions

What we did. In this paper, we provided a theoretical foundation for using negative examples in data processing, and we showed, on simplified toy examples, that the resulting algorithms indeed lead to more accurate models.

What still needs to be done. Now that the theoretical foundation has been formulated, we hope that the resulting algorithms and ideas will be applied to real-life problems.

Acknowledgements This work was supported in part by the National Science Foundation grants 1623190 (A Model of Change for Preparing a New Generation for Professional Practice in Computer Science) and HRD-1242122 (Cyber-ShARE Center of Excellence). The authors are thankful to Boris Kovalerchuk for his encouragement, and to the anonymous referees for valuable suggestions.

References

1. Belohlavek, R., Dauben, J.W., Klir, G.J.: Fuzzy Logic and Mathematics: A Historical Perspective. Oxford University Press, New York (2017)
2. J. Contreras, F. Zapata, O. Kosheleva, V. Kreinovich, and M. Ceberio, "Let us use negative examples in regression-type problems too", *Proceedings of the 24th International Conference on Information Visualisation IV'2020*, Vienna and Melbourne, September 7–11, 2020, pp. 296–300
3. Cormen, T.H., Leiserson, C.E., Rivest, R.L., Stein, C.: Introduction to Algorithms. MIT Press, Cambridge, MA (2009)
4. Jaulin, L., Kiefer, M., Dicrit, O., Walter, E.: Applied Interval Analysis. Springer, London (2001)
5. Klir, G., Yuan, B.: Fuzzy Sets and Fuzzy Logic. Prentice Hall, Upper Saddle River, New Jersey (1995)
6. Kovalerchuk, B.: Relationship between probability and possibility theiories. In: Kreinovich, V. (ed.) Uncertainty Modeling, pp. 97–122. Springer Verlag, Cham, Switzerland (2017)
7. Kreinovich, V., Lakeyev, A., Rohn, J., Kahl, P.: Computational Complexity and Feasibility of Data Processing and Interval Computations. Kluwer, Dordrecht (1998)
8. Luenberger, D.G., Ye, Y.: Linear and Nonlinear Programming. Springer, Cham, Switzerland (2016)
9. Mayer, G.: Interval Analysis and Automatic Result Verification. de Gruyter, Berlin (2017)
10. Mendel, J.M.: Uncertain Rule-Based Fuzzy Systems: Introduction and New Directions. Springer, Cham, Switzerland (2017)
11. Moore, R.E., Kearfott, R.B., Cloud, M.J.: Introduction to Interval Analysis. SIAM, Philadelphia (2009)
12. Nguyen, H.T., Kreinovich, V.: Nested intervals and sets: concepts, relations to fuzzy sets, and applications. In: Kearfott, R.B., Kreinovich, V. (eds.) Applications of Interval Computations, pp. 245–290. Kluwer, Dordrecht (1996)
13. Nguyen, H.T., Walker, C., Walker, E.A.: A First Course in Fuzzy Logic. Chapman and Hall/CRC, Boca Raton, Florida (2019)
14. Novák, V., Perfilieva, I., Močkoř, J.: Mathematical Principles of Fuzzy Logic. Kluwer, Boston, Dordrecht (1999)
15. Papadimitriou, C.: Computational Complexity. Addison-Wesley, Reading, Massachusetts (1994)
16. Rabinovich, S.G.: Measurement Errors and Uncertainties: Theory and Practice. Springer, New York (2005)
17. Zadeh, L.A.: Fuzzy sets. Information and Control **8**, 338–353 (1965)

Visualizing and Explaining Language Models

Adrian M. P. Braşoveanu and Răzvan Andonie

Abstract During the last decade, Natural Language Processing has become, after Computer Vision, the second field of Artificial Intelligence that was massively changed by the advent of Deep Learning. Regardless of the architecture, the language models of the day need to be able to process or generate text, as well as predict missing words, sentences or relations depending on the task. Due to their black-box nature, such models are difficult to interpret and explain to third parties. Visualization is often the bridge that language model designers use to explain their work, as the coloring of the salient words and phrases, clustering or neuron activations can be used to quickly understand the underlying models. This paper showcases the techniques used in some of the most popular Deep Learning for NLP visualizations, with a special focus on interpretability and explainability.

1 Introduction

Deep Learning (DL) models applied on texts need to cover the morphological, syntactic, semantic and pragmatic layers. Crafting networks that operate on so many levels is a challenging task due to the sparseness of the training data. Such networks have been traditionally called Language Models (LMs) [63]. The early iterations were based on statistical models [63], whereas the latest iterations use neural networks and embeddings. Current LMs are trained on large corpora and are generally sparse. Most current popular LMs are based on Transformer architectures [6].

A. M. P. Braşoveanu (✉)
Modul Technology GmbH, Am Kahlenberg 1, 1190 Vienna, Austria
e-mail: adrian.brasoveanu@modul.ac.at

A. M. P. Braşoveanu · R. Andonie
Transylvania University of Brasov, Bulevardul Eroilor 29, Braşov 500036, Romania

R. Andonie
Central Washington University, 400 E University Way, Ellensburg, WA 98926, USA
e-mail: Razvan.Andonie@cwu.edu

B. Kovalerchuk et al. (eds.), *Integrating Artificial Intelligence and Visualization for Visual Knowledge Discovery*, Studies in Computational Intelligence 1014,
https://doi.org/10.1007/978-3-030-93119-3_8

The first implementation of a Transformer network [85] proved that it was possible to design networks that achieve good results for Natural Language Processing (NLP) tasks with a set of multiple sequential attention layers. A Transformer contains a series of self-attention layers that are distributed through its various components. Self-attention is an attention mechanism that computes a representation of a sequence from a set of different positions of the same sequence. The Transformer model itself is simple and consists from pairs of encoders and decoders. Encoders encapsulate layers of self-attention coupled with feed-forward layers, whereas decoders encapsulate self-attention layers followed by encoder-decoder attention and feed-forward layers. The attention computation is done in parallel and the results are then combined. The result is termed a multi-head attention, and it provides the model with the ability to orchestrate information from different representation subspaces (e.g., multiple weight matrices) at various positions (e.g., different words in a sentence) [85]. Its outputs are fed either to other encoders or into decoders, depending on the architecture. There is no fixed number of encoders and decoders which can be included in this architecture, but they will typically be paired (e.g., 10 encoders and 10 decoders). In newer architectures, encoders and decoders can also be used for different tasks (e.g., encoder for Question Answering, and decoder for Text Comprehension) [65]. While the model was initially developed for machine translation tasks, it has been tested on multiple domains and was demonstrated to work well.

During the last three years, hundreds of papers and LMs inspired by Transformers were published, the best-known being BERT [14], RoBERTa [51], AlBERT [47], XLNet [103], DistilBERT [69], and Reformer [43]. Some of the most popular Transformer models are included in the Transformers library, maintained by HuggingFace [101].

Many of these models are complex and include significant architectural improvements compared to the early Transformer and BERT models. Explaining their information processing flow and results is therefore difficult, and a convenient and very actual approach is visualization. Our survey is focused on visualization techniques used to explain LMs. We investigate two large tool classes: (i) model-agnostic tools that can be used to explain BERT predictions; and (ii) custom visualizations that are focused only on explaining the inner workings of LMs based on neural networks. An early version of this survey was published a year ago, but it was focused only on visualizing Transformer networks [5]. We have since extended the material to include new articles about Transformer visualizations, other types of networks, as well as an extended section about model-agnostics AI libraries that are focused on interpretability and explainability for NLP.

In this survey, we look at the visualization of several types of LMs based on DL networks, review the basic charts and patterns present in them and try to understand the basic methodology that was used to produce these visual representations. The rest of the paper is organized as follows: Section 2 presents the motivation and methodology of this survey. Section 3 showcases the two classes of tools, whereas Section 4 discusses the various findings. The paper concludes with some thoughts on the future of this class of visualizations.

2 Background and Methodology

The need to quickly update NLP models in case of unforeseen events suggests that developers will be well-served by explainable AI and visualization libraries, especially since debugging Transformers is a complex task. Visualizations are particularly important, as they help us debug the various problems that such models exhibit and which can only be discovered through large-scale analyses.

Traditional visualization libraries are based on the classic grammar of graphics philosophy [100] which is focused on the idea that visualizations are compositional by design. They provide various *visualization primitives* like circles or squares and *a set of operations* that can be applied on top of these primitives to create more complex shapes or animations. Unfortunately, such traditional visualization libraries like D3.js [3], Vega [70] or Tableau,[1] do not offer specific functions for visualizing feature spaces, neural network layers or support for iterative design space exploration [61] when designing AI models. What this means is that for AI tasks, a lot of the functionality will have to be developed from scratch.

When visualizing more complex models like those built with Transformers, we typically need to understand all the facets of the problem, from the data and training procedure, to the input, network layers, weights or the outputs of the neural network. The outputs are the core of explainability, as people will not use the networks in commercial products if they can't explain how the outputs were obtained in the first place. What is also beneficial is to highlight the paths that lead to certain outputs, as this illuminates the features or parts of the models that may need to be changed to achieve the desired results. This can sometimes be accomplished by using model-agnostic tools specifically built for benchmarking or hyperparameter scoring, such as *Weights and Biases*. We include such tools in our survey if examples of how to use them for visualizing Transformers already exist, either in scientific papers or other types of media posts (Medium posts, GitHub, etc.).

The second big class of visualizations discussed in this article is, naturally, the class of visualizations specifically built around Transformers, either for explaining it (like ExBERT [35]), or for explaining certain model specific attributes (like embeddings or attention maps [86]).

We selected the libraries and visualizations presented here by reviewing the standard Computer Science (CS) publication libraries (e.g., IEEE, ACM, Elsevier, Springer, Wiley), but also online media posts (YouTube, Medium, GitHub and arXiv). In this extremely dynamic research field, some articles might be published on arXiv even up to a year before they are accepted for publication in a traditional conference or journal, time in which they might already garner hundreds of citations. The original BERT article [14] and also one of the first articles that used visualization to explain it [11] were cited over a hundred times before being published in conference proceedings.[2]

[1] www.tableau.com.

[2] Article [11] has garnered 149 citations at the moment of the submission, before being published in a conference or journal.

When testing new models, benchmarking and fine-tuning are the two operations where we might spend the most time, as even if the scores are good, we might want to try different hyperparameter settings (e.g., learning rate, number of epochs, batch size, etc) [22]. A *hyperparameter sweep* (or trial) is a central notion in both hyperparameter optimization and benchmarking. It involves running one or multiple models with different values for their hyperparameters. Since quite often the main goal behind running such sweeps is improving existing models, but it is not necessarily related to interpretability and explainability, we decided to include a minimal number of such libraries here.

TensorBoard[3] is a specialized dashboard deployed with Google TensorFlow that covers the basic visualization needs for ML experiments, from tracking, computing and visualizing metrics, to model profiling and embeddings. It is not necessarily a good tool for creating custom visualization or for explaining results, but it can be a good tool for improving accuracy. It is sometimes also used with TensorFlow's competing libraries like PyTorch or FastAI. *Neptune*[4] is an open-source ML benchmarking suite deployed for a variety of collaborative benchmarking tasks, including notebook tracking. *Sacred*[5] and *Comet.ml*[6] are Neptune alternatives that provide basic charting capabilities and dedicated dashboards. *Weights and Biases*[7] provides perhaps the largest sets of visualization and customization capabilities. It comes packed with advanced visualizations that include parallel coordinates [32], perhaps the best method to navigate hyperparameter sweeps. It is the easiest and the most agile solution to integrate with production code or Jupyter notebooks out of all the ones mentioned here. *Ray* [58], a distributed benchmarking framework that contains its own fine-tuning engine called *Tune* [50] is popular for optimizing Transformers.

Due to space limitations, we resume ourselves to discussing only the most interesting visualizations, especially in the model-agnostic visualization section, as otherwise this article could easily become an entire book.

3 Visualizing Language Models

Language models are difficult to train for a multitude of reasons, including (but not limited to) cost, time or carbon footprint [78]. Most of the LMs need to be trained on GPUs or TPU pods for days or weeks. Due to their generalization capabilities, such LMs can reliably estimate the actions for which they have a reasonable number of examples in their training datasets, whereas in cases with fewer examples they might overestimate the predicted actions therefore inserting some bias [72]. Debugging or retraining such models therefore becomes a necessity, even if the costs of such oper-

[3] https://www.tensorflow.org/tensorboard.

[4] https://neptune.ai/.

[5] https://github.com/IDSIA/sacred.

[6] https://www.comet.ml/site/.

[7] https://www.wandb.com/.

ations are still high. Visualization is just one of the methods that can help us explain such large LMs, especially since it is often combined with linguistics or statistics. Explaining the results in plain English should be what we are aiming for when we build new LMs, but this may sometimes require additional steps. An interpretation of the results, for example, would generally depend on the target domain (e.g., medicine, law, etc) [84], as in some cases a complex reasoning process (e.g., compliance with local or international regulations) may need to be applied before selecting the right words for an explanation. The visualization will essentially highlight the intermediary steps (e.g., the components that lead to a side effect for medication or a legal aspect in a certain jurisdiction) required to create a basic interpretation, and therefore it is often kept minimal and visualized features or processes are carefully selected. If it is easy to navigate the various information pathways and understand the results and their interpretations (e.g., where they may lead us), it may be safe to call the respective visualizations explainable.

Using visualization to explain the AI processes is an expanding research field. The main idea behind AI user interfaces should be to augment and expand user capabilities, rather than replace intelligence [31]. While not necessarily needed to understand the next section, several recent surveys about visualizations and DL can help provide additional context to the interested readers. We particularly recommend the following: the introduction on how Convolution Neural Networks "see" the world from [64], the discussion on visual interpretability from [106], and the discussion on the importance of visualizing features from [60].

3.1 Model-Agnostic Explainable AI Tools

Explainable[8] AI (XAI) is the key to enterprise adoption of the current wave of AI technologies, from vision to NLP and symbolic computation. An early XAI survey [34] describes methods through which visualizations can be turned into explanations for the AI models and goes on to define the terminology of the field. Early XAI libraries focused on visualizing ML features, whereas recent libraries are focused on visualizing embeddings, attention maps or various neural network layers [68].

Traditionally, the first step towards transparency was to describe the contribution of each feature to the final result [27]. This often lead to partial explanations of the results, as in reality if the models themselves were black-box, knowing the name of the features was not in itself enough. Output is definitely the most important part that we would like to explain, but not the only one. To create full explanations, we need to be able to explain the entire process from its input, to its various transformations (e.g., layers), training process and output. Explanations also need to be able to reflect state changes. For example, when computing Shapley values feature contributions are combined, and then a score that signifies the feature importance within that set of

[8] *Explainable* points to the idea of describing or explaining in an intuitive manner, via charts or tables, the prediction of an algorithm.

features is generated. If features are added or removed from this bundle, the Shapley value for a particular feature will change accordingly. This dynamic nature of the explainability is rarely explained, but it is one of the reason why visualizations in particular are a good fit for creating explanations in the first place.

Some early model-agnostic XAI libraries that were applied to NLP and Transformers visualizations include LIME [67] and SHapley Additive exPlanations (SHAP) [53]. The later was introduced to unify multiple explanation methods into a single model for interpreting predictions. Both SHAP and LIME can be used with classical ML libraries like scikit-learn and XGBoost, as well as with modern DL libraries like PyTorch or Keras/TensorFlow. SHAP provides visualizations for summary and dependency plots. Another XAI alternative to SHAP and LIME, ELI5 [19], is currently routinely used for explaining BERT predictions, and was found to be more secure in case of adversarial network attacks [74].

The visualizations created with LIME and SHAP are typically restrained to classic charts (e.g., line, bars or word clouds. The summary plots or interaction charts [53] from SHAP are relatively easy to understand, whereas the more complex force plot charts like feature impact [52] are not necessarily easy to use as they require a certain learning time. While the feature impact chart simply plots the expected feature impact with red (features with positive contribution to the prediction result) or blue (features with a likely negative contribution to the result) colors and should in theory be an easy-to-understand chart, there are no direct (e.g., in chart via a legend) explanations on how to interpret the start or end values, or what do the indicators placed on top of various components mean in some cases. The interpretation of such force plot charts is generally missing and people need to read additional documentation to understand the results. This is far from ideal, as, in our opinion, visualizations need to be self-explanatory.

It can be argued that explainability libraries like SHAP or LIME tend to focus on highlighting correlations or statistical effects rather than features, and are, therefore, less reliable than interpretable models which only showcase a list of features or algorithms that contributed to the results. We consider auditing to encompass both sides of the problem, as interpretability and explainability are the key towards understanding and clearing such LMs for deployment in real-world products. Keeping this in mind, we think that focusing on neural network visualization for NLP can only help in this process, as visualizations can help process, select and highlight the most important features included in such models.

Both SHAP and LIME were proven to be easily fooled with adversarial attacks [74]. The idea of deploying biased classifiers for tasks like credit rating, recommendation or search ranking sounds a bit counterintuitive because a single classifier should not be able to do much harm. However, given the fact that such large models typically end up being ensembles, one single classifier can actually lead to severe damage including wrong predictions, different sets of biases and ultimately even different outputs than the ones typically expected from the respective LMs. Not being able to correctly audit such models (e.g., investigate their output and the features that have contributed the most to it) can lead to problems with clients and regulatory agencies.

The list of attacks that can be perpetrated using LMs is extended every month, and therefore models may need to be periodically tested to assess their suitability for certain tasks. Some of the most robust attacks include: creating token sequences that act like universal adversarial triggers on specific target predictions when concatenated to any input from a dataset [96]; training data extraction attacks in which text sequences like public information or code are extracted from the training corpora and used to attack the trained language model [8]; spelling attacks in which random spellings for well-known words are generated through modifying the gradients during training [80]; hotflip attacks in which a gradient-based embeddings swap is performed to change classification results [16] or even textfooler attacks [37] in which multiple attributes like embeddings, part-of-speech matches or cosine similarities are used to perform a counter-fitted embeddings swap to fool untargeted classifiers and entailment relations. Possibilities to reuse part of the code of these attacks to create new attacks also exists if frameworks like TextAttack [59] or OpenAttack [105] are used. Sometimes these kinds of adversarial attacks can also be used to improve results, as demonstrated by improving aspect-based sentiment analysis tasks by creating artificial sentences [39] or by using various BERT attribution patterns (e.g., pruned self-attention heads from a certain task) as adversaries [30]. To understand these attacks, visualization can be a useful tool. For example, Chen et al. [10] showcases three types of attacks perpetrated at character, word and sentence level using visualizations built around various metrics like accuracy and successful attack ratios. The Interpret framework [95] uses saliency maps color highlights to showcase defenses against hotflip and untargeted classification attacks. A later publication then shows how almost all interpretations built with Interpret can be manipulated through gradient attacks (e.g., using some large gradients for irrelevant words) [98]. However, most of the visualization efforts have been focused on explaining the various neural network activations from Transformer networks, rather than on the various attacks that can fool Transformers, therefore the visualization of such attacks is a relatively nascent area.

An alternative approach to such explainability libraries that can easily pick up wrong signals could be to simply use models designed specifically to be interpretable. The Neural Additive Models (NAMs) combines the features of classic DNNs with the interpretability approach advocated by Generalized Additive Models (GAM) [2]. However, since such models are quite recent, their applicability to NLP has not yet been fully explored.

Interpretability and explainability are often used with interchangeable meaning. It has to be noted that quite often, interpretability has a domain-specific component [84], whereas explainability is a more general term. Explainability is the term preferred by Information Visualization designers and researchers, whereas interpretability is generally the term that is preferred by ML researchers, statisticians and mathematicians.

Many other explainable AI libraries use Shapley values for computing feature importance. However, in many cases we were only barely able to discover mentions of their usage for NLP (e.g., DeepExplain[9]), and therefore we decided not to include them in this survey.

3.2 *Visualizing Recurrent Neural Networks for NLP*

When visualizing LMs, it is best to start with the language resources used for their creation, from corpora to embeddings. To uncover biases in such large models we need to study gender differences, disciplines, languages, cultural context or regional and diachronic variations. A good method to include such information and compare resources for language variation is showcased in Fanhauser's work [20]. It uses a grids, heatmaps and word clouds to provide quick access to large amounts of data about English dialects. Another work uses scatter plots to visualize how and why large corpora differ [41]. Similar methods have later been used for visualizing large-scale embeddings.

Karpathy et al. [40] proposed a method through which to visually interpret the results of Recurrent Neural Networks (RNNs), Long Short-Term Meories (LSTMs) and Gated Recurrent Units (GRUs). He suggests that using interpretable activations can help navigate longer texts, whereas saturation plots would help showcase the gated units statistics.

Around the same time, an LM visualization survey [48] notes that classic Computer Vision visualization was focused on inverting representations, back-propagation and generating images from sketches, all techniques that work well for images. In NLP, however, it is important to focus on important keywords, composition and dimensional locality; especially since many of the words will depend on the context [48]. Important models will be able to capture this kind of information, and therefore it should be present in visualizations. The early saliency heatmaps clearly showcased these aspects for LSTM and Bi-Directional LSTMs [77]. By later adapting the ideas of first order saliency from Computer Vision, researchers were able to highlight intensification and negation, as well as differences between two sequences at various at consecutive time steps. Their saliency heatmap for SEQ2SEQ auto-encoders also works well for predicting corresponding tokens at each time step [54].

The main characteristic that connects these papers, as it can easily be observed in Table 1, regardless of the number of visualizations included in them, is the fact that they are focused on a interpretability and explainability of NLP models through visualization. We have analyzed the following characteristics:

- *Topic*—the main topic of the paper (e.g., attention, representation, information probing) followed by the papers in which this topic is addressed;

[9] https://github.com/marcoancona/DeepExplain.

- *Visualization Subject*—since visualizations included in these papers were focused on a large set of subjects from Transformer components (e.g., attention heads), to correlation between tasks (e.g., via Pearson correlation charts) or performance (e.g., accuracy or other metrics represented via line charts), we have decided to extract all these in a separate column to understand what kind of charts we might be interested in creating when exploring a certain topic.
- *Chart Type*—includes the various types of visual metaphors used for rendering the chosen subjects. Most of the chart types are classic (e.g., line, bar chart, t-SNE), very few being rebranded (e.g., attention maps are heatmaps) or actually new (e.g.,

Table 1 Articles focused on explaining RNN LMs through visualizations

Topic	Visualization subject	Chart type
(a) Special topics		
Predicting next word [56]	Syntactic heights head attentions	Parallel lines heatmaps
Emergence of units [48]	Activations of cells and gates connectivity	Line charts connectivity charts
(b) Hidden states		
Visualizations of representations [50]	Modification and negation clause composition first-derivative saliency	t-SNE heatmap saliency bars saliency grids
Hidden states semantics [73]	Predictive Semantic Encodings performance metrics	PSE charts Bar charts
Activation of RNNs [42]	Cells with interpretable activation	Saturation plots bar charts overlap charts
Hidden states LSTMVis [79]	Phrase selections hidden state patterns	Hidden pathways tables with cell activation PCA
Hidden states RNNVis [59]	Navigation sentence hidden state word	Control panel glyph-based chart state clusters word clusters
Hidden states ActiVis [40]	Navigation neuron activation instance selection	Model overview heatmaps
(c) Graph convolution		
Inductive text classification [107]	Attention performance metrics	Attention map line charts
Unsupervised domain adaptation [102]	Embeddings performance metrics	PCA line charts
GCN with label propagation [97]	Node embeddings performance	Graphs line charts

comparative attention graphs). The chart names need to give us a clear idea of what they represent.

A large class of visualizations is dedicated to the activation of neurons and the representation of hidden states, as it can easily be seen from Table 1.

LSTMVis [77] uses a large grid of sentences for matching the various state patterns that are then expanded through additional views in the same interface. The key innovation of visualizing hidden state changes keeps the focus on the right keywords, whereas the match views help enlighten particular cases. Many other visualizations similar to LSTMVis (e.g., RNNVis [57] or ActiVis [38]) follow the same template: a control panel is used for selecting the phrase or sentence, a middle view is focused on the word clusters or neuron activations, whereas the last view is typically a matrix view with highlighted cells which showcases the important words that are featured in the activated pathways. Such integrated views offer us holistic views of what these models can accomplish, except for the fact that they are rarely focused also on the corpora that was used for training. In time, the visualization designers started to include this information as well, as we will observe in the next section.

We considered that Graph Convolutional Networks are also worth exploring, but since there are entire libraries dedicated to this task which are not necessarily model-agnostic (e.g., PyTorch Geometric [21]) we have limited ourselves to including some papers that offer some classic visualizations that are typically included in such libraries (e.g., node embeddings, performance).

3.3 *Visualizing Transformers for NLP*

No other types of neural networks have led to such an increased demand for custom visualizations since the days of the Kohonen's Self-organizing Maps [45] or Manbelbrot's fractals [55] like the Transformers. When selecting Transformer visualization papers, we decided to focus on the most important topics related to interpretability and explanation. We have therefore eliminated papers that used only classic charts (e.g., bars, lines, pies). We decided to focus on the works that tried to visualize as many aspects of Transformer models as possible, from attention maps, to structural or informational probing, neural network layers, and multilingualism.

Many Transformer visualizations are focused on attention. While the attention mechanism is indeed important for the Transformer architecture, and it improves results for NLP tasks, they are not necessarily easy to interpret if sophisticated encoders are used [36]. Due to this fact, it is often not easy to test various explanations by simply modifying the weights and verifying if the outputs are also changed as a result of this. Alternative theories suggest that simply looking at information flows through such models is not enough, and that attention should only be used as explanation if certain conditions are met, e.g., if the weight distributions found via adversarial training do not perform well [99]. Since attention is central to Transformer models, many visualizations are rightly focused on this topic. The fact that such visualizations capture the dynamic nature of the output is not necessarily sufficient to consider them explanations. A good explanation needs to highlight the reasoning chain that lead to the particular output. This is the main reason why we have mainly looked for those visualizations that focus on multiple aspects of the network in this work.

The recent success of Transformers helped power many NLP tasks to the top of the leaderboards. BERT visualizations have focused on explaining these great results through visualizations, therefore highlighting: (i) the role of embeddings and relational dependencies within the Transformer learning processes [66]; (ii) the role of attention during pre-training or training (e.g., [79] or [86]) or (iii) the importance of various linguistic phenomena encoded in its language model like direct objects, noun modifiers, or possessive pronouns [11].

Current XAI methods for Transformer models have further developed and supported the idea that understanding the linguistic information which is encoded in the resulting models is key towards understanding the good performances in NLP tasks. Probing tasks [12] are simple classification problems focused on linguistic features designed to help explore embeddings and LMs. For example, by using structural probing [33], structured perceptron parsers [56]) or visualization (e.g., as demonstrated through BERT embeddings and attention layers visualizations like those from [11, 86]), one should be able to understand what kind of linguistic information is encoded into a Transformer model, but also what has changed since previous runs.

We have discovered two large classes of Transformer visualizations:

- *Focused*—visualizations centered on a single subject like attention. The papers themselves might present multiple visualizations, but these visualizations are not single tools.
- *Holistic*—visualizations or systems which seek to explain the entire Transformer model or lifecycle.

3.3.1 Focused Transformer Visualizations

The most important papers dedicated to focused visualizations are summarized in Table 2. We used the same conventions in this table like the ones applied in Table 1.

We can clearly distinguish several large topics in this group of focused papers: the relation between attention and model outputs (e.g., especially in [1, 36, 93, 99]), the analysis of captured linguistic information via probing (e.g., in [81, 90]), the interpretation of information interaction (e.g., in [30, 90]), and multilingualism (e.g., in [81] or [17]). In fact, if we look close to Table 1 we can distinguish 3 large classes of subjects: (a) attention; (b) hidden states and (c) structural or information probing. Papers that work on similar topics also tend to use the same kind of visual metaphors. This sometimes happens due to replication of a previous study (e.g., [99] replicates the experiments from [36] to prove that attention weights do not explain everything), whereas in other cases this happens because there is no need for more complicated visual metaphors (e.g., line charts are used in more than half of the papers to represent

Table 2 Articles focused on explaining transformer topics through visualizations

Topic	Visualization subject	Chart type
(a) Attention		
Attention explanation [38, 99]	Feature importance correlation permutation adversarial attention performance metrics	Kendall rank statistics scatterplots adversarial charts multiple line charts
Attention flow [1, 14]	Raw attention raw attention map comparative attention flows	Attention graph heatmap comparative attention graphs
Multi-head self-attention [93]	Layers attention for rare words dependency scores active heads	Importance charts heatmaps bar charts line charts
(b) Hidden states		
Intermediate layers [78]	Intermediate layers clustering	PCA
Information interactions interpretation [31]	Scoring attention information flow for tokens evaluation accuracy attention heads correlation	Heatmap attribution graphs line charts Pearson correlation charts
Causal mediation analysis [89]	Indirect effects effects comparison attention	Heatmaps line chart attention heads
Evolution of representations [90]	Token changes and influences distances between layers token occurences	Line charts line charts t-SNE clustering
(c) Probing		
Structural probes [83]	Summary statistics layer-wise performance predictions probing	Bar chart bar distribution chart multiple bar charts
Multilingual probes [16, 18]	Probing task positional embeddings performance metrics stability of training size	PCA cosine similarity matrices line charts bar charts
Information theoretic probing of classifiers [92]	Coding components performance metrics	Bar charts line charts
Psycholinguistics tests [26]	Comparing predictions performance metrics test content	Tables distribution charts line charts

performance). Besides the widespread use of the heatmaps that represent attention maps, one chart type that deserves to be highlighted in this category is the attention graph [30] which tracks the information flow between the input tokens for a given prediction.

One of the key methods used for explaining the output of LMs is called probing [18], used to analyze the linguistic information encoded in a fixed length vector.

Recent iterations like structural probes [33] evaluate if syntax trees are embedded in a network's word representation space. Identifying linear transformations is evidence for entire syntax trees embedded in the LM's vector geometry. This method works well for limited cases in which distances between words are known. The critics of probing argue that differences in probing accuracy between the various classifiers essentially render them unusable as they fail to distinguish between different representations, e.g., two LMs can end up having different linguistic representations even if based on the same initial BERT model. In such scenarios, one cannot compare the accuracy of the classifiers used to predict their labels. To counteract for such cases, a recent information-theoretic probing with minimum description length method was proposed [91]. The basic idea is that instead of predicting labels, the probe transmits data (a description) which is then evaluated based on the returned description's length. Such probes can be implemented on top of the classic structural probes, and are fairly stable. According to Pilault, classic structural probes may not be enough for complex tasks like summarization simply because it seems that increasing the number of random encoders provide significant performance boosts [62]. This suggests that the information-theoretic probes with minimum description length may be the better probes for this task, however this remains to be demonstrated by future experiments, as Voita's model was published after Pilault's article.

3.3.2 Holistic Transformer Visualizations

While in Sect. 3.2 several groups tried to expand their visualizations of hidden states to encompass the entirety of the model, they have rarely included the corpora provenance or an easy method to navigate it. Due to this aspect, in the case of Transformers, since such visualizations often included the whole lifecycle (e.g., including the corpora with known provenance), we decided to present them in a separate subsection.

Some of the most interesting tools or papers included in the category of *holistic visualizations* are compared in Table 3. These visualization systems typically integrate most of the components of a Transformer and provide detailed summaries of them. We have examined two large classes of attributes:

- *Components* represents the various components of the neural networks: from corpus, to embeddings, positional heads, attention maps or outputs.
- *Summary* includes the various types of views that offer us information about the state of a neuron or a layer, as well as overviews, statistics or details about the various errors encountered. Statistics might include different types of information: from correlations between *layers* or *neurons* to *statistical analyses of the results*. The *errors* column represents any error analysis method through which we can highlight where a particular error comes from (e.g., corpus, training procedure, layer, etc). While it can be argued that neuron or layer views should be included in the components section, the way these views are currently implemented suggests they are rather summaries, as neurons or layers can have different states.

Table 3 Comparison of holistic Transformer visualizations. Legend: Corpus (C); Embeddings (E); Heads (H); Attention Maps (A); outputs (out); errors (err); neuron (neu); layers (lay); overview (ove); statistics(sta)

Article	Components				Summary					
	C	E	H	A	Out	Err	Neu	Lay	Ove	Sta
BertViz [87]		✓	✓	✓			✓	✓	✓	✓
Clark [11]		✓	✓	✓		✓	✓	✓	✓	✓
VisBERT [82]		✓	✓	✓			✓	✓	✓	✓
ExBERT [35]	✓	✓	✓	✓	✓			✓	✓	✓
AttViz [73]	✓	✓	✓	✓	✓			✓	✓	✓
Vector norms [44]		✓	✓	✓				✓	✓	✓
Dictionary learning [104]		✓	✓	✓				✓	✓	✓

We have decided against including chart types in Table 3, as each visualization suite or paper included some novel visualization types besides attention maps (heatmaps), parallel coordinates or line and bar charts.

In our view, none of the examined visualization systems has yet managed to examine all the facets of the Transformers. This is perhaps because this area is relatively new and there is no consensus on what needs to be visualized. While it is quite obvious that individual neurons or attention maps (regardless of if they are averaged or not) are useful, and it is best to visualize them, the same cannot be said about the training corpora today, as only a few systems considered this aspect (e.g., [35, 73]). This is not really ideal, as lots of errors might simply come from a bad corpora, but researchers might simply not be aware of them [4]. Errors themselves are only seriously discussed in a single publication [11]. ExBERT [35] and AttViz [73] deserve a special mention here, as they combine different views on the corpus, embeddings and attention maps to provide a holistic image of a Transformer model.

A study that looks at the similarity and stability of neural representations in LMs and brains [83] shows that combining predictive modelling with Representation Similarity Analysis (RSA) techniques can yield promising results. This article deserves a special mention as it can be included in both focused and holistic visualizations. Their visualizations are rather basic in terms of design, but they contain lots of insights, for example one of the tables they produced showcases the RSA results for various layers of multiple models like BERT, Elmo and others. These kinds of analyses are rather new, and we hope they will become more common in the next years, as they might help us clarify which LMs are more similar to the human brains.

Table 4 Articles focused on explaining Language and Vision (LAV) models through visualizations

Domain	Topic	Chart type
Language and vision [8, 29]	Modality importance layer-level importance	Attention heatmaps line charts feature maps
Biomedical [51]	Attention maps	Head attention maps
VQA [24, 44]	Multimodal alignment training curves	Visual alignment charts line charts

3.3.3 Vision Transformers

Transformers have been applied in numerous fields due to their flexibility. Perhaps the most important application is the unification of vision and NLP. In the last two years, this has seen explosive growth. It would be impossible to cover all the various models in this article, and therefore we have selected only a few interesting topics related to this expansion. The one that comes to mind first is Visual Question Answering (VQA) [23], as this is perhaps the task that multimodal researchers have been trying to solve for decades. Vision Transformers offer an elegant, but expensive solution. Table 4 provides a short summary of interesting papers in this research direction.

4 Discussion

Multilingual models based on Transformer architectures are error-prone since they are trained on large collections of texts. Due to this training process, the extraction of semantics from language data often captures implicit biases hidden in languages themselves or contextual biases triggered by adaptation to various niches [108].

Besides the obvious questions of interpretability (e.g., which features are most important for the prediction, what are best-performing models for creating ensembles?) and explainability (e.g., what linguistic information is actually encoded in the model and why do random encoders perform better for summarization?), another important question is security (e.g., are these models stable enough/do they obtain the same result in any circumstance?). Through adversarial training [95] it is possible to remove some procedural biases from model-agnostic explainability libraries like LIME or SHAP, as well as well-known attack vectors (e.g., trigger words that might lead to a different language model output) [74]. It can be argued that probing tasks are really only good for simple tasks like NER or POS and for verifying if certain structures were encoded in a LM, but not for more complex tasks like summarization or word-level polysemy disambiguation [104]. Dictionary activation is a method that uses visualization heavily to peek into the activation of words and their senses. Such a method helps uncover both the various layers on which certain word senses are actually learned, and the contexts that might trigger these new senses.

There are several available options for understanding the inner workings, as well as the results produced by Transformer networks. Each of them have their advantages or shortcomings, briefly discusses in the following.

Model-agnostic tools like the XAI libraries or the hyperparameter optimization and benchmarking tools can be used with a variety of networks. Due to this, model-agnosticism the visualization skills learned while debugging a certain network (e.g., a Convolutional Neural Network) will be easily transferred to debugging and optimizing other networks (e.g., Recurrent Neural Networks). By building a transferable set of skills, users might be more reluctant to try model-specific approaches, like those from the second category discussed in this paper. Some of these model-agnostic tools might be more susceptible to various adversarial attacks [75], whereas some other tools might not provide us with sufficiently advanced visualizations to match our needs. If the users are already comfortable with some of these options, then they might well be their Swiss-Army knife for any scenario, whereas if they will need specific visualization scenarios (e.g., visualize a specific attention map), it is possible that they will eventually use the Transformer focused visualizations.

While visualization of RNNs started quite early in the DL era, it can be seen as a precursor to the Transformer visualization. Quite often, many of the topics that were important during this ear have remained important during the Transformer era (e.g., hidden states)

Some of the most useful tools discovered during this exploration include: visualization of attention maps (e.g., [11]) and embeddings [86], hidden states visualization (e.g., [77]), parallel coordinates plots [11], and the inclusion of corpus views from ExBERT [35].

Current generation of pre-trained LMs based on Transformers [101] was shown to be relatively good at picking up syntactic cues like noun modifiers, possessive pronouns, prepositions or co-referents [11] and semantic cues like entities and relations [29], but has not performed well at capturing different perspectives [9], global context [94] or relation extraction [24]. This may be because biases can be already included in the embeddings and later propagated to the downstream tasks [26].

The MIT Computation Psycholinguistics Laboratory created two useful tools for exploring LMs: the LM Zoo and the SyntaxGym [25]. The first allows users to install some classic LMs, whereas the SyntaxGym allows users to run pyscholinguistics tests and generate useful visualizations based on their results.

The two large classes of Transformer visualizations we examined (focused on explaining Transformer topics or holistic) are proof that the field is extremely dynamic. While many of the articles focused on explaining Transformer topics like attention or information probing tend to use classic statistical chart types (e.g., bar charts, line charts, PCA, or Pearson correlation charts), we do not consider this a bad thing as we are still in the exploration phase of this technology. Some of these articles also showcase new charts like attention graphs or attention maps.

The second class of visualizations includes tools like BertViz [86], AttViz [73], VisBERT [82] or ExBERT [35], that aim to visualize the entire lifecycle of a neural network from corpora and inputs to the model outputs mainly through following the information flow through the various components. They also offer detailed statistics for neurons or network layers. Since most of the models included in this category are rather new, it is expected that this class will expand in the next years.

One important thing to note about visualization methods is that they can easily be imported into other domains. The averaged attention heatmaps used by Vig in his causal mediation analysis for NLP [89], for example, were later reused for protein analysis in biology [88]. Similarly, attention maps [86] developed for BERT models are now used in a wide variety of disciplines, from vision and speech to biology or genetics.

The end goal of future visualization frameworks should be to visualize the entire lifecycle of the Transformer models, from inputs and data sources (e.g., training corpora), to embeddings or attention maps, and finally outputs. In the end, errors observed when creating such models can come from a variety of sources: from the text corpora, from some random network layer or even from some external Knowledge Graph that might feed some data into the model. Tracking such errors would be costly without visualizations.

5 Conclusion

While the current visualizations aspire to be model-agnostic, we think the directions opened by the various Transformer or RNN visualizations are worthy of expanding upon. In fact, since this is a ubiquitous architecture that has also branched from NLP into areas like semantic video processing, natural language understanding (e.g., speech, translation) and generation (e.g., text generation, music generation), the next generation XAI libraries will probably be built upon it.

Going beyond current visualizations that are model-agnostic, future frameworks will have to provide visualization components that focus on the important Transformer components like corpora, embeddings, attention heads or additional neural network layers that might be problem-specific. By focusing on the common components from larger architectures, it should be possible to enhance reusability. Other important features that should be included in future frameworks are the ability to summarize the model's state (e.g., through averaged attention heatmaps or similar visualization mechanisms) at various levels (e.g., neurons, layers, inputs and outputs), as well as the possibility to compare multiple settings for one or multiple models. It is important to note that most of the current visualizations seem to be designed to showcase methods and properties that were already known to belong to neural networks. It would be interesting to design visualizations that allow us to explore neural networks, that help us discover new properties. Such interactive exploration tools would significantly expand the role of visualization from communication to knowledge discovery.

One interesting direction is the automated development of model specific visualizations, as more complex neural networks might also include many specific components that cannot always be included into more general model agnostic frameworks.

Acknowledgements The research presented in this paper has been partially conducted within the EPOCH and GENTIO projects funded by the Austrian Federal Ministry for Climate Action, Environment, Energy, Mobility and Technology (BMK) via the ICT of the Future Program (GA No. 867551 and 873992).

References

1. Abnar, S., Zuidema, W.H.: Quantifying attention flow in transformers. In: Jurafsky, D., et al. (eds.) Proceedings of the 58th Annual Meeting of the Association for Computational Linguistics, ACL 2020, Online, July 5–10, 2020, pp. 4190–4197. Association for Computational Linguistics (2020). ISBN: 978-1-952148-25-5
2. Agarwal, R., et al.: Neural additive models: interpretable machine learning with neural nets (2020). CoRR arXiv: abs/2004.13912
3. Bostock, M., Ogievetsky, V., Heer, J.: D3 data-driven documents. IEEE Trans. Vis. Comput. Graph. **17**(12), 2301–2309 (2011). https://doi.org/10.1109/TVCG.2011.185
4. Brasoveanu, A., et al.: Framing named entity linking error types. In: Calzolari, N., et al. (eds.) Proceedings of the Eleventh International Conference on Language Resources and Evaluation, LREC 2018, Miyazaki, Japan, May 7–12, 2018. European Language Resources Association (ELRA) (2018)
5. Braşoveanu, A.M.P., Andonie, R.: Visualizing transformers for NLP: a brief survey. In: 2020 24th International Conference Information Visualisation (IV). IEEE. 2020, pp. 270–279. ISBN: 978-1-7281-9134-8. https://doi.org/10.1109/IV51561.2020
6. Brown, T.B., et al.: Language models are few-shot learners. In: Larochelle, H., et al. (eds.) Advances in Neural Information Processing Systems 33: Annual Conference on Neural Information Processing Systems 2020, NeurIPS 2020, December 6–12, 2020, virtual (2020)
7. Cao, J., et al.: Behind the scene: revealing the secrets of pre-trained vision-and-language models. In: Vedaldi, A., et al. (eds.) Computer Vision—ECCV 2020—16th European Conference, Glasgow, UK, August 23–28, 2020, Proceedings, Part VI. Lecture Notes in Computer Science, vol. 12351, pp. 565–580. Springer (2020). ISBN: 978-3-030-58538-9. https://doi.org/10.1007/978-3-030-58539-6_34
8. Carlini, N., et al.: Extracting training data from large language models (2020). arXiv: 2012.07805
9. Chen, S., et al.: Seeing things from a different angle: discovering diverse perspectives about claims. In: Burstein, J., Doran, C., Solorio, T. (eds.) Proceedings of the 2019 Conference of the North American Chapter of the Association for Computational Linguistics: Human Language Technologies, NAACL-HLT 2019, Minneapolis, MN, USA, June 2–7, 2019, Volume 1: Long and Short Papers, pp. 542–557. Association for Computational Linguistics (2019). ISBN: 978-1-950737-13-0. https://doi.org/10.18653/v1/n19-1053
10. Chen, X., et al.: BadNL: backdoor attacks against NLP models (2020). arXiv: 2006.01043
11. Clark, K., et al.: What does BERT look at? An analysis of BERT's attention (2019). arXiv: 1906.04341
12. Conneau, A., et al.: What you can cram into a single vector: probing sentence embeddings for linguistic properties. In: Gurevych, I., Miyao, Y. (eds.) Proceedings of the 56th Annual

Meeting of the Association for Computational Linguistics, ACL 2018, Melbourne, Australia, July 15–20, 2018, Volume 1: Long Papers, pp. 2126–2136. Association for Computational Linguistics (2018). ISBN: 978-1-948087-32-2. https://doi.org/10.18653/v1/P18-1198
13. DeRose, J.F., Wang, J., Berger, M.: Attention flows: analyzing and comparing attention mechanisms in language models. IEEE Trans. Vis. Comput. Graph. **27**(2), 1160–1170 (2021). https://doi.org/10.1109/TVCG.2020.3028976
14. Devlin, J., et al.: BERT: pre-training of deep bidirectional transformers for language understanding. In: Burstein, J., Doran, C., Solorio, T. (eds.) Proceedings of the 2019 Conference of the North American Chapter of the Association for Computational Linguistics: Human Language Technologies, NAACL-HLT 2019, Minneapolis, MN, USA, June 2–7, 2019, vol. 1 (Long and Short Papers), pp. 4171–4186. Association for Computational Linguistics (2019). ISBN: 978-1-950737-13-0. https://doi.org/10.18653/v1/n19-1423
15. Dufter, P., Schütze, H.: Identifying necessary elements for BERT's multilinguality (2020). arXiv: 2005.00396
16. Ebrahimi, J., et al.: HotFlip: white-box adversarial examples for text classification. In: Gurevych, I., Miyao, Y. (eds.) Proceedings of the 56th Annual Meeting of the Association for Computational Linguistics, ACL 2018, Melbourne, Australia, July 15–20, 2018, Volume 2: Short Papers, pp. 31–36. Association for Computational Linguistics (2018). ISBN: 978-1-948087-34-6. https://doi.org/10.18653/v1/P18-2006
17. Eger, S., Daxenberger, J., Gurevych, I.: How to probe sentence embeddings in low-resource languages: on structural design choices for probing task evaluation. In: Fernández, R., Linzen, T. (eds.) Proceedings of the 24th Conference on Computational Natural Language Learning, CoNLL 2020, Online, November 19–20, 2020, pp. 108–118. Association for Computational Linguistics (2020). https://doi.org/10.18653/v1/2020.conll-1.8
18. Ettinger, A., Elgohary, A., Resnik, P.: Probing for semantic evidence of composition by means of simple classification tasks. In: Proceedings of the 1st Workshop on Evaluating Vector-Space Representations for NLP, RepEval@ACL 2016, Berlin, Germany, August 2016. Association for Computational Linguistics, pp. 134–139 (2016). https://doi.org/10.18653/v1/W16-2524
19. Fan, A., et al.: ELI5: long form question answering. In: Korhonen, A., Traum, D.R., Marquez, L. (eds.) Proceedings of the 57th Conference of the Association for Computational Linguistics, ACL 2019, Florence, Italy, July 28–August 2, 2019, Volume 1: Long Papers, pp. 3558–3567. Association for Computational Linguistics (2019). ISBN: 978-1-950737-48-2. https://doi.org/10.18653/v1/p19-1346
20. Fankhauser, P., Knappen, J., Teich, E.: Exploring and visualizing variation in language resources. In: Calzolari, N., et al. (eds.) Proceedings of the Ninth International Conference on Language Resources and Evaluation, LREC 2014, Reykjavik, Iceland, May 26–31, 2014, pp. 4125–4128. European Language Resources Association (ELRA) (2014)
21. Fey, M., Eric Lenssen, J.: Fast graph representation learning with PyTorch geometric (2019). CoRR arXiv: abs/1903.02428
22. Florea, A., Andonie, R.: Weighted random search for hyper-parameter optimization. Int. J. Comput. Commun. Control **14**(2), 154-169 (2019). https://doi.org/10.15837/ijccc.2019.2.3514
23. Gan, Z., et al.: Large-scale adversarial training for vision-and-language representation learning. In: Larochelle, H., et al. (eds.) Advances in Neural Information Processing Systems 33: Annual Conference on Neural Information Processing Systems 2020, NeurIPS 2020, December 6–12, 2020, virtual (2020)
24. Gao, T., et al.: FewRel 2.0: towards more challenging few-shot relation classification. In: Inui, K., et al. (eds.) Proceedings of the 2019 Conference on Empirical Methods in Natural Language Processing and the 9th International Joint Conference on Natural Language Processing, EMNLP-IJCNLP 2019, Hong Kong, China, November 3–7, 2019, pp. 6249–6254. Association for Computational Linguistics (2019). ISBN: 978-1-950737-90-1. https://doi.org/10.18653/v1/D19-1649
25. Gauthier, J., et al.: SyntaxGym: an online platform for targeted evaluation of language models. In: Çelikyilmaz, A., Wen, T.-H. (eds.) Proceedings of the 58th Annual Meeting of the Association for Computational Linguistics: System Demonstrations, ACL 2020, Online, July 5–10,

2020, pp. 70–76. Association for Computational Linguistics (2020). ISBN: 978-1-952148-04-0. https://doi.org/10.18653/v1/2020.acl-demos.10
26. Gonen, H., Goldberg, Y.: Lipstick on a pig: debiasing methods cover up systematic gender biases in word embeddings but do not remove them. In: Axelrod, A., et al. (eds.) Proceedings of the 2019 Workshop on Widening NLP@ACL 2019, Florence, Italy, July 28, 2019, pp. 60–63. Association for Computational Linguistics (2019). ISBN: 978-1-950737-42-0
27. Guyon, I., Elisseeff, A.: An introduction to variable and feature selection. J. Mach. Learn. Res. **3**, 1157–1182 (2003)
28. Han, K., et al.: Transformer in transformer (2021). CoRR arXiv: 2103.00112
29. Han, X., et al.: OpenNRE: an open and extensible toolkit for neural relation extraction. In: Padó, S., Huang, R. (eds.) Proceedings of the 2019 Conference on Empirical Methods in Natural Language Processing and the 9th International Joint Conference on Natural Language Processing, EMNLP-IJCNLP 2019, Hong Kong, China, November 3–7, 2019—System Demonstrations pp. 169–174. Association for Computational Linguistics (2019). ISBN: 978-1-950737-92-5. https://doi.org/10.18653/v1/D19-3029
30. Hao, Y., et al.: Self-attention attribution: interpreting information interactions inside transformer (2020). CoRR arXiv: 2004.11207
31. Heer, J.: Agency plus automation: Designing artificial intelligence into interactive systems. Proc. Natl. Acad. Sci. USA **116**(6), 1844–1850 (2019). https://doi.org/10.1073/pnas.1807184115
32. Heinrich, J., Weiskopf, D.: Parallel coordinates for multidimensional data visualization: basic concepts. Comput. Sci. Eng. **17**(3), 70–76 (2015). https://doi.org/10.1109/MCSE.2015.55
33. Hewitt, J., Manning, C.D.: A structural probe for finding syntax in word representations. In: Burstein, J., Doran, C., Solorio, T. (eds.) Proceedings of the 2019 Conference of the North American Chapter of the Association for Computational Linguistics: Human Language Technologies, NAACL-HLT 2019, Minneapolis, MN, USA, June 2–7, 2019, Volume 1 (Long and Short Papers), pp. 4129–4138. Association for Computational Linguistics (2019). ISBN: 978-1-950737-13-0. https://doi.org/10.18653/v1/n19-1419
34. Hohman, F., et al.: Visual analytics in deep learning: an interrogative survey for the next frontiers. IEEE Trans. Vis. Comput. Graph. **25**(8), 2674–2693 (2019). https://doi.org/10.1109/TVCG.2018.2843369
35. Hoover, B., Strobelt, H., Gehrmann, S.: ex BERT: a visual analysis tool to explore learned representations in transformer models. In: Celikyilmaz, A., Wen, T.-H. (eds.) Proceedings of the 58th Annual Meeting of the Association for Computational Linguistics: System Demonstrations, ACL 2020, Online, July 5–10, 2020, pp. 187–196. Association for Computational Linguistics (2020). https://doi.org/10.18653/v1/2020.acl-demos.22
36. Jain, S., Wallace, B.C.: Attention is not explanation. In: Burstein, J., Doran, C., Solorio, T. (eds.) Proceedings of the 2019 Conference of the North American Chapter of the Association for Computational Linguistics: Human Language Technologies, NAACL-HLT 2019, Minneapolis, MN, USA, June 2–7, 2019, Volume 1 (Long and Short Papers), pp. 3543–3556. Association for Computational Linguistics (2019). ISBN: 978-1-950737-13-0. https://doi.org/10.18653/v1/n19-1357
37. Jin, D., et al.: Is BERT really robust? A strong baseline for natural language attack on text classification and entailment. In: The Thirty-Fourth AAAI Conference on Artificial Intelligence, AAAI 2020, The Thirty-Second Innovative Applications of Artificial Intelligence Conference, IAAI 2020, The Tenth AAAI Symposium on Educational Advances in Artificial Intelligence, EAAI 2020, New York, NY, USA, February 7–12, 2020, pp. 8018–8025. AAAI Press (2020). ISBN: 978-1-57735-823-7
38. Kahng, M., et al.: ActiVis: visual exploration of industry-scale deep neural network models. IEEE Trans. Vis. Comput. Graph. **24**(1), 88–97 (2018). https://doi.org/10.1109/TVCG.2017.2744718
39. Karimi, A., Rossi, L., Prati, A.: Adversarial training for aspect-based sentiment analysis with BERT. In: 25th International Conference on Pattern Recognition, ICPR 2020, Virtual Event/Milan, Italy, January 10–15, 2021, pp. 8797–8803. IEEE (2020). ISBN: 978-1-7281-8808-9. https://doi.org/10.1109/ICPR48806.2021.9412167

40. Karpathy, A., Johnson, J., Li, F.-F.: Visualizing and understanding recurrent networks (2015). CoRR arXiv: 1506.02078
41. Kessler, J.S.: Scattertext: a browser-based tool for visualizing how corpora differ. In: Bansal, M., Ji, H. (eds.) Proceedings of the 55th Annual Meeting of the Association for Computational Linguistics, ACL 2017, Vancouver, Canada, July 30–August 4, System Demonstrations, pp. 85–90. Association for Computational Linguistics (2017). https://doi.org/10.18653/v1/P17-4015
42. Kim, W., Son, B., Kim, I.: ViLT: vision-and-language transformer without convolution or region supervision. In: Meila, M., Zhang, T. (eds.) Proceedings of the 38th International Conference on Machine Learning, ICML 2021, 18–24 July 2021, Virtual Event, vol. 139, pp. 5583–5594. Proceedings of Machine Learning Research. PMLR (2021). http://proceedings.mlr.press/v139/kim21k.html
43. Kitaev, N., Kaiser, L., Levskaya, A.: Reformer: the efficient transformer. In: 8th International Conference on Learning Representations, ICLR 2020, Addis Ababa, Ethiopia, April 26–30, 2020. OpenReview.net (2020)
44. Kobayashi, G., et al.: Attention module is not only a weight: analyzing transformers with vector norms. (2020). CoRR arXiv: 2004.10102
45. Kohonen, T.: Self-organizing Maps. Springer Series in Information Sciences, vol. 30. Springer (1995). ISBN: 978-3-642-97612-4. https://doi.org/10.1007/978-3-642-97610-0
46. Lakretz, Y., et al.: The emergence of number and syntax units in LSTM language models. In: Burstein, J., Doran, C., Solorio, T. (eds.) Proceedings of the 2019 Conference of the North American Chapter of the Association for Computational Linguistics: Human Language Technologies, NAACL-HLT 2019, Minneapolis, MN, USA, June 2–7, 2019, Volume 1 (Long and Short Papers), pp. 11–20. Association for Computational Linguistics (2019). ISBN: 978-1-950737-13-0. https://doi.org/10.18653/v1/n19-1002
47. Lan, Z., et al.: ALBERT: A Lite BERT for self-supervised learning of language representations. In: 8th International Conference on Learning Representations, ICLR 2020, Addis Ababa, Ethiopia, April 26–30, 2020. OpenReview.net (2020)
48. Li, J., et al.: Visualizing and understanding neural models in NLP. In: Knight, K., Nenkova, A., Rambow, O. (eds.) NAACL HLT 2016, The 2016 Conference of the North American Chapter of the Association for Computational Linguistics: Human Language Technologies, San Diego California, USA, June 12–17, 2016, pp. 681–691. The Association for Computational Linguistics (2016). https://doi.org/10.18653/v1/n16-1082
49. Li, Y., Wang, H., Luo, Y.: A comparison of pre-trained vision-and-language models for multimodal representation learning across medical images and reports. In: Park, T., et al. (eds.) IEEE International Conference on Bioinformatics and Biomedicine, BIBM 2020, Virtual Event, South Korea, December 16–19, 2020, pp. 1999–2004. IEEE (2020). ISBN: 978-1-7281-6215-7. https://doi.org/10.1109/BIBM49941.2020.9313289
50. Liaw, R., et al. (eds.): Tune: a research platform for distributed model selection and training (2018). CoRR arXiv: 1807.05118
51. Liu, Y., et al.: RoBERTa: a Robustly Optimized BERT pretraining approach (2019). CoRR arXiv: 1907.11692
52. Lundberg, S.M., et al.: Explainable machine-learning predictions for the prevention of hypoxaemia during surgery. Nat. Biomed. Eng. **2**(10), 749–760 (2018)
53. Lundberg, S.M., Lee, S.-I.: A unified approach to interpreting model predictions. In: Guyon, I., et al. (eds.) Advances in Neural Information Processing Systems 30: Annual Conference on Neural Information Processing Systems 2017(4–9), December 2017, pp. 4765–4774. Long Beach, CA, USA (2017)
54. Luo, H., et al.: Improving neural language models by segmenting, attending, and predicting the future. In: Korhonen, A., Traum, D.R., Marquez, L. (eds.) Proceedings of the 57th Conference of the Association for Computational Linguistics, ACL 2019, Florence, Italy, July 28–August 2, 2019, Volume 1: Long Papers, pp. 1483–1493. Association for Computational Linguistics (2019). ISBN: 978-1-950737-48-2. https://doi.org/10.18653/v1/p19-1144
55. Mandelbrot, B.: Fractal Geometry of Nature. Freeman, W. H (1977)

56. Maudslay, R.H., et al.: A tale of a probe and a parser (2020). CoRR abs/2005.01641
57. Ming, Y., et al.: Understanding hidden memories of recurrent neural networks. In: Fisher, B.D., Liu, S., Schreck, T. (eds.) 12th IEEE Conference on Visual Analytics Science and Technology, IEEE VAST 2017, Phoenix, AZ, USA, October 3–6, 2017, pp. 13–24. IEEE Computer Society (2017). https://doi.org/10.1109/VAST.2017.8585721
58. Moritz, P., et al.: Ray: a distributed framework for emerging AI applications. In: Arpaci-Dusseau, A.C., Voelker, G. (eds.) 13th USENIX Symposium on Operating Systems Design and Implementation, OSDI 2018, Carlsbad, CA, USA, October 8–10, 2018, pp. 561–577. USENIX Association (2018)
59. Morris, J.X., et al.: TextAttack: a framework for adversarial attacks, data augmentation, and adversarial training in NLP. In: Liu, Q., Schlangen, D. (eds.) Proceedings of the 2020 Conference on Empirical Methods in Natural Language Processing: System Demonstrations, EMNLP 2020—Demos, Online, November 16–20, 2020, pp. 119–126. Association for Computational Linguistics (2020). ISBN: 978-1-952148-62-0. https://doi.org/10.18653/v1/2020.emnlp-demos.16
60. Nguyen, A., Yosinski, J., Clune, J.: Understanding neural networks via feature visualization: a survey. In: Samek, W., et al. (eds.) Explainable AI: Interpreting, Explaining and Visualizing Deep Learning. Lecture Notes in Computer Science, vol. 11700, pp. 55–76. Springer (2019). ISBN: 978-3-030-28953-9. https://doi.org/10.1007/978-3-030-28954-6_4
61. Park, D., et al.: ConceptVector: text visual analytics via interactive lexicon building using word embedding. IEEE Trans. Vis. Comput. Graph. **24**(1), 361–370 (2018). https://doi.org/10.1109/TVCG.2017.2744478
62. Pilault, J., Park, J., Pal, C.J.: On the impressive performance of randomly weighted encoders in summarization tasks. In: CoRR abs/2002.09084 (2020).https://arxiv.org/abs/2002.09084
63. Ponte, J.M., Bruce Croft, W.: A language modeling approach to information retrieval. In: Bruce Croft, W., et al. (eds.) SIGIR'98: Proceedings of the 21st Annual International ACM SIGIR Conference on Research and Development in Information Retrieval, August 24–28, 1998, Melbourne, Australia, pp. 275–281. ACM (1998). ISBN: 1-58113-015-5. https://doi.org/10.1145/290941.291008
64. Qin, Z., et al.: How convolutional neural networks see the world-a survey of convolutional neural network visualization methods. Math. Found. Comput. **1**(2), 149–180 (2018). https://doi.org/10.3934/mfc.2018008
65. Raghu, M., Schmidt, E.: A survey of deep learning for scientific discovery (2020). CoRR abs/2003.11755
66. Reif, E., et al.: Visualizing and measuring the geometry of BERT. In: Wallach, H.M., et al. (eds.) Advances in Neural Information Processing Systems 32: Annual Conference on Neural Information Processing Systems 2019, NeurIPS 2019(8–14), December 2019, pp. 8592–8600. Canada, Vancouver, BC (2019)
67. Ribeiro, M.T., Singh, S., Guestrin, C.: "Why Should I Trust You?": explaining the predictions of any classifier. In: Krishnapuram, B., et al. (eds.) Proceedings of the 22nd ACM SIGKDD International Conference on Knowledge Discovery and Data Mining, San Francisco, CA, USA, August 13–17, 2016, pp. 1135-1144. ACM (2016). ISBN: 978-1-4503-4232-2. https://doi.org/10.1145/2939672.2939778
68. Samek, W., et al. (eds.): Explainable AI: Interpreting, Explaining and Visualizing Deep Learning. Lecture Notes in Computer Science, vol. 11700. Springer (2019). ISBN: 978-3-030-28953-9. https://doi.org/10.1007/978-3-030-28954-6
69. Sanh, V., et al.: DistilBERT, a distilled version of BERT: smaller, faster, cheaper and lighter. (2019). CoRR arXiv: 1910.01108
70. Satyanarayan, A., et al.: Vega-Lite: a grammar of interactive graphics. IEEE Trans. Vis. Comput. Graph. **23**(1), 341–350 (2017). https://doi.org/10.1109/TVCG.2016.2599030
71. Sawatzky, L., Bergner, S., Popowich, F.: Visualizing RNN states with predictive semantic encodings. In: 30th IEEE Visualization Conference, IEEE VIS 2019—Short Papers, Vancouver, BC, Canada, October 20–25, 2019, pp. 156–160. IEEE (2019). ISBN: 978-1-7281-4941-7. https://doi.org/10.1109/VISUAL.2019.8933744

72. Shwartz, V., Choi, Y.: Do neural language models overcome reporting bias? In: Scott, D., Bel, N., Zong, C. (eds.) Proceedings of the 28th International Conference on Computational Linguistics, COLING 2020, Barcelona, Spain (Online), December 8–13, 2020, pp. 6863–6870. International Committee on Computational Linguistics (2020). https://doi.org/10.18653/v1/2020.coling-main.605
73. Skrlj, B., et al.: AttViz: online exploration of self-attention for transparent neural language modeling (2020). CoRR arXiv: 2005.05716
74. Slack, D., et al.: Fooling LIME and SHAP: adversarial attacks on post hoc explanation methods. In: Markham, A.N., et al. (eds.) AIES'20: AAAI/ACM Conference on AI, Ethics, and Society, New York, NY, USA, February 7–8, 2020, pp. 180–186. ACM (2020). ISBN: 978-1-4503-7110-0. https://doi.org/10.1145/3375627.3375830
75. Slack, D., et al.: How can we fool LIME and SHAP? Adversarial attacks on post hoc explanation methods (2019). CoRR arXiv: 1911.02508
76. Song, Y., et al.: Utilizing BERT intermediate layers for aspect based sentiment analysis and natural language inference (2020). CoRR arXiv: 2002.04815
77. Strobelt, H., et al.: LSTMVis: a tool for visual analysis of hidden state dynamics in recurrent neural networks. IEEE Trans. Vis. Comput. Graph. **24**(1), 667–676 (2018). https://doi.org/10.1109/TVCG.2017.2744158
78. Strubell, E., Ganesh, A., McCallum, A.: Energy and policy considerations for deep learning in NLP. In: Korhonen, A., Traum, D.R., Marquez, L. (eds.) Proceedings of the 57th Conference of the Association for Computational Linguistics, ACL 2019, Florence, Italy, July 28–August 2, 2019, Volume 1: Long Papers, pp. 3645–3650. Association for Computational Linguistics (2019). https://doi.org/10.18653/v1/p19-1355
79. Su, W., et al.: VL-BERT: pre-training of generic visual-linguistic representations. In: 8th International Conference on Learning Representations, ICLR 2020, Addis Ababa, Ethiopia, April 26–30, 2020. OpenReview.net (2020)
80. Sun, L., et al.: Adv-BERT: BERT is not robust on misspellings! Generating nature adversarial samples on BERT (2020). CoRR arXiv: 2003.04985
81. Tenney, I., Das, D., Pavlick, E.: BERT rediscovers the classical NLP pipeline. In: Korhonen, A., Traum, D.R., Marquez, L. (eds.) Proceedings of the 57th Conference of the Association for Computational Linguistics, ACL 2019, Florence, Italy, July 28–August 2, 2019, Volume 1: Long Papers, pp. 4593–4601. Association for Computational Linguistics (2019). ISBN: 978-1-950737-48-2. https://doi.org/10.18653/v1/p19-1452
82. van Aken, B., et al.: VisBERT: hidden-state visualizations for transformers. In: El Fallah Seghrouchni, A., et al. (eds.) Companion of the 2020 Web Conference 2020, Taipei, Taiwan, April 20–24, 2020, pp. 207–211. ACM (2020). ISBN: 978-1-4503-7024-0. https://doi.org/10.1145/3366424.3383542
83. van der Heijden, N., Abnar, S., Shutova, E.: A comparison of architectures and pretraining methods for contextualized multilingual word embeddings. In: The Thirty-Fourth AAAI Conference on Artificial Intelligence, AAAI 2020, The Thirty-Second Innovative Applications of Artificial Intelligence Conference, IAAI 2020, The Tenth AAAI Symposium on Educational Advances in Artificial Intelligence, EAAI 2020, New York, NY, USA, February 7–12, 2020, pp. 9090–9097. AAAI Press (2020). ISBN: 978-1-57735-823-7
84. Vellido, A.: The importance of interpretability and visualization in machine learning for applications in medicine and health care. Neural Comput. Appl. **32**(24), 18069–18083 (2020). https://doi.org/10.1007/s00521-019-04051-w
85. Vaswani, A., et al.: Attention is all you need. In: Guyon, I., et al. (eds.) Advances in Neural Information Processing Systems 30: Annual Conference on Neural Information Processing Systems 2017(4–9), December 2017, pp. 5998–6008. Long Beach, CA, USA (2017)
86. Vig, J.: A multiscale visualization of attention in the transformer model. In: Costa-jussà, M.R., Enrique Alfonseca, M.R. (eds.) Proceedings of the 57th Conference of the Association for Computational Linguistics, ACL 2019, Florence, Italy, July 28–August 2, 2019, Volume 3: System Demonstrations, pp. 37–42. Association for Computational Linguistics (2019). ISBN: 978-1-950737-49-9. https://doi.org/10.18653/v1/p19-3007

87. Vig, J.: Visualizing attention in transformer-based language representation models (2019). CoRR arXiv: 1904.02679
88. Vig, J., et al.: BERTology meets biology: interpreting attention in protein language models (2020). CoRR arXiv: 2006.15222
89. Vig, J., et al.: Causal mediation analysis for interpreting neural NLP: the case of gender bias (2020). CoRR arXiv: 2004.12265
90. Voita, E., Sennrich, R., Titov, I.: The bottom-up evolution of representations in the transformer: a study with machine translation and language modeling objectives. In: Inui, K., et al. (eds.) Proceedings of the 2019 Conference on Empirical Methods in Natural Language Processing and the 9th International Joint Conference on Natural Language Processing, EMNLP-IJCNLP 2019, Hong Kong, China, November 3–7, 2019, pp. 4395–4405. Association for Computational Linguistics (2019). ISBN: 978-1-950737-90-1. https://doi.org/10.18653/v1/D19-1448
91. Voita, E., Titov, I.: Information-theoretic probing with minimum description length. In: Webber, B. (eds.), Proceedings of the 2020 Conference on Empirical Methods in Natural Language Processing, EMNLP 2020, Online, November 16–20, 2020. Association for Computational Linguistics, pp. 183–196 (2020). https://doi.org/10.18653/v1/2020.emnlp-main.14
92. Voita, E., Titov, I.: Information-theoretic probing with minimum description length (2020). CoRR arXiv: 2003.12298
93. Voita, E., et al.: Analyzing multi-head self-attention: specialized heads do the heavy lifting, the rest can be pruned. In: Korhonen, A., Traum, D.R., Marquez, L. (eds.) Proceedings of the 57th Conference of the Association for Computational Linguistics, ACL 2019, Florence, Italy, July 28–August 2, 2019, Volume 1: Long Papers, pp. 5797–5808 Association for Computational Linguistics (2019). ISBN: 978-1-950737-48-2. https://doi.org/10.18653/v1/p19-1580
94. Wadden, D., et al.: Entity, relation, and event extraction with contextualized span representations. In: Inui, K., et al. (eds.) Proceedings of the 2019 Conference on Empirical Methods in Natural Language Processing and the 9th International Joint Conference on Natural Language Processing, EMNLP-IJCNLP 2019, Hong Kong, China, November 3–7, 2019, pp. 5783–5788. Association for Computational Linguistics (2019). ISBN: 978-1-950737-90-1. https://doi.org/10.18653/v1/D19-1585
95. Wallace, E., et al.: AllenNLP interpret: a framework for explaining predictions of NLP models. In: Padó, S., Huang, R. (eds.) Proceedings of the 2019 Conference on Empirical Methods in Natural Language Processing and the 9th International Joint Conference on Natural Language Processing, EMNLP-IJCNLP 2019, Hong Kong, China, November 3–7, 2019—System Demonstrations, pp. 7–12. Association for Computational Linguistics (2019). ISBN: 978-1-950737-92-5. https://doi.org/10.18653/v1/D19-3002
96. Wallace, E., et al.: Universal adversarial triggers for attacking and analyzing NLP. In: Inui, K., et al. (eds.) Proceedings of the 2019 Conference on Empirical Methods in Natural Language Processing and the 9th International Joint Conference on Natural Language Processing, EMNLP–IJCNLP 2019, Hong Kong, China, November 3–7, 2019, pp. 2153–2162. Association for Computational Linguistics (2019). ISBN: 978-1-950737-90-1. https://doi.org/10.18653/v1/D19-1221
97. Wang, H., Leskovec, J.: Unifying graph convolutional neural networks and label propagation (2020). arXiv: 2002.06755
98. Wang, J., et al.: Gradient-based analysis of NLP models is manipulable. In: Cohn, T., He, Y., Liu, Y. (eds.) Proceedings of the 2020 Conference on Empirical Methods in Natural Language Processing: Findings, EMNLP 2020, Online Event, 16–20 November 2020, pp. 247–258. Association for Computational Linguistics (2020). ISBN: 978-1-952148-90-3. https://doi.org/10.18653/v1/2020.findings-emnlp.24
99. Wiegreffe, S., Pinter, Y.: Attention is not explanation. In: Inui, K., et al. (eds.) Proceedings of the 2019 Conference on Empirical Methods in Natural Language Processing and the 9th International Joint Conference on Natural Language Processing, EMNLP–IJCNLP 2019, Hong Kong, China, November 3–7, 2019, pp. 11–20. Association for Computational Linguistics (2019). ISBN: 978-1-950737-90-1. https://doi.org/10.18653/v1/D19-1002

100. Wilkinson, L.: The Grammar of Graphics. Statistics and Computing, 2nd edn. Springer (2005). ISBN: 978-0-387-24544-7
101. Wolf, T., et al.: HuggingFace transformers: state-of-the-art natural language processing (2019). CoRR arXiv: 1910.03771
102. Wu, M., et al.: Unsupervised domain adaptive graph convolutional networks. In: Huang, Y., et al. (eds.) WWW'20: The Web Conference 2020, Taipei, Taiwan, April 20–24, 2020, pp. 1457–1467. ACM (2020). https://doi.org/10.1145/3366423.3380219
103. Yang, Z., et al.: XLNet: generalized autoregressive pretraining for language understanding. In: Wallach, H.M., et al. (eds.) Advances in Neural Information Processing Systems 32: Annual Conference on Neural Information Processing Systems 2019, NeurIPS 2019(8–14), December 2019, pp. 5754–5764. Canada, Vancouver, BC (2019)
104. Yun, Z., et al.: Transformer visualization via dictionary learning: contextualized embedding as a linear superposition of transformer factors (2021). CoRR arXiv: 2103.15949
105. Zeng, G., et al.: OpenAttack: an open-source textual adversarial attack toolkit (2020). CoRR arXiv: 2009.09191
106. Zhang, Q., Zhu, S.-C.: Visual interpretability for deep learning: a survey. Front. Inf. Technol. Electron. Eng. **19**(1), 27–39 (2018). https://doi.org/10.1631/FITEE.1700808
107. Zhang, Y., et al.: Every document owns its structure: inductive text classification via graph neural networks. In: Jurafsky, D., et al. (eds.) Proceedings of the 58th Annual Meeting of the Association for Computational Linguistics, ACL 2020, Online, July 5–10, 2020, pp. 334–339. Association for Computational Linguistics (2020). https://doi.org/10.18653/v1/2020.acl-main.31
108. Zhong, M. et al.: A closer look at data bias in neural extractive summarization models. In: CoRR abs/1909.13705 (2019). http://arxiv.org/abs/1909.13705

Transparent Clustering with Cyclic Probabilistic Causal Models

Evgenii E. Vityaev and Bayar Pak

Abstract In the previous work data clusters where discovered and visualized by causal models, used in cognitive science. Centers of clusters are presented by prototypes of clusters, formed by causal models, in accordance with the prototype theory of concepts, explored in cognitive science. In this work we describe the system of transparent analysis of such clasterization that bring the light to the interconnection between (1) set of objects with there characteristics (2) probabilistic causal relations between objects characteristics (3) causal models—fixpoints of probabilistic causal relations that form prototypes of clusters (4) clusters—set of objects that defined by prototypes. For that purpose we use a novel mathematical apparatus—probabilistic generalization of formal concepts—for discovering causal models via cyclical causal relations (fixpoints of causal relations). This approach is illustrated with a case study.

1 Introduction

In the previous work [1] data clusters was discovered and visualized of by causal models, used in cognitive science. Centers of clusters are presented by prototypes of clusters, formed by causal models, in accordance with the prototype theory of concepts, explored in cognitive science. This visualization of clusters is explainable and interpretable.

In this work we describe the system of transparent analysis of such clasterization that bring the light to the interconnection between (1) set of objects with there characteristics (2) probabilistic causal relations between objects characteristics (3) causal models—fixpoints of probabilistic causal relations that form prototypes of clusters (4) clusters—set of objects that defined by prototypes. For that purpose we use a novel

E. E. Vityaev (✉)
Sobolev Institute of Mathematics of SD RAS, 4 Acad. Koptyug avenue, Novosibirsk, Russia
e-mail: vityaev@math.nsc.ru

B. Pak
Novosibirsk State University, Novosibirsk, Russia

B. Kovalerchuk et al. (eds.), *Integrating Artificial Intelligence and Visualization for Visual Knowledge Discovery*, Studies in Computational Intelligence 1014,
https://doi.org/10.1007/978-3-030-93119-3_9

mathematical apparatus—probabilistic generalization of formal concepts—that discover causal models via cyclical causal relations (fixpoints of causal relations).

"Causal models" base on the probabilistic causality [2–4]. Probabilistic causality is a type of causality that can be discovered on data and it is not guaranteed that it is actually causality. Probabilistic causality is defined in terms of a set of objects features. Thus, it is a relative causality to this set of features. The proposed approach is base on such probabilistic causal relations between features.

In [5, 6], a probabilistic generalization of the Formal Concepts Analysis (FCA) was developed. In Sect. 5 a probabilistic formal concepts with negation are defined for the causal models representation via cyclical causal relations (fixpoints of causal relations). Also a statistical method for probabilistic formal concepts discovery is defined in Sect. 6 [6, 7].

The center of the cluster is formalized in accordance with the "prototype" theory of concepts presented in [1] and shortly described in the next section. This theory explored in cognitive science in terms of a correlational structure of perceived attributes. Traditionally correlational structure and causal models are described using Bayesian networks. However, Bayesian networks do not support cycles. We use probabilistic formal concepts for representation cyclic "causal models" and correlational structure via cyclical causal relations (fixpoints of causal relations) that form clusters and generate cluster prototypes.

Probabilistic generalization of formal concepts was obtained by generalization of the definition of formal concepts as fixpoints of implications. For this purpose implications were replaced by probabilistic causal relations. These relations satisfy some very strong condition of maximal specificity (Definition 18) and also Cartwright's [21] definition of causality relative to some background: every condition of the premise of the causal relation increases the conditional probability of this relation relative to other conditions of that premise as a background. Consistency of fixpoints for these causal relations was proved and probabilistic formal concepts was defined as fixpoints of these causal relations [6, 8].

Figuratively speaking, probabilistic formal concepts do not display data in all details as in photographs, but give an "idealized" representation of data. In [7] it was shown that these fixpoints model the process of perception.

In this work, probabilistic formal concepts are defined in terms of probability. The proof of their consistency as fixpoints of causal relationships is also carried out in terms of probability. To develop a clustering method (Sect. 6) that discover probabilistic formal concepts on data, it was necessary to use statistical estimations. But in this case, maximum specificity and consistency of fixpoints is no longer guaranteed. So, Sect. 6 provides a statistical approximation of both maximum specific causal relationships and fixpoints of these causal relationships. For this purpose, a special criterion of the maximal consistency of fixpoint for the statistically discovered maximum specific causal relationships is introduced. This criterion determines the meaning of the clustering method—maximal consistency of causal relationships for the prototypes characteristics.

The explainability of the resulting clusters follows from the explainability of causal relations as logical expressions and representation of clusters as fixpoints of these causal relations.

Thus, as a result of clustering we receive four types of entities: (1) set of objects with there characteristics, (2) probabilistic causal relations between objects characteristics, (3) causal models—fixpoints of probabilistic causal relations that form prototypes, (4) clusters—set of objects that defined by prototypes. How they are interconnected with each other?

For transparent analysis of these interconnection we developed a system that provide:

1. for each discovered probabilistic causal relations to "see" how it fulfilled on objects by highlighting causal relations characteristics among objects characteristics;
2. for each causal model, which is presented by the corresponding prototype characteristics (that form a fixpoint of probabilistic causal relations between these characteristics) to "see" all probabilistic causal relations that form this fixpoint. Moreover you can "see" the place of each probabilistic causal relation among prototypes characteristics by highlighting them. Fixpoint of the probabilistic causal relations form "idealized" representation of the objects of the class in form of its causal relational structure;
3. for each causal model and corresponding prototype to "see" all objects that produce this causal model by applying causal model formation procedure ("idealization" procedure see Definition 32) to the set of all object characteristics. These objects form a class of the corresponding causal model and prototype.
4. for each object of some class to "see" those characteristics that are included in the prototype characteristics of this class, and, therefore, essential for belonging of this object to this class—informative attributes of the object;
5. for each object of some class to "see" object characteristics, which are not included in the prototype characteristics of this class, and, therefore, not informative or even random relative to the causal relational structure of the prototype;
6. for each object of some class to "see" those object characteristics that are present, both in the object and in the prototype characteristics, but having different values. This means that this values was corrected during the causal model formation procedure (and, therefore, by the entire set of causal relations) and are possibly erroneously.

2 "Natural" Concepts

In works of Rosch [9–11] the principles of categorization of "natural" categories was formulated on the basis of the conducted experiments.

One of them is the "Perceived World Structure": "The second principle of categorization asserts that ...perceived world—is not an unstructured total set of equiprobable co-occurring attributes. Rather, the material objects of the world are perceived to possess ...*high correlational structure*. ...In short, combinations of what we perceive as the attributes of real objects do not occur uniformly. Some pairs, triples, etc., are quite probable, appearing in combination sometimes with one, sometimes another attribute; others are rare; others logically cannot or empirically do not occur".

Directly perceived objects (basic objects)—information-rich bundles of observed and functional properties that form a natural partition that creates a categorization. These bundles form "prototypes" of the clusters: "Categories can be viewed in terms of their clear cases if the perceiver places emphasis on the correlational structure of perceived attributes ...By prototypes of categories we have generally meant the clearest cases of category membership [10, 11]. Later on, the theory of "natural" concepts of Eleanor Rosch was called a "prototype" theory.

Further research has found that models based on features, similarities and prototypes are not sufficient to describe classes. Studies have shown that people's knowledge of categories isn't limited to the list of properties, but also and includes a rich set of causal relationships between these properties.

Robert Rehder formulated a causal-model theory: "people's intuitive theories about categories of objects consist of a model of the category in which both a category's features and the causal mechanisms among those features are explicitly represented ...and the degree of coherence of a set of features might be an important factor determining membership in a category" [12]. In the theory of causal models, the relationship of an object to a category is no longer based on a set of features and proximity based on features, but on the similarity of the generating causal mechanism: "Specifically, a to-be-classified object is considered a category member to the extent that its features were likely to have been generated by the category's causal laws" [12].

Some researchers have used Bayesian networks to represent causal models [14, 15]. However, these models cannot model cyclical causal relationships because Bayesian networks do not support cycles. In [13], Robert Rehder proposed a model of causal cycles based on the "disclosure" of causal graphical models (CGMs). The "disclosure" is done by creating a Bayesian network that deploys over time.

A cyclic causal models that we propose is directly based on cyclical causal relationships represented by probabilistic formal concepts.

3 Basics of Formal Concepts Analysis

Let's start with the main definitions of formal concepts analysis. The majority of them start with classical works on FCA [16, 17], others are taken from [18].

Definition 1 *A* formal context K *is a triple* $K = (G, M, I)$*, where* G *and* M*—arbitrary sets of objects and attributes, and* $I \subseteq G \times M$*—binary relation, expressing the belonging of an object to an attribute.*

Definition 2 *For* $A \subseteq G$, $B \subseteq M$

1. $A^{\uparrow} = \{m \in M \mid \forall g \in A, (g, m) \in I\}$
2. $B^{\downarrow} = \{g \in G \mid \forall m \in B, (g, m) \in I\}$.

Derivative operators $A^{\uparrow}$, $B^{\downarrow}$ link a subset of objects and attributes of a context. In the future we will also refer to both operators as $'$.

Definition 3 *A pair* (A, B)*—is a* formal concept, *if* $A^{\uparrow} = B$ *and* $B^{\downarrow} = A$.

Definition 4 *An* implication *is a pair of attribute sets* (B, C), $B, C \subseteq M$ *that is recorded as* $B \rightarrow C$. *It is true on the context* $K = (G, M, I)$, *if* $\forall g \in G(B \not\subseteq g' \vee C \subseteq g')$. *The set of all true implications on the context* K *is denoted as* $Imp(K)$.

Definition 5 *For any set* Imp *of implications we define a* direct inference operator f_{Imp}*:*

$$f_{Imp}(X) = X \cup \{C \mid B \subseteq X,\ B \rightarrow C \in Imp\}$$

Theorem 1 [16] For any subset $B \subseteq M$, $f_{Imp(K)}(B) = B \Leftrightarrow B'' = B$.

4 Probabilistic Logic in a Formal Context

We will present a fundamentally different way of the formal concepts definition, based on probabilistic implications. For this purpose, we redefine the context within the framework of logic. Further we will consider only finite contexts.

Definition 6 *For the finite context* $K = (G, M, I)$ *we define a* signature Ω_K *that contains predicate symbols* $m(x)$ *for each* $m \in M$. *For the signature* Ω_K *and the context* K, *considered as a model, we define the interpretation of predicate symbols as follows* $K \models m(x) \Leftrightarrow (x, m) \in I$.

Definition 7 *For the signature* Ω_K *let's define the following variant of the first order logic:*

1. X_K*—set of* variables*;*
2. At_K*—set of* atomic formulas *(atoms)* $m(x)$ *where* $m \in \Omega_K$ *and* $x \in X_K$*;*
3. L_K*—set of* literals, *including all atoms* $m(t)$ *and their negations* $\neg m(t)$*;*
4. Φ_K*—set of* formulas *defined inductively: every atom is a formula, for any* $\Phi, \Psi \in \Phi_K$ *following syntactic constructions* $\Phi \wedge \Psi$, $\Phi \vee \Psi$, $\Phi \rightarrow \Psi$, $\neg\Phi$ *are formulas.*

For a set of literals $L \subseteq L_K$ we define their *conjunction* as $\wedge L = \underset{P \in L}{\wedge} P$. In the same way $\neg L = \{\neg P \mid P \in L\}$.

Definition 8 *A set* $\{g\}, g \in G$, *together with signature* Ω_K *forms a* model K_g of this object. *The fact of truth of a formula* ϕ *on an object model* K_g *is written as* $g \models \phi \Leftrightarrow K_g \models \phi$.

Definition 9 *Let us define a* probability measure μ *on the set G, in the Kolmogorov sense. Then we can define a probability measure on the set of formulas as:*

$$\nu : \Phi_K \to [0, 1],\ \nu(\phi) = \mu(\{g \mid g \models \phi\});\ \mu(\{g\}) \neq 0,\ g \in G.$$

Definition 10 *A set M of literals is* ν-joint *if* $\nu(\wedge M) > 0$.

Further, the compatibility of the set of literals will be considered within the probability measure of the context, and in the absence of ambiguity, the symbol of the measure will be omitted.

Let us consider some subset of atoms $L \subseteq \mathrm{At}(K)$. The formula $m_1 \wedge m_2, \ \dots,\ \wedge m_k \to m = \wedge\{m_i\} \to m$ defines implication $(\{\mathrm{m_i}\}, \{\mathrm{m}\})$ on the context.

Let's define rules and probability of rules on the context.

Definition 11 *Let* $\{H_1, H_2, \ \dots, H_k, C\} \in L_K, C \notin \{H_1, H_2, \ \dots, H_k\}, k \geq 0$.

1. *The* rule $R = (H_1, H_2, \ \dots, H_k \to C)$ *is called implication* $(H_1 \wedge H_2 \wedge \ \dots \ \wedge H_k \to C)$;
2. *The* premise $R^{\leftarrow}$ of the rule R *is a set of literals* $\{\mathrm{H}_1, \mathrm{H}_2, \ \dots, \mathrm{H}_k\}$;
3. *The* conclusion of the rule *is* $R^{\rightarrow} = C$;
4. *The* length of the rule *is called the premise power* $|R^{\leftarrow}|$;
5. *If* $R_1^{\leftarrow} = R_2^{\leftarrow}$ *and* $R_1^{\rightarrow} = R_2^{\rightarrow}$, *then* $R_1 = R_2$.

Definition 12 *The* probability η *of the rule R is the value*

$$\eta(R) = \nu(R^{\rightarrow}|R^{\leftarrow}) = \frac{\nu(R^{\leftarrow} \wedge R^{\rightarrow})}{\nu(R^{\leftarrow})}.$$

If the rule denominator $\nu(\wedge R^{\leftarrow})$ is equal to 0, then the probability is not defined.

5 Consistency of Predictions and Probabilistic Formal Concepts

This section is compilation of works [6, 8, 16].

Definition 13 Prediction operator *by a set of rules* $\mathcal{R}$ *is*

$$\Pi_{\mathcal{R}}(L) = L \cup \{C \mid R \in \mathcal{R}, R^{\leftarrow} \subseteq L, R^{\rightarrow} = C\}.$$

Definition 14 *A* closure $\bar{L}$ *of the set of literals L is the smallest fixpoint of the prediction operator applied to L*

$$\bar{L} = \Pi_{\mathcal{R}}(\bar{L}) = \Pi_{\mathcal{R}}^{\infty}(L) = \bigcup_{k \in \mathbb{N}} \Pi_{\mathcal{R}}^{k}(L).$$

Definition 15 *A rule* R_1 *is a* subrule *of a rule* R_2 *if* $R_1^{\rightarrow} = R_2^{\rightarrow}$, $R_1^{\leftarrow} \subset R_2^{\leftarrow}$. *We denote this fact as* $R_1 \sqsubset R_2$.

Definition 16 *A rule* R_1 specifies a rule R_2, *or* $R_1 > R_2$ *if* $R_2 \sqsubset R_1$ *and* $\eta(R_1) > \eta(R_2)$.

Class $M_1(C)$ defined below contains only those rules R whose conditional probability is strictly greater than the probability of an unconditional rule $R_C = (\emptyset \Rightarrow C)$.

Definition 17 *Definition of the class* $M_1(C)$. *For* $R, R^{\rightarrow} = C,\ R \in M_1(C) \Leftrightarrow \eta(R) > \nu(R_C)$.

The next class $M_2(C)$ define the property of maximum specificity—the impossibility of increasing the conditional probability by refining the rule [8].

Definition 18 *Definition of the class* $M_2(C) : R \in M_2(C) \Leftrightarrow (R \in M_1(C))\ \&\ (R \sqsubset \tilde{R} \Rightarrow \eta(\tilde{R}) \leq \eta(R))$.

Definition 19 $R \in Imp(C) \Leftrightarrow (R^{\rightarrow} = C)\ \&\ (\eta(R) = 1)$.

Definition 20 *Define classes* M_1, M_2, *Imp by joining over all literals*

$$M_1 = \bigcup_{C \in Lit(K)} M_1(C);\ M_2 = \bigcup_{C \in Lit(K)} M_2(C);$$

$$Imp = \bigcup_{C \in Lit(K)} Imp(C).$$

The class *Imp* contains all precise implication and corresponds to the set $Imp(K)$ of Definition 4.

Definition 21 *A set of rules* $\mathcal{R}$ *is* exact, *if* $Imp \subset \mathcal{R}$.

Definition 22 System of rules *is any subset* $\mathcal{R} \subseteq M_2$.

A set of literals L is *consistent* if it does not simultaneously contain some atom C and its negation $\neg C$.

Theorem 2 (Joint predictions [6]). If L is joint, then $\Pi_{\mathcal{R}}(L)$ is also joint and consistent for any system of rules $\mathcal{R} \subseteq M_2$.

Definition 23 *A* probabilistic formal concept on the context K *is a pair* (A, B) *that meets the conditions:*

$$\Pi_{\mathcal{R}}(B) = B,\ A = \bigcup_{\Pi_{\mathcal{R}}(C)=B} C', \mathcal{R} \subseteq M_2.$$

To distinguish probabilistic formal concepts from classical formal concepts on the context K, we will call the latter simply formal concepts. The definition of a set A is also based on the following theorem linking probabilistic and simple formal concepts on context K.

Theorem 3 Let K be a formal context.

1. If (A, B) a formal concept on K, then there is a probabilistic formal concept (N, M) on K such that $A \subseteq N$.
2. If (N, M) is a probabilistic formal concept on K, then there is a family $\mathcal{C}$ of formal concepts on K, such that

$$\forall (A, B) \in \mathcal{C} \; (\Pi_{\mathcal{R}}(B) = M),$$
$$N = \bigcup_{(A,B)\in\mathcal{C}} A$$

6 Clustering by Statistical Discovery of Probabilistic Formal Concepts on Data

The procedure for obtaining all M_2 rules by exhaustive search is exponential, because the $O(2^{|M|})$ candidate rules (each atom or its negation can be included in the rule) are evaluated at $|G|$. Of course, this estimation is rather rough and may be improved by the following approximation of M_2-rules.

Definition 24 *R is a* probabilistic law, *if* $(\tilde{R} \sqsubset R) \Rightarrow (\tilde{R} < R)$.

This definition may be interpreted in terms of Cartwright's [21] definition of causality relative to some background. If the premise $R^{\leftarrow}$ of the rule R is a set of literals $\{H_1, H_2, \ldots, H_k\}$ and we consider this set as a background then every literal from that set is a cause of the conclusion of the rule $R^{\rightarrow}$ relative to that background, i.e.

$$\nu(R^{\rightarrow}/R^{\leftarrow}) > \nu(R^{\rightarrow}/(R^{\leftarrow}\backslash H))$$

for every $H \in \{H_1, H_2, \ldots, H_k\}$.

Definition 25 *A rule $\tilde{R}$ is* semantically derived *from the rule R, $R \rhd \tilde{R}$, if R and $\tilde{R}$—probabilistic law and $\tilde{R} > R$.*

Definition 26 *The probabilistic law R will be called* strongest *if there is no probabilistic law $\tilde{R}$ such as $(\tilde{R} > R)$.*

Definition 27 Semantic Probabilistic Inference (SPI) *of some strongest probabilistic law R_m is a sequence of probabilistic laws $R_0 \rhd R_1 \rhd R_2. \rhd R_m$, $R_0^{\leftarrow} = \emptyset$.*

Let us denote by $\mathcal{R}$ the set of all the strongest probabilistic laws obtained by various semantic probabilistic inferences.

In practical applications on data, we cannot assume that the probability is known. Then for determining a probabilistic inequalities used in the SPI definition we need to use some statistical criterion. For that purpose we use exact Fisher independence criterion with a confidence level α [22]. The resulting set of rules $\mathcal{R}_\alpha$ will already make contradicting predictions. Therefore, to approximate a prediction operator $\Pi_{\mathcal{R}}(L)$ by some other operator $\Upsilon_{R_\alpha}(L)$, it is necessary to introduce a further additional criterion of mutual consistency of predictions by rules $\mathcal{R}_\alpha$ on the set L.

Definition 28 *The rule $R \in \mathcal{R}_\alpha$ is* confirmed *on the set of letters L, if $R^{\leftarrow} \subset L$ and $R^{\rightarrow} \in L$. In this case, we write that $R \in \mathrm{Sat}(L) \subseteq \mathcal{R}_\alpha$.*

Definition 29 *The rule $R \in \mathcal{R}_\alpha$ is* refuted *on the set of letters L, if $R^{\leftarrow} \subset L$ and $R^{\rightarrow} \in \neg L$. In this case, we write that $R \in \mathrm{Fal}(L) \subseteq \mathcal{R}_\alpha$.*

Definition 30 *The* criterion of mutual consistency of predictions *by rules $\mathcal{R}_\alpha$ on the set L is the value*

$$\mathrm{Int}(L) = \sum_{R \in \mathrm{Sat}(L)} \gamma(R) - \sum_{R \in \mathrm{Fal}(L)} \gamma(R).$$

The choice of rule evaluation γ may depend on additional task specifics. In the experiments below, we are guided by Shannon's considerations as

$$\gamma(R) = -\log(1 + \epsilon - \eta(R)).$$

Definition 31 *Now we can define a* consistent predictions operator $\Upsilon_{R_\alpha}(L)$ *that changes a set of literals L by one element so as to strictly increase the consistency criterion:*

1. *For all $\phi \in L_K \backslash L$ calculate the effect of adding ϕ: $\Delta^+ = \mathrm{Int}(L \cup \{\phi\}) - \mathrm{Int}(L)$, Find ϕ^+ that maximaze Δ^+;*
2. *For all $\phi \in L$ calculate the effect of removing ϕ: $\Delta^- = \mathrm{Int}(L \backslash \{\phi\}) - \mathrm{Int}(L)$, Find ϕ^- that maximaze Δ^-;*
3. *The operator $\Upsilon_{R_\alpha}(L)$ adds a literal ϕ^+ to the set L if $\Delta^+ > 0$ and $\Delta^+ > \Delta^-$; the operator deletes a literal ϕ^- from L if $\Delta^- > 0$ and $\Delta^- > \Delta^+$. If $\Delta^- = \Delta^+$ and $\Delta^- > 0$, the operator deletes a literal ϕ^-;*
4. *If $\Delta^+ \leq 0$ and $\Delta^- \leq 0$, the operator returns L and we have a fixpoint.*

Definition 32 Clustering *of the objects of some context $K = (G, M, I)$ is the set of all fixpoints that can be obtained by application of the operator $\Upsilon_{R_\alpha}(L)$ to the set of literals $\{\phi \mid K_g \models \phi, \phi \in L_K\}$ generated by some object $g \in G$, where $\mathcal{R}_\alpha$ obtained by various semantic probabilistic inferences with a confidence level α on the context K.*

7 Experiments on the Digits Prototypes Formation

Let us illustrate the digit's prototypes formation on the partial digit's images. Let's encode digits from the Fig. 2 with 24 features, located as shown in Fig. 1a. Each feature has 7 values (see Fig. 2), where the value 1, designated by the little black box, is the empty (white square). Some lines of the digits are encoded twice, for example, the vertical line the digit 4 and 1 is encoded both by the vertical right line (feature 3) in cells 3, 7, 11, 15, 19, 23 and by the vertical left line (feature 5) in cells 4, 8, 12, 16, 20, 24.

For experiment 600 images were taken: 400 of them were randomly generated by a random choose of the feature value for each feature of the image. Another 200 images are 20 copies of each digit, in each of which two features are randomly deleted and replaced by the empty field. Thus, data does not have a complete image of any digit. Discover prototypes for these digits means a restoration of digit's images based on their incomplete images.

For that purpose, by a statistical method a set $\mathcal{R}_\alpha$ of strongest probabilistic laws (Defintion 26) with $\alpha = 0.0001$ was discovered on all 600 images. This set contained 135437 laws. By clustering of 200 images of digits with two randomly deleted feature exactly 10 fixpoints, which corresponds to 10 digits, were discovered as probabilistic formal concepts, using the set $\mathcal{R}_\alpha$. These fixpoints present prototypes of the digits images, because it was not known for the program how 600 images are generated and there are no digits images in the learning set. In Fig. 1b there are two examples of the digits restoration by clustering.

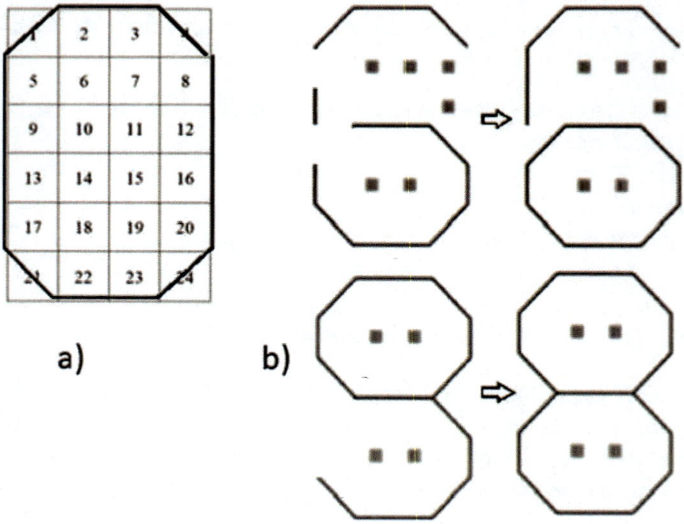

Fig. 1 Features numbering and digits restoration

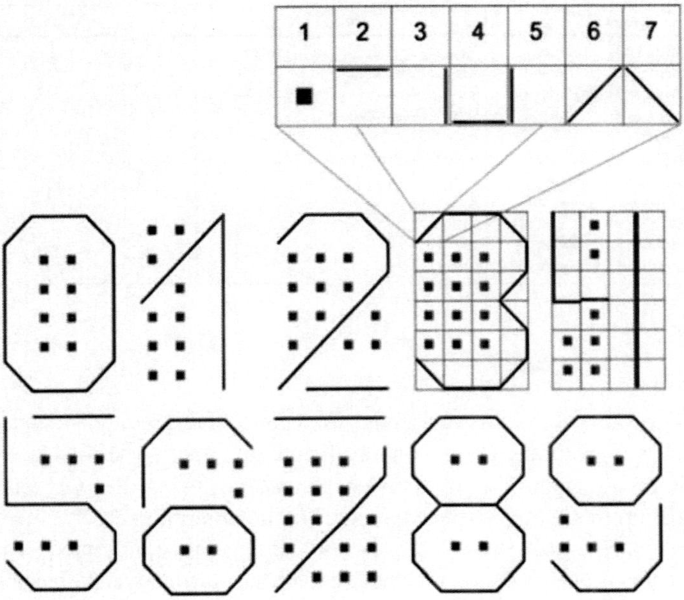

Fig. 2 Digits encoding

8 Transparent Analysis of the Relationships Between Objects, Rules and Classes

In this section the system of transparent analysis of relationships between objects, probabilistic causal relations, prototypes and clusters is described. It base on a modified interface of the Norton Commander system [23] that containing three windows: data—objects and their characteristics; causal relations with their characteristics and prototypes with their characteristics.

In the first window, loaded objects are shown in Fig. 3.

Objects			Rules		Classes	
Obj 1	Attribute 1	Value 1				
Obj 2	Attribute 2	Value 2				
Obj 3	Attribute 3	Value 3				
Obj 4	Attribute 4	Value 4				
Obj 5	Attribute 5	Value 5				
Obj 6	Attribute 6	Value 6				
Obj 7	Attribute 7	Value 7				
Obj 8	Attribute 8	Value 8				
Obj 9	Attribute 9	Value 9				
Obj 10	Attribute 10	Value 10				
Obj 11	Attribute 11	Value 11				

Fig. 3 Objects loading

Objects			Rules		Classes	
Obj 1	Attribute 1	Value 1	Rule 1	---- Predicted attribute ----		
Obj 2	Attribute 2	Value 2	Rule 2	**Attribute** **value**		
Obj 3	Attribute 3	Value 3	Rule 3	---------- premise ---------		
Obj 4	Attribute 4	Value 4	Rule 4	*Attribute 1* *Value 1*		
Obj 5	Attribute 5	Value 5	Rule 5	*Attribute 2* *Value 2*		
Obj 6	Attribute 6	Value 6	Rule 6	*Attribute 3* *Value 3*		
Obj 7	Attribute 7	Value 7	Rule 7			
Obj 8	Attribute 8	Value 8	Rule 8			
Obj 9	Attribute 9	Value 9	Rule 9			
Obj 10	Attribute 10	Value 10	Rule 10			
Obj 11	Attribute 11	Value 11	Rule 11			

Fig. 4 Rules loading

The first window (two left columns) display objects and their characteristics. You can load objects using the corresponding menu command. The loaded objects can be compared with each other by their characteristics, using the "Object Analysis" function. To do this, you can put the cursor on a certain object, and then click the menu item "Object Analysis". Then, after selecting some other object, you will see those characteristics that match the characteristics of the analyzed object. They will be highlighted by the bold text. Thus, you can "see" how objects differ from each other.

Then you can analyze probabilistic causal relations (see Fig. 4). They can be loaded if they already have been (at least partially) discovered using the menu item "Load relations". After loading, a list of causal relations with their characteristics will appear in the central window (third and fourth columns). Characteristics of the relation will appear in the second column of the window and include, starting from the top: the predicted characteristic and its value (consequence of the relation) and then attributes of the premise of the relation together with their values. The predicted characteristic is highlighted by the bold text and characteristics of the premise by italic text. If you select by cursor some rule in the first column, then its characteristics will appear in the second column.

If causal relations have not been previously discovered, you can discover them by clicking "Relations discovery" menu item. Each newly discovered relation appears in the second window. Rules discovery algorithm can be stopped in any time and results saved.

To analyze a certain relation, select it with the cursor and click the "Relation analysis" menu item. Then a list of objects will appear in the first window (see Fig. 5) for which the premise of the relation is fulfilled. If predicted characteristic of the relation is also true for some object, then it will be highlighted by bold text. If you select with the cursor some of the appeared object, then characteristics of the object included in the premise of the relation will be highlighted by the italic text and predicted characteristic of the relation will be highlighted by the bold text. Thus, you can "see" the relation behavior on all objects to which it applicable.

When objects and relations are loaded, it is possible to discover prototypes (causal models) and their characteristics. As well as for relations, already discovered (pos-

Objects		Rules		Classes	
Obj 1	Attribute 1 Value 1	Rule 1	---- Predicted attribute ----		
Obj 12	Attribute 2 Value 2	Rule 2	**Attribute value**		
Obj 31	**Attribute value**	Rule 3	--------- premise ---------		
Obj 35	Attribute 4 Value 4	Rule 4	*Attribute 1 Value 1*		
Obj 44	*Attribute 1 Value 1*	Rule 5	*Attribute 2 Value 2*		
Obj 61	Attribute 6 Value 6	Rule 6	*Attribute 3 Value 3*		
Obj 75	*Attribute 2 Value 2*	Rule 7			
Obj 77	Attribute 8 Value 8	Rule 8			
Obj 94	Attribute 9 Value 9	Rule 9			
Obj101	*Attribute 3 Value 3*	Rule 10			
Obj114	Attribute 11 Value 11	Rule 11			

Fig. 5 Rule analysis

Objects		Rules		Classes	
Obj 1	Attribute 1 Value 1	Rule 5	---- Predicted attribute ----	Class 1	Attribute 1 Value 1
Obj 14	*Attribute 2 Value 2*	Rule 21	**Attribute value**	Class 2	Attribute 2 Value 2
Obj 21	*Attribute 3 Value 3*	Rule 25	--------- premise ---------	Class 3	*Attribute 1 Value 1*
Obj 25	Attribute 4 Value 4	Rule 33	*Attribute 1 Value 1*	Class 4	Attribute 4 Value 4
Obj 34	*Attribute 5 Value 5*	Rule 36	*Attribute 2 Value 2*	Class 5	**Attribute Value**
Obj 41	*Attribute 6 Value 6*	Rule 41	*Attribute 3 Value 3*	Class 6	Attribute 6 Value 6
Obj 55	Attribute 7 Value 7	Rule 47		Class 7	*Attribute 3 Value 3*
Obj 67	*Attribute 8 Value 8*	Rule 48		Class 8	Attribute 8 Value 8
Obj 74	*Attribute 9 Value 9*	Rule 55		Class 9	*Attribute 2 Value 2*
Obj 85	Attribute 10 Value 10	Rule 59		Class10	Attribute 10 Value 10
Obj 90	*Attribute 11 Value 11*	Rule 61		Class11	Attribute 11 Value 11

Fig. 6 Classes analysis

sibly partially) prototypes can be loaded, or they can be discovered (see Fig. 6) by clicking the "Prototypes discovery" menu item. After loading or discover prototypes, a list of prototypes will appear in the third window along with their characteristics, located in the fifth and sixth columns.

For each prototype, you can see a set of relations that are fulfilled on it and, therefore, belong to the causal model of the prototype. To analyze a causal model of some prototype you need to select it by cursor (see Fig. 6) and click "Causal model analysis" menu item. Then in the second window all relations included in the fixed point of the causal model will appear (all relations the premise of which is fulfilled for the prototype characteristics) and in the first window all objects of the class corresponding to this prototype will appear (all objects that produce this causal model by applying causal model formation procedure ("idealization" procedure) to the set of objects characteristics). If, at the same time, in the second window, we select some relation, then characteristics of the relation will be highlighted among the characteristics of the prototype (predicted characteristic will be highlighted by the bold text and characteristics of the premise by the italic text). By moving cursor from relation to relation, you can see the entire structure of the causal model. This allows us to "see" how characteristics of the prototype mutually predict each other by relations of the second window that form a fixed point the causal model. If you select by cursor some object, then those attributes of the object that are among the attributes of the prototype will be shown by italic. By moving cursor from object to object you can see the appearance of the prototype characteristics among objects characteristics.

All this provide a transparent analysis described at the end of introduction.

9 Conclution

The initial idea of visualizing clusters centers as prototypes of clusters using a prototypical theory of categorization and causal models is experimentally confirmed. For this purpose a novel mathematical apparatus—probabilistic generalization of formal concepts for describing causal models via cyclical causal relations was developed. The experiment performed shows that clusters centers can be found even if they are absent in data. This approach can significantly reduce the volume of data by replacing clusters by their prototypes.

Acknowledgements The work is financially supported by the Russian Foundation for Basic Research 19-01-00331-a and also was carried out within the framework of the state contract of the Sobolev Institute of Mathematics (project no.0314-2019-0002) regarding theoretical results.

References

1. Vityaev, E.E., Pak, B.: Explainable rule-based clustering based on cyclic probabilistic causal models. In: Proceedings of the International Conference on Information Visualisation, September 2020, paper no. 9373131, pp. 307–312 (2020)
2. Suppes, P.: A Probabilistic Theory of Causality. North-Holland Publishing Company, Amsterdam (1970)
3. Pearl, J.: Causality: Models. Cambridge University Press, Reasoning and Inference (2000)
4. Hitchcock, C.: Probabilistic causation. In: Hájek, A., Hitchcock, C. (eds.) The Oxford Handbook of Probability and Philosophy. Accessed Sep 2016
5. Vityaev, E.E., Demin, A.V., Ponomariov, D.K.: Probabilistic generalization of formal concepts. Programming **38**(5), 18–34 (2012) (in Russian)
6. Vityaev, E.E., Martynovich, V.V.: Probabilistic formal concepts with negation. In: Voronkov, A., Virbitskaite, I. (eds.) Perspectives of System Informatics. LNCS, vol. 8974, pp. 385–399 (2015)
7. Vityaev, E.E., Neupokoev, N.V.: Formal model of perception and image as fix-point of anticipation. In: Approaches to Thinking Modeling (Collected under the editorship of V.G. Redko, Ph.D., m.D.). URSS Editorial-Al, Moscow, pp. 155–172 (2014) (in Russian)
8. Vityaev, E.E.: The logic of prediction. In: Proceedings of the 9th Asian Logic Conference Mathematical Logic in Asia (Novosibirsk, Russia, August 16–19, 2005), pp. 263–276. World Scientific, Singapore (2006)
9. Rosch, E.H.: Natural categories. Cognitive psychology **4**, 328–350 (1973)
10. Rosch, E., Lloyd, B.B. (eds.): Cognition and categorization, pp. 27–48. Lawrence Erlbaum, Hillsdale, NJ (1978)
11. Rosch, E.: Principles of categorization. In: Rosch, E., Lloyd, B.B. (eds.) Cognition and Categorization, pp. 27–48. Lawrence Erlbaum Associates, Publishers, Hillsdale (1978)
12. Rehder, B.: Categorization as causal reasoning. Cogn. Sci. **27**, 709–748 (2003)

13. Rehder, B., Martin, J.B.: Towards a generative model of causal cycles. In: 33rd Annual Meeting of the Cognitive Science Society, Boston, Massachusetts, USA, 20–23 July 2011, vol.1, pp. 2944–2949 (2011)
14. Cheng, P.: From covariation to causation: a causal power theory. Psychol. Rev. **104**, 367–405 (1997)
15. Griffiths, T.L., Tenenbaum, J.B.: Theory-based causal induction. Psychol. Rev. **116**(4), 661–716 (2009)
16. Ganter, B., Wille, R.: Formal Concept Analysis. Mathematical Foundations, Springer, Berlin-Heidelberg-New York (1999)
17. Ganter, B.: Formal Concept Analysis: Methods, and Applications in Computer Science. TU Dresden (2003)
18. Ganter, B., Obiedkov, S.: Implications in Triadic Formal Contexts. Springer, TU Dresden (2004)
19. Kuznetsov, S.O.: On stability of a formal concept. Ann. Math. Artif. Intell. **49**, 101–115 (2007)
20. Buzmakov, A., Kuznetsov, S., Napoli, A.: Concept stability as a tool for pattern selection. In: CEUR Workshop Proceedings, vol. 1257, ECAI 2014, pp. 51–58 (2014)
21. Cartwright, N.: Causal laws and effective strategies. Noûs **13**(4), 419–437 (1979)
22. Kendall, M.G., Stuart, A.: The Advanced Theory of Statistics. Volume 2, Inference and Relationship, pp. ix+676, 132s. Charles Griffin, London (1961)
23. Norton Commander. https://ru.wikipedia.org

Visualization and Self-Organising Maps for the Characterisation of Bank Clients

Catarina Maçãs, Evgheni Polisciuc, and Penousal Machado

Abstract The analysis and detection of fraudulent patterns in banking transactions are of most importance. However, it can be a laborious and time-consuming task. We propose a visualization tool—VaBank—to ease the analysis of banking transactions over time and enhance the detection of the transactions' topology and suspicious behaviours. To reduce the visualization space, we apply a time matrix that aggregates the transactions by time and amount values. Additionally, to provide a mechanism to characterises the different sub-sets of transactions and facilitate the distinction between common and uncommon transactions, we represent the transactions' topology through matrix and force-directed layouts. More specifically, we present: (i) a visual tool for the analysis of bank transactions; (ii) the characterisation of the transactions' topology through a self-organising algorithm; (iii) the visual representation of each transaction through a glyph technique; and (iv) the assessment of the tool effectiveness and efficiency through a user study and usage scenario.

Keywords Information visualization · Glyph · Finance · Profiling · Self-organising maps

---

This work is funded by national funds through the FCT—Foundation for Science and Technology, I.P., within the scope of the project CISUC—UID/CEC/00326/2020 and by European Social Fund, through the Regional Operational Program Centro 2020. The first author is funded by the Portuguese Foundation for Science and Technology (FCT), under the grant SFRH/BD/129481/2017 and by the project PS0738lLEXIA, under the grant 679746.

C. Maçãs (✉) · E. Polisciuc · P. Machado
Department of Informatics Engineering, University of Coimbra, Centre for Informatics and Systems of the University of Coimbra, Coimbra, Portugal
e-mail: cmacas@dei.uc.pt
URL: https://cdv.dei.uc.pt/authors/catarina-macas/

E. Polisciuc
e-mail: evgheni@dei.uc.pt

P. Machado
e-mail: machado@dei.uc.pt

B. Kovalerchuk et al. (eds.), *Integrating Artificial Intelligence and Visualization for Visual Knowledge Discovery*, Studies in Computational Intelligence 1014,
https://doi.org/10.1007/978-3-030-93119-3_10

1 Introduction

The analysis of financial data and the search for fraudulent activities may prevent possible future losses for institutions and their clients and, for this reason, it is a task of high importance. The management of fraud usually focuses on three main pillars: (i) detection, defined by a continuous monitoring system that measures and evaluates possible fraudulent activities; (ii) prevention, defined by a preventive method that creates barriers to fraud; and (iii) response, defined by a set of protocols that should be applied when fraud is detected [1]. In this work, we focus on the first pillar and develop a visualization tool that aims to facilitate the analysis of financial data and the detection of fraud.

Nowadays, experts in charge of fraud management support their analysis on tabular data, usually presented in the form of a spreadsheet and seldom supplemented with simple visualizations. With those methods, the inspection of irregularities and suspicious behaviours can be laborious, time-consuming, and arduous. Regarding the data to be analysed, it can be in a raw state or be the result of a previous analysis from Machine Learning (ML) systems trained to detect fraudulent behaviours. In both cases, experts' current tools may be of little use for the analysis and overview of such complex data. Additionally, as technology evolves and the techniques applied to detect fraud become publicly available, fraudsters adapt and modify their ways of acting [2]. This may prevent Machine Learning (ML) models from correctly detecting all fraudulent transactions and may lead to their incorrect classification. For this reason, investing only in ML algorithms for the detection of fraud can lead to undetected fraud cases.

To solve the aforementioned problems, especially the lack of tools to analyse both raw and pre-processed data, Information Visualization can be applied. Through visualization models that emphasise data patterns, it is possible to make fraud detection more reliable, effective, and efficient [2, 3]. Also, through the combination of computational means with our visual cognitive intelligence [4] and by enabling the detailed analysis of each suspicious behaviour that still needs to be carefully investigated, visualization can facilitate the analysis of financial data and reveal new undetected fraudulent patterns.

The visualization tool developed in this work is the result of a collaboration with Feedzai.[1] Feedzai is a world-leading company specialised in fraud prevention that owns a risk management platform powered by big data and ML. This platform is used mainly to identify fraudulent payment transactions and minimise risk in the financial industry (e.g., retail merchants and bank institutions). With their platform, Feedzai provides its clients with the possibility to analyse information and keep their customers' data and transactions safe. Also, Feedzai has its own fraud analysts whose goal is to provide a more detailed and humanised analysis of the data and detect unknown patterns of fraud. In this context, Feedzai highlights the need to design systems that can take advantage of the complementarity between humans and

[1] Feedzai (https://feedzai.com) is the market leader in fighting financial crime with Artificial Intelligence. One of their main products is an advanced risk management platform.

machines, to surpass the current limitations of humans and machines, while being provably beneficial [5, 6].

Feedzai's main goal is to provide its analysts with a visualization tool that enables the proper analysis and characterisation of different subsets of data—containing transactions made by several clients of a specific bank.[2] The analysts' level of expertise in Information Visualization is reduced and their experience in analysing fraud may vary. During their analysis process, there is a lack of tools to aid them. In fact, they only have at their disposal spreadsheet-style tools and a limited data analysis platform, which makes the analysis of fraud an overwhelming task. In this matter, the creation of our visualization tool—*VaBank*—is intended to hasten their analysis process by giving information about data relations, such as time intervals between events, similarity, and recurring patterns, and to provide an overall sense of scale in the financial time-oriented data. In the end, these findings may lead to the detection of suspicious behaviours and the finding of fraudulent activities, enabling the banks to take action.

More specifically, with VaBank, we aim to ease the analysis of the distribution of bank transactions over time, the detection of the main characteristics and topology of those transactions and, with this, aid in the detection of suspicious behaviours. These objectives were based on the goals of Feedzai's fraud analysts, which can be summarised as: (i) be able to inspect collections of transactions in a single place—usually, these collections are grouped by attributes, such as client ID or location IP; (ii) understand the overall behaviours of a set of transactions; and (iii) detect the most common types of transactions.

Our tool is implemented in Java and uses Processing[3] to render the visualization. VaBank is divided into two main areas, the visualization of the transactions' distribution over time and the visualization of the client's topology of transactions. In the latter, we apply a Self-Organising Map (SOM) algorithm to represent the topology of a subset of transactions, enable the detection of the most common type of transactions, and with this, characterise the client main behaviours. Other visual representations for multi-dimensional data were not considered as we aimed to create a coherent visual representation of the transactions between the two main areas described above. For example, if parallel-coordinates were to be used, the visualization in the first main area would augment in complexity, as multiple parallel-coordinates would be needed to represent every transaction at the different periods of time. The challenge to create a single representation led us to the definition of a glyph simple enough to be perceived at first glance, but with another level of detail for a more thorough analysis. Also, SOMs have already proved its usefulness and robustness for the analysis of large amounts of data [7]. The visualization of their results provide a visual summary of the data topology and can ease the interpretation of behaviours in a single image [8, 9]. We present the SOM's results through

[2] Note that due to the high sensitivity of the dataset, it was previously anonymised and encrypted, but retained the fraud patterns of the real datasets. This enables us to visually explore the data in real case scenarios, without compromising the users' anonymity.

[3] Processing is an open-source graphical library.

two visualization techniques: a matrix and force-directed projections. Both aim to represent the profiling of a group of transactions and enable the understanding of the characteristics of the most common transactions. Finally, the transaction history visualization provides a set of analytical features, enabling the analyst to navigate, explore, and analyse the sequence of transactions over time.

Our main contributions are (i) a user-centred visual tool, developed with the aid of fraud experts; (ii) a method that characterises the topology of the transaction through a SOM algorithm; (iii) the visual characterisation of transactions through complex glyphs; and (iv) a usage scenario and a user study that assess the tool effectiveness. Based on the analysts' feedback, we could conclude that our tool can improve substantially their line of work which currently involves the time-consuming analysis of spreadsheets.

The current article extends the work presented in [10] in two main aspects. First, we give a more detailed description of the dataset and its processing. Also, we extend our description of the tasks, the VaBank design, and how we performed the user testing with experts in fraud analysis. Second, we expand the validation of the VaBank tool with three usage scenarios. The usage scenarios goal is to highlight the efficiency and effectiveness of VaBank in enabling a detailed analysis of the transactions.

The remainder of the article is structured as follows. In Sect. 2, we present the related work on fraud visualization and self-organising algorithms. In Sect. 3, we introduce the dataset and how it was processed. In Sect. 4, we present the tasks and requirements of our tool and, in Sect. 5, we give a detailed description of the VaBank design. In Sect. 6, we describe three different usage scenarios and, in Sect. 7, we present a user testing with experts in fraud analysis. In Sect. 8, we discuss the results of the tests. Finally, in Sect. 9, we present our conclusions.

2 Related Work

The visual exploration of data has already proved its value concerning exploratory data analysis, as the user is directly involved in the exploration process, adjusting its goals along the analysis [11]. In this section we present the related work on the visualization of fraud in the finance domain. Additionally, we present the related work on Self-Organising Maps.

2.1 Visualization of Fraud in Finance

In what concerns the visual highlight of fraudulent activities, and regardless of the domain of application, the most common visualization techniques are line and bar charts and node-link diagrams. These techniques are used to represent changes over time, facilitate the comparison of categorical values, and represent networks and relationships, respectively. Focusing only on the financial domain, two surveys present a set of projects which apply techniques, such as parallel coordinate plots, scatterplots, and bar and line charts [12, 13]. For a more detailed description of the techniques and the different taxonomies, please refer to the works [12, 14].

For the representation of specific financial fraud patterns, six works can be found concerning the visualization of (i) *Stock Market Fraud*, which focus on the analysis of abnormal changes in stock market values along time [15, 16]; (ii) *Profile Analysis*, which focus on the analysis of personal bank transactions [17]; (iii) *Credit Card Fraud*, which focus on the analysis of improper use of credit cards [18]; and (iv) *Money Laundering*, which focus on the analysis of the network of transactions [19, 20]. From these, four projects [17–20] focus merely on the improvement of the respective automatic evaluation systems, not applying visualization for the manual analysis of fraud cases. Also, in the work of Sakoda et al. [18], they visualise directly the fraud labels given by the ML system, not giving further details of each transaction to enhance its analysis. Finally, from this subset, most tools apply more than one visualization technique in separate or multiple views.

From our research, we only found one visualization model related to the visualization of bank data. Wire Viz [21], is a coordinated visualization tool that aims to identify specific keywords within a set of transactions. Also, they apply different views to depict relationships over time. For example, they use a keyword network to represent relationships among keywords, a heatmap to show the relationships among accounts and keywords, and a time-series line chart to represent the transactions over time. Their goals are to give an overview of the data, provide the ability to aggregate and organise groups of transactions, and compare individual records [21].

With this research, we could conclude that the analysis of fraudulent activities through visualization is gaining popularity, but its use to detect specific types of fraud is uncommon. In the case of bank transactions, the only related work uses a different type of dataset, which contains transactions to and from other banks, whereas, in the dataset made available by Feedzai, we only have access to the transactions made from the accounts of a specific bank. With this dataset, we are not able to follow the connections between different transactions, being our main aim to characterise the transactions of specific clients that may be referred to as suspicious cases. We argue that to properly understand the behaviours of a certain client, a more detailed analysis of his/her patterns of transactions must be conducted, so it is possible to distinguish atypical and suspicious transactions from the common transactions.

2.2 *Self-Organising Maps*

Self-Organising Map (SOM) take advantage of artificial neural networks to map high-dimensional data onto a discretised low-dimensional grid [22]. Therefore, SOM is a method for dimensionality reduction that preserves topological and metric relationships of the input data. SOMs are a powerful tool for communicating complex, nonlinear relationships among high-dimensional data through simple graphical representations. Although there are multiple variants, the traditional SOM passes through different stages that affect the state of the network [22]. In the first, all neurons are initialised with random values. Then, for each datum of the training data input, the so-called Best Matching Unit (BMU) is defined. This is done by computing Euclidean distances to all the neurons and choosing the closest one. Finally, the weights of the BMU and the neighbour neurons are adjusted towards the input data, according to a Gaussian function—which shrinks with time. This process is then repeated for each input vector for a predefined number of cycles.

Since the present work deals with mixed data, we present SOM algorithms which work with that type of data. The topological self-organising algorithm for analysing mixed variables was proposed in [23], in which categorical data is encoded to binary variables. Also, the algorithm uses variable weights to adjust the relevance of each feature in the data. Hsu et al. [24, 25] proposed another method in which they use semantics between attributes to encode the distance hierarchy measure for categorical data. Similarly, the authors in [26] use semantic similarity inherent in categorical data to describe distance hierarchy by a value representation scheme. The authors in [27] use distance hierarchies to unify categorical and numerical values, and measure the distances in those hierarchies. Finally, in [28] a frequency-based distance measure was used for categorical data and a traditional Euclidean distance for continuous values.

Visualization of Self-Organising Maps The visualization of SOMs is typically concerned with the projection of neurons into a 2D/3D grid. The most common projection is the Unified Distance Matrix (U-Matrix), in which neurons are placed in a grid and the Euclidean distances between neighbouring neurons are represented through a greyscale colour palette. This visual mapping can be used in the detection of clusters [29, 30] or in the definition of thresholds [31]. Additionally, hexagonal grids [32] can also be used [33], increasing neighbourhood relations, although not always resulting in more detailed insights [33]. The results of SOMs have also been used as data inputs for other visualization models. In most cases, researchers used SOMs to define clusters or characterise different behaviours and then represent such groups in the visualization models. In [34], a 3D SOM was used to define clusters categorised visually with colour, which later is applied in geographic areas with different characteristics. In [35], SOM was also used to define clusters in data, and then those clusters were represented through various visualization models, such as parallel coordinates and Chernoff faces. In fact, the usage of Chernoff faces and glyphs, in general, was found in multiple works, which will be discussed in more detail later. Finally, in [36], the clusters resulting from the SOM algorithm were visualised through a two views visualization, consisting on the representation of the clusters on a map and in a temporal grid.

To improve the reading and understanding of each neuron, some works used more complex glyphs. In [37], the neurons are represented through a timeline, portraying the temporal profile of call logs, and, in the background, a circle is drawn with the size depending on the number of elements used to train each neuron. In [38], each neuron is represented by a squared glyph coloured according to the quantisation error and, inside each square, a line is drawn to represent a certain trajectory. In [39], the neurons are represented with a radar glyph which shows the consumption value of a specific product. Finally, in [40] a rose diagram is applied to represent the weights of each feature of the SOM.

Self-Organising Maps in Finance The application of SOM algorithms to analyse transactional data have been applied in a variety of projects. The majority of them apply SOMs to provide an analytical view on the financial market trajectories [38, 41, 42] and to analyse their stability and monitor multi-dimensional financial data [43]. Other works applied SOM to better comprehend the stock market dynamics[44] or to analyse the financial performance of companies [7].

3 Data Analysis and Preprocessing

We worked with an anonymised dataset that contains only the transactions generated by the clients of a certain bank—there is no data about the transactions that each client received. Each transaction of the dataset is characterised by attributes corresponding to the: client (e.g., ID, IBAN), location (e.g.,Client IP, Country IP), monetary amount (e.g., amount, currency), transaction (e.g., type, descriptor, fraud label), beneficiary details (e.g., IBAN), and date. Each transaction can be of two types: online, corresponding to regular transactions; and business, corresponding to business transactions. All clients can have transactions of both types. The transactions also have a descriptor, composed of two or three acronyms, that characterise the transaction according to (i) the interface used; (ii) the type of operation (e.g., national, international, loan); and (iii) whether it is for a new beneficiary or not. These characteristics must be known by the analysts, so they can be properly analysed. This task has a high level of difficulty as these descriptors can have different combinations. To enable a better understanding of the descriptor elements, we herein list them according to each type of transaction:

Business Transactions:

- Type of Interface: ATM Specific, Telephone, ATM, and Branch
- Type of Operations: Cash In and National
- Type of Beneficiary: New and Old.

Online Transactions:

- Type of Interface: Barc. Mobile, MBWay, Web, and App
- Type of Operations: Instant, International, National, Loan, Address change, and Agenda
- Type of Beneficiary: New and Old.

Additionally, all transactions are labelled by the bank as fraudulent or not. For this project, we group dynamically the transactions of a certain subset in different range scales in two axes: time and monetary amount. In terms of time granularity, the time axis can be divided in different ranges of days, being one day the smallest granule possible. Also, to be able to properly summarise the data, for each pair $[time, amount]$ we aggregate the transactions with the same characteristics (i.e., the same values for the attributes type, descriptor, and fraud label).

3.1 SOM Algorithm

We applied a variant of the Frequency Neuron Mixed Self-Organising Map (FMSOM), a SOM algorithm prepared to handle mixed data [28]. It consists of preserving the original algorithm for handling the numerical variables and extending the neuron prototype with a set of category frequency vectors. The algorithm follows the traditional *competition, cooperation* and *adaptation* process. Since we focus on the visualization tier of the SOM and not on the algorithm, any other method could be used. However, the Frequency Neuron Mixed Self-Organising Map (FMSOM) model allowed us to adapt it to define the dissimilarity between neurons, used in the visualization of the transaction's topology.

Features First, we extracted the features for each input raw data. In our project, 7 features and their types were identified: *amount*, *day* of week, *month* of the year, *year*, *time passed* since the last and until the next transactions (in milliseconds), *fraud*, *transaction type*, *operation type*, *beneficiary*, and *interface channel*. The later five features were briefly described in Sect. 3 and cannot be fully revealed due to the specificity and sensitivity of the dataset. The *amount* is the amount of money involved in the transaction. From the date of a transaction, we extract only the day of week $[1-7]$, the month of the year $[1-12]$, and the year. The features *time passed* since the last transaction and until the next transaction are previously calculated and are intended to capture the patterns of the transactional regularity.

Dissimilarity Metric We applied different measures to compute the distances between neurons. We applied the traditional Euclidean distance for continuous values and the measure based on probabilities (described in [28]) for categorical features. Ultimately, two types of dissimilarity measures were defined: one for the training of the SOM; another for the visualization.

Regarding the SOM domain, as in FMSOM [28], the dissimilarity measure between the neuron and the input feature vector consists of the following. Suppose that P is the number of input feature vectors $X_p = [x_{p1}, ..., x_{pF}]$, where F is the number of features in that vector. Also, suppose that n and k are the number of continuous and categorical features, respectively, where $[a_k^1, ..., a_k^r]$ is the set of categories of the k_{th} feature. Finally, suppose that the reference vector of the i_{th} neuron is $W_i = [W_{i1}, ..., W_{in}, W_{in+1}, ..., W_{iK}]$, where I is the number of the neurons in the network. With that said, the dissimilarity between an input vector and the reference vector of a neuron is defined as the sum of the numerical and categorical parts. The numerical part is calculated using Euclidean distance on normalised values. For the categorical dissimilarity measure the sum of the partial dissimilarities is calculated, i.e., the dissimilarity is measured as the probability of the reference vector not containing the category in the input vector. For more details on the FMSOM algorithm consult [28].

Regarding the visualization domain, the dissimilarity measure between two neurons is determined as follows. For the numerical part, the traditional Euclidean distance is applied $Dn(W_i, W_j) = \sqrt{\sum_{z=1}^{n}(W_{iz} - W_{jz})^2}$. For the categorical features the dissimilarity measure was defined as the Euclidean distance between the probabilities for each of the categories present in the reference vector $Dk(W_i, W_j) = \sqrt{\sum_{z=n}^{k}\sum_{m=1}^{r}(W_{iz}[a^m] - W_{jz}[a^m])}$. So, the final dissimilarity measure is given by $d(W_i, W_j) = Dn(W_i, W_j) + Dk(W_i, W_j)$.

4 Tasks and Requirements

From our collaboration with the fraud detection company, we were able to hold several meetings with their analysts, which aided us in better defining the domain and requirements for the analysis of the bank data. The analysts emphasised two main tasks: **[T1]** comprehend the transaction history; and **[T2]** detect the most common types of transactions. The latter is especially important as it enables the distinction between typical and atypical behaviours.

The analysts described their line of work, referring that their analysis usually starts by grouping the data by a specific attribute. Usually, they group the transactions by a specific client ID to better analyse and characterise the client's transactions. Then, they search for groups of transactions with similar characteristics, especially the ones labelled as fraud. This task is especially difficult using a spreadsheet, as if the transactions are not ordered, common attributes will not stand out. From all attributes, the analysts referred to the amount spent, type of transaction, and fraud label, as the most relevant. In the end, they referred to the importance of detecting similar transactions and identifying the profile of the client or a subset of transactions. Through our meetings, the analysts defined five requirements to which VaBank should comply:

R1 Search by field. The analysts usually sort the data by a certain field, such as client IBAN, client ID, or Country of IP, and analyse the transactions with common values on those fields. The creation of a mechanism that enables the analyst to easily select a field and choose a certain value of that field to group the transactions is of utmost importance. This will speed up the analysis process and ease the analysis of all transactions with common values;

R2 Distinguish amount values. When dealing with bank transactions, the transacted amount can be a sign of fraudulent activity, being transactions with high amounts, or above a certain threshold worth of a more detailed analysis. The visual sorting of the transactions by their amounts can enhance the detection of suspicious transactions;

R3 Distinguish transactions. By visually characterising each transaction, the analysts can more easily distinguish the transactions and focus their attention on the ones of the same type. With this, they can perceive the behaviours within the different types of transactions, facilitating the detection of atypical behaviours;

R4 Search common fields. When dealing with this data through spreadsheets, the analysts have difficulties in detecting transactions that share more than one attribute. This is of utmost importance when analysing fraudulent transactions which can share attributes with others. For this reason, it is important to implement a mechanism that enables the analyst to select an attribute and highlight all transactions with that same attribute.

R5 Detect typical transactions. Understanding the most common types of transactions can enhance the analysis of the data and aid the analyst in the detection of unusual transactions, which can be related to fraudulent behaviours. Hence, it is important to characterise the space and facilitate the detection of typical transactions in a certain subset of the data;

5 VaBank Design

The tool is divided into three views: the *Transactions History*; the *Transactions Topology*; and the *Transactions Relationships*. The first view (Fig. 1) aims to answer the task **[T1]** and arranges all transactions by time and amount. The last two views aim to answer the task **[T2]** and display the results of the SOM algorithm (see Sect. 3.1) in a grid and through a force-directed graph, respectively. With these views, for the same subset of the data, we aim to enable the analysis of the transactions by time (i.e., first view), and enable the analysis of their topology (i.e., second and third views). All three views share a common visual element, the transactions. We developed a glyph that serves to identify the type of transaction and its position in time. With this, we aim to facilitate the distinction between transactions with different characteristics and to provide coherence between views. In the following subsections, we present the design rationale of the glyph and the three views.

5.1 *Transaction Glyph*

To ease the distinction and visual characterisation of the transactions, we implemented a glyph **[R3]**. The glyphs are composed of three levels of visual detail. These levels were defined together with the company's analysts, according to the relevance of the types of attributes when analysing bank data. First, the analysts aimed to distinguish online transactions from business transactions. Then, the transaction amount and whether it was considered as fraud or not are analysed. These three characteristics represent the first level of visual impact. Then, the analysts want to drill down and distinguish between: inbound and outbound transactions; and new and old benefi-

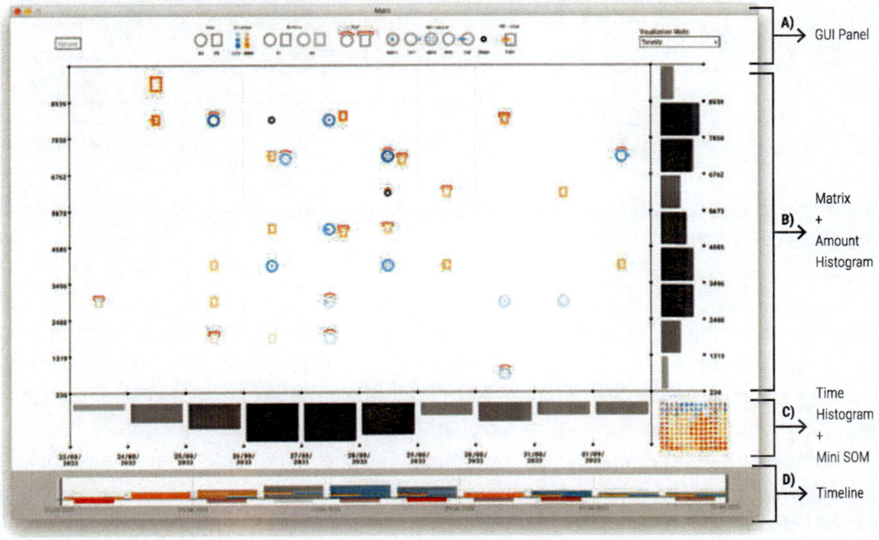

Fig. 1 Transaction history view and its components, from top to bottom: GUI panel (**a**); Matrix view and amount histogram (**b**); Time histogram and mini SOM (**c**); and Timeline (**d**)

ciaries. These characteristics represent the second level of visual impact. Finally, the time characterisation of each transaction and the interface with which the transaction was made were defined as less important than the described above. For this reason, they are grouped into the third level of visual impact and should have a lower visual impact.

As colour has a high impact on visualization [45], we apply colour to emphasise the characteristics of the first visual level. We apply different hues to the types of transaction: orange for business; blue for online. Additionally, we use different shapes to emphasise this distinction between transaction types—a rectangle for business; and a circle for online. Then, we use saturation to represent the amount: the brighter the colour, the higher the amount. As small differences in saturation would be imperceptible to the human eye, we defined three levels of saturation to distinguish: low, medium, and high amounts. These levels are computed as follows. We compute the average amount $\overline{x}$, define a window w, and if the value is: below $\overline{x} - w$, we consider the amount as low; between $\overline{x} - w$ and $\overline{x} + w$, the amount is medium; and higher than $\overline{x} + w$, the value is high. The window w is a percentage of the average value that was defined in collaboration with the company's analyst. Finally, to represent a fraudulent transaction, we place a red line above the main shape (see Fig. 2a). Note that these colours were not tested on colour blind people.

The transactions' shape is complemented with a set of symbols that represent the types of operation. They are divided according to the directionality of the transaction,

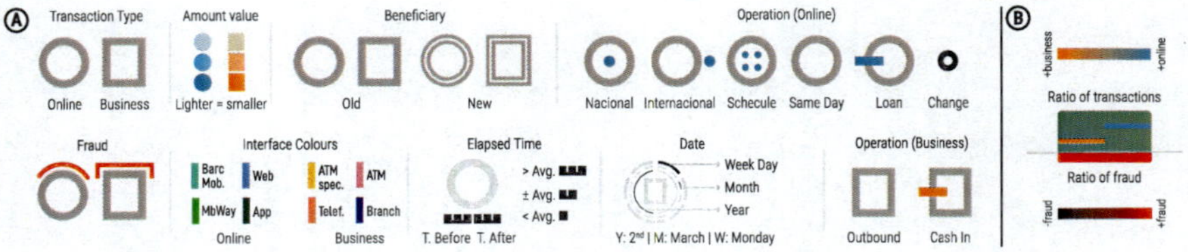

Fig. 2 Glyph elements that characterise each transaction (side **a**) and timeline bar composition and respective colour ranges (side **b**)

outbound or inbound.[4] The inbound is represented by the same symbol in online (i.e., Loan) and business (i.e., Cash In) transactions: a vertically centred horizontal rectangle positioned on the left. The outbound operations are represented as depicted in Fig. 2, in which the business transaction only have one type, and the online transactions have five. As the new beneficiary characteristic is a binary value, we represent transactions for new beneficiaries by dividing the stroke of the main shape in two. If the beneficiary is not new, no change is made (Fig. 2a).

For the third level, we represent the year, month, and day of the week of the transaction. Each time variable is represented by a ring with a different radius centred in the main shape, being the year the smallest ring, and the day of the week the biggest ring (Fig. 2). To distinguish periods of time, we divide the ring into 7 wedges, for the days of the week; 12 wedges for the months; and, for the years, in the total number of years in the dataset. All wedges are coloured in light grey, except the wedge that marks the period of the transaction, coloured in black. The day of the week has a thicker stroke, as the analysts referred it is the most important time variable. We also represent the elapsed time between the current transaction and the previous and following transactions. We apply an equal rationale to represent these two-time distances. As with the amount thresholds, we defined three levels of time distances that are computed in the same way. These three levels are represented as depicted in Fig. 2. Note that for the sake of simplicity this data was aggregated, even though in the SOM we use absolute values. Finally, the interface of the transaction is represented by filling the elapsed time's shape with the corresponding interface colour (Fig. 2a).

The glyphs used in the views concerning the SOM's result make use of all representations described above. However, in the Transaction History view, we only represent the first two levels of visual detail, as time is already being represented in the x-axis.

[4] Note that the inbound are transactions made only by the client, when asking for a loan (in online transactions) or when doing a deposit (in business transactions).

5.2 Transaction History View

In this view, there are a set of visualization models that display different data aggregations. The main representation, which occupies more canvas space, is the *Transaction Matrix* (Fig. 1b). It divides the space into different ranges of monetary values on the y-axis and temporal values on the x-axis **[R2]**. The transactions' glyphs are then distributed by the cells of the matrix, according to their date and amount. If more than one transaction with the same characteristics (defined in Sect. 5.1) occurs within the same cell, they are aggregated and its glyph grows in size. The placement within each cell is made through a circle packing algorithm which starts by placing the biggest glyph in the middle of the cell and the others around it.

In the bottom and right sides of the Transaction Matrix, histograms are drawn to show the total number of transactions per column and row, respectively (Fig. 1b and c). The histogram's bars are coloured according to the number of transactions: the darker the bar, the higher the number of transactions. Also, in the bottom right corner of the Transaction Matrix area, we draw a small matrix of glyphs that represents the result of a SOM algorithm, concerning three attributes: amount, transaction type, and fraud (Fig. 1c). With this, we aim to give a visual hint to the analyst about the distribution of the different transactions, enhancing the understanding of typical/atypical transactions **[R5]**.

At the bottom of the canvas, we placed an interactive timeline, so the analyst can select different periods of time (Fig. 1d). This timeline represents all time-span. To be able to represent all data in the timeline, we applied a hierarchical time aggregation algorithm that aggregates semantically the transactions according to the space of the timeline (see Sect. 5.2). The timeline is divided horizontally into equal sections, representing different periods of time with the same duration. Each section of the timeline is vertically divided into two parts.

In the upper part, we represent the number of transactions through a bar. To put it briefly, each bar is drawn as follows: (i) its height represents the total number of transactions; and (ii) its main colour is defined by a gradient between blue and orange—the bluer, the higher the number of online transactions, the more orange, the higher the number of business transactions (Fig. 2b). With this, we aim to represent which type of transaction occurs the most. To give a more detailed view, we also represent the quantity of each transaction type with two thin rectangles which are drawn inside the previous bar. They are placed horizontally according to the type of transaction—being the business one on the left and the online one on the right—and are placed vertically according to the percentage of occurrence. Additionally, they are coloured according to the transaction type.

In the bottom part of each timeline section, we place a rectangle with a predefined height. This rectangle is only visible if one or more fraudulent transactions occur. Then, it is coloured according to the percentage of fraudulent transactions in that specific period of time. The higher the number of fraudulent transactions, the brighter and redder the bar will be (Fig. 2b). If no fraud occurs, no bar is drawn.

Hierarchical Temporal Aggregation Fixed timelines can create multiple problems (see for example [46, 47]). For example, different time spans can result in either a tremendously cluttered timeline; a timeline with an uneven distribution of the time bars (e.g., one bar on the left and the other bars concentrated on the right); or a timeline that uses inefficiently the canvas space, due to the time granularity (e.g., one thin bar on the left and another on the right). With our algorithm, we intend to solve the problem of fixed timelines. The main goal is to allow the representation of any temporal range where the timeline would adapt its granularity and adjust the size of time bars.

Our adaptive timeline algorithm takes as arguments the available space for the timeline and the minimal width of a time bar. The algorithm follows an iterative top-down approach. We start at the biggest time unit existing in the computation systems (e.g., epoch), and descent, iterating over consecutive ISO time units (e.g., years, quarter years, months) until we find an optimal balance between the time granularity and the size of the time bars. The algorithm has to meet one single criterion that is tested at each temporal resolution. Consider T_i being the time tier currently evaluated, T_{min} and T_{max} being the minimal and maximal timestamp of the selected data subset, W_{min} being the minimal allowed width for the bars, and W_{total} being the width of the timeline. So, the criteria to determine the time resolution and the width of a bar is computed as follows: $W_{total}/T_{i+1}(T_{max} - T_{min}) < W_{min}$.

Note that we compute the width of bars at the $i + 1$ temporal tier. If the bar width at the next tier is smaller than W_{min} we stop, and the current tier is the one that we are looking for. The left part of the expression is the found width of bars.

Interaction To enable the analysts to analyse the transactions in more detail, we defined a simple set of interaction techniques. In the Transaction Matrix, the analyst can hover each glyph to see more details—Country IP, amount, beneficiary, and the number of transactions. If the analyst clicks on a glyph, these details are fixed in the canvas. By doing so, the analyst can interact with each one of the attributes. If the analyst clicks on an attribute, the transactions which share that same attribute will be highlighted with a black ring. With this, we aim to enhance the understanding of the transactions that may share the same suspicious attributes **[R4]**. Also, the user can interact with each bar of the histograms. By hovering a bar, the total number of transactions is shown and the analysts can more easily perceive the total number of transactions in a certain period of time (x-axis) or the total number of transactions within a certain range of monetary values (y-axis).

We also defined a set of interaction techniques for the timeline. The analyst can select different periods of time to visualise in the transaction matrix. To do so, the analyst must drag two vertical bars which are positioned in the leftmost and rightmost parts of the timeline area. By selecting a shorter period of time, the transaction matrix will be more detailed (i.e., the different periods of time in the x-axis will have shorter durations, being one day the shortest possible). With this, we aim to enable the analysts to see in more detail the distribution of the transactions over time. Also, the analyst can drag the selected to maintain the selected duration, but change the initial and final periods of time. Finally, the analyst can hover each bar of the timeline.

By doing so, a set of statistics are made available concerning the total number of transaction in that period of time, the start and end dates, the percentage of online and business transactions, and the percentage of fraud.

5.3 *Transaction Topology View*

In this view, we visualise the result of the SOM algorithm defined in Sect. 3.1. The SOM algorithm uses all transactions available of a predefined bank client. To visualise its result, we use the positions of the neurons in the SOMs matrix to distribute the glyphs on the canvas within a grid with the same number of columns and rows (Fig. 3, top). Also, and as referred previously, we use the three levels of visual detail to represent each neuron (see Sect. 5.1).

This approach enables the analyst to visualise the most common types of transactions through the analysis of the distribution of the different glyphs (representing the transaction's characteristics) in the matrix.[5] However, this view lacks a more detailed representation of the dataset, which could enable, for example, the representation of how many transactions are related to each neuron and which neuron is the most representative of the dataset. The latter task is especially difficult to achieve when more than one feature is being represented in the glyphs, as it can hinder the comparison between glyphs. To overcome this, we implemented a second approach, in which we place each neuron within a force-directed graph and represent their relations to the transactions. With this, we aim to achieve a better understanding of the client's profile.

5.4 *Transaction Relations View*

For the force-directed graph, neurons and sets of transactions are represented as nodes and are positioned within the canvas according to their dissimilarity measure: the similar two neurons are, the closer they will get (Fig. 3, bottom). The force-directed graph can be seen as a simplification of the SOM result produced and visualised in the Transaction Topology View. Our implementation of the graph is based on the *Force Atlas 2* algorithm [48]. All nodes have forces of repulsion towards each other so they do not overlap. However, only nodes whose dissimilarity is below a predefined threshold have forces of attraction. This makes similar nodes to get closer to each other, generating clusters defined by the SOM topology. Additionally, we added a gravitational force that pulls all nodes towards the centre of the canvas. The higher the number of connections between nodes, the higher the gravitational force. With

[5] Note that, as in the analysis of any SOM, the number of glyphs in the canvas is not representative of the number of transactions within the dataset.

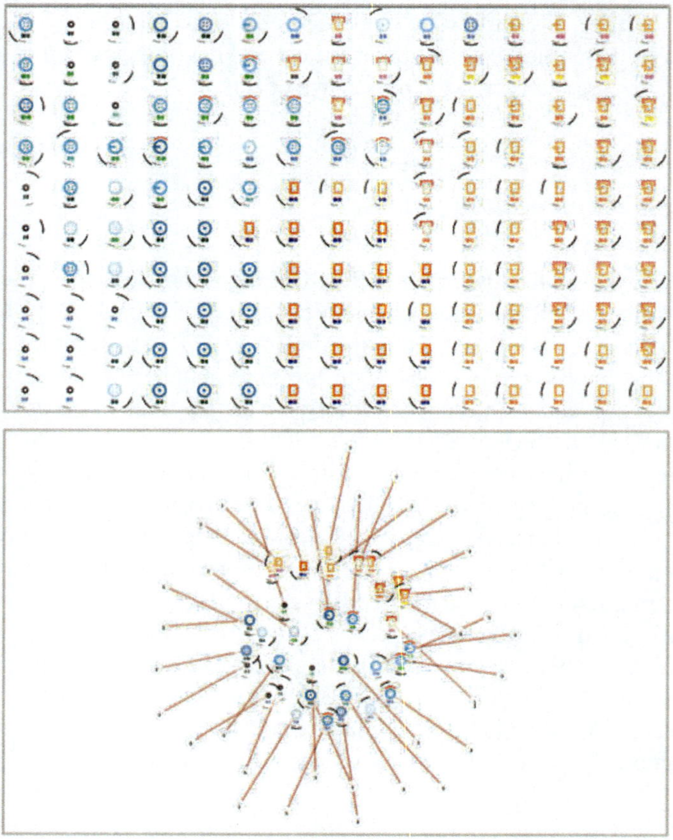

Fig. 3 Projections of the SOM results for the same bank client through the matrix projection (top) and force-directed graph (bottom). The self-organising map used in the matrix projection is generated with all transactions available from a certain client. The force-directed graph can be seen as a simplification of the matrix, as it only visualises the most representative neurons from the self-organising algorithm used in the matrix projection. Both projections aim to represent the most common transactions of a specific bank client and, therefore, characterise the client

this, clusters which are more representative of the dataset will be in the centre of the canvas, and the ones representing atypical transactions in the periphery.

To avoid clutter, only neurons selected as BMU in the training process of the SOM are represented. We opted to filter the neurons with this method, as the neurons that are selected as BMU are the ones which are more similar to the transactions within the dataset, and for this reason, are the ones which are more representative. Also, the transactions which have the same neuron as BMU are aggregated and this aggregation is represented with a node. These new nodes' forces of attraction are defined by their average force of attraction to other neurons.

The nodes have distinct representations. The neurons are represented with the glyphs described in Sect. 5.1. For the groups of transactions, we use a circular chart that represents the number of transactions by month of occurrence. This representation is intentionally simpler since our main goal is to give more visual impact to the result of the SOM. Also, if these nodes are connected to a certain neuron, it means they share similar characteristics, being redundant to use the glyphs approach.

We used lines to connect the nodes. These lines are coloured: (i) in red if they connect a node representing a group of transactions and their BMU neuron; (ii) in light grey, if they connect a group of transactions and other neurons which are also similar to them, but are not their BMU; and (iii) in blue, if they connect two similar neurons. These lines are represented to enhance the comprehension of the proximity of the nodes, but as they should have less visual emphasis, their opacity and thickness diminish according to the similarity values.

5.5 *Control Panel*

To enable a better transition between views we created a *Control Panel* (Fig. 1a). By clicking on the "Options" button, on the upper left corner, the *Options Panel* is shown, containing a list of all unique attributes of a predefined field—client ID. This list is scrollable and is sorted in an ascending way, according to the number of transactions of each client. Also, each row contains a set of statistics concerning the grouped transactions: the total number of transactions, the maximum, minimum, and average amount values, and the percentage of fraudulent transactions. On the *Options Panel*, the analyst can also access a list of fields and select a different one to group the transactions **[R1]**. On the upper right corner of the *Control Panel*, there is a dropdown that enables the analyst to change between the three views. Finally, in the middle of the *Control Panel*, a caption is shown to describe the glyphs that represent the transaction's characteristics (Fig. 1a). This caption is especially important due to the complexity of the glyphs: with it, the analyst can easily read the glyph without needing to memorise or search for the caption anywhere else.

6 Usage Scenario

In this section, we discuss three usage scenarios in which we analyse subsets of the dataset with fraudulent transactions. With this, we aim to highlight the efficiency and effectiveness of VaBank in enabling a detailed analysis of the data. In each scenario, we visualise the transactions made by a certain bank client in one month.[6] Due to the limited time range, all scenarios present a reduced number of transactions. However, we argue that this is not a limitation as our model is prepared to aggregate the data in different time ranges, and for this reason, a larger dataset would not add more difficulty to the analysis. Also, the number of transactions per period of time would

[6] This small temporal range is due to the limited accessibility to the data.

not change significantly, meaning that wider time spans would only result in bigger time ranges in the timeline. Nonetheless, with our timeline, the user can select smaller periods of time to reduce the time span being represented in the Transaction Matrix, enabling a more detailed analysis of each transaction.

6.1 Client A

In Fig. 4a, we can see the *Transactions History* View of *Client A*. We can instantly perceive, through the positioning of the first transactions, that the monetary value of those transactions is relatively low—concerning the other ranges of values visible in the y-axis. Also, by looking at the bar chart in the timeline, it is possible to understand that the transactions tend to occur periodically—there is an initial set of transactions, then no transactions are made in the following three days, then another set of transactions are made, and so on. On March 14, there was a business transaction with higher amount values that was marked by the bank as fraud. We can also see that Client A tried consecutively to make that type of transaction on the two following days with similar and smaller values, but got the same result, a fraud label by the bank. All of these transactions are Cash In operations, which means that the client attempted to add money to his/her account. Later, on March 2nd, we can see the same type of transactions with smaller values, however, this time they were not labelled as fraud. By looking at the small matrix—generated from the SOM—it is possible to see that the majority of the business transactions were considered fraudulent, especially the ones with high values.

When analysing the *Transactions Topology View* (Fig. 4b), we can verify the assumptions made previously and see that for the business transactions Client A used mainly the ATM interface (yellow) and for the online transactions the interface used was the Barc. Mobile. Also, we can see that the majority of the online transactions were of the national type and for new beneficiaries. Finally, by checking the *Transactions Relations View*, we can see these clear distinctions between online and business transactions (Fig. 5). In the cluster of business transactions, we can easily define two sub-clusters: the fraudulent transactions with high values and the ones with smaller values. Also, it is possible to see that the business transactions with low values were made on Tuesday, whereas the fraudulent ones occurred between Wednesday and Friday. For all these reasons, this client can be seen as suspicious.

6.2 Client B

In the second usage scenario, there were also visible fraudulent transactions (Fig. 6 (a)). Although the values are low, in comparison to the previous client, this client attempted several transactions of the business type with different amounts and aimed to add money to the account. When comparing these business transactions with the

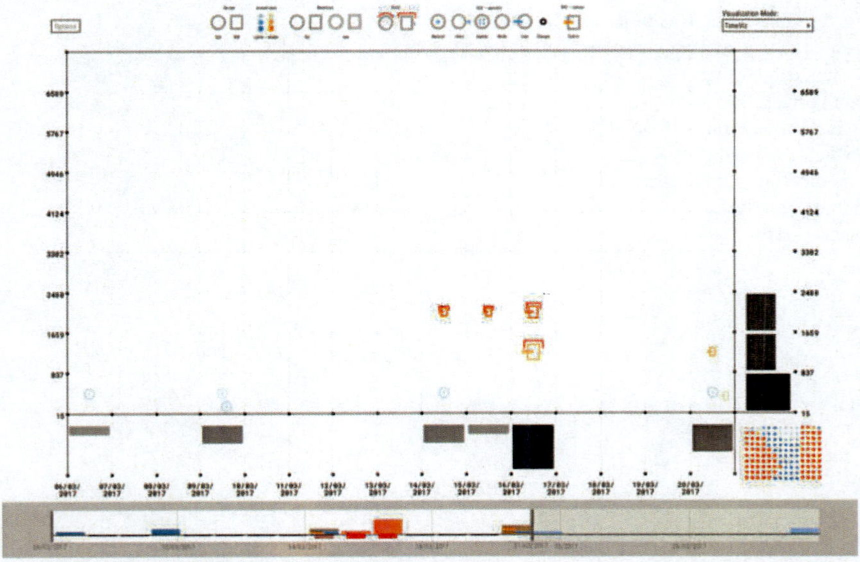

(a) Transaction History View. With this first view, it is possible to detect a group of business transactions

(b) Transactions Topology View. Through this view it is possible to perceive that the majority of the business transactions with higher values are considered as fraudulent.

Fig. 4 Two different views of Client A

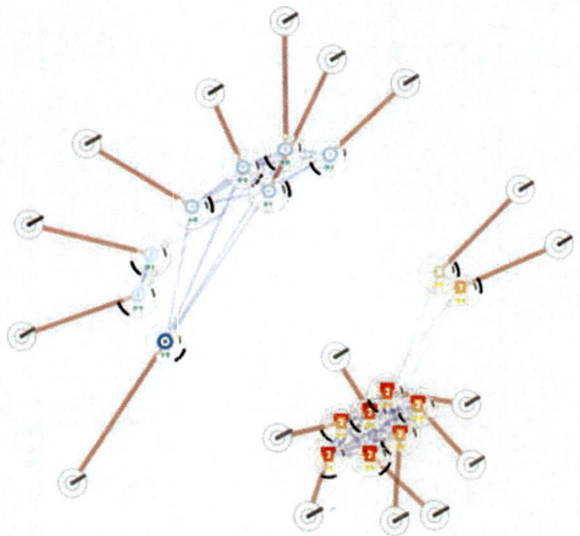

Fig. 5 Transaction Relation View of Client A. Two major clusters can be seen, separating online from business transactions. Also, the business cluster is subdivided into fraudulent and non-fraudulent transactions

rest of Client B transactions, which are usually placed below the €50 limit, the business transactions are of high value. Through the timeline, we can see that this client, after the peak of business transactions on March 10, made far fewer transactions. By looking at the small SOM matrix, we can see that this client behaves similarly to the previous one, as the majority of the business transactions are considered fraudulent and the online transactions are all of reduced value.

All the previous assertions can be verified with the *Transactions Topology View* (Fig. 6 (b)). The business transactions are, in the majority of the cases, considered as fraud, and the online transactions have in their majority smaller value ranges—in comparison to the business transactions. Additionally, the online transactions are divided into two subtypes: the ones of the agenda type and the others of the national type. With the aid of the *Transactions Relations View*, we can easily visualise these assumptions through the two well-defined clusters, one for the online transactions and the other for the business transactions (Fig. 7). Similarly to Client A, this client can be seen as suspicious.

6.3 Client C

When analysing the third client's data, we can see that it differs from the previous examples as the majority of the transactions are of the online type (Fig. 8 (a)). The values in these transactions fall in their majority in two different ranges: in the lowest range, from €30 to €500, and in the highest range, above €4200. Also in both ranges, there are fraudulent transactions of the online type. The fraudulent transactions are common in higher values, but not so common in lower ranges. In this case, we can

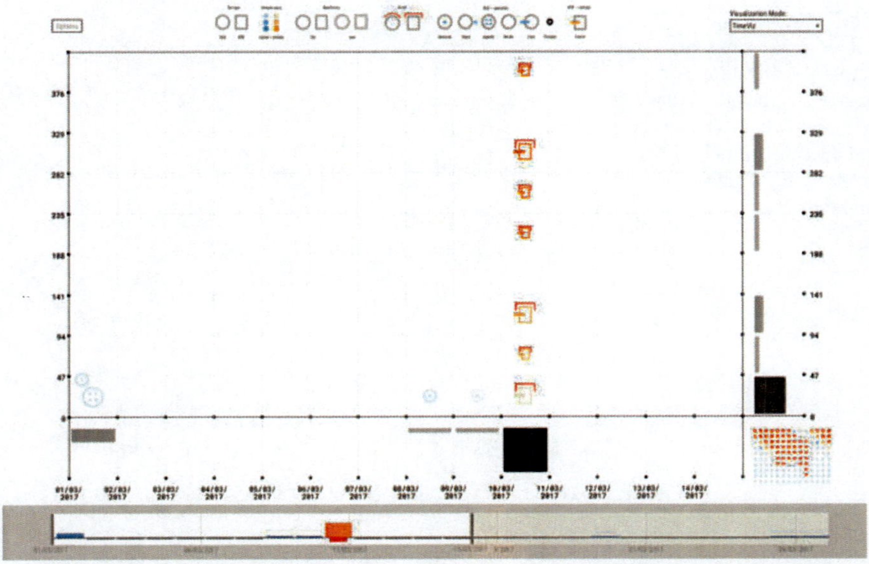

(a) Transaction History View. Similarly to Client A, Client B performs a set of fraudulent business transactions. This occurs in the same day with different amount ranges.

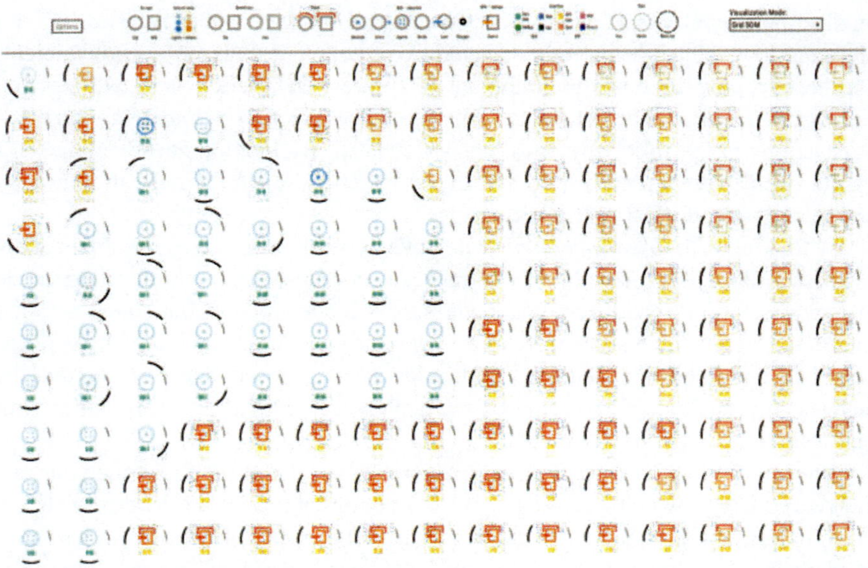

(b) Transaction Topology View. The majority of the neurons are of the business type and are considered to be fraudulent.

Fig. 6 Two different views of Client B

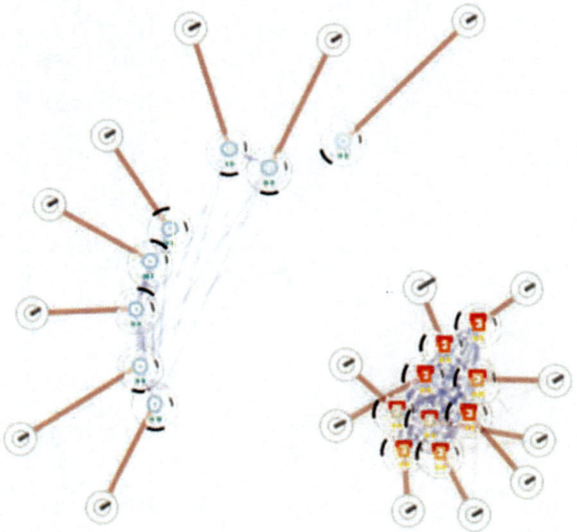

Fig. 7 Transaction Relations View of Client B. In this visualization, two clusters can be found, one of the fraudulent business type and the other of online transactions of small amounts

also see that, on March 5, there are fraudulent transactions in both ranges, which is uncommon. This client starts by doing an international transaction of low value and on the other day makes a few more national transactions to new beneficiaries of both high and low values, being only detected two fraudulent transactions out of eight. On the next day, there is a high amount of business transactions and two national transactions. This can be defined as a suspicious behaviour, since there are no more transactions in the following days, until March 26. By looking at the small SOM matrix, we can see a more varied SOM representation, in which fraud appears only on the national online transactions.

By looking at the *Transactions Topology View*, it is possible to see that the majority of the transactions have low-value ranges (Fig. 8b). Also, it is interesting to see that fraudulent transactions are made via Barc. Mobile and non-fraudulent national transactions are made via the web. This may indicate a breach in one of the applications and should be analysed in more detail. Also, it is possible to perceive that the business transactions of low ranges were made via ATM and the transactions with high values via Branch.

When analysing the *Transactions Relations View*, we can see three main clusters and two outliers—which are the business transactions of low and high ranges (Fig. 9). Also, in online transactions, we can see a distinction between fraudulent and non-fraudulent transactions. Among the non-fraudulent, we can distinguish four types of transactions: international, to new beneficiaries, national with low values, and national with high values. From this, we can refer that with the graph representation of the SOM's results, we were able to analyse more rapidly the different transactions and their relations.

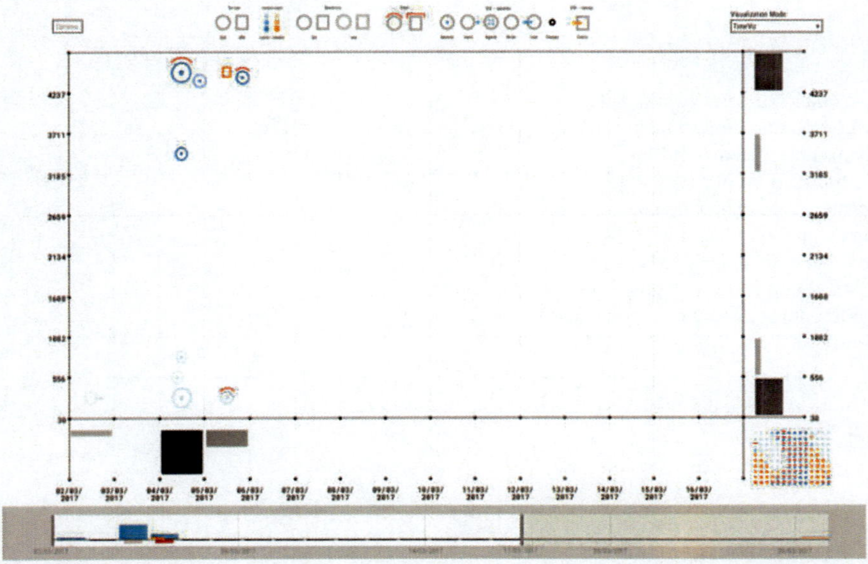

(a) Transaction Topology View. This client starts to perform a set of online transactions of small values and then performs a set of online transactions of high amounts, which are considered to be fraudulent.

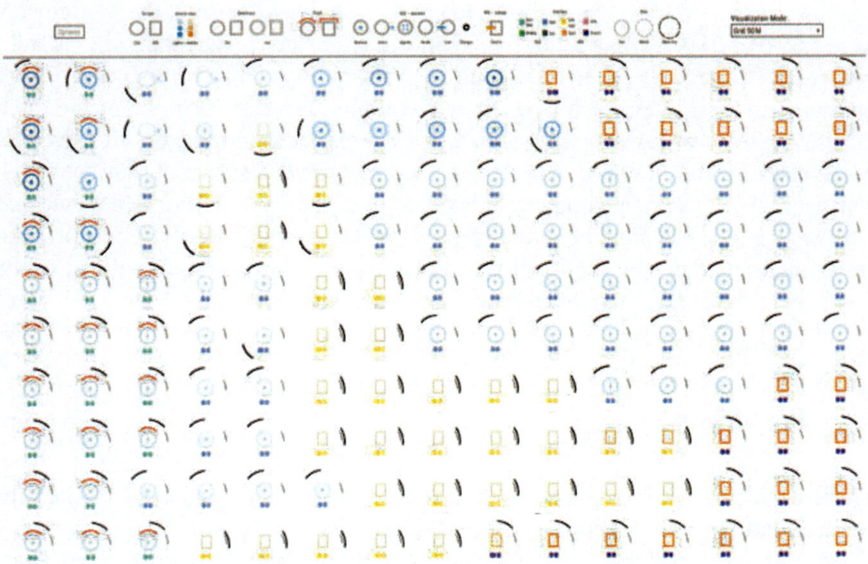

(b) Transaction Topology View. In this view, we can see a more varied transaction matrix, when comparing with Client A and Client B. This means that this client performs a more varied types of transactions.

Fig. 8 Two different views of Client C

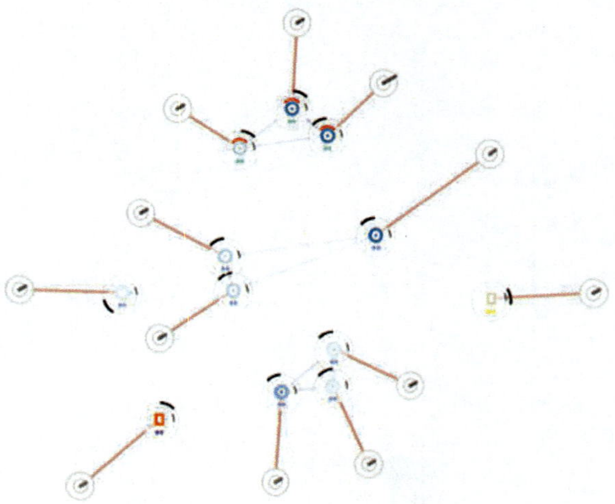

Fig. 9 Transaction Relations View of Client C. In this visualization, it is possible to perceive two types of transactions which can be considered as outliers, the transactions of the business type oh small and high amounts. Also, it is possible to distinguish two types of online transactions, the fraudulent, and the non-fraudulent

7 User Testing

To evaluate the tool's usefulness and effectiveness in the analysis of bank transactions, we performed user tests with a group of fraud analysts from the Feedzai company that were not present during the tool development. In this user testing, the participants were asked to (i) perform a set of specific tasks, (ii) to analyse the transactions from two clients through the interaction with the VaBank tool, and (iii) to give feedback on the aesthetics, interpretability, aid, and learning curve of each one of the three views. The tasks were defined to validate the models and determine the effectiveness of the visual encodings. The second part—the analysis of a subset of transactions—was defined to assess the complete functionality of the VaBank tool as a whole and whether the analysts were able to retrieve insights from the visualizations, proving is usefulness in the analysis of bank transactions. The third part was defined so the opinions of the analysts could be registered and analysed.

7.1 Participants

The user testing was performed by five fraud analysts. These analysts worked for Feedzai, but have no a-priori knowledge about the VaBank tool. On average, they worked in fraud analysis for five years, being the participant with the least years of experience, been working for three years, and the one with the most years of experience, been working with fraud for eight years. Also, three of the analysts had no experience with working with Information Visualization, and the other two had a reduced number of interactions with the field. Despite this being a reduced number of participants, this user testing aimed at understanding the impact of a tool such as VaBank in the analysis process of fraud experts—which are more used to deal with spreadsheets. For this reason, we believe that this number of participants was sufficient to fulfil the test requirements and provide a general sense of the VaBank impact on their analysis process.

7.2 Methodology

The tests were performed as follows: (i) we introduced the glyphs of the transactions, the views of the tool, and respective interaction mechanisms; (ii) we asked the analysts to perform 18 tasks concerning: the *Transactions History View* (6), the interpretability of the glyphs (4), the *Transactions Topology View* (4), and the *Transactions Relations View* (4); (iii) then, the analysts analysed two clients in terms of fraudulent behaviours; and (iv) the analysts were asked to give feedback on the models concerning aesthetics, interpretability, aid in the analysis, and learning curve. The second and third part of the tests were timed and, at the end of each task or analysis, the analysts were asked to rate the difficulty of the exercise and certainty of their answers on a scale from 1 to 5—from low to high, respectively.

The 18 tasks of the user testing were divided into 4 groups, depending on the component they aim to validate: **G1** Transaction History view; **G2** Transaction glyphs; **G3** SOM Matrix; and **G4** SOM Graph. In the Transactions History View, we tested the analysts' ability to comprehend temporal patterns and the transactions' distribution concerning time and amount values. In the views related to the SOM projections, we aimed to compare both views and perceive which one was more useful and efficient in solving tasks like counting clusters and identifying all glyphs from a certain attribute. For this reason, the tasks are equal for both views.

The third part of the test—which is concerned with the interaction with the VaBank tool and the analysis of two different clients' data—aims to understand the tool's usefulness and its ability in aiding the analysts to detect suspicious patterns and possible frauds. During the performance of this part of the test, the analysts were asked to explore and analyse the visualization, explain out loud what they were seeing at each moment of their exploration, and refer to whether the client was fraudulent, non-fraudulent, or suspicious.

The final part of the test was also intended to give to the analysts the opportunity to express their opinions on the tool. Although such feedback might be subjective, it is an indicator of the tool's impact within the analysts' workflow and can give clues on its effectiveness and efficiency.

All tests occurred in the same room within the Feedzai installations and were performed under the same conditions (i.e., the participants had access to the same computer and performed the test in the same sequence). We recorded the audio from each test so we could analyse each session afterwards.

7.3 Results

In Fig. 10, we summarise the results concerning difficulty, certainty, accuracy, and duration for each group of tasks. Hereafter, we further analyse each group of tasks, discuss the results from the third part of the test, and analyse the analysts' feedback.

Tasks Analysis The tasks related to the analysis of the *Transaction History View* (**G1**) and glyphs (**G2**) were the ones which arouse more difficulty. Nonetheless, all values are low, considering that on average the difficulty was no higher than two (i.e., the second-lowest level of difficulty). Regarding the Transaction History View, the analysts had more difficulties in interpreting the positioning of the glyphs in the grid and the histograms. For example, for the task "In which period of time the business transactions had the highest amount?", some analysts started to look at the histogram on the right, which gives the total number of transactions for each range of amount values. However, as this was the first question of the test, they were still assimilating all the information and rationale of the tool. The analysts also had some difficulty in interpreting the glyphs, which made their certainty to be lower than the other groups. Nonetheless, the certainty on average was no lower than four (i.e., the second-highest level of certainty). Also, an interesting point is that the accuracy of the analysts' answers for the glyphs tasks is higher than the accuracy for the Transaction History View tasks. With this, we could perceive that, as the glyphs were complex, the analysts were not certain if they were characterising all their attributes correctly, which caused the lower rates of certainty. Nonetheless, in the majority of the tasks related to the glyphs, their answers were accurate.

The groups of tasks related to the SOM analysis—Transaction Topology View and Transaction Relations View—took less time to perform (20 s, on average), had 100% of accuracy, and were the ones in which the analysts had more certainty in their answers and less difficulty in completing the tasks. Comparing both views, the Transaction Relations View (**G4**) had the lowest duration and the difficulty of completion was also considered low. This can be explained by the fact that, as the graph is less complex (has fewer glyphs), for the same tasks the analysts could analyse more quickly the glyphs and their relationships. These good results on both views

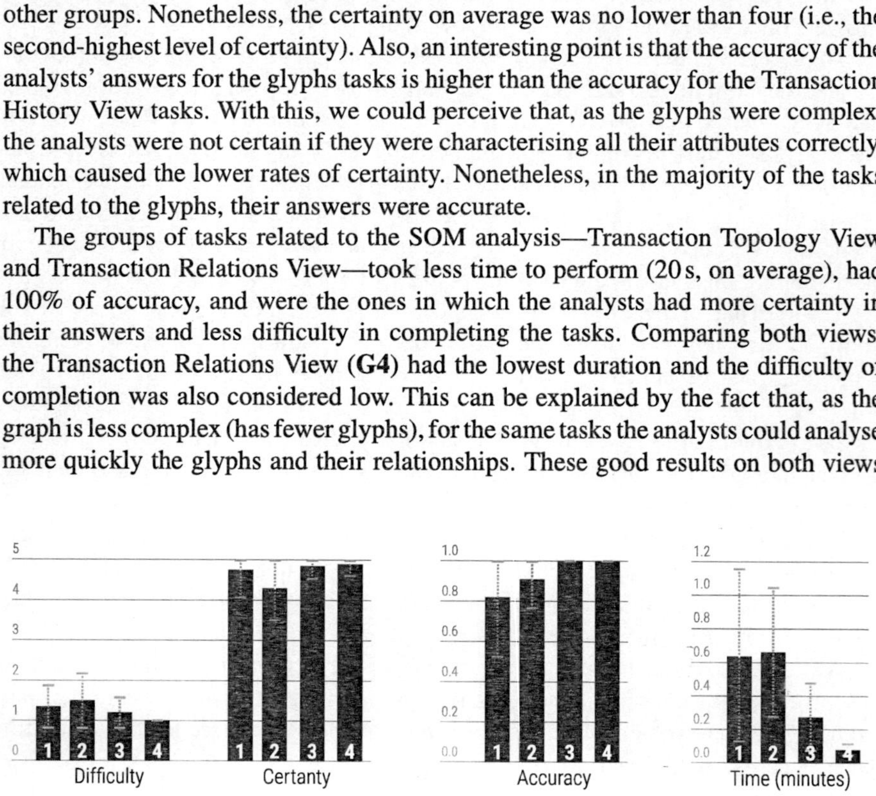

Fig. 10 Difficulty, certainty, accuracy and time values for the 4 tasks groups. In general, the difficulty of the tasks was considered to be low, and the certainty and accuracy are considered to be high. With regard to time, the majority of the tasks was completed in less than one minute

might also indicate the good acceptance of such models and the ease with which the analysts could interpret the topology of the transactions.

VaBank Analysis The third part of the test was concerned with the free exploration and analysis of the transactions of two clients. These clients have two different behaviours: Client A has a suspicious behaviour at the end of his/her data, and Client B commits fraud at the beginning of his/her data.

The majority of the analysts interacted with the tool in the same way. Hence, we hereafter summarise their interaction when analysing both clients—Client A and Client B. At the beginning of the Client A analysis, the analysts spotted no fraud in the first period range and all detected a weekly periodicity of online transactions of the agenda type with low values. One analyst started to be suspicious when perceived that there were 13 transactions to different new beneficiaries. Then, the analysts scrolled on the timeline to see the other transactions of that month. By doing so, the analysts understood that there was a disruption of the initial pattern. At the end of the month, the transactions had no pattern, being scattered along the last days, and the rate and value of the transactions increased. Through the interaction with the glyphs, one analyst noted that some transactions on the same day were made in different countries, which was considered suspicious. Also, by interacting with the glyphs, another analyst found that the beneficiary attribute changed in all transactions but the Internet Service Provider (ISP) was always the same. Additionally, after analysing the data through the Transaction Relations View and Transaction Topology View, one analyst referred that the suspicious behaviours were more evident in the Transaction History View than in the other views.

In summary, the majority of the analysts identified Client A as a suspicious case especially due to the pattern changes and the increase of amount and rate of transactions.

In regards Client B, the analysts directly detected the fraudulent activities through the glyphs. In this subset, the client asked for a loan of increased value that was considered as fraud and on the same day performed several online transactions of the agenda type, which were also considered as fraud. All analysts were intrigued by the fraudulent transactions, and all interacted with the glyph to try to understand what were the attributes of those transactions. By doing so, they could perceive that the transactions were made to different beneficiaries. The majority of the analysts referred to this type of behaviour as an external attack on a legitimate client account. Also, one analyst stated that it was also suspicious that Client B tried so many transactions of increased value, and then, after one week, made another transaction of a relatively small value. In summary, the majority of the analysts identified Client A as a suspicious case especially due to the pattern changes and the increase of amount and rate of transactions.

In summary, Client B was instantly classified as fraudulent, for his attempts of doing several transactions with high amounts for different accounts. Also, most analysts referred to Client B as an account that might have been hacked. As this client had few transactions, the analysts could see every transaction in the transaction matrix, without needing to interact with the timeline.

Feedback At the end of each test, the analysts rated each view in terms of aesthetics, interpretability, aid in the analysis, and learning curve. The Transaction History view got a higher rate in terms of aesthetics and aid. Additionally, it was defined as easier to interpret but had lower ratings in terms of the learning curve. This last rate may be caused by the complexity of this view, which included the histograms, the small SOM matrix, and the timeline.

Concerning the Transaction Topology View and the Transaction Relations View, the analysts took more time to complete the tasks with the first and rated it with higher values of difficulty. However, the Transaction Topology View was seen as a better aid for the analysis of the transaction patterns and was also defined as easier to learn, compared to the Transaction Relations View. In fact, the Transaction Relations View was the view with the lowest ratings in terms of aesthetic value and aid in the analysis of the data.

At the end of the tests, some analysts made some comments on the tool. They referred to the Transaction Topology View as a good auxiliary for their work and referred that with more practice the glyphs would get easier to read and interpret. One analyst also suggested a new positioning of the glyphs in the transaction matrix: to place them in each cell radially to represent the hours at which the transaction was made. At the end of each test, the analysts rated each view in terms of aesthetics, interpretability, aid in the analysis, and learning curve. The Transaction History view got a higher rate in terms of aesthetics and aid. Additionally, it was defined as easier to interpret but had lower ratings in terms of the learning curve. This last rate may be caused by the complexity of this view, which included the histograms, the small SOM matrix, and the timeline.

8 Discussion

Through our interaction with the fraud analysts, we were able to define the two main tasks to which VaBank should answer: to enable the visualization of the transactions over time and to enable their profile characterisation. The analysts also aided us in the definition of the specific requirements for the tool which allowed us to define the visualization models and interaction mechanisms. These first steps revealed to be important for the development of a tool to be used for the analysis and detection of suspicious behaviours in bank transactions.

Through the user testing, we could validate our tool in terms of efficiency, as most of the tasks were completed in reduced times—on average the tasks took less than 1.3 min to complete—and the exploration and analysis of the clients' data were also completed in a short time—took on average 4 min, which the analysts referred to as a good time for the analysis in comparison to their current tools. We could validate the tool in terms of effectiveness, as all analysts were able to complete correctly the tasks and also, through their interaction with VaBank, they were able to analyse the details of the transactions, their main characteristics, and detect suspicious behaviours.

With the analysis of the tasks' results, we could assess the interpretability of the visualization models. For example, we could understand that despite the complexity of the glyphs, the three levels of visual impact achieved their purpose, as the analysts could focus on the first level (the type of transaction, amount, and fraud) and with a more close reading analyse the operations types and the rest of the transaction's characteristics. Also, although during the execution of the tasks the Transaction History View was seen as the most difficult, after the interaction, the analysts found it to be easier to interact with. Also, through the analysts' feedback, we could understand that this view was well received by the analysts which defined it as a good auxiliary for their work. We could also perceive that, although the Transaction Relations View was faster to analyse, defined as easier to learn and in which the analysts were more certain about their answers, this view was also seen as less informative than the Transaction Topology View. The Transaction Topology View was seen by the analysts as a better aid for the analysis of the transaction patterns and was also defined as easier to learn.

With the analysis of the results of the second part, we could conclude that all analysts understood the tool's interaction mechanisms and all were able to interact properly with the tool. Also, the analysts took a small amount of time to analyse and perceive the types of behaviours of the analysed clients. With this, we can conclude that VaBank can aid in the detection of suspicious behaviours, which in turn, can improve the analysts' decisions.

Concerning the analysts' feedback, they stated that after the completion of the tasks they were more familiarised with the tool, and could easily use all interactive features. Additionally, they referred to the highlight of transactions with the same attribute as a good feature that was relevant for their line of work. This highlight aided in the creation of relationships between transactions and in the analysis of their attributes. Nonetheless, all visual elements were well received and understood. One analyst also referred that the timeline was an important asset as it enabled the visualization of different periods of time and the understanding of the types and amount of transactions on different time periods. Also, the representation and highlight of fraud were well understood by every analyst.

9 Conclusion

In this work, we explored different visual solutions for the representation of temporal patterns in the finance domain. We presented our design choices for VaBank, a visualization tool which aims to represent the typical behaviours and emphasise atypical ones in bank datasets. VaBank is a user-centred visualization tool developed in the context of a partnership with Feedzai—a Portuguese Fraud Detection Company—and intended to be implemented in the workflow of the company's fraud analysts. The company's main aim for the tool was that it could promote an efficient analysis of the dataset and could emphasise suspicious behaviours.

More specifically, we represented the temporal patterns of bank transaction data. With the collaboration with Feedzai, we were able to define the main requirements and tasks that would enable our tool to improve their analysis concerning their current tool—spreadsheets. Thus, our visualization models focus on: (i) the visual representation of the transaction characteristics through a glyph visualization; (ii) the temporal visualization of the transactions; (iii) the characterisation of the transactions topology through a Self-Organising Map (SOM) algorithm; and (iv) the projection of the SOM results into a matrix and a force-directed graph.

We validated and compared the different visualization components of the tool through formative and summative evaluations with experts in fraud detection. Through these tests, we could assess the effectiveness of the tool on the characterisation of the transactions. The analysts were able to properly analyse the visualization and detect different behaviours in different bank clients. In summary, the results showed that the tool was well received by the analysts and it could enhance their analysis, overpassing their current method—spreadsheets.

We contribute to the visualization domain in finance with a tool which focuses on the characterisation of bank transactions, on the representation of the topology of the transactions and, consequently, on the highlight of uncommon behaviours. By enabling in the same tool the visualization of the transactions along time—emphasising the ones with higher amounts—and their topology—emphasising the typical behaviours—, we were able to promote a better analysis of atypical transactions and suspicious behaviours. In conclusion, the presented work demonstrates that VaBank is effective and efficient for the analysis of bank data and in the detection of suspicious behaviours. Also, the characterisation of transactions with complex glyphs can aid in the understanding of transaction patterns and facilitate the analysis of the overall data.

References

1. Crain, M.A. et al.: Fraud prevention, detection, and response. In: Chap. 8, Essentials of Forensic Accounting, pp. 211–243. Wiley, Ltd (2017). ISBN:9781119449423. https://doi.org/10.1002/9781119449423.ch8
2. Bolton, R.J., Hand, D.J.: Statistical fraud detection: a review. In: Statistical Science, pp. 235–249 (2002)
3. Dilla, W.N., Raschke, R.L.: Data visualization for fraud detection: Practice implications and a call for future research. Int. J. Account. Inf. Syst. **16**, 1–22 (2015). ISSN:1467-0895. https://doi.org/10.1016/j.accinf.2015.01.001. http://www.sciencedirect.com/science/article/pii/S1467089515000020
4. Lemieux, V.L., et al.: Using visual analytics to enhance data exploration and knowledge discovery in financial systemic risk analysis: the multivariate density estimator. In: iConference 2014 Proceedings (2014)
5. Russell, S.: Human Compatible: Artificial Intelligence and the Problem of Control. Penguin (2019)
6. Mitchell, M.: Artificial Intelligence: A Guide for Thinking Humans. Penguin, UK (2019)

7. Eklund, T., et al.: Assessing the feasibility of using self-organizing maps for data mining financial information. In: Wrycza, S., (ed.), Proceedings of the 10th European Conference on Information Systems (ECIS) 2002, vol. 1. AIS (2002)
8. Kiang, M.Y., Kumar, A.: An evaluation of self-organizing map networks as a robust alternative to factor analysis in data mining applications. Inf. Syst. Res. **12**(2), 177–194 (2001). ISSN:10477047, 15265536. http://www.jstor.org/stable/23011078
9. Costea, A., et al.: Analyzing economical performance of central-East-European countries using neural networks and cluster analysis. In: Proceedings of the Fifth International Symposium on Economic Informatics, pp. 1006–1011. Bucharest, Romania (2001)
10. Maçãs, C., Polisciuc, E., Machado, P.: VaBank: visual analytics for banking transactions. In: 24th International Conference Information Visualisation, IV 2020, pp. 336–343. Melbourne, Australia (2020). https://doi.org/10.1109/IV51561.2020.00062
11. Keim, D.A.: Information visualization and visual data mining. IEEE Trans. Vis. Comput. Graph. **8**(1), 1–8 (2002)
12. Ko, S., et al.: A survey on visual analysis approaches for financial data. Comput. Graph. Forum **35**(3), 599–617 (2016). https://doi.org/10.1111/cgf.12931. eprint: https://onlinelibrary.wiley.com/doi/abs/10.1111/cgf.12931
13. Dumas, M., McGun, M.J., Lemieux, V.L.: Finance- vis. net-a visual survey of financial data visualizations. In: Poster Abstracts of IEEE Conference on Visualization, vol. 2 (2014)
14. Leite, R.A., et al.: Visual analytics for event detection: focusing on fraud. Vis. Inf. **2**(4), 198–212 (2018). ISSN:2468-502X. C. Maçãs, E. Polisciuc, P. Machado, https://doi.org/10.1016/j.visinf.2018.11.001. http://www.sciencedirect.com/science/article/pii/S2468502X18300548
15. Huang, M.L., Liang, J., Nguyen, Q.V.: A visualization approach for frauds detection in financial market. In: 2009 13th International Conference Information Visualisation. 2009, pp. 197–202. https://doi.org/10.1109/IV.2009.23
16. Kirkland, J.D., et al.: The NASD regulation advanced-detection system (ADS). AI Mag. **20**(1), 55 (1999). https://doi.org/10.1609/aimag.v20i1.1440. https://ojs.aaai.org/index.php/aimagazine/article/view/1440
17. Leite, R.A., et al.: Visual analytics for fraud detection: focusing on profile analysis. In: Proceedings of the Eurographics/IEEE VGTC Conference on Visualization: Posters. EuroVis '16, pp. 45–47. Eurographics Association, Groningen (2016)
18. Sakoda, C., et al.: Visualization for assisting rule de?nition tasks of credit card fraud detection systems. In: IIEEJ Image Electronics and Visual Computing Workshop (2010)
19. Didimo, W., et al.: An advanced network visualization system for financial crime detection. In: 2011 IEEE Pacific Visualization Symposium, pp. 203–210 (2011). https://doi.org/10.1109/PACIFICVIS.2011.5742391
20. Didimo, W., Liotta, G., Montecchiani, F.: Vis4AUI: visual analysis of banking activity networks. In: GRAPP/IVAPP, pp. 799–802 (2012)
21. Chang, R. et al.: WireVis: visualization of categorical, time-varying data from financial transactions. In: 2007 IEEE Symposium on Visual Analytics Science and Technology, pp. 155–162 (2007)
22. Kohonen, T.: The self-organizing map. Proc. IEEE **78**(9), 1464–1480 (1990). https://doi.org/10.1109/5.58325
23. Rogovschi, N., Lebbah, M., Bennani, Y.: A self-organizing map for mixed continuous and categorical data. Int. J. Comput. **10**(1), 24–32 (2011)
24. Hsu, C.-C., Lin, S.-H.: Visualized analysis of mixed numeric and categorical data via extended self-organizing map. IEEE Trans. Neural Netw. Learn. Syst. **23**(1), 72–86 (2012). ISSN:2162-237X (Print); 2162-237X (Linking). https://doi.org/10.1109/TNNLS.2011.2178323.
25. Hsu, C., Kung, C.: Incorporating unsupervised learning with self-organizing map for visualizing mixed data. In: 2013 Ninth International Conference on Natural Computation (ICNC), pp. 146–151 (2013). https://doi.org/10.1109/ICNC.2013.6817960.
26. Tai, W.-S., Hsu, C.-C.: Growing self-organizing map with cross insert for mixed-type data clustering. Appl. Soft Comput. **12**(9), 2856–2866 (2012). ISSN:1568-4946. https://doi.org/10.1016/j.asoc.2012.04.004. http://www.sciencedirect.com/science/article/pii/S1568494612001731

27. Hsu, C.-C.: Generalizing self-organizing map for categorical data. IEEE Trans. Neural Netw. **17**(2), 294–304 (2006). https://doi.org/10.1109/TNN.2005.863415
28. del Coso, C., et al.: Mixing numerical and categorical data in a self-organizing map by means of frequency neurons. Appl. Soft Comput. **36**, 246–254 (2015). ISSN:1568-4946. https://doi.org/10.1016/j.asoc.2015.06.058. http://www.sciencedirect.com/science/article/pii/S1568494615004512
29. Koua, E.L.: Using self-organizing maps for information visualization and knowledge discovery in complex geospatial datasets. In: Proceedings of 21st International Cartographic Renaissance (ICC), pp. 1694-1702 (2003)
30. Shen, Z., et al.: BiblioViz: a system for visualizing bibliography information. In: Proceedings of the 2006 Asia-Pacific Symposium on Information Visualisation, vol. 60, pp. 93–102. Australian Computer Society, Inc (2006)
31. Olszewski, D.: Fraud detection using self-organizing map visualizing the user profiles. Knowl.-Based Syst. **70**, 324–334 (2014). ISSN:0950-7051. https://doi.org/10.1016/j.knosys.2014.07.008
32. Milosevic, M., et al.: Visualization of trunk muscle synergies during sitting perturbations using self-organizing maps (SOM). IEEE Trans. Biomed. Eng. **59**(9), 2516–2523 (2012). ISSN:1558-2531 (Electronic); 0018-9294 (Linking). https://doi.org/10.1109/TBME.2012.2205577
33. Astudillo, C.A., Oommen, B.J.: Topology-oriented self-organizing maps: a survey. Pattern Anal. Appl. **17**(2), 223–248 (2014). https://doi.org/10.1007/s10044-014-0367-9
34. Gorricha, J.M.L., Lobo, V.J.A.S.: On the use of three-dimensional self-organizing maps for visualizing clusters in georeferenced data. In: Popovich, V.V., et al., (eds.), Information Fusion and Geographic Information Systems: Towards the Digital Ocean, pp. 61–75. Springer, Berlin (2011). ISBN:978-3-642-19766-6. https://doi.org/10.1007/978-3-642-19766-6_6
35. Morais, A.M.M., Quiles, M.G., Santos, R.D.C.: Icon and geometric data visualization with a self-organizing map grid. In: Murgante, B., et al., (eds.), Computational Science and Its Applications - ICCSA 2014, pp. 562–575. Springer International Publishing, Cham (2014). ISBN:978-3-319-09153-2
36. Andrienko, G., et al.: A framework for using self-organising maps to analyse spatio-temporal patterns, exemplified by analysis of mobile phone usage. J. Locat. Based Serv. **4**(3–4), 200–221 (2010). https://doi.org/10.1080/17489725.2010.532816
37. Furletti, B., et al.: Identifying users profiles from mobile calls habits. In: Proceedings of the ACM SIGKDD International Workshop on Urban Computing. UrbComp '12. Beijing, China: Association for Computing Ma-34 Catarina Maçãs, Evgheni Polisciuc, Penousal Machado chinery, 2012, pp. 17–24. ISBN:9781450315425. https://doi.org/10.1145/2346496.2346500
38. Schreck, T., et al.: Visual cluster analysis of trajectory data with interactive Kohonen Maps. In: 2008 IEEE Symposium on Visual Analytics Science and Technology, pp. 3–10 (2008). https://doi.org/10.1109/VAST.2008.4677350
39. Kameoka, Y., et al.: Customer segmentation and visualization by combination of self-organizing map and cluster analysis. In: 2015 13th International Conference on ICT and Knowledge Engineering (ICT Knowledge Engineering 2015), pp. 19-23 (2015). https://doi.org/10.1109/ICTKE.2015.7368465
40. Wehrens, R., Buydens, L.M.C.: Self- and super-organizing maps in R: the Kohonen package. J. Stat. Softw. **21**(5), 1–19 (2007). ISSN:1548-7660. https://doi.org/10.18637/jss.v021.i05. https://www.jstatsoft.org/v021/i05
41. Schreck, T., et al.: Trajectory-based visual analysis of large financial time series data. SIGKDD Explor. Newsl. **9**(2), 30–37 (2007). ISSN:1931-0145. https://doi.org/10.1145/1345448.1345454
42. Sarlin, P., Eklund, T.: Fuzzy clustering of the self-organizing map: some applications on financial time series. In: Laaksonen, J., Honkela, T., (eds.), Advances in Self-Organizing Maps, pp. 40–50. Springer, Berlin (2011). ISBN:978-3-642- 21566-7
43. Sarlin, P.: Sovereign debt monitor: a visual Self-organizing maps approach. In: 2011 IEEE Symposium on Computational Intelligence for Financial Engineering and Economics (CIFEr), pp. 1–8 (2011). https://doi.org/10.1109/CIFER.2011.5953556

44. Šimunić, K.: Visualization of stock market charts. In: In Proceedings from the 11th International Conference in Central Europe on Computer Graphics, Visualization and Computer Vision, Plzen-Bory (CZ) (2003)
45. Mackinlay, J.: Automating the design of graphical presentations of relational information. ACM Trans. Graph. **5**(2), 110–141 (1986). ISSN:0730-0301. https://doi.org/10.1145/22949.22950
46. Chang, R., et al.: Scalable and interactive visual analysis of financial wire transactions for fraud detection. Inf. Vis. **7**(1), 63–76 (2008). https://doi.org/10.1057/palgrave.ivs.9500172
47. Olsson, J., Boldt, M.: Computer forensic timeline visualization tool. In: Digital Investigation 6 (2009). The Proceedings of the Ninth Annual DFRWS Conference, pp. S78-S87. ISSN:1742-2876. https://doi.org/10.1016/j.diin.2009.06.008. http://www.sciencedirect.com/science/article/pii/S1742287609000425
48. Jacomy, M., et al.: ForceAtlas2, a continuous graph layout algo- rithm for handy network visualization designed for the gephi software. PLOS ONE **9**(6), 1–12 (2014). https://doi.org/10.1371/journal.pone.0098679

Augmented Classical Self-organizing Map for Visualization of Discrete Data with Density Scaling

Phillip C. S. R. Kilgore, Marjan Trutschl, Hyung W. Nam, Angela P. Cornelius, and Urška Cvek

Abstract In a previous publication, we introduced a method called hSOM to improve the comprehension of Kohonen's self-organizing map. Self-organizing maps are an unsupervised analogue of the artificial neural network which preserves the topology of its input space. It efficiently summaries multidimensional data, but is difficult to visualize in a manner that is accessible to those trying to interpret it. The hSOM method improves upon the classical visualization depicting a SOM by allowing for the proportion of each output node's instances of a discrete variable to be visualized, allowing distribution to be ascertained. This chapter extends that research by addressing visual noise that can arise out of dense hSOM visualizations and by adding an additional case study to evaluate hSOM's performance.

Keywords Self-organizing map · Visualization · Unsupervised learning · Histogram · Discrete · Neural network · Artificial intelligence · Machine learning

---

P. C. S. R. Kilgore (✉) · M. Trutschl · U. Cvek
Department of Computer Science, Louisiana State University Shreveport, One University Place, Shreveport, LA 71115, USA
e-mail: pkilgore@lsus.edu

M. Trutschl
e-mail: mtrutsch@lsus.edu

U. Cvek
e-mail: ucvek@lsus.edu

H. W. Nam
Department of Pharmacology, Toxicology and Neuroscience, Louisiana State University Health Sciences Center Shreveport, 1501 Kings Highway, Shreveport, LA 71103, USA
e-mail: hnam@lsuhsc.edu

A. P. Cornelius
Emergency Department, John Peter Smith Hospital, 1500 South Main Street, Fort Worth, TX 76014, USA
e-mail: angiepmd03@gmail.com

North Caddo Medical Center, Caddo Fire Districts 1 & 8, 815 South Pine Street, Vivian, LA 71082, USA

B. Kovalerchuk et al. (eds.), *Integrating Artificial Intelligence and Visualization for Visual Knowledge Discovery*, Studies in Computational Intelligence 1014,
https://doi.org/10.1007/978-3-030-93119-3_11

1 Introduction

Kohonen's self-organizing map (SOM) is an unsupervised neural network that is used for tasks such as dimensionality reduction and topological optimization of multidimensional data [1]. Although it has proven useful in making large, multidimensional data sets comprehensible as a form of dimensionality reduction [2–6], its underlying principles may be difficult to understand and may prove challenging to visualize. A classical visualization for the SOM exists, but serves mainly to depict the structure of the SOM in question rather than its relationship to data. Several approaches to this problem have been considered. They often involve either transforming the map [7] or clustering output nodes [8, 9].

In the original publication that this book chapter expands upon [10], we developed a method called histogram-SOM (hSOM) to improve the comprehension of the map in situ by utilizing embedded histograms to depict the distribution of a given discrete variable in each output node in the SOM. In SOMs, proximity between two output nodes indicates relative degree of distance between two output nodes; clusters of output nodes are related to each other because their weight vectors have comparatively small distances that indicate similarity. In some cases, it may be useful to know if a certain class is strongly represented in an output node. To test this notion, we implemented hSOM (as described below) and evaluated its output on three empirical data sets. We found that our method was able to highlight such instances, even for high-resolution SOMs.

This chapter retains much of the original text, but addresses a phenomenon where differences in output node density could create visual noise that confounds the interpretation of the hSOM visualization. New in this chapter is a proposed solution to this issue (Sect. 3.3) and the evaluation thereof (Sect. 5). We have also added an additional case study involving empirically gathered data concerning new SARS-CoV-2 (herein, COVID or COVID-19) cases in Louisiana during 2020. We have amended our discussion and conclusions to reflect these changes, which we conclude add greater utility to the hSOM method.

2 Background

Kohonen's SOM was pioneered by Teuvo Kohonen in his seminal 1982 paper [1], where it was presented as a method to preserve the topological features of complex, multidimensional data. SOMs may be understood as a grid of output nodes to which records from a data matrix are mapped, each consists of a weight vector. An output node's weight vector (i.e., codebook vector) is defined such that it is the same length and that it is constrained to the same space as the input records. For each record, there exists an output node called its *winning node* where the distance between the winning node's weight vector and the record in question is minimal. Records which share the same winning node have similar dimensional values to that node's weight vector and can thus be reasoned to be similar records. Let $\vec{x}$ and $\vec{w}$ be an input record and weight vector respectively; let $|\vec{x} - \vec{w}|$ be a metric and W be the set of weight vectors in the SOM. The winning node $\vec{u}$ is defined in (1). Thus, each output node may be treated as a bin in which records for which that node is a winning node are associated.

$$\vec{u} \equiv \arg\min_{\vec{w} \in W} |\vec{x} - \vec{w}| \tag{1}$$

The input topology is learned by adjusting the weight vectors in proximity of the record's winning node. A process to update the neighborhood about the winning node to conform to $\vec{x}$ to varying degrees, with the neighborhood function often being defined as a von Neumann Neighborhood. Let $\vec{p_u}$ and $\vec{p_w}$ be the positions in the grid of the winning node and an output node respectively. The von Neumann neighborhood about $\vec{u}$ is defined by u for the radius r by (2).

$$N_{vn}(r, W, \vec{p_u}) = \bigcup_{\vec{w} \in W} \vec{w} : |p_u - p_w| < r \tag{2}$$

The weight vector for each element of the neighborhood is adjusted using the corresponding node $\vec{w}$ neighborhood weight λ_w, which is typically a Gaussian function of the node's position, the neighborhood radius, and the current learning rate λ. The process of mapping and updating the neighborhood about the winning node for each input record is sometimes referred to as an *epoch*. The process typically terminates under one of two conditions: when a predetermined number of epochs have been executed, or when entropy in the SOM has been maximized beyond some threshold ϵ.

$$\lambda_w = \lambda * e^{-\frac{|u-w|}{r^2}} \tag{3}$$

$$\vec{w'} = \vec{w} + \lambda_w(\vec{x} - \vec{w}) \tag{4}$$

The resulting SOM has an important property. Because of the neighborhood update process, nearby output nodes will have similar weight vectors, making it possible to hypothesize that records mapped to nearby nodes will share the same degree of similarity. Generally speaking, proximity can now be used as a rough measure of similarity because of this; although the degree of difference can vary between adjacent output nodes, output nodes which are maximally distant will also maximize their dissimilarity.

Typically, the grid either uses a square [1] or hexagonal tiling [11, 12]; this naturally admits a planar topology, although it may also mapped to a cylinder or torus because these are homomorphisms of the plane. The latter topologies allow for the grid to "wrap" and can be achieved by gluing parallel edges of the grid together. The toroidial topology is sometimes used in cases where the boundary introduced by a planar topology cannot be assumed [13].

The SOM has been used in conjunction with other forms of visualization to address the problem of point occlusion caused by the limited dimensionality in traditional media [2–6]. However, it has also been used as a method of visualization in its own right. In its most basic form, the visualization consists of a grid representing the network's output nodes, although it has also become more common to embed a depiction of the output node's weight vector in each grid cell. Obayashi and Sasaski apply additional clustering to the output nodes in a form of meta-clustering to make similar nodes more visible[8]. Ultsch used a method called the U-matrix to depict distance between adjacent weight vectors, allowing for non-linearities in the map to be detected in the SOM [14].

The main benefit of these types of visualizations is that they convey the degree to which two output nodes are similar; however, they do not convey information respecting the records that map to them. When the number of records is less than the number of output nodes, it is inevitable that some output nodes will have no records mapped to them; this may be the case even if the number of records far exceeds the number of output nodes available. Often, it is useful to know which records map to the output nodes in the SOM as this can identify novel clusters that were not previously considered.

There have been several approaches for addressing this problem. The iNNfovis family of algorithms augments other forms of visualization to resolve point occlusion that may otherwise occur [2–6]. This method is useful in that it transforms the output nodes into the space of the visualization it is augmenting and may provide insight within it. In some cases, there may not be a more natural form of visualization for the input data than the SOM itself.

Merkl and Rauber have proposed several extensions to the classical SOM visualization which are worth noting in this discussion. One of their first was an adaptive coordinate where output nodes are repositioned according to their relative change in distance over time [7]. They have also approached this problem using smoothed data histograms (SDH) in order to transform the visualization into a contour map, allowing the SOM to be treated in a continuous (rather than discrete) manner [9]. Pampalk, Rauber, and Merkl argue that it is frequently more useful to consider groups of output nodes given that data may map well to more than just the winning node. Their exemplar uses a database of 359 musical pieces with 1200 features (corresponding to 1200 features based on frequency spectra) in a 14×10 and yields more granular clustering than either k-means or distance matrix methods.

3 The hSOM Approach

Unlike the above methods, the method that we propose is an extension to the classical SOM visualization which allows the constituents of the SOM's output nodes to be easily identified. We present hSOM, an extension of the classical SOM visualization which depicts the output nodes with histograms highlighting an additional class variable for specialization. In our model, the user is interested in the distribution of a particular discrete variable for each output node. We focus on the individual node rather than groups of them. As with Kohonen's SOM, hSOM is an unsupervised approach and requires the existence of a discrete variable that may either be intrinsic to the data or may be elicited by some external means. Its purpose is to highlight data distribution of discrete data that is present in each output node.

3.1 Rendering the Classical Visualization

In our model, we assume that each output node in the SOM maps to some logical grid S (which we call SOM space) and that it is visualized in the world space D. S is a multidimensional space with n_S dimensions and in R^{n_S}. D is an n-cubes with a number of dimensions n_S and n_D respectively. Let $d(\vec{s})$ define the mapping from S to an element of D, where $\vec{s}$ is an element of S such that each component is in the interval [0,1]. It is not necessary that $n_S = n_D$, although it is typically the case that $n_S \leq n_D$. Because of this (and because the mapping may vary based on tiling and dimensionality), we do not elaborate on the definition of $d(\vec{s})$; however, we do assume that D is a Euclidean space and that it is in R^{n_D}. In practice, n_D rarely exceeds 3, and frequently $n_D = 2$.

Let $\vec{g_u}(x)$ be the xth component of a coordinate vector $\vec{g_u}$ identifying the position in a grid for an output node $\vec{u} \in W, \forall g \in G$. Given an affine matrix A corresponding to a transformation into world space and centroid offset $\vec{k}$, the centroid of the resulting graphical depiction for the output node is located at $\vec{o_u}$ in world space (5). Assuming a rectangular tiling, $c(g_u) = g_u$; it may represent a more complex tiling, however.

$$\vec{o_u} = Ad(g_u) + \vec{k} \tag{5}$$

Because D is Euclidean and A is affine, we now define A and $\vec{k}$. Let $N \subseteq D$ and represent the region of D in which the output nodes shall be rendered with the least coordinate N_s, and let $||X||$ be a vector specifying the size of space X. Then A is defined by (6) and $\vec{k}$ by (7), where $\vec{x} \oplus \vec{y}$ represents the Hamadard product (i.e., componentwise or elementwise product) of $\vec{x}$ and $\vec{y}$ and $\vec{x}/\vec{y}$ denotes componentwise division. In other words, A is a translation to N_s, then scaling by $||N||$.

$$A = \begin{bmatrix} ||N||_1 & 0 & \cdots & 0 & N_{s(1)} \\ 0 & ||N||_2 & \cdots & 0 & N_{s(2)} \\ \vdots & \vdots & \ddots & \vdots & \vdots \\ 0 & 0 & \cdots & ||N||_{n_D} & N_{s(n_D)} \end{bmatrix} \tag{6}$$

$$\vec{k} = (N_e - N_s) \oplus 1/||G|| \tag{7}$$

From this region, a single output node reserves a region covering an n-cube of $2\vec{k}$. Typically, a visualization such as a parallel coordinates plot, star plot, or Nightingale's rose diagram (herein, "rose") is rendered to depict the contents of its weight vector. In our case, we have chosen the rose visualization. For simplicity, we will assume that this region covers a rectangle at this point, although a mapping to a 3-dimensional or higher space may be possible. Each component of the weight vector is represented by a sector of a circle ("petal"), where the radius of that circle is r and the radius of the sector depicts its value. First, the weight vector is min-max normalized (8); then, the radius for dimension x is calculated from that normalization (9).

$$z(\vec{u}) = \vec{u} \oplus 1/||W|| \tag{8}$$

$$r_x(\vec{u}) = z(\vec{u})_x \cdot r \tag{9}$$

Next, the start and end points for the sector must be determined. This is easily expressed in polar coordinates: if there are m organized dimensions in the SOM, then the portion of the circumference covered is $\tau/m = 2\pi/m$. We can define the sector in terms of three vertices: the circle's origin at (0,0), it's start vertex, and its end vertex. Given a dimension at index t, the sector sweeps through angles θ_t and θ_{t-1} (where $\theta_0 = 0$), where θ_t is defined by (10). Their corresponding coordinates are given by (11) and (12) for weight vector $\vec{u}$.

$$\theta_t = \frac{2\pi t}{m} \tag{10}$$

$$x_t(u) = o_{u(x)} + u_t r \cos\theta_t \tag{11}$$

$$r_t(u) = o_{u(y)} + u_t r \sin\theta_t \tag{12}$$

The value of r is unspecified here, but typically $0 < r < \min(\vec{k})$. It is common to label each sector within the rose using its own color so that each component of the weight vector may be easily identified. Additionally, it can be useful to be able to characterize the distance between individual output nodes. In order to visualize this, we use the method by Ultsch et al. [14] where the thickness of the edges constituting the boundaries between adjacent nodes (i.e., the internodal distance).

In order to depict the internodal distances, the distance matrix D between all output nodes' weight vectors is calculated, then lines of varying thickness are drawn (Lst. 1). For a planar SOM, it is convenient to draw only the upper and left boundaries for each output node because the distance will be undefined at the perimeter of the SOM. D may be normalized by greatest distance between adjacent nodes; the s parameter controls the maximum thickness of the boundary. The result is that adjacent output nodes with greater distance between their weight vectors will be depicted with a heavier border than those that have similar weight vectors.

```
void renderUltsch(W, D, s) {
    foreach (u in W) {
        x = o_u(x)
        y = o_u(y)
        # Render the upper boundary
        drawLine([0,0], [1,0], weight=D_(x,y-1) * s)
        # Render the left boundary
        drawLine([0,0], [0,1], weight=D_(x,y-1) * s)
    }
}
```

Listing 1: Pseudocode describing the rendering of the node boundaries to depict the internodal distance between adjacent nodes.

3.2 *Rendering the Histogram*

Up to this point, we have described how to render the classical depiction of a SOM with some common extensions (Fig. 1). This was done to establish a frame of reference for which we can express our contribution and is described herein. To assist in this, we have described the rendering of each output node in pseudocode (Lst. 2). Logically, there are four steps involved: calculate the centroid of the output node, render the histogram, render the weight vector plot, and perform any desired rendering after this.

```
void renderOutputNodes(W, N, H, X) {
    foreach (u in W) {
        N_u = (i ∈ N | o_u − k⃗ ≤ i ≤ o_u + k⃗)
        drawRect(N_u)
        renderHistogram(u, N_u, H, X)
        renderWeight(u, N_u)
    }
}
```

Listing 2: Pseudocode describing the rendering of the output nodes in hSOM which was run for 1000 epochs. Note that the histogram is rendered under the weight vector plot.

We have already approached the definition of `renderWeight` (the routine which renders the weight vector plot), so we now focus on the definition of `renderHistogram`. Let H be a discrete variable, X be the set of all input records, N_u be the region of N in which the output node will be rendered, and $X \mapsto \vec{u}$ denote the elements of X which map to the output node $\vec{u}$. Construct a support vector $\vec{c}$ such that c_i denotes the support of $x \in X \mapsto \vec{u} | x_h = i$, where $i \in H$. Then the frequency vector $\vec{f}$ is defined by (13).

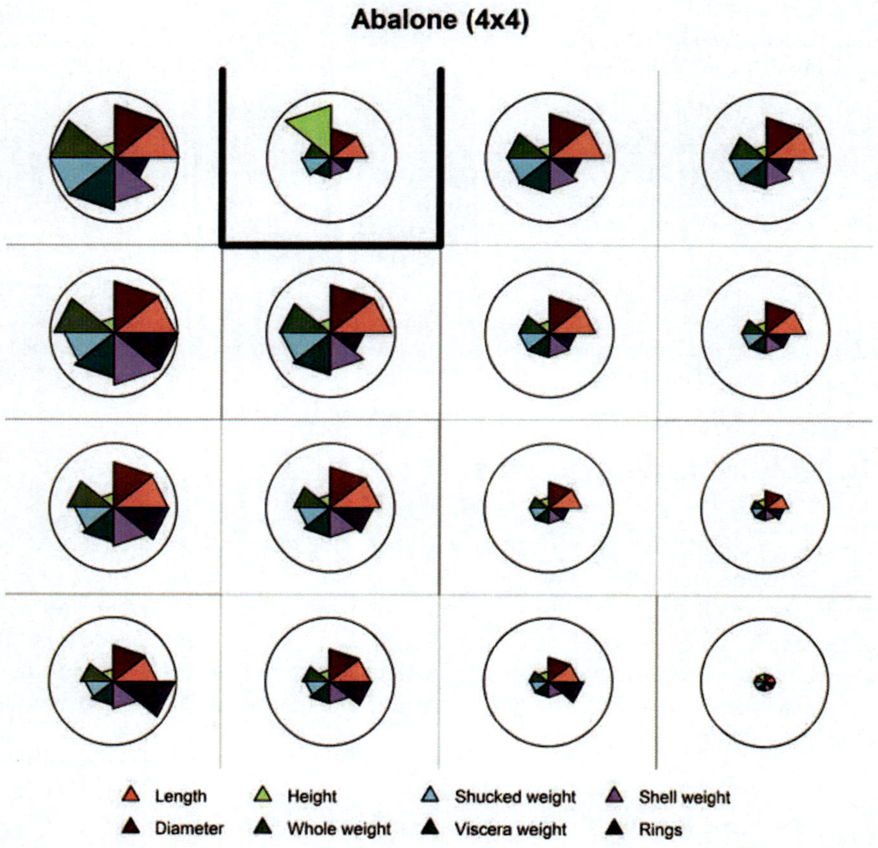

Fig. 1 A classical visualization of a 4 × 4 SOM using the methods described in Sect. 3.2. Note that only the structure of the SOM is depicted with this method

$$\vec{f} = \frac{\vec{c}}{|X \mapsto \vec{u}|} \tag{13}$$

To plot the histogram, we simply render filled rectangles for each $i \in H$ proportional to it's corresponding frequency in i. Here, we use `drawRect` and `fillRect` to denote routines which respectively draw the outline of and fill a rectangle that bounds N_u. To aid the reader comprehension of the plot, we also color the histogram slice (selected with a function named `colorOf`) and depict the count centered therein (using the routine `drawCount`). The resulting structure of the output node can be seen in Fig. 2d, and how it relates to others can be seen in Fig. 2a.

This visualization has two legends because there are two dimensions embedded within each output node: the histogram depicting its constituents (Fig. 2e) and weight vector plot (Fig. 2g). To aid viewers in interpreting the plot, we use different shapes

```
void renderHistogram(u, N_u, H, X) {
    l_0 = min(n_u)_x
    mapped = false;
    foreach (i in H) {
        l_1 = l_0 + f_i
        Δl = l_1 − l_0
        if (Δl ≤ 0) {
            fillRect((n ∈ N_u | l_0 ≤ n ≤ l_1), colorOf(i))
            drawCount(N_u, c_i, (l_1 − l_0)/2)
            l_0 = l_1
        }
    }
}
```

Listing 3: Pseudocode for plotting the histogram in an output node.

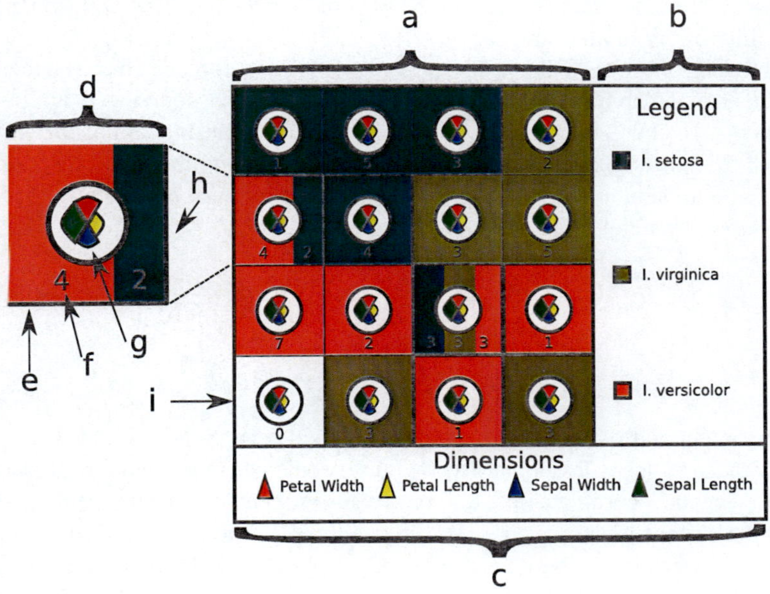

Fig. 2 A schematic mockup of an hSOM visualization to illustrate its components. **a** the output nodes. **b** the histogram legend. **c** the weight vector legend. **d** an individual output node, consisting of **e** zero or more histogram slices with **f** a corresponding record count for that slice and **g** the weight vector plot. **g** An internodal distance indicator depicting the distance between adjacent nodes' weight vectors. **i** An output node for which no records were mapped

and orientations to depict the graphs. For instance, we depict the histogram values using squares (Fig. 2b) and the organized dimensions using filled triangles to represent the rose's petals (Fig. 2c).

It is possible that an output node will have no records mapped to it. In this case, the output node has no special coloring and will appear empty (Fig. 2i). This is a consequence of the rendering algorithm (Lst. 3), which is designed to skip any element of H where the frequency is 0. This behavior is retained even of the event where no records map since it can easily be inferred that a lack of any mapped records was the cause.

3.3 Histogram Scaling

It may occasionally be the case that most of the records map to a small subset of the output nodes in the SOM. As previous described and as expected, the hSOM method will omit drawing histograms for output nodes with zero density; however, it cannot differentiate between two output nodes with non-zero density. Because of this, an output node for which most of the surrounding records proportionally map to will appear visually the same as one which has significantly less density, leading to the possibility of visual interference.

To account for this, we first calculate a count matrix C for which each element C_{yx} is defined as the number of records mapping to the output node located at position (x, y) (14). To obtain the relative density, we divide it by the maximum element of C to yield the density matrix D. This represents the height of the histogram. We define D_{min} to be some minimum threshold which the density must be; this so that records with extremely small density do not fail to render.

$$C_{yx} = ||\vec{x} \in X : \underset{\vec{w} \in W}{\arg\min} |\vec{x} - \vec{w}| = W_{yx}|| \tag{14}$$

$$D_{yx} = \max(C_{xy}/\max(C), D_{min}) \tag{15}$$

The output node with the maximum density will occupy its entire rendering area, but for those with any lesser density, we must decide where to anchor the histogram. Assume that the coordinate $(0.5, 0.5)$ represents the center of the rendering area, and $(1, 1)$ the upper-right corner (i.e., it is in quadrant I). The histogram has a "radius" $r_{yx} = D_{yx}$. Because the width is used to represent the distribution of the histogram, we need only consider the height. In this case, the center of a rectangle anchored to the top is located at $y = 1 - r$ and that anchored to the bottom is located at $y = 0 + r$, while a centered one is located at $y = 0.5$. Note that the center case is not dependent on radius; to calculate y_{yx}, we derive first derive a matrix of attachment factors A, such that when $a_{yx} = 1$, $y_{yx} = 1 - r$, and when $a_{yx} = 0$, $y_{yx} = 0 + r$. We do this by means of (16) and (17).

$$a'_{yx} = a_{yx} - 0.5 \tag{16}$$

$$y_{yx} = 0.5 + a'_{yx} - a'_{yx}r = a'_{yx}(0.5 - r) \tag{17}$$

The selection of a_{yx} probably depends on the topology; for toroidal or for some cylindrical topologies, there is always a node above or below another output node, so (18) will apply. In a planar topology, we assume that the top row of output nodes always attaches to its bottom neighbor and the bottom row to its top neighbor. We define (18) such that output nodes are "attached" to the neighbor with greater density; if the densities are equal, then the histogram is centered in the output node visualization.

$$a_{yx} = \begin{cases} 1 & C_{(y-1)x} < C_{(y+1)x} \\ 0.5 & C_{(y-1)x} = C_{(y+1)x} \\ 0 & C_{(y-1)x} > C_{(y+1)x} \end{cases} \quad (18)$$

The attachment mechanism exists because we hypothesize that users will be most interested in regions where density is maximized. Output nodes which are relatively sparse may represent outliers in isolation, while those nearby an occupied output node may be interpreted as a cluster of related output nodes. This creates visual whitespace which may make the boundaries of such clusters more apparent.

4 Methodology

We provide four examples: using a set of blood factors associated with alcoholism and using Warwick Nash's abalone set [15], COVID-related sampling from the State of Louisiana in March-April 2020 (COVID set), and a set containing myocardial infarction [16]. Both of these have several advantages with respect to this study: they were collected empirically and not designed for a machine learning study, both contain a mixture of numeric and categorical data, and both are multidimensional. The first data set (herein Alcoholism set) was collected by Dr. Hyung W. Nam's lab at LSU Health Shreveport and contains 12 metabolomic blood factors associated with 19 patients: 7 control patients and 12 patients with alcoholism. The blood factors are numeric measurements which are thought to be relevant to alcoholism. The hSOM approach was developed for this data and was meant to identify potential differences between patients.

Although the Alcoholism set is gathered empirically and used a large number of organized dimensions, it does not contain a large number of records and an alternative to test hSOM under these circumstances was sought. It is common to use Fisher's Iris set [17] for this purpose, but it has few organized dimensions and a small number of records to demonstrate the SOM at saturation. Instead, we used the Abalone set [15], which has over 4000 records and 8 numeric dimensions.

The COVID set consists of various lab samples of empirically-gathered 412 COVID-positive patients from a one month period and was collected by Dr. Angela Cornelius's team from LSU/Ochsner-affiliated sites Shreveport, LA and Monroe, LA. The subset of numeric and Boolean variables were selected as the input variables, while the histograms are derived from race. The myocardial infarction (MI) set consists of 1700 myocardial infarction events and contains 124 variables (which are mostly categorical). We organized on the numeric variables in the data and categorized the histogram on sex.

To inspect the visualization, we produced eight SOMs in total: for all sets, we produced a 4 × 4 and a 15 × 15 SOM, representing a total of 16 and 225 output nodes, respectively. This allowed us to investigate several questions related to the visualization. The 4 × 4 SOM was expected to give insight of how the hSOM behaves with relatively densely occupied output nodes. The number of available output nodes in the 4 × 4 SOM is two orders of magnitude less than the number of available records for Abalone set, and for the Alcoholism set, there are fewer output nodes than records. Conversely, the 15 × 15 SOM has about 11.8 times the available output nodes than the Alcoholism records and guarantees that some output nodes will be unoccupied. The 15 × 15 SOM also allowed us to inspect the visualization in conditions where visual acuity is reduced.

In cases where we inspected the histogram scaling method proposed above, we only used 15 × 15 SOMs. This was because 4 × 4 SOMs have insufficient resolution to exhibit the visual noise that this method addressed. Interpretation of the hSOM visualization remains the same, except that the scaled histogram will be anchored to the vertical neighbor with the least distance.

SOM processing was performed using the `kohonen` package for the R statistical language. All SOMs were processed using every available variable (except for the highlight variable H) and ran for 1000 epochs. The weight vectors were randomly initialized using a uniform distribution. The initial neighborhood radius was set to the width of the SOM and shrank linearly until it reached 0 at the final epoch. A rectangular, planar topology was used, Rendering the hSOM visualization was performed using custom code making use of R's core graphics library. We visualized the histogram according to disease state in the Alcoholism set, and sex in the Abalone set.

5 Discussion

One may see that even in the 4 × 4 SOM for the Alcoholism set (Fig. 3), there were output nodes with no records mapping to them. No output nodes had mixed classes (as depicted in the output nodes' histograms); however, four did have multiple records map ((row, column) = count): (1, 1) = 3, (1, 2) = 2, (1 ,4) = 2, and (4, 2) = 2. In the 15 × 15 SOM (Fig. 4), each record mapped to a unique output node. As expected with 19 records against a SOM with 225 nodes and the likelihood of adjacent output nodes having marginally different weight vectors, the alcoholism cohort is much more spread out, indicating that the patients involved have different metabolomic profiles (even within the same cohort). Nonetheless, the output nodes that they were mapped to may be interpreted as forming into a cluster of similar profiles and may benefit from a method such as the one proposed by Pampalk, Merkl, and Rauber [9].

There are potential drawbacks to this method. The real estate provided by N_u is usually small compared to the rest of the visualization, and the weight vector plot dominates each output node. For a sufficiently large number of elements from H, there may be an insufficient amount of space in the histogram to display them

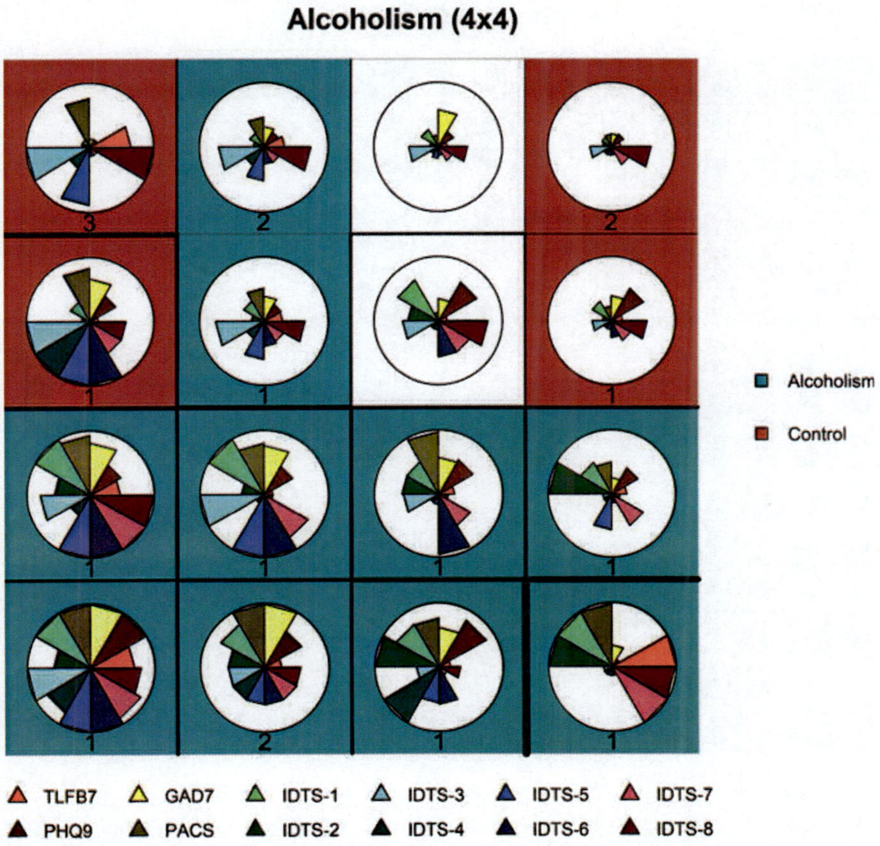

Fig. 3 A 4 × 4 SOM performed on Alcoholism set. Even though there were fewer output nodes than records, empty output nodes were still present

all. There are two cases to consider in this scenario: where the proportions in the histogram are skewed towards one element and when they have roughly equal proportions. In the former situation, this may be excusable since the most relevant information may be that the output node maps to mostly one element from H. In the latter situation, this may become more problematic since classes may be obscured. When the output nodes are rendered with insufficient resolution, it may be the case that even a histogram depicting relatively few elements would be difficult to interpret. We do not address a solution to this problem, but we do predict that these two cases may be differentiated on the basis of a measure like Shannon entropy (which should decrease as the proportions become skewed to one element).

This behavior in particular is exhibited in both the 4x4 (Fig. 5) and 15 × 15 (Fig. 6) SOMs for the Abalone set. In the 4 × 4 SOM, every output node is occupied by tens or hundreds of records. In most cases, the counts are comprehensible, but the output nodes in cells (1, 2), (1, 4), and (4, 1) show minimal occupancy by the I sex such

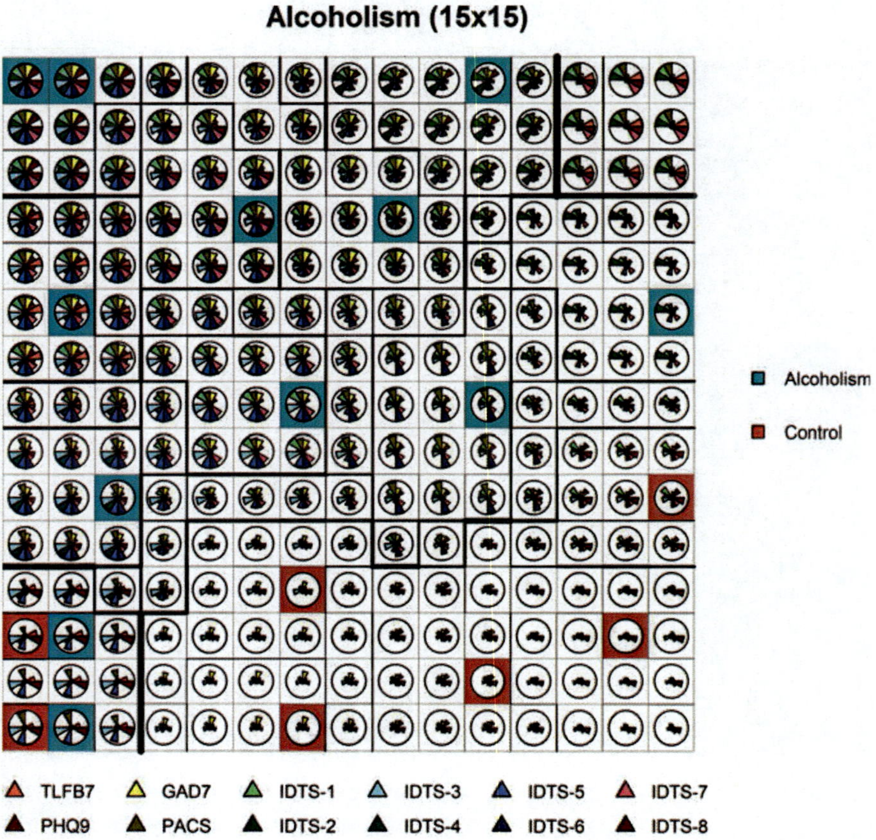

Fig. 4 A 15 × 15 SOM performed on Alcoholism set. The output nodes which each record maps to are dispersed, but still tend to form clusters

that it is barely perceptible. Even so, it is apparent that all output nodes in this SOM correspond to records of multiple sexes.

The 15 × 15 SOM demonstrates a worst case scenario for this visualization: low resolution with a large number of output nodes containing a mixture of records of varying classes. As in (Fig. 4), the counts become imperceptible; however, most output nodes also map to more than one sex and have visual interference from the weight vector plot. Some of these nodes have low information entropy and the color associated with the most common sex will dominate. For those with high entropy, their color is harder to perceive, and some ambiguity exists between the M and F classes as a result. Thus, a large cluster of nodes corresponding to the *I* sex is likely perceptible to the viewer. One way of interpreting this visualization is that the *M* and *F* sexes are much more similar to each other (at least in terms of the organized dimensions) than to the *I* sex.

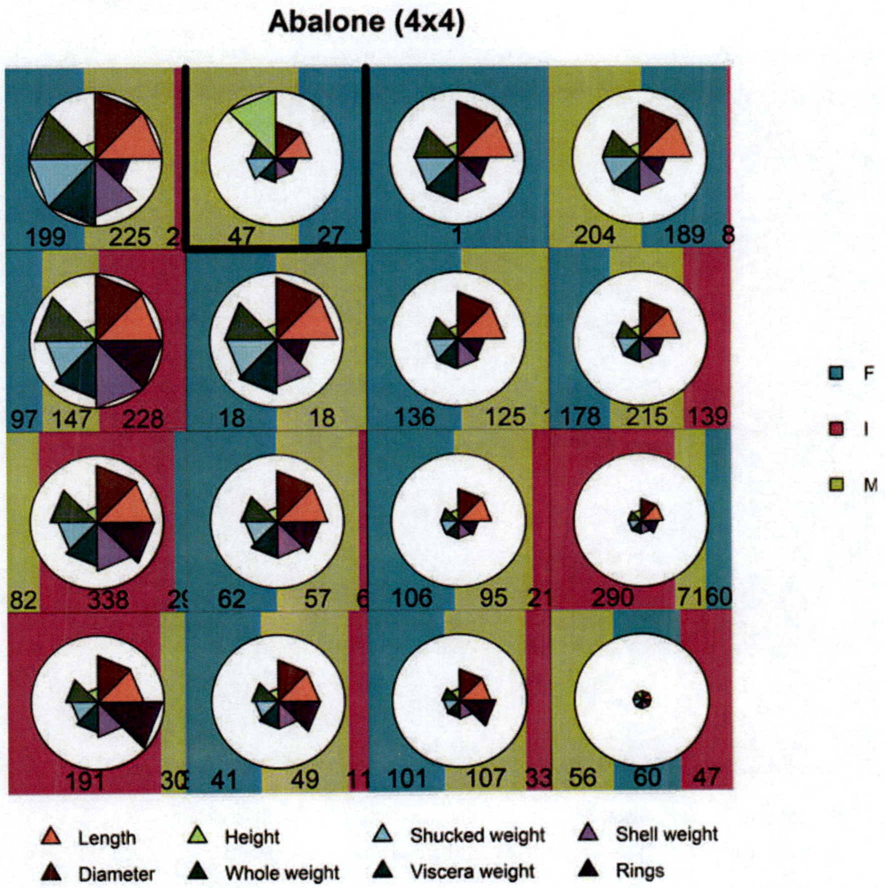

Fig. 5 A 4 × 4 SOM performed on Abalone set. This SOM has dense mapping to each output node, and no single node consists of a single group. It may be the cases that a larger SOM is needed to express this data or that not much separation exists between the classes involved

It is worth noting that many of the aforementioned weaknesses are also present in the classical visualization; it too, is susceptible to issues such as overplotting and visual interference. One strength here is that despite the visual interference involved, hSOM still conveys practical information about the record distribution according to the highlighted variable. With a greater number of output nodes, we hypothesize that cases where the histogram does not have sufficient resolution will result in a mixture of colors (similar to physical subpixel rendering on display surfaces such as color cathode-ray tube (CRT) and liquid crystal display (LCD) monitors).

However, one potential method of addressing this problem has already been presented: by using the aforementioned histogram scaling approach (Fig. 7). This version of the visualization tells us two things: that a large portion (perhaps the majority) are sparsely-mapped, and that there is more than one place where an output node is

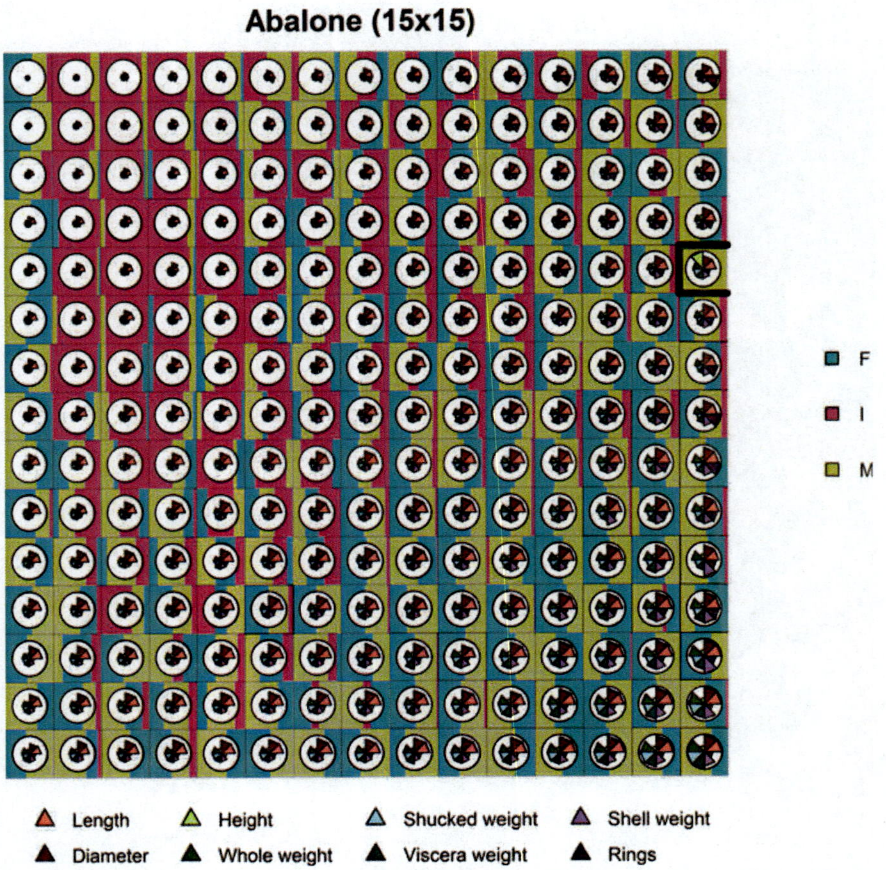

Fig. 6 A 15 × 15 SOM performed on Abalone data. As with (Fig. 5), each output node is densely mapped, but the insufficient resolution causes significant visual interference. Despite this, a large cluster mapping to the *I* sex can be identified

densely-mapped. As in (Fig. 6), we can see that the I class clusters in a particular region of the SOM and that the M and F classes appear to make up several smaller clusters; however, this fact is easier to resolve.

The justification for this approach is that when an output node is less densely mapped, it represents fewer of the nodes in the data. With respect to the visual clusters that are formed, these may represent extreme values that might be characterized as outliers or edge cases for the cluster in question. However, it is still important that we know that these output nodes have records mapped to them for some applications, particularly when we are interested in records that may represent outliers. In this case, we have used $D_{min} = 0.25$ so that the minimum reported density is visible.

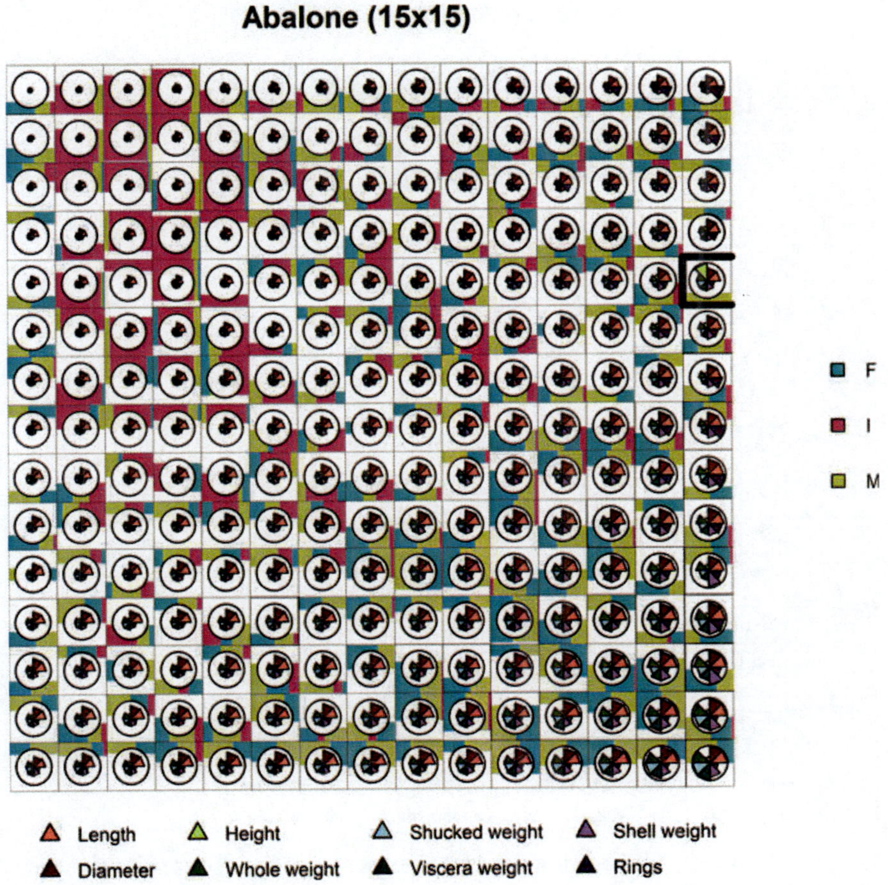

Fig. 7 A 15 × 15 SOM performed on Abalone data using histogram scaling. This method is similar to (Fig. 6), but the density of each output node is also depicted. The resulting whitespace resolves some of the visual interference of associated with the fact that every output node is mapped

In the COVID set SOM, the organized dimensions are primarily boolean values. One challenge here is that there is disagreement on how one should calculate distance between categorical variables [18]. In this instance, Boolean data is ordinal and has a total order; additionally, it can be converted into a continuous interpretation via fuzzy logic. This is what is done with this SOM; therefore, the SOM depicts partial truth values (Fig. 8).

Here, race is chosen as the highlight variable; however, in both Shreveport [19] and Monroe [20], the Black/African American population made up the majority of the population at the time of sampling, while the Asian and Native American population is underrepresented. This data is also collected from the Ochsner Health System, a non-profit institution. Because of these things, the Black/African American cohort represents nearly 75% of the input data.

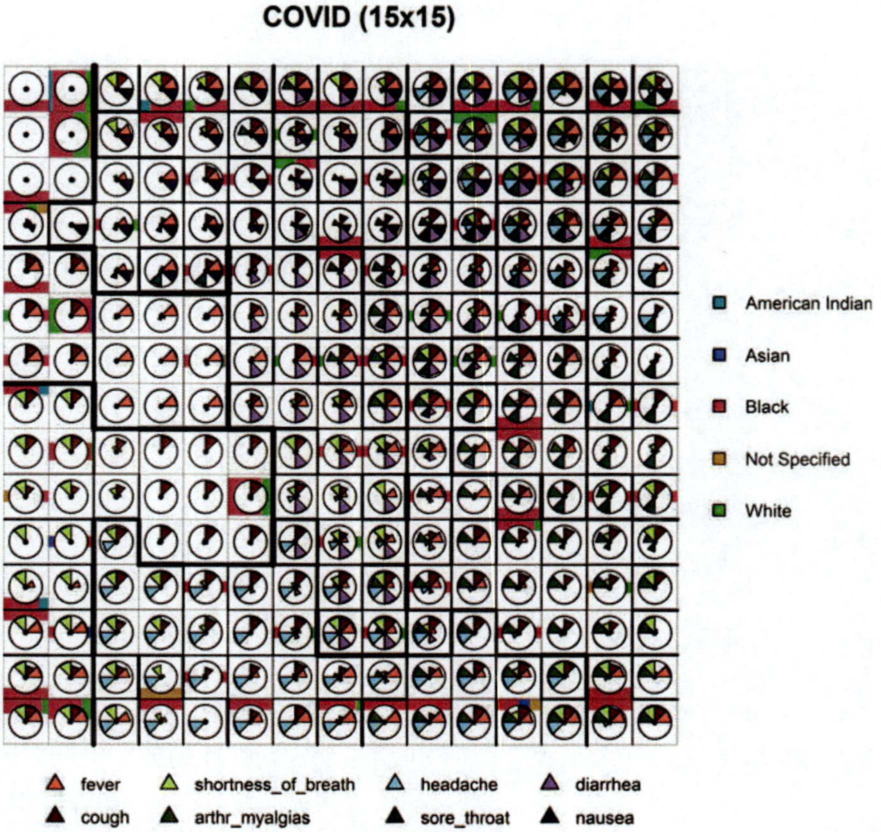

Fig. 8 A 15 × 15 SOM performed on COVID data using histogram scaling. The density of most output nodes are relatively small and do not form large clusters, indicating several subgroups

Because of this, we see several small clusters of mostly Black/African American, with small proportions of other races interspersed. We have also included an hSOM visualization where the records are highlighted by sex (Fig. 9). This was chosen because although females constitute the majority, the split is more even (58.6% vs. 41.4%). One may observe several small clusters consisting of mostly female patients.

The MI SOM is organized according to a selection of several potential predictors of myocardial infarction (Fig. 10). Here, we can see that nearly all output nodes are occupied; most output nodes are a mixture of categories; the Ultsch visualization would indicate that it is one large cluster with several interior clusters. However, the existence of several smaller, embedded clusters using the histogram scaling method can be inferred.

It may be possible to replace the histogram with another visualization, such as a tree map (for multiple dimensions) or a distribution curve (for numeric data). As this has not been tested, it is not commented on here to any great extent, but a brief

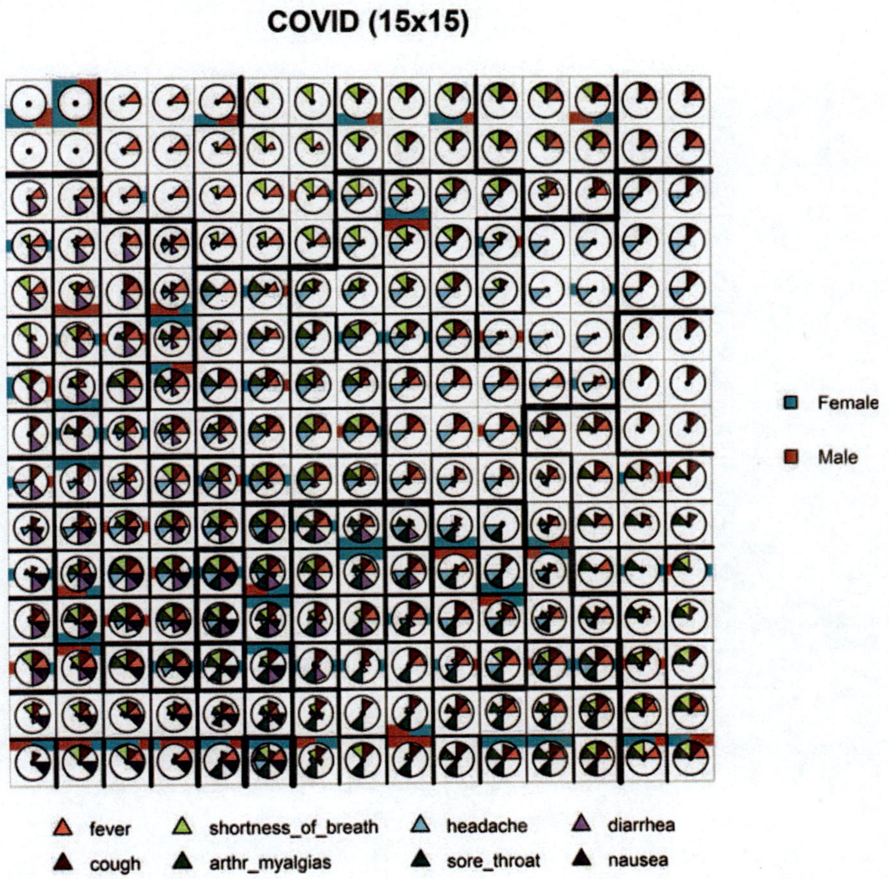

Fig. 9 A 15 × 15 SOM performed on COVID data using histogram scaling, this time highlighted by sex

discussion is warranted. On one hand, the histogram is essentially a one-dimensional visualization method and world space is typically multidimensional, so an argument can be made that this approach wastes available resources (particularly when world space is three-dimensional). A caveat to that approach is that the plot depicting the weight vector will likely occlude data. A possible solution may be to depict the weight vector some other way, but the weight vector needs to be conspicuous because it yields a great deal of information about the output node as a whole.

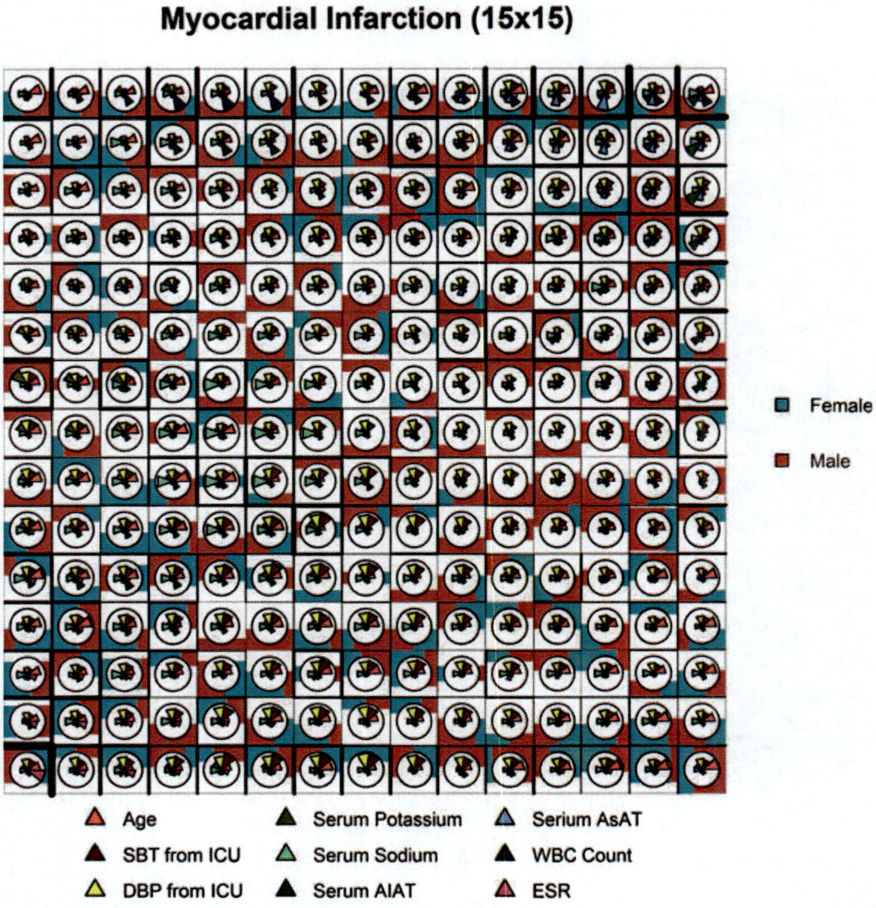

Fig. 10 A 15 × 15 SOM performed on myocardial infarction data using histogram scaling

6 Conclusions

The hSOM visualization removes one disadvantage associated with the classical SOM visualization: it not only allows for the data density of an output node to be ascertained, but it also shows the distribution of records in that node. This further increases the utility of the SOM because it provides specific information about what data each output node contains, rather than just the profile of its weight vector. Visual noise may occasionally become to great with hSOM; we address this issue through a scaling method.

In the future, we will investigate replacing the histogram using other visualization methods as described above. Preserving the ease at which an output node's weight vector can be inspected is a primary goal in this, but the remaining hypervolume in world space may be significant and can be utilized. There may also be ways of merging this approach with other visualization approaches to perceive some of the weaknesses identified in the Discussion section.

Acknowledgements Research reported in this manuscript was supported by the Noel Foundation and Louisiana Board of Regents Endowed Professor/Chair Programs: Abe Sadoff Distinguished Chair in Bioinformatics and Lisa Burke Bioinformatics Scholarship. Research work from Urska Cvek, Phillip Kilgore, and Marjan Trutschl reported in this manuscript was supported by an Institutional Development Award (IDeA) from the National Institute of General Medical Sciences of the National Institutes of Health under grant number P2O GM103424-19. Research work from Hyung Nam reported in this publication was supported by the NARSAD Young Investigator Award (26530) from Brain & Behavior Research Foundation and P20 GM1213-07. We thank Dr. Angela Cornelius and Dr. Mary Ann Edens and their team at Ochsner LSU Health Shreveport for sharing their COVID-19 dataset to be used in this study.

References

1. Kohonen, T.: Self-organized formation of topologically correct feature maps. Biol. Cybern. **43**(1), 59–69 (1982)
2. Kilgore, P.C.: Optimization of innfovis algorithms via concurrent programming. Ph.D. thesis, Louisiana State University in Shreveport (2012)
3. Kilgore, P.C., Trutschl, M., Cvek, U.: Parallel execution of self-organized visualization. In: Proceedings of the IASTED International Conference, vol. 201, pp. 1 (2013)
4. Trutschl, M., Grinstein, G., Cvek, U.: Intelligently resolving point occlusion. In: IEEE Symposium on Information Visualization 2003 (IEEE Cat. No. 03TH8714), pp. 131–136. IEEE (2003)
5. Trutschl, M., Kilgore, P.C., Cvek, U.: High-performance visualization of multi-dimensional gene expression data. In: 2012 Third International Conference on Networking and Computing, pp. 76–84. IEEE (2012)
6. Trutschl, M., Kilgore, P.C., Cvek, U.: Self-organization in parallel coordinates. In: International Conference on Artificial Neural Networks, pp. 351–358. Springer (2013)
7. Merkl, D., Rauber, A.: Alternative ways for cluster visualization in self-organizing maps. In: Workshop on Self-Organizing Maps, pp. 106–111. Citeseer (1997)
8. Obayashi, S., Sasaki, D.: Visualization and data mining of pareto solutions using self-organizing map. In: International Conference on Evolutionary Multi-Criterion Optimization, pp. 796–809. Springer (2003)
9. Pampalk, E., Rauber, A., Merkl, D.: Using smoothed data histograms for cluster visualization in self-organizing maps. In: International Conference on Artificial Neural Networks, pp. 871–876. Springer (2002)
10. Kilgore, P.C., Trutschl, M., Cvek, U., Nam, H.W.: hSOM: Visualizing self-organizing maps to accomodate categorical data. In: Proceedings of the 24th International Conference on Information Visualization, pp. 644–650 (2020)
11. Kohonen, T.: The self-organizing map. Proc. IEEE **78**(9), 1464–1480 (1990)
12. Vesanto, J., Himberg, J., Alhoniemi, E., Parhankangas, J., et al.: Self-organizing map in matlab: the som toolbox. In: Proceedings of the Matlab DSP Conference, vol. 99, pp. 16–17 (1999)
13. Mount, N.J., Weaver, D.: Self-organizing maps and boundary effects: quantifying the benefits of torus wrapping for mapping som trajectories. Pattern Anal. Appl. **14**(2), 139–148 (2011)
14. Ultsch, A.: Self-organizing neural networks for visualisation and classification. In: Information and Classification, pp. 307–313. Springer (1993)
15. Nash, W.J. et al.: The population biology of abalone (Haliotis species) in Tasmania. 1, blacktip abalone from the (h. rubra) from the north coast of Bass Strait. Technical Report 48, Sea Fisheries Division (1994). https://archive.ics.uci.edu/ml/datasets/Abalone

16. Golovenkin, S.E.: Myocardial infarction complications database. Technical Report 3, University of Leicester (2020)
17. Fisher, R.A.: The use of multiple measurements in taxonomic problems. Ann. Eugen. **7**(2), 179–188 (1936)
18. Boriah, S., Chandola, V., Kumar, V.: Similarity measures for categorical data: a comparative evaluation. In: Proceedings of the 2008 SIAM International Conference on Data Mining, pp. 243–254. SIAM (2008)
19. QuickFacts: Shreveport city, Louisiana. Technical Report 1, United States Census Bureau (2021). https://www.census.gov/quickfacts/fact/table/shreveportcitylouisiana/IPE120219
20. QuickFacts: Monroe City, Louisiana. Technical Report 1, United States Census Bureau (2021). https://www.census.gov/quickfacts/fact/table/monroecitylouisiana/IPE120219

Gragnostics: Evaluating Fast, Interpretable Structural Graph Features for Classification and Visual Analytics

Robert Gove

Abstract Graph comparison is important for many common analysis tasks, such as machine learning classifiers and visual analytic tools. Researchers have developed many graph comparison methods, such as graph-level statistics that are often not scalable, or machine learning methods that are often uninterpretable. To address the need for fast, interpretable graph comparison methods, this work proposes gragnostics, a set of fast, interpretable structural graph features. Gragnostics is comprised of 10 structural graph features that can each be computed in linear time. An evaluation shows that these features can discriminate graph topologies better than DDQC, an alternative set of features based on degree distribution. Example usage scenarios of Chiron, a visual analytic tool designed using gragnostics, show that gragnostics can be effective in a rank-by-feature framework. This book chapter also presents several new analyses: A deeper analysis showing relationships between features, and showing how individual features separate some graph classes; a new comparison of gragnostics and DDQC to graph kernels, showing that gragnostics has substantially faster runtime than graph kernels and better accuracy than DDQC and graph kernels; and additional depth and new figures illustrating gragnostics in the Chiron visual analytics usage scenarios.

1 Introduction

Comparing graphs, or networks, is a common analysis task. For example, to identify the characteristics of successful communities [53], or predict graph layout quality [33]. However, graph comparison algorithms are often slow or uninterpretable, and many have restrictions, e.g. that graphs be the same size or only have one connected component. Many visualization tools have used slower and/or less interpretable methods to compare graphs, such as graph features like betweenness centrality [18, 25, 34] or graph kernels like graphlet sampling [33].

R. Gove (✉)
Two Six Technologies, 901 N. Stuart Street, Suite 1000, Arlington, VA, USA
e-mail: robert.gove@twosixtech.com

B. Kovalerchuk et al. (eds.), *Integrating Artificial Intelligence and Visualization for Visual Knowledge Discovery*, Studies in Computational Intelligence 1014,
https://doi.org/10.1007/978-3-030-93119-3_12

Some graph comparison methods, such as graph kernels or calculating the Euclidean distance between adjacency matrices [4, 16], require that graphs satisfy certain properties, such as being composed of a single component, and some feature sets require that all graphs be the same size. This limits the utility and range of applications for these methods.

Data set size is constantly growing, which necessitates innovation to develop faster methods for comparing graphs. Prior work often relies on a mix of fast statistics, like number of vertices and edges, and also slow statistics, like graph diameter [34] that are typically implemented to run in $O(|V|^2 \log |V| + |V||E|)$ time [29].

Humans benefit from interpretability for a variety of reasons: interpretability supports goals such as fairness, failure diagnosis, etc. [40]; and interpretability aids the usability of systems that use these features. One element of machine learning interpretability is that users should be able to understand individual features used by the system [36]. System designers can aid this by using terminology already familiar to users, which can reduce cognitive load, speed up learnability, and improve re-learnability.

Gragnostics, a portmanteau of "Graph" and "Diagnostics," is designed to address these concerns. Gragnostics is a set of 10 graph-level structural features. Gragnostics mixes two new features, five existing features, and modifications to three existing features. The features have few restrictions compared to alternative graph comparison methods, which aids use in a variety of applications. The features can be computed quickly, in linear time. Finally, Gragnostics are carefully designed to capture common visual patterns in graphs, and therefore to be understandable by laypeople without a background in mathematics or network science.

In comparison to degree distribution quantification and classification (DDQC) and graphlet sampling graph kernels, gragnostics achieves notably higher classification accuracy. It is also 14% faster than graphlet sampling. Two example usage scenarios demonstrate that a rank-by-feature framework using gragnostics is an effective way to explore graphs with visual analytics.

This book chapter extends prior work [23] in the following ways: an extended discussion of related work, a deeper analysis of the relationships between gragnostic features, discussion of how individual features separate some graph classes, new analysis comparing gragnostics to graph kernels in a machine learning classification task, and more depth and new figures for the visual analytics example usage scenarios.

2 Background

A *graph* $G = (V, E)$ is a set of vertices V and a set of edges E, where if $e \in E$ then $e = (v, u)$ such that $v, u \in V$. In *undirected* graphs, (v, u) is the same as (u, v). *Simple* graphs have no duplicate edges or self-loops (e.g. (v, v)). $|V|$ and $|E|$ denote the number of vertices and edges, respectively. A graph $S = (V', E')$ is a *subgraph* of G if $V' \subseteq V$, $E' \subseteq E$, and if $(u, v) \in E'$ then $u, v \in V'$. The number of edges containing v is the *degree* and denoted $d(v)$.

A *path* in G is a distinct sequence of vertices from V such that every consecutive pair of vertices corresponds to an edge in E. A *component* of G is subset of vertices such that there is a path between every pair of vertices, but no other vertex can be added to the component and have a path connecting it to any other vertex in the component. A graph is *connected* if it has exactly one component, and a graph is *disconnected* if it has more than one component.

A *cycle* is a path where the first and last vertices are the same. A *tree* is a graph without cycles, and a *forest* is a graph with multiple components where each one is a tree. The *path graph*, denoted P_n, is a tree with n vertices such that two vertices have degree 1 and the other $n - 2$ vertices have degree 2. The *star graph*, denoted S_n, is a tree with n vertices such that one vertex has degree $n - 1$ and all other vertices have degree 1. The *complete graph*, denoted K_n, is a graph with n vertices where each vertex has an edge to the other $n - 1$ vertices. A *bridge* is an edge such that removing it would increase the number of components in the graph. Similarly, a *cut vertex* is a vertex such that removing it and all its edges would increase the number of components in the graph.

3 Related Work

This work builds on related work in finding relevant views in large data sets by identifying visual patterns, automatic techniques for comparing similarity of graphs, and interactive visualization tools for exploring graphs.

3.1 Data Features and Visual Patterns

Wilkinson et al. [63] introduced a graph-theoretic approach to calculating scagnostics (Scatterplot Diagnostics), which substantially reduced the running time of existing scagnostics. Scagnostics measure a scatterplot along nine interpretable characteristics. These are useful for sorting, filtering, or clustering scatterplots in order to identify visualizations that may reveal interesting relationships between variables. Similarly, Pixnostics [47] and Pargnostics [12] are approaches for pixel-oriented visualizations and parallel coordinate plots, respectively. Behrisch et al. proposed Magnostics, an approach to guide exploration of matrix visualizations of graph datasets [7]. Magnostics uses image descriptors of the matrix visualization to rank visualizations by visual patterns, cluster or search for visually similar matrices, or compare the quality of matrix ordering algorithms. Magnostics can only be used with matrix visualizations, and could be sensitive to the row and column orderings of the matrix. This is in contrast to gragnostics, which only uses the graph's topology and is independent of the visual representation of the graph.

The above techniques primarily operate on the visualization itself, and not on the underlying data. Seo and Shneiderman [48] propose the rank-by-feature framework where analysts can rank visualizations by statistical features of the underlying continuous-valued data, such as normality or correlation. One drawback is that many feature names are technical statistical terms that may be unfamiliar to novices, such as least squares error. Seo and Shneiderman's rank-by-feature approach is similar in concept to the approach of gragnostics, except that gragnostics are features of graph datasets instead of continuous-valued datasets, and gragnostics strives to use feature names that are familiar to users.

3.2 Comparing Graphs

Graph-level statistics can capture the relationship between graph-level structure and information flow in graphs. Himelboim et al. [26] discuss research that indicates high interconnectivity corresponds to shared knowledge, better information transmission, and the rate of information spread; clusters and modularity correspond to shared characteristics between vertices within a cluster; centralization corresponds to information flow dominated by a few individuals and vulnerable to disruption; and isolation corresponds to slow information flow and weak relationships. This supports (1) the idea that characterizing a graph's structure can lead to insights about the graph, and (2) the features should be understandable by humans so that analysts can understand, e.g., that two graphs are similar but differ in their clustering.

Many statistics exist for vertices and edges (see Newman [43] for an overview), and they have proven quite useful for comparing vertices and edges in visual analytic tools like SocialAction [45]. There are also graph-level statistics—for example, clustering coefficient and graph diameter—but some of them do not account for graphs with multiple components. A second problem with several of these graph-level statistics is that they are not interpretable by laypeople who do not have a background in graph theory. A third problem is that many graph-level statistics run in $O(|V|^2)$ time or slower, making them impractical for large graph datasets. Nonetheless, these statistics are sometimes used as features in machine learning applications [1, 34], as well as visualization tools [18, 45]. A different approach is to create graph-level statistics by averaging vertex- and edge-level statistics, such as closeness centrality, over a graph. There are several downsides: the average may not be easily interpretable, in most cases this would not describe the graph's higher-level topology, and the averages of two graphs might be the same despite having very different distributions (see Anscombe's Quartet [2]).

Several visualization tools use Euclidean distance on adjacency matrices to compute distances and cluster graphs [4, 16]. Although the systems appear useful, the feature extraction and distance computation is $O(|V|^2)$, and the features do not lend themselves to interpretable explanations of why two graphs are similar. This technique also requires graphs to have the same number of vertices, which reduces its generality.

Motif detection [25] and graph kernels [33] are related techniques for measuring similarities between graphs. However, the naive method is slow [25]. Random sampling techniques to compute motif (aka graphlet) frequencies can improve performance [50], but the features do not lend themselves to easy human interpretability. Furthermore, many graph kernels are not designed for graphs with multiple components [33]. Nonetheless, graph kernels have been successful in cases where explanation is not required [33].

Degree Distribution Quantification and Comparison (DDQC) [1] extracts features in linear time, but it is not designed for human interpretability. Furthermore, classification accuracy on real datasets has not been shown to be very high.

All of the above face at least one challenge with speed, interpretability, or requirements that graphs have only one component or be the same size. In contrast, gragnostics runs in $O(|V| + |E|)$ space and time, does not require familiarity with obscure graph theory terminology, and does not impose restrictions on graphs such as number of vertices or number of components.

3.3 Graph Visualization

Many algorithms provide scalable performance in computing graph layouts [3, 21, 22, 39, 55, 65]. This is an important area of research, but users need analytics and interactions that support the sensemaking process, and not just fast performance.

There are several works that are very effective at analyzing graphs with a few hundred to a few thousand vertices [6, 14, 15, 32, 45, 49, 52]. However, they are not designed to compare multiple graphs, and they have difficulty scaling up to support graphs with tens of thousands or hundreds of thousands of vertices for a variety of reasons: Insufficient rendering performance, unreadable graph layouts, or interaction techniques not suited to very large graphs. For example, ZAME [15] uses a pyramid hierarchy to speedup rendering, but it is limited to visualizing adjacency matrices, which are not appropriate for all analysis tasks [19, 20, 30]. Gephi has fast algorithms for computing layouts and good rendering performance, but limited interactivity for zooming, filtering, and exploring large graphs. In particular, its tools are designed to analyze vertices and edges, which can be challenging when there are hundreds of thousands of them. Similarly, SocialAction has a very effective interface designed specifically to rank and filter vertices and edges by their attributes and statistics. Users analyzing large graphs likely have questions that would not easily be answered with those tools, such as analyzing multivariate attributes [56], paths between vertices [44], or connections between clusters [24].

ManyNets [18] is a visualization tool for comparing multiple graphs using graph-level statistics in a sortable tabular display, and graphs can be selected to view in a node-link diagram. EgoNav [25] computes distances between ego networks using motif frequencies. Similarly, von Landesberger et al. [34] use a mix of graph-, vertex-, and edge-level statistics and motif frequencies to cluster multiple components of a graph. Unlike Chiron, the focus is on clustering, and not searching for graphs with specific topologies.

None of the above visualization tools used only linear-time graph features, so it is unknown whether linear-time features can yield the same level of insight. Furthermore, it is not known whether combining subgraph exploration with an overview of the entire graph provides benefits. As shown later, linear-time features can be effective, and an overview of the entire graph can provide useful context.

In contrast with many other visualization tools, Chiron is designed to enable users to explore a large graph by its subgraphs, which are believed to be units of interest in large graphs [24]. Chiron's focus+context design allows users to see the details of a subgraph while also seeing an overview of the entire large graph, which is believed to be useful even if low-level details are not visible [13]. Chiron integrates gragnostics, enabling users to search for graphs with specific, interpretable topological characteristics.

4 Topological Gragnostics

Gragnostics were designed to balance several constraints. First, gragnostics must scale to large graphs. To accomplish this, gragnostics was designed to be computed in $O(|V| + |E|)$ time. Second, the features must be comprehensible to analysts who are not experts in graph theory. To accomplish this, gragnostics correspond to topological characteristics described in plain language. This enables broad audiences to easily understand gragnostics. Third, gragnostics should not be constrained to certain types of graphs. To ensure this, gragnostics do not have restrictions on graph size or number of components.

The gragnostics features were selected by surveying the literature and identifying features for measuring graph similarity to many types of fundamental graphs, such as trees, stars, and complete graphs. Only features that can be calculated in linear time were selected. Gaps in the set of features were identified by looking for common graph motifs and fundamental graphs that were not modeled by these metrics, such as trees and lines. Additional features could detect cycle graphs, but algorithms to count cycles are computationally expensive, and therefore not included. This is an avenue for future research. However, as shown below, gragnostics achieves high accuracy without including this feature.

Of the proposed gragnostic features, two gragnostics are new methods to measure graph-level features: line and tree. Three are based on existing graph-level features, but with proposals to normalize them to improve human interpretability: bridge, disconnection, and constriction. The other five are existing graph-level features: Nodes (number of vertices), links (number of edges), density, isolation, and star. These features are chosen because they all describe distinct graph structure commonly discussed in network analysis (e.g. Himelboim et al. [26]). The new line and tree features were created because no known statistics exist to measure these structures on a [0, 1] scale. Figure 1 illustrates gragnostics with several graphs (excluding the nodes and links gragnostics).

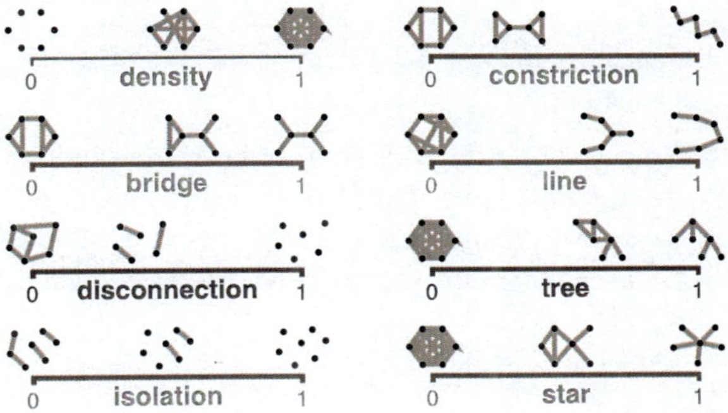

Fig. 1 Example graphs illustrating each gragnostic (excluding the nodes and links gragnostics)

The gragnostic feature names were chosen while considering understandability by laypeople. The names minimize the use of obscure graph theoretic terminology that is primarily only understood by graph theoreticians. Most names are in the list of 10,000 most commonly used words or in Basic English.[1] This helps ensure that the gragnostic feature names will be easily understood by users.

4.1 *Calculating the Features*

There are 10 gragnostics features. Each feature's range is [0, 1], where 0 is a low gragnostic value, and 1 is a high value. The gragnostics are presented below, along with possible interpretations. These features assume simple undirected graphs, but many gragnostics can be adapted to directed graphs.

Density This measures the interconnectivity of vertices within a graph. Higher density can indicate a higher rate of information flow. This uses the typical definition for undirected graphs:

$$\frac{2 \cdot |E|}{|V| \cdot (|V| - 1)} \tag{1}$$

This is minimized in a graph with no edges, and maximized in a graph where an edge connects every pair of vertices. Calculating density requires a simple count of the vertices and edges, and therefore runs in $O(|V| + |E|)$ time.

Bridge Edges are called *bridges* if their removal will disconnect the graph, analogous to a bridge that connects two cities. Bridge measures relationships that are required for information to flow in the graph. Graphs that have moderate values for bridge and density can indicate tight clustering. The bridge feature is calculated by

$$\frac{bridge(G)}{|V| - 1} \tag{2}$$

[1] https://github.com/first20hours/google-10000-english http://ogden.basic-english.org/words.html.

where $bridge(G)$ is the number of bridges in graph G. There are at most $|V| - 1$ bridges in a graph, which occurs if the graph is a tree. The number of bridges can be calculated using Tarjan's algorithm [54], which runs in $O(|V| + |E|)$ time.

Disconnection A high disconnection feature indicates that many vertices are disconnected from each other. This indicates little or no information flow, or broken or disrupted communication. Let C be the set of all maximally connected components of G. Then the disconnection gragnostic is defined by

$$\frac{|C| - 1}{|V| - 1} \tag{3}$$

This is minimized when the graph is a single component, and maximized when there are no edges in the graph, i.e. the graph is completely disconnected. If $|C| = 1$ then disconnection is defined as 0. C can be calculated by running breadth-first searches to identify each component [28], which runs in $O(|V| + |E|)$ time.

Isolation This describes the fraction of vertices in the graph that have no edges connecting them to other vertices. Low isolation indicates that there is some information flow, but there might or might not be clusters of high information flow. This is the definition given by Himelboim et al. [26]:

$$\frac{|\{v \in V : d(v) = 0\}|}{|V|} \tag{4}$$

This is 0 when $|E| = 0$, and 1 when every vertex has at least one edge (i.e. when $\forall v \in V, d(v) > 0$). High isolation can indicate slow information flow or weak relationships. The degree of each vertex can be calculated with a simple iteration over each edge to count the number of edges incident on each vertex. Therefore isolation can be calculated in $O(|V| + |E|)$ time.

Constriction The constriction gragnostic is similar to bridge. Constriction measures vertices that are required for information to flow in the graph. Graphs that have higher constriction and higher density can indicate tight clustering. Constriction is calculated by:

$$\frac{cut(G)}{|V| - 2} \tag{5}$$

where $cut(G)$ is the number of cut vertices in G. If $|V| < 3$ then this gragnostic is defined as 0. Calculating $cut(G)$ is very similar to calculating $bridge(G)$ [28], and therefore runs in $O(|V| + |E|)$ time.

Line (new) Measures how close a graph of $|V|$ vertices is to being the path graph P_n. Gragnostics uses the name "line" to avoid confusion with the other graph theory definition of path. Line measures the degree of sequential connections, such as hierarchies without branches or train stations without transfers. Let $\mathbf{D}$ be a vector of length

$|V|$ and let $\mathbf{D}_i, \mathbf{D}_j$ be elements of D where $\mathbf{D}_i = d(v_i), \mathbf{D}_j = d(v_j)$ for $v_i, v_j \in V$ such that if $\mathbf{D}_i = 1$ and $\mathbf{D}_j > 1$ then $i < j$. For example, if G is a graph with four vertices that have degrees 1, 2, 2, and 3, then both $\langle 1, 2, 2, 3\rangle$ and $\langle 1, 3, 2, 2\rangle$ are valid $\mathbf{D}$ vectors. Then the line gragnostic is calculated by

$$\sum_{i=1}^{|V|} \frac{l(i)}{|V|} \tag{6}$$

where

$$l(i) = \begin{cases} 1, & \text{if } \mathbf{D}_i = 1 \text{ and } i \leq 2, \text{ or if } \mathbf{D}_i = 2 \text{ and } i > 2 \\ 0, & \text{otherwise.} \end{cases} \tag{7}$$

In English, the line gragnostic is the fraction of vertices that have the correct degree of a path graph. This is derived from the fact that a path graph is a tree with two leaves, so a graph with $|V|$ vertices has 2 vertices with degree 1 and $|V| - 2$ vertices with degree 2. Creating $\mathbf{D}$ requires first calculating an array of degrees for each vertex, and then two linear-time loops over the array of degrees (one loop to find all degrees of value 1 and put them at the beginning of $\mathbf{D}_i$, and one loop to append the remaining degrees to $\mathbf{D}_i$). Therefore computing line runs in $O(|V| + |E|)$ time. If G has multiple components, then the line gragnostic is calculated as the arithmetic mean of the line feature of all components weighted by the number of vertices in each component.

Tree (new) Graphs that are tree-like can indicate hierarchy, dependency, or parent-child relationships. Trees are graphs with no cycles, and therefore a tree with $|V|$ vertices has $|V| - 1$ edges. Therefore, for a connected graph, we can calculate the tree gragnostic using the following definition:

$$1 - \frac{|E| - (|V| - 1)}{|V| \cdot (|V| - 1)/2 - (|V| - 1)} \tag{8}$$

where $|E| - (|V| - 1)$ is the number of edges needed to remove to make the graph a tree, and $|V| \cdot (|V| - 1)/2 - (|V| - 1)$ represents the maximum possible number of edges needed to remove to make the graph a tree (i.e. if G were a complete graph). If G is disconnected, then tree is calculated as the arithmetic mean of the tree feature of all connected components weighted by the number of vertices in each component. If $|V| \leq 2$ then the tree gragnostic is defined as 0. This gragnostic is a simple algebraic equation involving the count of the number of vertices and edges, and therefore it runs in $O(|V| + |E|)$ time.

Star The star gragnostic measures how much a graph with n vertices is like S_n, or the degree to which a single vertex is more central than the other vertices. Star measures hub-and-spoke relationships, or how much more central one vertex is than the others to the graph's information flow. This is measured using Freeman's degree centralization [17]:

$$\sum_{v \in V} \frac{d(v^*) - d(v)}{(|V| - 1)(|V| - 2)} \tag{9}$$

where v^* indicates the vertex with the largest degree. For the purposes of gragnostics, if $|V| \leq 2$ and then star is defined as 0. The star gragnostic is minimized when all vertices have the same degree. For a connected graph G, centralization is maximized when one vertex has degree $|V| - 1$ and all other vertices have degree 1, which is a star graph. If G has multiple components, then the star gragnostic is calculated as the arithmetic mean of the centralization of each component. This gragnostic is calculated by first calculating the degree of each vertex, and then computing the sum over all vertices, and therefore it runs in $O(|V| + |E|)$ time.

Nodes and Links Because two graphs may have similar gragnostic features but be significantly different in size, the number of vertices and edges is useful for discriminating graphs or finding graphs of a particular size. Gragnostics uses the name "nodes" for vertices, and the name "links" for edges because these terms are not based on graph theory. This makes the names more likely to be understood by analysts and users without a background in graph theory. The nodes gragnostic is simply $|V|$, and likewise links is $|E|$. These can be normalized to the range [0, 1] for a set of graphs by calculating $|V|/|V^*|$ and $|E|/|E^*|$, where V^* and E^* indicate the largest set of vertices and edges, respectively, from the set of graphs.

4.2 *Alternatives Calculation Methods*

There are other ways to calculate the above gragnostics. Line could be calculated by dividing the graph diameter by $|V| - 1$, but calculating diameter is slow (running time is $O(|V|^2 \log |V| + |V| \cdot |E|)$). Similarly, the largest normalized betweenness centrality could be used for the star gragnostic, but this also runs in $O(|V|^2 \log |V| + |V| \cdot |E|)$ time. The star gragnostic could be calculated using the star magnostic [7], but that is also slower, and introduces challenges with finding a good ordering for the matrix visualization. For some existing graph types, such as stars, trees, or complete graphs, we could use binary tests to decide if the graph matches the definition, but this can create large distances in the feature space between graphs that are very similar (e.g. if removing a single edge would make it a tree). Edit distance [10] could give a more nuanced notion of similarity, but this is also slow.

5 Evaluations

There are three evaluations that examine different aspects of gragnostics: (1) The independence of the features, (2) the effectiveness of gragnostics in a visual analytics setting, and (3) the effectiveness of gragnostics for visual analytic applications. The first two evaluations use a dataset of 92 graphs comprising the following seven classes of graphs:

Artificial The Barabási-Albert [5] and Watts-Strogatz algorithms [62] were designed to mimic the structure and degree distribution of real human interaction networks. Both algorithms take a parameter n, which is the number of vertices. Barabási-Albert takes two parameters, n and m (the number of edges that link a new vertex to already existing vertices). Six Barabási-Albert graphs were generated with $n \in \{250, 500, 1000\}$ and $m \in \{3, 4\}$. Watts-Strogatz takes three parameters: n, k (the number of neighbors for each vertex), and p (a "rewiring" probability). Six Watts-Strogatz graphs were generated with $n \in \{250, 500, 1000\}$, $k \in \{6, 8\}$, and $p = 0.5$.

Character Fictional character interaction networks from Anna Karenina [31], The Winter King (The Arthur Books #1) [27], David Copperfield [31], The Hobbit [27], Huckleberry Finn [31], Les Miserables [31], Star Wars 1–7,[2] and Storm of Swords [27].

Collaboration Research paper co-authorship graphs from Network Science [42], CHI,[3] International Symposium on Graph Drawing, International Conference on Machine Learning, KDD, Symposium on Discrete Algorithms, Transactions on Visualization and Computer Graphics, and arXiv (Astrophysics [41], High-Energy Theory [41], General Relativity and Quantum Cosmology [35], High Energy Physics - Phenomenology [35], High Energy Physics - Theory [35]).

Ego Ego networks of users from Facebook [37].

Geometric Graphs that exhibit regular geometric structure: cubical, dodecahedral, Frucht, 4×4 grid, 4D hypercube, icosahedral, octahedral, tetrahedral, S_6, S_{10}, Chvátal, Desargues, Heawood, Petersen, Pappus, a graph of three components where each is a K_3, and ladder graphs of length 4 and 6.

Software Software class dependency networks: jUnit [57], jMail [57], Flamingo [57], Jung [57], Colt [57], Org [57], JavaX [57], Guava [58], slucene [60], Weka [59], and sjbullet [61].

Subway Graphs of subway networks from cities in 2009 [46].

These graph classes were chosen because they are commonly analyzed in the literature. Common graphs of each class were added until each class had at least 10 graphs, while maintaining similar numbers of graphs in each class.

In the evaluations, the nodes and links gragnostics were scaled to the range [0, 1]. The evaluations began by extracting the 10 gragnostic features for all 92 graphs.

[2] https://github.com/evelinag/StarWars-social-network.

[3] http://gmap.cs.arizona.edu/datasets.

5.1 Are Gragnostic Features Sufficiently Independent?

This evaluation examines whether there are redundancies in the gragnostics features. We might expect correlation between: nodes and links, since graphs with more vertices can have more edges; between links and density, since graphs with more edges tend to become more dense; between isolation and disconnection, since a vertex with degree 0 forms its own component; between tree, star, and line, since stars and lines are special cases of trees; between bridge and constriction, since a bridge edge necessarily has two cut vertices; and negative correlation between density and tree, star, and line, since trees are not dense by definition. However, each feature measures a distinct topological characteristic, and therefore they will probably not be highly correlated in practice.

We compute the Pearson correlation between each feature using the 92 evaluation graphs described above to examine their correlation in a practical example. Since we are primarily concerned with the degree of correlation, and not necessarily with whether the correlation is positive or negative, this evaluation computes the absolute value of the correlation. This is shown in Fig. 2. The highest correlations are between the nodes and links features and disconnected and isolation features, which both have a value of 0.79. These two features are arguably important for differentiating graphs of different sizes that otherwise have similar topologies. Since all other correlations are smaller, we argue that the other features are sufficiently independent and should be included. With regards to the other feature correlations, note that our predictions about correlation were also correct for tree and density, bridge and constriction, and disconnection and isolation. However, our predictions were wrong about correlation between links and density; tree, star, and line; and density and star and line. The absolute value of all these correlations are below 0.5.

To examine the gragnostic feature relationships more closely Fig. 3 shows a scatterplot matrix of the 10 gragnostics for the 92 graphs. Most gragnostics appear uncorrelated, as in Fig. 2. Several scatterplots have outliers, indicating that the different gragnostics describe different characteristics, even if there are correlations. (For example, despite the correlation seen in this dataset, a hypothetical graph composed of 10 components where each was K_3 would have both a low tree and low density value. The tree value would be 0 and the density would be approximately 0.069.) In the scatterplot matrix we also see some class separation, even in 2 dimensions. For example, in the star versus density plot the geometric graphs are separated from the others; in the star versus bridge plot the ego graphs are separated; and in the constricted plots the subway graphs are separated.

5.2 Can Gragnostics Effectively Differentiate Graphs?

This evaluation examines the utility of gragnostics for clustering and machine learning classification. This evaluation compares a graph kernel [50] and Degree Distribution Quantification and Classification (DDQC) [1]—a set of graph-level features

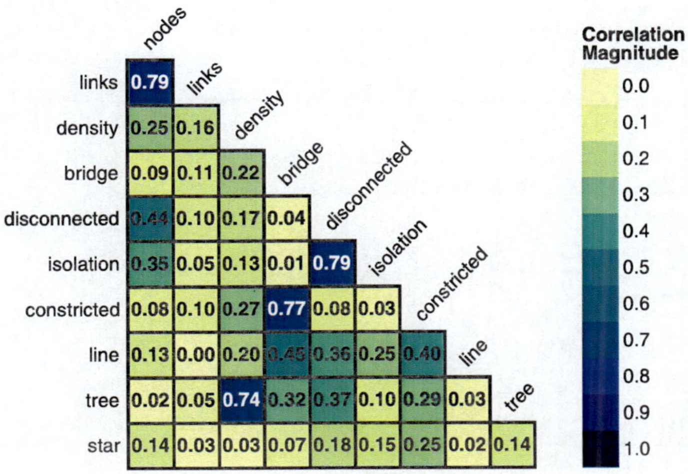

Fig. 2 The absolute correlation between each gragnostic on the 92 graphs. The nodes and links features and the disconnected and isolation features have the highest magnitude (0.79). Most features have low correlation

used for visualization [16]—to gragnostics. This evaluation does not include other feature sets because of known limitations: edge-based feature vectors require graphs to be the same size, and many other common graph statistics such as diameter or betweenness centrality run in $O(|V|^2)$ time or slower. Many graph kernels only work on graphs with one component, but graphlet sampling has been used successfully in other graph visual analytics scenarios [33] so we include it here. For these reasons, this evaluation only compares gragnostics, DDQC, and graphlet sampling. See Sect. 3 for further discussion of graph comparison methods.

Gragnostics and DDQC do not require parameterization, but graphlet sampling is parameterized by k, ϵ, and δ. The dimension of the graphlets is k, and ϵ and δ control the level of sampling. During experiments accuracy was highest with $k = 5$, $\epsilon = 0.05$, and $\delta = 0.05$ so those values were used for the results reported here. The evaluation uses the implementation provided by GraKeL [51].

The classification task is to predict a graph's class (e.g. artificial, character, etc.) using a k-nearest neighbors (KNN) classifier. Past research indicates different classes have different structure [1, 38]. The KNN classifier uses Euclidean distance to find the k nearest neighbor graphs, and uses inverse distance-weighted voting to predict the class from the nearest neighbors. (Note that a different type of classifier, such as a support vector machine with a Gaussian kernel, might perform better by identifying non-linear relationships between features. But this might require a larger dataset for training.) Each element in the graph kernel matrix K is converted to a distance matrix D using $D_{ij} = \sqrt{1 - K_{ij}}$ for each element i and j in the matrix.

This evaluation was run on a 2019 MacBook Pro, with a 2.4 GHz 8-Core Intel Core i9 processor and 32 GB of 2667 MHz DDR4 RAM. Table 2 describes the run

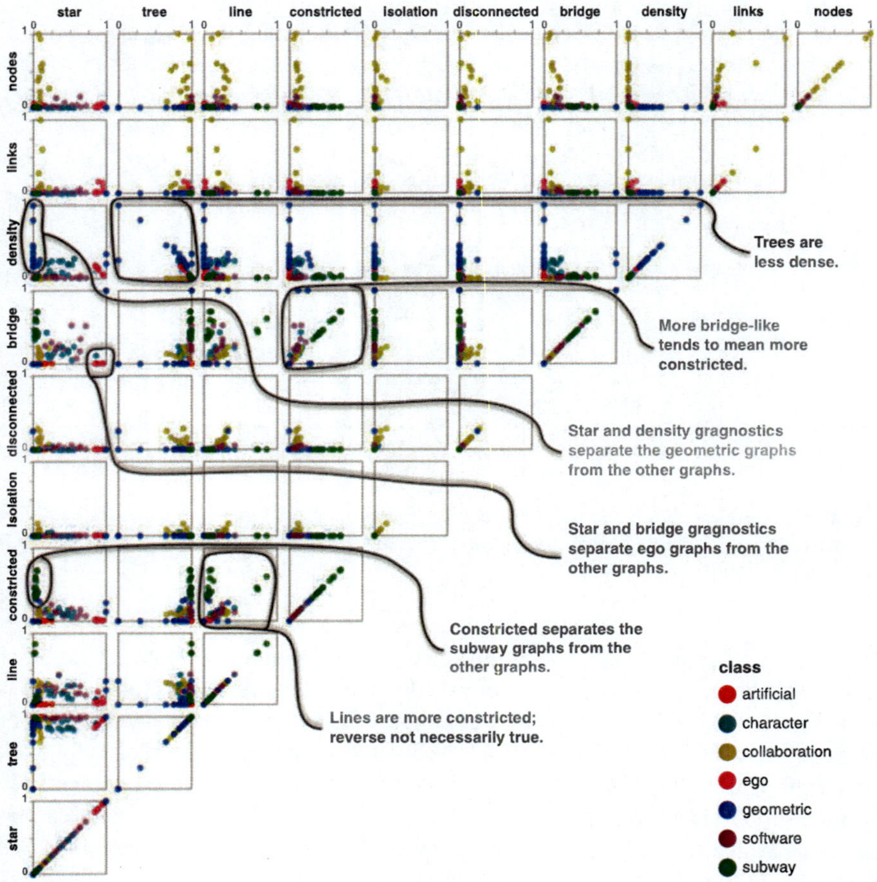

Fig. 3 A scatterplot matrix showing gragnostic values for 92 graphs

time for feature extraction on the 92 graphs. Gragnostics is slower than DDQC, but gragnostics is substantially faster than graphlet sampling.

The evaluation uses leave-one-out cross validation and KNN classifiers with k from 1 to 10 (10 is the number of graphs in the ego graph class, which is the smallest class). The evaluation metrics are accuracy (fraction of classifications that are correct), and precision@k (P@K, mean fraction of k nearest neighbors with the correct class). Figure 4 shows the accuracy and P@K for each value of k. Accuracy using gragnostics is high (0.90–0.93) and robust to the value of k. P@K of gragnostics is similarly high for small values of k. Figure 5 shows the 13 graphs that were misclassified along with the number of classifiers that misclassified them; the remaining 79 graphs were always correctly classified. Interestingly, we see that the artificial graphs are never misclassified. This is surprising because they are designed to generate graphs that mimic human social interaction graphs, and there are several in this dataset (collaboration graphs, character-interaction graphs, and ego social graphs).

Table 1 Summary of the 92 evaluation graphs: The number of graphs in each class, and the vertex and edge ranges

Graph class	Number	$\|V\|$	$\|E\|$
Artificial	12	250–1, 000	741–4, 000
Character	14	18–138	41–493
Collaboration	12	1, 001–20, 046	2, 627–198, 110
Ego	10	53–1, 035	198–30, 772
Geometric	18	4–20	5–32
Software	11	128–2, 956	310–10, 845
Subway	15	82–433	85–475

Table 2 Feature extraction time for the 92 graphs described in Table 1

Feature type	Calculation time
Gragnostics	16.8 s
DDQC	0.4 s
Graphlet sampling	119.0 s

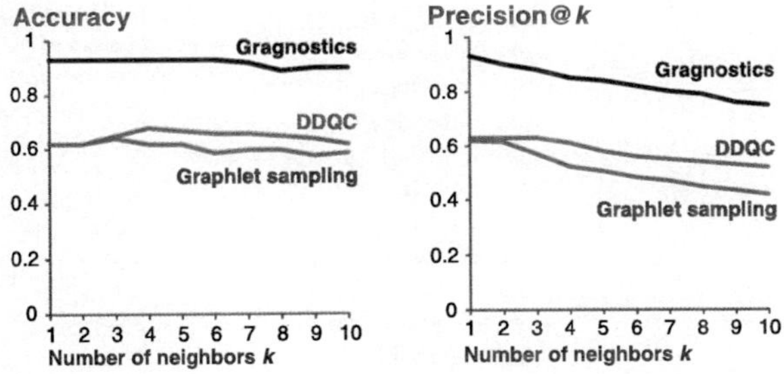

Fig. 4 Accuracy and precision@k for each tested value of k in the KNN classifiers using gragnostics, DDQC, and a graph kernel (graphlet sampling). Accuracy stays consistently high for gragnostics regardless of the number of neighbors k used. Gragnostics substantially outperforms DDQC and the graph kernel for both accuracy and precision@k

This indicates that the Barabási-Albert and Watts-Strogatz models do not capture some essential characteristics found in real and fictional human interaction graphs.

Accuracy and P@K for DDQC and graphlet sampling performed similarly to each other but considerably worse than gragnostics (see Fig. 4). In addition, using DDQC 41 graphs were misclassified at least once, and 23 were misclassified by all 10 KNN classifiers. With graphlet sampling 53 graphs were misclassified at least once, and 27 were misclassified by all 10 KNN classifiers. To test whether a different type of classifier would perform better than KNN, a support vector classifier ($C = 1$ performed the best) was trained on the graphlet sampling kernel matrix using a leave-

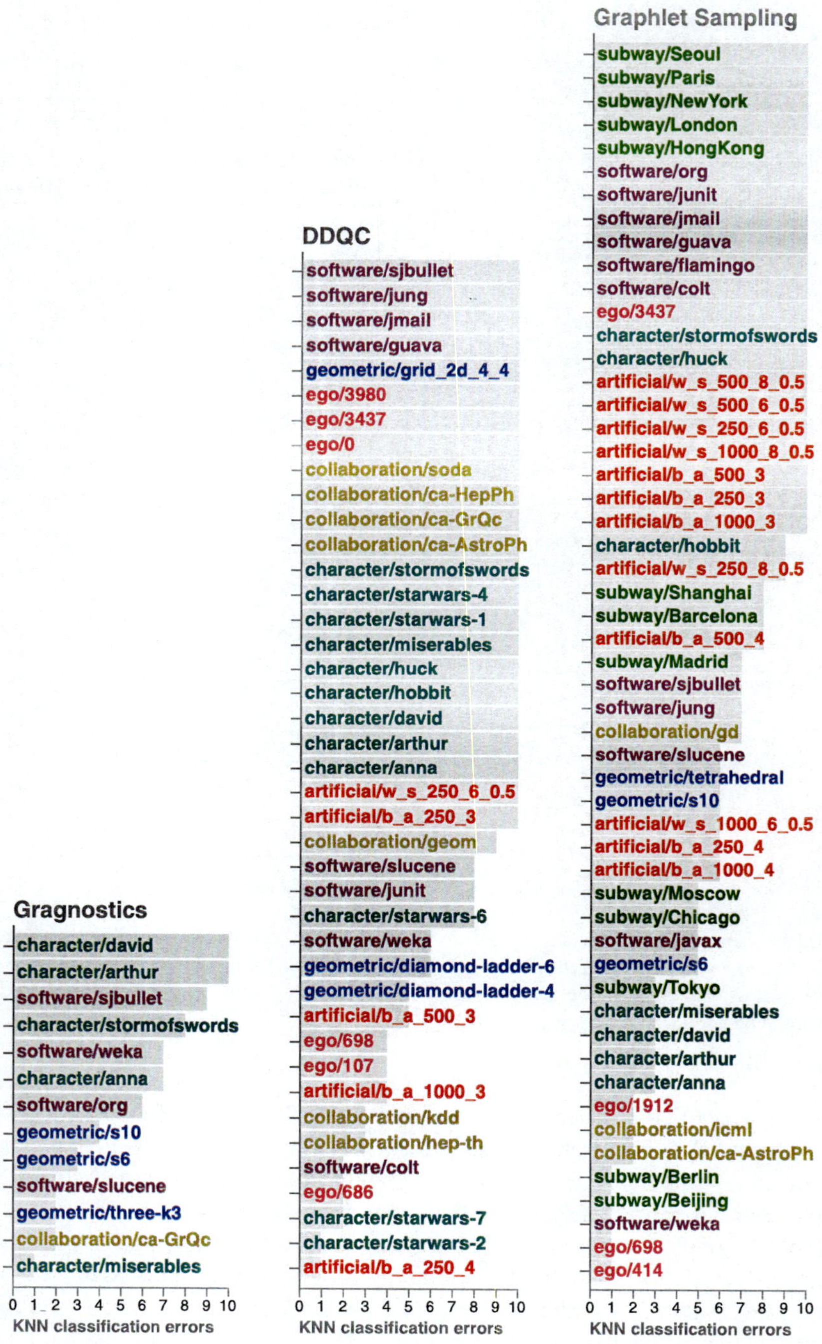

Fig. 5 The graphs that were misclassified by the KNN classifiers using gragnostics, DDQC, and a graph kernel (graphlet sampling). The length of each graph's bar indicates the number of classifiers that misclassified that graph. Graphs not shown for a featurization technique were always correctly classified when using those features

one-out cross validation and achieved 42.4% accuracy on the hold out graphs. This is notably worse than the accuracy of the KNN classifiers using graphlet sampling.

If we remove the four most highly correlated features from gragnostics (links, disconnected, bridge, and tree), we can test whether the extra features provide a useful increase in classification performance. By doing this and running the same experiment above using only the 6 features, the accuracy drops to 0.78-0.80, depending on the value of k. Similarly, the precision drops to 0.66-0.82. Additionally, 25 graphs were misclassified for at least one value of k, with 13 graphs always misclassified regardless of the value of k. Therefore, the extra four features appear to be beneficial for improving classifier performance.

This analysis indicates that graph-level features like gragnostics can be substantially faster than graph kernels and also achieve substantially better accuracy than DDQC or graph kernels. This is interesting because graph kernels have received considerable attention, but in this evaluation simpler methods that are more human-understandable are faster and achieve better accuracy.

Figure 6 shows a multi-dimensional scaling (MDS) plot [9] of the 92 graphs projected onto 2 dimensions from the 10-dimensional gragnostic feature space. We see

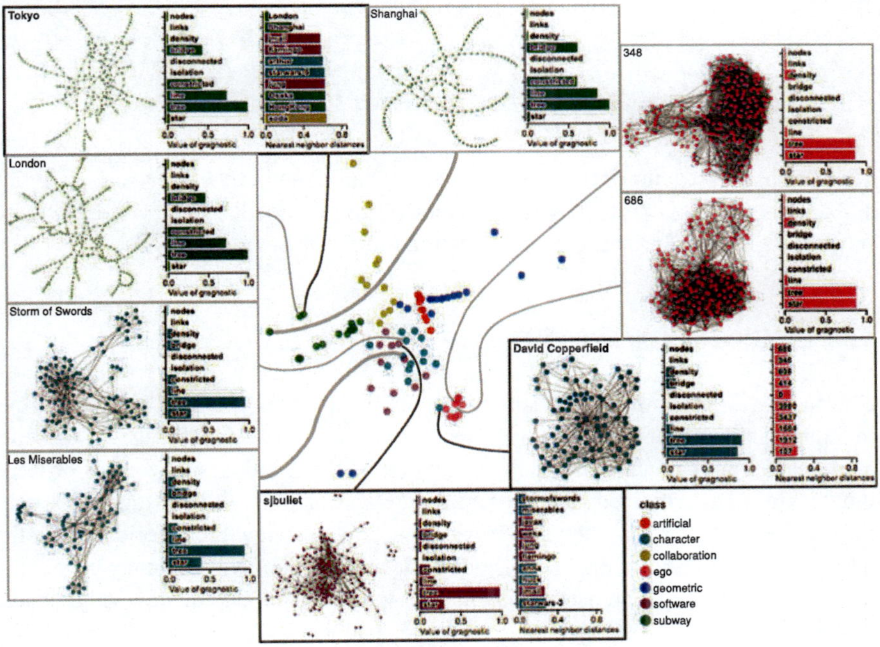

Fig. 6 A multi-dimensional scaling plot showing distances between 92 graphs using gragnostics. In general, there is good class separation, although the character and software classes overlap. The figure calls out Tokyo, David Copperfield, and sjbullet in **bold** and shows their gragnostic values and a bar chart of their 10 nearest neighbors and the distances to those neighbors. The two nearest neighbors are also called out for comparison, allowing us to see that the nearest neighbors do indeed have similar topological structure

good overall class separation, indicating that each graph class has different gragnostics values. We do see a few exceptions, which correspond to graphs misclassified by the KNN classifiers. For example, the David Copperfield graph's nearest neighbors are all Facebook ego networks. This classification makes sense if we consider that David Copperfield is often considered to be a semi-autobiographical novel about Charles Dickens, in essence making it an ego graph of the central character. The Les Miserables and Storm of Swords graphs are more typical of fictional character graphs. The software and the character classes overlap in the MDS plot. The sjbullet software graph's gragnostics are similar to the Storm of Swords character graph, although it is larger and less dense. We can visually confirm the topological similarity by comparing the two force-directed node-link diagrams. Finally, let's examine the Tokyo subway graph: its distance to the London subway graph is very short. Their gragnostics are nearly identical, and their force-directed node-link diagrams share the same visual structure. Meanwhile, the Tokyo graph's second nearest neighbor is the Shanghai subway graph, which is farther away than London. Shanghai has higher bridge, constriction, and line gragnostics. We can visually confirm this dissimilarity by looking at Shanghai's force-directed node-link diagram and noting that it has more bridge edges, it has more constriction points, and it is more line-like because more vertices have only two edges.

5.3 *Are Gragnostics Effective for Visual Analytics?*

This evaluation examines whether gragnostics can be effective for visual analytics applications. This is performed by implementing gragnostics in a rank-by-feature framework in the Chiron graph visualization tool.

5.3.1 Chiron System Description

Chiron is designed to explore subgraphs within a larger graph. Chiron has four main user interface components. (1) The *Detailed overview* panel (Fig. 7c) presents a zoomable and pannable overview of the graph. Node-link visualizations of large graphs are often criticized as being uninterpretable. Although they might not convey detailed information, they can provide a useful overview [13]. For example, the overview in Fig. 7 shows that there is one very large component and many very small components. This can be useful information to help users decide how to filter and select subgraphs. (2) The *Suggested subgraphs* panel (Fig. 7a) is a sortable and filterable list of all identified subgraphs of the graph. The panel shows the attributes and gragnostic feature values for each subgraph. (3) The *Settings panel* (Fig. 7b) controls filtering and rendering of the subgraphs. The panel has scented range sliders [64] that show users the distribution of subgraphs for each attribute and gragnostic feature. Users can also filter out clusters of similar subgraphs. (Clusters are determined by

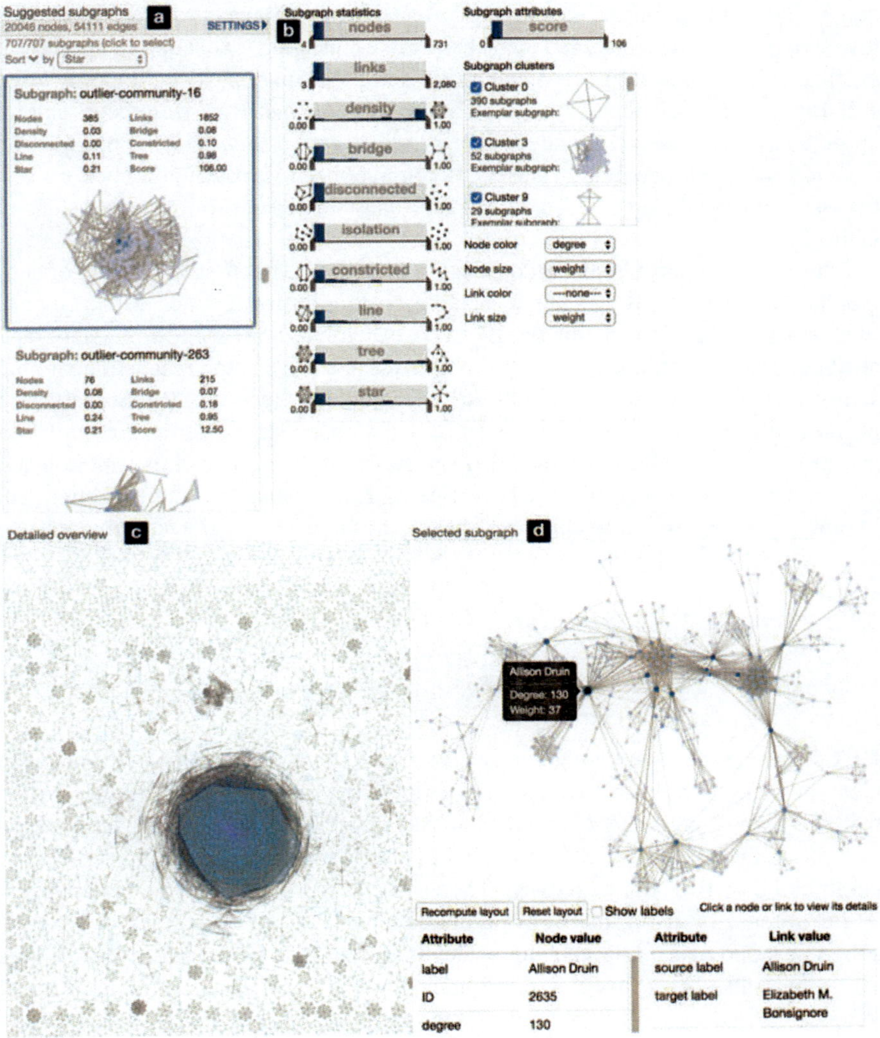

Fig. 7 Gragnostics can be used to identify patterns and features of subgraphs. Panel b shows the gragnostics and other subgraph attributes for the CHI dataset. The *Suggested subgraphs* panel **a** shows a list of subgraphs, the bottom left **c** is an overview of the entire graph, and on the bottom right **d** is a detailed view of one of the communities. This subgraph has a relatively high star feature, where a few authors tend to work with many other authors. In this dataset, this tends to indicate advisor-student relationships. Vertex color indicates degree, vertex and edge size indicate number of papers

clustering the subgraphs on the gragnostic features using k-means clustering. Chiron tests a range of values for k, and then chooses the clustering that achieves the highest average silhouette score.) (4) When a user selects a subgraph from the *Suggested subgraphs* panel, it is then highlighted in the *Detailed overview* with a convex hull and displayed for detailed analysis in the *Selected subgraph* panel (Fig. 7d). The *Selected subgraph* panel allows users to compute better layouts, and click on vertices and edges to see their details. Figure 7 shows a screenshot of these panels in Chiron.

Chiron can import Graphviz dot files. In the dot file, users specify a list of subgraphs that each vertex belongs to using a custom attribute called `subgraphs`. Each entry in the list contains the ID of the subgraph, and any optional subgraph attributes, such as outlier score. Chiron uses these attributes to create sliders in the settings panel for users to sort and filter the subgraphs (see Fig. 7b). Some examples of subgraphs are the graph's biconnected components or vertices grouped by an attribute. If users have not specified any subgraphs, Chiron can run the Louvain community detection algorithm [8] to create subgraphs when the dot file is imported. During import, Chiron also computes the gragnostic features for each subgraph.

5.3.2 Example Usage Scenarios

Below are two example analyses using gragnostics and Chiron's rank-by-feature functionality to analyze two graphs.

CHI This dataset is the co-authorship network of the Conference on Human Factors in Computing Systems from 1990 to 2014. Vertices represent authors, and edges denote that two authors co-authored at least one paper together. The `weight` attribute on vertices and edges indicates the number of papers written by an author (vertices) or co-authored by two authors (edges). There are 20,046 vertices and 54,111 edges.

Let's begin by using the Louvain community detection algorithm to generate subgraphs, and discarding all subgraphs with fewer than four vertices. This leaves 707 subgraphs. The overview (see Fig. 7) shows one large central component of vertices, and many small components on the periphery. Chiron's subgraph clustering functionality is especially useful here: Cluster 0 (see the settings panel in Fig. 7b) is composed only of small cliques. By deselecting this and other clusters of small, peripheral communities, we filter the suggested subgraphs down to 84 subgraphs. The subgraphs tend to have low density, and there are 54 subgraphs that have a tree feature greater than 0.9. In these communities, we see many vertices that connect local areas of density; indeed, many of the subgraphs have a relatively high star feature. This indicates that communities in the CHI research community are sparse, but there are a few key authors who connect the community (see the selected subgraph in Fig. 7d).

csrankings The csrankings dataset is a collaboration graph of faculty at top computer science schools based on publications in the most selective conferences.[4] The dataset

---

[4] http://csrankings.org/.

also includes the raw count of papers an author has published at top conferences, an adjusted count weighted by the number of co-authors, and the area of each publication (e.g. AI, visualization, HCI, etc.). A custom Python script parsed the dataset to create a Graphviz dot file to import into Chiron, and vertex positions were generated with a force-directed group-in-a-box layout [11]. Vertices represent authors, and edges represent a co-authorship relationship between authors. The resulting graph contains 5,995 vertices and 15,094 edges. (Some vertices are duplicates of the same author, such as in Fig. 8d.) Initially, vertices are grouped into subgraphs by the author's affiliation. In this grouping, there are 192 subgraphs.

After loading the dataset in Chiron, in the *Detailed overview* panel in Fig. 8c we see a large group of vertices that appears to be a component, with several isolated vertices (degree 0) on the perimeter. Next, let's adjust the filters to show only subgraphs with low star, density, and disconnection features. Following the typology from Himelboim et al. [26] this should show us CS departments at universities that are clustered, i.e. there are some groups of researchers who collaborate, but there is no collaboration between groups. We choose the disconnection feature instead of the isolation feature because we are looking for subgraphs with many components, and not necessarily many isolated researchers who do not collaborate with anyone. Sorting the subgraphs by descending disconnection feature reveals several CS departments that are clustered. For example, in the University of Pittsburgh subgraph (Fig. 9a), we find five authors publishing work in AI, NLP, and machine learning, but only two of them have collaborated, producing one paper in this dataset. In contrast, there is a cluster of six authors whose work centers around architecture, design automation, software engineering, embedded systems, high performance computing, communication, and data management. There is also a third cluster who have published mostly in data management.

Next, let's discard the subgraphs grouped by affiliation and run the Louvain community detection algorithm on the graph. This gives us 1092 subgraphs, with 932 that are only a single vertex, and 107 that only have two vertices. We discard these, and import the remaining 53 subgraphs. Filtering by the number of affiliations in each subgraph, we find 12 subgraphs that only have one affiliation. All of these have either three or four vertices. The subgraph with the highest number of affiliations has 132 affiliations.

Next, we reset the filters and filter out subgraphs with small numbers of vertices, and sort by the star feature. We see a few subgraphs like the one in Fig. 9b, which has one vertex from one university that is a cut vertex connecting two researchers from a different university. This is David R. Cheriton, who is a professor at Stanford University, but he is known for his strong ties to Waterloo. Sorting by star in descending order, we see a common pattern where authors from the same institution collaborate, and one or two of those authors will then collaborate with members of another institution (see Fig. 9c).

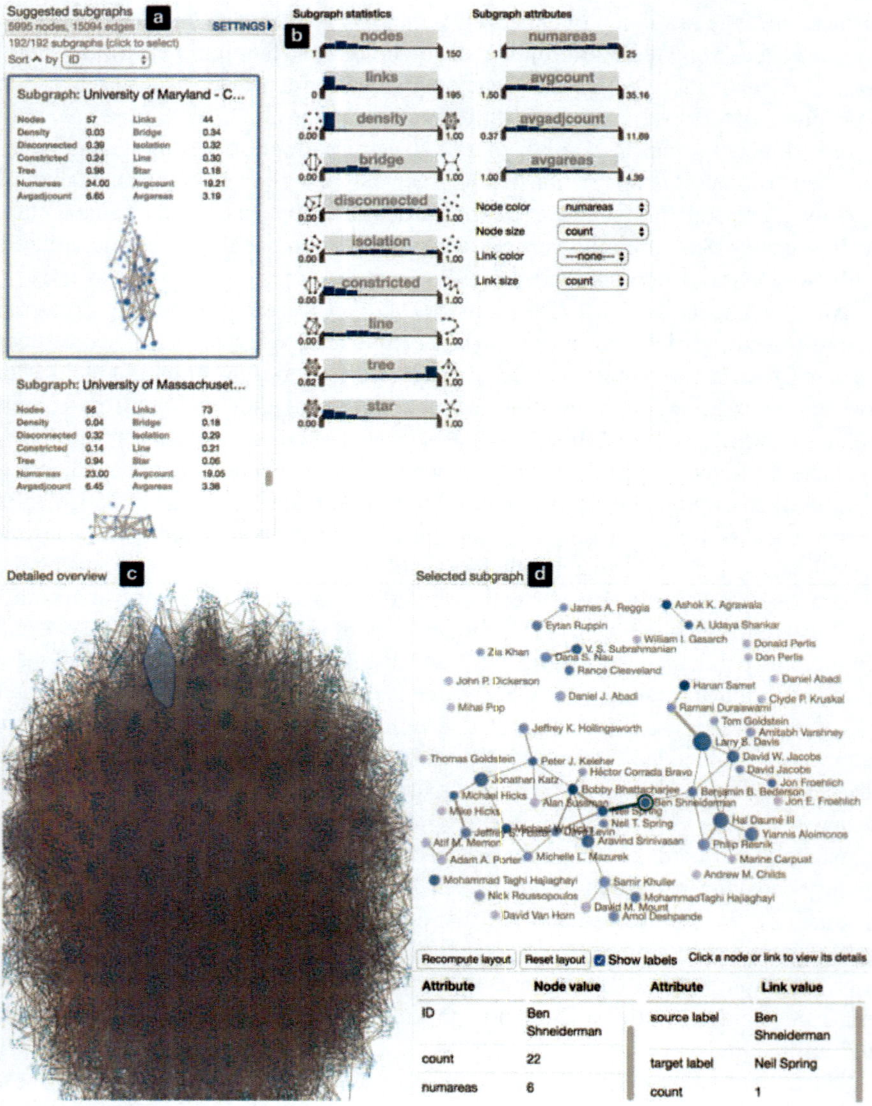

Fig. 8 Gragnostics can be used to identify patterns and features of subgraphs in Chiron. On the top **b** is the distribution of gragnostics and other subgraph attributes for the csrankings dataset. The *Suggested subgraphs* panel **a** shows a list of subgraphs, the bottom left **c** is an overview of the entire graph, and on the bottom right **d** is a detailed view of the University of Maryland subgraph

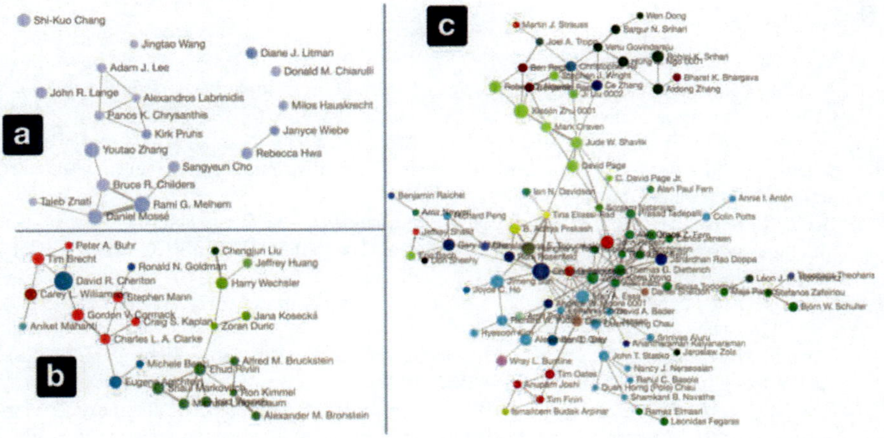

Fig. 9 Subgraphs from csrankings. In the U. of Pittsburgh collaboration subgraph **a** vertex color is the count of papers published, and vertex and edge size are the number of areas. In **b** and **c** vertex color indicates affiliation, vertex size is the number of areas, and edge size is the count of papers. In **b** and **c**, intra-institution collaboration is more common than inter-institution, but interestingly David R. Cheriton connects researchers from the U. of Waterloo

6 Discussion

This chapter introduced two graph-level features (tree and line) and combined them with eight existing graph-level features to form gragnostics, a set of $O(|V| + |E|)$ human-understandable features for comparing graphs. This chapter evaluated gragnostics for graph clustering, classification, and visual analytics.

This chapter shows that gragnostics are a capable tool for classifying graphs, understanding the classifications, and finding interesting subgraphs within a larger graph. A byproduct of this research shows that artificial and geometric graphs are easily distinguishable from real-world graphs, suggesting that researchers should carefully choose datasets for research studies. Gragnostics-based classifiers outperformed DDQC-based and graph kernel classifiers. Like Scagnostics, some gragnostics are correlated; however, in practice the features are sufficiently uncorrelated such that removing them degrades classification performance.

We found Chiron's gragnostics-based rank-by-feature framework quite useful for exploring the subgraphs in the csrankings and CHI datasets, and it guided us to insights about specific subgraphs as well as patterns across many subgraphs. The ability to sort and cluster subgraphs proved useful, replicating the findings of other researchers [18, 25, 34].

References

1. Aliakbary, S., Habibi, J., Movaghar, A.: Feature extraction from degree distribution for comparison and analysis of complex networks. Comput. J. **58**(9), 2079–2091 (2015)
2. Anscombe, F.J.: Graphs in statistical analysis. Am. Stat. **27**(1), 17–21 (1973)
3. Arleo, A., Didimo, W., Liotta, G., Montecchiani, F.: A distributed multilevel force-directed algorithm. In: Graph Drawing, pp. 3–17 (2016)
4. Bach, B., Shi, C., Heulot, N., Madhyastha, T., Grabowski, T., Dragicevic, P.: Time curves: Folding time to visualize patterns of temporal evolution in data. IEEE TVCG **22**(1), 559–568 (2016)
5. Barabási, A.L., Albert, R.: Emergence of scaling in random networks. Science **286**(5439), 509–512 (1999)
6. Bastian, M., Heymann, S., Jacomy, M.: Gephi: an open source software for exploring and manipulating networks. ICWSM **8**, 361–362 (2009)
7. Behrisch, M., Bach, B., Hund, M., Delz, M., Von Rüden, L., Fekete, J.D., Schreck, T.: Magnostics: image-based search of interesting matrix views for guided network exploration. IEEE TVCG **23**(1), 31–40 (2017)
8. Blondel, V.D., Guillaume, J.L., Lambiotte, R., Lefebvre, E.: Fast unfolding of communities in large networks. J. Stat. Mech.-Theory **2008**(10), P10008 (2008)
9. Borg, I., Groenen, P.J.: Modern Multidimensional Scaling: Theory and Applications. Springer Science & Business Media (2005)
10. Bunke, H., Allermann, G.: Inexact graph matching for structural pattern recognition. Pattern Recogn. Lett. **1**(4), 245–253 (1983)
11. Chaturvedi, S., Dunne, C., Ashktorab, Z., Zachariah, R., Shneiderman, B.: Group-in-a-box meta-layouts for topological clusters and attribute-based groups: space-efficient visualizations of network communities and their ties. CGF **33**(8), 52–68 (2014)
12. Dasgupta, A., Kosara, R.: Pargnostics: screen-space metrics for parallel coordinates. IEEE TVCG **16**(6), 1017–1026 (2010)
13. Eades, P., Hong, S.H., Klein, K., Nguyen, A.: Shape-based quality metrics for large graph visualization. In: Graph Drawing, pp. 502–514. Springer (2015)
14. Ellson, J., Gansner, E., Koutsofios, L., North, S.C., Woodhull, G.: Graphviz-open source graph drawing tools. In: Graph Drawing, pp. 483–484 (2002)
15. Elmqvist, N., Do, T.N., Goodell, H., Nathalie, H., Fekete, J.D.: ZAME: interactive large-scale graph visualization. In: Proceedings of the PacificVis, pp. 215–222 (2008)
16. van den Elzen, S., Holten, D., Blaas, J., van Wijk, J.J.: Reducing snapshots to points: a visual analytics approach to dynamic network exploration. IEEE TVCG **22**(1), 1–10 (2016)
17. Freeman, L.C.: Centrality in social networks conceptual clarification. Soc. Netw. **1**(3), 215–239 (1978)
18. Freire, M., Plaisant, C., Shneiderman, B., Golbeck, J.: ManyNets: an interface for multiple network analysis and visualization. In: CHI, pp. 213–222 (2010)
19. Ghoniem, M., Fekete, J.D., Castagliola, P.: A comparison of the readability of graphs using node-link and matrix-based representations. In: IEEE Information Visualization, pp. 17–24 (2004)
20. Ghoniem, M., Fekete, J.D., Castagliola, P.: On the readability of graphs using node-link and matrix-based representations: a controlled experiment and statistical analysis. Inf. Vis. **4**(2), 114–135 (2005)
21. Gove, R.: It pays to be lazy: reusing force approximations to compute better graph layouts faster. In: 11th Forum Media Technology, pp. 43–51 (2018)
22. Gove, R.: A random sampling O(n) force-calculation algorithm for graph layouts. Comput. Graph. Forum **38**(3) (2019)
23. Gove, R.: Gragnostics: Fast, interpretable features for comparing graphs. In: 2019 23rd International Conference Information Visualisation (IV), pp. 201–209. IEEE (2019)

24. Guerra-Gomez, J., Wilson, A., Liuy, J., Daviesz, D., Jarvis, P., Bier, E.: Network explorer: design, implementation, and real world deployment of a large network visualization tool. In: Proceedings of the AVI, pp. 108–111 (2016)
25. Harrigan, M., Archambault, D., Cunningham, P., Hurley, N.: Egonav: exploring networks through egocentric spatializations. In: Proceedings of the AVI, pp. 563–570. ACM (2012)
26. Himelboim, I., Smith, M.A., Rainie, L., Shneiderman, B., Espina, C.: Classifying Twitter topic-networks using social network analysis. Soc. Media + Soc. **3**(1), 1–13 (2017)
27. Holanda, A.J., Matias, M., Ferreira, S.M.S.P., Benevides, G.M.L., Kinouchi, O.: Character networks and book genre classification. ArXiv e-prints (2017)
28. Hopcroft, J., Tarjan, R.: Algorithm 447: efficient algorithms for graph manipulation. Commun. ACM **16**(6), 372–378 (1973)
29. Johnson, D.B.: Efficient algorithms for shortest paths in sparse networks. J. ACM (JACM) **24**(1), 1–13 (1977)
30. Keller, R., Eckert, C.M., Clarkson, P.J.: Matrices or node-link diagrams: Which visual representation is better for visualising connectivity models? Inf. Vis. **5**(1), 62–76 (2006)
31. Knuth, D.E.: The Stanford Graph Base: A Platform for Combinatorial Computing, 1st edn. ACM Press (1994)
32. Krzywinski, M., Birol, I., Jones, S.J.M., Marra, M.A.: Hive plots-rational approach to visualizing networks. Brief. Bioinform. **13**(5), 627–644 (2012)
33. Kwon, O.H., Crnovrsanin, T., Ma, K.L.: What would a graph look like in this layout? A machine learning approach to large graph visualization. IEEE TVCG **24**(1) (2018)
34. von Landesberger, T., Gorner, M., Schreck, T.: Visual analysis of graphs with multiple connected components. In: Proceedings of the VAST, pp. 155–162. IEEE (2009)
35. Leskovec, J., Kleinberg, J., Faloutsos, C.: Graph evolution: densification and shrinking diameters. ACM Trans. Knowl. Discov. Data **1**(1) (2007)
36. Lipton, Z.C.: The mythos of model interpretability. ACM Queue **16**(3), 30 (2018)
37. McAuley, J., Leskovec, J.: Learning to discover social circles in ego networks. In: Proceedings of the NIPS, pp. 539–547. Curran Associates Inc. (2012)
38. Milo, R., Itzkovitz, S., Kashtan, N., Levitt, R., Shen-Orr, S.: Superfamilies of evolved and designed networks. Science **303**(5663), 1538–1542 (2004)
39. Muelder, C., Kwan-Liu, Ma.: Rapid graph layout using space filling curves. IEEE TVCG **14**(6), 1301–1308 (2008)
40. Murdoch, W.J., Singh, C., Kumbier, K., Abbasi-Asl, R., Yu, B.: Definitions, methods, and applications in interpretable machine learning. Proc. Natl. Acad. Sci. **116**(44), 22071–22080 (2019)
41. Newman, M.E.J.: The structure of scientific collaboration networks. Proc. Natl. Acad. Sci. **98**(2), 404–409 (2001)
42. Newman, M.E.J.: Finding community structure in networks using the eigenvectors of matrices. Phys. Rev. E **74**(3), 036104 (2006)
43. Newman, M.E.J.: Networks: An Introduction. Oxford University Press (2010)
44. Partl, C., Gratzl, S., Streit, M., Wassermann, A.M., Pfister, H., Schmalstieg, D., Lex, A.: Pathfinder: visual analysis of paths in graphs. Comput. Graph. Forum **35**(3), 71–80 (2016)
45. Perer, A., Shneiderman, B.: Balancing systematic and flexible exploration of social networks. IEEE TVCG **12**(5), 693–700 (2006)
46. Roth, C., Kang, S.M., Batty, M., Barthelemy, M.: A long-time limit for world subway networks. J. R. Soc. Interface (2012)
47. Schneidewind, J., Sips, M., Keim, D.A.: Pixnostics: Towards measuring the value of visualization. In: Proc. VAST, pp. 199–206 (2006)
48. Seo, J., Shneiderman, B.: A rank-by-feature framework for interactive exploration of multidimensional data. Inf. Vis. **4**(2), 99–113 (2005)
49. Shannon, P., Markiel, A., Ozier, O., Baliga, N., Wang, J., Ramage, D., Amin, N., Schwikowski, B., Ideker, T.: Cytoscape: a software environment for integrated models of biomolecular interaction networks. Genome Res. **13**(11), 2498–504 (2003). https://doi.org/10.1101/gr.1239303.metabolite

50. Shervashidze, N., Vishwanathan, S., Petri, T., Mehlhorn, K., Borgwardt, K.: Efficient graphlet kernels for large graph comparison. Art. Int. Stat. 488–495 (2009)
51. Siglidis, G., Nikolentzos, G., Limnios, S., Giatsidis, C., Skianis, K., Vazirgiannis, M.: Grakel: a graph kernel library in python. J. Mach. Learn. Res. **21**(54), 1–5 (2020)
52. Smith, M., Ceni, A., Milic-Frayling, N., Shneiderman, B., Mendes Rodrigues, E., Leskovec, J., Dunne, C.: NodeXL: A Free and Open Network Overview, Discovery and Exploration Add-in for Excel 2007/2010/2013/2016. Social Media Research Foundation (2010). https://nodexl.codeplex.com/
53. Sopan, A., Rey, P., Shneiderman, B.: The dynamics of web-based community safety groups: Lessons learned from the nation of neighbors. IEEE Signal Process. Mag. **30**(6), 157–162 (2013)
54. Tarjan, R.: A note on finding the bridges of a graph. Inf. Process. Lett. **113**(7), 241–244 (1974)
55. Tikhonova, A., Ma, K.l.: A scalable parallel force-directed graph layout algorithm. In: Eurographics Symposium on Parallel Graphics and Visualization, pp. 25–32 (2008)
56. Van Den Elzen, S., Van Wijk, J.J.: Multivariate network exploration and presentation: from detail to overview via selections and aggregations. IEEE TVCG **20**(12), 2310–2319 (2014)
57. Šubelj, L., Bajec, M.: Community structure of complex software systems: analysis and applications. Phys. A Stat. Mech. Appl. **390**(16), 2968–2975 (2011)
58. Šubelj, L., Bajec, M.: Clustering assortativity, communities and functional modules in real-world networks. ArXiv e-prints (2012)
59. Šubelj, L., Bajec, M.: Software systems through complex networks science: review, analysis and applications. In: Proceedings of the KDD Workshop on Software Mining, pp. 9–16 (2012)
60. Šubelj, L., Bajec, M., Blagus, N.: Group extraction for real-world networks: The case of communities, modules, and hubs and spokes. In: Proceedings of the International Conference on Network Science, pp. 152–153 (2013)
61. Šubelj, L., Žitnik, S., Blagus, N., Bajec, M.: Node mixing and group structure of complex software networks. Adv. Complex Syst. **17**(7), 1450022 (2014)
62. Watts, D.J., Strogatz, S.H.: Collective dynamics of 'small-world' networks. Nature **393**(6684), 440–442 (1998)
63. Wilkinson, L., Anand, A., Grossman, R.: Graph-theoretic scagnostics. In: Proceedings of the IEEE Information Visualization, pp. 157–164 (2005)
64. Willett, W., Heer, J., Agrawala, M.: Scented widgets: improving navigation cues with embedded visualizations. IEEE TVCG **13**(6), 1129–1136 (2007)
65. Yunis, E., Yokota, R., Ahmadia, A.: Scalable force directed graph layout algorithms using fast multipole methods. In: Proceedings of the ISPDC, pp. 180–187 (2012)

VisIRML: Visualization with an Interactive Information Retrieval and Machine Learning Classifier

Craig Hagerman, Richard Brath, and Scott Langevin

Abstract VisIRML is a visual analytic system to classify and display unstructured data. Subject matter experts define topics by iteratively training a machine learning (ML) classifier by labeling of sample articles facilitated via information retrieval (IR) query expansion—i.e. semi-supervised machine learning. The resulting classifier produces high quality labels better than comparable semi-supervised learning techniques. While multiple visualization approaches were considered to depict these articles, users exhibited a strong preference for a map-based representation.

Keywords Machine learning · Text classifiers · Active learning · Semi-supervised learning · Information visualization

1 Introduction

In this chapter we present VisIRML (Visualization with an Interactive Information Retrieval and Machine Learning Classifier), which provides a platform for the classification and display of unstructured data to be used to extend an existing ambient visualization system.

There are many pre-existing visualization techniques which depict structured data, for example, as scatterplots, bar charts, maps, distributions and so forth. Unstructured data can potentially be represented in these visualizations; however, some structure must be extracted from these documents in order to display them with these techniques.

C. Hagerman · R. Brath (✉) · S. Langevin
Uncharted Software Inc, Toronto, VIC, Canada
e-mail: rbrath@unchartedsoftware.com

C. Hagerman
e-mail: chagerman@unchartedsoftware.com

S. Langevin
e-mail: slangevin@unchartedsoftware.com

B. Kovalerchuk et al. (eds.), *Integrating Artificial Intelligence and Visualization for Visual Knowledge Discovery*, Studies in Computational Intelligence 1014,
https://doi.org/10.1007/978-3-030-93119-3_13

Ambient visualizations convey timely information in the periphery of human attention [1]. Such visualizations provide an abstract summary of real-time information and usually have a low level of interaction, or a lack of interaction. As a result, the data must be a form that facilitates updates. The new data for VisIRML behind our ambient visualization is unstructured text documents. The objective was to extend the existing system to depict unstructured financial data, such as news articles, research documents, call transcripts and other documents. How to provide structure to such unstructured data was a key challenge to this project.

Most of these documents are not categorized by subject matter (i.e. unlabeled). In some cases, the documents may have tags, but the tags may not be relevant to the topics of interest. For example, news articles may be pre-tagged with labels such as politics, business, sports, etc., whereas the desired tags may be technology, health care, energy, real-estate, etc. Thus, to visualize these articles by topic, it is necessary to create topic labels for each article.

Manual labeling of documents might be an option for a small corpus of unstructured data, but was not feasible for our purposes. One of our data sources is GDELT, an open-data service which collects real-time news and provides more than 70,000 articles per day (gdeltproject.org). That quantity of data makes manual labeling impractical and also highlights the fact that for some data sources the data is constantly being updated. For these reasons an automated solution is necessary.

Automated labeling of text documents could be done by a machine learning (ML) classifier using *supervised learning*. The supervised aspect involves training a model on a preferably large set of pre-existing labeled data in order to learn patterns within the data. However, there was no existing labeled data available for the topics of interest (or the labels were not the desired categories of interest). The lack of training data is a common problem in Natural Language Processing (NLP), or, more generally, for machine-learning. *Unsupervised machine learning* can automatically segment data into groups, but those groups will not align with the desired topics, thus unsupervised approach is not feasible.

This project thus required a system for subject matter experts (SMEs) to:

1. Provide human-assisted labeling of unstructured data
2. Assist with rapid construction of a machine learning classifier
3. Aid the workflow to construct, view, validate and interact with these classifiers

The goal is for a visual analytic system combining visualization (Vis), information retrieval (IR) and machine learning classifiers (ML)—i.e. VisIRML. Our contribution is this system [2], which we have now further extended with a variety of document previews.

Visualization, IR and ML classification approaches are mature areas of research. Ambient visualizations are not artworks subject to a wide range of interpretation: these visualizations must be decipherable [3, 4], and they need to engage attention [5]. The design and content of ambient visualizations varies widely. Figure 1 shows example ambient visualizations, such as Viegas' Artifacts of the Presence Era [6]; Lozano-Hemmer's Pulse [7]; and a visualization of real-time stock market data on the NASDAQ MarketSite Tower [8].

Fig. 1 Example ambient visualizations, which do not necessarily need user interaction to update the data and display

Traditional IR focuses on ranking documents by relevance to a query [9]. This largely involves matching query keywords with the terms found in each document. This is often a simplistic search that will not return results for documents with alternate spellings, or similar words or related concepts that do not contain any of the search terms. This is often enough to retrieve a set of topically relevant documents. However this is suboptimal to return all documents of interest for a particular topic.

Using IR requires trial-and-error query formulation and inspection. Instead, supervised machine-learning can improve relevant results. ML automates categorization, ranking and filtering of documents into classes of interest based on prior labeled examples. However, a challenge with ML classifiers is the significant time and effort required to create a high quality training data set (i.e. labeled data) in order for ML to construct a high quality classifier.

Semi-supervised ML is an iterative process that can be used to drastically reduce the time and manual effort required to label data. This is accomplished by using human judgement to manually label a small amount of data to bootstrap the learning process. This small amount of data is used to train a supervised ML classifier, which is then used to label unlabeled data. The effectiveness of the resulting classifier relies on the quality of the small human-labeled data. The accuracy and efficiency of this labeling is aided by combining the human labeling through iterative interactive queries (i.e. Active Learning) in a visual environment.

Our VisIRML enables a SME to construct domain-specific document classifiers with minimal effort and no knowledge of machine learning. The SME provides guidance for the Active Learning classifier and can check the performance improvement with each iteration. This approach reduces the time required to label the corpus at the same time ensuring that training data is representative of the classification task.

2 Background

Our goal was to create an ambient visualization displaying articles of interest. For example, the display should focus on the most important global business and financial related news stories but filter out sports, culture and other non-hard news topics. We solve this by using an iterative Active Learning approach to first build up a sufficiently large labeled corpus.

Active Learning is a machine learning approach that involves a learning algorithm iteratively asking a user to make decisions and then using that human guidance to refine its pattern recognition. This is most commonly used for semi-supervised labeling of unlabeled data. In such an application the end goal is to classify and label all data.

The Active Learning algorithm starts by querying the user about a small number of samples. It then uses those few labels to train a rudimentary classifier to identify the 'decision boundary' between classes (Fig. 2 center). Since this is based on very few samples much of the data will be mislabeled. The Active Learning algorithm then queries the user to label more data points near the decision boundary (Fig. 2 right). This process repeats interactively with each iteration resulting in refinements to the classifier and more accurate auto-labeling. This method allows a user to accurately label a large dataset after only a few such iterations.

2.1 *Challenges with Text Classifiers*

Machine learning classification can be used for many tasks where one wants to be able to classify documents. For example an email client may include a classifier to classify whether or not emails are likely to be spam. Labeled training data is used for the classifier to learn the statistical relationship between input variables and the class.

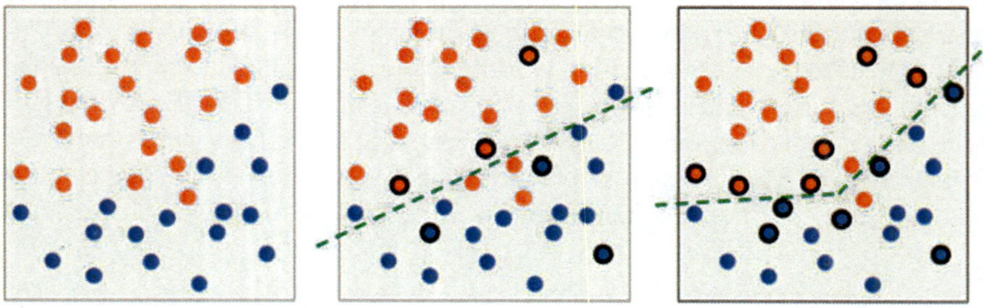

Fig. 2 Conceptual example of semi-supervised learning. Left is all documents shown as dots in two categories (blue vs. orange). Center shows a small subset of training documents (circled in black), used to train a classifier, shown as a line splitting the dots into two groups, which some erroneous dots on either side of the line. Right shows classifier after successive iterations splitting the groups more accurately

For example, in binary classification, examples are needed for both the positive and negative classes of interest. This labeled data can then be used to train the machine learning algorithm to model the classes.

The process of labeling data is prone to many issues. (1) The use of humans to label data for specialized subjects, such as financial articles, cannot be done with generalists or low-cost labor: expensive subject matter experts are required to create correct labels relevant to the domain. (2) Labeling a large dataset is prohibitive in terms of time: there are too many documents, such as the GDELT example discussed earlier. (3) Articles may be long, and some topics may occur in only a portion of the article—the time to review even individual documents is too high. (4) Data science based approaches can be used to automate the task; however, subject matter experts typically do not have data science skills [10]. (5) There are not easy-to-use interfaces to streamline these tasks or may not be suitable for the type of document or the domain. (6) Sampling based approaches are problematic, as it is not always obvious if the sample is representative of the distributions in the underlying areas of interest.

Labeled data is foundational to supervised machine learning. Thus, it is important to create a good subset of labeled data in order to improve the classifier's performance [11].

2.2 *Reducing Effort for Text Classification*

A simple manual workflow for annotating documents with labels would use a document viewer, such as PDF, to review documents; and a worksheet to record the appropriate label(s). The result would then be a set of labeled documents suitable for machine learning. However, the process is tedious, error-prone, and difficult to scale. Instead, a tool that can simplify and automate the process as much as possible is desirable for the users.

Such a tool should make it easy to review the relevant parts of documents for the users; as well as reducing the number of documents need to be reviewed. One approach is to build an interactive Active Learning system, such as the examples shown by Burr in [12]. This can be further streamlined, in the case of binary classifiers, to label only positive matches. An approach to semi-supervised learning that uses only positive examples to build a binary classifier is shown by Li and Liu in [13].

We improve on Li and Liu's approach with several enhancements:

1. *Query Expansion*: We perform query expansion by using a word embedding model together with edit distance similarity. This finds both semantically and context similar terms, but also alternate spellings.
2. *Iterative Interface*: We use a visual process to build the classifier which allows the expert to preview results.
3. *Document Review based on Model*: We use the text classifier results in successive iterations to return matched documents for review and guidance by the experts to improve labeling of borderline documents.

2.3 Visualizing Classified Text

The target use of the classified results is an ambient visualization system, such as the example shown in Fig. 3. The target system has six animated visualizations (e.g. bar chart, line chart, distribution, scatterplot, map). Each visualization provides an initial overview of the dataset, animates to specific subset of data points, animates an observation such as a callout of the largest values [14], and simple interactions to further explore the data (if a user is nearby and chooses to interact with the visualization). This approached is based on the the martini glass structure of narrative visualization outlined by Segel and Heer [15]. For example, a distribution of 100 companies starts by showing all companies, followed by an animation to companies within one sector, followed by animated text call-out(s) overlaid to indicate specific observations. Any nearby user has the ability to tap and show any specific value. If no user interaction occurs, the visualization will automatically animate to a different visualization with a different live dataset.

Each visualization type has unique animations and interactions. The primary use of the visualizations is for the ambient visualization of quantitative data. However, for this project the goals were to reuse these visualizations for animated visualization of textual data; and to use them to facilitate the text classification process.

Fig. 3 Snapshot of the ambient visualization, depicting a stacked distribution of current data

3 Technical Approach

In our approach, the subject matter experts label a corpus in an iterative process guiding the construction of a machine learning text classifier. The result is that each article in the corpus is labeled positive or negative for zero or more topics, with positive topics to appear within the visualization. Our approach also adds in techniques from information retrieval and visualization to facilitate the process.

3.1 Workflow Overview

The VisIRML process starts with a keyword search. This is followed by query exploration where the information retrieval process suggests semantically similar terms; and provides example documents matching these terms. The subject matter expert can then reject terms or groups of documents whether from their original query or the suggestions. Otherwise the expanded terms and documents are implicitly taken to be acceptable positive examples. The process then labels all documents based on the user's approvals with a *positive* label. All other documents in the corpus are then implicitly assigned a *negative* label. These labeled documents are then used to train a supervised classifier, after which the model is applied to the entire corpus to re-label according to the classifier predictions. It should be noted that this process results in a great deal of false positives and false negatives initially. However when repeated through successive iterations the resulting labels more and more closely align with ground truth. A visualization may be used to assess the results.

Our approach will improve document classification with several techniques: (1) the use of *Query Expansion* to suggest potential additional query terms to use for labeling using alternate spellings, semantic and context similarity of terms and phrases; (2) an iterative process that builds the classifier and allows the subject matter expert to preview results; (3) using these interactively defined text classifiers to search document repositories for positively matched documents, for review and validation by subject matter expert; and (4) using visualizations to aid assessment and refinement of the classifier.

3.2 Data

In order to validate our approach, we use an existing labeled dataset of news articles: the Reuters-21578 text collection [16]. This is a high-quality labeled news dataset that has been used in NLP research for 30 + years. This allows us to compare the results of our Active Learning labels with the given labels to measure how close we approach the ground-truth.

For our intended application, we use two data sources. One is a commercial news data source and the other is the open-source news data source, GDELT [17]. GDELT provides machine extracted events and metadata (including headline and URL) from world news sources. From GDELT we can extract top events, then with successive steps we can fetch the associated article text and metadata. As business news articles follow an inverted pyramid style of writing, we retained only the initial paragraph(s) as these contained the key information of the story. We collected tens of thousands of news articles over the course of a year.

Using GDELT to collect news events presents additional issues. Worldwide there is a constant stream of news stories being published. The GDELT API makes an update available every 15 min. Thus, given a trained text classifier we can fetch the most recent update every 15 min, label stories with the text classifier and use select candidates in the ambient visualization. The fact that the text corpus is being continuously updated means that we must also be cognizant of data drift and the effect it has on the model. That is, over time the input data being classified will resemble the training data less and less. Thus over time the classifier will be expected to lose accuracy to correctly identify news stories of interest. Moreover, there is the related issue that there are likely to be new topics of interest we would like to highlight in the ambient visualization which did not exist when the classifier was first trained (e.g. the emergence of the term "Covid"). Thus, for both of these reasons it is necessary to update the model every so often, to make sure that it addresses data drift as well as to identify and handle new topic classes of interest. This partly involves the same issues as above—how to classify unlabeled text. Therefore we can re-use the same iterative Active Learning approach to update and expand the text classifier to ensure it remains current.

3.3 Language Model

Statistical modeling, such as the text classifier we undertake here, requires that input data be encoded as numeric values. There are various options for how to encode text ranging from a simple bag-of-words count vector to a Term Frequency -Inverse Document Frequency (TF-IDF) vector [18]. We use TF-IDF as part of an ensemble approach to computing document similarity.

An encoding approach that allows for both flexibility and statistical power is to use a word embedding model. Such a model creates a high-dimensional vector space with each unique word in the corpus assigned a vector, such that words sharing a similar context have vectors that are close together. This approach allows for both a more compact representation of texts and more robust comparative properties.

We use word embeddings derived from a Word2Vec model [19] for use in the term similarity service and for document similarity. The Word2vec software library can generate word embeddings from a given corpus. These embeddings are used by the similarity service to identify semantically similar and related terms and documents.

This is augmented by word and phrase level similarity provided by Elasticsearch edit distance similarity [20]. This similarity service can identify that for a given search term such as *barrel* the words *barrels,* and *barren* are similar based on edit distance.

Word2Vec is a Natural Language Processing algorithm that learns associations between words in a corpus. It creates *word embeddings* (vectors) which are coordinates within a multi-dimensional space such that similar and related words are in similar areas of the vector space. This is a very effective way to create a language model, but requires a large corpus in order to build an accurate model.

For document-level similarity, we use three different text vectorization approaches: term frequencyș—inverse document frequency (TF-IDF), averaging the word embedding vectors of each word in a document, and Doc2vec [21]. Doc2vec is analogous to word embeddings, but represents documents rather than words as a vector.

3.4 Search Keywords and Query Expansion

The first step for the subject matter expert is to create and explore their query to find a set of positive documents for their target label. Exploration is critical to help the user find alternate spellings, misspellings, similar phrases, as well as the ability to exclude alternative meanings that might be associated with keywords (such as homonyms, idioms, proper nouns, etc.).

Query Expansion can be, and often is, incorporated into search engines to improve recall and identify the largest set of candidate documents. In the present case we allow the subject matter expert to interact with Query Expansion results (selecting or deselecting suggestions) since the purpose here is not just to improve the recall of information retrieval but also the precision.

Query exploration helps the expert find and assess potential positive documents, by: (1) improving discoverability across the differences in articles; (2) reducing cognitive effort by reducing reliance on recall as suggested terms can be recognized as similar (recognition vs. recall); (3) easy interaction to include or exclude relevant suggested terms; and, (4) easy interactions to label various subsets of data with a few clicks.

On the server side, a similarity service takes the initial keywords and make appropriate calls to subsystems. Elasticsearch is used to find common and alternate misspellings of terms and phrases using the built-in completion suggester [20]. ElasticSearch is used as part of the Query Exploration service. ElasticSearch is a search engine that provides very fast keyword search. It also incorporates fuzzy searching and can thus return results for misspelled or alternately spelled words, morphologically related words and phrases that contain gaps. We leverage that built-in capability to expand query terms by identifying related query terms with morphological or phrasal similarity.

The Word2Vec model is used to find related terms [21]. Domain-specific vocabulary is handled by training the Word2Vec model using terms and phrases from a representative set of domain articles. However, if a term does not appear in the training corpus it will not be recognized when the language model is used for similarity. For this reason, when the term is missing from the trained Word2Vec model, the service will fallback to use a pretrained Google News word embedding model.

To incorporate domain-specific vocabulary into the semantic similarity service, Word2vec models are trained for both terms and phrases using a representative document corpus. If a query term is not part of the domain vocabulary, the service falls back to a pretrained Google News word embeddings model to incorporate a generic vocabulary.

The subject matter expert receives these suggestions as they are typing in the query (i.e. autocomplete) or listed after the query is entered. Figure 4 shows an example snapshot of the user interface. At the top left the user defines the label e.g. "crude" and adds keywords in the search bar—in this example, *oil*, *opec*, *crude*, *barrel*. These are retained as easily deletable tags immediately below the search bar. Below this is a list of terms including additional suggestions from the Query Expansion, such as additional terms (e.g. *gas*) as well as potential alternate spellings (e.g. *barred*). For each of these, the number of documents is provided as well, giving a sense of breadth and depth of these terms in the article repository.

3.5 *Refine*

Feature selection is a phase of a machine learning pipeline that involves identifying and choosing input variables that have the highest correlation to the output variable. This process is crucial in building a robust, highly predictive classifier. In the present context, the selection of query terms guides the features selection of the classification model to associate keywords with labels.

Within the user interface, as shown in Fig. 4, the left-hand side below the search field lists terms, any of which can be expanded. In the figure, the keyword *barrel* is expanded. Note that under the keyword barrel, four example articles are shown. The groups may include alternate spellings nominated by Query Expansion. In Fig. 4, under *barrel*, the groupings include *barred* and *barren*.

For each article, counts of similar articles are shown on the left. Document similarity via the Doc2vec model, is used to find and group similar articles – for example groups associated with the keyword *oil* may differentiate semantically different types of oil such as crude oil, vegetable oil, lubricating oil, and so on. Approximate Nearest Neighbors (ANN) [22] is used to quickly find similar documents. ANN is a proximity search method that trades off increased speed and memory savings for identifying approximate matches. The proximate documents that ANN identifies are then labeled according to the input documents and added to the training corpus.

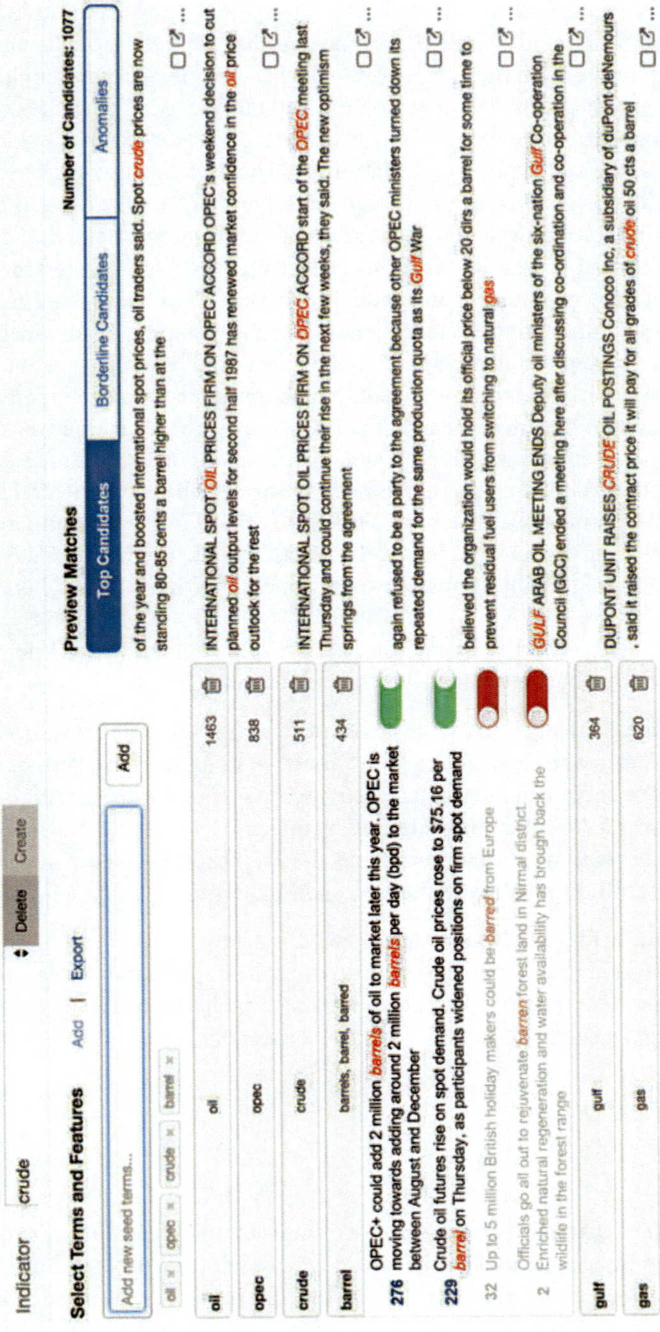

Fig. 4 Snapshot of the user interface for the label "crude", with keywords, Query Expansion, grouped articles (left side) and article previews (right side)

Within each group of articles, a snippet of text is shown from the highest-ranking article including the keyword(s) highlighted in red. This keywords-in-context (KWIC [23]), help the user understand how the keywords are used being used within a group.

In this newer version of VisIRML, we have further included a preview panel on the right side of the user interface. The preview panel provides a list of sample articles matching the current query set or any article group that the user has clicked on. Each of these articles are also shown as snippets, enabling the user to rapidly skim many of the articles associated with a particular group. The user can also click to read the full article on the original source webpage, as some articles may contain minimal text and contain more information in the original form, for example, as a charts or infographics not visible with keyword-in-context. In Machine Learning in general, user trust in automated systems is an ongoing concern. This preview panel with many snippets assists the user in their understanding of the grouping as provided by the underlying service and results in higher confidence of the labeling decision.

Beside each group, is an on/off switch, which the subject matter expert can use to label the group of documents as either positive or negative. The user can skim the exemplar snippet, the preview snippets, and link to the original article. The user may determine that the group is not useful for determining the label and turn off that group so that it is labeled negative when the classifier is built. In the example in Fig. 4, the terms *barred* and *barren* are clearly erroneous, and the user has excluded them. Thus, with each click, a subject matter expert can easily skim and label hundreds of articles.

As such, we have created a user interface pattern following the Visual Information-Seeking Mantra by Shneiderman [24]:

- *Overview first*: where we show the keywords and frequencies in a vertical list
- *Zoom and filter*: where our interactions allow us to expand any item on the list allowing us to zoom and inspect the groupings associated with that term; and the switches allow us to filter out irrelevant documents.
- *Details on demand*: were we provide a detailed right side preview panel allowing the user to scroll through many snippets and link to source articles.

Starting with a few seed query terms, within a few minutes, the user is able to quickly explore the relevant subject matter and label thousands of articles from the corpus as positive training examples. Explicitly excluded articles and unlabeled articles are used as negative examples when the classifier is built.

3.6 Iterate

A subject matter expert can use the above workflow to quickly explore and partition a text corpus by entering search terms, selecting or deselecting expanded query term, view some documents and select or deselect those for training. This is likely to result in an acceptable data set for training a classifier, but not one that will be highly accurate. The real benefit from the above workflow comes from iterating over it multiple times to produce a labeled dataset with accurate labels.

In successive iterations, the model can label data and then these model labels can be used in the presentation of snippets in the user interface. Many classification algorithms create a quantitative metric in addition to the binary class. Ordering matches by the quantitative metric is useful to better understand the characteristics of the documents which the model scores highly and which are borderline. The preview panel aids this process as the user can show a list of sample documents which score highly from the classifier as shown under "Top Candidates". The "Borderline Candidates" shows matches near the binary threshold. Close inspection of documents near the threshold can help fine-tune the labels as the user can then interactively relabel some of the documents, e.g. rejecting false positives.

3.7 Classifier

Our classifier is based on the semi-supervised learning approach from Li and Liu [10]. It uses the positive labeled examples and considers all other examples as negative. Initial versions were implemented in Python using Gensim and Scikit-learn machine learning libraries. To create a faster, more scalable version, these were reimplemented in Apache Spark—which distributes the data and computation across a compute cluster. Within Spark, we use the MLlib library which includes functionality for common Natural Language Processing (NLP) tasks, for example, text pre-processing, document vectorization, and machine learning algorithms such as classifiers [25]. Whereas Li and Liu use only a TF-IDF vectorization model, our ensemble approach includes Word2vec and Doc2vec which improves the quality of our models, discussed below.

3.8 Visualization

In our ambient visualization system, we had two visualizations that seemed relevant to visualization of news articles: a map and a scatterplot. In Fig. 5, positive articles for the classifier are displayed on the map. All matching articles are plotted as tiny dots on the base map, note the tiny bright blue dots on the US east coast and the east coast of Australia. From these, the top ranked articles are displayed as bars on a map, bar-height corresponding to an article metric (e.g. rank, number of sources, etc.). Then, within each geographic area, the headlines for the top articles are displayed above the map with leader lines to the article. Animations step through the articles corresponding to each bar, with the pop-up showing country (as a flag), article title, lead photo (if any), opening paragraph, and newly added links to the full story as a QR-code so that a person walking by the screen can access the original story on their mobile device, to read the full article at a later time if desired.

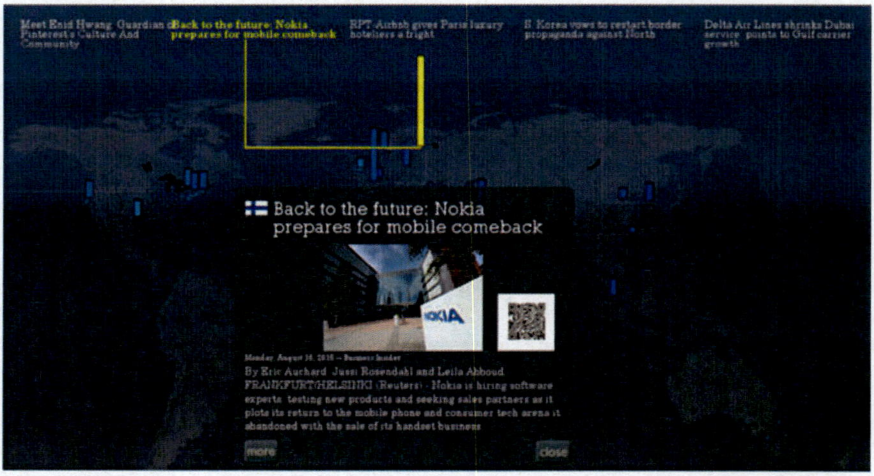

Fig. 5 Snapshot of map view of news articles in ambient mode; including dots on map, bars for top articles, headlines per region, and an animated popup of one story with a QR link to the story

Alternatively, our ambient visualization system includes a scatterplot, which can display dots, labels, or phrases—which we repurposed to display headlines. These marks can be color-coded to indicate different classifiers (e.g. oil, corn, technology) or other metadata (e.g. country, sector). The axes of the scatterplot can be configured in different ways. Explicit axes encode data, such as sentiment on the x-axis, number of sources on the y-axis, and recency on the z-axis (the z-axis is perpendicular to the screen, so that older stories are further back and smaller), as shown in the example in Fig. 6. The coordinate space can alternatively be configured with multi-dimensional reduction, such as a principal component analysis (PCA) or positions can be randomized (like a word cloud).

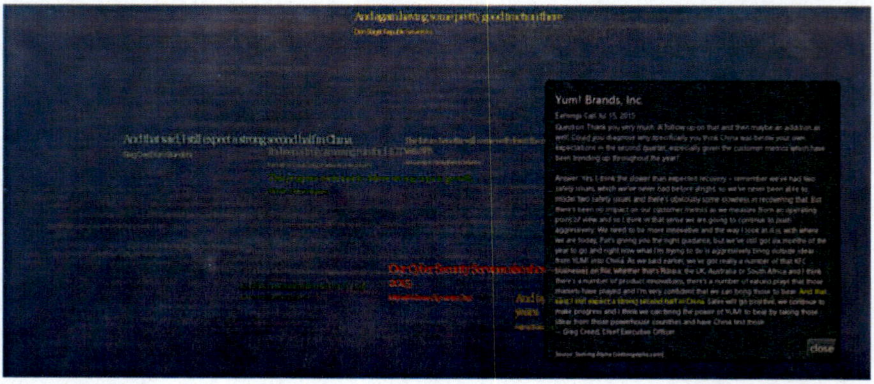

Fig. 6 Scatterplot with a word cloud of headlines. Headlines animate between foreground and background, and an animated pop-up shows story details

The visualizations can be useful to the subject matter expert at the time of building the labels. For example, the map facilitates differentiating news articles in different parts of the world: stories about oil will be different in OPEC countries as opposed to countries highly dependent on imported oil. In the scatterplot view, for example, the color-coding can provide insights into oil based on the point of view of different sectors such as energy producers (e.g. oil companies) or energy users such as transportation (e.g. airlines). The subject matter expert can preview the content and can use their pre-existing real-world knowledge of locations and industries to validate the quality of the labeling and remove the need for reliance on machine learning experts.

4 Results

The system has been evaluated with multiple approaches and revised. Evaluation includes: (1) the quality of our machine learning approach was compared to benchmarks in Table 1 below; (2) the quality of our machine learning approach was compared to a small sample of commercial news data feed; (3) end-user feedback on the system provided for a number of enhancements, including the preview panel with additional modes described below; and (4) end-user feedback on the query expansion and the classifier.

Table 1 F1 score results

Topic label	Svm baseline	Li and liu results	Spark implementation	Improved implementation
Acq	0.94	0.905	0.939	0.898
Corn	0.9	0.635	0.611	0.753
Crude	0.89	0.811	0.89	0.871
Earn	0.98	0.886	0.865	0.893
Grain	0.95	0.903	0.911	0.923
Interesr	0.78	0.614	0.68	0.769
Moneyfx	0.75	0.764	0.777	0.776
Ship	0.86	0.829	0.774	0.848
Trade	0.76	0.728	0.714	0.775
Wheat	0.92	0.779	0.738	0.764
Average	**0.873**	**0.7854**	**0.7899**	**0.827**

4.1 Classifier Performance Versus Benchmark

To validate our machine learning approach, we compare our results to the ground truth and other systems performance on the *Reuters-21578* news text dataset. The top ten largest categories were measured. In our system we used query exploration to find additional terms for the target labels, from which we selected the features and documents to label. Then the machine learning classifier was built against this labeled data (the unlabeled data for a given label being negative).

Table 1 shows a summary of the performance of our approach. The first column show the topic labels. The remaining columns show the performance of different systems versus the ground-truth (i.e. labeled topics in the source dataset). Performance metrics are F1 scores, a common measure for the performance of a binary classifier which combines precision and recall scores.

We present results for a baseline supervised classifier, our implementation of the algorithm in Li and Liu, and our improved version of the Li and Liu algorithm. Table 1 shows F1 scores for each variation. The F1 score, or F measure, is a common measurement of binary classifier performance that equally weights precision and recall scores.

The SVM Baseline column shows the results of classification based on the given labels in the data, then applied to a withheld dataset. This baseline represents the performance of classifiers where all the training data is correctly labeled and SVM performs well in all categories with an average of 0.873 with half above 0.9.

The following columns then show results for semi-supervised machine learning, where users interactively label a subset of positive examples. Li and Liu represents the starting point for semi-supervised learning, with an average F1 of 0.7854; well below the SVM average of 0.873. Our Spark based implementation of Li and Liu's approach resulted in similar scores to Li and Liu, with an average of 0.7899.

The final column shows the results of our improved implementation, using the ensemble query expansion and workflow. We see a significant increase in F1 to 0.827—halving the gap between Li and Liu and the SVM Baseline. Also note that in two categories, our approach marginally outperformed SVM (*moneyfx* and *trade*). We hypothesize that this may be the result of occasional erroneous labels in the ground-truth data: we have found in real-world labeled news data that some articles may be labeled with a topic even though the article is primarily about some other topic with only a marginal mention of the particular topic. The inclusion of these articles could then increase the error of the SVM scores.

4.2 Machine Learning Versus News Tags

We compared the results of automated classification and labeling from our system to ground truth labels attached to news articles provided by another system that uses a rule-based approach to categorize and tag articles. Examining stories with conflicting tags applied by our approach versus the other system revealed that the other system contained both false positives and false negatives in tags.

4.3 User Feedback

The overall prototype system was reviewed with end-users. Issues were raised by stakeholders included:

- *Set and forget*. Subject matter experts were interested in setting up stable topics in a one-time configuration rather than revising topics. While flexibility to revise was an important system criteria, once defined topics were rarely revised.
- *Emerging topics*. Managers liked that new classifiers could be easily created, and envisioned use for emerging topics (e.g. *Covid*). We prototyped functionality to detect new and trending terms. New terms are identified by tracking word frequencies over time and triggering thresholds.
- *Preview with borderline and anomalous articles*. The singular summary keyword-in-context was not sufficient to gain trust in the use of labeling. In this updated version of VisIRML, we added the preview panel to see snippets from more sample articles. This can be used to see top matches as shown in Fig. 4.

  Borderline articles: The article preview panel can also be used to view borderline matches when using the classifier – these are articles that score near the threshold on the binary classifier. The user can then modify the label of these candidate stories to help improve the quality of the classifier.

  Anomalous articles. We noticed a variety of anomalous news articles in real-world data as we worked with real-world data in the system. There are articles with no body text (headline only); a headline where the story is a table of data; non-English articles (which were outside our scope); artifacts such as non-printable characters or HTML codes; sentences missing spaces between words and so on. We added a tab intended for reviewing these articles which had issues with parsing or length, although this remains unimplemented as of time of writing (i.e. if these anomalous articles are by default considered unlabeled and therefore negative, we deem that OK for our intended use as we don't want to see these kinds of articles in our ambient displays).
- *Relevance*. Subject matter experts are acutely focused on story relevance – a story of local relevance is likely less important than a global story. While meta-data such as "number of sources" might be useful for relevance, local stories might gain international attention, for example, for a famous person or humorous anecdote. We added scoring for different news sources to help define relevancy.
- *Interestingness*. Correctly classified, highly relevant documents may still be uninteresting. For example, there are many government statistical reports. Some of these will have unanticipated information and have high interest. Some will have no unexpected information and these are not interesting. This was not addressed.

- *Sentiment blocklisting.* Management did not want negative content about their organization or their close customers. A simple blocklist was implemented.
- *Map.* Invariably, users liked map-based representation of news. The map provides global orientation and global context. The map aids users by: 1) personal relevance (e.g. look at countries of personal interest); 2) orientation in that region is representative of different roles with regards to a topic – for example, Middle-East countries are oil producers, Japan and Western Europe are importers; and 3) the global map helps users align their real-world expectations to the data. For example, technology stories may not be expected in some regions.
- *Scatterplot.* User reactions to the scatterplot were mixed. In general, users wanted the encoding between the data and visualization to be immediately and easily decodable. Essentially, for an ambient visualization, users wanted a representation that required very low cognitive effort to understand. As such, multidimensional reduction is too difficult to explain and comprehend, and it was dismissed by users. We believe multidimensional reduction visualization will be useful for labeling, but given the strong negative reaction, it has not been implemented yet.

  A scatterplot using explicit axes of sentiment, number of sources and recency (similar to Fig. 6) was liked by a small subset of users. Logically, the configuration was understandable, but it did not engage with them like the map. We hypothesize that there are fewer affordances to understand the representation (whereas a map is immediately recognized) and fewer opportunities to compare their real-world knowledge to the data in the visual representation (whereas a map provides immediate geographical associations). Possibly with practice a novel scatterplot of headlines might become familiar.

  Interestingly, the word cloud variant of news headlines was also rejected. The organization previously had some missteps with ambient visualization. Prior ambient visualizations at the firm were data-driven visualizations but unfamiliar representations with non-standard layouts and strong graphic design elements. The resulting visualization was not decodable to the target community. In effect, the result was perceived as information art instead of information visualization [26] and dismissed.
- *Data quality*: Open source news data (e.g. GDELT) returns a high volume of news articles, but many of these are local, or irrelevant, or low quality. High-quality sources are expensive.

4.4 Discussion

What started as a simple system to label news shows that users have many additional requirements beyond simple labels, such as relevance, interestingness and sentiment. These requirements did not become articulated until after the initial system was completed and running with real data, even though these users were engaged in defining the initial requirements and reviewing interim results.

We did not anticipate the level of rejection of scatterplots. We hypothesize that this is related to the ease of decoding. In Munzner's nested model of visualization design and validation [27], this is an error at the abstraction level. This mismatch does not align with definitions of ambient visualizations that might prioritize aesthetics over data [26, 28]. This mismatch was apparent with the users' prior issues with earlier ambient visualizations, which we tried to address through highly-logical encodings (a defined set of axes), yet still missed on the ease of decoding.

The grouping of articles by similarity was an ongoing experimentation throughout the development of the system. The grouping helps users label many similar documents at once. We experimented with Doc2vec vectors and ANN search to find similar articles but while the approach shows promise we found the results inconsistent between various data sources (i.e. open source GDELT or commercial sources). We hypothesize that the effectiveness will increase with larger datasets. We also experimented with Latent Dirichlet Allocation (LDA) topic modeling. This approach will find clusters of documents and extract a list of keywords characterizing the cluster. However, this approach did not align with the user needs for pre-defined topics. We hypothesize that LDA could aid query expansion with suggestions of related terms.

5 Conclusions

VisIRML is a system for the classification of news articles by subject matter experts for use in an ambient visualization system. We believe the approach is generalizable to a wider variety of unstructured data and we have applied components of VisIRML to classification of Internet advertisements and of social media.

VisIRML also revealed unexpected results in working with real users and real data—additional features and metadata beyond the initial labeling were required to make a usable system; and the visualization in the end-use ambient system needed to be easily understood.

Future work should include extending the scatterplot visualizations to aid the review of the documents during the labeling stage. We also considered a means for measuring interestingness. We captured interactions on the visualization, either as direct taps on the visualization, or through a QR-code associated with a document. We hypothesize that these interactions are useful to identify the most interesting stories and potentially usable to model an interestingness score.

Additional approaches to grouping the data during the feature selection stage, such as ANN search and non-parametric topic modeling, could lead to greater human-guided classification and labeling efficiencies. Another area of future work could include incorporating a label recommendation algorithm derived from a term-document bipartite graph [29, 30] or ontological guidance [31, 32].

References

1. Moere, A.V.: Towards designing persuasive ambient visualization. In: Issues in the Design and Evaluation of Ambient Information Systems Workshop. Citeseer (2007)
2. Hagerman, C., Brath, R., Langevin, S.: Visual analytic system for subject matter expert document tagging using information retrieval and semi-supervised machine learning. In: 2019 23rd International Conference Information Visualisation (IV), 2019, pp. 234–240 (2019). https://doi.org/10.1109/IV.2019.00047
3. Lang, A.: Aesthetics in Information Visualization, Same Book As Above
4. Kosara, R.: Visualization criticism-the missing link between information visualization and art. In: 2007 11th International Conference Information Visualization (IV'07), pp. 631–636. IEEE (2007)
5. Bafadikanya, B.: Attractive visualization. In: Trends in information visualization. In: Baur, D., Sedlmair, M., Wimmer, R., Chen, Y.-X., Streng, S., Boring, S., De Luca, A., Butz, A., (eds.), Technical ReportLMU-MI-2010-1, Apr. 2010. ISSN 1862-5207. University of Munich, Department of Computer Science, Media Informatics Group (2010)
6. Viegas, F.: Artifacts of the Presence Era, flickr.com, CC-BY-2.0 by Viegas (2009)
7. Lozano-Hemmer, R.: Pulse. Flickr.com. CC-BY-SA-2.0 by Anokarina (2019)
8. The Nasdaq Stock Market, Inc. NASDAQ MarketSite Tower. © Copyright 2000 reprinted with the permission of The Nasdaq Stock Market, Inc., Photo credit: Peter Aaron/ Esto
9. Manning, C.D., Raghavan, P., Schütze, H.: Introduction to Information Retrieval. Cambridge University Press (2009)
10. Pustejovsky, J., Stubbs, A.: Natural Language Annotation for Machine Learning. O'Reilly Media (2012)
11. Pujara, J., London, B., Getoor, L.: Reducing label cost by combining feature labels and crowdsourcing. In: Proceedings of the 28th International Conference on Machine Learning (2011)
12. Settles, B.: Active Learning Literature Survey. University of Wisconsin, Madison (2010)
13. Li, X., Liu, B.: Learning to classify texts using positive and unlabeled data. In: Proceedings of the 18th International Joint Conference on Artificial Intelligence (2003)
14. Brath, R., Matusiak, M.: Automated annotations. In: An IEEE VIS Workshop on Visualization for Communication (VisComm) (2018)
15. Edward Segel and Jeffrey Heer: Narrative visualization: telling stories with data. IEEE Trans. Visual Comput. Graph. **16**(6), 1139–1148 (2010)
16. UCI Machine Learning Repository: Reuters-21578 Text Categorization Collection Data Set. archive.ics.uci.edu, 2016. https://archive.ics.uci.edu/ml/datasets/Reuters-21578+Text+Categorization+Collection.
17. Leetaru, K., Schrodt, P.A.: Gdelt: Global data on events, location, and tone, 1979–2012. In: ISA Annual Convention, vol. 2, no. 4. Citeseer (2013)
18. Ramos, J.: Using tf-idf to determine word relevance in document queries. In: Proceedings of the First Instructional Conference on Machine Learning 2003 Dec 3, vol. 242, No. 1, pp. 29–48 (2003)
19. Mikolov, T., Chen, K., Corrado, G., Dean, J.: Efficient Estimation of Word Representations in Vector Space. arXiv:1301.3781
20. Completion Suggester Elasticsearch Reference. Elastic.co, 2016. https://www.elastic.co/guide/en/elasticsearch/reference/2.1/search-suggesters-completion.html
21. Dai, A.M., Olah, C., Le, Q.V.: Document embedding with paragraph vectors (2015). arXiv preprint arXiv:1507.07998
22. Avarikioti, G., Emiris, I.Z., Psarros, I., Samaras, G.: Practical linear-space Approximate Near Neighbors in high dimension (2016). arXiv preprint arXiv:1612.07405
23. Hearst, M.: Search User Interfaces. Cambridge University Press (2009)
24. Shneiderman, B.: The eyes have it: a task by data type taxonomy for information visualizations. In: The Craft of Information Visualization 2003 Jan 1, pp. 364–371. Morgan Kaufmann

25. Meng, X., Bradley, J., Yavuz, B., Sparks, E., Venkataraman, S., Liu, D., Freeman, J., Tsai, D.B., Amde, M., Owen, S., Xin, D., Xin, R., Franklin, M.J., Zadeh, R., Zaharia, M., Talwalkar, A.: MLlib: machine learning in apache spark. J. Mach. Learn. Res. **17**(1), 1235–1241 (2016)
26. Lau, A., Moere, A.V.: Towards a model of information aesthetics in information visualization. In: 2007 11th International Conference Information Visualization (IV'07). IEEE (2007)
27. Munzner, T.: A nested model for visualization design and validation. IEEE Trans. Visual Comput. Graph. **15**(6), 921–928 (2009)
28. Pousman, Z., Stasko, J., Mateas, M.: Casual information visualization: Depictions of data in everyday life. IEEE Trans. Visual. Comput. Graph. **13**(6), 1145–1152 (2007)
29. Song, Y.: Automatic tag recommendation algorithms for social recommender systems. ACM Trans. Comput. Logic (2008)
30. Guan, Z., Wang, C., Bu, J., Chen, C., Yang, K., Cai, D., He, X.: Document recommendation in social tagging services. In: Proceedings of the 19th International Conference on World Wide Web (WWW '10), pp. 391–400. ACM, New York (2010)
31. Ha-Thuc, V., Mejova, Y., Harris, C., Srinivasan, P.: News event modeling and tracking in the social web with ontological guidance. 2010 IEEE Fourth International Conference on Semantic Computing, pp. 414–419 (2010)
32. Ha-Thuc, V., Renders, J.M.: Large-scale hierarchical text classification without labeled data. In: Proceedings of the Fourth ACM International Conference on Web Search and Data Mining (2011)

Visual Analytics of Hierarchical and Network Timeseries Models

David Jonker, Richard Brath, and Scott Langevin

Abstract Confidence in timeseries models can be gained through visual analytics that represent the many aspects of the models including, the timeseries data (input, predicted, intermediate factors), model structure, model behavior, model sensitivity and model quality in one holistic application. We show examples ranging from simplistic prototypes of financial ratios, to nowcasting and economic forecasting, and massive transaction analysis. The approach is perceptually scalable to the exploration of large-scale structures with millions of nodes by visually representing many node characteristics; on-demand navigation through sub-graphs; hierarchical clustering of nodes; and aggregation of links and nodes. These visual analytics allow expert users to compare the many aspects of the model to their real-world knowledge helping them gain an understanding of the model and ultimately build confidence.

Keywords Timeseries model · Timeseries visualization · Factor model

1 Understanding Timeseries Models

There are relationships between sets of timeseries data: for example, employment impacts GDP, rainfall affects crop yields, and so on. Modeling related timeseries is a common task in many domains and frequent in financial services. They are broadly applicable to diverse analytical tasks; for example, aggregate models such as the Consumer Price Index, or decomposition, such as geographical decomposition of public opinion for a political party. In financial markets, timeseries models are common:

D. Jonker · R. Brath (✉) · S. Langevin
Uncharted Software, Toronto, Victoria, Canada
e-mail: rbrath@unchartedsoftware.com

D. Jonker
e-mail: djonker@unchartedsoftware.com

S. Langevin
e-mail: slangevin@unchartedsoftware.com

B. Kovalerchuk et al. (eds.), *Integrating Artificial Intelligence and Visualization for Visual Knowledge Discovery*, Studies in Computational Intelligence 1014,
https://doi.org/10.1007/978-3-030-93119-3_14

- *Financial sums and ratios.* Many financial ratios such as Return On Equity (ROE) are a chain of simple arithmetic combinations deriving from widely available fundamental timeseries.
- *Multi-regression models.* Given that the price of many financial instruments is correlated to other financial instruments, expert users may create multiple regressions on the fly for some small subset of chosen series.
- *Risk models.* Some risk models are linear models often viewed over time. For example, the Altman Z-Score is a measure of credit risk based on a number of reported data values, linearly combined into features and then into a risk score; and users often view these as timeseries to assess change in probably of default.
- *Factor and attribution models.* Factor models are used extensively in the construction and analysis of portfolios. It is not uncommon to find implementations with hundreds or thousands of factors; any of which a user may want to view as a timeseries to assess the factor trend.
- *Anti-money laundering models.* Money laundering may involve transaction flows between related entities. Models may be used to flag suspicious accounts and transactions. Humans need to review temporal flows across networks.
- *Causal models.* Financial research reports may use causal models, for example, using weather models to predict crop yields.
- *Nowcasting models.* Key economic metrics, such as GDP, can be forecast based on underlying components which are released more frequently (such as employment, CPI, etc.)
- *Semi-automated models.* Wherein a user provides datasets and machine learning approaches automate feature selection and model selection. The user then needs to understand the result, including the model building decisions and model output.

While the relationships between these timeseries can be modeled by experts, a significant challenge for the use of these models is the communication and understanding of these models by downstream consumers. In our experience, these downstream users are reluctant to use a model that they do not comprehend. Explanation is required. Explanation in the form of documentation outlining the technical approach and the equations is insufficient. Model documentation (1) assumes analysts are familiar with statistical or quantitative formulae; (2) requires analysts to cross-reference between model results and documentation rather than direct interaction with the model; and (3) requires the reader to relate between abstract examples described in the documentation to the specific concrete scenarios and data that they currently are using in the interface.

Understanding a model goes beyond confidence in the forecast variables. Using a model only for an immediate predictive point value is highly useful for answering *what* (e.g. what is the value of GDP going to be?, what is the probability of default of Microsoft?). Beyond the *what*, there is value in conceptualization, validation and internalization of the model. That is, users may be unable to construct an internal mental model thereby limiting their ability to reason with the model. When a user has internalized the model, they are better able to leverage the model to answer questions such as *why* and *what-if* (e.g. why did this outcome occur, or what-if this scenario occurs).

Visualizing causality, visual inference and explainable predictions are a top unsolved visualization problem [1]. What is needed is increased transparency in a visual format that facilitates ease of deep understanding of models (e.g. the Explainable AI challenge articulated by [2]). This challenge includes aspects such as model structure, model behavior, model quality, and model change over time in a holistic approach to facilitate rapid internalization of the model. Our prior contribution in this area is a visual analytics approach to facilitate deep understanding of hierarchical timeseries models [3], which we further extend here with contributions showing additional applications including extensions to scale the approach to networks (not just hierarchies); and to scale the approach to larger models with millions of nodes.

2 Visual Analytics and Timeseries Models Background

Visual analytics is "the science of analytical reasoning facilitated by interactive visual interfaces" [4]. Analytical reasoning with timeseries is multi-faceted—there are many aspects that can be made explicit with visual analytics:

1. *Input timeseries data.* The ground truth is input timeseries data associated with each entity, and the visual depiction of these is paramount to financial analysts who visually analyze timeseries throughout the day.
2. *Modelled timeseries data.* The predicted timeseries outputs are based on the model and the input timeseries, and these must be visually explicit in order to see the prediction.
3. *Model behavior*. The relationship between the multiple input timeseries and output timeseries is necessary to visually assess which inputs are most impactful on the outputs.
4. *Model sensitivity*. Models may be non-linear: understanding the conditions that lead to unexpected behavior may be critical in abnormal markets.
5. *The model structure* is the set relationships between in the input timeseries, intermediary timeseries and output timeseries—which can be thought of as a hierarchy (for some models) and more generally as a network.
6. *Changes in model structure.* Relationships change over time, which may be implicit in the relationships or can potentially be made explicit, for example, as dynamic graphs.
7. *Model quality*. Models are approximations—indication of difference between the model and actual may be important.
8. *Commentary and Narrative.* Models may be interactive or otherwise create updates, and acquire a life of their own in the user community. Ability to annotate any of the above aspects can aid recording thoughts and communicating insights.

An ideal visual analytics system bringing all the above together can create a deep understanding of a model leading to improved decision making [5].

2.1 Timeseries

While it may be feasible to represent timeseries in different ways, such as discrete events, calendars and so forth, within the financial domain, timeseries are almost universally quantitative data varying over time, and as such are depicted as line charts. A capital markets expert at one of the largest firms providing data and analytical software estimates that as much as 90% of analysis involves timeseries, which includes the depiction of one or more of these timeseries as line charts. The behavior of one timeseries is readily visible as anomalies, trends, changes, gaps, reversals and other phenomena are readily perceived as slopes, highs, lows, change in line direction and so on. The CMT Association is an example of a professional organization devoted to the analysis of timeseries data as line charts (and related charts such as candle charts and associated analytics). As timeseries are critically important, line charts are a key element required in any timeseries model visual analytics.

2.2 Model Structure

In most timeseries models there is an intrinsic structure between the input timeseries and the output timeseries. This structure may be explicit, such as a hierarchy in an attribution model or a directed acyclic graph in a causal model or neural network. This structure could be implicit: for example, a flat multi-regression model may be better explained by grouping more strongly correlated input variables—which forms a hierarchical grouping.

Making the model structure explicit and visual can help the analyst understand how the inputs are configured, how they are connected to each other and their relative weights. Prior work in data visualization has focused on depicting the structure of models by explicitly representing the connections as node-link diagrams, for example in biology [6]; neuroscience [7]; network anomalies [8]; and visualization [9].

Explicit representation of connections as node-link diagrams are popular in graph visualization tools such as Gephi, yEd, Cytoscape, etc., [10]) as shown in Fig. 1 left and middle, with small markers for nodes, lines (or arrows) for links, and short text labels for annotating the nodes. Hierarchies can also be presented with space filling layouts, such as treemaps or hierarchical pie charts, as shown in Fig. 1 right.

However, all of these hierarchical layouts are limited in the amount of information each visual element can convey, and each relies on cumbersome drill-downs to see more detailed sets of attributes associated with each element. For example, in all the hierarchies shown in Fig. 1, nodes only visualize three data attributes using color, size and a textual label. A review of the 300 + tree visualizations at treevis.net shows that almost all techniques focus primarily on the depiction of hierarchical structure with the possible addition of a few metrics—not the tens to hundreds of datapoints in timeseries. It is difficult to show a long timeseries—Elmqvist and Tsigas [11] use animation, however a transient display relies on short-term memory

Fig. 1 Popular graph visualization tools showing hierarchies: a force-directed graph with Gephi (left); a radial layout graph in yEd (center) and a hierarchical pie chart with D3.js (right). Visualizations focus on the structure, and typically only show minimal data per node, such as size, color and label

making it cognitively difficult to compare across periods (e.g. see Tufte [12], or Larkin and Simon [13]). These hierarchical layouts do not provide affordances for manipulation of input/output values. Finally, these layouts do not provide a good means for embedding commentary directly into the visual representation, thereby requiring cross-referencing to other blocks of prose.

2.3 *Model Behavior*

In financial services, analysts often review two or more timeseries (e.g. see Chapter 22, Dual Y Axes Charts Defended: Case Studies, Domain Analysis and a Method). Even without a model, a visual comparison between the timeseries aids comparison of relative movement between timeseries: do they move together? which one moves first? when do they stop moving together?

Explicit depictions of one or more timeseries allows users to understand the past (*what happened*) and future prediction (*what will happen*). Future prediction may include the forecast predicted by the model; known upcoming events (e.g. earnings release); or trend projection. Analysts also want to uncover *why* (e.g. why a particular anomaly, gap, reversal, etc., occurred). Understanding the *why* helps the analyst assess, generalize, predict and act on similar patterns in the future—i.e. to formulate hypotheses or to create and internalize a mental model. This requires interactive visualization techniques to explore underlying causes, for example, in stock market data this may include timeseries of indexes, peers, economics, fundamentals, earnings, news, social media, business activity and so on [14, 15].

One approach to visualize model behavior is to visualize only the model inputs and the model outputs, such as Krause et al. [16]. This conveniently bypasses the need to visualize model structure and thus becomes a general visualization approach applicable across a wide range of models. User interactions, such as selection, are then used to associate input(s) with output(s). However, this approach is difficult to scale to models with hundreds or thousands of inputs. Furthermore, the model structure remains opaque, requiring the user to infer potential model structure based only on what can be observed, for example, perturbing an input and noticing what changes in the output. Other approaches focus on explicitly visualizing internal or hidden nodes, such as Strobelt et al. [17].

With timeseries models, the analyst is interested in understanding the relationship between features in the timeseries (trends, reversals, anomalies, etc.), how these are reflected across model stages, and the predictive output. Predictions must be accompanied by an understanding of how and why in order to leverage the considerable human abilities of the decision-maker (Jonker [18]). Without this transparency of the model and behavior, the decision-maker is required to place wholesale trust in the accuracy and totality of the models. The goal is for decision-makers is to understand how the relationships work, and it is this working understanding that enables them to take action.

Prior authors have used different methods to combine timeseries with hierarchies. For example, Chintalapani [19] provides a single panel which shows a timeseries for a selected node on a hierarchy. Hao et al. [20] create a treepmap, wherein each rectangle depicts a timeseries; however, it is difficult to compare timeseries as each rectangle is a different size with a different aspect ratio. Schreck et al. [21] remedy the variable sizes so that timeseries are consistently sized thereby supporting comparison; however, the hierarchy may not be explicitly visible nor are intermediate aggregations depicted, thus the timeseries decomposition is not available. Fischer et al.'s novel ClockMap [22] depicts a hierarchy of nodes as circles (similar to Fig. 1 left), then encodes the timeseries radially around the node using color: while this does depict comparable timeseries at each node, color variation has low accuracy compared to size (e.g. see [23, 24]). Furthermore, representing timeseries radially by color is unfamiliar to expert users, who generally expect to see timeseries as familiar line charts with time on the x-axis and the measured value(s) on the y-axis.

2.4 Model Sensitivity

Graphs and timeseries only indirectly depict the effects that occur across complex networks, such as amplification, dampening, transience, timing and degree of effect (Yao [25]). Causality may be linear, non-linear, or probabilistic with varying levels of likelihood and varying levels of lag. There are may other aspects of causality to consider, such as characteristics associated with each causal node, and those characteristics may aid reasoning about potential effects (Wright and Kapler [26]).

2.5 *Model Quality*

Quantitative measures of model quality, such as forecast error, mean absolute error, and root mean squared error can provide value by comparing models to known ground truth. Model drift is also an issue: some features become more important, some irrelevant, relationships shift, and so on. Participants in capital markets discuss "correlation breakdowns", wherein pricing and valuation may be highly correlated to some timeseries for a period of time, then shifting to correlation to another timeseries. For example, shipping prices may be correlated to fuel price (when fuel prices are high), supply and demand (when shipping capacity is deficient), port lag times (when ports are congested), tariffs (when government regulations change, such as Brexit), and so on. Models may need to be updated to reflect these new changes. Comparing models quantitatively against new ground truth is important for ongoing model maintenance.

2.6 *Commentary*

The addition of prose to the visualization is important, especially if the user community includes novice users and casual consumers who interact with the model infrequently. Explanation is becoming more important in modeling, machine learning and artificial intelligence, as seen in the popular press (e.g. [27, 28]); narrative infographic explainers on news media sites (e.g. [29]) and now more generally data-driven storytelling [30]; national research programs (e.g. [31]); and the increasing in popularity of data science notebooks such as *Jupyter* and *ObserservableHQ* which allow for code, models, visualizations and narrative blocks to be intermingled into a narrative explanation of data science models. There are many different kinds of commentary that could be relevant, such as references to data sources and calculations; user authored notes; semi-automated or fully automated natural language generation (NLG) to highlight or explain insights.

3 Visual Analytics of Timeseries Models

The above requirements are extensive and we prioritized a subset. In various proofs of concepts we initially focused on timeseries representation together with hierarchical model structure. Then, in a broader application for financial nowcasting, we extend these ideas with a more scalable interactive workflow, model manipulation and commentary. Most recently, in a money flow analysis application, we go further, to extend these approaches beyond hierarchies to address networks and graphs; and to address greater scale.

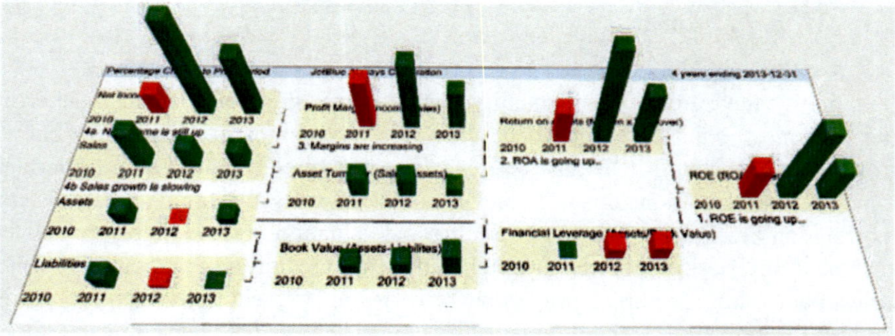

Fig. 2 Quick prototype showing decomposition of financial ratios for JetBlue Airways. The visualization starts on the left with timeseries from the company annual reports (such as sales, assets and liabilities) and then derived through successive ratios into Return on Equity (ROE), shown at right

3.1 *Timeseries and Model Structure POC*

To explore concepts, we created a quick prototype with real data in Microsoft Excel combining both timeseries (as small sets of vertical bars) and model structure (as nodes containing raw data on the left, connecting to successive derived ratios progressing to the right) as shown in Fig. 2. The analyst can visually trace which variables have a similar correlated trend to the output variable. In the example showing financial metrics for JetBlue Airways, increasing *ROE* (Return on Equity—a measure of financial efficiency shown at the right of the diagram) can be visually traced through the preceding ratios to the left which show similar trend: to *Return on Assets*, to *Profit Margin*, and to *Net Income*. The use of 3D, Excel, perspective text were not useful, although the primary goal combining timeseries and model structure were promising.

3.2 *Financial Nowcasting Visual Analytics*

From several early prototypes, we incorporated what we learned into our first timeseries model decomposition visual analytics platform called Prism which has since been used for several applications, including economics, social sciences and financial nowcasting. Incorporating lessons learned from several such prototypes and domains, we have since designed and implemented more advanced and comprehensive solutions for a variety of needs such as *nowcasting*. The platform is scalable to more complex models, for use in interactive, collaborative, multi-user, web-based environments.

3.2.1 Nowcasting Overview

Nowcasting seeks to predict financially significant values in the very near term (under 3 months), such as global economic metrics (e.g. Gross Domestic Product), company finances (e.g. quarterly revenue), or merger activity. Accurate prediction of these values in advance of their official publication provides an opportunity for quantitative financial professionals to make a profit on the anticipated stock price movement which will occur immediately after the official publication becomes broadly available to everyone. Hedge funds use nowcasting algorithms to create estimates before official filings in their trading strategies, for example:

- Prescription trends to estimate pharmaceutical company revenue (e.g. IQVIA)
- Daily visit counts from mobile phone data to estimate retail sales activity (e.g. PlaceIQ)
- Satellite imagery can be used to measure container traffic at ports to estimate imports and exports; oil storage levels at refineries; or the nighttime illumination to estimate economic activity. (e.g. go.spaceknow.com/africa-lights-index)
- Tariff data can be used to used to measure shipments imported by a retail company and estimate sales (e.g. panjiva.com)
- Email volumes can be used to estimate sales transactions for e-commerce sites
- Social media can be used to measure company sentiment and estimate impact to stock prices
- Movement of corporate jets can provide insights into sales and mergers (e.g. quandl.com)
- Employee satisfaction to measure potential employee turnover and estimate increased costs

For the purposes of showing our nowcast platform, we use a GDP nowcasting model. The Federal Reserve Bank of Atlanta publishes a United States GDP nowcasting model [32], with regular updates on their website.[1]

3.2.2 GDP Nowcasting Hierarchical Analysis

Our GDP nowcast visualization example shown in Fig. 3 starts with a global overview. The overview provides a GDP marker per country: the bubble indicates growth (by bubble size) and change to prior period (by ring color indicating the amount of growth or shrinkage in the most recent period). Interactive buttons allow the user to change the historic time period, forecast time period (outlook), and indicator (i.e. the metric, here showing GDP with a legend top left). Prose includes instructions (top) and analyst authored commentary (bottom left).

The analyst can click a country, in this example, United States (Fig. 4). As the map is no longer relevant, it animates to a thumbnail to the top left (clicking acts like a back button, returning to the full map). The timeseries chart shows two lines: the officially

[1] www.frbatlanta.org/cqer/research/gdpnow.aspx

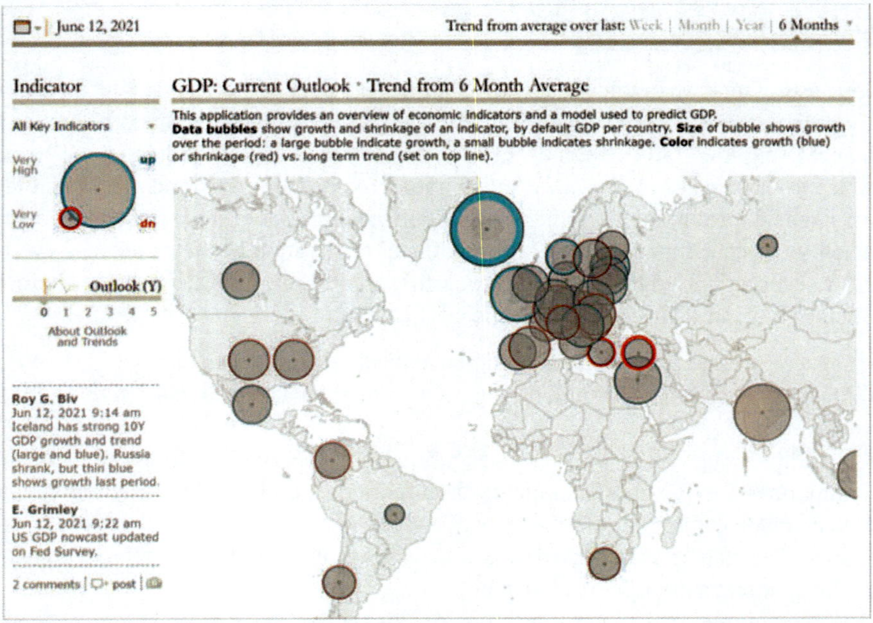

Fig. 3 Nowcast overview with zoomable map, legend, introductory text and analyst commentary. The viewer can skim global metrics geographically to see anomalies, for example, rapid growth in Iceland (big and blue); or rapid shrinking in Turkey (small with thick red ring)

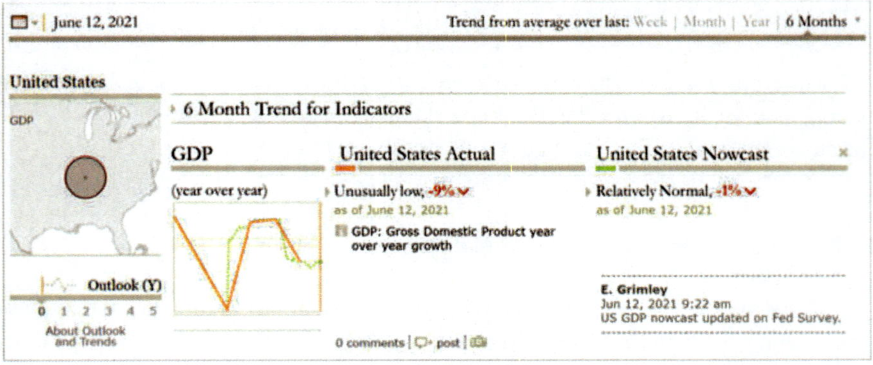

Fig. 4 Nowcast focused on USA. The yellow line is actual GDP, the green line is the nowcast. The green line extends further than the yellow, showing the GDP nowcast estimate

released timeseries values are shown as a solid amber line and nowcast as a green dotted line. Note that overlaying multiple timeseries within the same chart is critical to fine-grain comparison in financial services [15], something not considered in prior hierarchical timeseries analysis. In the timeseries chart, the yellow line represents the actual GDP values and visibly starts high on the left of the chart, significantly drops to a low value in middle, then rebounds. The green line represents the the nowcast value. Note from the low point near midchart, the nowcast captures most of this rebound almost immediately after the low, thereby providing significant advance indication that the drop was a one-time anomaly. For the current time period, the green line extends beyond the last real observation of actual GDP in the yellow line, thereby showing the current nowcast estimate for GDP.

The analyst can then drill-down to show the detailed structure and component timeseries which comprise the model as shown in Fig. 5 left. The overall GDP nowcast chart animates to the top right, with successive layers of decomposition to the left. The raw input timeseries data is shown in the leftmost column (e.g. New Orders Manufacturing, Production Price Index) and intermediate factors in center column (e.g. Production Indicators, Trade, Consumer). Note that underlying input timeseries are represented with solid amber lines whereas the intermediate factors are shown as dotted lines. These factors are synthetic modeled values and are differentiated by dotted lines to indicate that they are estimates rather than ground truth. The analyst can visually trace through the factors to see which raw timeseries are impacting the model most. In this example, there is a recent drop in GDP, and a corresponding recent drop in trade, shown both with the descending yellow line and the large red circle on the last observation.

There can be many input timeseries, more than fit on the screen at once. The synthetic factors in the center column provide a view of intermediate aggregates. Any of these factors can be clicked to further drill-down to the underlying input variables to that factor. In Fig. 5 right, the factor "Production Indicators" has been

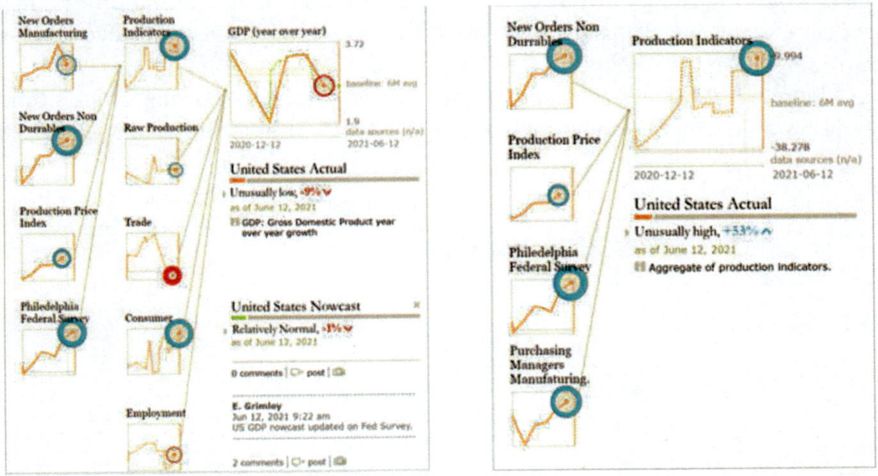

Fig. 5 Nowcast showing timeseries decomposition. Left image: Overall GDP is selected showing the output on the right, synthetic factors in the center, and raw input timseries on the left. Right image: the synthetic factor *Production Indicators* has been chosen and shows the underlying raw timeseries. In this case, all underlying production indicators have increased, all contributing to the net increase in the production factor

selected and promoted to the large visual in the upper right corner with the underlying constituent timeseries on the left.

The analyst can also create *what-if* scenarios. The most recent data points for the input variables can be selected, which then displays interactive arrows adjacent to the data point. These arrows can be dragged up and down which will then cause the model to recompute and the display to update. Multiple variables can be changed to create hypothetical scenarios to assess the potential impact of changes in the world according to the model (e.g. the impact to GDP of due to Brexit; Covid; or global trade slowdown due to supply chain congestion). The analyst not only sees the change in the model results; they can compare that to their real-world knowledge to potentially question the model; and they can uncover new potential scenarios and outcomes previously unconsidered. The latter is particularly important, as future financial crises are often different than prior crises—thus, historical data does not contain the relevant scenario to current conditions and the expert must explicitly create new scenarios with inputs updated to reflect plausible conditions.

3.3 Economics Model

The same application has been used in other domains such as economics, commodities and social data. Figures 6 and 7 show an example of the application applied to economics data regarding the global steel industry. The image in Fig. 6 left shows a marker per country as a clover, with the top petal indicating cost drivers, the left petal indicating supply drivers and the right petal indicating demand drivers: note the large supply and demand growth in China (lower blue petals), while overall costs are going down in all countries (red top petals).

Figure 6 right shows selection of one country—China—showing China's timeseries in yellow for demand, supply and cost, along with associated commentary per

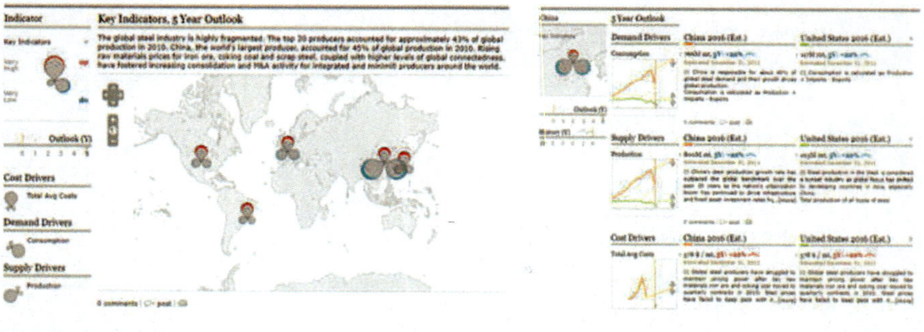

Fig. 6 Similar version of the application applied to economics data showing factors impacting the global steel industry. Left shows key drivers for major steel producing nations; right shows comparison of these driver variables for China (yellow) vs. United States (green)

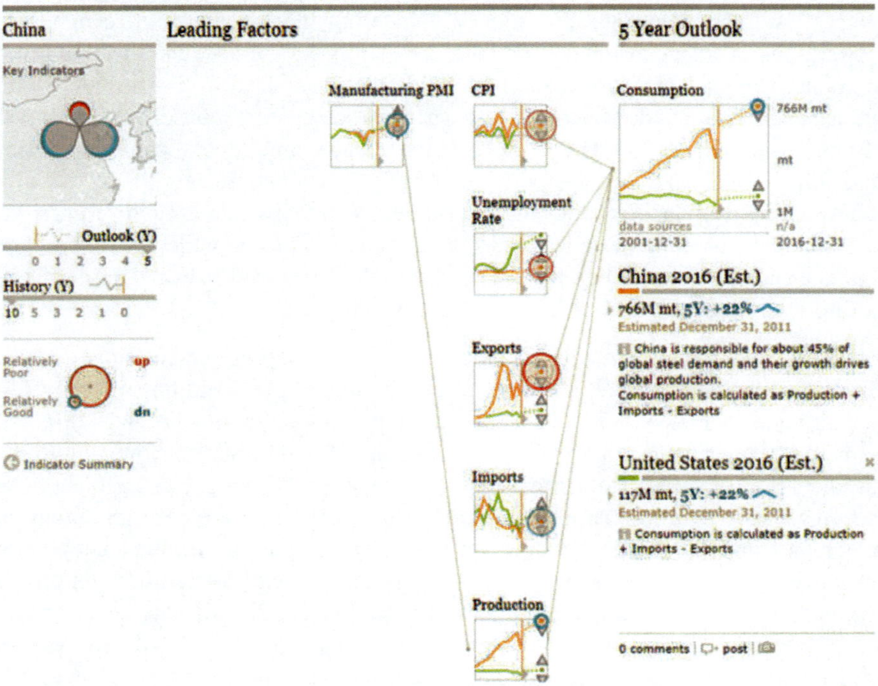

Fig. 7 Timeseries decomposition of the consumption driver for China and United States

each. A comparison to the US has been added as an overlay in green. The big downward drop in demand and supply correspond to the global financial crisis in 2008, for which the forecast model in this example predicts a strong recovery in 2009.

In Fig. 7, the consumption metric has been selected (shown on the right), economic factors (center, such as CPI, unemployment, exports, imports, etc.), and underlying input timeseries far left (e.g. Manufacturing PMI). In this example, it can be seen that net consumption goes up slightly (small blue circle), which can be traced to an increase in production and imports, but a reduction in exports.

3.4 Financial Fraud Model

The prior examples were small, deterministic models, with tens to hundreds of timeseries and strictly hierarchical models. We have since extended these techniques to larger, more complex timeseries models. Challenges with large networks typically include visibility and usability when too many nodes and too many edges make an unintelligible tangled diagram. This is further complicated by the need to understand temporal patterns and additional account characteristics.

For example, fraud and anti-money laundering models may utilize hundreds of millions of financial transaction data across millions of accounts resulting in a complex web of connections. Analytic models may be used to flag accounts with issues such as anomalous activity or suspect patterns. In response, we built upon the prior approaches to create a much more scalable and generalizable timeseries modeling visual analytics system. We achieve perceptual scalability by visually representing many node characteristics; on-demand navigation through sub-graphs; hierarchical clustering of nodes; and aggregation of links and nodes. A snapshot of the interface is shown in Fig. 8 showing a subset of data from a loan dataset of 150 million financial transactions across two million accounts.

Nodes as data-dense Cards: Nodes (i.e. accounts) are represented as square cards. Each card includes a label, a timeseries as a histogram, and additional icons indicating node characteristics, such as account type, region, and model flags (e.g. red pins, yield signs, stop signs, etc.). The histogram indicates money flowing into the account (i.e. deposits) as positive bars over time, and money flowing out of the account (i.e. withdrawals) as negative bars. At top left is a card for Farrah Sorenson, a person from USA who is lending out money, which can be seen as a couple big negative bars, and is receiving monthly payments in return, as can be seen on the regular small positive bars. (Note names have been replaced with generated fictional

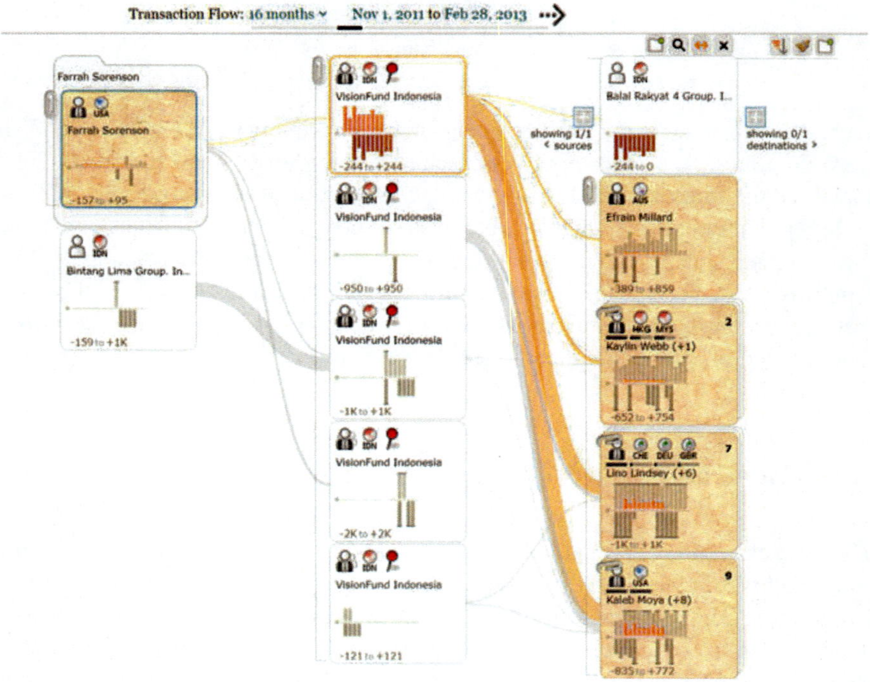

Fig. 8 View showing money flow between groups of accounts, each account showing timeseries of deposits and withdrawals

names.) Beneath Farrah is another account, Bintang Lima Group, which has received one large payment (i.e. a loan) and provided 4 smaller outgoing payments (i.e. monthly payments). The use of cards provides flexibility to represent a variety of node characteristics beyond the timeseries: there are multiple different flag types, background color, card outline color, annotations, counts and so on that can be added to cards.

Graph Network: Lines between nodes indicate money flow, with line thickness indicating the total dollar amounts. Here the viewer has expanded the flows from Farrah and Bintang Lima Group to 5 different accounts shown in the center column—all associated with VisionFund Indonesia. VisionFund Indonesia is an intermediary administrator providing funds to the loan applicants and then facilitating the monthly repayments back to the providers. All the payments from those five VisionFund Indonesia accounts are then provided to the accounts shown in the third column to the right. In this snapshot, the user has highlighted the top card in the center column, thereby highlighting all money flows into and out of that account in amber, as well as highlight the corresponding transaction within the histograms of any connected cards. Thus, the user can see (1) the amount of money flowing between accounts as line thickness, (2) the proportion of money in each account flowing to the selected account, as shown by the proportion of the bar highlight, and (3) the trend over time for those flows, as shown in the bar sequence.

Graph Navigation: Note that flows are not hierarchical, rather there can be many crossing flows. As these graphs can become large very quickly, interactions provide for expanding and managing the flows of interest to the current investigation. To manage the proportion of the network shown on the screen out of millions of accounts, the user can expand/collapse connections and add/remove cards as needed. Above the top right card, Balai Rakyat 4 Group, a number of buttons can be seen. To the left and right of the card are plus buttons, allowing the user to expand and show additional columns of nodes connected to that node. Above the node are a series of buttons providing operations such as grouping selected accounts, searching for related accounts, highlighting all transactions to/from the account, deleting the account, sorting the column (e.g. so that the largest dollar volume cards are at the top, or the risky accounts are at the top), and so on.

Hierarchical Clustering: Note on the third column the paper-clips on the top left corners of some cards. The paper-clips are indicators of groups of cards, and clicking on the paper-clip will expand/contract the group. Groups can have subgroups allowing for the display and interaction with large hierarchies of groups. The system provides automatic clustering whenever the user expands a link, so that hundreds or thousands of cards do not appear individually. In the right most column, the bottom 3 cards indicate groups, via the paper-clip across the top of the card, via the card stack, and via the count in the top right corner indicating the number of cards. Also, the name of the account holder on the card indicates the largest account holder within the card stack. The histogram, icons, etc., represent summaries of the underlying characteristics. For example, for the card Kaylin Webb (+1), the tiny horizontal bars under the regions indicate the proportion of accounts from HKG (Hong Kong) and from MYS (Malaysia). Users can explicitly create and manage groups by simply dragging cards or card stacks into file folders for custom clusters relevant to an investigation, and then "sweep away" the remaining irrelevant nodes.

With the above techniques, we achieve perceptual scalability through the display and management of relevant subgraphs while rolling up and exposing the relevant temporal patterns and account characteristics. In practice, users can effectively navigate, manipulate and analyze thousands or tens of thousands of nodes out of graphs with millions of nodes, and the application has been deployed to users at multiple financial institutions.

4 Discussion and Future Work

The ongoing evolution of our timeseries visual analytics has been driven in response to user feedback over the last 20 years. Through close cooperation with these experts in iterative agile development we have had the opportunity to make many small tweaks to visual representations, for example, to make the visual representations work effectively at small scales. More importantly, with our long engagements with expert users, we have gained an appreciation for the tacit knowledge of the data gained by users with many years of experience in the field. Expert users have deep understanding of financial markets and benefit from representations that reflect their real-work knowledge. Explicit representation of the timeseries means that experts can relate features in the timeseries to their knowledge, building confidence in the model. For example, users may recognize the sharp drop in 2008 in Fig. 6 as corresponding to the financial crisis at that time. Similarly, users can interact with model inputs or model structure to explore characteristics of the model. This allows analysts to create what-if scenarios to explore potential outcomes. It also allows them to construct prior conditions that they know to see how well the model represents those prior conditions—which allows them to test the model and build confidence. We believe that this association between model structure, behavior and data to real-world expert knowledge confirms our belief that model internalization is a key requirement to successful use and deployment of models into real-world applications.

With our newer implementation, we are also able address a variety of shortcomings in the earlier systems. For example, we can directly show the relative influence of inputs over time: selecting one account and seeing the proportion overlaid on the next timeseries gives an indication of the impact of the prior node. We provide both automated and user-generated hierarchical grouping, and we continue to evolve additional approaches for graph clustering based on our work in large scale graph decomposition and visualization (Jonker et al. [33]). We are also investigating techniques for user recomposition of the model structure—that is, using the interface as a means to author the model, not just represent the model (Kapler et al. [34]).

Future challenges include introducing model quality, feature weights and changes over time while retaining the ease of use and ease of exploration to achieve better deep understanding. Significant extensions to causal models are ongoing.

References

1. Chen, C.: Top 10 unsolved information visualization problems. IEEE Comput. Graph. Appl. **25**(4), 12–16 (2005)
2. Gunning, D.: Explainable Artificial Intelligence (XAI). https://www.darpa.mil/attachments/DARPA-BAA-16-53.pdf. Last accessed 16 Oct 2018
3. Jonker, D., Brath, R., Langevin, S.: Industry-driven visual analytics for understanding financial timeseries models. In: 2019 23rd International Conference Information Visualisation (IV) 2019 Jul 2, pp. 210–215. IEEE (2019)
4. Cook, K.A., Thomas, J.J.: Illuminating the path: the research and development agenda for visual analytics. Technical report, Pacific Northwest National Lab (PNNL), Richland, WA (United States) (2005)
5. Endsley, M.R., Connors, E.S.: Situation awareness: state of the art. In: Power and Energy Society General Meeting-Conversion and Delivery of Electrical Energy in the 21st Century, 2008, pp. 1–4. IEEE (2008)
6. Opgen-Rhein, R., Strimmer, K.: From correlation to causation networks: a simple approximate learning algorithm and its application to high-dimensional plant gene expression data. BMC Syst. Biol. **1**(1), 37 (2007)
7. Seth, A.K.: A Matlab toolbox for granger causal connectivity analysis. J. Neurosci. Methods, **186**(2), 262–273 (2010)
8. Zhang, H., Sun, M., Yao, D., North, C.: Visualizing traffic causality for analyzing network anomalies. In: Proceedings of the 2015 ACM International Workshop on International Workshop on Security and Privacy Analytics, pp. 37–42. ACM (2015)
9. Kadaba, N., Irani, P., Leboe, J.: Visualizing causal semantics using animations. IEEE Trans. Visual Comput. Graph. **13**(6), 1254–1261 (2007)
10. Brath, R., Jonker, D.: Graph Analysis and Visualization. Wiley (2015)
11. Elmqvist, N., Tsigas, P.: Growing squares: animated visualization of causal relations. In: Proceedings of the 2003 ACM Symposium On Software Visualization, pp. 17–ff. ACM (2003)
12. Tufte, E.: The Visual Display of Quantitative Information. Graphics Press (1983)
13. Larkin, J.H., Simon, H.A.: Why a diagram is (sometimes) worth ten thousand words. Cogn. Sci. **11**(1), 65–100 (1987)
14. Sorenson, E., Brath, R.: Financial visualization case study: correlating financial timeseries and discrete events to support investment decisions. In: 2013 17th International Conference Information Visualisation (IV), pp. 232–238. IEEE (2013)
15. Brath, R., Hagerman, C., Sorenson, E.: Why two Y-axes (Y2Y): a case study for visual correlation with dual axes. In: 2020 24th International Conference Information Visualisation (IV). IEEE (2020)
16. Krause, J., Perer, A., Bertini, E.: Using Visual Analytics to Interpret Predictive Machine Learning Models (2016). arXiv:1606.05685
17. Strobelt, H., Gehrmann, S., Pfister, H., Rush, A.M., et al.: Visual analysis of hidden state dynamics in recurrent neural networks. CoRR (2016). arXiv:abs/1606.07461
18. Jonker, D.: Linked visible behaviors: a system for exploring causal influence. In: AHFE Cross-Cultural-Decision-Making (CCDM) conference, (2012)
19. Chintalapani, G.: Temporal Treemaps for Visualizing Time Series Data. Diss (2004)
20. Hao, M.C., Dayal, U., Keim, D.A., Schreck, T.: Importance-driven visualization layouts for large time series data. In: IEEE Symposium on Information Visualization, INFOVIS 2005, pp. 203–210. IEEE (2005)

21. Schreck, T., Keim, D., Mansmann, F.: Regular treemap layouts for visual analysis of hierarchical data. In: Proceedings of the 22nd Spring Conference on Computer Graphics. IEEE (2006)
22. Fischer, F., Fuchs, J., Mansmann, F.: ClockMap: enhancing circular treemaps with temporal glyphs for time-series data. In: EuroVis (Short Papers) (2012)
23. Ware, C.: Information Visualization: Perception for Design. Morgan Kaufmann (2019)
24. Munzner, T.: Visualization Analysis and Design. CRC press (2014)
25. Yao, M.: Visualizing Causality in Context Using Animation. Simon Fraser University (2007)
26. Wright, W., Kapler, T.: Challenges in visualizing complex causality characteristics. In: Proceedings of the IEEE Pacific Visualization, 2018. IEEE (2018)
27. O'Neil, C.: Weapons of Math Destruction. Crown Books (2016)
28. D'ignazio, C., Klein, L.F.: Data Feminism. MIT Press (2020)
29. Segel, E., Heer, J.: Narrative visualization: telling stories with data. IEEE Trans. Vis. Compu. Graph. **16**(6), 1139–1148 (2010)
30. Riche, N.H., Hurter, C., Diakopoulos, N., Carpendale, S.: Data Driven Storytelling. A K Peters and CRC Press (2018)
31. Turek, M.: Explainable artificial intelligence. DARPA (2016). https://www.darpa.mil/program/explainable-artificial-intelligence
32. Patrick, H.: GDPNow: a model for GDP "Nowcasting". Federal Reserve Bank of Atlanta (2014). https://www.frbatlanta.org/-/media/documents/research/publications/wp/2014/wp1407.pdf
33. Jonker, D., Langevin, S., Giesbrecht, D., Crouch, M., Kronenfeld, N.: Graph mapping: multi-scale community visualization of massive graph data. Inf. Vis. **16**(3), 190–204 (2017)
34. Kapler, T., Gray, D., Vasquez, H., Wright, W.: Causeworks: a framework for transforming user hypotheses into a computational causal model. In: International Conference on Information Visualization Theory and Applications (IVAPP), (2021)

Integrated Systems and Case Studies

ML Approach to Predict Air Quality Using Sensor and Road Traffic Data

Nuno Datia, M. P. M. Pato, Ruben Taborda, and João Moura Pires

Abstract Air quality is an important issue that impacts who live in cities. During COVID-19 pandemic, it becomes clear that low air quality can increase the effects of the disease. It is very important that in this era of Big Data and Smart Cities we use technology to address health issues like air quality problems. Often, air quality is monitored using data collected using fixed selected stations in a region. Such approach only gives us a global notion of the air quality, but do not support a fine-grained comprehension about spots distant from the collector's stations, specially in residential urban places. In this paper, we propose a visual analytics solution that provides city council decision-makers an interactive dashboard that displays air pollution data at multiple spatial resolutions, that uses real and predicted data. The real air quality data is collected using low-cost portable sensors, and it is combined with other environmental contextual data, namely road traffic mobility data. Estimated air quality data is obtained using a machine learning regression model, that is integrated into the interactive dashboard. The visual analytics solution was designed with the city council decision-makers in mind, providing a clutter-free interactive exploration tool that enables those users to improve the quality of life in the city, focusing on one of the most important cities' health quality key issues.

N. Datia (✉)
NOVA LINCS & Future Internet Technologies, ISEL - Instituto Superior de Engenharia de Lisboa, Instituto Politécnico de Lisboa, Lisbon, Portugal
e-mail: datia@isel.ipl.pt

M. P. M. Pato
LASIGE & IBEB & Future Internet Technologies, ISEL - Instituto Superior de Engenharia de Lisboa, Instituto Politécnico de Lisboa, Lisbon, Portugal
e-mail: matilde.pato@isel.pt

R. Taborda
ISEL - Instituto Superior de Engenharia de Lisboa, Instituto Politécnico de Lisboa, Lisbon, Portugal
e-mail: A39134@alunos.isel.pt

J. M. Pires
NOVA LINCS, FCT, Universidade NOVA de Lisboa, Lisbon, Portugal
e-mail: jmp@fct.unl.pt

B. Kovalerchuk et al. (eds.), *Integrating Artificial Intelligence and Visualization for Visual Knowledge Discovery*, Studies in Computational Intelligence 1014,
https://doi.org/10.1007/978-3-030-93119-3_15

1 Introduction

Visualization tools has foundational components—the visual representations—based on combination of different visual encoding like length, position, size, colour saturation and others. Both data and visual encoding are designed to develop interactive visualization tools [3, 41]. Such tools allow to specify direct mappings between the data and the visual representation by a user without much programming experience.

Lisbon's city council is using an urban data platform (PGIL) to manage the city, with many real-time integrated data streams, as well as dashboards to support decision-making [30]. City planners have the responsibility to accurately and realistically propose sustainable alternative to their citizens. The ability to select and present information to support decision-making is an important part of such process. The importance of computer visualization for planning, lies in its potential for improving the quality of decision-making. City council requires an interactive visualization dashboard to address air quality issues, that can be integrated into the PGIL, supporting multiple levels of detail (LoD). However, at the time of this writing, there is no such dashboard, making it difficult to act in situations of large concentration of air pollutant. This work intends to present a solution, which can scale, and will enable an interactive visualization of particulate matter (PM) distributions throughout the city.

The Expert Panel highlighted that Lisbon was the first capital in Europe to sign the New Covenant of Mayors for Climate Change and Energy in 2016 [9]. In 2020, Lisbon makes commitments[1] for the future, to

- reduce in 60% of carbon dioxide (CO_2) emissions until 2030,
- achieve carbon neutrality until 2050,
- become resilience to climate change,
- achieve air quality targets (PM, NOx, SOL_2, O_3) for 2025, among others [8].

At the time of writing, Lisbon council has no fine-grained detailed data about air pollution. The official values came from the Portuguese Ambient Agency (APA), that collects data in five fixed stations.[2] Lisbon has an airport inside the city boundaries, are frequently visited by several boat cruises, and has a handful of highways ways that cross the city, which makes it difficult to get a real picture of Lisbon's air quality using the official data. It is known that those infrastructures are an important contribution to the city's air pollution [24, 26, 29].

Air pollution has a huge impact on the health of the population [31]. Data from World Health Organization (WHO) states that air pollution is a critical risk factor for non-communicable diseases, causing about 24% of deaths from cardiovascular disease, 25% from stroke, 43% for chronic obstructive pulmonary disease (COPD) and 29% associated with lung cancer. Exposure to PM is particularly dangerous to health. These particles, less than 2.5μm (PM2.5) in diameter and less than 10μm

[1] Published on the web-page of the Lisbon Municipal Assembly on 15/07/2016, https://lisboagreencapital2020.com/en/commitment/.

[2] More details at https://airindex.eea.europa.eu/.

(PM10), are considered one of the sixth leading causes of premature death in South-east Asia [4]. In particular, PM2.5 easily penetrate the lungs, irritating and corroding the alveoli, causing breathing difficulties and various lung diseases [44].

In this paper we address some of the missing analytic capability of PGIL. Our contributions are:

- Proposing an interactive dashboard with multi-level spatial resolution, to analyse air quality data;
- Integrating a predictive model, seamlessly, on the dashboard to support city council power users on their analysis.

The rest of the paper is structured as follows. Section 2 reviews current state of the art on visual analytics, including visualization of events in multiple LoD, usage of machine learning and interaction. Section 3 details the problem and introduces the proposed architecture of the solution. Next, we describe the data that were used, both for visualization and for predictive modelling (Sect. 4). Section 5 details the creation and usage of the predictive model, and how it was integrated in the visualization solution. The interactive dashboard is described on Sects. 6, and 7 shows the dash-board evaluation and, finally, we draw some conclusions and point out directions for future work (Sect. 8).

2 Related Work

The previous decade brought us many systems that assist users to visualize their data for a broad range of problems. We have selected some to highlight the different paths and solution taken [2, 3, 7, 34, 41, 47]. Despite their different approaches, all of them suggest alternative views that combine visual encoding, data attributes and data transformations based on user-specified period of interest. As such, visualization developers use elementary graphical units called visual encoding to map multivariate data to graphical representation to put it into the context. We will look at the related work considering their use of:

- multiple levels of detail (LoD) in both space and time;
- machine learning to "fill in the blanks";
- map based visualizations;
- interactivity.

2.1 Multiple LoDs

What patterns we can "see" on data highly depends on the level on which we are observing the data. In the one hand, if we look too "close", we may miss patterns, since we pass beyond the cycle of repetition. On the other hand, looking "too far" needs a data aggregation, that, depending on the function, may soften patterns making them less clear [35]. Choosing the correct level of detail that suits a specific application is key in visual analytics. Works like Silva et al. [34] help us to select the proper LoD where a spatial, temporal, or spatio-temporal pattern emerge. Albino et al. [2], besides the analysis at different LoDs, also shows the impact of choosing different combinations of attenuation and accumulation functions to model the phenomena impact. Those functions act as an aggregation that models the effect of the observed phenomena, for visual analytic purposes. Taborda et al. [38] has an initial implementation of an interactive dashboard to analyse air pollution, based on collected air quality data.

2.2 Machine Learning

The usage of machine learning (ML) is key to support users making their analysis, as data grow in size, complexity and rate of production. ML not only is capable to model data and provide users new visualization scenarios, but can also leverage the data to automate and support their exploration. Bouali et al. [3] use an interactive genetic algorithm that, based on a set of visualizations mapping heuristics, and a specific data mining task chosen by the user (a domain expert), assists the user improving the visualization mapping between the data and the final visualization. Deng et al. [7] employ ML to detect pollution propagation patterns. Those patterns are integrated in a tool that, using graph visualizations, enables users to capture and interpret propagation patterns of air pollution. Using spatial clustering of pollution data, Zhou et al. [47] proposed a visual analytics system that reveals the structure of the data using Voronoi diagrams and hierarchical clustering.

2.3 Map Based Visualizations

Communicate visually to the user the space where events happen helps the analysis and contributes to a better understanding of what is depicted. Generally, most of the works that uses base maps allow spatial exploration using zoom, pan, and adjusting the viewport to relevant areas of analysis. However, some works (e.g. [2, 34, 38]) use maps and allow an aggregation/desegregation of the depicted data based on the available spatial LoD. Other works (e.g. [5, 14, 42, 47]) use maps, supporting zooming to regions of interest, but shows data at the same LoD. Despite the zoom level, maps helps passing the message to the users [18], and end up as an essential visualization tool for searching and exploring the depicted events in context—in this case, the spatial context.

2.4 Interactive Dahsboards

In its seminal work "The eyes have it" [33], Ben Shneiderman enunciates the Visual Information-Seeking Mantra: "overview first, zoom and filter, then details on demand". Implicitly, to achieve such mantra, interaction is key. Silva et al. [34] uses interaction to change the LoD, to link temporal and spatial representations of the events, towards a visual exploration of data. On dashboards concerning air pollution visualization [25, 42] only a limited interaction is available to the user. Most of the time, such interactivity is restricted to pan and zoom. Although in [42] it is possible to get more details once a air quality measurement station is selected, we end-up with static graphs resuming the behaviour of some metrics in the previous day.

2.5 Summary

From the visualization solutions that has air pollution as its main driver, we can point out some missing aspects. First, it is not possible to browse historical data using specific dates or interval of dates. Some websites present air quality information, most of them specific to a country or a region (e.g. [1, 11]), and static, with no interaction, presenting only the past data. This makes difficult the comparison between periods. Besides, the visualizations do not show in context external factors that contribute to the level of air pollution, even thou many of that information is publicly available. The other missing aspect is the lack of multiple levels of detail. Air pollution information is generally presented at a fixed detail, despite the selected zoom level is generally fixed (e.g. [1]). The visualizations that exist in the literature do not seem to take into account specific target users; they assume to be a tool for the public, and not specialist and decision-making users. Finally, the use of machine learning integrated in the visualization is still not fully explored. Although some works predict the air pollution for a given region [1], they are not fully integrated.

In this work, we try to address all the mention issues, presenting an interactive dashboard that depicts measured and predicted air pollution data using two levels of detail, with possible pollution causes visualized into context (the traffic jams).

3 Problem Description

In a smart city environment, besides pollution, there are others problems that share similar issues for which we can foresee similar solutions. Smart city data come from multiple, heterogeneous, data sources [17], belonging to different operators and having different access policies. For those related to the work described in this article, we can highlight different temporal and spatial grains. Thus, for any solution using smart city data, we must choose a proper temporal and spatial grain to support all the

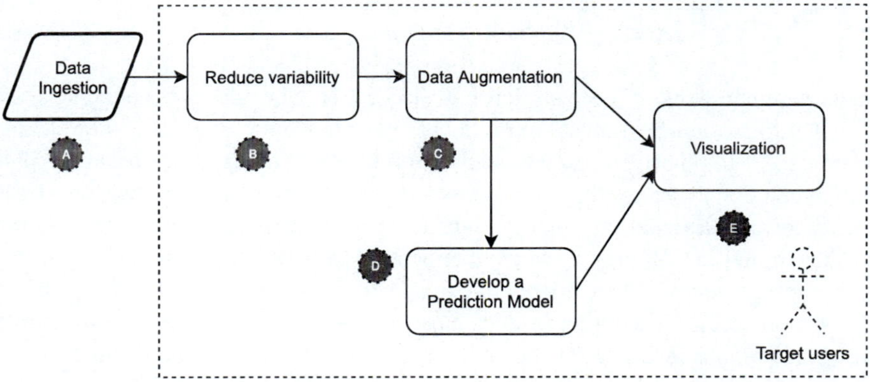

Fig. 1 High-level description of the approach taken to address the problem

analysis, including visualizations and the development of predictive models. Figure 1 depicts a high-level solution to get insights and knowledge from pollution data, that can be applied to other problems having the same restrictions. The first aspect to highlight is the dependence of the solution on the end users and their requirements. After the data ingestion (Fig. 1 Ⓐ), all the aspects of the solution must be user driven. We are supporting those users for decision-making based on what is presented to them. However, the decision based on historical data must report to the same points in space and time. For example, if the spatial coordinates suffer from jitter caused by the Global Navigation Satellite Systems (GNSS) attached to the data source, care must be taken to "correct" them for the solution. Thus, it is crucial to enable a mechanism that maintains the same spatio-temporal points (Fig. 1 Ⓑ). The data manipulation, either by transforming the ingested data individually, by joining different datasets, solving missing data, or by generating new variables should be focused on helping the visualization and the prediction, always maintaining things intelligible to the end-users (Fig. 1 Ⓒ). The prediction model (Fig. 1 Ⓓ) should provide good accuracy and be kept up-to date, to deal with the concept shift and concept drift [40]. It is important to provide a dependable solution. And estimations need to be clear presented to the users using the proper visualization paradigms, guaranteeing that data is displayed in context and that real and predictive information are shown seemingly together (Fig. 1 Ⓔ). This may include the explanation of the prediction in terms of visual artefacts. The process identified in Fig. 1 it is continuous.

In the next subsections we will focus and detail the pollution data issues, the dashboard target users and propose an architecture for the solution.

3.1 Problem Statement

It is known [36] that air quality can vary in relatively close places, for example, giving different wind conditions [43]. Lisbon's air quality it is generally good [27], but episodes of high pollution level still exists for some regions. PGIL has not real-time input data stream of air quality data, nor it has an interactive dashboard that can display the (few) available data.

3.2 The Target Users

Dashboards are an efficient way to communicate with users, improving decision-making activities [39, 46]. In particular, if they enable interaction, users can explore data and get new insights for a given context. The work described in this article aims to support city council employees, specially those working in the CGIUL—a group that is responsible, among other things, to curate data in the Lisbon's city council. Those users interact, in a daily basis, with Lisbon's urban data platform PGIL, to deal with aspects of urban governance, aiming at improving citizens' life [23]. Those aspects include environmental issues, ranging from noise problems to urban waste management. At the time of writing, there is no integrated dashboard whose focus is on air pollution monitoring. CGIUL users have a good spatial knowledge of the city, are aware of the official air pollution data and know much air pollution spots. However, they lack deep knowledge of the real (actual) situation at different spots in the city, as they are not covered by air quality measurement stations. Thus, the dashboard should give an overview of the air pollution inside the city boundaries, enabling exploration of pollution spots on demand. Therefore, the dashboard is a tool that will help those users to deliver a better risk management and prevention in the air pollution area.

3.3 Proposed System Architecture

The interactive dashboard is the visible interface of a more complex system, illustrated in Fig. 2. The system architecture is designed as a "ready to deploy" prototype that shows freshly collected data in context on a dashboard. The external data sources, represented by Fig. 2Ⓐ, are queried on a regular basis and data is ingested into to the system (Fig. 2Ⓑ) and stored in a spatial database, a PostgreSQL with PostGIS extension (Fig. 2Ⓒ). For sake of completeness, the air pollution data is served by a REST API. The value obtained from the API is aggregated using a 7-day sliding average, emphasising the trend of variation in PM2.5. Thus, the minimum granularity of the available data is the day, and it is not possible to disaggregate these values, e.g., for an hour. Thus, for each cell, for each day, data is collected making a HTTP GET like

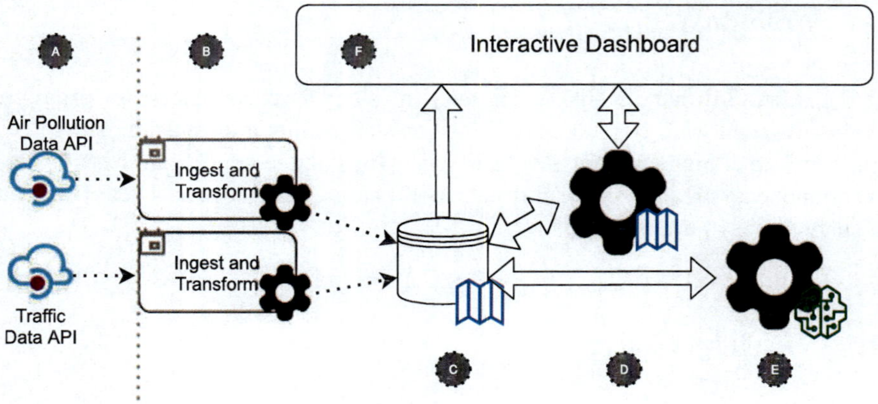

Fig. 2 Architecture of the solution supporting the interactive dashboard

http://localhost:8080/pollution?lat=38.799899&lon=-9.129835. There is no drastic changes to the fetched data, just parsing the JSON representation, particularly of the georeferenced data that are converted to the spatial types available in the database. Regarding the traffic data, and since traffic, has its own variations as a consequence of commuting movements, known as "rush hours", it is necessary to collect data several times a day, e.g. each 10m. This is the granularity with which the data is stored. However, for the purposes of joint analysis between traffic data and pollution, it was necessary to perform, after the ingestion, an aggregation to the same granularity of the pollutant PM2.5 data, that is, making the average of the day.

The transformed data is used by the map service, represented by Fig. 2Ⓓ, to serve specific pre-calculated data to enhance the response of the Interactive dashboard (Fig. 2Ⓕ). Finally, Fig. 2Ⓔ illustrates the integration of the predictive model.

4 The Data

In this section we will describe the data used in this work, prior to any transformation necessary to the end usage, either to present it on a dashboard, or to develop a prediction model using machine learning algorithms. As presented in Fig. 2, we assume that all data is made available through a WEB API. The details about authentication is out of the scope of this article.

4.1 The Air Pollution Data

To get a better knowledge of the air pollution inside Lisbon, our region of interest, there is a need to collect data in more places than those where it is actually done. However, classical air pollution measurement stations have a high cost [15], turning it very unlikely that the city council will spread them all over the city. A midway solution seems to be the installation of hundreds of small stationary station. However, it is currently feasible to collect pollution data with portable devices and low cost sensors (e.g. based on laser technology, like the Grove HM-3301 laser dust sensor[3]), that enables a real-time fine-grained data collection, that are not possible otherwise [16, 21]. Using a portable air quality sensor, data can be though as a function of time and location (a 2D point consisted by a longitude and latitude). To enable data comparison and detection of spatial patterns through time, it is necessary that the historical data report to the same 2D point. Thus, locations need to be aligned with a fixed reference. To achieve this, a 100m cell grid covering the area of Lisbon city was created. Every air quality measure that falls inside a grid cell will be reported using the coordinates of the centre of that cell (centroid), as the portable device has a GNSS sensor. Figure 3 illustrates the grid on the map. The number of relevant cells were reduced, after the intersection between the grid and the city limit's polygon. In the end, air quality can be reported for about ten thousand cells. These number of cells allows fine grain air pollution data, that is economical not viable to get with fixed stations. Table 1 illustrates the data used in this exploratory study. The `keycol` column identifies one cell inside the grid. The `value` column reports to PM2.5 values in $\mu g/m^3$. Other air quality parameters are possible thou. The data model is simplified down to a single table with the relevant information, for the sake of presentation. Other tables exist, namely, data about the bounding box associated with each cell.

Data was collected for some locations on the city. Thus, the reported data is sparse, both in space and time. For that reason, values were averaged for each day, for each cell. This procedure is needed also in a production scenario, where collected data can have fluctuations over the time. However, data validity and accuracy is outside the scope of this study and must be done in a wider time frame.

4.2 Road Traffic Data

The gathering of the pollutant PM2.5 on a daily basis, can be paired with the collection of traffic information. The idea is to correlate PM2.5 and jams, since the traffic in the city of Lisbon [19] is one of the main causes of the increased level of air pollution. The road traffic data source is Waze, a turn-by-turn navigation application with user-submitted travel times and route detail. The data is supplied by EMEL, a Portuguese

[3] See https://www.seeedstudio.com/Grove-Laser-PM2-5-Sensor-HM3301.html.

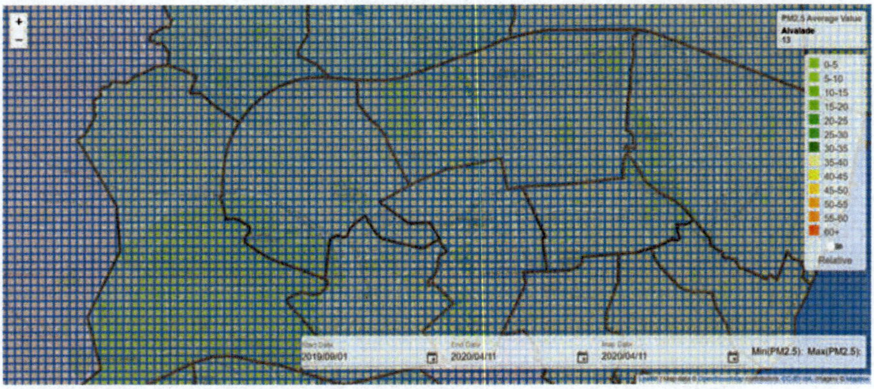

Fig. 3 Illustration of the logical grid used to divided the region of interest, Lisbon, into 100m × 100m cells to where the air pollution data will be reported, geolocated by the centroid of each cell. The grey lines represent administrative region's boundaries, in this case the parishes

Table 1 Excerpt of the air quality data, for a given day. The **keycol** column contains the identification of the cell inside the grid

Keycol	Location	Timestamp	Value
8881	(38.7567, −9.1205)	2020-02-26	11
8882	(38.7567, −9.1194)	2020-02-26	12
8883	(38.7567, −9.1182)	2020-02-26	15

public mobility company, through a WEB API.[4] Each request returns a valid GeoJson document, illustrated in Table 2, containing a set of active jams.

In the next section, we will cover the construction of the predictive model and how it is integrated in the interactive dashboard.

5 Air Pollution Predictive Model

Despite the many roots of urban air pollution, human activity and, in particular, road traffic is still the main source of particles generation [13]. For the predictive model developed in this work, the traffic data to induce the values of PM2.5 in a certain area is considered.

[4] See https://emel.city-platform.com/opendata/.

Table 2 Description of the columns used to report a traffic jam

Column	Description
Country	Code country represented as an ISO 3166-1 code
City	Name of the city or state
Level	Five values representing the jam' severity (0 = free, 5 = highly jam street)
Length	Length of the jam, in meters
Turn_type	Type of the turn for a segment (e.g. allowed turns at each end of the segment)
Uuid	Unique identifier
End_node	Nearest exit at the end of the jam
Speed	Average speed on the jam segment, in meters per second
Road_type	Road type where the jam exists (e.g. Primary Street)
Delay	Delay time in seconds (−1 = totally jammed)
Street	Name of the street
Pub_millis	Publication time in Unix time
Bbox	Jam segment bounding box

5.1 Dataset

The collected data stored in the spatial database went through a sequence of changes until the creation of the dataset used to train the pollution predictive model. Having as input the air pollution data and the traffic data, each one is transformed prior their integration, in a process commonly refer to as data fusion. Figure 4 illustrates the macro steps used in that process. First, the air pollution data is joined with grid

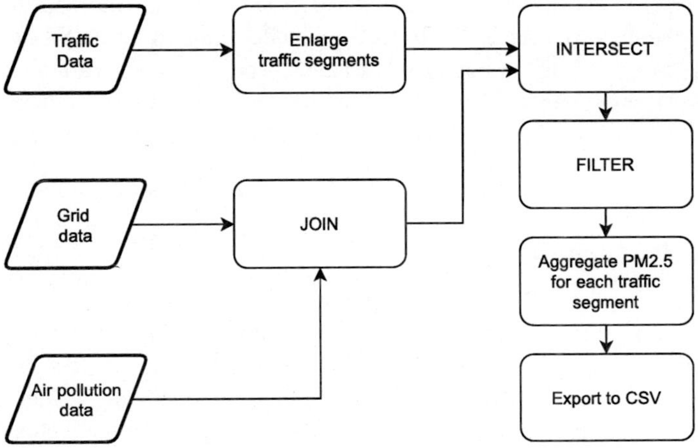

Fig. 4 Dataset creation process

(a) Section with road traffic (original) (b) Section with road traffic (enlarged)

Fig. 5 Representation of the data augmentation, considering one segment of road traffic represented as is (left) and after the augmentation algorithm (right)

data to get the spatial attributes necessary to intersect with the traffic data. However, traffic data is reported by segments of different lengths, as shown in Table 2. Since the way of knowing the value of the pollutant for a specific date, in a line segment (e.g. street, intersection, motorway) with traffic is achieved by intersecting that line with the grid of the map, it was decided to enlarge the area of this segment. The rational is the following. Many streets fall close to the boundary of multiple cells having a set of different PM2.5 values. Thus, turning a line into a rectangle like shape will cover a larger area and intersect the adjacent cells with different pollutant values. Therefore, the dataset that was used to create the predictive model takes into account calculated geographic data that represents the line segment extended by about 20 m. Figure 5 illustrates the result of the augmentation procedure. It was done using the spatial functions available on Postgis, including the `ST_Buffer`. As can be seen, both Fig. 5a, b represent the same segment, with the second covering a larger spatial area.

The intersection between traffic and air quality data will generate a set of PM2.5 values for each traffic segment than need to be aggregated, producing a single pollution value for each segment. We have used the average as an aggregation function. There is also a filtering step, to remove unwanted columns for the modelling. In the end, data is exported to CSV, to be used by the machine learning tool. Table 3 illustrates the exported data. The traffic related columns (the first 5) are described in Table 2; those are the independent variables. The dependent variable is labelled `pm2.5` and represents the 24h moving average of particles concentration, measured in $\mu g/m^3$. For this work, we have used 1 month of data, from February 2020.

5.2 *Model Creation*

To create the predictive model, the H20.ai open-source [12], a Java-based software for data modelling and general computing is used. The H20 is tested in a single machine, using a single node, and is manipulated pragmatically using the Python module. Given the problem at hand, we have used a regression, given the dependent variable, that is a supervised learning task. The H2O's AutoML [6] for automating the machine learning workflow is applied as a started point and prior analysis of the data.

Table 3 Sample of the dataset used to build the predictive model

Speed	Road_type	Level	Length	Delay	pm2_5
0	7	5	106	−1	24
2.198	2	4	483	161	20
0	2	5	285	−1	13
1.855	1	3	232	86	25
5.342	6	3	1064	144	21
0	7	5	170	−1	24
3.494	1	3	459	81	21
1.433	1	2	279	101	23
0	2	5	111	−1	12
0	7	5	56	−1	25
0	7	5	99	−1	25
4.177	2	2	660	83	23

The AutoML includes automatic training and tuning of many models within a user-specified time-limit. Among the models used by AutoML, we can find the *Stacked Ensembles* method, that is a supervised ensemble machine learning algorithm that finds the optimal combination of a collection of prediction algorithms (e.g. *Random Forest* and *Gradient Boosting Machine* (GBM)) using a process called stacking (a kind of boosting technique [28]). The *Stacked Ensembles* supports regression and is based on all previously trained models. Other variation used by AutoML is the best model of each family—an ensemble considering the best models achieved in each individual ML algorithm alone. In most cases, this variation will be the top performing model in the AutoML leaderboard.

The AutoML is used with the default parametrization. We randomly split the input dataset into training and testing, using $\frac{2}{3}$ and $\frac{1}{3}$ respectively, and denoted as `trainSet` and `validSet`. The number of folds for k-fold cross validation is equal to 5. The output variable is pm2_5. Table 4 resumes the 4 best models, as well as their individual performance measures. Since regression refers to predictive modelling problems that estimates a numeric value, the metrics for regression involve calculating an error score to summarize the predictive skill of a model. In this way, the chosen measures to assess the model are the *Mean Squared Error* (MSE), *Root Mean Squared Error* (RMSE), *R-Squared*(R^2), *Mean Absolute Error* (MAE), and *Root Mean Squared Log Error* (RMSLE). Best results for each measure are highlighted using bold typeface font.

As we can see, the best model was achieved when AutoML used *Stacked Ensembles*, in a combination of all individual models, including several *XGBoost* and GLM models. The *StackedEnsemble_AllModels* presents a good R^2 for this domain, and small generalisation errors. For example, MAE give us a magnitude of error around $\pm 2\mu g/m^3$, on average. Since the air quality is reported using the European Air Quality Index [10], the predictive values are aligned with the reported severeness

Table 4 Best 4 Models ordered by the (mean residual deviance), from best to worst. This measured is used by AutoML as the default way of ranking the models in the leaderboard

Validation measures	MSE	RMSE	R^2	MAE	RMSLE
StackedEnsemble_AllModels	**8.3544**	**2.8904**	**0.7093**	1.9208	**0.1551**
StackedEnsemble_BestOfFamily	8.3602	2.8914	0.7091	1.9189	0.1553
XGBoost_grid__1	8.3801	2.8948	0.7084	**1.9178**	0.1556
XGBoost_2	8.6609	2.9429	0.6987	1.9674	0.1580

of the PM2.5. Thus, the usage of this model in the dashboard give users the correct perception of the pollution value.

5.3 *Model Deployment*

H20 can export the models to be integrated into an external Java application, for real time prediction. In our case, we have used a MOJO model, that stands for Model Object, Optimized. MOJO models support both the dataset used and the AutoML models, and is reported[5] to be 20–25 times smaller in disk space and 10–40 times faster from cold start, than previous exported versions. The integration of the predictive model (please refer to Fig. 2Ⓔ) is done using the spatial database to store the predictions. To achieved this, a Java service has been developed to:

- use the MOJO model to make the PM2.5 predictions;
- use the spatial database to read and transform new road traffic data, pass it to the model and store PM2.5 predictions.

The service is responsible to deploy the predictions for each cell into the spatial database. The service runs at a schedule time, on a daily basis. No real-time prediction is made. Although technically feasible, it would have no impact on the prediction values, since the source data, once retrieved for a given day, do not change. Besides, the model was built using PM2.5 with a day granularity. Thus, predictions should follow this granularity. Finally, pre-calculating the predictions simplify the server side architecture that serves the map, making no difference (besides the semantics) between real and predictive data. Therefore, the predictions are calculated offline, using today's data to predict the next day PM2.5 value. Figure 6 illustrated the steps carried on by the Java service.

[5] See http://docs.h2o.ai/h2o/latest-stable/h2o-docs/mojo-quickstart.html.

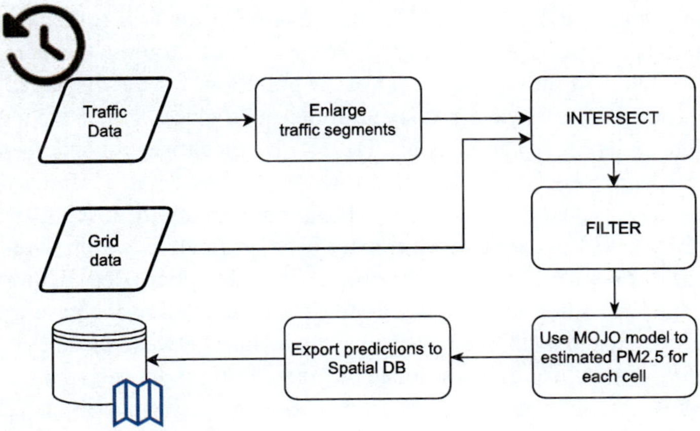

Fig. 6 Steps to integrate the air pollution predictions into the solution. It as procedure that is schedule to occur in a daily basis

6 Interactive Dashboard

The design of the dashboard simplify the information presented to the user. Citing DJ. Mayhew, "We want powerful functionality, but a simple, clear interface" [20]. Following the categorisation present in [46], we consider the dashboard in terms of:

Purpose : Performance Monitoring;
Users : Spatial/Simple/Feeling/Judgement;
Design Features : Single page/Map/2.5D/Drill down & drill up;
Decision Making : Consistency/Speed/Prediction.

As much as possible, the design has followed the Visual Information Seeking Mantra [32]: Overview first, zoom and filter, then details on demand. Figure 7 illustrates the dashboard, highlighting the areas where users can see the pollution information, interact and receive the visual feedback of their interaction. The interface also follows some of the desired features from Human Computer Interaction (HCI), namely, it: *(i)* has affordances (e.g. Fig. 7Ⓓ); *(ii)* provides feedback on the user's action (e.g. clicking on a region changes the summary information, Fig. 7Ⓑ); *(iii)* prevents users from making errors (e.g. selecting dates are constrained by data pick controls and a data range definition, Fig. 7Ⓓ). The dashboard is responsive React single page application, communicating with a Spring server side server. As it can be seen, the result is a clutter free context map,[6] where the colour is used to discriminate areas, grading them according their air pollution value. Figure 7Ⓐ shows the map area, where the Regions of Interest (ROI) are presented. Users can see further details by stimulating a ROI, either by hover it, or by a mouse click. Clicking on a ROI selects

[6] The map in the dashboard was implemented using the open-source JavaScript library Leaflet (see https://leafletjs.com/).

that region (illustrated in Fig. 7Ⓕ). The information of the ROI is then displayed in the dashboard area identified by Fig. 7Ⓑ. For sake of illustration, that area indicates the parish name, and the average level of air pollution, for the displayed day. The detailed information is always displayed in the same place, to achieve a consistent visual feedback. The area identified as Fig. 7Ⓒ give the proper context to identify the level of pollution in each ROI. The colour range used by default follows the palette used in the EU Air Quality Index (EAQI) [10], representing the level of air pollution of each ROI. The EAQI uses 6 discriminator bands, namely: 1. Good (Green) 2. Fair 3. Moderate 4. Poor (Yellow) 5. Very poor 6. Extremely poor (Red). However, since the air quality in Lisbon is generally good, it was decided to increase the number of tones (2/1 ratio) in each band for better distinction between close EAQI values. For example, it were used several tones of green to highlight the small differences between ROI, even if they fall in the same band (of the six possible). When there is no data for a given ROI, we use no colour (transparent) to inform users that for a given day no data is available, as can be seen at Fig. 7Ⓖ. To select the period of analysis, users can pick up three dates, in the area identified as Fig. 7Ⓓ, with the following semantics: 1. The first and last dates are use to settle the temporal range of analysis; 2. The middle date is used to set day whose PM2.5 data is displayed on the map. The interval dates are used to calculate the maximum and minimum values of the PM2.5, and then display it at the area Fig. 7Ⓔ. The idea is to enable a quick visual exploration of different days inside an interval, maintaining the pollution limit values (min. and max.) fixed for that interval. The rational is the following. Sometimes there are incidents that raises the PM2.5 values, which are keep high for a while and decreasing rapidly towards "normal" values for the area [45]. Finally, users can change between historical and current values of PM2.5, and between predictive values using the top bar (Fig. 7Ⓖ). There is a visual feedback, displayed in blue, to indicate the selected visualization (real or predicted).

Besides the colour palette used in EAQI, we provide an alternative way of colouring ROI, by enabling a normalised feature button at the bottom of the colour legend (Fig. 7Ⓒ). The normalisation is used to compare ROIs, despite their values fall inside the EAQI bands. Using the maximum PM2.5 value for the depicted day, colours are mapped into the 13 colour palette, using a even distribution between 0 and the maximum PM2.5 of the day. The result is depicted at Fig. 8. Again, this feature is inline with one of the purposes of the dashboard—A simple and quick judgement by the user's how air pollution is spread, in a given day, among different city locations. As we can see, top left ROIs, coloured in reddish tones, presented higher pollution levels than the ROI at the centre, painted in greenish tones. The ROI have a direct relation with administrative areas, since usually people think and reasoning that way, specially the target users. Two levels of detail (LoD) were considered for users to explore the data. One is the administrative divisions of city in parishes. The other is a statistical region that have a good socio-demographic characterisation—the statistical subsection [22]. This LoD can be latter used to study the affected population by low air quality events.[7] Figure 9a, b illustrates these two levels of detail, respectively.

[7] This is something out of the scope of the article.

Fig. 7 User interface of the interactive dashboard. The highlighted areas represent key components for the interaction and to get visual feedback based on the interaction, namely: **A** the map area; **B** place to show detailed information for the selected area; **C** colour legend; **D** dashboard's time selector; **E** Minimum and Maximum PM2.5 values for the selected time range; **F** selecting a zone; **H** alternate between real data and predictive data. Do note that traffic data is not shown in this example

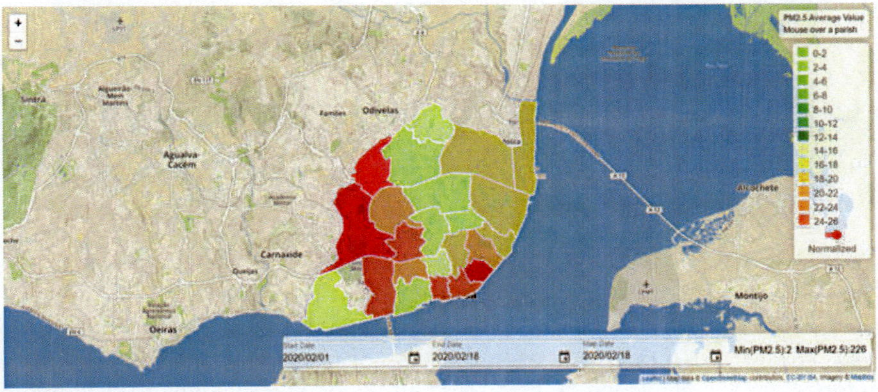

Fig. 8 Illustration of the map using a relative representation of the data on the map. The colour palette is always the same, but relative PM values highlights the difference between regions on the map, enabling a direct comparison between them

The switch between both levels of detail are done manipulating the zoom of the map. However, when the user double click on a ROI at the parish level, the dashboard is automatically centred on the ROI, selecting the lowest LoD, the statistical subsection.

Figure 7 G let users alternate between current real data (*Map*) and predicted data (*Preview Map*). Thus, users need to explicitly change their view to access predicted data, making them aware of the data that is depicted. However, the integration of the predictive model do not change the way users interact in the dashboard. It maintains the same information, the same LoDs and the same options to colour the air pollution

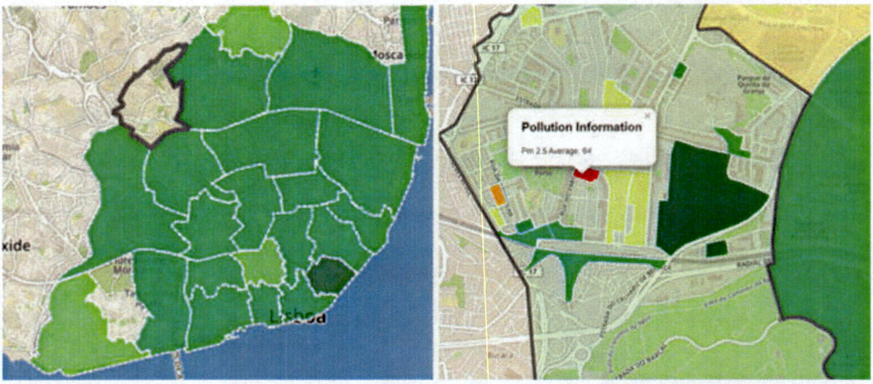

(a) Representation of the first level of detail — 24 parishes of Lisbon.

(b) Representation of the second level of detail — statistical subsection. The detail is for the selected parish.

Fig. 9 Levels of detail available for analysis

Fig. 10 Illustration of traffic congestion display, to be used as contextual information

data. However, users can see on top of today's context data, including road traffic information, the prediction of tomorrow's air pollution levels. Figure 11 illustrates the dashboard view for the predictive air pollution is put in context with the road traffic data. The traffic information, gathered from Waze data, was inserted at the dashboard as a contextual information. The jam level, ranging from 0 to 5, where 0 indicates no jam and 5 is totally jammed, was average for each day and displayed as a dashed line. The 0 was represent as a long spaced dashed line, decreasing the space between dashes until the level 5 is represented as a straight line. We avoid colours in the representation to lower the clutter and reduce the visual information overflow. The result is displayed in Fig. 10, and can be seen in context on Fig. 11. Do notice that, at low levels of detail the difference between level of jam is not noticeable. We have decided use the contextual information only on the preview map.

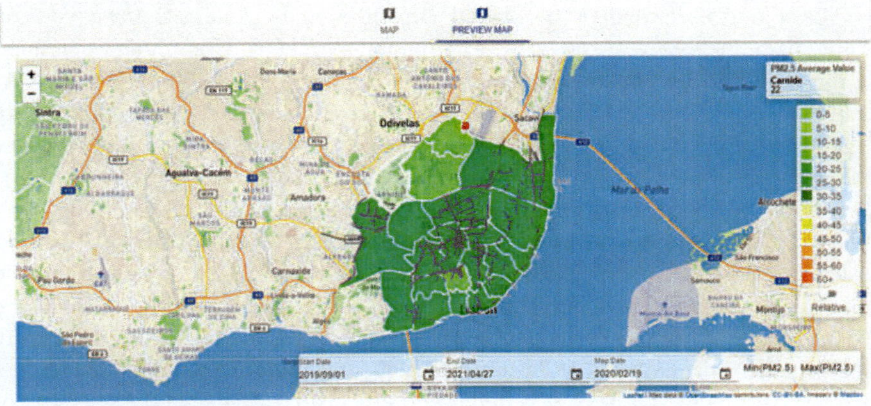

Fig. 11 Complete information made available on the dashboard. The air pollution data is predicted using the machine learning model

7 Assessment and Discussion

The dashboard was assessed, at an early stage of development, by a small number of end users (5 users and an advisor to a CML councillor) belonging to the city council's CGIUL group. The assessment's goals is to detect usability and interface issues. Thus, the number of users are enough to spot 85% of the user interface (UI) problems [37], since the variance for individual users is low. The users were selected among those whose responsibility is to follow the pollution levels in the city and that were involved in previous settlement of fixed air pollution collection spots. For the purpose of this work, they are considered to be domain experts. Only one of the users presented on the test had took part in a previous kick-off meeting of this work. The assessment was made at the city council's facilities, using a large monitor (65"), whose characteristics are close enough to those of the monitors used by end users in real situations.

The test was carried out as follows:

1. Presentation of the dashboard to the users, focused on the goals and design issues;
2. Live demonstration of dashboard's interaction;
3. Collection of feedback from the end users.

The first thing to notice is the dashboard adapt seamlessly to the large screen, causing an effect on the users, as they see how the pollution spread across the city, for places where, at the moment, no official data is available. Everyone has an empiric idea on the air quality, but is different seeing it on the screen. Some higher pollution spots caused curiosity in some users that wanted to drill down to the statistical subsection to see what area is causing the higher values. Changing between parish level and statistical subsection level in the ROI seems to be enough, as no users suggested other level of detail.

The colours used seems to be aligned with the users judgement. Being experts on the domain, they were aware of the existing index (EAQI). Even thou the number of tonalities for each band is different in the dashboard, the users understand correctly the level of PM reported. Changing for relative colouring (see Fig. 8) were not straightforward for users to understand. The meaning and the purpose of this new colouring needed an extra explanation, after which the idea was grasp by the users. Regarding the selection of the temporal range, displayed on the dashboard (Fig. 7Ⓓ), there were found some issues. The prototype presented to the users had the following order on the dates: 1. Start Date; 2. Map Date; 3. End Date. The rational was that the map date, that mandates what is shown, is one inside an interval of dates. Despite the explanation that one day at the time is displayed on the map, the users consistently understand the displayed PM2.5 was a representation of the selected temporal range. It was clear that the UI need to be redesign in this part, to separate the selection of the data to be displayed from the selection of the temporal range used to established a background knowledge to compare with. Nevertheless, users found interesting and usable the comparison between a day and an interval, even thou they want more information besides the minimum and maximum PM2.5 values. For example, they wanted to see a visual display of the evolution of the data for each day of the range.

Other aspects that was pointed out to be a subject of change in the UI is the interaction with the statistics of the temporal interval (Fig. 7Ⓔ). The prototype, at the time of the assessment, do not offer users any interaction on this part, although we used similar representation where such interaction it is possible (Fig. 7Ⓓ). Thus, besides the addition of a visual representation of the PM2.5 evolution, when Fig. 7Ⓔ is clicked, users want to see where it happens. Thus, in the current prototype, when a user click on the `min` or `max`, the visualization adapts and put the map at finest level of detail (statistical subsection) where the `min` or `max` value, respectively, have occurred. When questioned if the represented day should remain the same, even if the `min` or `max` occurred in a different day inside the temporal range, the answers were not conclusive. Finally, during the interaction and exploration of the map, the users often stop in areas where the expected PM2.5 level is different from their preconceived notion of pollution level for that spot. Being not only domain experts, they also live in Lisbon, making them aware of the space where the pollution is reported. We found that some contextual information (e.g. traffic congestion) may help users to convey some reason behind these pollution variations and deviations. This contextual information, to be informative, should not clutter the dashboard.

As a result of the UI initial assessment, all the changes were incorporated on the current version of the dashboard. However, it is worth to mention that since we are using predictions to estimate values in context, we must assure dependable predictions. Even if the predictive results deviate from the user's expectation, the UI should provide the reason why those values are different, engaging the users on the system and enabling to turn information into knowledge. We foresee every future system needs that uses ML need to provide explanations to support the results.

8 Conclusion and Future Work

This work presents a prototype of a dashboard visualization for air quality data, used as a tool to analyse the evolution of air pollution inside the Lisbon's city boundaries. It enables the exploration of pollution spots on demand. The target users are city council's employees responsible for monitoring many aspects of the city's daily life, including the air pollution evolution. The dashboard visualization, a map based iterative tool, is a clutter-free web based React responsive application that provides a quick overview of the air pollution levels in two levels of details, known by the end users. The integration of regression model to estimate the pollution, with estimated data put into context, opens new ways to manage the city regarding the expected pollution levels. With the first assessment of the dashboard, it was possible to conclude that it has a good approval by the end users, where most of the information was easily grasped. The reported issues were corrected during development. The solution can be extend to other sensor data sharing the same requirements.

For future work, the dashboard can include other historical information, namely, the history of PM2.5 and other pollutants in each ROI. The changes must produce a clutter-free dashboard, that is aligned with the objective on CGIUL users—an interactive visualization dashboard focused on pollution monitoring. The regression model must be improved, by including more contextual information, e.g. meteorological data, and to explore other ways to resume the road traffic data. Finally, it must be explored the inclusion into a single dashboard, real and estimated data, where users can see where estimations are being used and can have a notion of how good those estimations are.

Acknowledgements We would like to thank the CGIUL team for its support under the "Data at the service of Lisbon" Protocol, and for their assessment of the first prototype of the air quality monitoring dashboard. This work is supported by LASIGE (UIDB/00408/2020) and by NOVA LINCS (UIDB/04516/2020) with the financial support of FCT- Fundação para a Ciência e a Tecnologia, through national funds.

References

1. AIRPARIF. Air Quality Forecast. online: https://www.airparif.asso.fr/en/# (2020)
2. Albino, C., Pires, J.M., Datia, N., Silva, R.A., Santos, M.Y.: Aa-maps—attenuation and accumulation maps for spatio-temporal event visualisation. In: 2017 21st International Conference Information Visualisation (IV), pp. 292–295 (2017)
3. Bouali, F., Guettala, A., Venturini, G.: VizAssist: an interactive user assistant for visual data mining. Vis. Comput. **32**(11), 1447–1463 (2016)
4. Chowdhury, S., Dey, S., Estimate adjusted for baseline mortality: Cause-specific premature death from ambient PM2.5 exposure in India. Environ. Int. **91**, 283–290 (2016)
5. Trafair consortium. Trafair air quality dashboard. https://trafair.eu/airquality/ (2020)
6. Anamaria Crisan and Brittany Fiore-Gartland. Fits and starts: Enterprise use of automl and the role of humans in the loop. arXiv preprint arXiv:2101.04296 (2021)

7. Deng, Z., Weng, D., Chen, J., Liu, R., Wang, Z., Bao, J., Zheng, Y., Wu, Y.: Airvis: visual analytics of air pollution propagation. IEEE Trans. Visu. Comput. Graph. **26**(1), 800–810 (2020)
8. European Comission. Covenat of Mayors for Climate and Energy. https://www.eumayors.eu/en (2020)
9. European Comission. European Green Capital Award Winner. https://ec.europa.eu/environment/europeangreencapital/lisbon-is-the-2020-european-green-capital-award-winner/ (2020)
10. European commission. European Air Quality Index. https://www.eea.europa.eu/themes/air/explore-air-pollution-data (2020)
11. FCT-NOVA. Air Quality Forecast. http://www.prevqualar.org/homepage.action (2020)
12. H2O.ai. H20
13. Harrison, R.M., Van Tuan, V., Jafar, H., Shi, Z.: More mileage in reducing urban air pollution from road traffic. Environ. Int. **149**, 106329 (2021)
14. King's College London. London Air. https://www.londonair.org.uk/LondonAir/nowcast.asx (2020)
15. Kumar, P., Morawska, L., Martani, C., Biskos, G., Neophytou, M., Di Sabatino, S., Bell, M., Norford, L., Britter, R.: The rise of low-cost sensing for managing air pollution in cities. Environ. Int. **75**, 199–205 (2015)
16. Li, Z., Che, W., Christopher Frey, H., Lau, A.K.H., Lin, C.: Characterization of PM2.5 exposure concentration in transport microenvironments using portable monitors. Environ. Pollut. **228**, 433–442 (2017)
17. Liu, X., Heller, A., Nielsen, P.S.: Citiesdata: a smart city data management framework. Knowl. Inform. Syst. **53**(3), 699–722 (2017)
18. MacEachren, A.M.: How Maps Work: Representation, Visualization, and Design. Guilford Press (2004)
19. Martins, A., Cerqueira, M., Ferreira, F., Borrego, C., Amorim, J.H.: Lisbon air quality: evaluating traffic hot-spots. Int. J. Environ. Pollut. **39**(3–4), 306–320 (2009)
20. Mayhew, D.J.: The usability engineering lifecycle. In: CHI'99 Extended Abstracts on Human Factors in Computing Systems, pp. 147–148 (1999)
21. McKercher, G.R., Salmond, J.A., Vanos, J.K.: Characteristics and applications of small, portable gaseous air pollution monitors. Environ. Pollut. **223**, 102–110 (2017)
22. Nation Statistic Institute. Statistical subsection. https://censos.ine.pt/xportal/xmain?xpid=CENSOS&xpgid=censos_subseccao (2020)
23. Open & Agile Smart Cities. Lisbon's Urban Data Platform - PGIL. https://oascities.org/lisbons-bet-on-urban-data-platform/ (2020)
24. Pérez, N., Pey, J., Cusack, M., Reche, C., Querol, X., Alastuey, A., Viana, M.: Variability of particle number, black carbon, and pm10, pm2. 5, and pm1 levels and speciation: influence of road traffic emissions on urban air quality. Aerosol Sci. Technol. **44**(7), 487–499 (2010)
25. Paris Air Quality. Paris air quality. https://capgeo.sig.paris.fr/Apps/QualiteAirParis/
26. Ruiz-Guerra, I., Molina-Moreno, V., Cortés-García, F.J., Núñez-Cacho, P.: Prediction of the impact on air quality of the cities receiving cruise tourism: the case of the port of barcelona. Heliyon **5**(3), e01280 (2019)
27. Santos, F.M., Gémez-Losada, Á., Pires, J.C.M.: Impact of the implementation of lisbon low emission zone on air quality. J. Hazard. Mater. **365**, 632–641 (2019)
28. Schapire, R.E.: A brief introduction to boosting. In: Ijcai, vol. 99, pp. 1401–1406. Citeseer (1999)
29. Schlenker, W., Walker, W.R.: Airports, air pollution, and contemporaneous health. Rev. Econ. Stud. **83**(2), 768–809 (2016)
30. Serrador, A., Tremoceiro, J., Cota, N., Cruz, N., Datia, N.: iLX—a success case in public tender methodology. In: ProjMAN 2018—International Conference on Project MANagement (2018)
31. Serviço Nacional de Saúde. WHO: Air Pollution. https://www.sns.gov.pt/noticias/2018/05/02/oms-poluicao-atmosferica/

32. Shneiderman, B.: The eyes have it: a task by data type taxonomy for information visualizations. In: Proceedings 1996 IEEE Symposium on Visual Languages, pp. 336–343. IEEE (1996)
33. Shneiderman, B.: The eyes have it: a task by data type taxonomy for information visualizations. In: The Craft of Information Visualization, pp. 364–371. Elsevier (2003)
34. Silva, R.A., Pires, J.M., Datia, N., Santos, M.Y., Martins, B., Birra, F.: Visual analytics for spatiotemporal events. Multimedia Tools Appl. **78**(23), 32805–32847 (2019)
35. Silva, R.A., Pires, J.M., Santos, M.Y., Datia, N.: Enhancing exploratory analysis by summarizing spatiotemporal events across multiple levels of detail. In: Sarjakoski, T., Santos, M.Y., Sarjakoski, L.T. (eds.) Geospatial Data in a Changing World, pp. 219–238. Springer International Publishing, Cham (2016)
36. Sorte, S., Arunachalam, S., Naess, B., Seppanen, C., Rodrigues, V., Valencia, A., Borrego, C., Monteiro, A.: Assessment of source contribution to air quality in an urban area close to a harbor: case-study in porto, portugal. Sci. Total Environ. **662**, 347–360 (2019)
37. Spool, J., Schroeder, W.: Testing web sites: five users is nowhere near enough. In: CHI'01 Extended Abstracts on Human Factors in Computing Systems, pp. 285–286 (2001)
38. Taborda, R., Datia, N., Pato, M.P.M., Pires, J.M.: Exploring air quality using a multiple spatial resolution dashboard—a case study in lisbon. In: 2020 24th International Conference Information Visualisation (IV), pp. 140–145 (2020)
39. Vilarinho, S., Lopes, I., Sousa, S.: Developing dashboards for smes to improve performance of productive equipment and processes. J. Industr. Inform. Integr. **12**, 13–22 (2018)
40. Webb, G.I., Lee, L.K., Goethals, B., Petitjean, F.: Analyzing concept drift and shift from sample data. Data Mining Knowl. Dis. **32**(5), 1179–1199 (2018)
41. Wongsuphasawat, K., Moritz, D., Anand, A., Mackinlay, J., Howe, B., Heer, J.: Voyager: exploratory analysis via faceted browsing of visualization recommendations. IEEE Trans. Vis. Comput. Graph. **22**(1), 649–658 (2015)
42. World Air Quality Index project. World Air Quality Index. https://waqi.info/ (2020)
43. Xie, J., Liao, Z., Fang, X., Xinqi, X., Yu Wang, Yu., Zhang, J.L., Fan, S., Wang, B.: The characteristics of hourly wind field and its impacts on air quality in the pearl river delta region during 2013–2017. Atmoph. Res. **227**, 112–124 (2019)
44. Xing, Y.F., Xu, Y.H., Shi, M.H., Lian, Y.X.: The impact of PM2.5 on the human respiratory system. J. Thoracic Dis. **8**(1), E69–74 (2016)
45. Jianming, X., Chang, L., Yuanhao, Q., Yan, F., Wang, F., Qingyan, F., The meteorological modulation on PM2.5 interannual oscillation during, : to 2015 in Shanghai, China. Sci. Total Environ. **572**(1138–1149), 2016 (2013)
46. Yigitbasioglu, O.M., Velcu, O.: A review of dashboards in performance management: implications for design and research. Int. J. Accoun. Inform. Syst. **13**(1), 41–59 (2012)
47. Zhou, Z., Ye, Z., Liu, Y., Liu, F., Tao, Y., Su, W.: Visual analytics for spatial clusters of air-quality data. IEEE Comput. Graph. Appl. **37**(5), 98–105 (2017)

Context-Aware Diagnosis in Smart Manufacturing: TAOISM, An Industry 4.0-Ready Visual Analytics Model

Lukas Kaupp, Kawa Nazemi, and Bernhard Humm

Abstract The integration of cyber-physical systems accelerates Industry 4.0. Smart factories become more and more complex, with novel connections, relationships, and dependencies. Consequently, complexity also rises with the vast amount of data. While acquiring data from all the involved systems and protocols remains challenging, the assessment and reasoning of information are complex for tasks like fault detection and diagnosis. Furthermore, through the risen complexity of smart manufacturing, the diagnosis process relies even more on the current situation, the context. Current Visual Analytics models prevail only a vague definition of context. This chapter presents an updated and extended version of the TAOISM Visual Analytics model based on our previous work. The model defines the context in smart manufacturing that enables context-aware diagnosis and analysis. Additionally, we extend our model in contrast to our previous work with context hierarchies, an applied use case on open-source data, transformation strategies, an algorithm to acquire context information automatically and present a concept of context-based information aggregation as well as a test of context-aware diagnosis with latest advances in neural networks. We fuse methodologies, algorithms, and specifications of both vital research fields, Visual Analytics and Smart Manufacturing, together with our previous findings to build a living Visual Analytics model open for future research.

L. Kaupp (✉) · B. Humm
Department of Computer Science, Darmstadt University of Applied Sciences, Haardtring 100, 64295 Darmstadt, Germany
e-mail: lukas.kaupp@h-da.de

B. Humm
e-mail: bernhard.humm@h-da.de

K. Nazemi
Darmstadt University of Applied Sciences, Human-Computer Interaction and Visual Analytics Research Group, Haardtring 100, 64295 Darmstadt, Germany
e-mail: kawa.nazemi@h-da.de

B. Kovalerchuk et al. (eds.), *Integrating Artificial Intelligence and Visualization for Visual Knowledge Discovery*, Studies in Computational Intelligence 1014,
https://doi.org/10.1007/978-3-030-93119-3_16

1 Introduction

Accelerated through Industry 4.0, the production industry experiences a significant change towards full automation. This change creates new challenges along the way. During the transformation, the side-by-side existence of legacy brownfield systems and cyber-physical systems form a complex and highly interconnected environment that produces large amounts of data. Additionally, the increased interconnectivity leads to novel relationships and dependencies between machinery, which play a crucial role in, for example, contextual faults [1]. Outliers as early traces of faults are challenging to recognize in a smart factory's high-density tempo-spatial data. Any kind of diagnosis in such a complex environment is time-consuming [2]. Visual Knowledge Discovery (VKD) and Artificial Intelligence (AI), both part of Visual Analytics (VA), can mitigate the complexity that arises from the increased interconnectivity and speed up the rectification of a fault in a smart factory [3]. Thereby, VA [4–9] as a holistic compound of algorithms from data pre-processing, over model engineering (machine learning), to assistive visualizations and the knowledge generation thrive the purpose of a Visual Knowledge Discovery process, to explore knowledge visually. Latest advances towards context-awareness by Zhou et al. [10], Wu et al. [11], Filz et al. [12] emphasize the need for a generalizable VA model for Smart Manufacturing (SM).

We proposed the VA model TAOISM (contex**t**-**a**ware diagn**o**sis **i**n **s**mart **m**anufacturing) in [3] to support the diagnosis process. Our approach takes the complex environment into account. It reflects the novel dependencies between cyber-physical systems in contexts (aggregated snapshots of the current situation) to accelerate the early detection of anomalous events for reducing the time during diagnosis through intelligent visualizations of contexts and outliers. In this book chapter, we present an extended version of this work [3], joint together with our latest findings around the reference implementation of our VA model.

Before we defined the context term mathematically in the VA model TAOISM for SM, the context was a vague periphrasis, where the domain expert came into play [6] and tried to deduct new insights. Our TAOISM model defines each area of a VA model. After the overview of the proposed TAOISM model, we start with a definition of data and potential data sources. Next, we provide the context definition and the first set of models within models. Additionally, we name requirements for visualizations in SM that arise from our model and provide a draft of visualizations. To complete our model, we identify and define the main tasks in SM that are significant for our model. We complement our previous more theoretic-oriented work with the automatic extraction and application of context in context-hierarchies. With the context-aware diagnosis, we can show in experiments that with context, prediction on our CONTEXT dataset [1] is more accurate even with small data and allows an improvement in outlier detection. Since this chapter is an extended version of our previous work [3], it may share identical sections, paragraphs, definitions, or descriptions. The enhancements in this section lead to new scientific contributions. As our previous work included a three-fold contribution (1–3), we provide in this chapter four more contributions (4–7) that enhance the previous work significantly. (1) We provide an overview about current trends and tendencies for VA and SM; (2) we formalize, construct and compose the context-aware diagnosis in SM through our model and (3) we propose a model that combines VA and SM through the definition of context and the context creation process. Additionally, (4) we introduce context-hierarchies to form a defined deduction environment, (5) provide the first algorithm for context building, (6) added a concept of context-based information aggregation, and (7) facilitate the automatically extracted context in prediction for more appropriate outlier detection.

2 Related Work

Our TAOISM model concerns multiple areas. For a novel VA model, we review the current state of the general VA models that build the foundation of our TAOISM model. Furthermore, first advances exist in SM that try to contextualize SM processes involving human perception or computational models. Next, multiple visualizations systems cover similar steps to achieve similar goals in SM that have traces of an underlying VA model. We cover these systems to connect the traces and develop a concrete VA model for SM. In the end, we additional cover graph neural networks (GNNs) as one first possible approach to employ context practically.

2.1 General Visual Analytic Models

Visual Analytics is a vibrant area of research and has been the foundation for the creation of multiple models over the past years [13, 14]. Keim et al. [4] with their model provoke one of the first general approaches, which was detailed later [5]. As an extension, Sacha et al. define the knowledge generation process [6] and human cognitive activities. Another well thought-out extension, given by Andrienko et al., who characterize the outcome of Keim's VA model [9]. We use Keim's Visual Analytics model as a foundation, specify and refine it for the SM domain. Additionally, we add the first draft of the context definition and how the context can be used in the SM domain. Context is a rather vague term that was already mentioned by Sacha et al. [6], where a domain expert is mandatory. As a consequence, contexts are enclosed by a domain for our model in the SM domain. Besides Sacha et al. [6], there are also data-driven approaches to enlighten the context term. Ceneda et al. [8] characterize guidance within Visual Analytics to complement Keim's model. Munzner [15] provides a framework to specify tasks as a tuple of action and target. We use their framework to characterize our Visual Analytics model.

2.2 *Context-Awareness in Smart Manufacturing*

The first step towards context-awareness was done by Emmanouilidis et al. [16]. Their conceptual model integrates the knowledge of domain experts as a single entity. Their ideas to contextualize machinery symptoms by integrating human perception are an inspiration and can be seen as an early predecessor of the proposed model in this paper. In addition, to context-awareness Zhou et al. [10] define a novel situational awareness model incorporating qualitative (temperature sensor data) and quantitative (temperature zones with boundaries) measures. Their situational awareness model is split into an index part (rules for temperature zones) and a computational model that utilizes the measurements; in their case, temperature. In addition, the computational model deducts a value from multiple measurements, which represents the state of the production line (low, guarded, elevated, high, severe). To the best of our knowledge, that is the first model that formalizes a context in an SM process and calculates the severity of that context. We adapt the principle to take not only the production line itself as the single source of information but the surrounding variables as well. In contrast to Zhou et al. [10] that relies only on temperature data and cannot be applied in scenarios with complex multivariate data, we are aware of that situation and employ a data transformation step even to cover complex cases. Additionally, we leverage other production-related systems (e.g., MES, ERP) to enhance the scope of context beyond the production line. After the acceptance of our TAOISM VA model publication [3], Filz et al. [12] published a product state propagation concept, which is similar to our context concept, whereas their concept can be referred to as a product-centered context to identify the malicious process that leads to a faulty product. Unlike their product state propagation, our context is used to identify faults, dependencies, and event chains in the production line, CPSs, and their communication. Therefore, with our approach, we want to identify faults in the production line and its involved sub-process, rather than identify the processes leading to the faulty product with an otherwise normal working factory. Nevertheless, Filz et al. [12] also encourage the analysis of the manufacturing system as a whole, instead of the analysis of the single and isolated processes, in favor of the risen complexity in SM.

2.3 *Visualization for Maintenance and Production*

The SM domain is challenging to assess due to concerns about security, rights of intellectual properties, or data sovereignty. Nevertheless, Zhou et al. managed to publish a thorough survey of current visualizations in this domain [17]. They structured the visualizations using the concepts of creation and replacement. Visualizations in the replacement concept free people from tedious work through the implacement of intelligent devices (e.g., replacing monitoring personnel through online fault diagnosis) or virtualize dangerous work environments where people can learn the needed skills [17]. Creation encompasses the design phase (creation of products), the production phase (ideology to physical forms), the testing phase (guarantee established standards), the service phase (insights from usage) [17]. In order to gain insights into manufacturing data, Xu et al. [18] propose a combination of an extended Marey-graph and a station graph to exploit production flow and spatial awareness and provide insights to uncover anomalies. Jo et al. [19] provide an aggregated view of ongoing tasks in the production line with an extended Gantt-chart. Where Post et al. [20] use flow, workload, and stacked graph to provide a user-guided visual analysis of a production line. Most impacts for the proposed model had the work of Arbesser et al. [21] and Zhou et al. [10]. Arbesser et al. developed a visual data quality assessment for time series data with integrated plausibility checks. Plausibility checks are simple rules that apply to given meta-information (e.g., sensor type, position) and observe, for example out of range values. Thus, these are similar to our foreknown case models, which can be initialized before installation, based on manufacturers' cyber-physical system (CPS) documentation. Their well-thought overview contains information with different levels of granularity (from overview to detail) with the employment of the checks to color the severity of the anomaly based on the checks. Wu et al. [11] set up a novel VA pipeline to manually combine and pick features for the machine learning models and visualize the effectiveness in a training set view to act accordingly if results do not match observations. Additionally, they added a system overview with an extended theme river and a radial layout with a multifaceted pane for details on demand. Where Wu et al. preference is more manually configured, is our model more driven by automation protocols such as the de facto communication standard OPC UA [22]. OPC UA comes with machine models in place, which combine sensors to groups for different aspects of the CPS. The manual selection of individual sensor values for the generation of feature vectors is inefficient. Our small learning smart factory already has about 17.000 values to choose from [23]. Additionally, the determination of machine learning can be an issue in hybrid scenarios, where the smart factory already has statistical models in place. Our model solves these issues by also including statistical models.

2.4 Methods to Learn Context

One possibility to represent the context and the context hierarchies is a graph. Therefore, for our reference implementation, we focus on novel techniques capable of learning our graph-based context. The latest innovations in Graph Neural Networks (GNNs) allow graph learning and graph prediction, which enables context-based prediction to improve outlier detection and context-aware diagnosis. Outlier detection is a stepping stone in our reference implementation of our TAOISM VA model that automatically triggers notifications in our visualizations. GNNs were developed by Defferrard et al. [24] using a new efficient CNN architecture to learn graph-based information. Since their invention, new categories of GNNs are presented, such as Graph Convolutional Networks (GCNs) [25] and Graph Attention Networks (GATs) [26]. GNNs are a vital research field and recently receive growing attention from the research community in surveys [27–29], especially in time series forecasting and anomaly detection [30–34]. Wu et al. simplify the computation effort and reduce the training effort on GNNs [35]. First, advances exist to use GNNs with the Internet of Things (IoT) [36], which hints, through the usage in the IoT domain, faster convergence of the neural network and a better prediction also in a SM scenario with IIoT and CPSs. However, to the best of our knowledge, no prediction or outlier detection with GNNs is done in SM, especially no context-based prediction using GNNs, as we propose in this paper.

3 TAOISM VA Model

Our proposed TAOISM VA model (Fig. 1) consists of four main layers (data, models, visualization, knowledge) and a cross-sector meta-layer (trigger). Originating in the established general VA model by Keim et al. [4], depends each layer on another with multiple cross-sector relationships. All layers are involved in providing

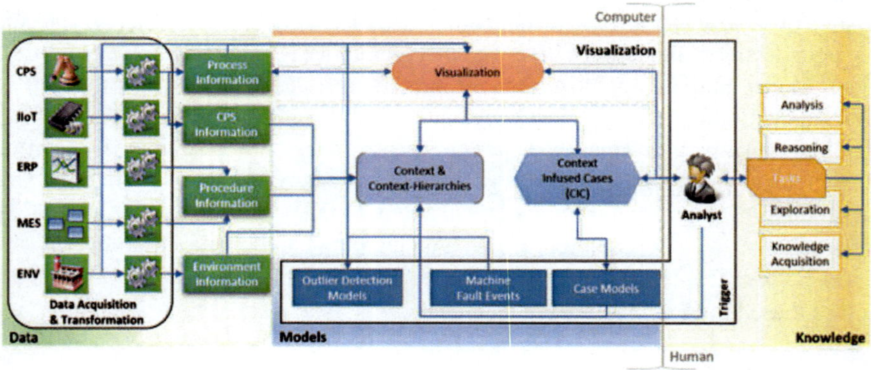

Fig. 1 Proposed contextualized TAOISM VA model for Smart Manufacturing embedded in the established general VA model [4]. The four main areas align with the general VA model, extended with a new meta-layer. (1) Data encompasses all accessible information within a SM-environment. Visualized information sources should be seen as an example and may vary. (2) Models comprise either active models (triggers) or passively build models (context-related). Context hierarchies represent layered aggregated contexts to describe inter context dependencies and relationships, with impact e.g. on the UI. Context Infused Cases (CIC) are generated by human, and stored and used actively again. (3) Visualization assists the users in their tasks and help to narrow down the underlying reason for an event. (4) Knowledge is where the users build hypotheses based on new insights, driven by their daily tasks. We also identified four main tasks within the production phase (concept of creation), which was presented by Zhou et al. [17]. (5) Finally, we extend the model with a cross-sector meta-layer (the trigger) used to combine all instances which can enforce a context creation build, such as machine entities (machine fault events, case models), observer models (outlier detection) or the user himself. Given instances are examples and may vary. Updated version of our previous revision [3]

a context-aware diagnosis. Starting at the data layer, which acquires and transforms the data and provides information. The models' layer observes the available information and triggers a context fetch (snapshot of the current situation) to construct *Context Infused Cases* (CICs). CICs are built with human perception by transforming presented information into a case with a description and a derived error model. Context hierarchies may be extracted at the beginning, but also continuously, to shape information aggregation in visualizations. For example, fuse extracted information automatically together visually and present them in one view if the information belongs to the same context hierarchy. Next, the visualization plays a crucial role in providing a meaningful representation of the data collected, supporting the tasks performed by the analyst, learning, exploring, analyzing, and reasoning. Knowledge is the area where the insights thrive new hypotheses and vice-versa. We identified four tasks, knowledge acquisition, exploration, analysis, and reason, as part of the production phase, in the concept of creation, which was presented by Zhou et al. [17]. We use Munzner's Framework [15] to characterize our TAOISM VA model. A smart factory is a complex environment that yields vast amounts of data [1, 2, 37]. As a matter of fact, it is impossible to capture all data in real-time concerning, e.g., low-bus speeds, where the actual process instructions and safety features have to be delivered as well. For this reason, we employ the asynchronous process of context creation. The process of context creation fetches the data in question one by one accordingly to a defined time frame without overloading the system. As a result, the analyst can reason over the adjacent current values and the current situation. We introduce the cross-sector meta-layer (trigger), where all instances are combined that can start the context creation process. Before introducing each layer in detail, we name the involved entities, state the triggers, describe the human part in the equation, and provide an example use case of error inference with our TAOISM VA model. Our TAOISM VA Model is based upon knowledge found through our various preliminary studies [2, 38].

3.1 Involved Entities

Multiple entities are involved in an SM-environment; tasks as error reasoning or predictive maintenance are complicated to fulfill in such an environment. Through the mass amount of log data [2] it is hard to gain insights and find the underlying fault for an occurring error. Schriegel et al. [22, 39] have analyzed the standard industry 4.0 architecture with its entities on a technical level. We align with their actor definition but use the bundle of sensors, PLCs, and SCADA-Systems as a synonym for a CPS. We also add the human as an equal-worth entity and add IIoT as a synonym for small cheap devices or device bundles within the smart factory (see Fig. 2). Our definition of actors is as follows:

- **CPS**. Cyber-physical systems are the source of process information. *Process Information* summarizes all information about the working process or the information

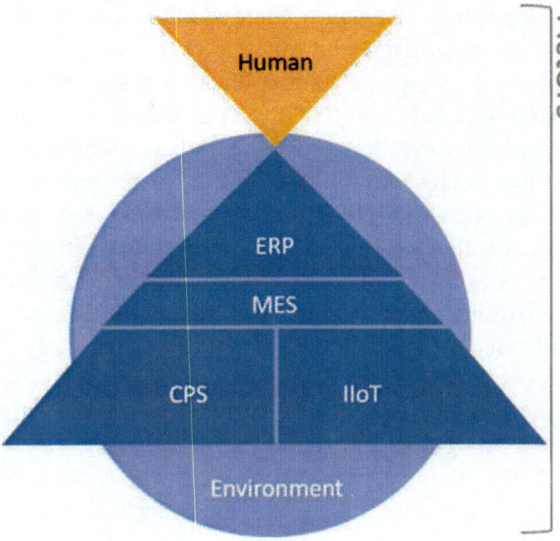

Fig. 2 Our actor definition for SM, redefined and adapted from the automation pyramid [22, 39]

which each machinery provides about the manufacturing process itself. This information may also thrive information about the health status of the machinery and its components or not (considered here as *CPS information*).

- **IIoT**. The Industrial Internet of Things encompasses devices that enhance current non-smart machines with the capabilities of a CPS–alternatively, used as pre- and post-processing units to gather additional information about the production process and its CPSs. As well as the CPS, the IIoT devices deliver *Process Information* or *CPS information.*
- **ERP**. Enterprise Resource Planning software contains information about the overall production capacity and provides higher information in terms of planned output, real output, estimated revenue, etc. This information becomes valuable in reasoning, where reasoning is implied to visualize a first estimation of the failure costs. The ERP system provides *Procedure Information.*
- **MES**. Manufacturing Execution System holds information about the overall manufacturing process. Which product is manufactured at which machinery and where the process will be continued, what steps are left for the finished product. This information is useful because of the visible chain of commands (manufacturing protocol) and the possibility of backtracking the relationships. Consequently, the MES also provides *Procedure Information.*
- **Environment**. The environment of the manufacturing process itself is not getting much attention from the research community yet. Nevertheless, in some cases, even in a smart factory, parts of the process chain maybe fail. The injection modeling process is vulnerable to environmental influences, and they are recognized as uncontrollable factors [40–42]. Such information can be gathered through environmental sensors, providing *Environment Information.*

- **Human**. The personnel involved in the manufacturing process, such as operation, maintenance, repair, or monitoring. The human provides information about certain situations, such as conditions, concerns, or upcoming faults. With this situation awareness, the human ties the loosely coupled ends of information provided through the visualization.

Each of the involved entities provides information that is useful for a context-aware diagnosis. The sources of information may vary from the technical side from factory to factory and should be taken as an extensible list, with a bottom line for discussion.

3.2 *Triggers*

Triggers in the cross-sector meta-layer observe the provided information automatically or manually and start the context creation process if the observations are subject to suspicion. We define four triggers:

- **Outlier Detection**. The unsupervised outlier detection, which consists of statistical and machine learning methods, will give a hint on abnormal behavior through an event as previously proposed [2]. These models can be re-trained either by reinforcement learning or by annual, monthly, or weekly wither manual or automatic updates to cover the latest developments within the surveyed CPSs.
- **Machine Fault Events**. In critical cases, the machine itself will throw an event on faulty components.
- **Case Models**. A case contains known issues enriched with context information and a pre-built error model that activates the trigger for observation.
- **Human**. In rare cases, where none of the automatic triggers will be activated, the human can force a context creation process if the observations look suspicious from his perspective. A typical situation is that quality assessment reported faulty products while each system operates normally.

Each context creation process ends in a derived view of the current situation presented to the user. The visualization in multiple views presents the aggregated interconnected blended information in an understandable sustainable manner that assists the personnel in the situation.

3.3 *The Human as Valuable Source of Expertise*

The role of the human is vital and considered as human-interactive systems within cyber-physical systems [43]. Therefore, the concepts in SM can be combined with Visual Analytics, where the human is also indispensable [4–9, 44]. The human interprets the presented information and draws a conclusion towards the goal of a task, e.g., take error inference (analysis task). The presented information encompasses basic process information, paired with procedure information, CPS information, and enriched with higher information from ERP- and MES-systems. Furthermore, the presented information is used to create a context-infused case (CIC). We define a CIC as a case of abnormal behavior that contains the annotated context information and where the human imposes an error model through, e.g., a rule-based approach to trigger the case. It differs from case models introduced earlier. The CIC is based on a formerly unknown issue and incorporates more information about the state (context of the production process and a human-validated error model). In comparison, case models are initialized with known machinery problems that operate context-independent referring maintenance schedules or CPS documentation. These simple cases are supplied at the installation step of the machinery. Therefore, the CICs are the contextual opposite. As a result, CICs can be trained and learned in production, compromising information relationships between on-site machinery and interactions to fuel the error model. In addition, these CICs are composable, learnable, and transferable to other production systems, enabling a smarter start of similar production lines. As error inference is only one analysis task, we also identified knowledge acquisition, exploration, and reasoning as primary tasks within a SM production setup. For Munzner [15] a task can be separated into an action and a target (goal). Countless combinations exist in Munzner's Framework, e.g., between the analysis of a current situation and the investigation of all data, attributes, network data, or spatial data using a defined search or query. The integration of Munzner's framework enables our model to be extensible for new tasks in the future. We emphasize the current task list as a first draft and may subject to change in further research.

3.4 Use-Case: Contextual Faults in the Smart-Factory

Together with our previous publication, we describe the general error inference procedure with our TAOISM VA model [3]. The procedure can be summarized as follows:

1. A suspicious observation triggers the asynchronous context creation process.
2. The context creation process gathers the data from all sources and presents the contextual information around the observation.
3. An analyst is now given access to the computed visualizations.
4. Exploratory search is assisted using self-search, annotation, and boundary employment.
5. Through the connection of multiple annotations, a custom error model can be built, and the model is saved as a CIC. CICs are composable, can be altered and enriched with further information, such as similarity or correlation measurements, to build a better fitting error model.

In order to concertize our work and have a test scenario for our TAOISM VA model, we recently presented the CONTEXT dataset [37], which contains contextual faults in a smart factory. Contextual faults are a new class of faults that need contextual knowledge to reason the fault's origin. A contextual fault cannot be reasoned with a single incident or simple rule and may contain multiple observations from different CPSs until the reason for the issue is obtained. In our CONTEXT dataset, we describe the following faults as a test scenario for our TAOISM VA model:

- *Missing Pressure*. The pressure system will begin to fail. Some CPSs in the smart factory slow down in production, unnoticed from all employed safety and maintenance functions. If the contextual fault can be recognized, a production slowdown and possible downtime (worst-case scenario) can be prevented.
- *Shuttle Dropout*. A monorail shuttle system delivers the product between stations. Under normal operation, two shuttles are employed. If a shuttle drop out happened, the shuttle stays, e.g., in the manual inspection bay, and the production output is cut in half. The production line operates normally, and as such, no fault is thrown. Prevention of this contextual fault keeps the production output on a high level and mitigates losses.
- *Missing Parts*. During assembly, the product starts missing parts, e.g., the part that fell of the shuttle. The robot's assembly routine works normally. As such, the robot assembles nothing, leading to a faulty product during quality inspection. If the contextual fault is hindered, the faulty tagged parts could be reused in the production process. Consequently, minimize the waste, environmental impact and at the same time reduce the costs.

We also employed sensing units (IIoT) to gather additional information about the production process to describe the faults in more detail. In order to succeed in detecting the aforementioned contextual faults, good outlier detection is critical to detect unforeseen incidents. On top of that, it also needs some intelligent form of information aggregation to allow the user to differentiate unusual internal behavior from a regular operation and not overwhelm the user with information. The first is achieved with context-based prediction and outlier detection, and the latter is done with context hierarchies. Both will be part of the reference implementation of the TAOISM VA model.

4 Data

Many sources for data exist for Smart Factories (see Sect. 3.1). Schriegel et al. [22, 39] structure the entities within an automation pyramid from a few instances such as ERP to many instances such as sensors. Applied to our previous definition, we change the pyramid to ERP, MES, CPS, IIoT, and at the bottom sensors (see Fig. 2). Starting at the bottom line of the pyramid, the following section outlines the available data types in smart factoring environments. Our CONTEXT dataset [1] describes all available data types briefly. The dataset contains numeric and categorical values and includes data structures like arrays (see Table 1). We also describe the testbed in detail [1]. The Smart Factory consists of high-bay storage, a six-axis robot for assembly, a pneumatic press, an inspection unit proofing optical and weight parameters,

Table 1 Datatypes of all available 33.089 OPC UA variables with examples [1]

Data type	Node count	Example values
Boolean	17874	True, "[False,]"
Byte	4918	11
ByteString	406	b'\ xff...'
DateTime	5	2020-07-09 16:05:11.795000
Double	1	0.0
Float	1248	3.1233999729156494
Int16	3003	4
Int32	1822	16
Int64	6	2103635700381566
SByte	34	"[32, 32,..]"
String	93	V3.0, "['+40,,', '', '', '', '']"
UInt16	2655	2
UInt32	1024	3

and an electrical inspection unit, and a shuttle system that interconnects everything. That testbed produces complex log data that is hard to analyze, and the published CONTEXT dataset consists of *Process Information*, *CPS Information* and *Environment Information*. We conduct a study with more system data included, such as data from ERP, MES to also include *Procedure Information*. The following table shows all available datatypes throughout the OPC UA models with example values (see Table 1).

The examples contain numeric and categorical values and include data structures like arrays. This further increases the complexity of possible values and makes a transformation step mandatory (as shown in Fig. 1). An additional layer of complexity is that OPC UA has 25 data types [45] that can be arranged in arrays, structures, and unions (see Table 1), which can also be extended in the future. Each CPS is typically bundled with an OPC UA model, which is shipped with the machinery itself. The model is a list of references, which hold information about the integrated sensors and routines. Each reference can be subscribed to in order to retrieve changing values. A PLC as part of a CPS, for example, has access to multiple sensors and has software routines, which can also emit messages. In CONTEXT we provide also access to *Environmental Information* through our developed sensing units (see Table 2).

In order to handle such a variety of data types, we already published some transformation steps [2, 38]. In order to access the information, we follow these steps:

- Parse incoming data into a standardized format
- Use complex event processing (CEP) to infer higher-level events
- Transform events and values to specific formats for the used algorithms

Especially, the last step is mandatory in order to utilize multiple algorithms, from rule-based approaches to neural networks. In our study [2] we had to transform

Table 2 Datatypes of sensing units with examples [1]

Columns	Data type	Example values
Datetime	DateTime	2020/07/01 16:15:40:647
TCAp, sGRP	Integer	7, 0
TSL_IR, TSL_Full, TSL_Vis, TSL_LUX, BMP_TempC, BMP_Pa, BMP_AltM, MPU_AccelXMss, MPU_AccelYMss, MPU_AccelZMss, MPU_GyroXRads, MPU_GyroYRads, MPU_GyroZRads, MPU_MagXuT, MPU_MagYuT, MPU_MagZuT, MPU_TempC	Float	41.00, 256.00, 215.00, 29.47, 27.57, 99359.20, 87.82, 5.72, −6.28, −5.15, −0.00, −0.01, 0.00, 21.58, 29.34, −28.94, 30.40

incoming data to a numerical format. Furthermore, the challenge was to maintain the data characteristics within the new format. Multiple transformation strategies are necessary for multiple algorithms that differ based on the input types of the involved algorithms. As an example, we proposed the following strategy to convert log events to a numerical format in order to use the data with an autoencoder neural network [2]:

- *Temporal Values*. We use temporal values to set the order of the sequence within the event stream. After sorting, the timestamps get deleted to reduce the feature space.
- *Categorical Values*. We vectorize all categorical values using one-hot encoding [46]. The encoding process enlarges the spatial size of each logline for each categorical value found in the log data. For log lines with a high number of categorical values, another encoding should be chosen, such as a lower-dimensional target embedding [47].
- *Numerical Values*. All numerical values get scaled between zero and one, so scaled values will not collide with the vectorized categorical values and achieve balanced inweights between categorical and numerical values.
- *Messages*. Each message gets automatically generated by the software to ensure human readability and interpretability of each logline. For the deduplication of information, we prune the messages.

All strategies have in common are that a transformation step needs to maintain the characteristics of a dataset. The disjoint communication between systems (different bus systems, parallel communication) is another major issue in SM. Therefore, also a strategy is mandatory to align the various pooled information to find outliers validly. We proposed a nïave strategy [1] to align communication by previously known activity windows of the CPSs in the smart factory. A more general approach would be the use of dynamic time warping (DTW) [48] to compute distances between activities

Algorithm 1: Determine order of CPSs based on data activity

```
Data:
windowed_datastream ws = W^{DxT}
model_names mn = getDistinctOPCUAModelNames()
acivities[mn] a = countSignalsPerModelInWindows(mn, ws)
peak_signals ps = zScoreRunningAlgorithmForPeakDetection[49](a)
orderdedModels o[mn] = Array()
possibleOrders o_List = Array()
Result: N-gram model with the probabilistic order of CPSs

while p in ps do
  if p[mn] equals 1 then
    if o is empty then
        o = 0
    end if
    new_postion i = max(o) + 1
    if i > len(mn) then
      /* all CPSs have a position, retrieve the order CPS1->CPS2->CPS3 */
      o_List += getOrderOfModelsAsString(o)
      buildRunningNGramModel(o_List)
    end if
    if not model_name n in o then
      o[n] = i
    else
      prev_val s = o[n]
      o[n] = max(o)
      while model_name n2 in mn do
        if n not equal n2 then
          if o[n2] > s then
              o[n2] = o[n2] - 1
          end if
      end while
    end if
  end if
end while
```

Fig. 3 CPS order algorithm

and align the time-series correspondingly. For that reason, our TAOISM VA model adds a transformation step (Fig. 1) for each data source. In order to automatically compute context hierarchies, information about the relationships and dependencies between the involved CPSs is needed. Therefore, we propose the following algorithm to determine order within CPSs automatically through OPC UA model activity in the OPC UA communication (see Fig. 3).

The algorithm describes the extraction of the order of the stations in the running smart factory. As a precondition, access to the OPC UA models needs to be provided, as well as the communication within the smart factory. Assuming that access is given, the stream W^{DxT} (matrix W with column length D with multiple rows over time T) is windowed. Furthermore, in each window, the activity of each model mn will be summarized to get the activities per window per model. Now, it is possible to use the z-score peak detection algorithm [49] to transform the activity signals into digital signals between $-1...+1$. The reduced peak signals can now be used to determine the order of the incoming signals and are used to order the CPSs. Over a row with all models $o[mn]$, each station gets positioned using numbers, starting from 0 to $len(mn)$, with every new peak, the model in the row gets a numbered position and gets placed at the end of the chain. If all stations received a position, the row is saved as a order string, each additional peak will reassemble the order of the complete row, and a new order string is computed. After a while, a corpus of order strings is built that is converted into an N-gram model. Furthermore, the N-gram model is now a probabilistic representation of the reduced data activity and CPS order in the manufacturing system. As a result, the N-gram model can be given a model and predict the most likely next model in data activity. Additionally, the input and output can now again be used to predict the next station, so on and so forth. In the end, the most likely order of CPSs is given in the SM environment. Through the automatically analyzed data activity, semantic relationships and dependencies between employed models are unveiled that beneficially can be exploited programmatically. Consequently, the new information is used to construct context hierarchies and help to set values, events, and incidents into perspective.

One limitation of the proposed approach is the variance in data activity, which leads to a wrong order in involved CPSs if the observation time of the data activity is too low. The longer the observation period, the better the automatically detected order

$$P = \{t \in T_O\}, T_O \subseteq T_A \quad (1)$$

$$W_n = \{t \in T_A \mid p \in P \wedge x \in X \wedge p_n - x \leq t \leq p_n + x\}, \\ W_n \subseteq W, n \in [1, |P|] \subseteq \mathbb{N} \quad (2)$$

$$C_n = \{pi_m, ci_m, pri_m, e_m \mid pi \in PI \wedge ci \in CI \wedge pri \in PRI \wedge e \in Env \wedge \\ \min(W_n) \leq m \leq \max(W_n)\}, C_n \subseteq C \quad (3)$$

will become. Another limitation is that our algorithm assumes that the production line is cyclic. If a station works in parallel, the computed order will always toggle the position of the stations in parallel. Lastly, a limitation exists because the algorithm assumes that all given models are involved production CPSs. As a result, models, e.g., from the management station (not directly involved in the production process), will toggle throughout the order because they are always active and, e.g., retrieve the result of each production step. A countermeasure against the limitations is the retraction of the toggling CPSs and presenting the order to the personnel to manually order the CPSs in question.

Concluding this section, we provided an overview of the information sources, their types and gave examples in the facilitation of our CONTEXT dataset [1]. Additionally, we presented a set of data transformation strategies from previous studies [2, 38]. As a quintessence, the data has to be transformed per algorithm that usage is planned. Lastly, we propose a novel order algorithm to extract semantic relationships and dependencies between CPSs. The algorithm provides novel information that is used to construct context hierarchies (Sect. 5). Furthermore, the complexity of the presented information adds additional requirements towards visualizations, in terms of complexity reduction and information highlighting, where context hierarchies can be employed, e.g., to automatically aggregate information on each hierarchy layer.

5 Models

Our TAOISM VA model consists of five models: outlier detection (Eq. (4)), machine fault events (Eq. (5)), case models (Eq. (7)), context and context hierarchies models (Eq. (3)) and Context Infused Cases (CIC, Eq. (8)). As denoted, the named models should be taken as an example and are subject to further research. Additional models may be added in the future. The following section outlines the formal definition of each of our employed models. In addition, the formal definition helps to differentiate between the individual models and enables models, algorithms, and methods that have already been published to be classified, assigned, or segregated to our models.

We differentiate two types of models either active models, which cause a context creation, or passive models, which will be built automatically or with user interaction. Active models that actively trigger a context creation are the outlier detection models, machine fault events, and case models.

The outlier detection models (OD) are a composition of different algorithms (A), classical approaches such as ARIMA [50] or neural networks such as autoencoder [51]. These methods can also be applied in ensembles to cover the weakness of one algorithm with the strength of another [52, 53].

$$OD = \{a_1, a_2, ..., a_n \mid a \in A\} \tag{4}$$

Nowadays, most machinery has some sort of report system that automatically reports faulty components or throws events on incoming issues. Consequently, the machine fault events (MF) consist of multiple events (E) with a mapping function ($F(CO)$: $(co_1, .., co_n) \rightarrow E$), where components ($CO$) trigger the events directly.

$$MF = \{(e_1, f_1), (e_2, f_2), .., (e_n, f_n)) \mid e \in E \wedge f \in F\} \tag{5}$$

The machine fault event model cover those trivial cases and help the professional by starting the fetch of context related data for the visualization after an event is caught.

The case models (CM) build the bridge between the initial setup and the operation phase of the CPS. Meanwhile, the setup of the CPS standard cases (see Eq. (7)) that were known upfront are employed in the case models. A basic case (CA) consists of an Error Model (EM) and a description (D).

$$CA = \{D, EM\} \tag{6}$$

$$CM = \{ca_1, ca_2, .., ca_n \mid ca \in CA\} \tag{7}$$

Those cases can encompass rule-based approaches that incorporate CPS-related logic. We already published a rule-based approach to fuse documentation and incoming machinery events [38].

The context model (C) (see Eq. (3)) fuses process information (PI), CPS information (CI), procedure information (PRI) and environment information (Env) together utilizing one or more windows (W). The fuse points (P) are the timestamps of the outliers (T_O), which are a subset of all available timestamps (T_A). A window (W) spans around an outlier timestamp (P) and configurable range (X). Furthermore, the earliest and latest timestamp will then be used to build the context and fetch the data within the interval. As a result, the context can be seen as a current snapshot or joint of the situation that will provide a fine-grained overview around a timestamp of a suspicious observation.

In our previous publication [3] we also proposed that a context can consists of multiple other contexts (see Eq. (3)) but do not specify the usage. Now, we clarify this point and adhere to our context hierarchies. Context hierarchies are a virtual construct independent from physical hard- and software of the involved machinery and systems to describe dependencies and relationships. Figure 4 exemplary shows three context levels, the machinery context, the prepost production context, and the production context. Furthermore, the three contexts should be seen as an extensible list rather than a hard border or requirement. Additionally, the shown contexts allow to define virtual areas of coherence where multiple machineries are grouped. The context focus enables a further abstraction of underlying possible data sources that mitigate complexity in the observation of different models at the same time. For instance, if only a specific part of the smart factory acts suspiciously, the context can be set around the involved entities. Besides the need for fewer data observations, contexts are virtual chambers, where faults can be analyzed on different layers with different granularity. For example, an outlier in the machinery layer cannot be referenced within the machinery context, but maybe together with other layers in the prepost production layer. A context can also be seen as a vertical zoom through the smart factory that can be shifted horizontally.

- *Machinery Context*. The Machinery or CPS context encompasses all data sources that belong to the CPS and involved foreign entities (e.g., IIoT). Additionally,

Fig. 4 Context hierarchies exemplary spanning three layers. Different layers exists from machinery context to subsequent process to production context

the CPS context can further be divided between software- and hardware-related data and accessibility. Software-related data is split among, e.g., an OPC UA model, which encompasses data the manufacturer pass to the outside. Internal software routines can also be a source of faults and mostly have their own logging mechanism. Hardware-related data are, e.g., sensor or actuator data, and their values are made available, e.g., through the OPC UA model. Nevertheless, internal sensors exist that do not pass any information to the outer world. These kinds of sensors are employed for redundancy and only activated if the others malfunction or are employed for internal surveillance to assure quality in operation. These sensors are an example of machinery hidden data that not even the monitoring systems of the smart factory retrieve. In order to mitigate the impact of hidden data on our analyses, we employed IIoT in our published recordings [1]. Therefore, the machinery context could also be referred to as a three-dimensional information sphere around each CPS which also encompasses, e.g., IIoT as an external signal transmitter.

- *Prepost production Context*. Similar to the CPS context, the pre and post production context form a three-dimensional information sphere around the working station and also includes the forthcoming and upcoming process. Therefore, the context encloses the CPS context of each machinery and the additional communication.
- *Production Context*. Another context layer is the production context. The final layer includes all CPS contexts, the communication, and external systems. Furthermore, we define the external system as systems that are not physically involved in the production process. Such systems are the enterprise resource planning (ERP), the manufacturing execution system (MES), and the environment. In ERP, the planning and commercial evaluation is done, and the system contains high-level information (e.g., planned output, factory schedule, cost factors). Additionally, the MES collects all information about the current factory job (e.g., the plan, the status, fuse information [e.g., CPS, communication, safety]). In the end, the environment has parameters about the environmental factors (e.g., humidity, light, pressure).

A fault may appear between and within a context, and multiple contexts may need to reason over the fault (Fig. 4). In order to build the context hierarchies, information about the positioning of the CPSs and their interaction is needed. Either the information is added manually, e.g., by the personnel, or as we proposed, automatically using data activity (Sect. 4). Is the chain of interaction and the order of the CPSs known clusters are build top-down. Furthermore, CPSs are the starting point, and each successor context is broadened by one upcoming and forthcoming process until the context hierarchy encompasses all CPSs and systems. Therefore, we correspondingly update our CIC definition. A CIC is a combination of both the context and context hierarchies (C_n) and the case (CA), with additional historical information (H):

$$CIC = \{C_n, CA, H\} \tag{8}$$

The process information is part of the context (C_n), the analyst (Fig. 1) employs domain knowledge within the exploration task to find patterns that can be connected to cover the case and build an error model. Additionally, it is possible to add historical data to strengthen the error model and refine the trigger. After everything is in place, the CIC is saved to the case models as an additional extended case. The now annotated information, current and historical data is also used for training of the different outlier detection algorithms, e.g., also to cover noisy cases.

6 Visualization

We integrate multiple layers in our TAOISM VA model (Fig. 1) that affect the visualization, e.g., data or models. Each layer has several implications that result in requirements for the visualizations. The following section outlines the requirements for visualizations that intend to use our TAOISM VA model. Additionally, we add the first draft of three user interfaces (overview, configuration, analysis), which arose from the given requirements. These drafts depend on each other and are ordered from overview to detail to provide a broad first hint about the impact between the relationships of the requirements. The key requirements that emerge from our model are:

- **Unified System Integration (1)**. Through the extendable automated transformation process, we provide a way to interconnect and integrate new system components (e.g., sensors). The visualization must be extendable and universal to adapt to new elements and provide an overview of the involved systems (CPSs, Env, ERP, MES, etc.).
- **Configurability (2)**. Our model is configurable. We put the analyst in charge of changing the transformation process and the models (outlier detection, case, machine events). Additionally, the analyst can limit the information provided by the different visualizations. Consequently, the visualization has to allow a modular configuration of each component within the model and provide an ability to slice the granularity of information.
- **Surveillance of Mass Information (3)**. We do not hide information in our model. We employ ways to provide the analyst with distilled information with details on demand. As a result, the visualization has to implement intelligent aggregation strategies to cope with the vast amount of information.
- **Highlight important Information (4)**. Our visualizations highlight outliers and use context information to project and compose different data sources to provide deeper insights. Therefore, the visualization must contain several ways to highlight information without overwhelming and interrupting human perception.
- **Context Creation and User Integration (5)**. Upon event notification, the build context information is automatically provided and visualized. The analyst can compose problem-related data, enrich the data by rules, and infuse historical data to enhance old cases or create new cases. For this reason, the visualization has to allow the user to mark important information and chain the found information and multiple observations together to generate new error models.

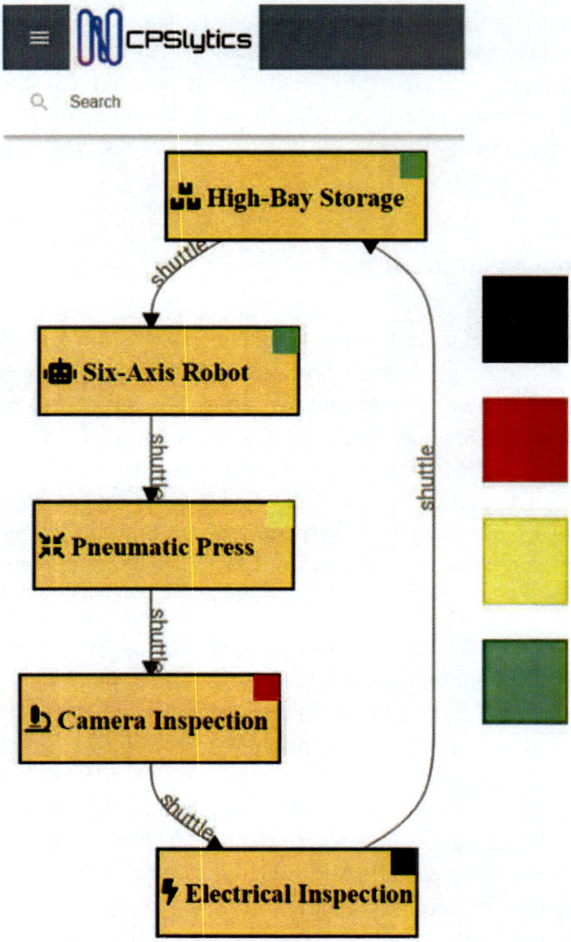

Fig. 5 Generated overview based on the incoming Information, validated using profound OPC UA models (left). Additionally, the current severity levels (right) [3]

The first overview draft (Fig. 5) shows a fusion of different hard- and software models and is part of the unified system integration (1). A SM production line consists of different CPSs and data sources (Sect. 4). We can acquire many data through OPC UA and other manufacturing protocols. OPC UA machine models provide information about all the available data, from process data to a single sensor. Each available model is used to automatically generate an overview of the production line (Fig. 5). The draft in Fig. 5 shows the output of the generation process. Currently, we are in a transition between loosely coupled interconnected production plants towards the fully automated smart factory. In this hybrid state, there is also the need to add production plants manually to the shown graph in Fig. 5. For this reason, the overview is editable.

The generation process of the overview draft is started after one or more OPC UA endpoints are added through the UI or automatically found due to used standard

OPC UA ports (spared to save space). Each found model is drawn as an orange square containing its OPC UA namespace to cover its name. An ontology is used to find an icon fitting the name for each model. The icons are interchangeable and are in place in favor of distinctness and perception. An icon together with a name is more recognizable as a single string. In addition, the visual layout supports the perception of the process sequence. The more advanced part of the visualization is the automatic information flow annotation. The algorithm, as we propose in Sect. 4 is used to annotate the information flow (arrow direction). Every automatism has its flaws so that the user can change the automatically build graph afterward. Each visualized node has an additional field in the right upper corner. Here, the current health status of the plant is visualized. The health status of the machine aggregates the different information sources to their highest level. That square creates a space for a visual placement of the results through a context algorithm, such as the situation value formula invented by Zhou et al. [10]. In our model, the status contains information about current machine load, process flow, outlier models and prediction models, case models, and machine fault events. Each event occurs in different severity levels, visualized from green (everything ok) to black (faulty machine).

Figure 5 shows also the different severity levels. Thereby, green indicates a normal running system, yellow stands for the most severe situation, and red is completely wrong if nothing is changed. The scale ends with black if the production line is forced to pause and the worst case situation has occurred, e.g., a machine fault. The elements within the overview are movable to ensure that it is adjustable to visually map the outline of the production line into the dashboard.

Next, we propose a visualization draft for the configurability (2) of the different transformation steps and the configuration of different algorithms. We name our approach visual orchestration of methods because we align different transformation steps with parametrization of different data models (outlier detection, case, machine fault events). Figure 6 shows the configuration dashboard. The view is split into two areas, the available nodes, and the configuration view. Each available node can be used to create a configuration graph. All found sensors or different layers of machine abstraction delivered through OPC UA can be selected in a source node. The views (Figs. 5 and 6) depend heavily on each other. Every time a node is added on the overview page or the configuration layer, the models are updated. After a source node is added, one of the information sources Sect. 4 can be selected. Multiple nodes can be joined together within a transformation node. A menu emerges on a node hover (Fig. 6) to visualize the output of the given node, e.g., the transformation step, to either inspect or explore the transformed data. As shown, different visualization techniques are available to the user (the selected node is colored orange). A click on edit enables the analyst to tweak the parameters for actual transformation steps.

It is shown that, e.g., each outlier detection model has its node, where again the parameters of the given model can be tweaked through the UI and the domain expert. Each node is mapped directly to an output node, which results in a new visualization in the downstream visualization views of our model. Multiple outputs of different transformation steps or models can be joined to operate, e.g., on a preprocessed data stream. Some used algorithms may process massive amounts of data slower, which

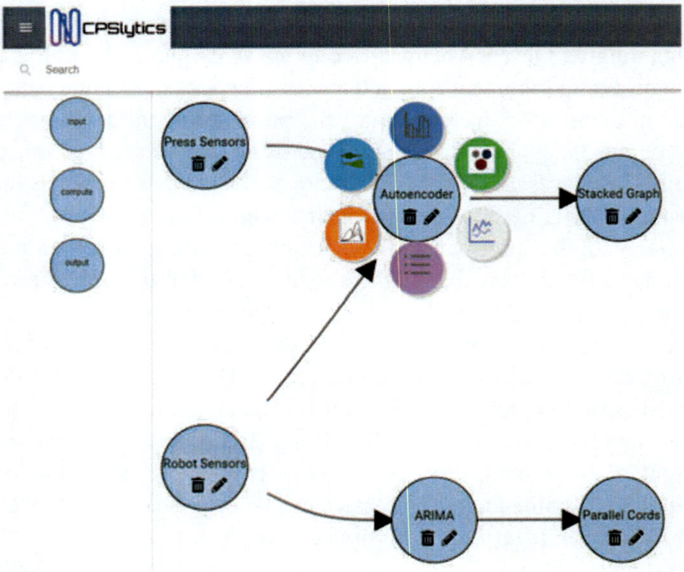

Fig. 6 Visual orchestration of methods: Dashboard configuration view [3]

would delay the analysis. A reduced data stream resolves such a bottleneck. Each configured pipeline can be stored by a name. This enables the analysis of different user-created pipelines. Clustering between simple and advanced users or domain experts is examined in a future study. The study will encompass if AI models can assist the basic user in building an optimal pipeline for a given problem. As a result, automated prefabrication of a pipeline based on user feedback may be possible in the future.

Figure 7 shows our concept for information aggregation using context hierarchies (from production to machinery). The concept is embedded between the general overview (Fig. 5) and the detailed system performance overview (Fig. 8). Furthermore, the view serves the surveillance and analytical purpose and shows high-level aggregated information in a dense pixel visualization. An outlier is colored in red and fused with neighboring pixels to reflect a hot spot dynamic. Each cell contains aggregated information from the current day (past 24h) over a certain history H (e.g., one week). Each layer from machinery context (top) to production context (bottom) contains more dense contextual information. Vice-versa, the information gets more aggregated from the production context layer towards the machinery context layer. The status square in the general overview (see Fig. 5) is the final aggregation that reflects the current context status. As a result, the value of the context algorithm gets normalized to match one of the mentioned severity levels. Currently, there are open points for discussion concerning the computational model (behind the aggregation), the visual alignment and visual aggregation of neighboring contexts, the integration of underlying context (e.g., bottom to mid), and the chosen information signals

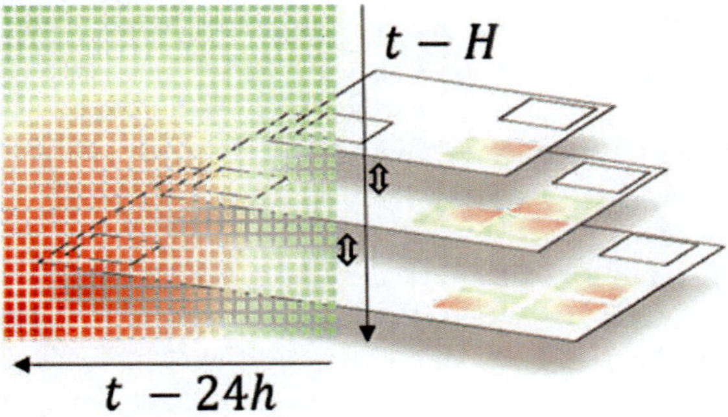

Fig. 7 Concept of information aggregation with context hierarchies using dense pixel visualization

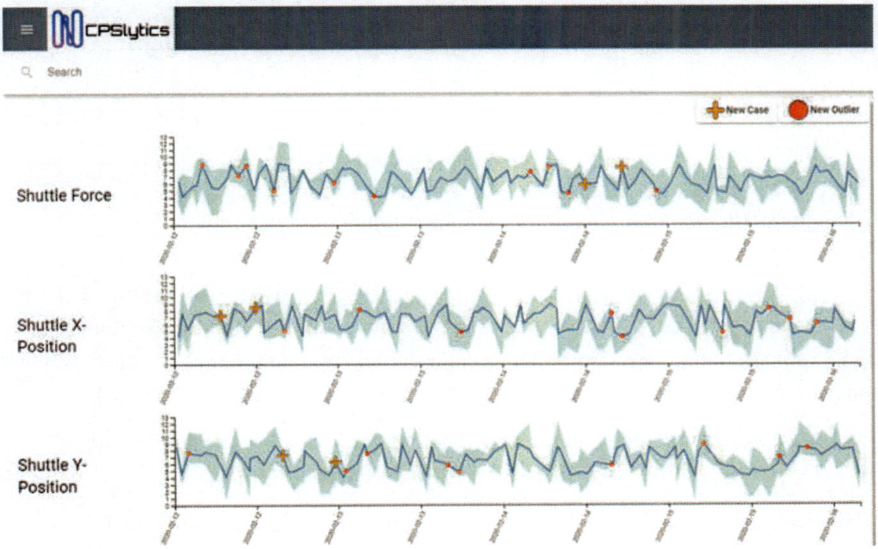

Fig. 8 Distilled system performance zoomed overview backed by predictions with marked outliers and triggered cases [3]

that get aggregated. We conduct two studies concerning the automatic extraction of production-relevant variables and a visualization-focused publication that considers and clarifies the open points.

The last visualization draft is the system performance overview (Fig. 8), which serves as the gateway to observe mass information (3), highlight important information (4), and be the port for context creation and user interaction (5). Each output element of the previous visualization layer is given its visualization, where the output is visualized, e.g., as a graph. Hereby, the visualization is based on both rules and prediction.

The estimation of the current trend or value range is visualized as bright blue, and for each point, a min and max value is shown, leading to an advanced river chart.

Circles represent reported outliers, and the color reflects its severity or probability, e.g., in the case of a neural network. Cases that are triggered are visualized with a plus, again colored to their severity. Furthermore, the context is visualized with a click on either a case or an outlier. On the downstream view (ditched in favor of the other views), the analyst can explore the underlying context and mark suspicious observations. For this reason, the analyst can mark areas on the graph with a plus for a case or a circle to explicitly save the area for the training process of one of the employed supervised outlier detection algorithms. Numerous pluses and circles can be connected to create fuzzy rule chains to trigger the case. In addition, the analyst describes each case and sets a proper name. All data, including the context, is stored in a database as an annotated corpus for the unsupervised outlier detection algorithms. The database leads to an adaptable system that may become better over time to detect different quirks of the CPS. As a result of centralized storage, the database, with stored contexts and cases, can be interchanged with other production lines with similar machines. In return, a database migration may provide a way to integrate a new production line in the future.

7 Knowledge

The term knowledge is used ambiguously. Therefore we define this term in this section. We use Munzner's work [15] as the foundation for formalizing and characterizing tasks. Munzner formalizes tasks as action and target [15]. Accordingly, we classify for our model within SM a task as an action (A) and a target (Tr):

$$T = (A, Tr) \tag{9}$$

These actions are used to analyze, search or query data. This can be done for all data, attributes, network, or spatial data to achieve different goals such as trend or outlier detection. The formal definition (Eq. (9)) may lead to a fine-grained task definition in future. In the meanwhile, we use the work of Zhou et al. [10] on visualizations in SM to illustrate the most widespread tasks and include them in our model. For this, we identified four main tasks for our model:

- **Knowledge Acquisition**. Knowledge acquisition is the task where the user gets familiar with the SM environment. It is mandatory to acquire an understanding of the complex CPSs to handle upcoming incidents. The task is essential for trainee programs, advisory, or recovery from a crisis to mitigate losses in domain knowledge through employee exchange. We assist the knowledge acquisition due to our prefabricated case models with descriptions and different levels of information granularity to assess the systems and their internals.
- **Exploration**. Exploration is defined as a task after the user gets familiar with the system through knowledge acquisition. The exploration process leads to new defined insights and correlations without a certain predefined task in the user's

mind. Therewith is a self-driven knowledge acquisition process without the main goal. Exploration is part of the day-to-day work in the form of surveillance and maintenance of the production line. The exploration process is mainly supported through our interactive visualizations.

- **Analysis**. Analysis is the process of investigating deeper insights, in particular, to detect a certain pattern or solve a complex analytical task, e.g., detecting and solving an outlier. We enable the analysts to investigate outliers and search the system for possible correlations. Outlier Detection and Predictive Maintenance are part of the analysis task. Already acquired knowledge is used to examine the system to search for possible reasons or causing effects.
- **Reasoning**. Finally, the reasoning task is the inference of a reason for a specific issue. It builds upon the acquired knowledge and vast exploration during an analysis. We support the analysts to reason outliers providing the tooling for recording, annotation, and interaction with the data and the process throughout the visualization. The CIC creation and a Fault Recovery is part of the reasoning task. Our proposed VA model provides the toolchain to put data into perspective.

The Knowledge Acquisition task is a prerequisite to the other task in order. CPSs are complex interconnected systems with networks and process schedules. Furthermore, the knowledge about the systems, system architecture, and common events has to be acquired upfront to use exploration, analysis, and reasoning. Our model provides the tools to assess the system simplified. In addition, the case models database provides the opportunity to browse through common, already known events. A trainee, professional, or analyst who wants to get familiar with the CPSs starts browsing the delivered data. They want to get to know the topology of the CPSs and the paths of the information flow. We introduced a general process overview (Fig. 5), the concept of information aggregation (Fig. 7) and the configuration dashboard (Fig. 6) to visualize the information flow in the system. In addition, information are delivered in different zoom levels (Figs. 7 and 8). The visualization based on our concept of information aggregation with context hierarchies (Fig. 7) allows the identification of unknown patterns through the hot spot identification. The aggregation can be detailed through the different layers of contexts in the hierarchy up to a side-by-side comparison between sensor values in the detail view (Fig. 8).

In the exploration task, the users are already familiar with the CPSs and the user interfaces. They want to discover more aspects of the production process. Meanwhile, the exploration tasks the analyst wants to discover new insights through exploration. The case models (Fig. 1) provide information in a centralized point. The search field (Fig. 5) helps to query the database of recent or historically known cases. If available, all cases are delivered with more information such as contexts (process information, procedure information, etc., and links to the context hierarchy). The provided information is visualized in already known views (detail view, Fig. 8) in order to keep the user's perception consistent. Now, after a basic understanding of the production process is established, attributes are additionally in focus. The analyst is able to survey different distortions, extremes or similarities (Fig. 8). The obtained knowledge is then utilized in the analysis task. An analysis of an outlier or a new fault is a complex task. Our model supports these tasks by providing the visualizations to annotate, record, and derive data from locating or identifying useful information in that scenario. In Sect. 6 we describe an analysis scenario. The analysts can review data and annotate outliers or cases to the found suspicious observations. Furthermore, the saved context and its values are transferred to the case models database. Different kinds of algorithms derive the data in order to observe, e.g., minima or maxima. Additionally, estimations are also visualized in the performance overview (Fig. 8). The analyst is also enabled to configure each algorithm for the analysis task (Fig. 6). Mature processes can be edited and altered to get extended and varied to match new conditions of historical events. As a result, the analyst is capable of the final task, the reasoning.

Besides the analysis tasks, the reasoning task involves more data. New insights and hypotheses have to be proven in order to get the underlying reason. Hawkins postulates a definition that an outlier is "an observation which deviates so much from other observations as to arouse suspicions that it was generated by a different mechanism" [54, p. 1]. The reasoning task tries to find these different mechanisms and provide a kind of rule that can be recognized–no matter whether it is just an outlier, a machine event, or a faulty component.

8 Context-Aware Diagnosis

We refer to diagnosis that includes the current situation, the context or context hierarchies, in the analysis as context-aware. Formerly we showed a context-based information aggregation concept. Now we elaborate on using context in algorithms to receive better results in the diagnosis process. One part of the process is the unsupervised outlier detection to start a context fetch and an analysis. For example, outlier detection uses, e.g., prediction to judge if an adjacency value is outlying by a variance measure between prediction and value. Here, context can beneficial used to better detect outliers through a better prediction on minor data using GNNs. A smart factory is a complex environment with complex dependencies, but for the chapter, we use a simple case from our CONTEXT dataset [1].

We show the benefits of context on a simple example using data from the sensor groups (sGRP=0) in our employed sensing units across the smart factory. Figure 9 exemplary shows the single acceleration value of the x-axis (of sensor group 0) of the robot with sixteen assemblies of an electronic relay. Furthermore, the data is taken from the subset *ground truth* of our CONTEXT dataset. For our test we train five deep learning models (feed forward, convolutional, long short-term memory, graph convolutional and graph transformer based neural networks). All networks are build using the same configuration (Fig. 11) and trained on all sensor values of sensor group 0 over all sensing units in parallel to predict 60 steps ahead. Therefore, we test a multi-input multi-output scenario. Consequently, our test differs from normal test scenarios where only a single time-series is learned. All models are trained with

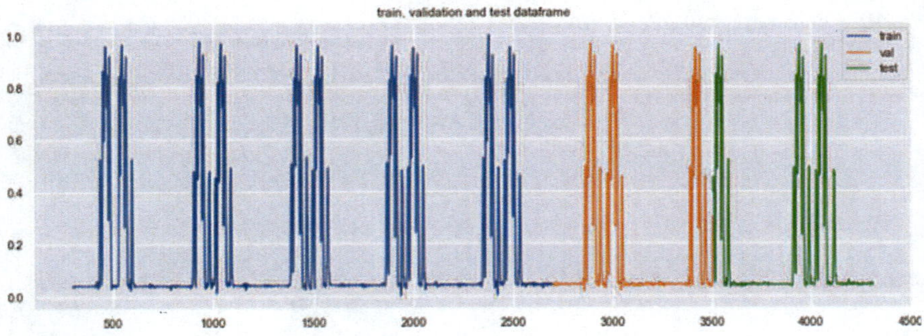

Fig. 9 Exemplary sensor value of robot in motion (sGRP=0, X-Axis) that shows sixteen assemblies of electronic relays

a polynomial decay learning rate (from 0.01 to 0.00001). We standardize (Eq. (10)) the data before training between range (r) 0 and 1 with:

$$\mathbf{x}_{std} = \left(\frac{\mathbf{x} - \mathbf{x}_{min}}{\mathbf{x}_{max} - \mathbf{x}_{min}}\right) \cdot (\mathbf{r}_{max} - \mathbf{r}_{min}) + \mathbf{r}_{min} \tag{10}$$

$$\widetilde{\mathbf{x}}_{sin/cos} = \begin{cases} \sin \\ \cos \end{cases} \left(\frac{2 \cdot \pi \cdot \mathbf{x}}{\max(\mathbf{x})}\right) \tag{11}$$

We additionally transform the timestamps using *sin-cos*-transformation (Eq. (11)) [55] and create two additional time-based periodic features that the models are able to learn. Afterward, the time series is resampled to one second. As a result, our dataset contains around 4.300 equidistant points for training. Additionally, we prepared a context graph that describes the different stations and their connectivity in the smart factory for our graph-based neural networks (Fig. 10). Moreover, the shuttle passes all stations and consequently has a connection to all nodes in the graph. Furthermore, the context graph can be built with our proposed algorithm (Fig. 3) either half-automatically with human-interaction or fully automatically if the mentioned limitations are solved in the future. Therefore, graph-based neural networks can use context native and have a theoretical advancement over standard neural networks (Fig. 11).

Now, the data is windowed and split among training, validation, and test data with a 60%, 20%, 20% ratio (Fig. 9). We train 100 epochs with a batch size of 32 on a Ryzen 9 3950X (16C/32T, 4.7 Ghz) with 32 GB DDR4-3600 RAM. Moreover, we elaborate on how good a neural network can become in the reconstruction of its inputs under the same constraints (architecture, learning rate, batch size) and time (epochs). The lower the reconstruction error, the better is the neural network reconstruction, the better the predictions will become. Table 3 shows the results of the trained neural network models on the training, the validation, and the test data. All networks differ in their computational complexity. Therefore, a simple feed-forward

Fig. 10 Context-graph of the smart factory

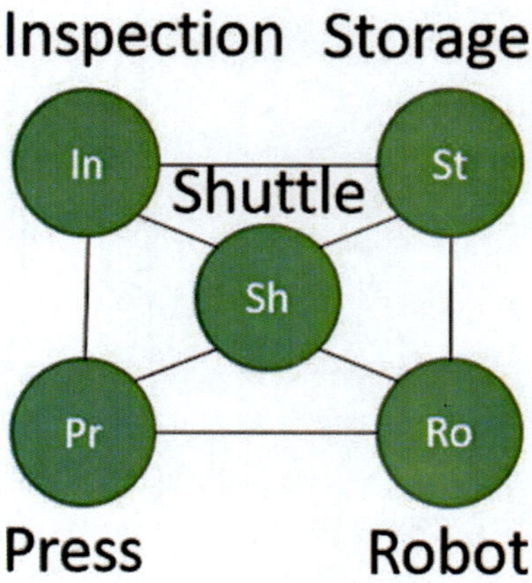

Table 3 Results. Mean absolute error (MAE) in reconstruction training, validation and test data. Lower is better. Best result per column is bold

Network	Training	Validation	Test	Training time (100 Epochs)
FF-NN	0.1042	0.1094	0.1186	**1:40 min**
C-NN	0.1039	0.1108	0.1186	15:00 min
LSTM-NN	0.0919	0.0993	0.0993	36:40 min
GC-NN	0.1448	0.1462	0.1671	3:20 min
GTrans-NN	**0.0610**	**0.0717**	**0.0899**	6:40 min

neural network has minor hyperparameters than a convolutional or LSTM network. As a result, the training time varies during the 100 epochs. The FF neural network is trained in 1:40 min, whereas the LSTM neural network needs 36:40 min to finalize 100 epochs. The more complex the neural network gets (more hyperparameters to tune), the better the actual reconstruction is and the lower the mean absolute error. The results begin at the Table 3 with the FF-NN, followed by the CNN and the LSTM-NN. Surprisingly, the graph convolutional neural network (GC-NN), with its complexity between CNN and LSTM, reaches the worst scores in our comparison, although it has the advantage of the context graph. Furthermore, the graph-based neural networks need minor training time for their more complex counterparts. Significantly, the GTrans-NN profit from the context graph and has the 3rd best training time while maintaining the lowest mean absolute error and get the best predictions. Figure 12 shows a chosen example of a single sensor value, with the input (blue), the actual path (green) as labels, and the predictions (orange) after training. Again for an explanation, the neural networks are trained on all sensor values at once. Nevertheless, the LSTM-NN learned the trend of a single sensor (top), whereas the GTrans-NN (bottom) has already learned the curves after the same time in training.

We are aware that with different network architectures, different parameters (e.g., learning rates, batch sizes), or a longer training time, the networks may have achieved

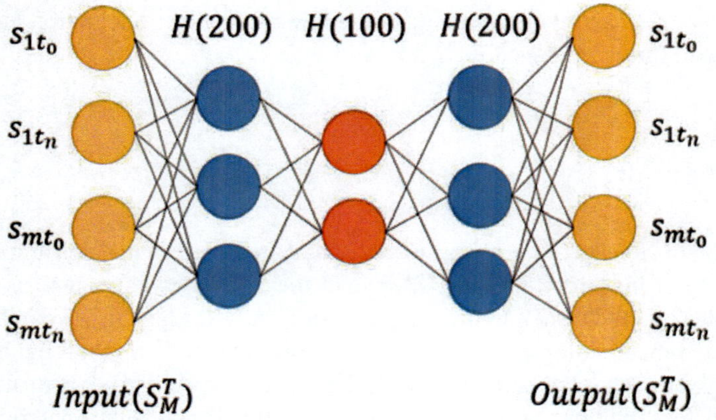

Fig. 11 Neural network configuration

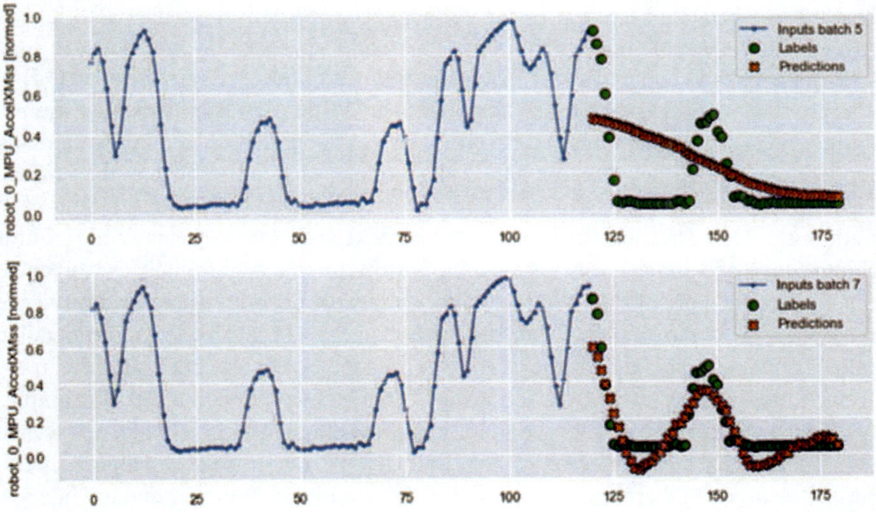

Fig. 12 LSTM-NN prediction results (top), GTrans-NN prediction results (bottom) on a random chosen sample

more accurate predictions. Nevertheless, an equal comparison needs a similar defined test scenario with the same parameters. Nevertheless, we could show that context-based prediction is possible with graph-based neural networks. Additionally, we were able to show that graph-based neural networks need minor training time and, in the case of the graph transformer neural network, allow for better predictions. Therefore, graph neural networks are one way to integrate our context for a better prediction while foremost receive better results on minor training time. Especially in the SM domain, where complexity and amount of data will rise in the future, graph-based neural networks can be one stepping stone towards context integration and should be evaluated further to unveil their full potential.

9 Conclusion

This chapter extended our previous work [3] and presented an updated and enhanced version of our industry 4.0-ready VA model for context-aware diagnosis in Smart Manufacturing (TAOISM). In order to form our TAOISM model, we combined and summarized methodologies, algorithms, and specifications. We thoroughly defined and explained each section of our TAOISM model. We specified context and the process of context creation and gave examples of the usage in a Visual Analytics system. Additionally, we named possible information sources and models for context-aware diagnosis in Visual Analytics. Furthermore, we presented requirements for visualizations and provided the first set of visualizations. Moreover, studies are planned to evaluate the visual orchestration of the method's approach. We further identified four main tasks in context-aware diagnosis and used Munzner's Framework [15] to classify those. To the best of our knowledge, no Visual Analytics model or system has provided such a classification before. We refer to our TAOISM model as a living model, open for further research and extension to open opportunities in the future.

With this in mind, we have extended our previous work [3], both top-down and bottom-up. Our extension focused on the context and context usage in algorithms (bottom-up) and impacts on visualizations (top-down). Additionally, we updated our former theoretical data section with our newest accomplishments around the recordings of contextual faults in the smart factory [1] and added more detailed transformation strategies. Furthermore, we broaden our context definition with the former implicit, now an explicit, definition of context hierarchies backed by examples of our smart factory. Moreover, we added a first algorithm to automatically extract and construct context hierarchies using the activity in OPC UA models. Our algorithm can further be used to determine the position of the machinery in the production line and determine the order of the internal hard and software routines, which are represented as subscriptional OPC UA model attributes. This consequently allows a new level of context extraction in a smart factory that can be used in visualizations or algorithms. On top, we released a concept of context-based information aggregation based on context and context hierarchies for our upcoming smart factory visualization. At the same time, we elaborate our context and context hierarchies in context-based prediction with graph-based neural networks in a multi-input multi-output scenario, which significantly lower the reconstruction error, improve prediction, and at the same time, lower training time. We trained five deep learning models, a simple feedforward, a convolutional, a long-short term memory, and two graph-based neural networks for our elaboration. The graph-based transformer neural network profits the most from our provided context-graph and surpasses the tested traditional neural networks. Additionally, we are the first who use graph neural networks and context-based prediction on a production-equal smart factory dataset. Moreover, our results pinpoint the usefulness of graph-based neural networks for context usage in a smart factory environment, which we plan to evaluate further in the future.

We plan to evaluate our TAOISM model in a reference implementation. Consequently, we use our TAOISM model as a foundation and join our different experiments and implementation of the data, the models, the visualizations and use the tasks in the knowledge section as a mental model during the design phase. Additionally, we use the concept of context-based information aggregation in an upcoming visualization, as well as our novel evaluated context-based graph transformer-based neural networks in outlier detection. Both theoretical and application benefits are joined to form a reference implementation that is then tested against our already published CONTEXT dataset [1].

To conclude our work, we provided a novel Visual Analytics model for context-aware diagnosis in Smart Manufacturing. We proposed an updated revision of our TAOISM model based on our previous work [3]. Furthermore, we extended our model and showed the impacts of context and context hierarchies in different parts of our TAOISM model (visualization and context-based prediction), which will be further evaluated in a reference implementation. Nevertheless, first hints towards the usefulness of context and our VA model could be shown. Despite the reference implementation, our model should be seen as a living model open to future research.

Acknowledgements This work was conducted within the research group on Human-Computer Interaction and Visual Analytics (https://vis.h-da.de). The Research Center for Applied Informatics supported the presentation of this work. We want to thank Prof. Dr. Stephan Simons for the permission to use the learning smart factory of the Darmstadt University. We thank our student, Arash Javanmard, for the implementation of our context-based prediction and outlier detection.

References

1. Kaupp, L., Webert, H., Nazemi, K., Humm, B., Simons, S.: Context: an industry 4.0 dataset of contextual faults in a smart factory. Procedia Comput. Sci. **180**, 492–501 (2021)
2. Kaupp, L., Beez, U., Hülsmann, J., Humm, B.G.: Outlier detection in temporal spatial log data using autoencoder for industry 4.0. In: Macintyre, J., Iliadis, L., Maglogiannis, I., Jayne, C. (Eds.) Engineering Applications of Neural Networks, Series Communications in Computer and Information Science, vol. 1000, pp. 55–65. Springer, Cham (2019)
3. Kaupp, L., Nazemi, K., Humm, B.: "An industry 4.0-ready visual analytics model for context-aware diagnosis in smart manufacturing. In: 24th International Conference Information Visualisation (IV). IEEE vol. 2020, pp. 350–359 (2020)
4. Keim, D., Andrienko, G., Fekete, J.-D., Görg, C., Kohlhammer, J., Melançon, G.: Visual analytics: Definition, process, and challenges. In: Hutchison, D., Fekete, J.-D., Kanade, T., Kerren, A., Kittler, J., Kleinberg, J.M., Mattern, F., Mitchell, J.C., Naor, M., Nierstrasz, O., North, C. (Eds.) Information Visualization, ser. Lecture Notes in Computer Science, vol. 4950, pp. 154–175. Springer, Berlin, Heidelberg
5. Keim, D. (ed.): Mastering the information age: Solving problems with visual analytics. Goslar, Eurographics Association (2010)
6. Sacha, D., Stoffel, A., Stoffel, F., Kwon, B.C., Ellis, G., Keim, D.A.: Knowledge generation model for visual analytics. IEEE Trans. Vis. Comput. Graph. **20**(12), 1604–1613 (2014)

7. Sacha, D., Senaratne, H., Kwon, B.C., Ellis, G., Keim, D.A.: The role of uncertainty, awareness, and trust in visual analytics. IEEE Trans. Vis. Comput. Graph. **22**(1), 240–249 (2016)
8. Ceneda, D., Gschwandtner, T., May, T., Miksch, S., Schulz, H.-J., Streit, M., Tominski, C.: Characterizing guidance in visual analytics. IEEE Trans. Vis. Comput. Graph. **23**(1), 111–120 (2017)
9. Andrienko, N., Lammarsch, T., Andrienko, G., Fuchs, G., Keim, D., Miksch, S., Rind, A.: Viewing visual analytics as model building. Comput. Graph. Forum **37**(6), 275–299 (2018)
10. Zhou, F., Lin, X., Luo, X., Zhao, Y., Chen, Y., Chen, N., Gui, W.: Visually enhanced situation awareness for complex manufacturing facility monitoring in smart factories. J. Vis. Lang. Comput. **44**, 58–69 (2018). http://www.sciencedirect.com/science/article/pii/S1045926X17301829
11. Wu, W., Zheng, Y., Chen, K., Wang, X., Cao, N.: "A visual analytics approach for equipment condition monitoring in smart factories of process industry. In IEEE Pacific Visualization Symposium. Piscataway, NJ: IEEE, pp. 140–149 (2018)
12. Filz, M.-A., Gellrich, S., Herrmann, C., Thiede, S.: Data-driven analysis of product state propagation in manufacturing systems using visual analytics and machine learning. Procedia CIRP **93**, 449–454 (2020)
13. Nazemi, K., Breyer, M., Burkhardt, D., Stab, C., Kohlhammer, J.: Semavis: a new approach for visualizing semantic information. In: Wahlster, W., Grallert, H.-J., Wess, S., Friedrich, H., Widenka, T. (eds.) Towards the Internet of Services: The THESEUS Research Program, ser. Cognitive Technologies, vol. 255, pp. 191–202. Springer International Publishing, Cham (2014)
14. Nazemi, K.: Adaptive Semantics Visualization, vol. 646. Springer International Publishing, Cham (2016)
15. Munzner, T.: Visualization Analysis & Design, ser. AK Peters Visualization series. CRC Press, Boca Raton, FL (2015)
16. Emmanouilidis, C., Pistofidis, P., Fournaris, A., Bevilacqua, M., Durazo-Cardenas, I., Botsaris, P.N., Katsouros, V., Koulamas, C., Starr, A.G.: Context-based and human-centred information fusion in diagnostics. IFAC-PapersOnLine **49**(28), 220–225 (2016)
17. Zhou, F., Lin, X., Liu, C., Zhao, Y., Xu, P., Ren, L., Xue, T., Ren, L.: A survey of visualization for smart manufacturing. J. Vis. **22**(2), 419–435 (2019)
18. Xu, P., Mei, H., Ren, L., Chen, W.: Vidx: Visual diagnostics of assembly line performance in smart factories. IEEE Trans. Vis. Comput. Graph. **23**(1), 291–300 (2017)
19. Jo, J., Huh, J., Park, J., Kim, B., Seo, J.: Livegantt: interactively visualizing a large manufacturing schedule. IEEE Trans. Vis. Comput. Graph. **20**(12), 2329–2338 (2014)
20. Post, T., Ilsen, R., Hamann, B., Hagen, H., Aurich, J.C.: User-guided visual analysis of cyber-physical production systems. J. Comput. Inform. Sci. Eng. **17**(2), 9 (2017)
21. Arbesser, C., Spechtenhauser, F., Mühlbacher, T., Piringer, H.: Visplause: visual data quality assessment of many time series using plausibility checks. IEEE Trans. Vis. Comput. Graph. **23**(1), 641–650 (2017)
22. Bruckner, D., Stanica, M.-P., Blair, R., Schriegel, S., Kehrer, S., Seewald, M., Sauter, T.: An introduction to opc ua tsn for industrial communication systems. Proc. IEEE **107**(6), 1121–1131 (2019)
23. Kaupp, L., Humm, B.G., Nazemi, K., Simons, S.: Raw opc-ua dataset of a working industry 4.0 smart factory (h-da autfab). https://doi.org/10.5281/ZENODO.3709619 (2020)
24. Defferrard, M., Bresson, X., Vandergheynst, P.: Convolutional neural networks on graphs with fast localized spectral filtering. In: Lee, D., Sugiyama, M., Luxburg, U., Guyon, I., Garnett, R. (Eds.) Advances in Neural Information Processing Systems, vol. 29. Curran Associates, Inc. (2016). https://proceedings.neurips.cc/paper/2016/file/04df4d434d481c5bb723be1b6df1ee65-Paper.pdf
25. Kipf, T.N., Welling, M.: Semi-supervised classification with graph convolutional networks. http://arxiv.org/pdf/1609.02907v4
26. Veličković, P., Casanova, A., Lio, P., Cucurull, G., Romero, A., Bengio, Y.: Graph attention networks

27. Zhou, J., Cui, G., Zhang, Z., Yang, C., Liu, Z., Wang, L., Li, C., Sun, M.: Graph neural networks: a review of methods and applications. https://arxiv.org/pdf/1812.08434
28. Zhou, Y., Zheng, H., Huang, X.: Graph neural networks: taxonomy, advances and trends. https://arxiv.org/pdf/2012.08752
29. Wu, Z., Pan, S., Chen, F., Long, G., Zhang, C., Yu, P.S.: A comprehensive survey on graph neural networks. IEEE Trans. Neural Netw. Learn. Syst. **32**(1), 4–24 (2021)
30. Guo, S., Lin, Y., Feng, N., Song, C., Wan, H.: Attention based spatial-temporal graph convolutional networks for traffic flow forecasting. Proc. AAAI Conf. Artif. Intell. **33**, 922–929 (2019)
31. Deng, A., Hooi, B.: Graph neural network-based anomaly detection in multivariate time series (2020)
32. Cao, D., Wang, Y., Duan, J., Zhang, C., Zhu, X., Huang, C., Tong, Y., Xu, B., Bai, J., Tong, J., Zhang, Q.: Spectral temporal graph neural network for multivariate time-series forecasting. In: NeurIPS (2020)
33. Wu, Z., Pan, S., Long, G., Jiang, J. Chang, X., Zhang, C.: Connecting the dots: multivariate time series forecasting with graph neural networks. In: Gupta, R. (Ed.) Proceedings of the 26th ACM SIGKDD International Conference on Knowledge Discovery & Data Mining, ser. ACM Digital Library, New York, NY, United States: Association for Computing Machinery, pp. 753–763 (2020)
34. Zhao, H., Wang, Y., Duan, J., Huang,C., Cao, D., Tong, Y., Xu, B., Bai, J., Tong, J., Zhang, Q.: Multivariate time-series anomaly detection via graph attention network. In: 2020 IEEE International Conference on Data Mining (ICDM). IEEE, 11/17/2020 - 11/20/2020, pp. 841–850
35. Wu, F., Souza, A., Zhang, T., Fifty, C., Yu, T., Weinberger, K.: Simplifying graph convolutional networks. In: Chaudhuri, K., Salakhutdinov, R. (Eds.) Proceedings of the 36th International Conference on Machine Learning, ser. Proceedings of Machine Learning Research, vol. 97. PMLR, pp. 6861–6871 (2019). http://proceedings.mlr.press/v97/wu19e.html
36. Zhang, W., Zhang, Y., Xu, L., Zhou, J., Liu, Y., Gu, M., Liu, X., Yang, S.: Modeling iot equipment with graph neural networks. IEEE Access **7**, 32–754–32 764 (2019)
37. Humm, B.G., Bense, H., Fuchs, M., Gernhardt, B., Hemmje, M., Hoppe, T., Kaupp, L., Lothary, S., Schäfer, K.-U., Thull, B., Vogel, T., Wenning, R.: Machine intelligence today: applications, methodology, and technology. Informatik Spektrum (2021)
38. Beez, U., Kaupp, L., Deuschel, T., Humm, B.G., Schumann, F., Bock, J., Hülsmann, J.: Context-aware documentation in the smart factory. In: Hoppe, T., Humm, B., Reibold, A. (eds.) Semantic Applications, vol. 23, pp. 163–180. Springer, Berlin (2018)
39. Schriegel, S., Kobzan, T., Jasperneite, J.: Investigation on a distributed sdn control plane architecture for heterogeneous time sensitive networks. In: WFCS 2018. Piscataway, NJ: IEEE, 2018, pp. 1–10. https://www.nist.gov/system/files/documents/el/CPS-WorkshopReport-1-30-13-Final.pdf
40. Pledger, M.J.: Utilising measurable uncontrollable factors in parameter design to optimise the response (1998)
41. He, S.-G., Li, L., Qi, E.-S.: Study on the Quality Improvement of Injection Molding in LED Packaging Processes based on DOE and Data Mining: WiCom 2007; pp. 21–25: [Shanghai, China, p. 2007. IEEE Operations Center, Piscataway, NJ (2007)
42. Modrak, V., Mandulak, J.: Exploration of impact of technological parameters on surface gloss of plastic parts. Procedia CIRP **12**, 504–509 (2013)
43. NIST: Foundations for innovation in cyber-physical systems: Workshop report (2013)
44. Andrienko, N., Andrienko, G.: A visual analytics framework for spatio-temporal analysis and modelling. Data Min. Knowl. Disc. **27**(1), 55–83 (2013)
45. OPC Foundation, Opc-ua datatype xml schema file (2020). https://opcfoundation.org/UA/schemas/1.04/Opc.Ua.Types.bsd
46. Qu, Y., Fang, B., Zhang, W., Tang, R., Niu, M., Guo, H., Yu, Y., He, X.: Product-based neural networks for user response prediction over multi-field categorical data. ACM Trans. Inform. Syst. **37**(1), 1–35 (2018)

47. Rodríguez, P., Bautista, M.A., Gonzàlez, J., Escalera, S.: Beyond one-hot encoding: lower dimensional target embedding. Image Vis. Comput. **75**, 21–31 (2018)
48. Salvador, S., Chan, P.: Toward accurate dynamic time warping in linear time and space. Intell. Data Anal. **11**(5), 561–580 (2007)
49. van Brakel, J.: Robust peak detection algorithm using z-scores, Stack Overflow (2014). https://stackoverflow.com/questions/22583391/peak-signal-detection-in-realtime-timeseries-data/22640362#22640362
50. Box, G.E.P., Pierce, D.A.: Distribution of residual autocorrelations in autoregressive-integrated moving average time series models. J. Amer. Stat. Assoc. **65**(332), 1509–1526 (1970)
51. Hinton, G.E., Zemel, R.S.: Autoencoders, minimum description length and helmholtz free energy. In: Proceedings of the 6th International Conference on Neural Information Processing Systems, ser. NIPS'93, pp. 3–10. Morgan Kaufmann Publishers Inc., San Francisco, CA, USA (1993)
52. Opitz, D., Maclin, R.: Popular ensemble methods: an empirical study. J. Artif. Intell. Res. **11**, 169–198 (1999)
53. Chen, J., Sathe, S., Aggarwal, C., Turaga, D.: Outlier detection with autoencoder ensembles. In: Chawla, N., Wang, W. (Eds.) Proceedings of the 2017 SIAM International Conference on Data Mining [S.l.]: Society for Industrial and Applied Mathematics, pp. 90–98 (2017)
54. Hawkins, D.M.: Introduction. In: Hawkins, D.M. (Ed.), Identification of Outliers, ser. Monographs on Applied Probability and Statistics, pp. 1–12. Springer, Dordrecht (1980)
55. Petneházi, G.: Recurrent neural networks for time series forecasting. https://arxiv.org/pdf/1901.00069

Visual Discovery of Malware Patterns in Android Apps

Paolo Buono and Fabrizio Balducci

Abstract The diffusion of smartphones and the availability of powerful and cheap connections allow people to access heterogeneous information and data anytime and anywhere. In this scenario billions of mobile users as well as billions of under-protected IoT devices have high risk of being the target of malware, cybercrime and attacks. This work introduces visualization techniques applied to software apps installed on Android devices using features generated by mobile security detection tools through static security analysis. The aim of this work is to help common people and skilled analysts to quickly identify anomalous and malicious software on mobile devices. The visual findings are reached through text, tree and other techniques. An app inspection tool is also provided and its usability has been evaluated with an experimental study with ten participants.

1 Introduction

With the widespread of smart devices and large mobile broadband, the use of web services and applications has become essential for most of the population, especially young people. For example, educational learning can take several forms, either online or offline, including flipped classroom schemes, MOOCs, e-books, certifications and online courses, serious games [1–3]. In this way, smartphones increasingly become an attractive target for online criminals.

The main reason for this phenomenon is that (smart)phones and tablets have increasingly become equipped with multi-tasking processors and high-bandwidth connectivity that allow interfacing with various peripherals and sensors. In many

P. Buono (✉) · F. Balducci
Computer Science Department, University of Bari A. Moro, Via Orabona 4, Bari, Italy
e-mail: paolo.buono@uniba.it

F. Balducci
e-mail: fabrizio.balducci@uniba.it

B. Kovalerchuk et al. (eds.), *Integrating Artificial Intelligence and Visualization for Visual Knowledge Discovery*, Studies in Computational Intelligence 1014,
https://doi.org/10.1007/978-3-030-93119-3_17

countries such devices provide payment apps and platforms to manage money credits in phone wallets, together with valuable personal information and credit card information.

According to Gartner[1] more than 1.5 billion mobile phones were bought in 2019 while IDC[2] reports that Android market is about 86.1% and 13.9% relates to Apple's operating system. Considering this context, cybercriminals are developing sophisticated attacks which are excellent at stealing valuable personal data or extorting money. With complex frameworks like Android mobile weaknesses are increasing as well as the vulnerabilities in app stores, where users browse, purchase, and remotely install apps. There are, for example, malwares that steal browser cookies from the infected desktop computer to pretend to be the user, in order to remotely install apps onto the victims' phones and tablets without their explicit consent.

In 2017 more than 300 Google Play Store apps were discovered to contain malware that officially were common apps, such as storage managers, video players, and ringtones (about 70,000 Android devices affected); when installed, in the app background they would take over the user's Android device, making it part of a giant botnet, called *WireX*,[3] able to launch distributed Denial-of-Service attacks from more than 100 countries.

The *OSX.Wirelurker* uses private APIs to manage sensitive functionalities as well as *YiSpecter*, consisting of four different modules signed with enterprise certificates, that spreads devices by hijacking traffic from nationwide ISPs, a Social Network Service worm on Windows and an offline app installation from a Command-and-Control server.

Moreover, about 256 iOS apps were removed from the Apple App Store[4] since they used third-party advertising technology from the *Youmi company*.[5] This software library was used to steal personal information like installed applications list, platform serial number, the Apple ID email address associated with the iOS device while *XcodeGhost* was discovered in a malicious variant of the Apple's Integrated Development Environment *Xcode* so that iOS developers were unknowingly inserting malicious code into their iOS apps. Symantec also detected the Android version of Youmi, Android, Youmi.[6]

Regarding the *usable security* topic, malwares deceive users when entering their banking credentials into a fake login page on top of the real banking apps as done by *Android.FakeLogin*.[7] The ransomware *ScarePakage*[8] mimics Google's design style

[1] Market Share: PCs, Ultramobiles and Mobile Phones, https://www.gartner.com/en/newsroom/press-releases/2021-02-22-4q20-smartphone-market-share-release, last visit: 2020-03-21.

[2] https://www.idc.com/promo/smartphone-market-share/os, last visit: 2020-03-21.

[3] Wirex, New Jersey Cybersecurity, https://bit.ly/2YUdzE4, last visit: 2019-03-21.

[4] Apple Threat Landscape, Symantec, https://symc.ly/2wtwSb2, last visit: 2019-03-21.

[5] youmi-ioslib, https://github.com/youmi/ios-sdk, last visit: 2019-03-21.

[6] Android.Youmi, Symantec, https://www.symantec.com, last visit: 2019-03-21.

[7] Android.FakeLogin, Symantec (2015), https://www.symantec.com/security-center/writeup/2015-102108-5457-99, last visit: 2019-03-21.

[8] Android.Locker, ESET (2014), https://www.virusradar.com, last visit: 2019-03-21.

to display fake FBI warnings that appear legitimate and intimidating. Static Analysis [4] and Artificial Intelligence [5] offer methods to prevent and recognize cybercrime activities removing the human from log file analysis as done through deep learning techniques [6] with Generative models [7]. The diffusion of embedded and portable communication devices on modern vehicles also entails security risks [8, 9]. Android malware bypass signature-based security software by obfuscating code while others, before the attack, check if they are running on real devices or sandboxes used by security researchers, making harder for analysts to recognize them. Such problems extend also to domains such as Internet Of Things (IoT) [10], that involves smart homes, recent cars, smart TVs, medical devices, embedded devices, etc., and also education [11].

Many clues and insights hidden into mobile devices can be investigated by analyzing log file since Android apps create them for debugging facilities and to notify the system all the app activities. This work introduces a series of visualization techniques that may help to discover and understand patterns hidden in the Android log files produced by devices that contain suspicious activities and possible malware, regardless to specific (obfuscated or plain) verbose and plain texts that without such visual facilities may distract the analyst.

This paper is a substantial revision and improvement of the paper [12]. The text has been thoroughly reworded and new content have been added. Specifically, the App Inspection Tool and the list of relevant features have been added. The tool adds novel visualization technique that increases the analytical possibilities because it is able to analyze multiple apps and it can be used by a non-expert. A usability test has been performed on the tool in order to inform the reader about the usability of the tool.

The paper is constituted as the following: related work in the field of log file visualization are in Sect. 2 while Sect. 3 introduces the proposed method that involves five visualization techniques on an existing dataset containing log files coming from more than 128.000 Android apps. Section 4 introduces the graph inspection tool followed by its experimental evaluation in Sect. 5. Finally, Sect. 6 shows conclusions and future works.

2 Related Work

Malware analysts still manually inspect suspicious elements by heuristic rules dissecting software pieces and looking for malware evidences in the code, in fact software applications generate log files to communicate system status reports, hardware faults or software errors that can be used to check the status of their work. A common problem with such text files is to separate and locate critical messages among the standard ones which flows from routine communications. Moreover, in mobile devices logs are also linked to data coming from sensors and real time events.

The visualization of log files, considering their nature of multi-source and multi-format textual data, has become a research topic in the scientific literature, mostly in the network/system security area, where scientists made efforts to help analyst in order to transform them into machine-readable files [13]. An example of such amount of textual data is in E-learning systems that accumulate log information on students' activities like learning practices, reading and writing exercises, test sessions and various tasks in real or virtual environments with peers [14]. The analysis of log data may improve education for both instructors and students. In fact by analyzing data about students' activities, the instructors can better organize their lessons and material.

There are two main goals that sometimes overlap in visualizations, i.e. explaining data to solve specific problems and exploring large datasets for better understand the nature and characteristics of such information. Visualizations that are meant to drive the user along defined paths are *explanatory* and may offer many perspectives on data by comparing multiple datasets at once.

See Soft is one of the earliest techniques used to visually analyze text log files through lines [15] while *parallel coordinates*, *scatter plots* and hierarchical visualization (in particular Treemaps) are used for firewall logs to perform computer network security analysis. A classification of visualization techniques in five categories (textual, parallel visualization, hierarchical, 3D, other) is provided by Zhang et al. [16].

Regarding large-scale systems, for intrusion and misuse detection it is used visual representation of network and system activity with the aim to detect traffic anomalies [17]. Xydas et al. [18] propose a 3D Graph Visualization for intrusion detection to process web server log files through a reduction method based on frequencies. The *Colored graphs* have been used to visualize real-time traffic and detect internal fraud in systems with pair of entities (e.g., employees and clients) and periodic activities with techniques like spirals, layered diagrams and stacked bar plots [19].

Lee et al. studied efficient visualizations of security log files [20] providing a solution based on *D3.js* for analyzing Advanced Persistent Threat attacks using time-based graph visualization, *Sankey diagrams* or circular hierarchical representation, with other components like multi-level monitoring graph, for easily check malicious packets in the network.

Considering the mobile context, Shen and Ma [21] developed a visual analytics tool to combine social and spatial information in one heterogeneous network with the aim to analyze people behavior (individual and in group) exploiting data in their mobile devices and their locations. In addition to networks and time plots, the *Behavior Rings* techniques is a compact radial representation of individual and group behaviors that has been introduced to allow the comparison of behavior pattern.

Lahmadi et al. [22] studied the visualization of logs and network flow data in Android mobile applications collecting data through Hadoop database and analyzing them with *Elasticsearch* engine while Somarriba et al. [23] describe a framework to monitor and visualize Android applications anomalous function, tracing restricted API functions used at runtime by using dendrograms.

3 Materials and Method

In 2014 Arp et al. [24] proposed a method for Android malware detection aimed to identify malicious applications directly on the device. It employs a a wide static code analysis gathering sets of information and features that can be embedded in a joint vector space as feature vectors to represent relevant patterns useful to automatically identify malware [25].

3.1 The Dataset

As a case study, DREBIN Android malware dataset is analyzed using a pipeline of visualization techniques with the aim to explore log files. Figure 1 shows an excerpt of log file that contains 4 permissions requests, 2 activities, 8 URL requests and 3 system calls.

The dataset is composed of more than 123,000 logs extracted from normal apps, over 5,500 logs extracted from malicious apps and contains the set of malware family names [26, 27]. In order to assess the quality of the dataset, it can be calculated its *Entropy* and the non-randomness of data considering $p(x_i)$ the probability of character number i appearing in the string, I the information function of an event i with probability p_i, b=2 the log base, the alphabet of symbols and their frequencies. The *Shannon Entropy* is calculated using Eq. 1:

```
permission::android.permission.ACCESS_CACHE_FILESYSTEM
activity::GameAlertDialog
url::http://client.go360days.com/report/return_push.do
url::http://client.go360days.com/report/return alert.do
real_permission::android.permission.ACCESS_NETWORK_STATE
call::getDeviceId
url::http://client.go360days.com/request/config.do
call::Execution of external commands
permission::com.android.launcher.permission.UNINSTALL_SHORTCUT
call::Cipher(AES/CBC/PKCS5Padding)
activity::.ImageTestActivity
call::getSubscriberId
url::http://client.go360days.com/report/down_apk.do
url::http://client.go360days.com/report/return_config.do
url::http://client.go360days.com/request/push.do
url::http://client.go360days.com/request/index.do
permission::android.permission.SET_WALLPAPER
url::http://10.0.0.172:80
```

Fig. 1 An excerpt from the DREBIN-generated log file dataset

$$H(X) = \sum_{i=1}^{n} p(x_i) I(x_i) = \sum_{i=1}^{n} p(x_i) \log_b \frac{1}{p(x_i)}$$
$$= -\sum_{i=1}^{n} p(x_i) \log_b p(x_i) \quad (1)$$

It results that at least 5 bits per symbol are needed to encode the information in binary form ($H(X) = 5.13846$) and at least 15380 bits are needed to encode the string optimally: the dataset is encoded with 25744 bits (the file also has header containing metadata). The *metric entropy* coming from Shannon entropy divided by dataset string length is 0.0016 and such low value assures that data are not random.

3.2 App Code Features

The static code analysis made on DREBIN dataset allows to identify information about the Android app code structure like:

- **Hardware Components**: access to the camera, touchscreen, GPS, network module;
- **Requested Permissions**: e.g., *SEND_SMS* permission to send premium SMS messages;
- **App Components**: activities, services, content providers and broadcast receivers, e.g., several variants of *DroidKungFu* family share the name of specific servers (*com:8511/search/*) [28];
- **Intents**: used to share data between different components and apps. An example of an intent message involved in malware is *BOOT_COMPLETED* that triggers malicious activity after the device reboot [29].

Android app bytecode can be disassembled, in order to gather information about API calls and data used in the application and to recover additional features like:

- **Restricted API Calls**: possible malicious behavior when restricted calls are used without asking the users for the required permissions; it may indicate that the malware is trying a root exploit to overcome the limits imposed by the Android system;
- **Used Permissions**: the extracted set of calls is used as the ground to understand the subset of permissions requested and permissions actually used. It can be useful the method introduced in Felt et al. [30] where API calls and permissions are matched to have a wide view on the application behaviour since multiple calls can be protected by a single permission. For example, *sendMultipartTextMessage()* and *sendTextMessage()* both require that the SEND_SMS permission is granted to the app;
- **Suspicious API Calls**: there are known API calls like *getDeviceId()*, *getSubscriberId()*, *setWifiEnabled()*, *execHttpRequest()*, *sendTextMessage()*, *Runtime.exec()* and *Cipher.getInstance()* that allow access to sensitive data or resources and that are frequently found in malware samples, so it becomes useful to extract and create a set of such features;

- **Network Addresses**: connections might be involved in botnets since malware regularly establishes network connections to retrieve commands or send data collected from the device. In this way all IP addresses, hostnames and URLs found in the disassembled code are included in this set of features

3.3 Visualization Pipeline

Initially, the DREBIN-generated log files were considered as a case study. Such dataset have no temporal data, thus the proposed pipeline involves the following visualization techniques:

- DocuBurst, to extract hierarchy between words [31];
- Word Tree, to show words concordance [32];
- Word Cloud, to highlight feature frequency [33];
- Textexture [34], to visualize most frequent features as a graph representation.

Circular visualization techniques has been proved as effective for people that are aware of their use [35]. The dataset contains hierarchical data, thus a feasible visualization technique that shows hierarchical in a circular fashion is *DocuBurst*. DocuBurst combines word frequency with WordNet lexical database to create a visualization that also reflects semantic contents. Figure 2 shows the *term entities*, each represented as single log entry (line). The components of the log entries appear as slices of concentric circles and each circle represents a hierarchy level.

Figure 2 reveals at least three levels of children nodes depth providing clues about the presence of main features that influence the others. The color intensity associated

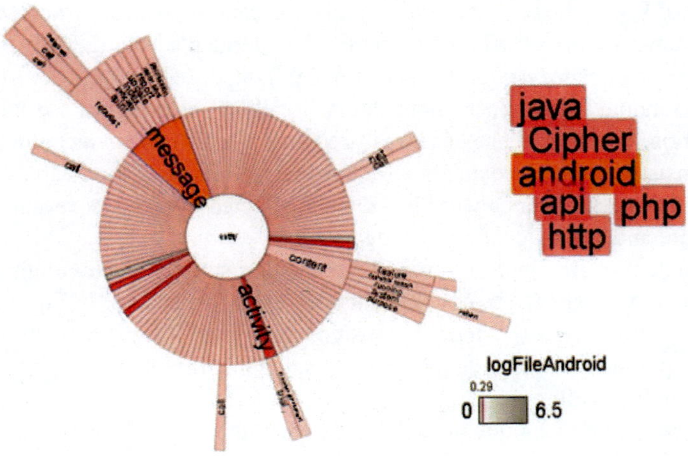

Fig. 2 Hierarchical visualization using *DocuBurst* with zoom filtering on the word *android*; in the bottom-right corner the score-bar represent how strong is the word occurrence in the document

Fig. 3 The DocuBurst output showing occurrence of the term *android* in a log file

to the slice reflects the frequency of a specific word in the collection while the size visual information reflects the number of its children. From the items on the top left of Fig. 2 it emerges that in general frequent entries are all at the first level with a set of related concepts denoted by the cluster terms *message*, *content* and *activity*.

The visual pattern of *message* entry means that in the log this entry has great variety featuring more depth than others, similarly to *content* on the right of the DocuBurst visualization while the term *activity* contains only two children. Considering the right part of Fig. 2, the term *android* appears in a central position; when clicking on it the system provides details according to Shneiderman's Mantra [36] as shown in Fig. 3 where all occurrence of the term are highlighted.

After noticing a few clues about the API calls topology from the hierarchical visualization, the log file has been analyzed with the *Word Tree* technique (Fig. 4), which consists in a visual search tool for unstructured text that shows all the different contexts in which appears a word or sentence selected by the user revealing details on structure and words location in a log entry.

The context is arranged in a tree-like branching structure to highlight recurrent sentences with the aim to trace API functions used at run-time. In Fig. 4 segments that connect nodes identify the relationships. However, it must be considered that zooming operations could hide relevant items and the filtering ones might affect the topology understanding.

The *WordCloud* technique provides hints about the most relevant items in a visualization in which colors and sizes of terms depicted together represent potential patterns in malware behavior, as shown in Fig. 5. However, a drawback of this technique is the changing positions of items at each run of the algorithm. While it can

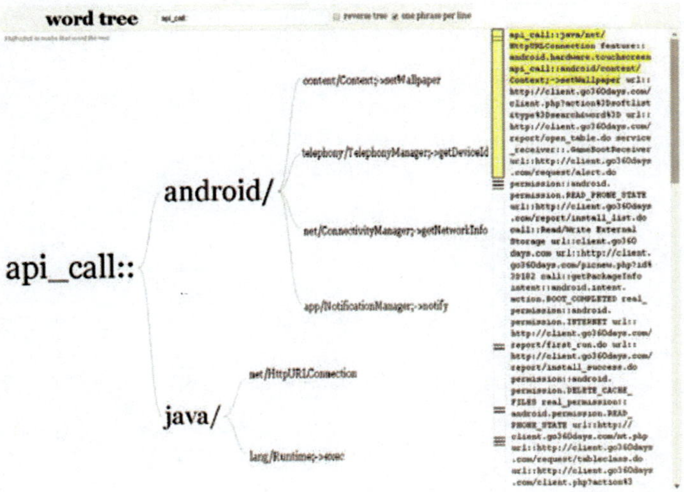

Fig. 4 A WordTree visualization used to trace API function calls used at runtime. The path of the API call that sets the wallpaper is highlighted

Fig. 5 The WordCloud output about frequent features where *go360days* appears a prominent string

be considered a problem, the aim of word cloud is to provide item distributions at glance, in order to allow the user to spot patterns.

According to Fig. 5, the most interesting features results are:

- *External*
- *STATE*
- *url::permission::android.permission.ACCESS*
- *api*
- *url::http://client.go360days.com/report/return*
- *url::http://client.go360days.com/client.php*

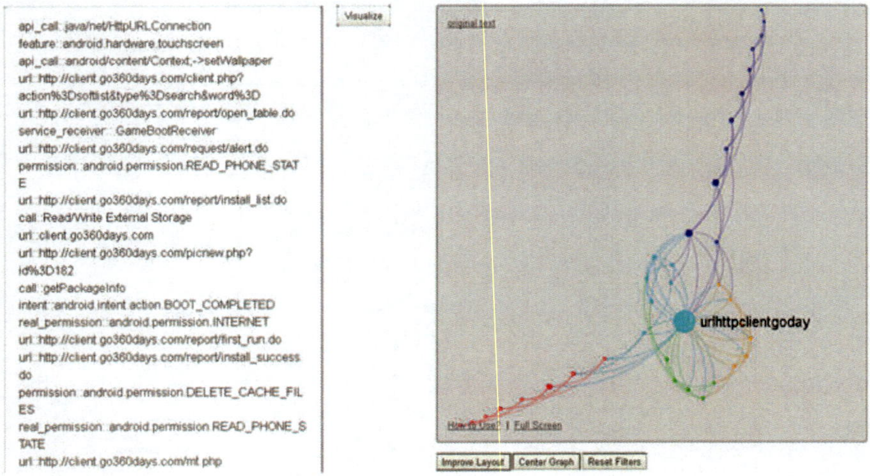

Fig. 6 Node-link visualization showing the topology of a polymorphic malware

- *permission::android.permission.READ*
- *action*
- *real*

Malware presence can be identified by using https://client.go360days.com as keyword in a search, however without additional information like topology and patterns is still hard to find more explicit behaviors. Thus, the next goal in the log visualization pipeline is to highlight internal feature relationships, for example using *Textexture*, a visualization tool that arranges text as semantic networks based on *Gephi* [37] where nodes of the same color represent similar features. Figure 6 shows the core of the malware in red node labeled *urlhttpclientgoday* (aggregation of terms in the string https://client.go360days.com previously seen).

The node-link graph visualization created with *Textexture* helps revealing the topology of a malware providing the analyst with hints on the search for similar visual patterns. Moreover, the allowed interactive actions result useful to manipulate the graph and make more visible patterns that may be hidden in a complex structure.

In Fig. 7 there are details about the light green "right wing" which has the most prominent features detected on external device disk and read/write operations, including installations.

Figure 8 shows a portion of the malware structure and, specifically, the details about the "left wing" of the generated graph (previously seen in Fig. 6) where only red-colored features related to graphical permission and app status check methods are shown.

Figure 9 shows a polygonal visualization called *Polysingularity* provided by *Textexture*: the intentional node-link malware representation appears as a specific signature and it shows the actual pattern of malware behaviour, in fact it depicts part

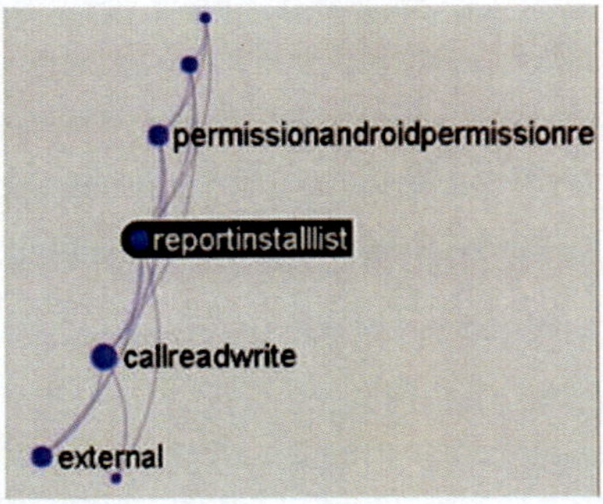

Fig. 7 Right side of the https://client.go360days connections in Fig. 6 where external disk access and read/write operations are highlighted

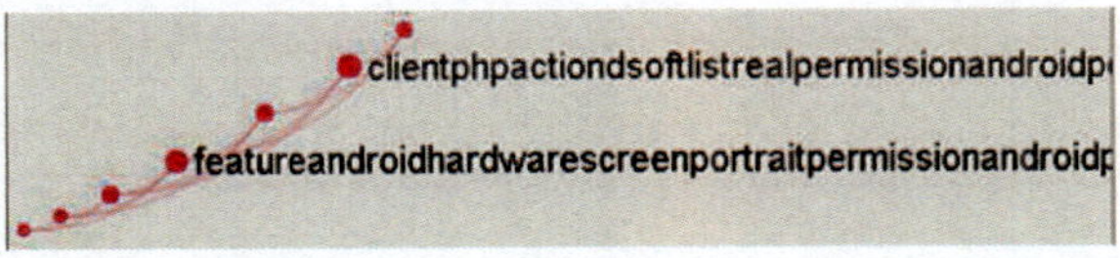

Fig. 8 Left side of the https://client.go360days connections in Fig. 6 where real and screen portrait permission requests are highlighted

Fig. 9 The polygonal view of the analyzed malware provided by Textexture visualization

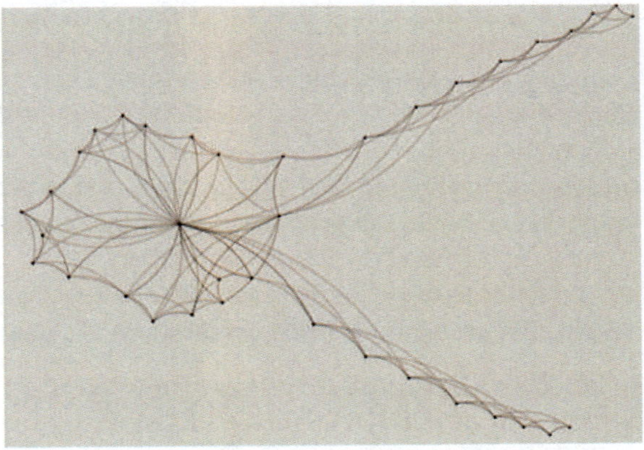

of the *GinMaster* Android trojanized family, gone through three generations since it was first found in August 2011 [38].

This malware family has been created in mainland China and has more than 13,000 known variants; new variants of *GinMaster* can successfully avoid detection by mobile anti-virus software through polymorphic techniques to hide malicious code, obfuscating class names for each infected object, and randomizing package names and self-signed certificates for applications.

4 App Inspection Tool

The analysis described so far requires an expert because it is needed to be quite expert of the internal components of Android. A fast inspection of single or grouped apps can be performed, with the aim to allow even unskilled cybersecurity people to understand the meaning of the obtained information and data from Android apps code.

Automated tools, such as *AndroBugs* are capable to provide various informative records about an Android app code structure and quality. In fact, when launched, 25 fields are provided for each record which includes: apk id, package name, platform, targetSdk, minSdk, package version name, file SHA1 / SHA256 / SHA512, details, analyze engine build, analyze status, etc.

Although all this data information are useful, it is very interesting the *details* vector field containing all the checks carried out together with the related results. There are several fields inside the details (one for each security check performed by AndroBugs) and, for each of them, additional fields are:

- **count**: number of times this leak / permission is detected
- **title**: the issue identification
- **vector details**: files and lines where the issue was found
- **summary**: short description of the problem
- **severity level**: severity of the problem (Info, Notice, Warning, Critical)

In order to develop the interactive visualization for the inspection tool, *Vega-lite* library has been employed and a dashboard to display and interact with selected information has been created. The data that were deemed relevant for a security analysis, in addition to the package name, the target and minimum SDK version, concern some of the fields contained within the details attribute mentioned above. The features have been divided in two groups, namely those that indicate characteristics related to the *used permissions* and those that indicate *security-related aspects*.

The first macro group (used permissions) is composed of:

- EXTERNAL STORAGE: permissions to access external storage
- USE PERMISSION SYSTEM APP: permissions that belong to the group *system*
- PERMISSION GROUP EMPTY VALUE: if the app does not require any permission
- FILE DELETE: permissions that uses the *file.delete()* function
- PERMISSION IMPLICIT SERVICE: if the app declares implicit services
- PERMISSION EXPORTED GOOGLE: if the app declares exported components (such as activities and services)
- PERMISSION INTENT FILTER MISCONFIG: errors in the configuration of intent filters
- PERMISSION DANGEROUS: dangerous permissions
- PERMISSION PROVIDER IMPLICIT EXPORTED: if the app declares implicit providers of type *exported*
- PERMISSION EXPORTED: if the app declares explicit providers of type *exported*

- INTERNET USE PERMISSION: if the app requires internet access

Although these fields refer to the use of permissions and app components, they are relevant from a security point of view as their incorrect use can lead to the exposure of sensitive data to other (malicious) applications.

In the second macro group (security-related), the following fields have been selected:

- SHARED USER ID: the app uses a shared user ID (which allows two applications of the same developer to access each other's data)
- SENSITIVE SMS: the app uses SMS (which can be intercepted and are not considered a secure method of communication)
- DYNAMIC CODE LOADING: the app loads some code dynamically (vulnerable to code injection)
- WEBVIEW JS ENABLED: wether *JavaScript* is enabled in a *WebView* (may be the cause of Cross Site Scripting vulnerabilities)
- SENSITIVE DEVICE ID: the app requires the device IMEI (unique identifier of the phone that would be not used and shared)
- ALLOW BACKUP: the app allows backup through Google services (it should be disabled for sensitive data)
- MODE WORLD READABLE OR MODE WORLD WRITEABLE: if the app writes files in a readable or writable mode (which makes that data exposed to anyone on the device)
- SSL CN1: if the hostname is correctly validated
- SSL CN2: if the app considers valid any hostname verifier
- SSL CN3: the app uses the *getInsecure* method to verify a hostname
- SSL WEBVIEW: the app is vulnerable to MITM (man in the middle) attacks in its WebViews
- SSL X509: the app uses an SSL certificate
- HACKER BASE64 STRING DECODE: if there are strings coded in Base64 (which are sometimes incorrectly used as a form of encryption to hide information)
- DEBUGGABLE: if the app has the debug flag set to true and therefore is executable in debug mode
- COMMAND: whether the app uses the function *exec("")* that can be exploited to run arbitrary code
- WEBVIEW ALLOW FILE ACCESS: there are WebView that access to files
- COMMAND MAYBE SYSTEM: the app requires root privileges or checks for their presence
- SSL URLS NOT IN HTTPS: checks for URLs without HTTPS secure protocol
- WEBVIEW RCE: whether the app is vulnerable to *Remote Command Execution* in a WebView
- HACKER DB KEY: any "hard-coded" access keys to databases (i.e. written statically within the code)

Starting from such information, a tool for the visual inspection through representing all the encountered issues in the Android app(s) under analysis can be employed.

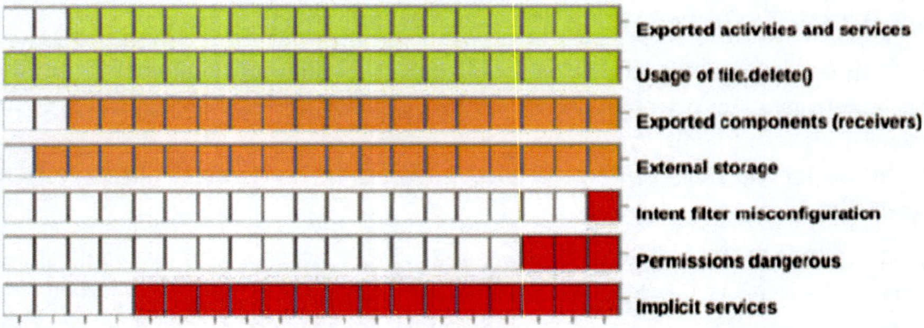

Fig. 10 Issues related to permissions found in the applications, ordered by increasing severity

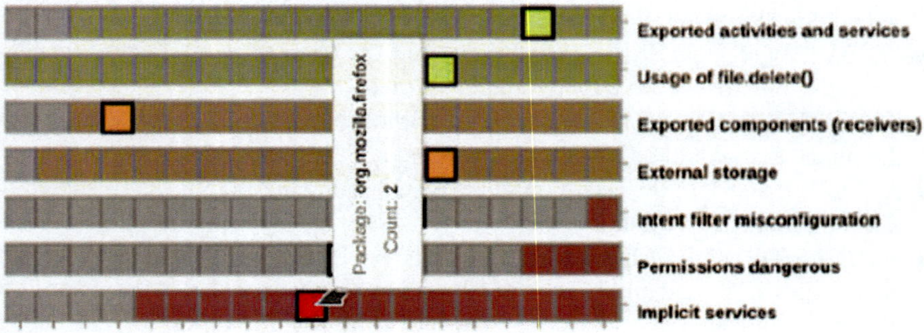

Fig. 11 Vulnerabilities details about a selected APK

As visible in Fig. 10, each square represents each analyzed APK; the color varies according to the severity "level" previously seen i.e. red if *Critical*, orange if *Warning*, yellow if *Notice* and white if *Info* (which indicates that this evidence has not been detected for an APK). On the right, a short description of the vulnerability is provided so it is possible to understand how many APKs are affected by one of them. This visualization technique is generated both for the set of features concerning permissions and for that concerning the security features of the application.

One or more square can be selected or highlighted using the mouse. Hovering the mouse over a square the tool shows the name of the package and additional information such as the number of implicit services inside the APK (see Fig. 11).

A variant of the previous visualization is a matrix that shows individual vulnerabilities for each app (Fig. 12) that is useful to analyze one or more apps allowing to understand, both, apps having a specific type of vulnerability and vulnerabilities of a specific app. The visualization presents the interactive characteristics described above (selection and mouse hovering) and the selection made on this propagates on the other visualizations, highlighting only the data to which a user is interested.

The last visualization technique proposed is a *tornado chart*, which represents the distribution of the minimum and the target Android SDK version (Fig. 13). The y-axis lists each APK under analysis to which a horizontal bar is associated. The

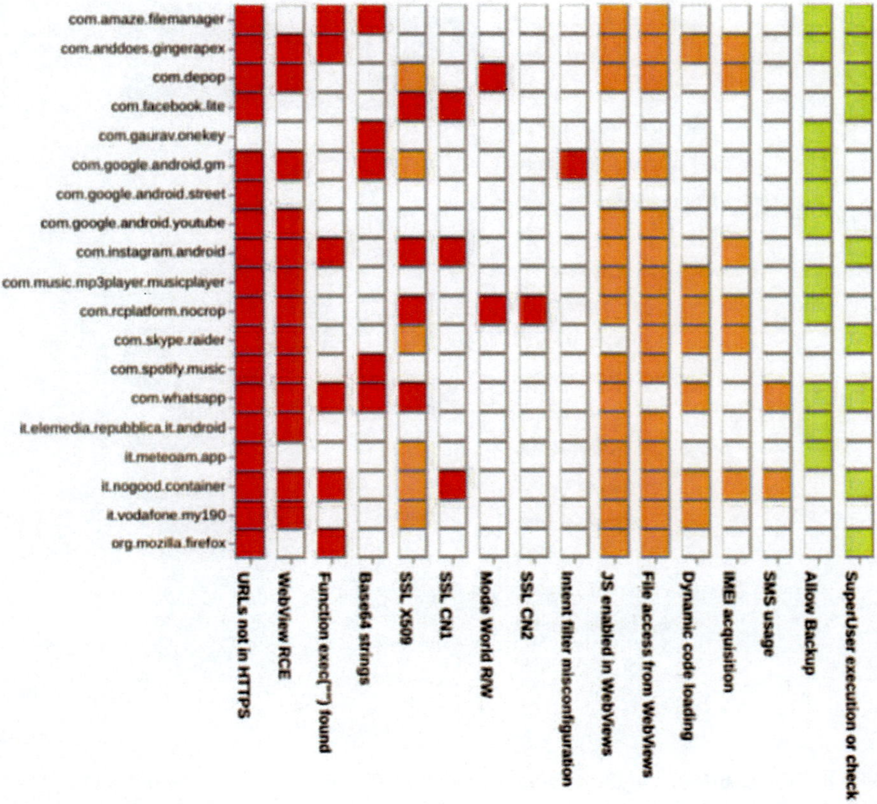

Fig. 12 The interactive matrix graph with the correspondence between application and vulnerabilities

width of a blue bar illustrates the range of the supported Android SDKs of the APK. Conversely, it is possible to select a specific SDK version and easily check from which APK it is supported drawing a vertical line and observing its intersection with the blue bars. Finally, when placing the mouse pointer on a bar a popup that summarizes the minSDK, maxSDK and APK name values appears. Such information can reveal the level of "health" of a mobile phone, because more recent Android SDK represent higher levels of security.

5 Usability Test

A small usability experiment was carried out to verify the effectiveness and how the developed tool was perceived as usable by common users. The setup consists in a virtual machine equipped with a python-notebook (a web application that allows

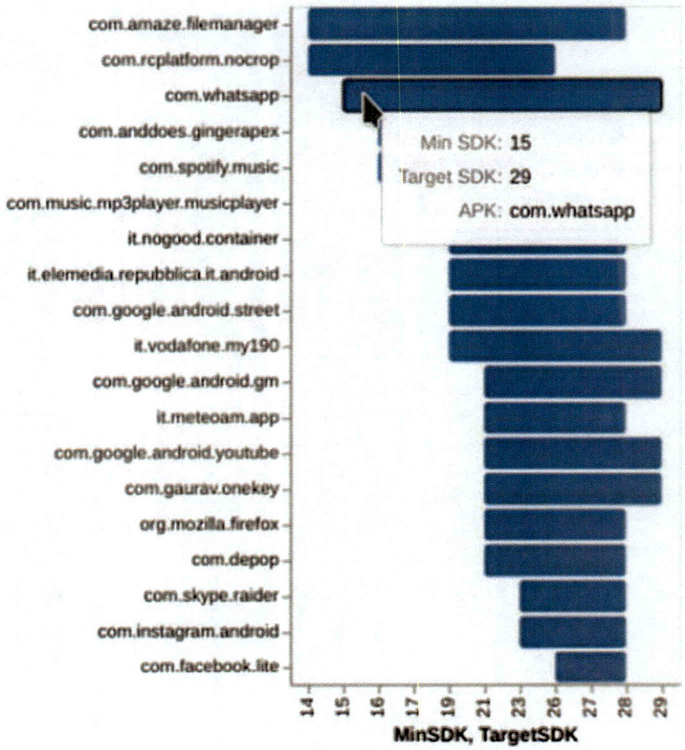

Fig. 13 The distribution of the minimum and target SDK among the analyzed APKs through the tornado chart visualization

to create interactive documents featuring live code and visualizations). Instructions about the use of the tool were provided in a text file that participants had to read and follow.

To improve the interaction with the proposed visualizations the User Interface was provided with four buttons, shown in Fig. 14. Two buttons allow the user to change the item sorting (less or most critical first) while the other two buttons allow to decide whether to show all the values or only the most meaningful ones.

Because of the pandemic constraints an online test was performed, asking the participant to use screen sharing software. Subjects were remotely monitored during the execution of the proposed tasks while the correct answers, the execution time and the actions performed were annotated. The test involved 10 subjects aged between 20 and 50 years that were divided in two groups of five people (each composed respectively of four men and one woman) and performed the test individually trying to execute 10 tasks that were randomly submitted. All the participants have a medium-high knowledge of the Android architecture and the security features present in it. The proposed tasks are reported in the following:

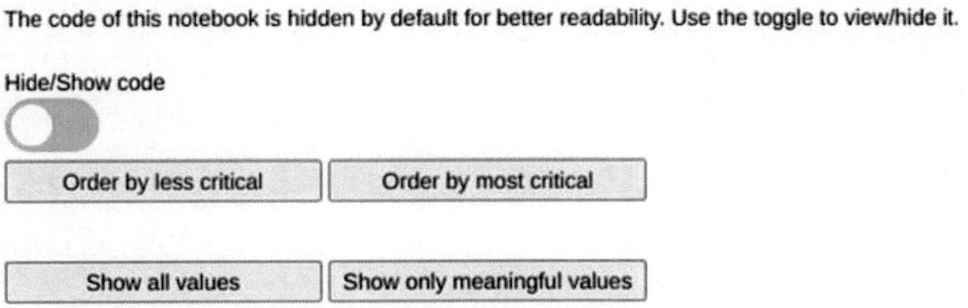

Fig. 14 The interactive buttons that allow users to interact with the visualizations performed by the app inspection tool

1. Find the most frequent critical vulnerability related to permissions in the sample of analyzed apps
2. Find the most frequent critical security vulnerability in the sample of analyzed app
3. Indicate the app with the most critical security vulnerabilities
4. Indicate the app with the most critical permissions related vulnerabilities
5. Sort the vulnerabilities from least significant to most significant
6. Indicate which are the non-significant fields relating to permissions (those for which no vulnerabilities were found in the sample)
7. Indicate which are the non-significant fields relating to security (those for which no vulnerabilities were found in the sample)
8. Indicate one of the vulnerabilities highlighted for the *com.facebook.lite* app with critical level
9. Indicate which app supports the older version of Android (minSdk)
10. Indicate which app has the least critical vulnerabilities related both to permissions that to security

Of the tasks listed above, those from 5 to 7 required interaction with the tool via the buttons on the top of the python-notebook in order to be performed. From this session it was noted that, after an initial disorientation due to the lack of knowledge on the tool, the candidates were able to carry out most of the tasks in an acceptable time.

From the evaluation comes that 67% of tasks were performed successfully and each subject successfully performed an average of 7 tasks (ranging from subjects-2 and subject-7 with only 3 tasks to subjects-3 and subject-4 with 10 tasks) while the average duration of a single experimental session was about 13 min.

The tasks performed successfully by all subjects were task-2 and task-8 (with an average execution time of about 1:20 min for the first and about 1 min for the second) followed by task-9 (9 successes) and task-1 (8 successes). Conversely, the tasks with multiple incorrect or incomplete executions are task-6 and task-7 (only 3 successes) followed by task-5 (5 successes).

Table 1 Results of the usability test. Each cell of the Table reports two values: "Time" indicates the performing duration of the task in 'minutes:seconds' format while "Succ." indicates the success (Y) or the failure (N) of the task execution

Task	Subj. 1 Time-Succ.	Sub. 2 Time-Succ.	Subj. 3 Time-Succ.	Subj. 4 Time-Succ.	Subj. 5 Time-Succ.	Subj. 6 Time-Succ.	Subj. 7 Time-Succ.	Subj. 8 Time-Succ.	Subj. 9 Time-Succ.	Subj. 10 Time-Succ.
1	1:32-Y	1:32-N	3:30-Y	1:01-Y	0:47-Y	1:17-Y	1:05-N	0:46-Y	0:51-Y	1:02-Y
2	1:40-Y	1:03-Y	0:55-Y	0:58-Y	1:41-Y	3:46-Y	0:20-Y	0:29-Y	1:20-Y	0:40-Y
3	1:21-N	2:03-N	0:50-Y	1:20-Y	1:27-N	1:33-Y	1:48-Y	0:34-N	1:09-Y	1:41-Y
4	0:41-Y	1:52-N	2:01-Y	0:55-Y	4:10-N	0:41-Y	1:02-N	0:36-N	1:18-Y	0:33-Y
5	0:38-Y	1:32-N	1:52-Y	0:37-Y	1:30-Y	1:30-N	1:06-N	2:51-Y	3:11-N	1:34-N
6	0:45-N	2:03-N	2:00-Y	2:35-Y	2:09-Y	1:47-N	0:55-N	0:51-N	1:40-N	2:10-N
7	0:52-N	1:58-N	1:31-Y	1:20-Y	7:10-Y	1:02-N	1:00-N	0:20-N	0:05-N	2:30-N
8	0:49-Y	2:47-Y	0:40-Y	1:18-Y	0:54-Y	0:36-Y	0:49-Y	0:33-Y	0:39-Y	0:28-Y
9	1:01-Y	0:55-Y	0:38-Y	1:10-Y	1:35-Y	2:04-Y	0:49-N	0:29-Y	0:36-Y	0:35-Y
10	1:24-Y	1:14-N	0:44-Y	0:48-Y	0:20-N	5:01-Y	1:21-N	1:35-Y	1:25-Y	0:40-Y
Tot.	10.43-7	16:59-3	14:41-10	12:02-10	21.43-7	18.17-7	10:15-3	9:04-6	12:14-7	1.53-7

Considering the provided results emerges that the greatest difficulties in using the tool are related to isolate non-significant fields, i.e. permissions (task-6) and security (task-7) vulnerabilities not found in the sample, and to sorting vulnerabilities from least significant to most significant (task-5) (Table 1).

6 Conclusions and Future Work

This paper presented a working pipeline that takes advantage of visualization techniques on Android log files to identify interesting patterns and malware behaviors.

The visualization of security log files is still an open challenge. In this work we found that it is possible to reveal malware families by observing graph topology patterns. The choice of the best visualizations is another challenge, because the user, the task and the context should be taken into account [39]. In cybersecurity context, analysts are used to interact with text commands on terminals [40] and so visualization tools can help them to increase the speed to recognize interesting patterns. Moreover, a software tool to inspect Android apps has been provided and tested with users. The test showed how average skilled people can understand the number of weaknesses and their severity.

The present work provides an overview that can help creating hypothesis, and navigation strategies in the data space allowing people to think about different strategies for the analysis of mobile malware. Intended users can be security tool developer, but also people using Android apps in different contexts such as in mobile learning. The visual interaction may lead to detect malwares by observing topological patterns that can be identified and classified into "visual taxonomies" revealing common characteristics and allowing pro-active protection. One of the goals is to find stealth malware or obfuscate them to trace it back to families and find patterns that are invariant with respect to obfuscation methods.

In future work, the presented tool and visualization techniques will be extended to log files produced by heterogeneous sources, such as IoT devices with the aim to investigate peculiar visual patterns. The goal could slightly change from searching for malware families to analyzing behavioural patterns. Another direction is to address advanced interaction between devices [41] that can lead to menu interaction styles to perform advanced queries.

Acknowledgements The authors thank Pietro Carella for the early contribution of this work and Vincenzo Nigro for his help in the implementation of the app inspection tool and the successive evaluation.

References

1. Bitonto, P.D., Roselli, T., Rossano, V., Frezza, E., Piccinno, E.: An educational game to learn type 1 diabetes management. In: Proceedings of the 18th International Conference on Distributed Multimedia Systems, DMS 2012, August 9-11, 2012, Eden Roc Renaissance, Miami Beach, FL, USA, pp. 139–143. Knowledge Systems Institute (2012)
2. Balducci, F., Buono, P.: Building a qualified annotation dataset for skin lesion analysis trough gamification. In: Catarci, T., Norman, K.L., Mecella, M., (eds.), Proceedings of the 2018 International Conference on Advanced Visual Interfaces, AVI 2018, Castiglione della Pescaia, Italy, May 29 - June 01, 2018, pp. 36:1–36:5. ACM (2018). https://doi.org/10.1145/3206505.3206555
3. Benzi, F., Cabitza, F., Fogli, D., Lanzilotti, R., Piccinno, A.: Gamification techniques for rule management in ambient intelligence. In: de Ruyter, B.E.R., Kameas, A., Chatzimisios, P., Mavrommati, I. (eds.), Ambient Intelligence - 12th European Conference, AmI 2015, Athens, Greece, November 11-13, 2015, Proceedings, Series. Lecture Notes in Computer Science, vol. 9425, pp. 353–356. Springer (2015). https://doi.org/10.1007/978-3-319-26005-1_25
4. Karim, A., Salleh, R., Shah, S.A.A.: Dedroid: a mobile botnet detection approach based on static analysis. In: 2015 IEEE 12th International Conference on Ubiquitous Intelligence and Computing and 2015 IEEE 12th International Conference on Autonomic and Trusted Computing and 2015 IEEE 15th Intl Conf on Scalable Computing and Communications and Its Associated Workshops (UIC-ATC-ScalCom), pp. 1327–1332 (2015)
5. Chakraborty, T., Pierazzi, F., Subrahmanian, V.S.: Ec2: Ensemble clustering and classification for predicting android malware families. IEEE Trans. Depend. Sec. Comput. **17**(2), 262–277 (2020)
6. Sharif, A., Nauman, M.: Function identification in android binaries with deep learning. In: Seventh International Symposium on Computing and Networking (CANDAR), pp. 92–101. IEEE (2019)
7. Chen, Y.-M., Yang, C.-H., Chen, G.-C.: Using generative adversarial networks for data augmentation in android malware detection. In: 2021 IEEE Conference on Dependable and Secure Computing (DSC), pp. 1–8. IEEE (2021)

8. Barletta, V.S., Caivano, D., Nannavecchia, A., Scalera, M.: Intrusion detection for in-vehicle communication networks: an unsupervised kohonen som approach. Fut. Internet **12**(7), 119 (2020)
9. Barletta, V.S., Caivano, D., Nannavecchia, A., Scalera, M.: A kohonen som architecture for intrusion detection on in-vehicle communication networks. Appl. Sci. **10**(15), 5062 (2020)
10. Caivano, D., Fogli, D., Lanzilotti, R., Piccinno, A., Cassano, F.: Supporting end users to control their smart home: design implications from a literature review and an empirical investigation. J. Syst. Softw. **144**, 295–313 (2018). https://doi.org/10.1016/j.jss.2018.06.035
11. Bevanda, V., Azemovic, J., Music, D.: Privacy preserving in elearning environment (case of modeling hippocratic database structure). In: Fourth Balkan Conference in Informatics, vol. 2009, 47–52 (2009)
12. Buono, P., Carella, P.: Towards secure mobile learning. visual discovery of malware patterns in android apps. In: 23rd International Conference Information Visualisation (IV), vol. 2019, pp. 364–369. IEEE (2019)
13. Kandel, S., Heer, J., Plaisant, C., Kennedy, J., van Ham, F., Riche, N.H., Weaver, C., Lee, B., Brodbeck, D., Buono, P.: Research directions in data wrangling: visuatizations and transformations for usable and credible data. Inf. Vis. **10**(4), 271–288 (2011)
14. Benito, J.C., García-Peñalvo, F.J., Therón, R., Maderuelo, C., Pérez-Blanco, J.S., Zazo, H., Martín-Suárez, A.: Using software architectures to retrieve interaction information in elearning environments. In: 2014 International Symposium on Computers in Education (SIIE), pp. 117–120 (2014)
15. Eick, S.G., Nelson, M.C., Schmidt, J.D.: Graphical analysis of computer log files. Commun. ACM **37**(12), 50–56 (1994)
16. Zhang, Y., Xiao, Y., Chen, M., Zhang, J., Deng, H.: A survey of security visualization for computer network logs. Secur. Commun. Netw. **5**(4), 404–421 (2011)
17. Erbacher, R.F., Walker, K.L., Frincke, D.A.: Intrusion and misuse detection in large-scale systems. IEEE Comput. Graphics Appl. **22**(1), 38–47 (2002)
18. Xydas, I., Miaoulis, G., Bonnefoi, P.-F., Plemenos, D., Ghazanfarpour, D.: 3d graph visualization prototype system for intrusion detection: a surveillance aid to security analysts. In: Handbook of Graph Drawing and Visualization (2006)
19. Argyriou, E.N., Sotiraki, A.A., Symvonis, A.: Occupational fraud detection through visualization. In: IEEE International Conference on Intelligence and Security Informatics, vol. 2013, pp. 4–6 (2013)
20. Lee, J., Jeon, J., Lee, C., Lee, J., Cho, J., Lee, K.: A study on efficient log visualization using d3 component against apt: How to visualize security logs efficiently? In: 2016 International Conference on Platform Technology and Service (PlatCon), pp. 1–6 (2016)
21. Shen, Z., Ma, K.: Mobivis: a visualization system for exploring mobile data. In: IEEE Pacific Visualization Symposium, vol. 2008, pp. 175–182 (2008)
22. Lahmadi, A., Beck, F., Finickel, E., Festor, O.: A platform for the analysis and visualization of network flow data of android environments. In: IFIP/IEEE International Symposium on Integrated Network Management (IM), vol. 2015, pp. 1129–1130 (2015)
23. Somarriba, O., Zurutuza, U., Uribeetxeberria, R., Delosières, L., Nadjm-Tehrani, S.: Detection and visualization of android malware behavior. In: JECE, vol. 2016 (2016)
24. Arp, D., Spreitzenbarth, M., Hübner, M., Gascon, H., Rieck, K.: Drebin: effective and explainable detection of android malware in your pocket. In: Symposium on Network and Distributed System Security (NDSS), vol. 02 (2014)
25. Canbek, G., Sagiroglu, S., Taskaya Temizel, T.: New techniques in profiling big datasets for machine learning with a concise review of android mobile malware datasets. In: International Congress on Big Data. Deep Learning and Fighting Cyber Terrorism (IBIGDELFT), vol. 2018, pp. 117–121 (2018)
26. Jiang, J., Li, S., Yu, M., Li, G., Liu, C., Chen, K., Liu, H., Huang, W.: Android malware family classification based on sensitive opcode sequence. In: IEEE Symposium on Computers and Communications (ISCC), vol. 2019, pp. 1–7 (2019)

27. Zhang, Y., Feng, C., Huang, L., Ye, C., Weng, L.: Detection of android malicious family based on manifest information. In: 2020 15th International Conference on Computer Science Education (ICCSE), pp. 202–205 (2020)
28. Jiang, X.: Security alert: new droidkungfu variant again! found in alternative android markets (2011). http://www.csc.ncsu.edu/faculty/jiang/DroidKungFu3/
29. Zhou, Y., Jiang, X.: Dissecting android malware: characterization and evolution. IEEE Symp. Secur. Privacy **2012**, 95–109 (2012)
30. Felt, A.P., Chin, E., Hanna, S., Song, D., Wagner, D.: Android permissions demystified. In: Proceedings of the 18th ACM Conference on Computer and Communications Security, Ser. CCS '11, pp. 627–638. ACM, New York (2011)
31. Collins, C., Carpendale, S., Penn, G.: Docuburst: visualizing document content using language structure. In: Proceedings of the 11th Eurographics / IEEE - VGTC Conference on Visualization, Series EuroVis'09, pp. 1039–1046. Chichester, UK: The Eurographs Association & Wiley, Ltd (2009)
32. Wattenberg, M., Viégas, F.B.: The word tree, an interactive visual concordance. IEEE Trans. Visual Comput. Graph. **14**(6), 1221–1228 (2008)
33. IBM.: (2016) Word-cloud generator. https://www-01.ibm.com/marketing/iwm/iwm/web/preLogin.do?source=AW-0VW
34. Nodus.: Textexture - visualize text network (2012). https://noduslabs.com/radar/textexture-visualize-text-network/
35. Buono, P., Costabile, M., Lanzilotti, R.: A circular visualization of people's activities in distributed teams. J. Vis. Lang. Comput. **25**(6), 903–911 (2014)
36. Shneiderman, B.: A grander goal: a thousand-fold increase in human capabilities. Educom Rev. **32**, 4–10 (1997)
37. Bastian, M., Heymann, S., Jacomy, M.: Gephi: an open source software for exploring and manipulating networks (2009)
38. Yu, R.: Ginmaster: a case study in android malware. In: Proceedings of Virus Bulletin Conference, pp. 92–104 (2013)
39. Ardito, C. Buono, P., Costabile, M., Lanzilotti, R.: Systematic inspection of information visualization systems. In: Proceedings of BELIV'06: BEyond Time and Errors - Novel EvaLuation Methods for Information Visualization. A Workshop of the AVI 2006 International Working Conference (2006)
40. Costabile, M., Buono, P.: Principles for Human-Centred Design of IR Interfaces. Lecture Notes in Computer Science (including LNAI and LNBI), LNCS, vol. 7757, pp. 28–47 (2013)
41. Desolda, G., Ardito, C., Jetter, H.-C., Lanzilotti, R.: Exploring spatially-aware cross-device interaction techniques for mobile collaborative sensemaking. Int. J. Hum Comput Stud. **122**, 1–20 (2019)

Integrating Visual Exploration and Direct Editing of Multivariate Graphs

Philip Berger, Heidrun Schumann, and Christian Tominski

Abstract A central concern of analyzing multivariate graphs is to study the relation between the graph structure and its multivariate attributes. During the analysis, it can also be relevant to edit the graph data, for example, to correct identified errors, update outdated information, or to experiment with *what-if* scenarios, that is, to study the influence of certain attribute values on the graph. To facilitate both, the visual exploration and direct editing of multivariate graphs, we propose a novel interactive visualization approach. The core idea is to show the graph structure and calculated attribute similarity in an integrated fashion as a matrix. A table can be attached to the matrix on demand to visualize the graph's multivariate attributes. To support the visual comparison of structure and attributes at different levels, several mechanisms are provided, including matrix reordering, selection and emphasis of subsets, rearrangement of sub-matrices, and column rotation for detailed comparison. Integrated into the visualization are interaction techniques that enable users to directly edit the graph data and observe the resulting changes on the fly. Overall, we present a novel integrated approach to explore relations between structure and attributes, to edit the graph data, and to investigate changes in the characteristics of their relationships. To demonstrate the utility of the presented solution, we apply it to explore and edit structure and attributes of a network of soccer players.

P. Berger (✉) · H. Schumann · C. Tominski
University of Rostock, Rostock, Germany
e-mail: philip.berger@uni-rostock.de

H. Schumann
e-mail: heidrun.schumann@uni-rostock.de

C. Tominski
e-mail: christian.tominski@uni-rostock.de

B. Kovalerchuk et al. (eds.), *Integrating Artificial Intelligence and Visualization for Visual Knowledge Discovery*, Studies in Computational Intelligence 1014,
https://doi.org/10.1007/978-3-030-93119-3_18

1 Introduction

Multivariate graphs comprise two key aspects: the structure as defined by nodes and edges, and the multivariate data attributes associated with them. Multivariate graphs are relevant in various domains. For example, social scientists study groups of people and their social media behavior. Sports analysts investigate team compositions based on player affiliation, relationships within teams, and individual performances.

In general, the analysis of multivariate graphs aims at understanding the graph's structure and the graph's attributes [1, 2]. Particularly interesting for the visual analysis are the relationships between structure and attributes. For example, given a subset of nodes being *similar* in their attributes, do they exhibit *similar* structural properties? Or, given a certain substructure of the graph, do the nodes in that substructure exhibit *similar* attribute values? Or, given two *similar* substructures, are their associated attributes *similar* as well?

Similarity obviously plays a central role in this context. It is relevant at different levels. At the level of individual data elements, it is of interest whether nodes or edges are similar. At the level of subsets, it is interesting to see whether local parts of the data exhibit similar characteristics. Finally, at the global level, we are interested in the overall similarities between structure and attributes.

In addition to such exploratory analysis, the editing or wrangling of the data becomes increasingly important [3, 4]. Multivariate graphs can become subject to editing operations, for example, when errors or outdated information are found during the data exploration. Besides correcting errors, data editing also facilitates what-if analyses, that is, to test how data characteristics change when certain attribute values are present or are varied in the data.

For example, in sports networks, updates are required when players' performances change over the course of a season due to injuries or completed training sessions, or when their affiliations change due to transfers between clubs. Furthermore, exploring created what-if scenarios can be used in preparation for upcoming matches and seasons. For example, different training scenarios and the resulting improvements of players can be taken into account to plan ahead substitutions within the team or transfers between different clubs.

The visual analysis of the relationships between structure and attributes in such what-if scenarios is of particular interest. To experiment with what-if scenarios, it is necessary to edit the graph data and to analyze the resulting effects. Consequently, not only the editing is important, but also the visual feedback of editing effects, so that users can identify and investigate data changes in the multivariate graph. Similarity plays an important role here as well. For example, given a substructure or subset of nodes in the graph, is the edited version similar to the original? Or, given two similar substructures before editing, are they still similar to each other afterwards? Or, given two subsets of nodes, how much must the attributes of one subset change to be similar to the other?

On the global level it is of interest whether changes between the relationships of structure and attributes occurred at all. At the level of subsets, it can be interesting to see whether relationships of local parts of the data changed their characteristics. Finally, at the level of individual elements, it is of interest how nodes or edges have changed in detail.

This chapter presents a novel.[1] integrated visualization approach to support both, the visual exploration of relations in multivariate graphs and the direct editing to correct and update the data as well as to perform what-if analyses. To provide a similarity-oriented overview of the data, structure and calculated attribute similarity are visualized in an integrated fashion as a matrix-based representation. On-demand details are available in a table-based representation showing the multivariate attribute values. Flexible interactive mechanisms allow users to dynamically adapt and rearrange the visualization so that selected parts of the data can be compared visually with ease. This includes the reordering of the matrix and the table, the extraction of subsets, and the dynamic alignment of rows and columns for detailed comparison of individual data elements. Additional visual editing facilities enable users to manipulate the multivariate graph data directly in the visualization. On-demand side-by-side arrangements and a difference view provide the necessary visual feedback so that changes in the relationships between structure and attributes resulting from edit operations can comprehended.

As an illustrating example, we explore and edit a network of soccer players. We will demonstrate how the introduced approach can help users to identify and investigate relations between the network structure and the characteristics of the players. The direct editing facilities will be applied to experiment with a what-if analysis where we investigate changes in the characteristics of relationships between players.

2 Related Work

Before describing the integrated exploration and editing approach in detail, we will first look at related work in the context of multivariate graph visualization and direct data editing.

Multivariate Graph Visualizations Visualizing multivariate graphs requires showing the multivariate attributes and the graph structure. The general visualization literature provides a wealth of techniques for multivariate attributes [6]. The most frequently applied techniques are parallel coordinates and scatter plot matrices [7]. Graph structures are typically visualized as node-link diagrams or matrix representations. While node-link diagrams require appropriate layout algorithms [8], matrix representations rely on suitable ordering mechanisms for the matrix [9].

Visualization approaches for multivariate graphs typically combine ideas from multivariate data visualization and graph visualization. Both Kerren et al. [10]

[1] Note that this chapter is an extended version of a previous paper [5] While the original paper focused solely on the visual exploration of multivariate graphs, this extended chapter adds the direct editing part.

and Nobre et al. [2] provide an overview, including multi-view and single-view approaches.

Multi-view approaches show the different aspects of multivariate graphs in separate, but linked views. For example, Shannon et al. [11] combine node-link diagrams and parallel coordinates. Lex et al. [12] use a combination of parallel coordinates, heat map, and a bucket technique to visualize pathways. In related works, Partl et al. [13, 14] visualize the graph structure via node-link diagrams, while attributes are shown in a tabular visualization.

Single-view approaches show structure and attributes in an integrated fashion. That is, aspects of attributes are integrated in the visualization of the structure or the other way around. For example, Cao et al. [15] use node-link diagrams where nodes are represented as glyphs to encode multivariate attributes. Van den Elzen and Van Wijk [16] aggregate selected nodes and show within them the value distributions of node attributes. Attribute-driven approaches arrange the graph structure according to the underlying attribute values. For example, Shneiderman and Aris [17] group nodes based on attributes and show the structure-forming edges only for selected parts of the data. In follow-up work, Rodrigues et al. [18] improve the readability of the graph structure. Wattenberg [19] uses a grid-based layout of aggregated nodes and edges to show relations between attributes and edges. Bezerianos et al. [20] and Eichner et al. [21] use bi-variate scatter plots and superimpose the graph structure. Major and Basole's Graphicle [22] supports dynamically switching between layouts that facilitate structure-related and attribute-related analysis. The technique by Nobre et al. [23] juxtaposes a filtered graph structure and a table of attribute values.

The existing approaches have different pros and cons. Providing multiple views allows for using the most appropriate visualizations for structure and attributes. However, structure and attributes are shown in different parts of the display, which requires additional mental effort to integrate the two aspects. The simultaneous display of structure and attributes in a single view does not have this drawback. Relations between structure and attributes can be discerned for individual data elements and for local subsets. Yet, the increased information density can lead to increased cognitive load, which can make it more difficult to study relations on a global scale.

Editing Multivariate Graphs While graph visualization can reveal the need to edit a graph, the actual editing is mostly carried out in external representations. This is often done via alphanumeric inputs in text editors or spreadsheets. For example, the graph visualization tools Gephi [24] and Tulip [25] provide means to edit the graph data in separate tabular views. This separation of exploring the data in one view and editing the data in another view requires additional screen space, leads to unwanted attention switches, and can make it more difficult for users to see the effects of edit operations.

Following Baudel's direct manipulation principle [3], previous works have proposed different approaches to manipulate graph data in the visualization. For example, Eichner et al. [21] edit node attributes by moving the nodes in a 2D-coordinate system with an overlaid node-link diagram. To edit the structure of graphs in node-link representations, specialized lens tools can be utilized [26]. For matrix based representations, direct editing techniques mostly focus on changing the structure by (un)marking cells to create or delete edges [27, 28]. The recent *Responsive Matrix Cells* support a seamless transition from exploration to editing of multivariate graphs [29]. To this end, selected regions of the matrix can be enlarged to embed additional interactive views that enable the exploration of details as well as the editing of data values.

Editing graph data directly in the visualization has advantages compared to editing in external representations. Data can be edited directly in the view in which the need for edits surfaced. Moreover, editing effects are immediately reflected in the visualization which helps users understand the consequences of the data manipulation. For multiple successive edit operations, however, additional visual feedback is necessary to communicate not only individual value changes but also the global effect on the data as a whole. This also involves providing the user with mechanisms to compare how the data looked like before and after editing.

In summary, exploring and editing multivariate graphs often relies on manually combining several views and tools, and therefore, remains challenging. The approach to be introduced next aims to integrate visualization and editing facilities to make exploring and editing the multivariate graph easier.

3 Requirements and Approach Outline

An integrated approach to exploring and editing multivariate graphs has to fulfill the following requirements:

R1 Show similarity: Similarities between structure and attributes should be identifiable easily.

R2 Integrate structure and attributes: Graph structure and attributes should be displayed in the same view.

R3 Enable drill-down: The user should be enabled to concentrate on subsets and individual data elements.

R4 Support exploration: Interaction techniques need to be provided for exploring structure and attributes.

R5 Support direct editing: Interaction facilities to edit the multivariate graph should be provided directly in the visualization.

R6 Communicate editing effects: The new state of edited data should be immediately reflected in the visualization and be comparable to previous states.

Addressing these requirements, the first step is to calculate the attribute-wise node similarity. Based on that, an integrated view of structure and attribute similarity is provided by means of a novel form of **matrix representation**. The similarity values are color-coded in the matrix along with the usual display of the graph structure. From the matrix overview, users can drill down into details as provided in a **table representation** on demand. The table visualizes the multivariate attributes in detail to make clear to users how the underlying data values contribute to the calculated similarity. Further drill-down and exploration is facilitated through **dynamic adaptations and rearrangements** of the aforementioned visual representations. Direct manipulation interactions support the **editing of multivariate graphs** directly in the matrix and the table representation. A dedicated visual encoding is employed to help users understand and compare the induced **changes in the relationships** between structure and attributes caused by editing operations.

In the next section, we describe the integrated matrix visualization and the on-demand table visualization in detail. After that, we explain the developed interaction techniques for data exploration and editing.

4 Integrated Visualization of Graph Structure and Attributes

As indicated, two types of visual representations are combined, a matrix representation and a table representation. Both will be explained in more detail in the following.

Matrix Representation Matrix representations have proven to be useful for communicating the structure of graphs [30]. In a regular matrix visualization, rows and columns correspond to graph nodes, and a matrix cell is marked if an edge exists between the row-node and the column-node. The novel idea is to extend matrix representations such that the similarity of node attributes can be visualized along with the graph structure.

To this end, we use the upper and lower triangular parts of the matrix for different purposes, as illustrated in Fig. 1. One triangular part encodes the graph structure as usual by marking the edges. Color coding is used to represent edge weights. The darker the color, the higher is the weight. The other triangular part does not show the edges, but instead visualizes attribute similarity of the nodes.

In a first step, the pair-wise similarity of the nodes needs to be calculated. There are many different ways of calculating multivariate similarity. For the purpose of illustration, it is sufficient to implement a basic approach that uses the Euclidean distance of the underlying attribute values. Where appropriate, additional derived attributes can be included, for example, graph theoretical measures such as degree, betweenness, or centrality. Note that specific applications might need different, potentially more complex methods, such as projection techniques [31] or self-organizing maps [32].

Once computed, the attribute similarity of the nodes is visualized in the matrix by color-coding. That is, the color of cell $c_{i,j}$ represents the attribute-wise similarity of the i-th and j-th nodes. Depending on the calculated similarity, different color scales can be used for the visualization. If the similarity is in the interval $[0, 1]$ (0 for *not similar* and 1 for *similar*), then a sequential color scale (e.g., from ColorBrewer [33]) is appropriate. In this context, darker cells indicate *similar* nodes. If the similarity is calculated in the interval $[-1, 1]$ (-1 for *dis-similar*, 0 for *not similar*, and 1 for *similar*), diverging color scales would be appropriate. This similarity-based color

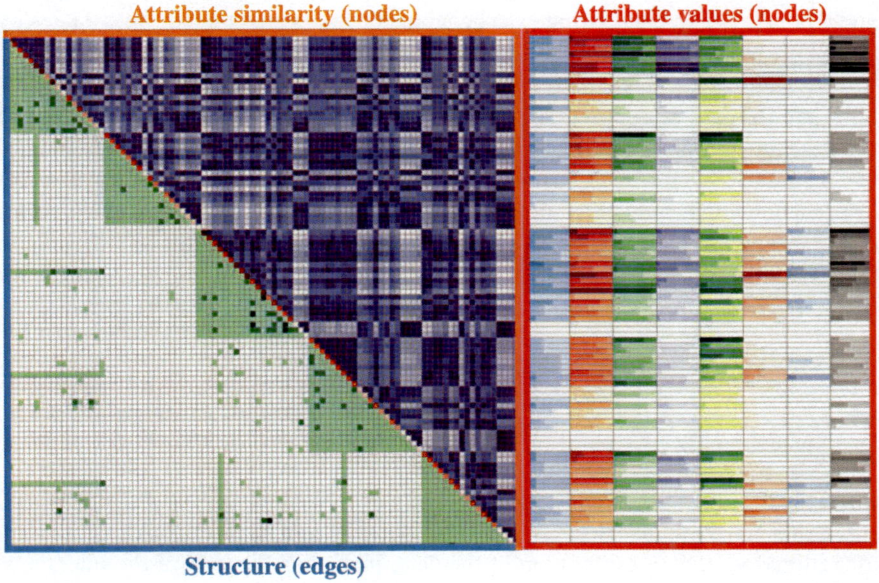

Fig. 1 Integrated matrix representation with attached table representation. The lower triangular matrix visualizes the edges that constitute the graph structure. The upper triangular matrix visualizes the pair-wise similarity of nodes with respect to their multivariate attributes. The main diagonal color-codes a selected node attribute. The table representation depicts all node attributes in detail

coding in the upper triangular matrix should be clearly distinguishable from the regular color-coding of edge weights in the lower triangular matrix.

Fulfilling requirement R1, the proposed matrix visualization facilitates the investigation of similarities in a multivariate graph. Moreover, the matrix integrates structural characteristics and attribute similarity in a single coherent view as demanded by requirement R2. Relations between structure and attributes can be discerned by comparing cells or regions in the upper diagonal matrix with the corresponding cells or regions in the lower diagonal matrix and vice versa.

Finally, it is also possible to utilize the matrix cells in the main diagonal. Usually, these cells represent loop edges, that is, edges that connect a node to itself. For graphs without loop edges, the main diagonal can alternatively be used to visualize properties of the nodes directly, for example, by color-coding a selected node attribute, as illustrated in Fig. 1.

Table Representation While the matrix representation allows users to see which nodes are similar, it is not possible to understand *why* nodes are considered similar. In fact, the calculation of similarity values is a kind of data abstraction that condenses down the multivariate attributes of two nodes to a single similarity value.

To enable users to understand why data elements are similar, an on-demand table visualization can be attached to the matrix as illustrated in Fig. 1. For each node respectively row in the matrix, there is a corresponding table row visualizing the node's multivariate data attributes. The table cells visualize the attribute values by means of two-tone pseudo coloring. This visual encoding combines the two visual variables of *color* and *length* to provide at the same time an overview and the possibility to read data values precisely [34].

Attaching the table to the matrix establishes a direct link between similarity values and their underlying attribute values. With the help of the table representation, users can drill down and study the similarity of nodes in detail. Yet, this is only a first means to address requirement R3. In the next section, we describe interactive mechanisms that enable users to further drill-down into details to explore the data for relationships between structure and attributes.

5 Interactive Data Exploration

The visualization introduced so far provides an overview of graph structure, attribute similarity, and attribute values. As formulated in requirement R4, it is also necessary to facilitate a detailed exploration of the data, including support for the interactive visual comparison of different subsets of the data. This involves the selection of subsets and the dynamic adaptation and rearrangement of the visualization to suit the comparison task.

Similarity Selection and Matrix Reordering The first step in any comparison task is to identify interesting subsets to be compared [35]. For the overview matrix, the differentiability of such interesting subsets depends on two aspects: the attributes used for the similarity calculation and ordering of rows and columns [9].

To focus on different subsets of attributes, users can select the ones to be taken into account when calculating the attribute similarity. Attributes can be included and excluded by clicking the corresponding column headers in the table visualization, which causes the similarity visualization in the upper triangular matrix to update on the fly.

To reveal interesting subsets in both parts of the overview matrix, it can be sorted by (i) structural characteristics, (ii) attribute similarity, and (iii) attribute values. Depending on the applied strategy, different patterns can be made visible, as shown in Fig. 2. By ordering the rows and columns based on structural characteristics, such as node degree or betweenness, different aspects of the graph structure can be emphasized. For example, the lower triangular matrix in Fig. 2a clearly shows cliques as green triangular patterns. Sorting the matrix based on the attribute similarity can be helpful to reveal groups of similar and dis-similar nodes. One can sort either by the similarity of all pairs of nodes or by the accumulated similarity of nodes. Figure 2b shows an example with two discernible groups after a pair-wise sorting according to the similarity with respect to a selected node. Another strategy is to sort rows and columns based on the node attribute values. This is facilitated via the column headers in the table very much like in common spreadsheet applications. After clicking a header, table and matrix are sorted by the values in the chosen

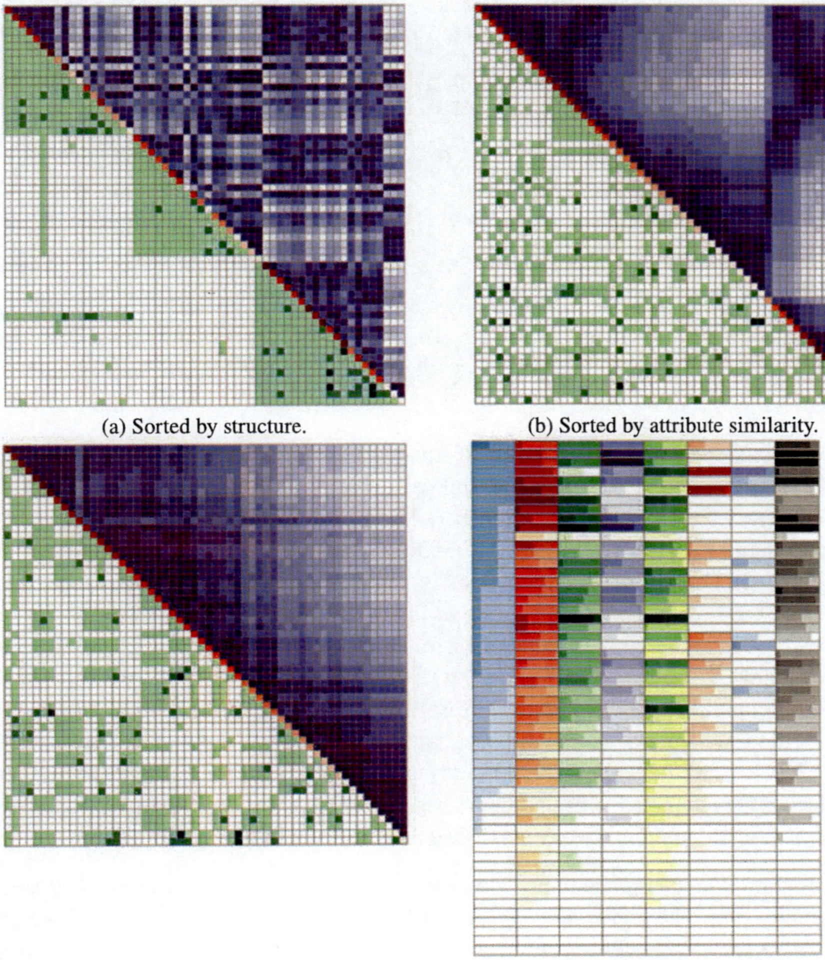

(a) Sorted by structure.

(b) Sorted by attribute similarity.

(c) Sorted by attribute values in the first table column.

Fig. 2 Integrated visualization of structure and attribute similarity

column. This can lead to interesting patterns. In Fig. 2c, a single node, distinguished by a differently colored horizontal and vertical line, stands out in the first third of the matrix.

Subset Selection and Emphasis As mentioned before, relations between structure and attributes can exist on different levels of the graph. The matrix and the table allow users to discover relations on a global level. To find local patterns, it is necessary to focus the exploration on subsets of the graph. To this end, we integrate an interactive highlight and filter technique.

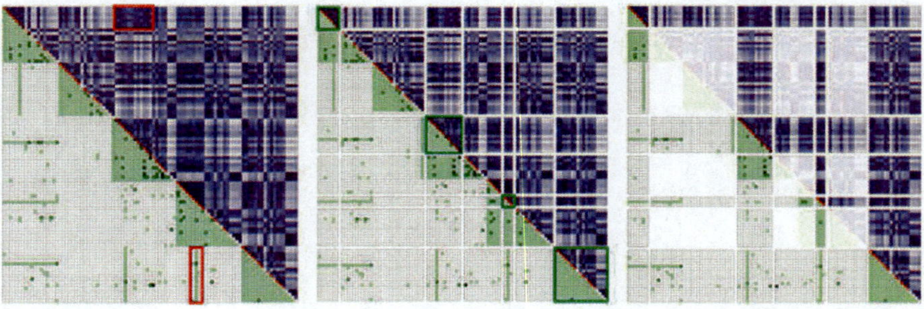

(a) Selected patterns. (b) Emphasis by inserting gaps. (c) Emphasis by filtering.

Fig. 3 Interactive selection and visual emphasis of selected data subsets

The first step is to select subsets of nodes that should be investigated in more detail. The selection is done by marking cells of the matrix or the table. Figure 3a illustrates two local patterns marked by red frames in the matrix. Marking cells in either triangular part of the matrix automatically selects the corresponding cells in the other triangular part as well. After the selection, the corresponding nodes are determined, indicated as green frames in Fig. 3b. By marking rows in the table or cells of the matrix' main diagonal, all associated nodes are selected. The selected subsets are then to be emphasized.

For a brief contemplation about the selected subsets, a rather subtle highlighting based on dynamically inserting gaps into the visual representations is sufficient. Figure 3b illustrates how these gaps divide the matrix into sub-matrices. For a stronger emphasis of the selected subsets, it makes sense to filter out those parts of the visualization that do not contain any selected nodes. This is illustrated in Fig. 3c.

Sub-matrix Rearrangement Emphasizing selected parts of the data already supports their comparison. Yet, the subsets could still be far apart in the matrix, which unnecessarily complicates the comparison. Therefore, it makes sense to dynamically rearrange the selected sub-matrices for a closer inspection. Side-by-side layouts are well suited for such comparison tasks [36].

To dynamically set up side-by-side layouts, sub-matrices can be detached from the main matrix. This effectively creates a hybrid NodeTrix layout where selected subsets are shown as separate matrices being connected via links [37]. Figure 4 shows an example with three detached sub-matrices, where two of them also show their associated attribute table. Once the sub-matrices have been studied in detail, they can be integrated back into the main matrix.

Column Rotation The techniques described so far allow users to study the data at a global level and at the level of local subsets. Finally, the analysis of individual data elements needs to be facilitated, where support for the detailed comparison of structure and attribute similarity of individual nodes is particularly important.

The matrix already visualizes the necessary information, and the selection mechanism can be used to emphasize an individual node as illustrated in Fig. 5a and b.

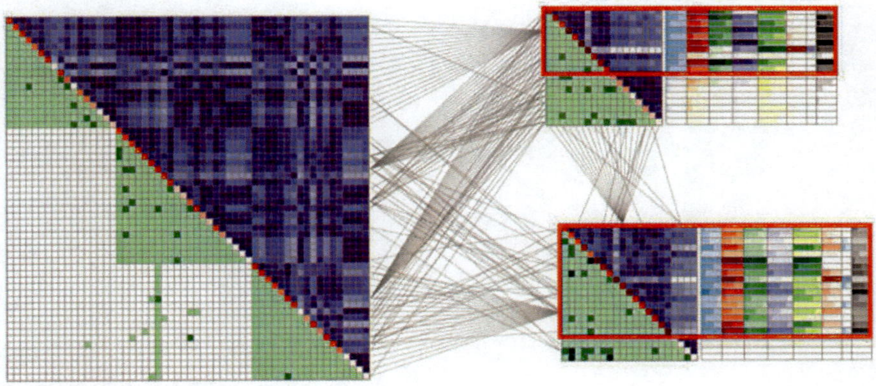

Fig. 4 Rearrangement of submatrices for side-by-side comparison

However, the separated partial rows and columns in Fig. 5b form ⌝ and ⌞ shapes along which data values must be read from the visual representation. This hampers an effective comparison of structure and attribute similarity. Ideally, structure and attribute similarity would be arranged side-by-side like so =. Therefore, the matrix column for a selected node can be dynamically rotated to create a side-by-side view of its structure and attribute similarity on the fly. As shown in Fig. 5c, the partial columns are rotated to align them with the partial rows. As a result, we obtain the two complete rows in Fig. 5d, one showing the similarity values and the other showing the structural information. To make the dynamic rearrangement easier to comprehend, it is carried out in a smooth animation.

In summary, choosing attributes for the similarity calculation, reordering, selection and emphasis, sub-matrix rearrangement, and column rotation provide the user with the interactive tools needed to explore and compare the relation between structure and attributes in multivariate graphs in detail. Next, we discuss how the multivariate graph data can be edited directly within the described matrix and table visualizations.

6 Interactive Data Editing

The interactive mechanisms introduced so far provide means to explore the multivariate graph. Additionally, addressing requirement R5, it is necessary to facilitate the editing of the graph data directly in the visualization. Considering requirement R6, appropriate visual feedback must be provided to communicate differences and similarities of the same data before and after editing. This also involves the investigation of whether relationships between structure and attributes have become more similar or dis-similar and what caused these changes in detail.

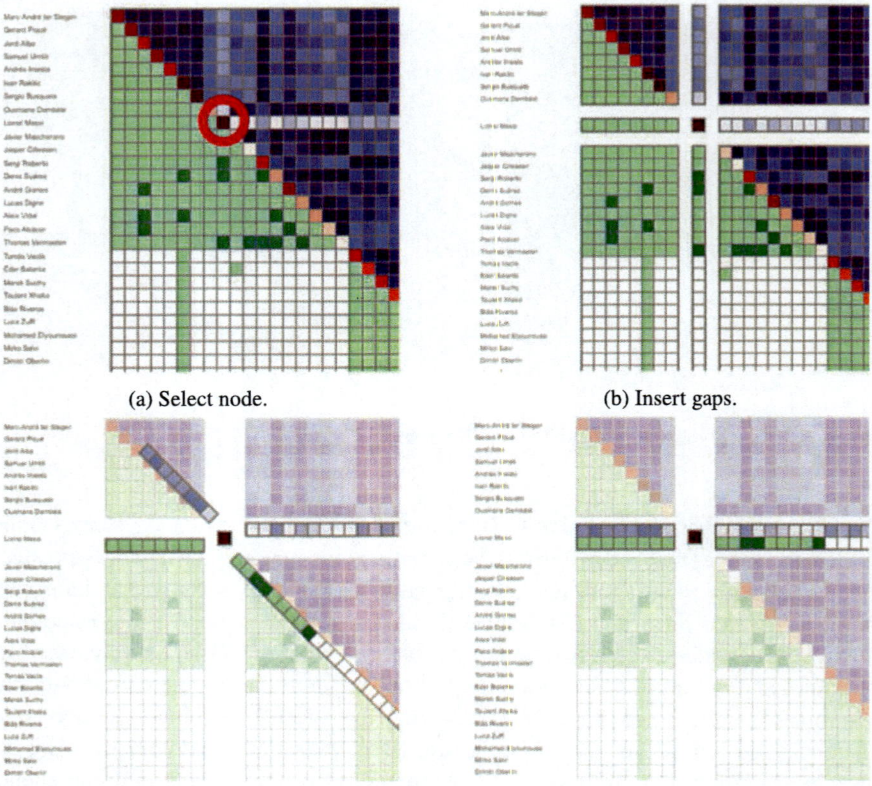

(a) Select node. (b) Insert gaps.

(c) Rotate partial columns. (d) Aligned view.

Fig. 5 Dynamic rearrangement of matrix cells to create an aligned view for the detailed inspection of structure and attribute similarity

6.1 Direct Editing Interaction

In general, edit operations can target the structure or the attributes of the graph. With our overview matrix and on-demand table representation, we have an appropriate basis for directly manipulating these targets.

Editing the Structure The lower triangular matrix facilitates editing the graph structure using different approaches from the literature [27, 28]. The edit operations are carried out directly on the matrix cells. Edges can be added or deleted by simply clicking on the corresponding matrix cell. This is useful for editing individual edges, but can be time-consuming and cumbersome when multiple edges needs to be changed. Therefore, a drag operation can be performed along rows or columns to quickly insert or remove multiple edges. With click and drag operations it is possible to directly edit the graph structure. However, the small size of the matrix cells can make it difficult to editing of edges precisely. To reduce the risk of adding or removing edges acci-

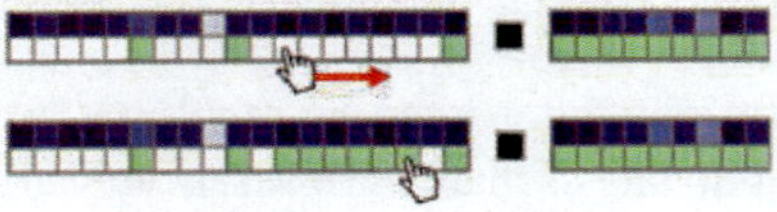

(a) Structure edit in column.

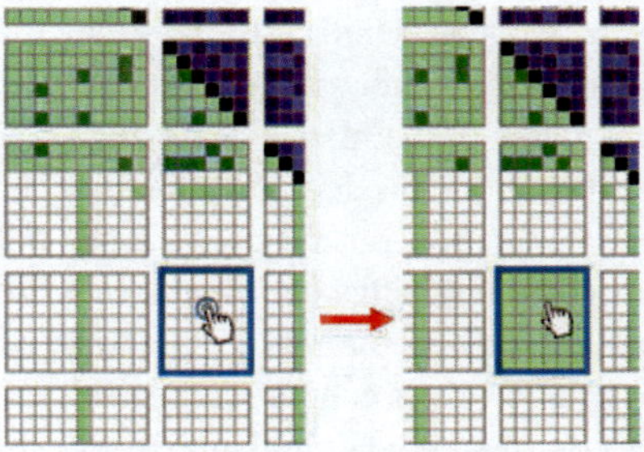

(b) Structure edit in selection.

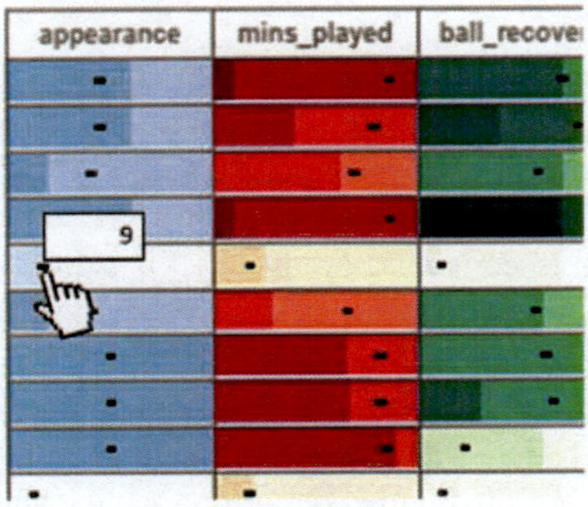

(c) Discrete attribute edit.

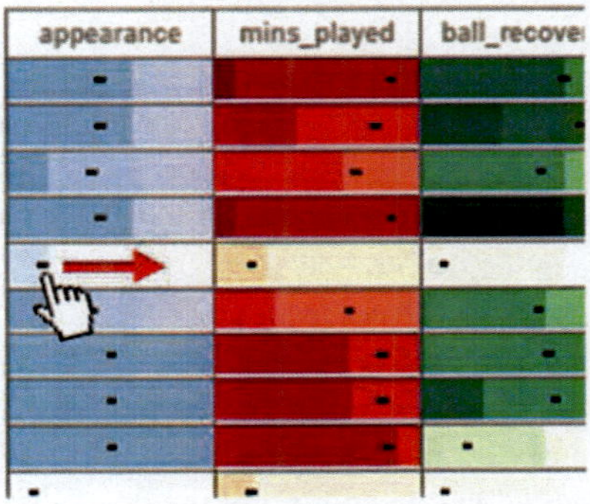

(d) Continuous attribute edit.

Fig. 6 Manipulation of the multivariate graph data through direct editing

dentally during the editing process, the previously introduced dynamic adaptation and rearrangement interactions can be employed. They are particularly helpful for adjusting multiple edges simultaneously.

The *column rotation* creates a dedicated view for selected nodes by aligning their partial rows and columns in a double row. With such a dynamically created alignment the drag operation can be constrained to the horizontal direction (see Fig. 6a), which makes the interaction easier to perform for the user. The gaps inserted between the aligned rows and the rest of the matrix further prevent unintentional selection of other cells.

The *selection and emphasis* technique can be used to facilitate the editing of blocks consisting of multiple cells. Additional interaction shortcuts are provided for adding and deleting edges based on entire blocks. A double-click inside a block in the lower triangular matrix connects the corresponding nodes and creates a clique (see Fig. 6b), whereas a long-press within the block deletes all edges.

Editing the Attributes In contrast to the lower triangular matrix, the cells in the upper triangular matrix correspond to multiple entities. In fact, each cell condenses the attribute values of pairs of nodes to a single similarity value. Here, the direct visual editing is more difficult. Previous work has demonstrated that it is possible to directly edit individual attribute values by embedding dedicated visual representations into selected matrix cells [29]. However, this approach requires focus+context distortions of the overall matrix, and hence, makes it difficult for users to interpret the visual feedback in the matrix as the data changes. Therefore, we propose editing attribute values in the on-demand table, which works without distorting the matrix.

Depending on the user's task, different edit interactions can be employed for manipulating attribute values. In correction and update scenarios, the user typically already knows the exact value to be entered. In this case, it is necessary that specific attribute values can be set precisely. Such *discrete edits* of attribute values are easily accomplished via alphanumeric keyboard input. For this purpose, a corresponding input field is provided when the user clicks and selects a table cell as illustrated in Fig. 6c.

However, in what-if scenarios no exact values are given a priori, but a whole range of values may need to be scanned before a concrete value is set. In such scenarios, typing individual values in a discrete fashion is impractical. Therefore, what-if scenarios are facilitated by *continuous edit* operations. To this end, interaction handles become available in the individual table cells as soon as a cell is selected by a user. To edit an attribute value, the action handles can be dragged along the horizontal axis, as shown in Fig. 6d.

What-if scenarios often require editing multiple elements in the graph. For this purpose, multiple selected nodes and their attribute values can be simultaneously edited in one column of the on-demand table. We distinguish between *absolute* and *relative* value changes. For absolute changes, all attribute values are set to the same value, either by entering a specific value via keyboard or by dragging an action handler, where a single handler is taken as a proxy for all selected nodes. In contrast, relative change means that each attribute value is shifted based on the relative movement of a single proxy action handler.

Visual Feedback for Data Edits Appropriate visual feedback is required to support users in understanding the effects and consequences of edit operations. Therefore, the matrix visualization and the adjacent table visualization are updated on the fly during the editing process. The *selection and emphasis* technique and *column rotation* can be used to set apart edited elements and subsets from the unedited parts of the graph, which makes it easier for the user to focus on those parts of the visualization that are effected by edit operations.

On-the-fly updates of the visualization and the possibility to focus on edited data help users understand individual edit operations, for example, to realize the influence of individual attributes in the similarity calculation. However, these means address requirement R6 only partially. The reason is that data editing may consist of multiple operations carried out over an extended period of time. In such cases, it can be difficult for users to keep all edits and data changes in mind. Therefore, we incorporate mechanisms that enable users to compare data changes in more detail.

Understanding edit effects in retrospect involves the comprehension of: (i) which aspects of the multivariate graph are affected, (ii) how the aspects have changed (e.g., stayed similar or became dis-similar) and (iii) what the changes are in detail (i.e., the concrete value changes). In order to help users to better comprehend edit effects, it

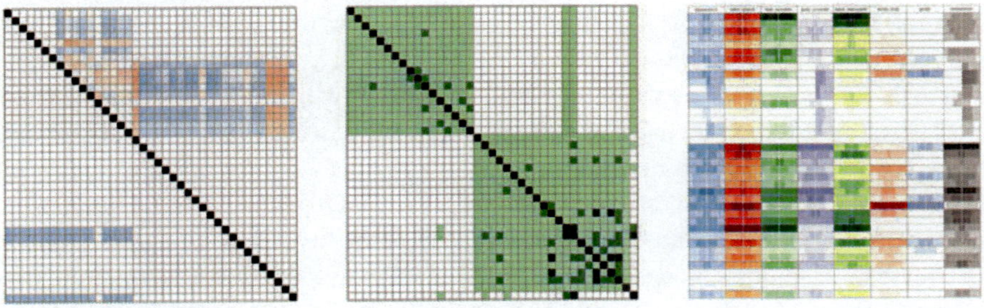

(a) Difference Encoding. (b) Mirrored Structure. (c) Attribute Comparison.

Fig. 7 Visual feedback for data edits through dynamic adaptation of the overview matrix and on-demand table

must be possible to compare how the data looked like before the edits and how they look after the edits. For this purpose, we propose additional dynamic adaptations and rearrangements of the visualization.

The first step is to identify which subsets of the multivariate graph are affected by edit operations. This step can be supported by adapting the visual encoding of the matrix More concretely, a dedicated difference encoding can be activated to enable users to compare different version of the multivariate graph.

The difference encoding visualizes the resulting data changes of each graph element and replaces the regular encoding of structure and node attribute similarity. Changes in the pair-wise node attribute similarity are visualized in the upper triangular matrix and edge weight changes are shown in the lower triangular matrix. To reveal data changes in both parts in a similar manner, we compute the differences in the interval $[-1, 1]$. In the upper triangular matrix, -1 means a pair of nodes has become more dis-similar, 0 means no change has occurred in terms of node similarity, and 1 means the two nodes got more similar. In the lower triangular matrix, -1 indicates a decrease in edge weight (or deletion of an edge), 0 means no change, and 1 indicates an increase of edge weight (or the creation of an edge). When the difference encoding is active, the matrix cells are color-coded with an appropriate diverging color scale, as illustrated in Fig. 7a. For a better distinction, this color scale is visually different from the regular encoding.

By integrating the additional difference encoding, the overview matrix can emphasize which aspects of the graph are affected by edit operations and also present a first impression of how the individual graph elements have changed. The next step is to investigate the graph elements in more detail to determine what in particular has been edited.

It would be ideal if the original and edited graph elements themselves could be displayed side-by-side for comparison. One option would be to show multiple attributes superimposed in a matrix cell. However, due to the limited space and color blending it would be difficult to make a comparison [35]. Therefore, we take advantage of the two-fold nature of our matrix. To compare edit effects in the structure, we rearrange the matrix so that the structure is mirrored in the upper triangular matrix. The lower triangular matrix represents the edges before editing and the upper triangular matrix after editing. This creates a side-by-side view similar to a regular matrix representation with directed edges, as shown in Fig. 7b. In concert with the previously presented interactive exploration techniques (see Sect. 5), changes in the graph structure can be explored and compared in detail.

To enable users to comprehend changes of the attributes in more detail, a side-by-side view in the on-demand table representation is created. Depending on whether an overview of multiple nodes or a detailed view of individual nodes is required, two variants are suitable. For an overview, the table cells are split vertically to display pairs of attribute values in a mirror-like manner, illustrated in Fig. 7c. The attribute values before the edits are displayed starting from the center to the left and those after the edits from the center to the right. Along the individual table columns, changes in an attribute can be compared for multiple nodes.

For a detailed view on individual nodes, a horizontal split of the table cells and a vertical arrangement of the attribute values on top of each other would be more suitable. In this case, the upper bar displays the attribute value before the edits and the lower after them. For each node, the individual value changes in their attributes can be compared along the table rows.

7 Use Case

To demonstrate our approach, we use a dataset of soccer players of 16 clubs participating in the Champions League season 2017/18. Nodes of the network are players with attributes such as the number of matches played, the number of minutes played, pass accuracy, ball recoveries, and many more defensive and offensive characteristics. Edges exist between two players if they have played for the same club at some time during their career. The edge weight corresponds to the number of clubs that two players have in common.

7.1 Exploration of Relations Between Structure and Attributes

The goal of the analysis of this multivariate network is to find out whether patterns occurring in the structure are related to the attributes, or vice versa. Of particular interest are the team compositions and individual player performances that stand out from others. To this end, multiple clubs have to be compared to each other before they are studied more closely.

To identify structural patterns, the matrix is sorted based on the current club affiliation. This reveals triangular cliques in the lower part of the matrix containing players in the same club. Figure 8 shows an extract with three clubs. Darker cells indicate

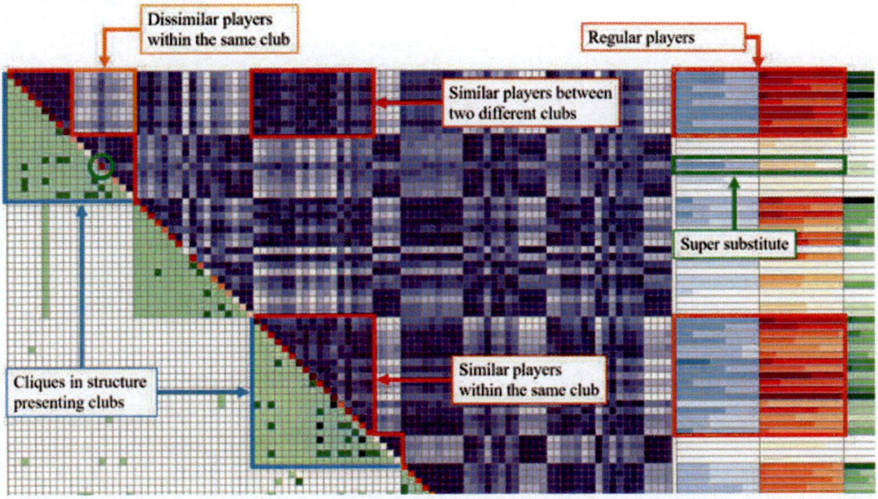

Fig. 8 Extract of the soccer network ordered according to team affiliation. Cliques in the lower matrix depict different soccer clubs. The upper triangular matrix shows the similarity between players based on their matches and minutes played

that players have played for the same club before in their career, which indicates similarities in their transfer history. The colored cells outside of the triangular groups represent former affiliations of the players besides their current club. If those cells form horizontal or vertical lines, the corresponding player has been a former member of the club whose triangular clique is parallel to the line.

To identify relations between club affiliation and team composition, we select the attributes *matches played* and *minutes played* in the attribute table. As a result, the upper triangular matrix shows the similarity of players with respect to these two attributes. In Fig. 8, we can see that for the two clubs at the top left and in the center, their triangular cliques in the lower triangular matrix are each mirrored by two smaller triangular patterns of similar players in the upper triangular matrix. Moreover, between the darker triangles of similar players, there are rectangular regions with brighter colors, indicating dissimilarity between the two groups. In correspondence with the table view, these patterns capture groups of regular players with many matches and minutes played (upper similarity triangles) and substitute players with much less time on the pitch (lower similarity triangles). A frequent substitute player with many matches but just a few minutes played, a so-called super substitute, stands out in the matrix, as indicated in Fig. 8.

Note that for the third club (bottom right in our example), the separation of frequent players and substitute players is much less pronounced. Instead of two triangles flanking a rectangular region, we rather see one large triangle. This indicates that the third club regularly rotates the starting lineup and substitute players. The structural part of the matrix also suggests that the players of the third club have stronger connections, that is, they have played for the same clubs.

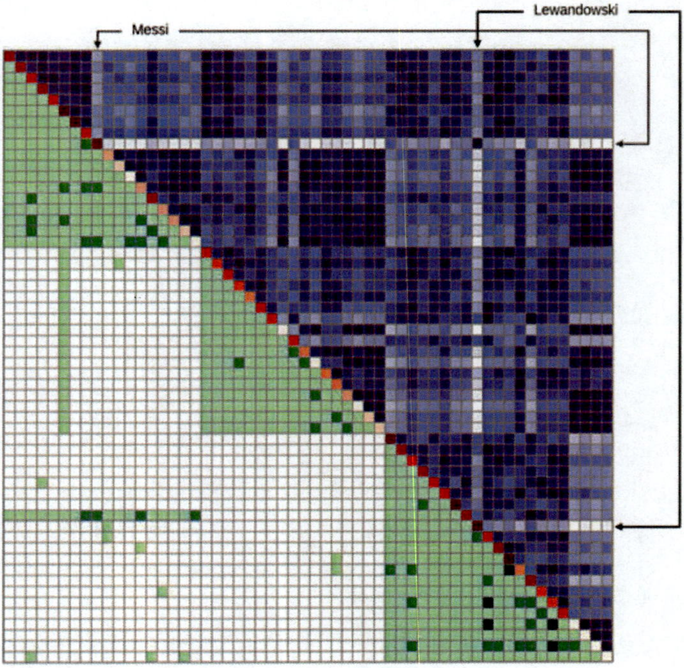

Fig. 9 Overview matrix with two players standing out

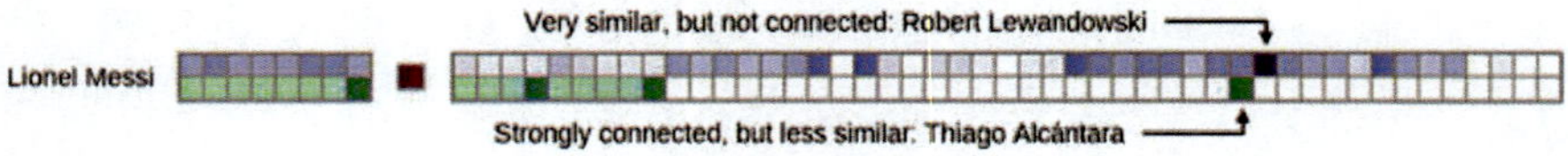

Fig. 10 Detail view of player Lionel Messi. Top row: Pair-wise similarity to other players. Bottom row: Connection to other players

For a detailed comparison, Fig. 4 shows the first and the third club detached from the main matrix. Yet, the players' offensive and defensive attributes in the linked tables show no discernible differences in the performance of the starting lineups of both clubs.

To investigate key offensive players in the dataset the upper triangular matrix in Fig. 9 displays the similarities of players with respect to shot goals, attempts at goal, successful passes, and failed passes. Two players appear to be rather dissimilar to the rest of the players as indicated by brighter rows and columns. The players are Lionel Messi and Robert Lewandowski.

In order to study Messi in detail, the matrix cells for the node are dynamically rotated. In Fig. 10, one can see that Messi is dissimilar to all players who played for the same clubs as he did. There is only one player being similar, indicated by a dark cell in the similarity row, which is Lewandowski, but we can see that Messi and he have never played for the same club. Yet, immediately next to Lewandowski is a player with whom Messi already played together, presented by a dark cell in the structure row. This player is Thiago Alcántara.

Taken together, the findings presented in this section demonstrate how well our techniques support the visual exploration of relations between structure and attributes in multivariate graphs. Starting with sorting the integrated matrix, users can apply different interaction techniques to study selected parts in detail. This allows users to quickly identify key characteristics of interesting data elements and to compare them with other parts of the data.

Besides exploring the multivariate graph and its characteristics, it is also important to be able to edit it and observe the resulting changes. To this end, we present a use case of a what-if analysis in the next section.

7.2 *Integrating Exploration and Editing for What-If Analysis*

As an example of a what-if scenario, we consider the second and the third club from the previous analysis (center and bottom of the lower triangular matrix in Fig. 9). Let's call them C2 and C3. Our goal is to prepare C2 for the new season by making the players of this club more similar to the players and style of play of C3. To this end, suitable transfer players have to be added to C2 and player attribute values have to be edited. Afterwards, the editing effects need to be investigated to decide if the required changes are feasible in preparation for the new season.

To identify existing similarities in the play style between the clubs, we look at players who participated in a lot of matches and their performances. For this purpose, we select the attributes matches played, minutes played, pass accuracy and ball possession, and display them in the upper triangular matrix. In Fig. 11a, we can see that this reveals dissimilarities between two major groups in the clubs.

In order to reduce the differences between the clubs, the first step is to find suitable transfer players. They should have high pass accuracy and ball possession, and preferably have already links to our club C2 so that they can quickly fit in. The lower triangular matrix and the on-demand table in Fig. 11a reveal three possible candidates. It becomes apparent that two candidates are from C3 and one from another club (bottom row).

To focus on the relevant subgraph and to facilitate editing, the two clubs and the third transfer candidate are rearranged in a new submatrix. In Fig. 11, clubs are distinguishable by the triangular cliques in the lower part of the matrix. C2 at the top left, C3 at the bottom and the unaffiliated possible third transfer player is positioned at the bottom right.

At the start of the editing process, new edges are added between C2 and the possible transfer players. This is done by a double click in the corresponding selection of transfer players and club C2, as illustrated in Fig. 11b [1]. Next, the players from the dissimilar group in C2 are gradually edited until they appear similar enough to the players in C3. Figure 11b [2] highlights the relevant attributes. By simply dragging

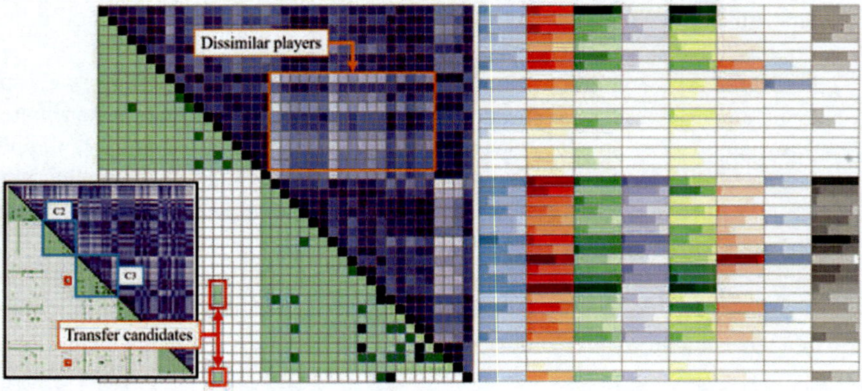

(a) An overview of the graph and the selection of the relevant submatrix.

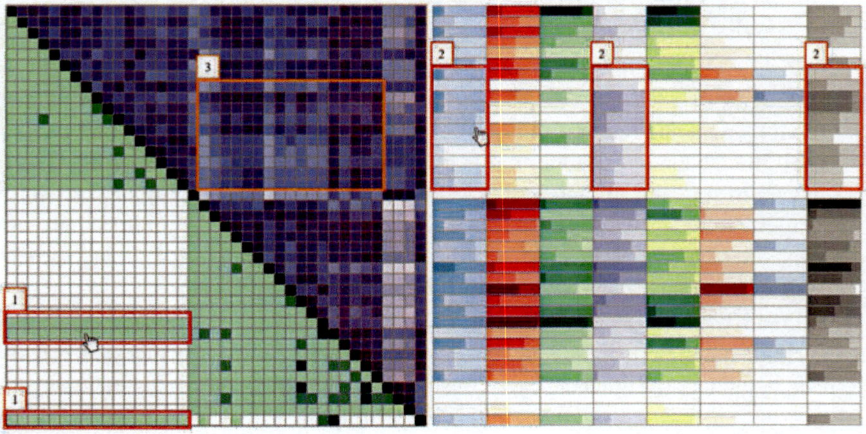

(b) Editing the structure (1) and attributes (2) of the submatrix.

Fig. 11 Submatrix rearrangement of an extract from the soccer network with associated attribute table, before and after editing the data

the action handler in the table cells, value ranges for the first attributes and players can be skimmed quickly. While the similarity part of the matrix is updated on the fly, desired value settings for the attributes can be found. To speed up the editing process, these settings can be used to directly enter values via keyboard or to edit multiple players at once. This process is repeated until the attributes in the table and the similarity part of the matrix show the desired values. To check this, the multivariate network can be explored anytime, as described in the previous use case. Figure 11b 3 shows the overall result of the editing steps.

To test the feasibility of the editing operations with respect to the goal stated before, the difference encoding is activated (Fig. 12a). The blue cells in the lower triangular matrix indicate the newly added edges. In the upper triangular matrix, red cells indicate that players have become more dissimilar and blue indicates an

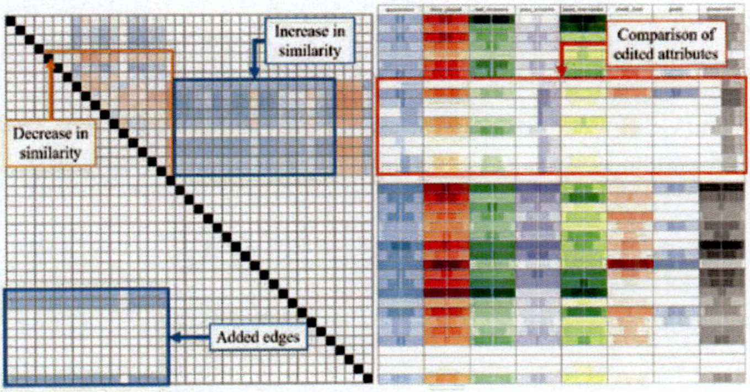

(a) Difference encoding to compare changes in the data.

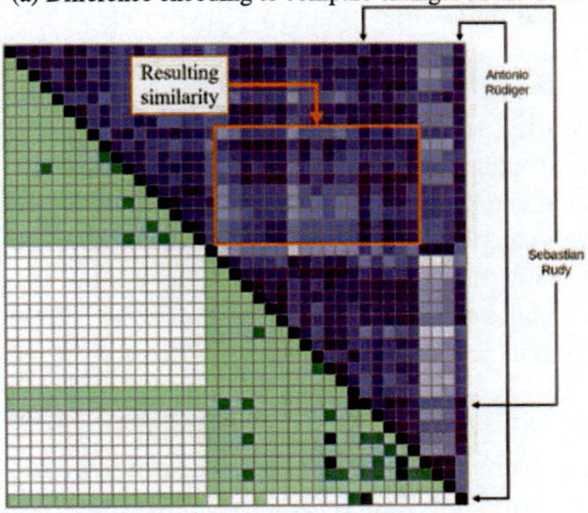

(b) Resulting similarity after editing.

Fig. 12 Visual feedback of the editing changes

increase in similarity. Figure 12a, shows that the players of C2 have become more dissimilar to each other. However, the blue cells show that players have become more similar between clubs, which is in line with our goal. In order to investigate the edit effects in detail, the split table cells are activated to directly compare the attribute values before and after the edit operations. The resulting similarity between C2 and C3 is highlighted in Fig. 12b. Furthermore, it can be seen from the updated similarities that the second and third transfer candidates fit the new setup of C2 best. Overall, they are very similar to all the players from C2. These players are Antonio Radiger and Sebastian Rudy. As a result, C2 becomes more similar to C3 due to the performed edits of the individual player values and the integration of the potential transfer players. Based on the values shown in Fig. 11b, the coach can now decide if the players' development can be achieved during the preparation for the new seasons or if the what-if scenario needs further tuning.

This walk-through illustrated how what-if scenarios can be supported through directly editing the data in the matrix and the table representation. The corresponding effects on the data are made comprehensible via dedicated visual and interactive means. Overall, the case studies showed how users can utilize our techniques to explore multivariate graphs and to edit them, and that it is possible to seamlessly switch between exploration and editing phases.

8 Conclusion and Future Work

In this chapter, we presented a novel approach to integrating the visual exploration and the direct editing of multivariate graphs. The core idea was to display structural aspects and multivariate attribute similarity in an integrated fashion as a matrix, and to provide details of the attributes in an on-demand table. Several interactive mechanisms can be used to adapt and rearrange the visualization to facilitate the comparisons of both structure and attributes at different levels of granularity. We think that such an integrated and flexible visualization of structure, attribute similarity, and multivariate attributes can provide new insight into the relation between structure and attributes in multivariate graphs. On top of that, direct editing facilities are integrated in the visualization. These are useful for data corrections and what-if analyses. The resulting editing effects can be investigated with the help of additional visual and interactive means.

In the future, the utility of the presented techniques can be improved further. For example, tracing paths through the graph by jumping from one matrix cell to another is known to be difficult [38]. To better support path-related exploration tasks one could further strengthen the interplay of matrix and node-link representations. The NodeTrix layout provides a good starting point for such investigations. Another promising research direction would be to improve the dynamic rearrangements to facilitate comparison tasks. So far, the rearrangements are triggered manually by the user. An alternative could be to automatically create suitable arrangements based on the calculated similarity of data elements [39]. This would also require improving the calculation of similarities. It is already possible for users to choose the attributes to be considered for the similarity calculation. But this is only a first step. Additional methods and control mechanisms would be needed for a flexible task-based calculation of similarities.

Another aspect that should be considered in the future is the scalability of the presented approach. In general, matrix visualizations of graphs with more than a couple of 100 nodes are difficult to handle. The NodeTrix layout and the flexible arrangements of submatrices are a first step in addressing this challenge. Another option would be to hierarchically aggregate matrix cells to meta nodes to create a zoomable multi-scale matrix [40].

It is further important to address scalability in terms of the number of attributes. The similarity encoding is but a first step in reducing the amount of information displayed to the users at once. An interesting future direction could be the integration of additional dimensional reduction techniques, for example PCA [41] or t-SNE [42] and their incorporation into the visualization. One option would be to combine manual and automatic approaches. For example, using machine learning capabilities to automatically make suggestions to the users based on their manual selections. These suggestions could then be used to support users in refining existing arrangements or creating new ones [43].

Future work should also consider improving the direct data editing. For example, a more comprehensive history mechanism is needed to not only compare the last edit operations but arbitrary states of the graph. For longer and more extensive analyses, data provenance features are particularly useful [44–46]. For the interaction itself, it would be interesting to study modern interaction modalities such as touch, pen, or speech input, which might help improving precision and simplify editing operations. In addition, the utility of the proposed concepts and techniques in this chapter need to be evaluated through user studies in the future.

References

1. Pretorius, J., Purchase, H.C., Stasko, J.T.: Tasks for multivariate network analysis. In: Multivariate Network Visualization, pp. 77–95. Springer International Publishing (2014)
2. Nobre, C., Meyer, M., Streit, M., Lex, A.: The State of the art in visualizing multivariate networks. Comput. Graph. Forum **38**(3), 807–832 (2019)
3. Baudel, T.: From information visualization to direct manipulation: extending a generic visualization framework for the interactive editing of large datasets. In: Proceedings of ACM Symposium on User Interface Software and Technology. ACM (2006)
4. Kandel, S., Heer, J., Plaisant, C., Kennedy, J., Van Ham, F., Riche, N.H., Weaver, C., Lee, B., Brodbeck, D., Buono, P.: Research Directions in data wrangling: visualizations and transformations for usable and credible data. Inf. Vis. **10**(4), 271–288 (2011)
5. Berger, P., Schumann, H., Tominski, C.: Visually exploring relations between structure and attributes in multivariate graphs. In: Proceedings of IEEE International Conference on Information Visualization. IEEE (2019)
6. Tominski, C.: Schumann, H.: Interactive Visual Data Analysis. CRC Press, AK Peters Visualization Series (2020)
7. Ward, M.O., Grinstein, G., Keim, D.: Interactive Data Visualization: Foundations, Techniques, and Applications, 2 edn. A K Peters/CRC Press (2015)
8. Tamassia, R. (ed.): Handbook of Graph Drawing and Visualization. CRC Press (2013)
9. Behrisch, M., Bach, B., Henry Riche, N., Schreck, T., Fekete, J.D.: Matrix reordering methods for table and network visualization. Comput. Graph. Forum **35**(3), 693–716 (2016)
10. Kerren, A., Purchase, H.C., Ward, M.O. (eds.): Multivariate Network Visualization. Springer International Publishing (2014)
11. Shannon, R., Holland, T., Quigley, A.: Multivariate Graph Drawing Using Parallel Coordinate Visualisations. University College Dublin, School of Computer Science and Informatics (2008)
12. Lex, A., Streit, M., Kruijff, E., Schmalstieg, D.: Caleydo: design and evaluation of a visual analysis framework for gene expression data in its biological context. In: Proceedings of IEEE Pacific Symposium on Visualization, pp. 57–64 (2010)
13. Partl, C., Lex, A., Streit, M., Kalkofen, D., Kashofer, K., Schmalstieg, D.: enRoute: dynamic path extraction from biological pathway maps for in-depth experimental data analysis. In: Proceedings of IEEE Symposium on Biological Data Visualization, pp. 107–114. IEEE (2012)
14. Partl, C., Gratzl, S., Streit, M., Wassermann, A.M., Pfister, H., Schmalstieg, D., Lex, A.: Pathfinder: visual analysis of paths in graphs. Comput. Graph Forum **35**(3), 71–80 (2016)

15. Cao, N., Lin, Y.R., Li, L., Tong, H.: g-Miner: interactive visual group mining on multivariate graphs. In: Proceedings of ACM Conference on Human Factors in Computing Systems, pp. 279–288. ACM (2015)
16. Van den Elzen, S., Van Wijk, J.J.: Multivariate network exploration and presentation: from detail to overview via selections and aggregations. IEEE Trans. Vis. Comput. Graph. **20**(12), 2310–2319 (2014)
17. Shneiderman, B., Aris, A.: Network visualization by semantic substrates. IEEE Trans. Vis. Comput. Graph. **12**(5), 733–740 (2006)
18. Rodrigues, E.M., Milic-Frayling, N., Smith, M., Shneiderman, B., Hansen, D.: Group-in-a-box layout for multi-faceted analysis of communities. In: Proceedings of IEEE Third International Conference on Privacy, Security, Risk and Trust and 2011 IEEE Third International Conference on Social Computing, pp. 354–361. IEEE (2011)
19. Wattenberg, M.: Visual exploration of multivariate graphs. In: Proceedings of ACM Conference on Human Factors in Computing Systems, pp. 811–819. ACM (2006)
20. Bezerianos, A., Chevalier, F., Dragicevic, P., Elmqvist, N., Fekete, J.D.: Graphdice: a system for exploring multivariate social networks. Comput. Graph. Forum **29**(3), 863–872 (2010)
21. Eichner, C., Gladisch, S., Schumann, H., Tominski, C.: Direct visual editing of node attributes in graphs. Informatics **3**(4), 17 (2016)
22. Major, T., Basole, R.C.: Graphicle: exploring units, networks, and context in a blended visualization approach. IEEE Trans. Vis. Comput. Graph. **25**(1), 576–585 (2019)
23. Nobre, C., Streit, M., Lex, A.: Juniper: a tree+table approach to multivariate graph visualization. IEEE Trans. Vis. Comput. Graph. **25**(1), 544–554 (2019)
24. Bastian, M., Heymann, S., Jacomy, M.: An Open Source Software for Exploring and Manipulating Networks. The AAAI Press (2009)
25. Auber, D.: Tulip—a huge graph visualization framework. In: Graph Drawing Software, pp. 105–126. Springer International Publishing (2004)
26. Gladisch, S., Schumann, H., Ernst, M., Füllen, G., Tominski, C.: Semi-automatic editing of graphs with customized layouts. Comput. Graph. Forum **33**(3), 381–390 (2014)
27. Gladisch, S., Schumann, H., Loboschik, M., Tominski, C.: Toward using matrix visualizations for graph editing. Poster at IEEE Conference on Information Visualization (2015)
28. Kister, U., Klamka, K., Tominski, C.: Dachselt, R: GraSp: combining spatially-aware mobile devices and a display wall for graph visualization and interaction. Comput. Graph. Forum **36**, 503–514 (2017)
29. Horak, T., Berger, P., Schumann, H., Dachselt, R., Tominski, C.: Responsive matrix cells: a focus+context approach for exploring and editing multivariate graphs. IEEE Trans. Vis. Comput. Graph. **27**(2), 1644–1654 (2021)
30. Ghoniem, M., Fekete, J.D., Castagliola, P.: A comparison of the readability of graphs using node-link and matrix-based representations. In: Proceedings of IEEE Symposium on Information Visualization (2004)
31. Behrisch, M., Davey, J., Fischer, F., Thonnard, O., Schreck, T., Keim, D., Kohlhammer, J.: Visual analysis of sets of heterogeneous matrices using projection-based distance functions and semantic zoom. Comput. Graph. Forum **33**(3), 411–420 (2014)
32. Kohonen, T.: Self-organizing Maps. Springer Science & Business Media (2012)
33. Harrower, M.A., Brewer, C.A.: ColorBrewer.org: an online tool for selecting color schemes for maps. Cartographic J. **40**(1), 27–37 (2003)
34. John, M., Tominski, C., Schumann, H.: Visual and analytical extensions for the table lens. In: Proceedings of SPIE Conference on Visualization and Data Analysis, pp. 680907–1–680907–12. SPIE (2008)
35. Tominski, C., Forsell, C., Johansson, J.: Interaction Support for Visual Comparison Inspired by Natural Behavior. IEEE Trans. Vis. Comput. Graph. **18**(12), 2719–2728 (2012)
36. Gleicher, M., Albers, D., Walker, R., Jusufi, I., Hansen, C.D., Roberts, J.C.: Visual comparison for information visualization. Inf. Vis. **10**(4), 289–309 (2011)
37. Henry, N., Fekete, J.D., McGuffin, M.J.: NodeTrix: a hybrid visualization of social networks. IEEE Trans. Vis. Comput. Graph. **13**(6), 1302–1309 (2007)

38. Henry, N., Fekete, J.D.: Matlink: enhanced matrix visualization for analyzing social networks. In: Proceedings of IFIP Conference on Human-Computer Interaction, pp. 288–302. Springer International Publishing (2007)
39. Tominski, C.: CompaRing: reducing costs of visual comparison. In: Short Paper Proceedings of the IEEE VGTC/Eurographics Conference on Visualization. Eurographics Association (2016)
40. Abello, J., Van Ham, F.: Matrix zoom: a visual interface to semi-external graphs. In: Proceedings of IEEE Symposium on Information Visualization, pp. 183–190 (2004)
41. Jolliffe, I.T.: Principal Components in Regression Analysis, pp. 129–155. Springer International Publishing, Cham, Switzerland (1986)
42. Van der Maaten, L., Hinton, G.: Visualizing Data using t-SNE. J. Mach. Learn. Res. **9**(86), 2579–2605 (2008)
43. Chegini, M., Bernard, J., Berger, P., Sourin, A., Andrews, K., Schreck, T.: Interactive labelling of a multivariate dataset for supervised machine learning using linked visualisations, clustering, and active learning. Vis. Inf. **3**(1), 9–17 (2019)
44. Kreuseler, M., Nocke, T., Schumann, H.: A history mechanism for visual data mining. In: Proceedings of IEEE Symposium on Information Visualization, pp. 49–56. IEEE (2004)
45. Mathisen, A., Horak, T., Klokmose, C.N., Grønbæk, K., Elmqvist, N.: Integrating data-driven reporting in collaborative visual analytics. Comput. Graph. Forum, InsideInsights (2019)
46. Nancel, M., Cockburn, A.: Causality: a conceptual model of interaction history. In: Proceedings of ACM Conference on Human Factors in Computing Systems, pp. 1777–1786. ACM (2014)

Real-Time Visual Analytics for Air Quality

Chiara Bachechi, Laura Po, and Federico Desimoni

Abstract Raise collective awareness about the daily levels of humans exposure to toxic chemicals in the air is of great significance in motivating citizen to act and embrace a more sustainable life style. For this reason, Public Administrations are involved in effectively monitoring urban air quality with high-resolution and provide understandable visualization of the air quality conditions in their cities. Moreover, collecting data for a long period can help to estimate the impact of the policies adopted to reduce air pollutant concentration in the air. The easiest and most cost-effective way to monitor air quality is by employing low-cost sensors distributed in urban areas. These sensors generate a real-time data stream that needs elaboration to generate adequate visualizations. The TRAFAIR Air Quality dashboard proposed in this paper is a web application to inform citizens and decision-makers on the current, past, and future air quality conditions of three European cities: Modena, Santiago de Compostela, and Zaragoza. Air quality data are multidimensional observations update in real-time. Moreover, each observation has both space and a time reference. Interpolation techniques are employed to generate space-continuous visualizations that estimate the concentration of the pollutants where sensors are not available. The TRAFAIR project consists of a chain of simulation models that estimates the levels of NO and $NO2$ for up to 2 days. Furthermore, new future air quality scenarios evaluating the impact on air quality according to changes in urban traffic can be explored. All these processes generate heterogeneous data: coming from different sources, some continuous and others discrete in the space-time domain, some historical and others in real-time. The dashboard provides a unique environment where all these data and the derived statistics can be observed and understood.

C. Bachechi (✉) · L. Po · F. Desimoni
Enzo Ferrari Engineering Department, University of Modena and Reggio Emilia, Modena, Italy
e-mail: chiara.bachechi@unimore.it

L. Po
e-mail: laura.po@unimore.it

F. Desimoni
e-mail: federico.desimoni@unimore.it

B. Kovalerchuk et al. (eds.), *Integrating Artificial Intelligence and Visualization for Visual Knowledge Discovery*, Studies in Computational Intelligence 1014,
https://doi.org/10.1007/978-3-030-93119-3_19

1 Introduction

The Word Health Organization (WHO) air quality guidelines updates [1] state that the reduction of air pollution levels can significantly decrease the burden of disease from heart disease, lung cancer, stroke, and both chronic and acute respiratory diseases (e.g. asthma). People already spend a lot of time deciding what to eat or drink to be healthy, they should also consider that the quality of the air they breathe strongly influences their health and lifestyle.

For this reason, public administrations and citizens need systems able to let them visualize the air pollutants concentration in order to change their behaviour and motivate the adoption of effective countermeasures. The TRAFAIR project[1] is a European project that aims to understand traffic flow in order to improve air quality. TRAFAIR monitors both urban air quality and road traffic in real-time and employs a chain of simulation models to predict urban air pollutant concentrations for the next two days. In the scope of the TRAFAIR project, a network of low-cost air quality sensors has been installed in 6 European cities: Modena, Florence, Livorno, Pisa, Santiago de Compostela, and Zaragoza. The measured pollutants are: CO, NO_2, NO, and O_3. These substances are recognized as toxic and the air is considered polluted if their concentration is higher than a legal limit for a certain period. Thus, to identify the level of pollution in a specific area the concentration of the pollutants should be monitored continuously and compared with the European standards.

This paper describes the TRAFAIR Air Quality dashboard implemented to share and visualize air quality data collected and generated during the TRAFAIR project. The dashboard is now employed by public administration members and citizens in the city of Modena, Zaragoza, and Santiago de Compostela.

Spatiotemporal data are difficult to visualize. Adequate visualization techniques are required to cope with data that have both spatial and temporal dimensions. Moreover, the majority of the available visualization needs to be frequently updated when new data are available. The dashboard communicates information about real-time, statistics, and trends of air quality conditions in a city through dynamic and/or interactive graphs and maps.

The dashboard was developed to manage a large amount of data coming in real-time from different sensors and characterized by both spatial and temporal dimensions. Moreover, it can handle complex data such as series of maps obtained from models or interpolation processes, and it is scalable and easily adaptable to different cities.

This work extends a previous work [2], where we mainly refer to the visualizations available in the city of Modena. In this version, the dashboard has been extended to Zaragoza and Santiago de Compostela adapting some views and demonstrating the versatility of the proposed solution. The related work section has been enriched with some comparisons with similar web services. Section 3 has been extended and revise adding a description of the different types of geodata, outlining the chain of simulation models in the TRAFAIR project, and describing the characteristics of

[1] https://trafair.eu.

air quality monitoring sensors. Moreover, new visualizations have been described in Sect. 5: air quality forecasts, historical interpolation and forecasted maps, and future scenarios.

The rest of the paper is structured as follows. In Sect. 2, a literature review on air quality data visualization is provided and, Sect. 3 provides some background knowledge to better understand the air pollution monitoring system and the chain of models used to generate forecasts. In Sect. 4, the system architecture that enables the visualization is discussed. Then, in Sect. 5, dashboard properties, charts, and visualizations are presented.

Finally, the results displayed in the dashboard are commented and future works are described in Sect. 6.

2 Related Work

Smart city data are generally communicated and shared with citizens and public administration members through dashboards. When dealing with air quality information, the main goal is to communicate where the hot spots are located. Air pollutants concentration data are difficult to interpret and their visual analytic can significantly influence the citizens' perception of the air quality. Visualizing gradually varied colour maps on Google Earth has a stronger expression than the original data table as proved for the city of BeiJing in [3]. In general, AQ maps provide an efficient way to investigate and understand the current status of air quality and to identify spatial-temporal patterns of air quality [4]. In [5], spatiotemporal data of weather and air quality have been compared with the incidence of COVID-19 in Spain to understand possible correlations. In [6], an interactive web-based geo-visual analytics platform allows the user to explore the results of the co-clustering analysis of their spatiotemporal data. Users can upload their data setting the co-clustering parameters and then the data are processed with a co-clustering algorithm considering both the spatial and the temporal dimensions of the dataset. Coordinated Multiple Views (CMV) are employed to display different aspects of the uploaded data: a geographical map, a linear timeline, a heatmap, and adjacent small multiples maps representing changes associated with timestamps. CMV is an efficient solution to help the user exploring spatiotemporal data.

City dashboards are analysed and compared in [7] with a focus on geo-visual analytics. The authors define the design principles for the communication of real-time time series. The main challenges of real-time are change-blindness and communication of spatiotemporal variability. When observing a map or a graph updated in real-time the user may not see the changes that occur if the changes are too fast or too slow. Moreover, spatiotemporal data can vary in both dimensions and the user should have the possibility to observe the changes in both perspectives. In the development of the TRAFAIR dashboard, these challenges and the proposed principles have been taken into account.

Geospatial dashboards are classified as a web-based interactive interface supported by a platform combining: mapping, spatial analysis, and visualization. In [8], 3 types of dashboards are defined:

- Operational dashboards: that use indicators to provide descriptive measurements of smart cities,
- Analytical dashboards: that are based on data inferred from geospatial data using spatial analytic and are used as a diagnostic method for smart cities,
- Strategical dashboards: that can predict, estimate and visualize possible future outcomes.

The TRAFAIR Air Quality dashboard is a unique platform designed to be operational, analytical, and strategical in a unique solution.

2.1 Air Quality Applications

The urge to monitor air quality has become more pressing throughout the population that more and more platforms are being developed and made accessible. The most well-known are: AQICN,[2] the platform of the World Air Quality Index project, a non-profit project started in 2007 to promote air pollution awareness for citizens and provide unified and worldwide air quality information, IQAir AirVisual Platform[3] a worldwide monitoring platform that brings together data collected by governments, companies, and individuals around the world; Air Portal[4] of the European Space Agency(ESA), a platform to collect, analyze, and display local air quality data that also combines satellite data, regional air quality forecasts, land use information, and local monitoring data to make accurate air quality predictions in urban environments; Airqoon[5] a hyperlocal air pollution management that performs visualization and mining of AQ, and designs action plans. Moreover, many companies that sell air quality sensors also developed and provided mobile or web applications for monitoring, like Plume Labs,[6] AirCasting.[7] We compared the main features of those applications (as shown in Table 1) to identified the most common and used functions and visualizations.

---

[2] aqicn.org.

[3] https://www.iqair.com.

[4] https://air-portal.nl/.

[5] https://www.airqoon.com/.

[6] https://plumelabs.com/.

[7] https://www.habitatmap.org/aircasting.

Table 1 Features comparison among Air Quality applications

	AQICN	IQAir AirVisual	AIR-PORTAL	AIRQOON	PLUME LABS[a]	AIRVISUAL[a]
Gas monitoring	x		x		x	
PM monitoring	x	x	x	x	x	
AQI visualization	x	x	x		x	x
Historical data	x	x	x			
AQ interactive maps		x	x	x		
AQ forecast		x	x	x	x	x
Personalization option						x
Critical activities					x	x

[Legend:[a]The mobile version of those applications]

3 Background

This section provides basic information on what we mean for geospatial data and how air quality data are generated within the TRAFAIR project before being visualized through the dashboard. Firstly an introduction on the main aspects of geospatial and spatiotemporal data is provided. Then, the TRAFAIR project is described. Subsequently, the air quality sensors that have been used and deployed and the models and methodologies to create urban air quality maps are introduced.

3.1 Geospatial and Spatiotemporal Data

Geospatial data are information that refers to a precise position on earth. Spatial data are generally related to natural and man-made features whose size is between 1m and 10km. The earth's surface is composed of different elements that can be represented as geometries or images. For this reason, there are two main types of geospatial data: vectors and rasters. Vectors are the geometrical representation of objects with a precise location on earth. The simpler geometry is a point associated with its coordinates in a given reference system, then there are lines (collection of points), polygons (collection of lines), multipoligon (collection of polygons), and other more complex geometries (e.g. collection of different geometries, triangulated irregular network, and polyhedral surfaces). Vectors represent the real world in a digital and deterministic manner: as a space populated by features each one with its geometry. For this reason, vectors are used to describe discrete geographic objects like ways, rivers, and buildings. Phenomenons affected by a geographical variation cannot be easily represented as vectors (vectorized). As described in [9], to represent variables that are continuous in space the study area is divided into smaller rectangles and the geographical variation can be represented by recording the local pattern of the variable over each rectangle. The resulting matrix of observations is called a raster data structure. More variables can be observed in the same rectangular area generating a raster composed of more layers (one for each variable) called 'bands'. Raster data are generally in modified images formats (e.g. GeoTIFFs, IMG, and JPEG 2000). Each rectangle is a 'cell' and is commonly referred to as a 'pixel'. Digital images, categories of land cover, weather conditions are typically stored as raster data. Mobile sensors and ubiquitous positioning devices have generated a new type of geospatial data: spatiotemporal data. Spatiotemporal data have both spatial and temporal dimensions. Both vector and raster data can have a temporal dimension. There are several different types of spatiotemporal data: geolocated time series, map associated with a timestamp, and trajectories. Measurements provided by a sensor installed at a certain location generate a geolocated time series: a sequence of values over time associated with a location on earth [10]. A geolocated time series has a fixed position in space and variations in the time dimension. Maps associated with a timestamp are data that cover a wide area of space and are associated with a fixed time instant. The time is fixed or aggregated and the variations are represented in the spatial dimension. The evolution of the observed phenomenon over time can be observed creating a collection of maps of subsequent instants. Finally, trajectories represent phenomena where both position and time change together, each timestamp is associated with a new position. Trajectories are used to represent the movement of objects (e.g. vehicles, people, and particles). Mobile sensors generate trajectories of values, each measurement is associated with a time-position couple. In our use case, sensors can be moved around the city but they have a fixed position when collecting measurements. For this reason, we have not managed trajectory data. However, the information collected by each air quality sensor are geolocated time series and we generate a big amount of maps associated with a timestamp in the process of real-time and historical air quality monitoring (Fig. 1).

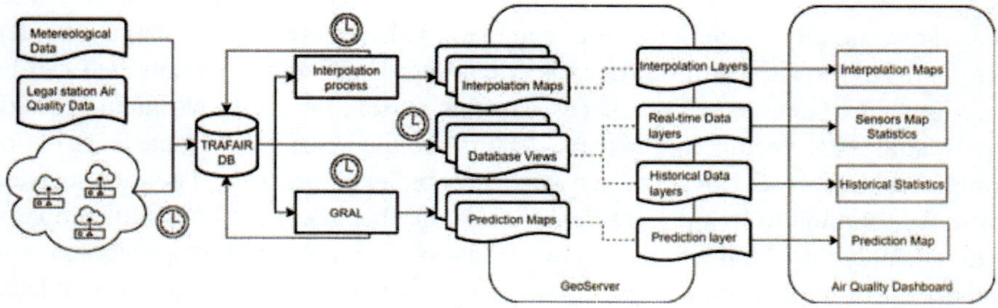

Fig. 1 The architecture of the TRAFAIR Air Quality dashboard

3.2 *TRAFAIR*

The Trafair project[8] [11] brings together 10 partners from two European countries (Italy and Spain) to develop innovative and sustainable services combining air quality, weather conditions, and traffic flows data to produce new information for the benefit of citizens and government decision-makers.

TRAFAIR raises awareness among citizens and public administrations about the air quality within an urban environment and the pollution caused by traffic. The project aims at monitoring air quality by using sensors in 6 cities and making air quality predictions thanks to simulation models. The two main goals of the project are:

1. monitoring urban air quality by using sensors in 6 European cities: Zaragoza (600,000 inhabitants), Florence (382,000), Modena (185,000), Livorno (160,000), Santiago de Compostela (95,000) and Pisa (90,000);
2. making urban air quality predictions thanks to simulation models based on the weather forecast and traffic flows (that are simulated thanks to traffic models [12]).

Monitoring air quality means setting up a network of low-cost sensors spread within the city [13] to monitor levels of pollutions in areas that are not covered by the legal air quality stations. Predicting urban air quality is possible thanks to a chain of simulation models. The traffic flows are simulated from the real measurements supplied by traffic sensors (as described in Sect. 3.2.1), then the emissions are calculated by taking into account the vehicle fleet in the city. Finally, the air quality predictions are calculated using an air pollution dispersion model (Sect. 3.2.2) taking into account the emissions, building shapes, and the weather forecasts.

3.2.1 Traffic Model

Traffic models are promising tools that can simulate the movement of vehicles in the streets starting from different sources of traffic data. Two different traffic models have been employed in TRAFAIR:

1. SUMO (Simulation for Urban MObility) open-source simulation model.
2. The WiP traffic model developed by the Department of Information Engineering of the University of Florence [14].

SUMO[9] is a micro-simulation model for road traffic. In micro-simulation models, each vehicle is simulated as a singular entity with its direction, trip, and speed. SUMO can be configured in different ways to accept several data sources [15]. The WiP model instead is a macro-simulation model based on differential equations and physical constraints applied to a detailed street graph. A stochastic learning approach is adopted to estimate the road-sections capacities at each time slot of the day. In the

[8] https://trafair.eu/.

[9] https://sumo.dlr.de.

city of Modena and Santiago de Compostela, induction loop detectors were installed under the surface of the street; this kind of traffic sensor can count vehicles that drive above them. Moreover, in the case of the city of Modena, they can also estimate the vehicle's average speed. Generally, these sensors collect an observation every minute. Otherwise, in Zaragoza only historical traffic data coming from mobile devices were available. In Modena (as described in [12]) and Zaragoza, the SUMO model was employed to perform daily simulations starting from the available traffic data. Then, the traffic of an average weekday of each month of the year was estimated considering historical simulations. While in Santiago de Compostela a similar approach was adopted employing the WiP model.

3.2.2 Emission Evaluation and Pollutant Dispersion Model

Once the estimation of average traffic flow for each day of the week and month of the year have been generated, they are employed as a forecast for the traffic flow of tomorrow and the day after tomorrow. Then, considering the number of vehicles in each lane for each hour of the day the emissions are estimated through the VERT (Vehicular Emission from Road Traffic) model described in [16]. This emission model aims at estimating vehicular emissions basing its calculation on the latest emission factors suggested by the European Environmental Agency in 2018. VERT estimates NOx exhaust emissions and directly performing cold and hot emissions estimation from the number of vehicles simulated by a traffic model within a road network, in a specific time step. Finally, to predict the air quality conditions for the next 48 h, we exploited the capabilities of the Graz Lagrangian Model (GRAL) [17], an open-source simulation software. The GRAL model simulates how the particles emitted by vehicles move in the air considering the weather conditions, the winds, and the shape of buildings. Moreover, additional emission sources (e.g. domestic heating) are included in the input data [18].

The GRAL model runs every day and, for each run, it generates 48 GeoTIFFs, one for each forecasted hour; each GeoTIFF represents the forecasted concentration of NOx in the urban area at a specific hour.

GRAL dispersion model was employed also to simulate in different seasons weekdays and holidays the effect on the air quality of a vehicle fleet composed differently (i.e. with more hybrid or electric vehicles). Emissions have been evaluated from the average weekday and holiday traffic flow of each season generating 24 different scenarios in each city (4 seasons, 2 day types, 3 vehicle fleets).

3.3 *The Sensor Network*

In the city of Modena, within the TRAFAIR project [11], 13 low-cost air quality sensors have been installed: 12 Decentlab Air Cubes, and one Libelium Smart Environment PRO. These sensors are connected to the municipality LoRaWAN network

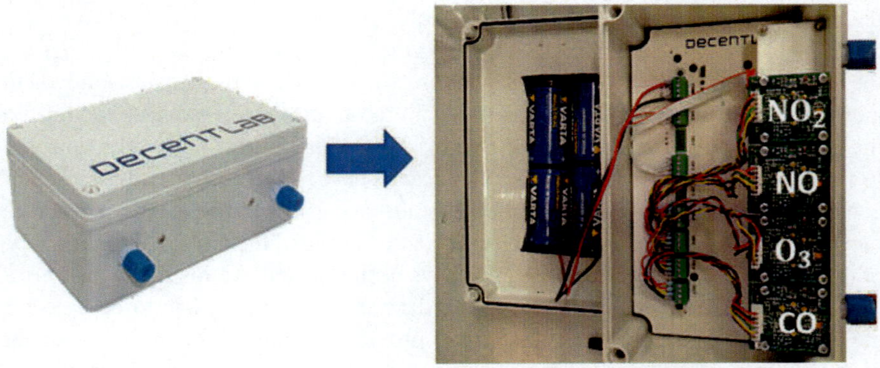

Fig. 2 An AQ device (on the left) and the inside content(on the right): 4 cells/sensors for measuring the level of 4 air pollutants (NO, NO_2, CO, and O_3 in this case)

and send values of CO, NO_2, NO, and O_3 every 2 min. Similarly, in Zaragoza 10 Decentlab air cubes have been installed. In Santiago De Compostela 3 Decentlab air cubes,[10] 3 Libelium Smart Environment PRO,[11] and a Kunak[12] sensor have been installed.

The low-cost devices are boxes with different sensors placed inside. Each sensor, also called 'cell', is devoted to the measurement of a specific pollutant. Figure 2 shows a Decentlab sensor[13] on the left and its interior view on the right. Inside the box, there is a sensor for air temperature and humidity, and 4 AQ sensors for NO, NO_2, CO, and O_3.

During their life cycle, the sensors are moved in different locations around the city; but, since the measurements of low-cost sensors are not highly accurate, they need to be periodically calibrated. The calibration phase is a period of the sensor's lifecycle that requires its collocation near an air quality legal station. The legal stations provide very accurate measurements about the concentration of pollutants in the air. In Modena, there are two legal stations managed by ARPAE, the environmental agency of the Emilia-Romagna region. One station is a background station located inside a green area, the other is a traffic station located near a highly congested road. In Zaragoza, there are eight air quality legal stations belonging to three different types: moderate traffic (3), intense traffic (3), background (2). In Santiago de Compostela, two ISO standard air quality stations are available, a background station and a traffic station, both managed by the local meteorological agency Meteogalicia. During the calibration process, machine learning algorithms (e.g. Random Forest, Support Vector Machines) or deep learning models (e.g. Multilayer perceptron, Long Short Term Memory) are employed to learn the association between the raw data measured by

10 https://www.decentlab.com/products/air-quality-station-no2-no-co-ox-for-lorawan.

11 https://www.libelium.com/iot-products/plug-sense/.

12 https://www.kunak.es/en/smart-environment/urban-air-quality/.

13 https://tinyurl.com/yunuaryt.

the sensor and the real concentrations registered by the legal station in the same location. Once the models are trained on the data collected during the calibration period, they can be employed to generate the calibrated value when the sensor is moved in a different location. Every 6 months, the calibration process is replicated to ensure the quality of the observations produced by the sensors. In this phase, the sensor is in the *calibration* status. Once calibrated, each sensor is able to provide a value of concentration for every pollutant. Once the calibration phase finishes, the sensors are moved from the legal station to different locations around the city. Within the scope of the project, some locations of interest have been selected. At this stage, the sensor is in the *running* status and the collected data are accurate enough to be used for the estimation of air pollution. The sensors provide raw milliVolt measurements for each pollutant every 2 min. Those values are stored in the project database and are converted in micrograms (milligrams for the CO) per cubic centimeter every 10 min interval. The concentration values are stored in the project database too for further trends and statistical analysis.

3.4 Air Quality Maps

Public administrators need to know the position and status of the installed sensors, monitor real-time air quality conditions, visualize statistics concerning the air quality conditions in the urban area, and quickly identify if there are hot spots, i.e. areas with a high concentration of pollutants. Each air quality sensor provides the four pollutants concentrations in the location where it is placed and, by spatially interpolating the values of all the sensors, it is possible to estimate the pollutant concentration in the whole urban area. Thus, an R script was implemented to produce GeoTIFF files interpolating the pointwise concentrations using Inverse Distance Weighted(IDW). As described in [19], IDW is a deterministic (non-geostatistical) estimation method where values at unmeasured points are determined by a linear combination of values at nearby measured points. Value at location $x*$ is evaluated as:

$$x^* = \frac{w_1 X_1 + w_2 X_2 + ... + w_n X_n}{w_1 + w_2 + ... + w_n}$$

where $x*$ is the value to predict and w_i is the weight of the sampled point X_i. Weights are evaluated as the inverse of the distance between the location to predict and the sampled data point i to the power of p:

$$w_i = \frac{1}{d^p_{i-x*}}$$

The rate at which the weights decrease is dependent on the value of p. As p increases, the weights for distant points decrease rapidly. If the value of p is very high, only the immediate surrounding points will influence the prediction. In our use case, we employ $p = 2$. Other interpolation techniques such as IDW with p equal to 1, nearest neighbourhood, ordinary Kriging, and thin plate spline, have been considered and tested. Both IDW and Nearest Neighbourhood provided good performances. We chose to use IDW and to produce a new map every time new calibrated values are available i.e. every 10 min.

In order to present the air pollutant concentrations to diverse audiences, we decide to share numeric data and to display them using three colour scales: the one used by the ARPAE in Modena, a customized colour scale named TRAFAIR in Zaragoza and Santiago de Compostela, and the one proposed by the European Environmental Agency (EEA) for all the cities.

4 Architecture and Technologies

The main purpose of the TRAFAIR dashboard was to support the public administration in decision-making activities. Public administration members were involved in the deployment of the dashboard and contribute to defining the requirements:

- **R0** automatically updating visualization of real-time measurements of low-cost sensors in the city,
- **R1** visualization of sensor positions and their status,
- **R2** statistical information about historical measurements;
- **R3** visualization of the average day of the week trends and other aggregations in space and time;
- **R4** visualization of the current concentration of pollutants in the city area updated every 10 min;
- **R5** comparison between the different points of observation;
- **R6** visualization of the forecast of NOx concentration in the whole city for today and tomorrow;
- **R7** visualization and download of historical forecasting maps;
- **R8** visualization of historical maps based on the daily average concentration measured by low-cost sensors;
- **R9** comparison between the NOx concentration generated by different scenarios;

Moreover, since the number of urban sensors is destined to increase scalability must be ensured. Scalability is the ability to add or remove sensor devices without affecting the system's availability. Figure 1 illustrates the architecture of the dashboard. The data are stored in the TRAFAIR DB. The information contained in the database is exposed through GeoServer over the Internet and used by the TRAFAIR Air Quality dashboard.

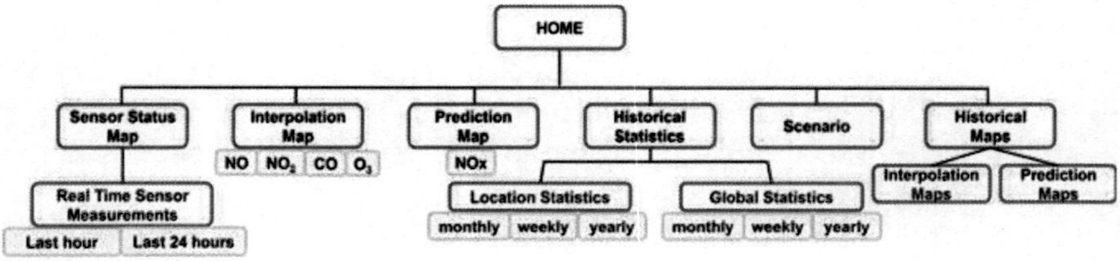

Fig. 3 Site Map of the TRAFAIR Air Quality dashboard

4.1 TRAFAIR Database

The TRAFAIR architecture relies on a PostgreSQL database. For efficient management of time series and spatial data, the database has been equipped with PostGIS and timescale extensions. The spatial extension ensures the correct management of both vector data (e.g. the position of sensors) and raster data (e.g. interpolation maps). Its main role is the collection of the stream of data generated by the low-cost air quality sensors. In order to preserve the complete history of the measurements, the database contains basic information about the sensors themselves. Then, every time the sensors are moved, new data related to their current positions and status are inserted. Each sensor has its own life-cycle and for every instant in time the physical sensor is associated with a position in space and one of the following status: *running*, *calibration*, *broken*, and *offline*. When the sensor is collecting data it is in the *running* status; While, when it is collecting data but it is located near a legal station for the calibration process its status is *calibration*; Otherwise, when the sensor is not able to collect data or the collected data are not reliable the sensor is in the *offline* or *broken* status.

Each sensor measures pollutants concentrations and sends data packages every 2 min. The data are sent over a LoRaWAN network and are captured by different gateways. Once the gateway receives a data package, it redirects the data to the TRAFAIR database. In particular, the sensors are able to capture the concentrations of CO, NO_2, NO, and O_3, humidity, and temperature and send to the database the intensity of the voltage generated during the analysis. Therefore, for each observation captured by the sensor, we store the measured voltages of each gas (in millivolts), the temperature, the humidity, and the battery voltage. As indicated in Sect. 3, each sensor needs to be calibrated to collect reliable data. Alongside the measurement of our sensors, we collect the air quality data generated by the legal air quality stations and run a calibration algorithm to associate the measured voltages to the air quality data collected from the legal station. The parameters of the calibration algorithm and the data used during this calibration phase are also kept in the DB. Once the calibration step is completed, the data incoming from the sensors, are converted from millivolts to concentration (milligrams for cubic meter) and then stored in the database.

Beside the main schema containing the core data, the database contains other schemas. In order to store the view that have been generated for the creation of visualizations in the dashboard, we created the *webapp* schema. While, the *open$_d$ata* schema was created to store the data that are published as open data, and the *mobile* schema for the visualizations of the mobile apps.

4.2 GeoServer

To share geospatial data in an easy and open source solution, GeoServer GeoServer[14] [20] offers a platform that allows to upload and publish several types of data sources. GeoServer accepts group of files or single files in raster or vector format, tables hosted by a PosgreSQL database, and even images. For each data source a store is generated. Then, from each store, different layers can be created. Layers enables the handling of temporal and spatial data using standard OGC protocols:

1. the WFS (Web Feature Service),
2. the WCS (Web Coverage Service),
3. the WMS (Web Map Service).

The WFS standard allow exposing spatial and temporal data as GML, KML, JSON, CSV, XML, and other data formats; This protocol was exploited to expose the data for drawing the graphics presented in Sect. 4.3. The WCS standard manage "coverages". A "coverage" refers to objects covering a geographical area like a set of points, a regular grid of points, or a set of segmented curves. Finally, the WMS standard is employed for generating server-side maps that are sent to the client as regular images or gif files. This solution is adopted to avoid sending fine-grained data to the client; The interpolation maps described in Sect. 5.4 and the prediction maps generated by the air pollution dispersion model described in Sect. 5.5 have been exposes through the WMS. Moreover, the choice to deploy an instance of GeoServer to share the data was motivated by the need of a higher security level. GeoServer allows separating the access to the database from the access to the generated visualizations. Furthermore, GeoServer helps to easily manage the publication of spatial and temporal data. Another important reason that motivate the adoption of GeoServer was the need of a common platform with the same structure. Since the data structure deployed in the 3 cities is slightly different, to manage this heterogeneity of data the same layers were generated in GeoServer. Moreover, GeoServer we exploited the GeoServer's ImageMosaic plugin that allows the creation of a mosaic from a set of GeoTIFFs (georeferenced rasters).

The mosaic is a set of geospatially rectified images related spatially or temporally to each other. For example, an ImageMosaic data store was generated to group the GeoTIFFs prediction map of several time slots into a unique element. In the ImageMosaic data store the images can be associated to a timesamp allowing to

[14] http://geoserver.org/.

investigate the evolution over time of the represented phenomenon. This solution is suitable for the maps associated with a timestamp described in Sect. 3.1. Furthermore, other services like WPS (Web Processing Service) or WMTS (Web Map Tile Service), and other plugins can be integrated into GeoServer if needed.

Since TRAFAIR project requires the generation of several maps and visualizations that need to be updated frequently we exploited the GeoServer API system to automatically generate the layers required for exposing the database views. A detailed description of the ingestion of layers in GeoServer and the publication through OGC services of air quality open data is depicted in [21].

For handling sensor observations, status and positions, and statistical data about sensors or locations, 9 views in the web app schema and 9 layers in GeoServer have been created. These layers consist of 3 Real-time Data layers and 6 Historical Data layers as shown in Fig. 1. A dashboard requires a quick response time; thus, some views have been materialized. As a result, the average response time decreases from 10 to 0,7 s. The map layers are different and requires some additional explanation about the structure of the file and the process that generates it. The interpolation map layers are collections of the GeoTIFFs file created by the interpolation process. These raster files are unique but are composed of 4 bands, one for each pollutant. Each layer is associated with one of the band and with a defined style to visualize the concentrations of a pollutant as a coloured map. The prediction map layer and the layers of the scenarios are created exploiting the ImageMosaic plugin to merge the 48 GeoTIFFs generated by GRAL into an ImageMosaic data store. One layer is generated for each forecasts and each possible scenario.

The content of the Historical data layers and Prediction layer is updated daily; moreover, the Interpolation layers and real-time data layers are refreshed every 10min.

4.3 Dashboard

The dashboard is a web application based on Angular 7 design framework described in [22] and writted in Typescript language. Two different libraries are employed for the graph visualization: D3 [23] and Chartjs [24]. Instead, in order to create maps, Open Layer[15] library is used. The TRAFAIR dashboard is a single-page client applications, Angular was chosen for its implementation due to its modular structure that allows the reuse of code. The Angular framework is based the NgModules i.e. building blocks of related codes. NgModules offers the compilation context for 'components'. In Angular, a component is a collection of related screen elements that compose a single view.

Besides, 'services' provides the functionalities that are shared between different components. They are also employed to share information and data between components. In our implementation, each view has its corresponding component.

[15] https://openlayers.org/.

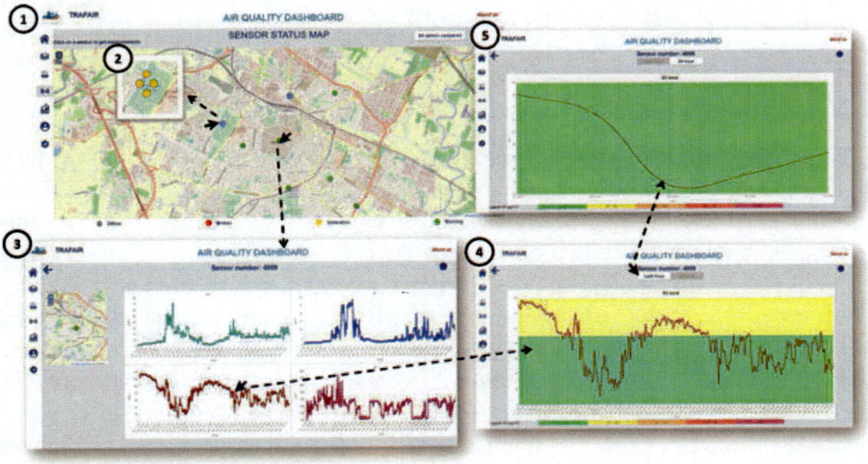

Fig. 4 Views accessible from the sensor map view: (1). A sensor cluster when clicked is converted into several single clickable sensors (2). Clicking a sensor the last 24 h measurements for the four pollutants are visualized (3). Zoomed view of a single graph with colour scale indicating the level of pollutant concentration

Component can be in a parent child relationship when the child component is a part of the parent component, in this case we have a single view represented by the parent component and the child component inside it. Parent and child components can share data through the '@input' decorator, without using services. We employ child components in some views where graphs change accordingly to some option selected by the user in a form defined in the parent component.

Moreover, several services have been defined to share information between components that are not in a parent-child relationship. Services support both the retrieval and the modification of the information.

The main services implemented in TRAFAIR Air Quality dashboard, as shown in Fig. 1, are:

- the authentication service: to share and modify data concerning the user session.
- the city selection service: to share and modify data about the selected city (our web application supports three cities)
- the language selection service: to share and modify data about the selected language (our web application supports three languages)
- the GeoServer service: to define functionalities related to interrogations to obtain data using the GeoServer API.
- the threshold service: to share and modify the selected colour scale thresholds used to generate graphics and maps colours.

5 TRAFAIR Air Quality Dashboard

Within the TRAFAIR project described in Sect. 3.2, a suite of monitoring tools for public administration was realized.

The TRAFAIR Air Quality Dashboard[16] [2] allows them to visualize air quality conditions in the cities. Another dashboard was realized, SenseBoard [25] an environmental expert-oriented dashboard for monitoring the status of the sensors and the calibration. These two dashboards have different purposes. SenseBoard is for technical experts; however, the TRAFAIR Air Quality dashboard is created for a non-expert user that needs to monitor quickly transitioning data. The dashboard enables the analysis and the estimation of the diffusion of pollutants in the urban area of all three cities. The user can get an overview of the air quality conditions in the cities: tracking and comparing over time and space pollutants concentration, viewing the real-time air quality condition here-and-now. Time series and geospatial data are displayed through interactive graphs and maps. The graphics are dynamically updated when new data are released. The user can interact with several visualizations selecting, filtering, and querying data, zooming in/out, panning, and overlaying. Since the dashboard was initially conceived for public administrations to help them in decision making, not all the visualized data are public and can be visualized by a not-logged user. Administration members should log in to access this private data. City maps are obtained from Open Street Map (OSM),[17] the data visualized and the map layers are generated querying GeoServer configured layers.

In order to respect the principles introduced for the visualization of real-time data in [7], the user needs a time reference to easily gasps changes in the visualization. For this reason, timers inserted inside the Typescript code of the web application views are updated with the same frequency. When the timer expires a new request is sent to the GeoServer layer and then the view is updated, and the timer is restarted. A label in the view advises the user about the time stamp the visualization refers to. The colours significantly help to convey the criticality of air quality conditions and the definition and choice of the colour scale can influence the perception of the user. The functionalities related to the colour scale are described in Sect. 5.1. As can be seen in Fig. 3, the site map is composed of 6 branches: the sensor status map (Sect. 5.2), the interpolation maps (described in Sect. 5.4), the prediction map (described in Sect. 5.5, the historical statistics (described in Sect. 5.3, the Scenario view (described in Sect. 5.6), and the historical maps (described in Sect. 5.7).

[16] The dashboard is available online at https://trafair.eu/airquality.

[17] https://www.openstreetmap.org/.

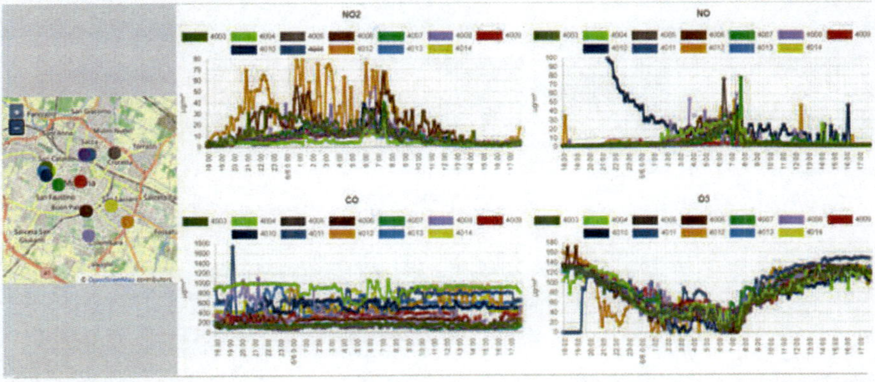

Fig. 5 The view that allows comparing the last 24 h measurements of the four pollutants concentration of all the available sensors

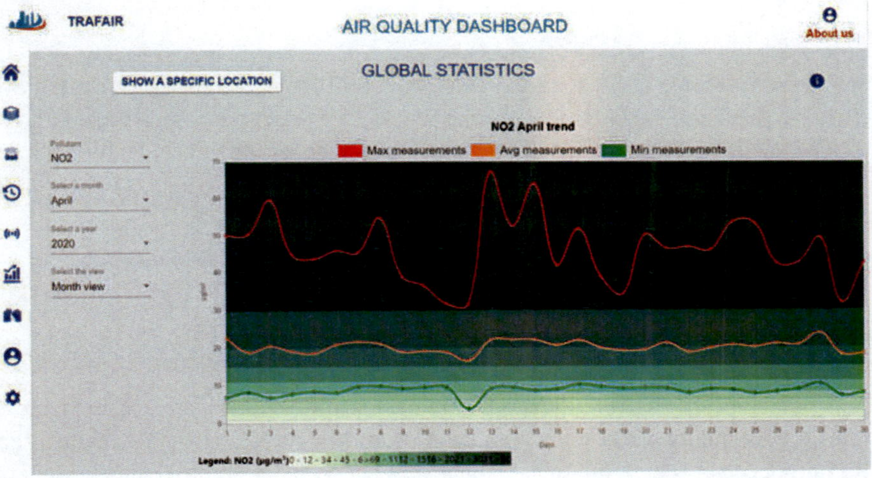

Fig. 6 NO_2 concentration trend for the whole urban area of Zaragoza in April 2020. The background colours are based on the TRAFAIR scale

5.1 Colour Scale

Environmental experts involved in the TRAFAIR project decide to adopt the EEA colour scale. This colour scale was defined by the European Environmental Agency[18] that defines the threshold and the colours for each pollutant. Moreover, in Modena, a different colour scale was defined and used by the ARPAE agency. The environmental experts' team also define a new colour scale that helps to visualize better small changes in pollutants concentration. This colour scale was defined for the cities of Santiago de Compostela and Zaragoza. Thus, the user can choose between ARPAE

[18] https://www.eea.europa.eu/.

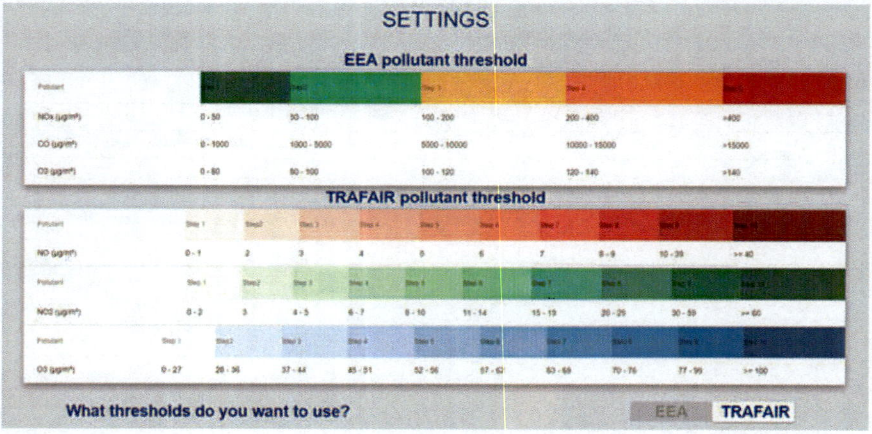

Fig. 7 The settings page of the dashboard where the user can select the colour scale

or EEA colour scale in Modena and TRAFAIR or EEA colour scale in the other two cities. In the TRAFIAR Air Quality dashboard, the setting-up page (Fig. 7) allows the user to choose between the available colour scales for the selected city. Once the user chooses a colour scale, the Threshold service will provide this information to all the views in the application.

5.2 *Sensor Status and Real-Time Measurements*

This branch of the TRAFAIR Air Quality dashboard displays the measurements of low-cost air quality sensors and their position and status. To satisfy the requirement **R1** introduced in Sect. 4, the 'sensor map' view shows the position of each sensor and its status(see Fig. 4-1). The status of the sensor is represented through the colour of its marker on the map. Besides, the sensor position is obtained from the TRAFAIR database and saved in the GeoServer instance as a layer queried with the WFS standard. During the calibration phase, sensors are placed near the legal station and it may happen that several sensors are in the same position. When sensors are placed in the same location, to allow the correct visualization of all the sensors, the map initially shows a big spot displaying the number of sensors located in the specific position. Then, the user can click on the big spot to see the status of every single sensor separately (see Fig. 4-2). To obtain this visualization we modified the Open Layer library adding some additional functions to manage clusters of sensors.

To satisfy requirement **R0**, by clicking on the marker of a sensor in the map the user can open a new view that displays the last 24 h measurements of the 4 pollutants of the selected sensor (see Fig. 4-3).

The time series of sensor measurements is associated with a position on earth (the position of the sensor); thus, it is a geolocated time series (Sect. 3.1). To display a time series of observations with a frequency of 10 min in a linear representation a great display space is required in order to produce a comprehensible visualization. Therefore, if the user wants to see more clearly the values of a specific pollutant by clicking on the corresponding graph a zoomed visualization will appear. In this zoomed visualization, the background of the graph is coloured accordingly with the selected colour scale (ARPAE or TRAFAIR and EEA). The user can select the colour scale on the setting-up page of the dashboard. The legend behind the graph describes the meaning of the colours and helps the user to identify values that are critical and may require his attention. Public administration members give us feedbacks about this visualization underlining the need for a view that allows them to easily compare the measurement of different sensors located in different points of the city. Therefore, an additional view was created to satisfy their requirement. By clicking on the 'all sensor compared' button in the top-right of the sensor status map, a new page is opened where the sensors' time series are overlaid in a single chart, one for each pollutant. In this way, the last 24 h measurements of the sensors can be compared; Moreover, since the view contains a map with the position of each sensor (see Fig. 5), the areas of the city with the highest pollutant concentrations can be easily identified. In order to associate to each curve the correct position, the colour of the curve is the same as the sensor marker in the map. These graphs are interactive, by clicking on the label of a sensor the user can remove its measurements and decide the set of sensors that he wants to compare. The contemporary use of maps and charts allows visualizing the geolocated time series together with their location, combining the spatial and the temporal dimensions in a single visualization.

Every 10 min, the graphics are automatically refreshed.

5.3 Historical Statistics

The requirements **R2** and **R3** underlines the necessity of a statistical overview of historical data. During the TRAFAIR project, all the collected data are stored in the TRAFAIR database and uploaded in GeoServer. A longitudinal archive of data is generated. Historical data can be examined over different time frames, such as a week, month, and year. Moreover, archival data are used to detect trends and provide contextual information for a better understanding of current data. The view is composed of several drop-down menu windows that allow defining the year, the month, the day, and the time aggregation. The user can interact with the view changing the set-up to visualize different graphs. Moreover, two different types of statistics are provided: global statistics, and location statistics. Global statistics are evaluated on the whole urban area considering all the available sensors observations. They provide an overview of the air quality conditions in the city. Location statistics are accessible through the button 'show a specific location' in the top left of the global statistics view. A map that shows all the available locations to select will appear and by

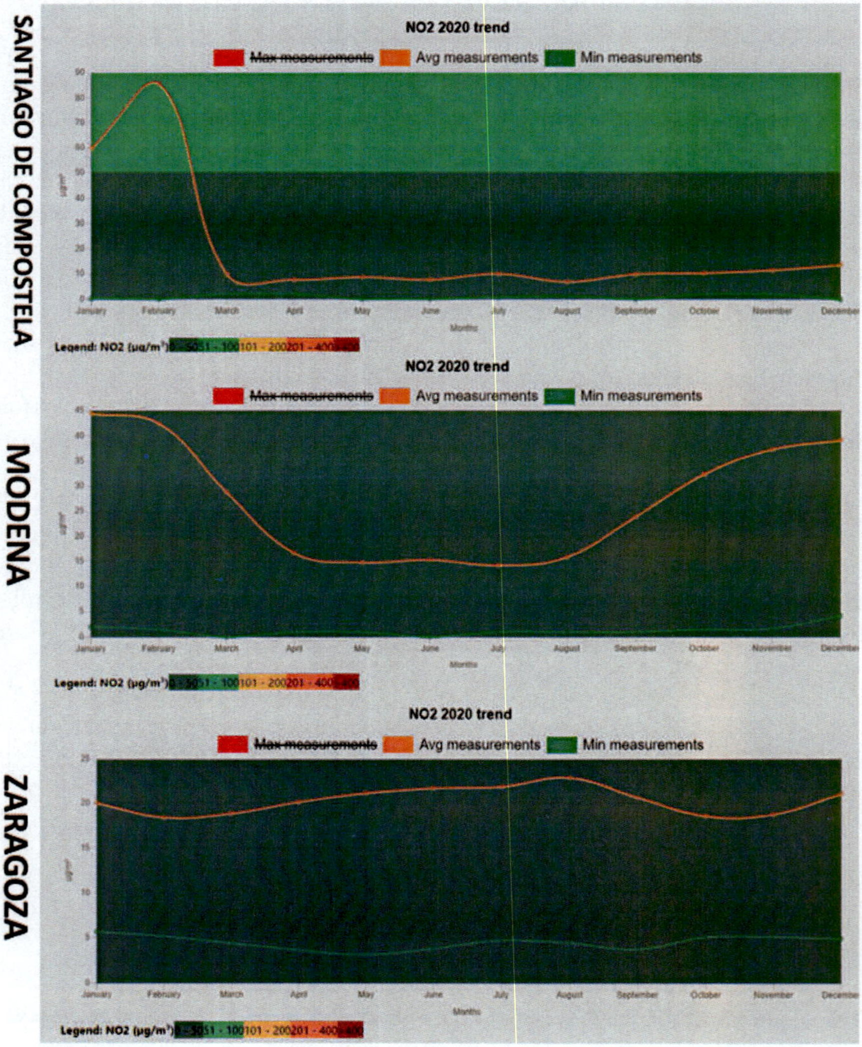

Fig. 8 The three year views of NO_2 concentration in 2020 for the cities of Santiago de Compostela, Modena, and Zaragoza

clicking on the location marker the user can visualize the selected location statistics. Location statistics are evaluated individually for each location where sensors have been placed during the project. Since sensors are moved around the city, they can be placed in the same position simultaneously or in different periods. However, the measured pollutant concentration does not depend on the physical sensor but strongly depends on the location. Thus, the location statistics are evaluated considering all the observations registered in the selected period by all the sensors that were placed in that location. The global and location statistics views are very similar, the main difference is that in the location statistics view the name of the location is at the top of the view. Three different graphics are provided: the month graph, the year graph, and the week graph.

The month graph is available for each month of each year. Once selected the month, the year, and the pollutant from the drop-down menu windows, a graph that shows the maximum, minimum, and average time series of values of pollutant concentration measured in every day of the selected month is generated. The user by clicking on the name of the curve can remove it from the chart. The background colours indicate the level of concentration of the pollutant accordingly to the selected colour scale.

Figure 6 display the global month view of NO_2 for April 2020 in the city of Zaragoza. The selected colour scale is the TRAFAIR colour scale and maximum values are in the higher colour class.

Figure 9 shows the same graph in the same location displaying the monthly trend of $O3$ in June 2020 in the city of Modena but with two different scales. As can be seen, the adoption of a different colour scale can significantly influence the perception of the user. The visualization on the top adopts the EEA scale and gives the impression of very high and dangerous values of concentration at the end of the month. The visualization with ARAPE scale instead associates an orange colour to these values giving the impression of less severe air quality conditions.

For each year in which the data are available, the user can select the year and the pollutant, and then the visualized year graph shows the time series obtained from maximum, minimum, and average values of the selected pollutant concentrations measured for every month of the selected year. The maximum, minimum, and average

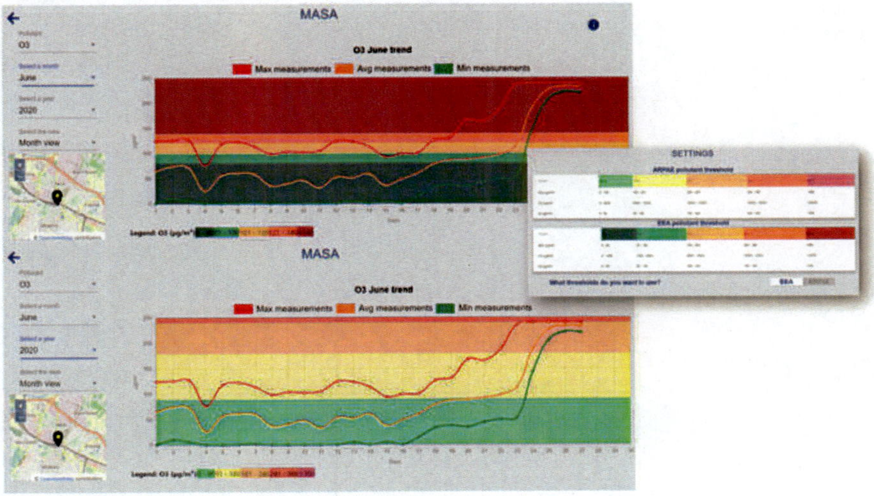

Fig. 9 Two month views for $O3$ in June 2020 in the city of Modena: the top view uses ARPAE scale and the lower EEA scale. The figure shows also the settings page where scales can be modified

curves can be removed to ensure a better visualization, and the background is coloured according to the selected colour scale. For example, in Fig. 8, the three cities' year views for the NO_2 concentration are displayed. The selected colour scale is the same in the three views, the EEA scale. The curve of the maximum values has been removed to guarantee a focus on the average yearly trend. Santiago de Compostela has higher values in January and February and then the concentration drops to a very low value for the rest of the year. In the city of Modena, the situation is different, we have lower values in the initial part of the year compared with Santiago de Compostela; However, the values in the rest of the year are higher. Zaragoza has a very constant trend during the year and the NO_2 concentration values are generally lower than in Modena.

Finally, the last graph is the week view. For each month, the graph displays the average concentration of the mean weekdays. In order to evaluate the mean hourly concentration trend, for each pollutant and each day of the week, all the available observations in the selected month of the selected day of the week in each hour are averaged. Moreover, to ensure an easy comparison between the trend of different days of the week, their curves are displayed together in the same graph. The user can also click on the day of the week name to remove it from the graph if necessary. The background of the graph is coloured according to the selected colour scale. For example, in Fig. 10, the curves of average days of the week trend for the O_3 pollutant on February 2020 are displayed. The graph is a location statistics view of the city of Zaragoza, the position of the location is indicated in the map on the left corner. In this graph, the curves have all a similar trend lower values at night and higher during the evening hours. However, Wednesday and Thursday had a higher peak than the

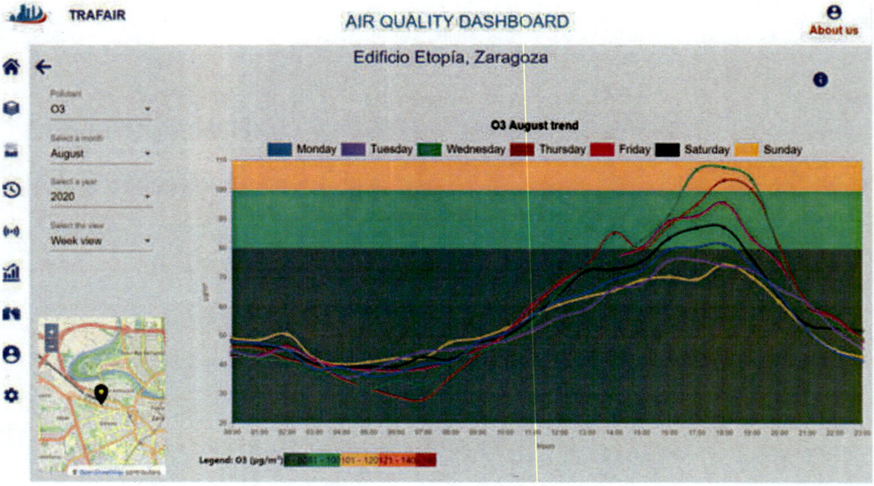

Fig. 10 The average weekdays view for O_3 in August 2020 in Edificio Etiopìa, Zaragoza. The colours of the background are based on the EEA scale

other days of the week, and Saturday and Sunday instead show higher values during night and morning hours.

5.4 *Interpolation Maps*

The interpolation maps view of the dashboard satisfy the requirement **R4**; the semi-real-time interpolation maps are obtained querying GeoServer layers. As displayed in Fig. 11, the interpolation maps view consists of a coloured map with 4 buttons on the top that can be clicked to switch between one pollutant to the others available. The coloured interpolation map obtained as described in Sect. 3.4 is overlayed on the OSM city map of the area. The style of the map depends on the selected colour scale. This is possible because interpolation maps are saved as raster data. Rasters are generally used to represent space-continuous phenomenons like pollutants concentration. Two different colour scales can generate a very different map and can be used to communicate diverse information. For example, in Fig. 12, the same raster map is visualized with two different colour scale. The maps refer to Santiago de Compostela and the top-view shows the interpolation map obtained with the EEA colour scale; The bottom-view, indeed, displays the same interpolation map obtained setting the TRAFAIR customized colour scale. The TRAFAIR colour scale can help to highlight differences in the area since there are more colour bands each one covering a smaller interval of concentration values. Besides, the EEA based map only shows a lighter green area that has a higher value than the rest of the city, while the one based on the TRAFAIR colour scale helps to identify two areas with higher values in a darker shade of blue and one area with a lower concentration of O_3 highlighted in white.

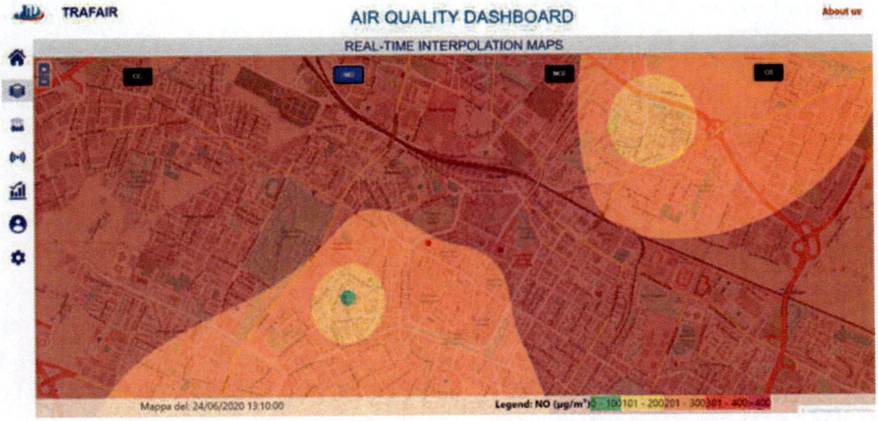

Fig. 11 The interpolation map of *NO* of 24th June 2020 for the city of Modena at 1:10 pm UTC. The colours are based on the EEA scale

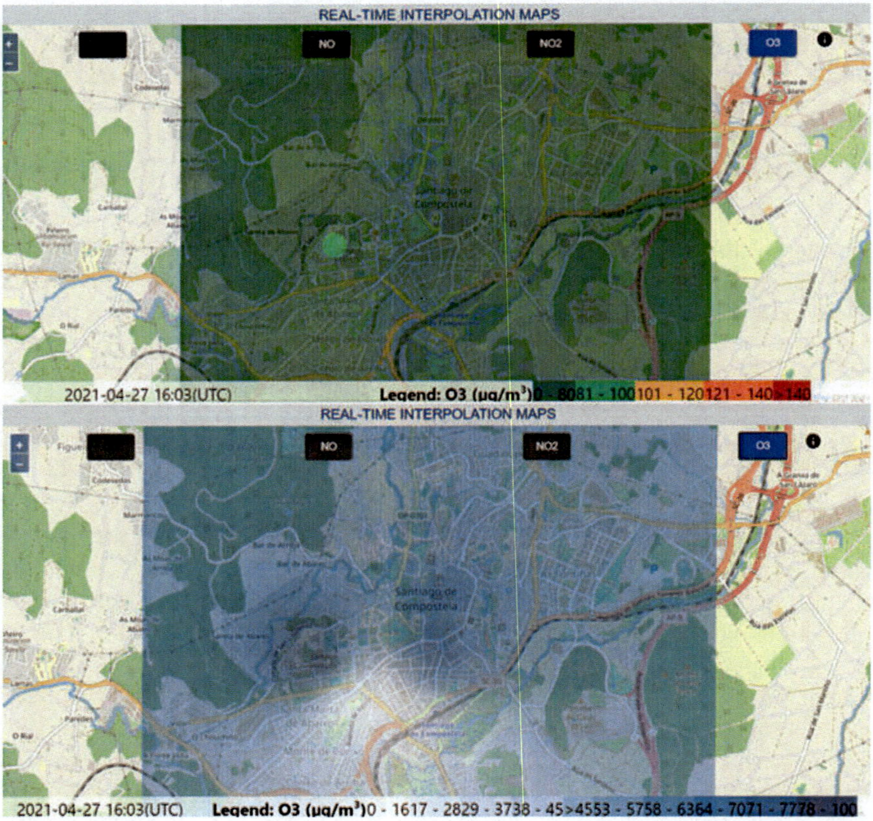

Fig. 12 O_3 interpolation map of Santiago de Compostela on the 27th of April 2021 at 16:03 UTC. The colour scale employed to represent the raster data is the EEA in the top figure, and the TRAFAIR customized colour scale in the bottom figure

In the left bottom of the view, the timestamp associated with the map is displayed. The maps are updated automatically every 10 min.

5.5 Prediction Maps

The requirement **R6** is satisfied by the prediction map view. As described in Sect. 3, data coming from traffic sensors are used to feed a traffic model, then emissions are evaluated and the GRAL dispersion model simulates the movement of the particles and generates 48 images, for each hour of the forecast. These images are saved as raster data that contain information about the forecast of NOx concentration for each 4X4 meters square of the city area. Moreover, when these images are uploaded in GeoServer they are associated with the hour of the day their forecast refers to. Images

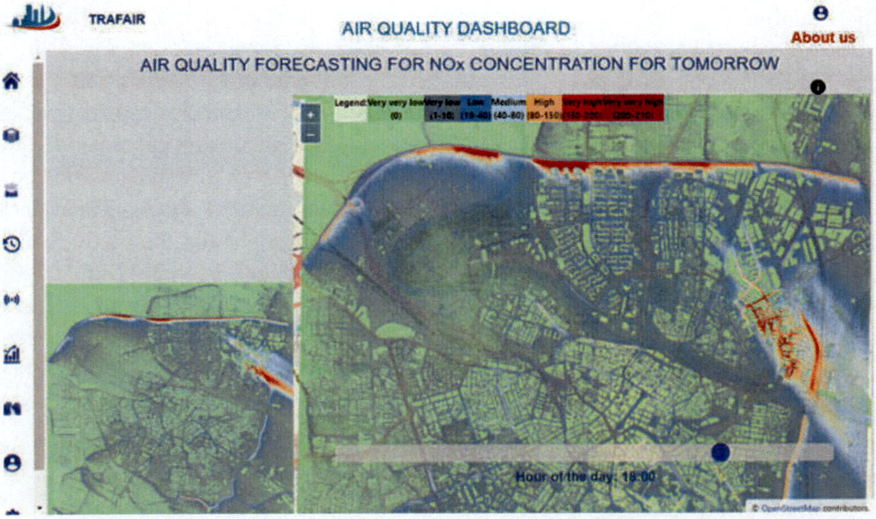

Fig. 13 Prediction view that displays the air quality forecast for today (2021-04-28). The map displays the air quality conditions predicted for 6 PM for the city of Zaragoza

generated with the same simulation are saved in the same Image Mosaic collection in GeoServer (as described in Sect. 4.2). The ImageMosaic plugin allows to see the evolution of images over time, the collection of images can be queried to obtain a GIF or a sequence of ordinated images. As can be seen in Fig. 13, the view displays a GIF summarizing the evolution of today's air quality forecast. Moreover, the city map is coloured accordingly to the concentration of NOX particles forecasted by the air pollutant dispersion model. The slide bar can be used to move through the time axis and see the evolution of the air conditions during the day. Furthermore, users can interact with the map zooming and moving in the area. In this case, the legend is qualitative because we are communicating a forecast that can contain an error. By clicking on the button 'Tomorrow prevision' the view showing the prevision for tomorrow is displayed.

5.6 *New Scenarios*

The **R9** requirement asks to compare *NOx* concentrations in different scenarios. In the Scenario view, the season, the day type (weekday or holiday), and the composition of the vehicle fleet should be selected through a drop-down menu. Whit the collaboration of the public administration members, we define a set of possible scenarios and this view allows comparing the predicted air quality conditions. The sustainability plan inspired the definition of the vehicle fleet options that are different for each city. For example, in the city of Modena, the three possible options are: (i) CURRENT:

the actual vehicle fleet composition with a majority of petrol and diesel alimented vehicles, (ii) PAIR 2020: the vehicle fleet inspired by the Integrated Air Plan of the Emilia Romagna region obtained increasing the number of electric and hybrid vehicles, and (iii) PUMS 2030: the future vehicle fleet that the city council wishes to reach in 2030, inspired by the Urban Sustainable Mobility Plan, characterized by a majority of electric and hybrid vehicles. The scenario view shows the composition of the vehicle fleets in a table beneath the drop-down menu windows. Then, once the options have been selected, by clicking on the 'ready' button the prediction of the possible concentration of NOx with the selected vehicle fleet and Similarly to the 'Prediction view', the user can select the hour of the day using the slide bar, and the GIF shows the evolution of the concentration of NO_x during the day. This view allows users to observe the differences in different periods of the year, the difference between weekdays and holidays, the effect of a different vehicle fleet composition on the urban air quality (Fig. 14).

This view can be used to see the differences between the period of the year, weekdays, and holidays and the effect of a change in the composition of the vehicle fleet.

5.7 *Historical Maps*

Comparing real-time maps with the ones obtained in the past days can help to better understand the current air quality values in the period context. For this reason, the dashboard allows the user to visualize maps of past days. As required by **R7** and **R8**, two main categories of maps are available: Historical Interpolation maps and Historical Prediction maps.

The Historical interpolation maps view enables the users to visualize interpolation maps regarding the past days. These interpolation maps are obtained with a spatial interpolation of the average daily values in each sensor position. The day can be selected from the calendar on the left side of the view (Fig. 15). When a day is selected it will be inserted into the history searches so that the user can easily compare several days moving from one to the other. This view shows the average air quality conditions in the whole urban area and enables citizens and public administrations to compare different days of the week and different areas of the city.

The Historical Prevision view allows the user to select a day in the past and see the forecast of NO_x concentration that was produced for that day using the GRAL model. Since GRAL generates 48 h previsions, for each day, two different forecasts are available: the prevision produced the day before and the prevision produced two days before. Generally, the prevision produced the day before should be the most reliable since the weather forecasts used to simulate the particles' movements are more precise. The user can visualize both of them by clicking on the 'show forecast evaluated the day before' button. An example of this view is displayed in Fig. 16. The user can also download and save locally an image of the current map he/she is visualizing. Moving the mouse cursor on the map a label appears with the predicted NO_x concentration value in the pointed position.

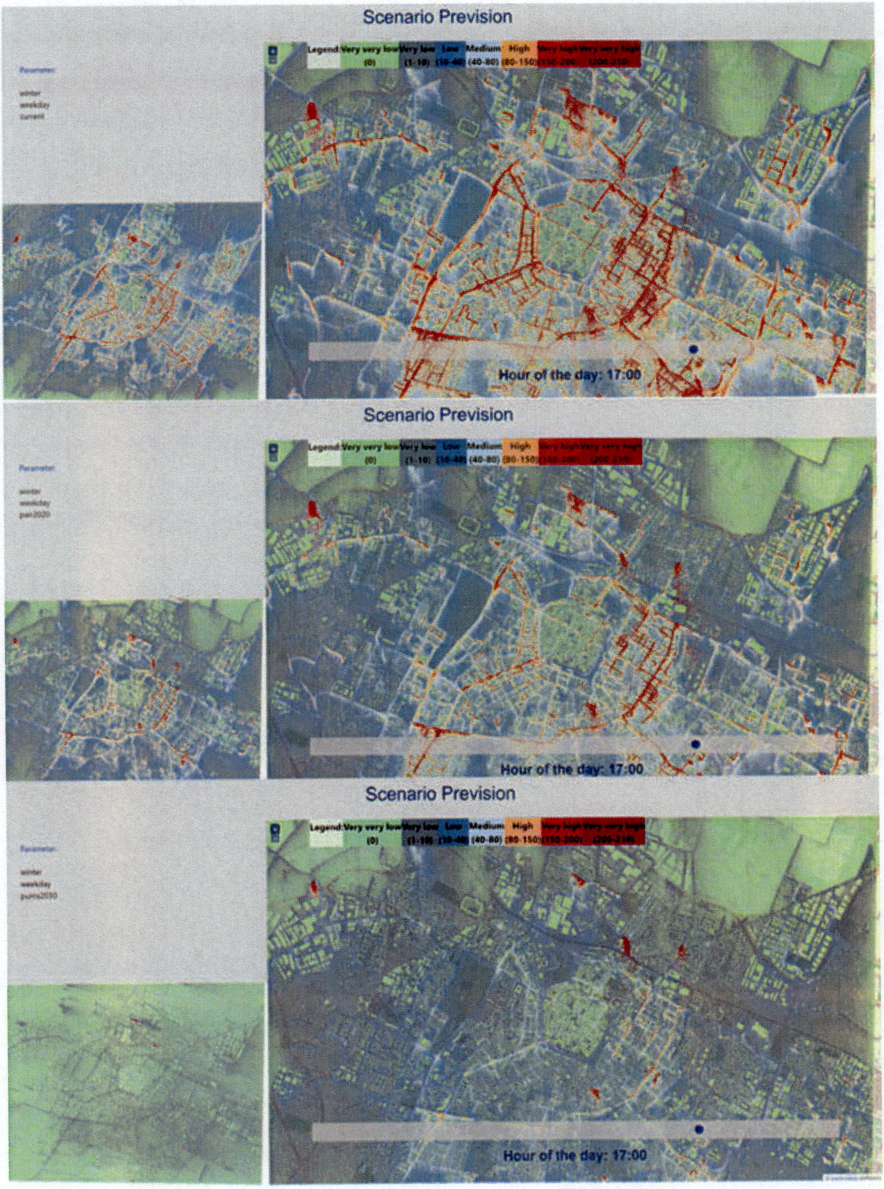

Fig. 14 Comparison of winter weekdays in 3 different scenarios: with current vehicles fleet (top), with Pair2020 vehicles fleet (middle), and with PUMS 2030 vehicles fleet (bottom).The image refers to the city of Modena

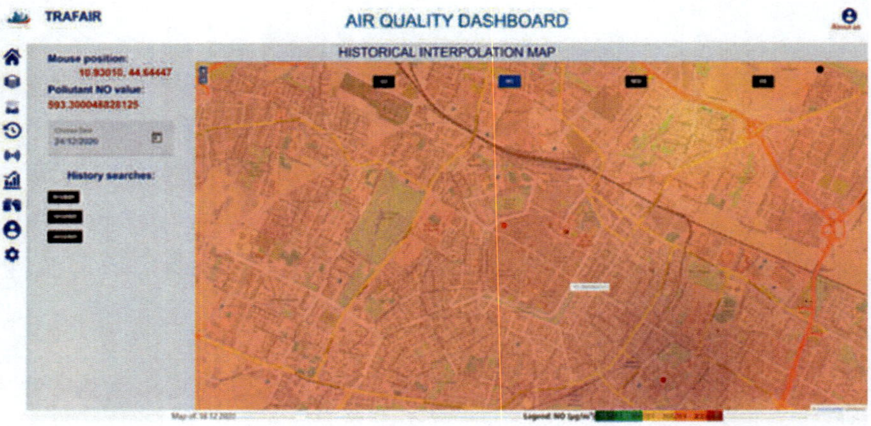

Fig. 15 Historical view of the average daily interpolation map for NO of 17th January 2021 in the city of Modena. The user can move the mouse on the map to see the value of average pollutant concentration in the pointed position

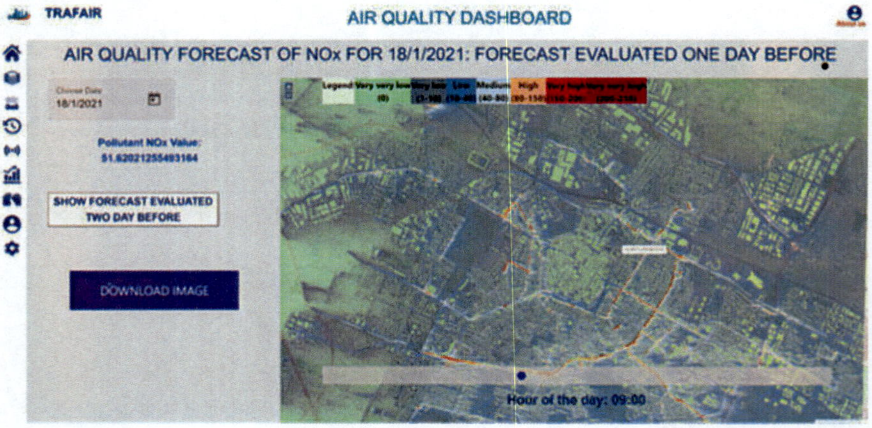

Fig. 16 Air quality forecast for NOx for 18th January 2021 at 09:00 AM in the city of Modena

6 Conclusion

We presented the TRAFAIR Air Quality dashboard that has been realized within the TRAFAIR project to display real-time, forecast, and statistical air quality observations through graphics, timelines, and maps. It allows decision-makers to monitor air conditions in the city of Modena and analyze trends regarding pollutant concentration. The architecture of the web application enables scalability allowing us to insert new sensors, move sensors around the city in new positions, and collect statistics regarding more months and years. We verify the replicability in different cities of the proposed solution: the dashboard is now in use in the cities of Modena, Santiago de Compostela, and Zaragoza. Through the use of GeoServer, data can be queried as coming from an API with a REST request. This allows us to manage a big amount of data structuring them in different layers and enabling different services on the same storage to obtain images, maps, or numerical values. The response time is reduced by the use of materialized views and automatized processes to update them.

We intend to integrate into the dashboard a combined visualization of air pollution dispersion model results and measured data to provide feedback on the effectiveness of the predictions to refine the model. Moreover, a future improvement of the dashboard may include the measurement of Particulate Matter (PM) in order to provide an overall air quality index that describes the air quality condition in the cities. This improvement requires additional sensors able to measure PM. Thanks to the experience gained in the context of the visualization of air quality sensor and model data, we create a dashboard that displays traffic data: both data coming from the sensors available in a smart city and data produced by traffic models [13, 26]. We are working on some views that allow visualizing both air quality and traffic conditions to better understand the impact of vehicles emission on the quality of the air in the urban context, an example of this views is included in the TRAFAIR Traffic dashboard described in [27].

Acknowledgements This research was developed in the scope of the TRAFAIR project 2017-EU-IA-0167), a European project co-financed by the Connecting Europe Facility of the European Union. The contents of this publication are the sole responsibility of its authors and do not necessarily reflect the opinion of the European Union. We kindly thank the City of Modena and Lepida S.c.p.A that both contribute to the management of sensors. Moreover, we acknowledge the important contribution of the LARMA team of the "Enzo Ferrari" Engineering Department for their contribution in installing, maintaining, and calibrating the air quality sensors in the city of Modena. We would like to thanks Jose R.R. Viqueira and Raquel Trillo Lado for contributing to making the dashboard also work in the city of Santago de Compostela and Zaragoza. Finally, we kindly regard Filippo Monelli and Giulio Querzoli for their help with software implementation.

References

1. Krzyzanowski, M., Cohen, A.: Update of who air quality guidelines. Air Quality Atmos. Health **1**, 7–13 (06 2008)
2. Bachechi, C., Desimoni, F., Po, L., Casas, D.M.: Visual analytics for spatio-temporal air quality data. In: Banissi, E., Khosrow-shahi, F., Ursyn, A., Bannatyne, M.W.M., Pires, J.M., Datia, N., Nazemi, K., Kovalerchuk, B., Counsell, J., Agapiou, A., Vrcelj, Z., Chau, H., Li, M., Nagy, G., Laing, R., Francese, R., Sarfraz, M., Bouali, F., Venturini, G., Trutschl, M., Cvek, U., Müller, H., Nakayama, M., Temperini, M., Mascio, T.D., Sciarrone, F., Rossano, V., Dörner, R., Caruccio, L., Vitiello, A., Huang, W., Risi, M., Erra, U., Andonie, R., Ahmad, M.A., Figueiras, A., Mabakane, M.S. (eds.) 24th International Conference on Information Visualisation, IV 2020, Melbourne, Australia, September 7–11, 2020, pp. 460–466. IEEE (2020). https://doi.org/10.1109/IV51561.2020.00080
3. Chen, P.: Visualization of real-time monitoring datagraphic of urban environmental quality. EURASIP J. Image Video Process. **2019**, 42 (2019). https://doi.org/10.1186/s13640-019-0443-6

4. Zhou, M., Wang, R., Mai, S., Tian, J.: Spatial and temporal patterns of air quality in the three economic zones of China. J. Maps **12**, no. sup1, 156–162 (2016). https://doi.org/10.1080/17445647.2016.1187095
5. Martorell-Marugãn, J., Villatoro-García, J.A., García-Moreno, A., López-Domínguez, R., Requena, F., Merelo, J.J., Lacasaña, M., de Dios Luna, J., Díaz-Mochón, J.J., Lorente, J.A., Carmona-Sãez, P.: Datac: a visual analytics platform to explore climate and air quality indicators associated with the covid-19 pandemic in Spain. Sci. Total Environ. **750**, 141424 (2021). https://www.sciencedirect.com/science/article/pii/S0048969720349536
6. Wu, X., Poorthuis, A., Zurita-Milla, R., Kraak, M.: An interactive web-based geovisual analytics platform for co-clustering spatio-temporal data. Comput. Geosci. **137**, 104420 (2020). https://doi.org/10.1016/j.cageo.2020.104420
7. Stehle, S., Kitchin, R.: Real-time and archival data visualisation techniques in city dashboards. Int. J. Geogr. Inf. Sci. **34**(2), 344–366 (2020). https://doi.org/10.1080/13658816.2019.1594823
8. Jing, C., Du, M., Li, S., Liu, S.: Geospatial dashboards for monitoring smart city performance. Sustainability **11**, 5648 (10 2019)
9. Guptill, S.: Spatial data. In: Smelser, N.J., Baltes, P.B. (Eds.) International Encyclopedia of the Social and Behavioral Sciences, pp. 14 775–14 778. Pergamon, Oxford (2001). https://www.sciencedirect.com/science/article/pii/B0080430767025080
10. Chatzigeorgakidis, G., Patroumpas, K., Skoutas, D., Athanasiou, S., Skiadopoulos, S.: Scalable hybrid similarity join over geolocated time series. In: Proceedings of the 26th ACM SIGSPATIAL International Conference on Advances in Geographic Information Systems, ser. SIGSPATIAL'18, pp. 119–128. Association for Computing Machinery, New York, NY, USA (2018). https://doi.org/10.1145/3274895.3274949
11. Po, L., Rollo, F., Viqueira, J.R.R., Lado, R.T., Bigi, A., López, J.C., Paolucci, M., Nesi, P.: Trafair: Understanding traffic flow to improve air quality. In: 2019 IEEE International Smart Cities Conference, ISC2 2019, Casablanca, Morocco, October 14–17, 2019, pp. 36–43 (2019)
12. Bachechi, C., Po, L.: Implementing an urban dynamic traffic model. In: Barnaghi, P.M., Gottlob, G., Manolopoulos, Y., Tzouramanis, T., Vakali, A. (eds.) 2019 IEEE/WIC/ACM International Conference on Web Intelligence, WI 2019, Thessaloniki, Greece, October 14–17, 2019, pp. 312–316. ACM (2019). https://doi.org/10.1145/3350546.3352537
13. Po, L., Rollo, F., Bachechi, C., Corni, A.: From sensors data to urban traffic flow analysis. In: 2019 IEEE International Smart Cities Conference, ISC2 2019, Casablanca, Morocco, October 14–17, 2019, pp. 478–485 (2019). https://doi.org/10.1109/ISC246665.2019.9071639
14. Bellini, P., Bilotta, S., Nesi, P., Paolucci, M., Soderi, M.: "Wip: Traffic flow reconstruction from scattered data. In: IEEE International Conference on Smart Computing (SMARTCOMP) **2018**, 264–266 (2018)
15. López, P.Á., Behrisch, M., Bieker-Walz, L., Erdmann, J., Flötteröd, Y., Hilbrich, R., Lücken, L., Rummel, J., Wagner, P., WieBner, E.: Microscopic traffic simulation using SUMO. In: 21st International Conference on Intelligent Transportation Systems, ITSC 2018, Maui, HI, USA, November 4–7, 2018, pp. 2575–2582. IEEE (2018)
16. Veratti, G.: The development of a building-resolved air quality forecast system by a multi-scale model approach and its application to Modena Urban area. Dissertation, University of Modena and Reggio Emilia (2020). https://iris.unimore.it/retrieve/handle/11380/1200723/261485/
17. Bigi, A., Veratti, G., Fabbi, S., Po, L., Ghermandi, G.: Forecast of the impact by local emissions at an urban micro scale by the combination of lagrangian modelling and low cost sensing technology: the trafair project. In: 19th International Conference on Harmonisation within Atmospheric Dispersion Modelling for Regulatory Purposes, Harmo 2019 (2019)
18. Fabbi, S., Asaro, S., Bigi, A., Teggi, S., Ghermandi, G.: Impact of vehicular emissions in an urban area of the PO valley by microscale simulation with the GRAL dispersion model. In: IOP Conference Series: Earth and Environmental Science, vol. 296, p. 012006 (07 2019)
19. Li, J., Heap, A.D.: Spatial interpolation methods applied in the environmental sciences: a review. Environ. Modell. Softw. **53**, 173–189 (2014). http://www.sciencedirect.com/science/article/pii/S1364815213003113
20. Iacovella, S.: GeoServer Cookbook. Packt Publishing (2014)

21. Nogueras-Iso, J., Ochoa-Ortiz, H., Janez, M.A., Viqueira, J.R.R., Po, L., Trillo-Lado, R.: Automatic publication of open data from ogc services: the use case of trafair project. In: 2020, Manuscript Submitted for Publication in The Twelfth International Conference on Advanced Geographic Information Systems, Applications, and Services GEOProcessing (2020)
22. Jain, N., Bhansali, A., Mehta, D.: Angularjs: a modern mvc framework in javascript. J. Global Res. Comput. Sci. **5**(12), 17–23 (2014)
23. Bostock, M., Ogievetsky, V., Heer, J.: D3 data-driven documents. IEEE Trans. Vis. Comput. Graph. **17**(12), 2301–2309 (Dec 2011). https://doi.org/10.1109/TVCG.2011.185
24. Chart.js | open source html5 charts for your website (2017). http://www.chartjs.org#Dataviz. Accessed 15 July 2020
25. Rollo, F., Po, L.: Senseboard: sensor monitoring for air quality experts. In: Costa, C., Pitoura, E., (eds.) Proceedings of the Workshops of the EDBT/ICDT 2021 Joint Conference, Nicosia, Cyprus, March 23, 2021, Ser. CEUR Workshop Proceedings, vol. 2841. CEUR-WS.org (2021). http://ceur-ws.org/Vol-2841/BigVis_3.pdf
26. Bachechi, C., Rollo, F., Desimoni, F., Po, L.: Using real sensors data to calibrate a traffic model for the city of Modena. In: Intelligent Human Systems Integration 2020-Proceedings of the 3rd International Conference on Intelligent Human Systems Integration (IHSI 2020), February 19–21, 2020, Modena, Italy, pp. 468–473 (2020). https://doi.org/10.1007/978-3-030-39512-4_73
27. Bachechi, C., Po, L., Rollo, F.: Big data analytics and visualization in traffic monitoring. Big Data Res. **27**, 100292 (2020). https://www.sciencedirect.com/science/article/pii/S221457962100109X. https://doi.org/10.1016/j.bdr.2021.100292. ISSN 2214-5796

Using Hybrid Scatterplots for Visualizing Multi-dimensional Data

Quang Vinh Nguyen, Mao Lin Huang, and Simeon Simoff

Abstract Scatterplot visualization techniques are known as a useful method that shows the correlations of variables on the axes, as well as revealing patterns or abnormalities in the multidimensional data sets. They are often used in the early stage of the exploratory analysis. Scatterplot techniques have the drawback that they are not quite effective in showing a high number of dimensions where each plot in two-dimensional space can only present a pair-wise of two variables on the x-axis and y-axis. Scatterplot matrices and multiple scatterplots provide more plots that show more pair-wise variables, yet also compromise the space due to the space division for the plots. This chapter presents a comprehensive review of multi-dimensional visualization methods. We introduce a hybrid model to support multidimensional data visualization from which we present a hybrid scatterplots visualization to enable the greater capability of individual scatterplots in showing more information. Particularly, we integrate star plots with scatterplots for showing the selected attributes on each item for better comparison among and within individual items, while using scatterplots to show the correlation among the data items. We also demonstrate the effectiveness of this hybrid method through two case studies.

Q. V. Nguyen (✉) · S. Simoff
MARCS Institute and School of Computer, Data and Mathematical Sciences, Western Sydney University, Penrith, Australia
e-mail: Q.Nguyen@westernsydney.edu.au

S. Simoff
e-mail: S.Simoff@westernsydney.edu.au

M. L. Huang
School of Software, University of Technology, Sydney, Australia
e-mail: Mao.Huang@uts.edu.au

B. Kovalerchuk et al. (eds.), *Integrating Artificial Intelligence and Visualization for Visual Knowledge Discovery*, Studies in Computational Intelligence 1014,
https://doi.org/10.1007/978-3-030-93119-3_20

1 Introduction

The advancement in information technology has generated an abundance of data that are usually in tabular forms or called multidimensional or multivariate data, such as database tables and spreadsheets. Unfortunately, such data normally have several dimensions which are much more than a human perception capability in both volume and dimensional information. Data should be represented in low-dimensional space for better human comprehension [1].

Visualization plays an important role as visual aids for decision making by presenting multidimensional data in readable forms, otherwise, it is difficult to extract and understand the data and the patterns. There are two ways to present the multidimensional data, including raw and derived data. The raw data visualization can be used directly to the original data with a low number of dimensions. For high dimensional data, dimensionality reduction or projection methods to reduce the number of dimensions, or relationship extraction methods to create hierarchies or graphs are usually applied before the visualization [1, 2].

1.1 Dimensionality Reduction Methods

Dimensionality reduction methods project multidimensional data to lower dimensions, usually to two-dimensional (2D) or three-dimensional (3D) spaces. These projection methods are great to reduce the dimensional complexity to a manageable number for further statistical analysis and visualization. This process also helps to reveal patterns and identify relevant subpopulations and outliers in the data. The dimensionality reduction methods include linear and non-linear approaches.

Linear projection methods, such as Principal Component Analysis and Linear Discriminant Analysis [3] have been popularly recognized as a basic yet effective method that can linearly transform data into low dimensional space while preserving well the variance of the data. However, the linear projection approach does not usually capture well non-linear structures that consist of arbitrarily shaped clusters or curved manifolds which are crucial in many applications, such as flow cytometry data [4].

The non-linear dimensionality reduction approach has gained popularity in data analysis due to its ability to generate a more meaningful organization of subpopulations including preserving the neighborhood information. Such methods in this approach can use non-linearize a linear dimensionality reduction method, such as Kernel Principal Component Analysis [5] and Multidimensional Scaling [6], or use manifold based methods, such as Laplacian Eigenmaps [7], Locally Linear Embedding [8], or based on other approaches, such as t-distributed Stochastic Neighbor Embedding (t-SNE) [9], and Uniform Manifold Approximation and Projection (UMAP) [10]. However, non-linear projection methods also introduce the complexity in the projection processes that make it difficult to understand the projection process to create trust in the outputs [11].

Dimensionality reduction methods can also integrate with feature selection methods, such as Random Forests [12] or artificial neural network models, such as [13] when handling data sets with a very large number of dimensions, such as genomic data sets. This process selects a handful number of dimensions or reducing irrelevant and redundant features prior to the projection process.

1.2 Multidimensional Data Visualization

There are several visualization techniques of such multivariate or multidimensional data, classified as geometric methods and iconographic displays, such as in [1, 14]. The methods can be roughly classified into two main streams, including geometric and glyph visualizations. Other methods can also create a structure of an image where the features are embedded in displays of other features, such as *Dimensional stacking* [15] and *Trellis display* [16].

1.3 Geometric Visualizations

Geometric visualizations use axes to present multidimensional points following geometric shapes or orientations. This type of visualization is popularly used in practice to analyze the correlation among data items, as well as reveal patterns and (ab)normality in the data sets. Typical techniques in this approach are *Scatterplots* and *Scatterplot matrices*, *Multiline graphs*, *Parallel coordinates*, and *Star plots*. Instead of using traditional horizontal and vertical axes in *Scatterplots* and *Scatterplot matrices*, *Parallel coordinates* or *Star plot* methods arrange the axes vertically or circularly.

Parallel coordinates methods plot the data points on the vertical axes and connect them by straight lines or curves (see Fig. 1). *Parallel coordinates* can also be linked with another type of visualizations, such as scatterplot [17] at each axis, to show more information, yet this approach could also create a discontinuity in the views.

On the other hand, *Star plot* or also called radar chart represents data on the axes arranged radially and originated from the same central point [18] (see Fig. 2). Sangli et al. adapted Star plot for visualizing data sets with a high number of dimensions [19]. The enhancement features include (i) drawing one star per data item for overlapped star plots, (ii) shift the origin away from the center point and (iii) partition the disc into multiple concentric circular regions each representing a portion of attributes or dimensions. However, this new visualization also creates the discontinuity problem when comparing variables on axes at different concentric circles.

The compactness of Star plot visualization makes it an excellent choice for displaying a small multivariate data set with less than a few hundred points and with a small number of dimensions. This method is useful for reviewing variables with similar, low or high values, or if there are any outlines among the variables.

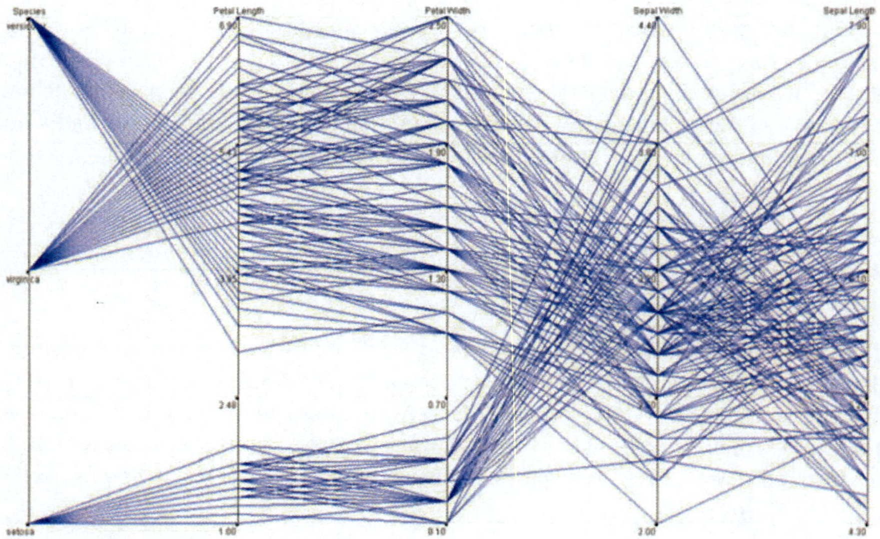

Fig. 1 An example of Parallel coordinates visualisation showing the classical Irish Flower data set (retrieved from https://archive.ics.uci.edu/ml/datasets/Iris). It is hard to see the correlation of sepal's and petal's width and length of the species in this visualisations

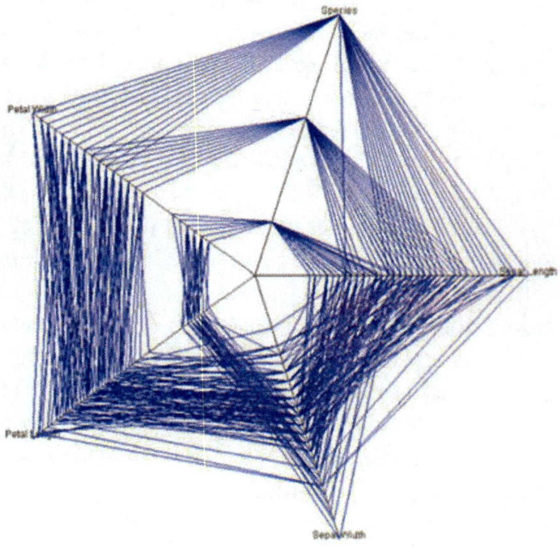

Fig. 2 An example of *Star plot* visualisation showing the same classical Irish Flower data set. It is hard to see the correlation of sepal's and petal's width and length of the species in this visualisations

However, Star plot visualization is not efficient to display large or high dimensionality data sets due to the high density in the display or densely packed radial axes, especially in the central area.

1.4 Glyph Visualizations

Glyph or icongraphic methods present multidimensional data on each object using glyph attributes A glyph contains the graphical entity of selected K dimensions as a subset of all N dimensions ($K \in N$). The glyphs might also integrate geometrical attributes (shape, size, orientation, position, and direction) and appearance attributes (color, texture, and transparency).

Sample glyph techniques are such as *Chernoff faces* [20] and *Star glyphs* [21]. *Chernoff faces* method maps multidimensional values to the facial features, such as the face shapes, the eyes, the eye browses, the width of a nose, etc. *Star glyphs* on the other hand display the variables using circular coordinates in each individual.

Follow the guidelines for layout strategy in [2] and extended from our work published in the 24th International Conference Information Visualisation [22], this chapter utilizes the compactness of the glyphs (such as Star glyphs) for conveying more multidimensional information on multiple scatterplots in addition to popular visual attributes, such as size, color, and shading. The proposed method enables greater flexibility and the ability to show more information of each individual on the contextual Scatterplot views. We use Scatterplots to show the correlation among the data items at the selected dimensions whilst our method can also represent the selected attributes of each item with a star plots. By showing additional information for each individual item, users can gain a better insight for the item as well as compare the additional attributes among them. And by maintaining the overall view of the entire data items with Scatterplots, users can also view the correlation and overall pattern of the mapping variables. The motivation of our approach is also based on the recent study which indicates that the presentation and the effective use of the radial layout thumbnails worked better than those thumbnails using the Cartesian layout [23]. We also demonstrate the effectiveness of our proposed method via various case studies.

2 Scatterplot Visualizations

One of the versatile, polymorphic, and generally useful methods for visualizing data is scatterplots. A scatterplot can show two or three variables on the 2D or 3D axes. Visual properties of items are also usually used to represent additional values. Scatterplot visualizations are very useful in the early stages of analysis where they can be used to show correlation and patterns in low-dimensional data [24, 25] as well as provide a snapshot of a large amount of data [26]. They are more effective than landscape visualizations for both visual search [27] and visual memory [28], especially when studying the correlation between two variables. Scatterplots have been extensively studied for both non-dimensional reduction [29] and dimensional reduction data [26].

However, the effectiveness of these elements deteriorates when the number of dimensions increase, as well as the number of data points. In addition, a single scatterplot does not handle well data sets with a high number of dimensions due to the limited mapping dimensions in the plot. Figure 3 presents an example of the visualization scatterplots with the classical Irish Flower data set. We can show all four attributes septal length, septal width, petal length, and petal width on the axes together with the species as color, which otherwise could not be possible on a single scatterplot. This visualization indicates three distinctive groups of the three species based on the width and length of the sepals and petals. In addition, the overplotting and overlapping of data points may also hinder the accuracy of the extracted information in the single scatterplot [30].

Scatterplot matrices extend the simple scatterplots that show all the pairwise scatterplots of attributes on a single matrix formation. Given a dataset with N dimensions, the scatterplot matrix provides N × N scatterplot panels in N rows and N columns. Having all attributes mapped in this formation means it is easier to scan horizontally and vertically to assess the differences in the relationships between multiple variables and compare the dimensions [26]. However, repeated comparison pairs at the upper- and lower halves, as well as unused diagonal panels, results in much redundancy within matrix displays (see Fig. 4).

Modern Scatterplot matrices may utilize the diagonal plots for presenting histograms or other information (such as in [31]), and integrate interactive techniques, including linking, brushing, and zooming to highlight items across multiple scatterplots or enlarge a focused plot.

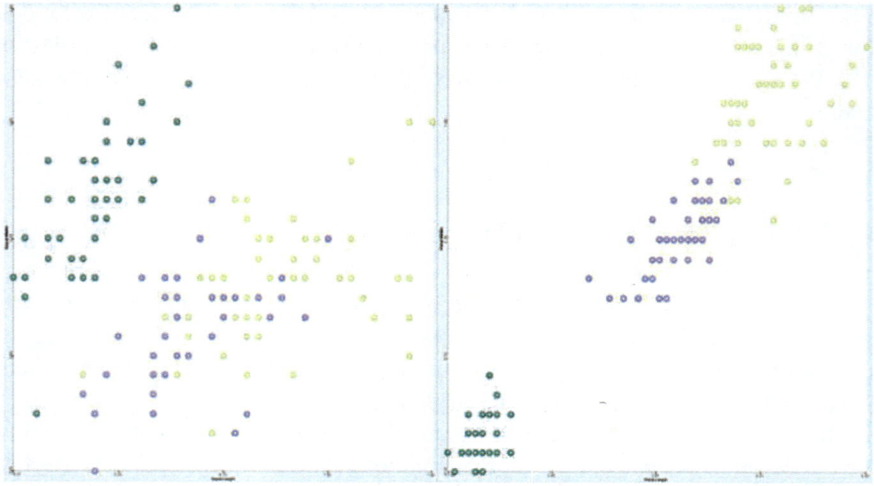

Fig. 3 An example of two individual scatterplots showing the classical Irish Flower data set. To visualize this data sets with four quantitative variables (septal length, septal width, petal length, and petal width) and one categorical variable (species), we need at least two scatterplots to show the pairwise comparison of the four quantitative variables as well as using colors to present the species

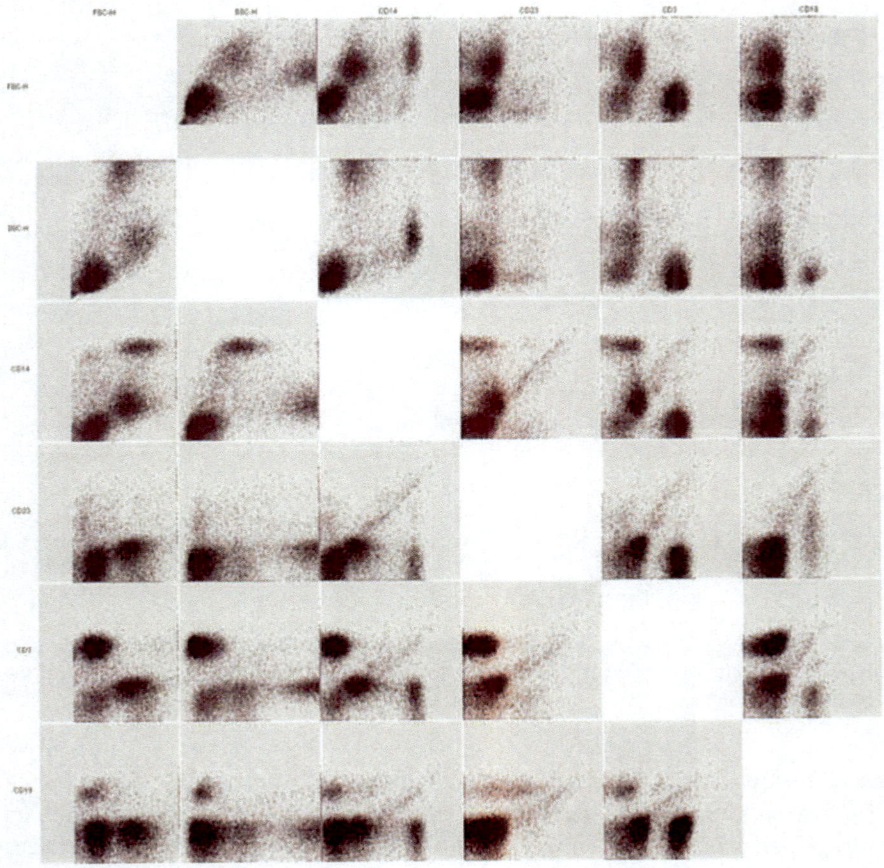

Fig. 4 An example of a scatterplots matrix on a flow cytometry data with 6 dimensions, each cell represents as a dot-point. While the visualization shows all dimensions in a pairwise manner in the plots, the display space of each plot is small for detail information it also creates a redundancy when showing all uninterested dimensions

To overcome the limitation of single scatterplot and scatterplot matrices, *Linkable Scatterplots* technique was introduced in [32–34]. This method provides a flexible way to specify the scatterplot panels over the user interface so that it can be arranged in a way that best meets users' data exploration goals. Depending on the nature of the data being represented and the intentions of the user, Linkable Scatterplots can provide more plot panels than the single scatterplot and reduce the unnecessary plot panels—a major problem in the scatterplot matrix methods, one which may contribute to cognitive overload.

Figure 5 is an example of *linkable scatterplots* for the Automobile data set with four panels. We map the quantitative dimensions or variables on the axes of each plot as well as the common visual mappings on categorical variables including "made" and "engine types" to color and shape respectively. The figure illustrates clearly

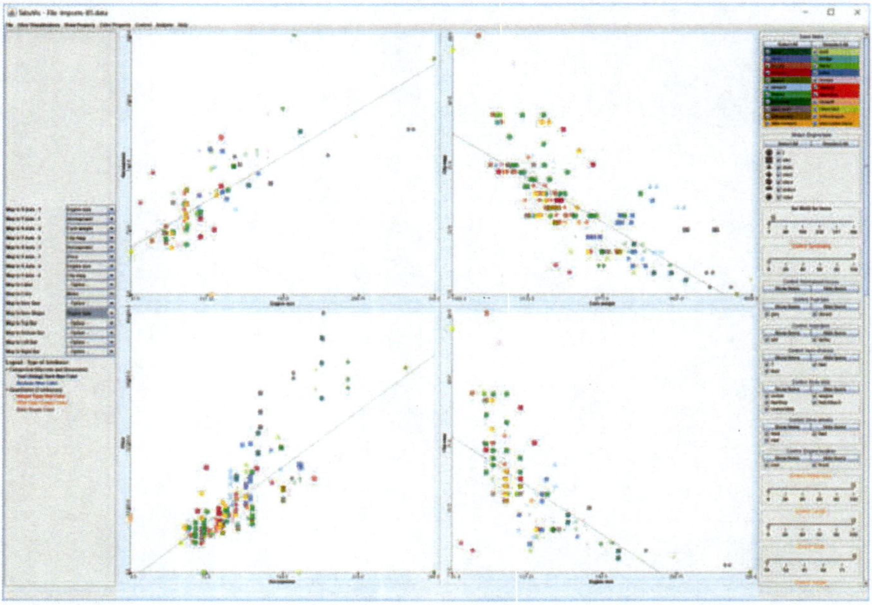

Fig. 5 An example of linkable scatterplots' visualization 4 panels of the Automobile data set (retrieved from https://archive.ics.uci.edu/ml/datasets/Automobile) showing pairwise plots of the necessary dimensions

the correlation among the mapping variables particularly *horsepower* versus *engine size*, *city miles per gallon* versus *curb weight*, *price* versus *horsepower*, and *city mileage per gallon* versus *engine size*. For example, cars with larger engine sizes linearly produce more horsepower, use more petrol per mile, and also relatively more expensive.

We use *linkable scatterplots* in the visualization thanks to its robustness and superiority for display the information in comparison with other scatterplot methods [35]. The study shows that the visual analysis with multiple scatterplots (e.g. linkable scatterplots) are better sequential scatterplots (e.g. single scatterplot) and simultaneous scatterplots (e.g. scatterplot matrix) in exploring multivariate data. Particularly, linkable scatterplots method is received higher accuracy and the other two methods. And this method also is the most preferable method with positively experienced technique in the study.

3 Hybrid Model

Extending the visual information communication model in [36], we propose a hybrid layout model for multidimensional data visualization (see Fig. 6). The model comprises three main components, including data processing, visualization, and interaction.

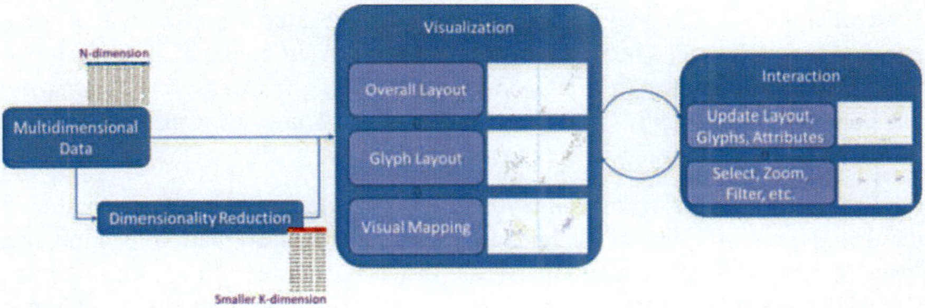

Fig. 6 The hybrid model for multidimensional data visualization

3.1 *Data Processing*

The first and the important step is to collect, clean, integrate and normalize the data to ensure minimal variations and errors prior to the visualization. Although the output data from the above processes can be used directly in the visualization, it can be inefficient to visualize data sets with a high number of dimensions. For such data sets, it is important to identify important features and reduce the number of dimensions.

Feature selection methods select a handful number of most useful dimensions, according to the target of interest. The feature selection can calculate the score of all features using a machine learning model such as *Random Forest* [37], or select the features based on the highest scores or in a greedy fashion. Feature selection methods for high dimensional classification data are benchmarked in the study in [38].

Dimensionality reduction methods reduce the number of dimensions by projecting them into a low dimensional space, usually in two-dimension or three-dimension. Methods in this direction can use linear project (such as *Principal Component Analysis*) or non-linear projection (such as *Kernel Principal Component Analysis* [39], *Self-Organizing Map* [40], *Laplacian Eigenmap* [41], *Locally Linear Embedding* [8], or popularly in biology field *t-Distributed Stochastic Neighbor Embedding* [42] and *Uniform Manifold Approximation and Projection* [43]).

3.2 *Visualization*

Visualization plays an important role to present multidimensional data in a clear and interpretable way, where we can interact and explore the information to gain insight. It is not likely possible to construct a one-type-fit-all and easy-to-understand visualization to cover the range of data. A visualization can include three components, including overall layout, glyph layout, and visual mapping. The use of both global and individual layouts can complement each other so that the former can show the correlation among the data items while the latter can provide additional information on data items for item-to-item comparison.

Overall layout—we first lay out the data items to reveal the overall insight information and the correlations between the data points. Layout methods for multidimensional data as discussed in the previous section, including *Scatterplots* and *Scatterplot matrices, Multiline graphs, Parallel coordinates*, and *Star plots*. Our prototype uses Scatterplots to generate the layouts for the data items.

Glyph layout—While the overall layout presents a useful overview presentation of the data, glyphs are popular methods to provide further information for individual items thanks to their visual compactness. Instead of representing each point as a simple dot from the overall layout, glyphs can also be used to represent additional information or dimensions at the data items. Among the glyphs methods discussed in the previous section, we use star plots to present additional dimensional information at each data item.

Visual mapping—visual mappings such as colors, shading, and size can be applied to the glyphs to provide additional information to each data item.

3.3 Interaction

Interaction is useful to allow a user to explore visualization and provide different views to support further the discovery and the insight. The interaction can also integrate the users' knowledge and preference from which they can evaluate, refine, and go beyond the current and previous iterations.

The interaction can apply to both the overall layout and the glyph layout. For example, the users can change the overall layouts based on the new dimension mappings or new layout methods to suit the visual interest or exploration requirement. The individual glyph layout can also be updated including changing the set of dimensions used in the glyphs or the glyph layout methods. Different visual styles and dimensional mappings can also during the interaction.

The users can also interact with the visualizations in different ways, in both global layout and individual items, depending on the exploratory tasks following popular Shneiderman's Mantra "overview first, zoom and filter, then details on demand" [44]. To support this, individual and multi-item selections are available to the users where the selected items can also be highlighted across multiple plot panels for further actions, such as zooming, masking, or filtering. Item details can also be provided via tool-tips when required.

4 Hybrid Visualization

Follow the hybrid model, our visualization provides a seamless view from a global overview of all data items to a glyphs view for each item, as well as a comprehensive interaction and visual mappings to enable the analysis. Our hybrid visualization was built on the *Linkable Scatterplots* platform [32]. The platform provides a flexible environment, where the analysts can choose the number of plot panels, mapping on axes as well as visual attributes in the visualization, such as shape, size, color, and visual bars. With the capability of showing selected variables concurrently, it is more effective to compare the correlation of more variables within the limited space than a single scatterplot while reducing the unnecessary and crowded presentation of all information as scatterplot matrices. Interaction such as linking, brushing, and zooming among the plots are also supported in the visualization (see Fig. 5).

We utilize the effectiveness of linkable scatterplots to provide an overview of the data to reveal the correlation and patterns in the data for multiple selected variables. To enable the further presentation of more variables, we represent each data item as a single star plot instead of a simple point or icon. The axes on the star plots can be customized so that they can represent the limited information of the selected variables without overwhelming the viewers. The arrangement and visual attributes such as color and size of individual star plot items follow the existing scatterplots visualization.

Figure 7 shows an example of visualization of the same Irish Flower dataset where the flowers are arranged in order and all five attributes of each item are presented as a Star plot. Red, green, and blue colors are representing Virginica, Setosa, and

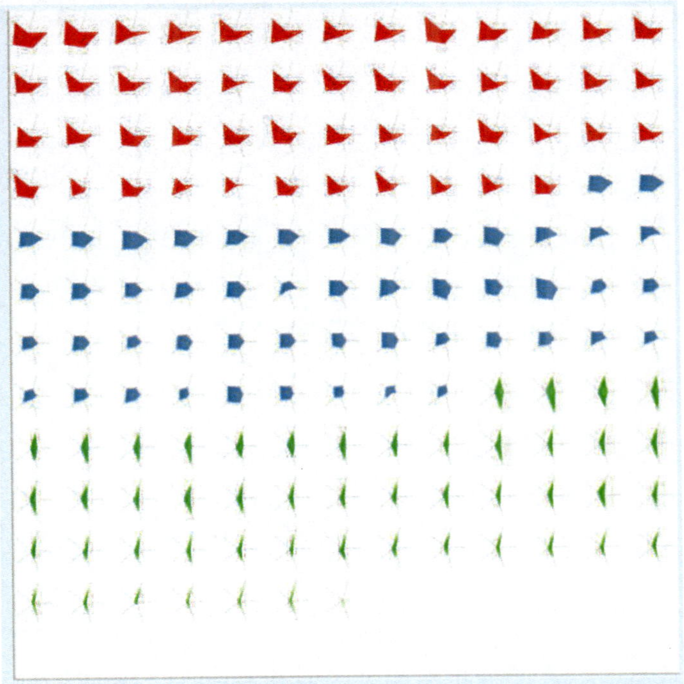

Fig. 7 An example of visualization with the Star plots showing all five attributes of the Irish Flower dataset while the items are positioned in order and sorted by the species

Versicolor species respectively. We can identify the property among the Star plot items, such as polygonal shapes are similar for the items of the same species and these shapes are different between species, such as Versicolor samples (blue color) are more round, Setosa samples (green color) are skinny. However, it is quite challenging to reveal a correlation among the species as shown in the Scatterplots.

Figure 8 is the visualization of the same Irish Flower dataset where the star plots are distributed based on Scatterplot visualization of Sepal length (X-axis) and Sepal width (Y-Axis). When combining with the Scatterplot, we can reveal much better the pattern or correlation among the selected variables in addition to the detailed information at each individual plot. Particularly, this figure shows clearly the distribution of species where Setosa samples (lower Sepal length and higher Sepal width) are clearly different from Virginica and Versicolor. The Sepal property of the two latter species is reasonably distinguishable yet as clear as the former one.

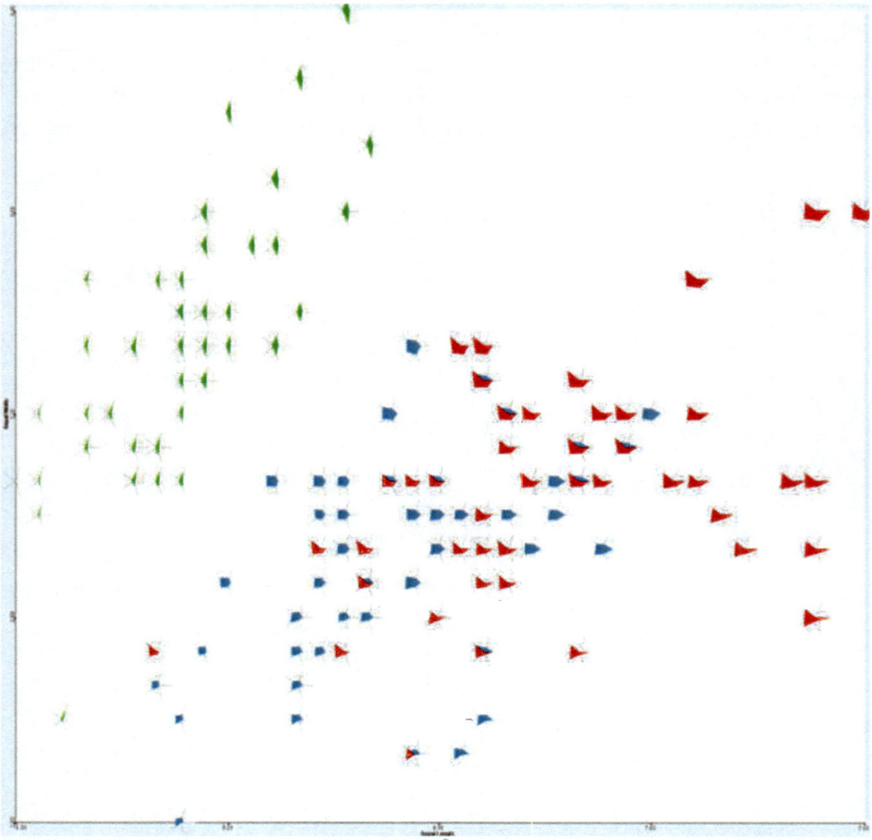

Fig. 8 An example of visualization with the star plots showing all five attributes of the Irish Flower dataset where the items are positioned by Sepal-length (X-axis) and Sepal-width (Y-axis)

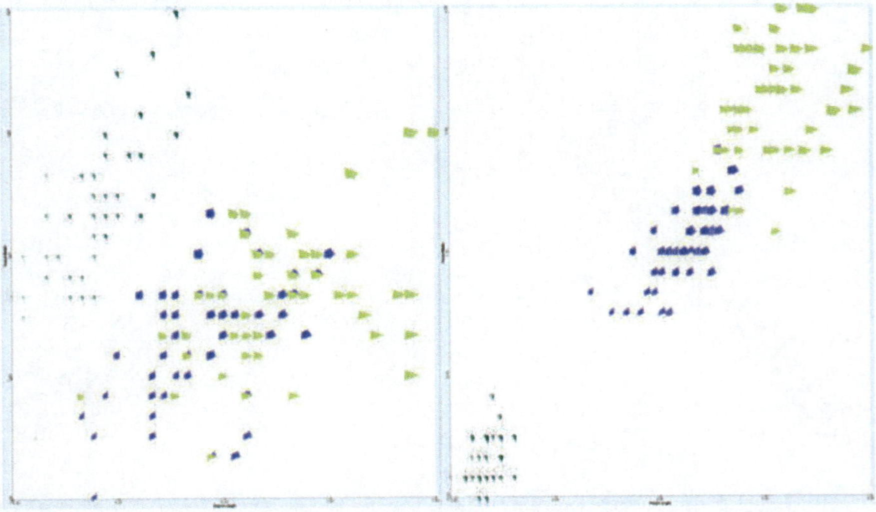

Fig. 9 An example of visualization with the star plots showing all five attributes of the Irish Flower data set where two scatterplots panels are used to show the correlations of Sepal-length versus Sepal-width and Petal-length versus Petal-width and different visual attributes are used

Figure 9 shows an alternative view of the hybrid visualization when two scatterplots are used to illustrate better the correlations among flower species in the data sets. The glyphs illustrate further the property of the flower species when we can compare them side by side in a compact way.

5 Case Studies

Our preliminary experiments show that the hybrid technique using Scatterplots and Star plots may not effective when showing a large number of data items. This is due to the over-crowded and overlapping views when showing several large star plots items concurrently on limited-size panels. Figure 10 illustrates an example of the hybrid visualization on a large data set with 1,380 items showing cigarette consumption in The United States for the period 1963 to 1992 (data retrieved from https://vincentarelbundock.github.io/Rdatasets/datasets.html). The two scatter plots indicate quite clearly the correlation and the pattern of the selected pairwise variables, including *price per pack of cigarettes* versus *minimum price in adjoining states per pack of cigarettes* (left panel) and *price per pack of cigarettes* versus *cigarette sales in packs per capita* (right panel). Unfortunately, the overlapping and overcrowded make it very hard to identify and compare other attributes in the individual star plots. Filtering and interaction are necessary to simplify the visualization and to focus on a smaller number of selected items.

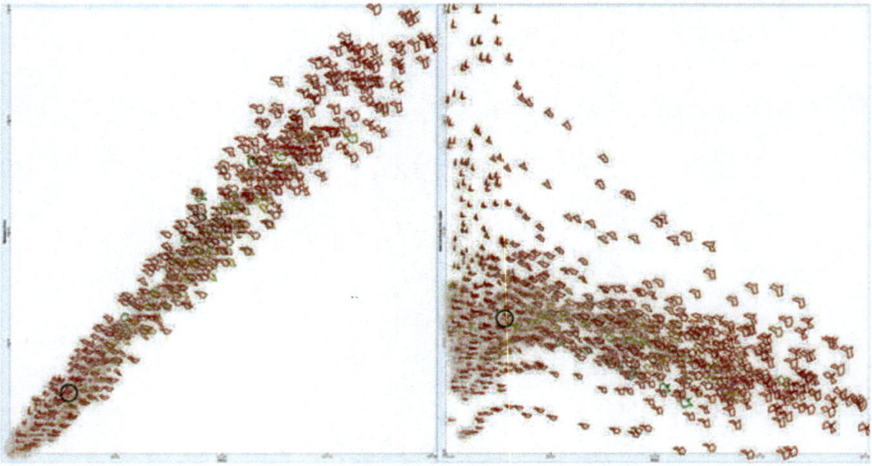

Fig. 10 A hybrid visualization showing the large number of data items of the scatterplots which make it really difficult to perceive the information from the individual start plots

Nevertheless, the hybrid method can provide a useful complementing view when the users wish to see further information for comparison on individual items that cannot be shown by the Scatterplots during the exploration and interaction. We demonstrate the effectiveness of the new hybrid technique via two case studies, including health and world data sets.

5.1 Case Study 1—Health Data Set

Using the studies and the Acute Lymphoblastic Leukaemia (ALL) data in [45, 46], we wish to explore whether the biomedical variables could have any impact on the survival rate of the disease. Using the genomic and biomedical profiles of 100 paediatric B-cell ALL patients treated at the Children's Hospital at Westmead were generated using Affymetrix expression microarray [46]. The automated genomic data analytics methods were applied to the very high dimensional data, such as Feature Selection with as Random Forests [12] and Dimensionality Reduction Methods (as described in the previous section), to create the similarity space in three-dimensional space. Two patients are located closely together if their genes are similar and otherwise are far from each other if their genomic data are different.

Figure 11 shows a scatterplot visualization of the patient populations. The display reveals the two groups of patients who were relapsed (red color and mostly on the top-left side) and who were survived (green color and mostly on the right side). While the scatterplot suggests an interesting genomic property of the two groups, this visualization does not provide additional biomedical information to the patients

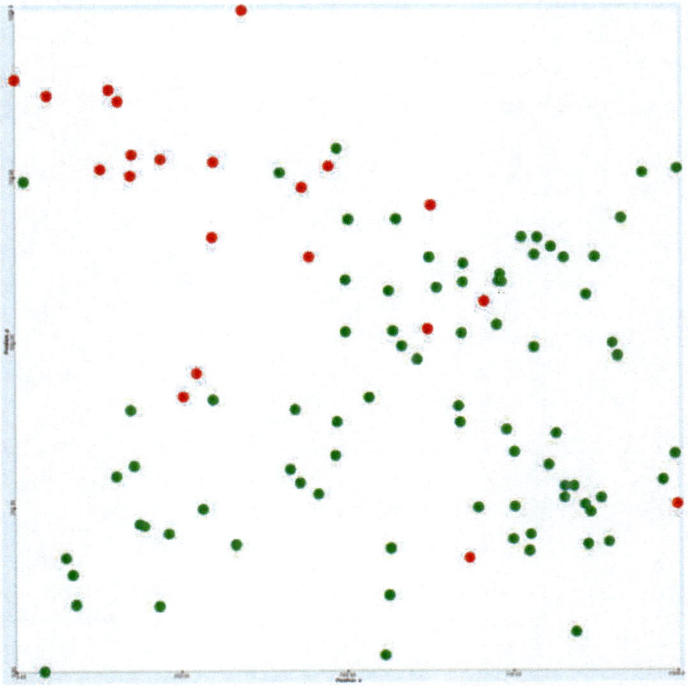

Fig. 11 A visualization of the patient population in a two-dimensional scatterplot showing their genomic similarity. The red dots are the relapsed patients and the green dots are the survived patients. The visualization reveals the potential of genomic difference among the two groups where the relapsed patients are mainly located at the top-left side of the visualization

for comparing the patient's property. To enable the greater ability to show more biomedical data, we use star plots to represent the selected variables on the items.

Figure 12 shows the visualization where star plots present the *gender, phenotype, immunophenotype, treatment protocol, risk strategy,* and *treatment method* for each individual patient. We can see from the *Star plot* glyphs on the horizontal axis, for example, most of the relapsed patients are males (in-ward mark on the axis).

Figure 13a illustrates another example when we use two scatterplots to show the patient cohort where the x-axes are mapped to the same dimension and the y-axes are mapped to two different dimensions. Star plots present the biological values, including *Ph_rearr, MI_Af4, E2a_pbx1, Cns involvement*, and *treatment protocol*. The visualization reveals useful information for further comparison among the patients, such as those patients who are marked with arrows are quite different from their neighbors (see Fig. 13a). We do further exploration by filtering out the treatment protocol to show only the patients in *Study 8* protocol (Fig. 13b). The visualization indicates *Study 8* is the dominant protocol, and unfortunately, most of the relapsed patients are also in this protocol.

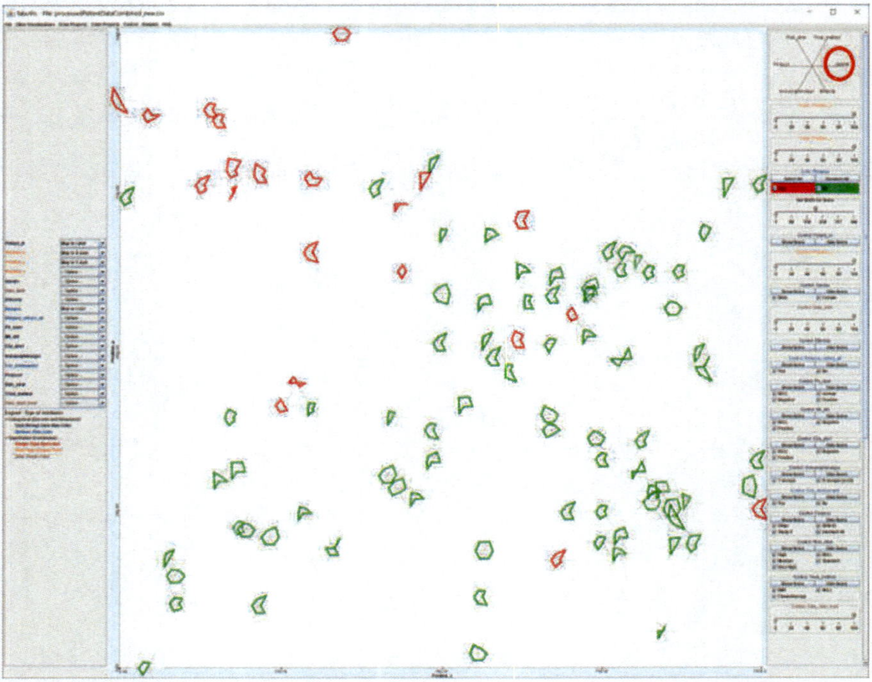

Fig. 12 A visualization of the patient population when we use star plots to show additional information, including gender, ethnicity, immunophenotype, treatment protocol, risk value, and treatment method

5.2 *Case Study 2—World Data Set*

We apply our hybrid visualization for analyzing the world data set (retrieved from https://www.worlddata.info/downloads/ which was compiled from files at http://gsociology.icaap.org/dataupload.html). The data contain several quantitative and categorical variables in relation to the countries, such as *population, area, population density, net migration, infant mortality, gross domestic product per capita, literacy, birth-rate, death-rate, agriculture,* and *industry.*

Using two scatterplots, we can show the correlation of *gross domestic product per capita* or *GDP* (x-axis) versus *literacy* (y-axis) and *birth-rate* (x-axis) versus *dead-rate* (y-axis) for all countries (see Fig. 14). The visualization indicates the interesting correlation of *GDP* versus *literacy* and *birth-rate* versus *dead-rate* for the countries. This scatterplot visualization improves dramatically when we add visual attributes into the data items, including colors for region and size of population (see Fig. 15).

Although using scatterplots with visual attributes on each item can provide a couple of extra dimensions compared to the former one, it is not possible to gain further information from the visualization due to the limited visual mapping capability.

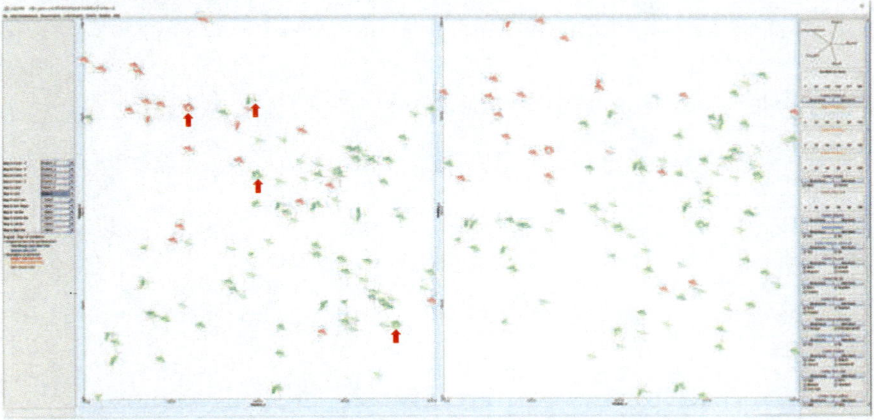

a

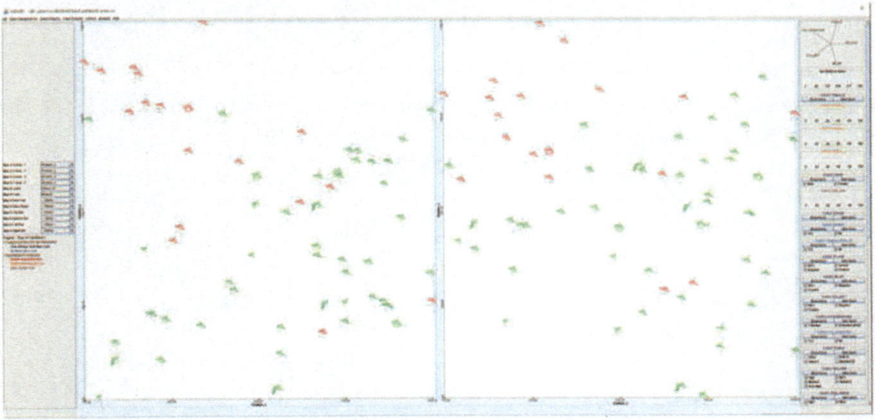

b

Fig. 13 A visualization of the patient population using two scatterplots to show the additional dimension on the global correlation view among the patients. We use star plots to show additional biological values, including Ph_rearr, MI_Af4, E2a_pbx1, Cns involvement, and protocol. Figure a shows the full patient cohort while Figure b shows the group of patients under Study 8 protocol

We extend the Scatterplot visualization further using Star plots to review additional variables where it reveals a much richer property of each item for better comparison among the data items (see Fig. 16). Five additional variables that we selected in the visualization including *area, net migration, phones (per 1000), agriculture,* and *industry.* While the Star plots can give additional information in each item, the crowded view in the hybrid visualization could reduce the readability, especially when showing a high number of items with smaller sizes and overlapping. This issue can be overcome with the interaction when we just select a small number of items for better item-to-item comparison among them. Figure 17 shows the view of the visualization at an interaction stage set when selecting 22 countries with the largest population. And Fig. 18 presents a combined view where the countries are laying in

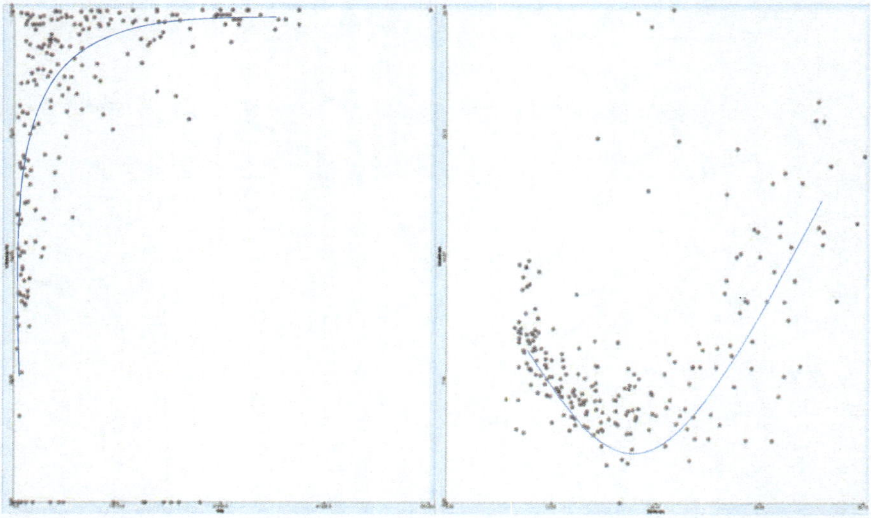

Fig. 14 An example of Scatterplots showing the non-linear correlation of *gross domestic product per capita* versus *literacy* and *birth rate* versus *death rate* variables

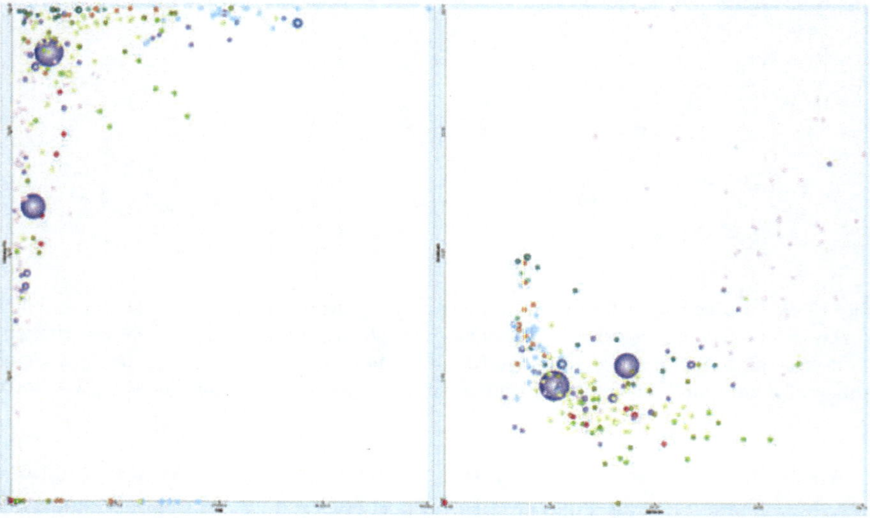

Fig. 15 An example of Scatterplots showing the same layout and data in which visual attributes are used to display more information, where the size of each item is proportional to the country's population and the color shows regions of the countries

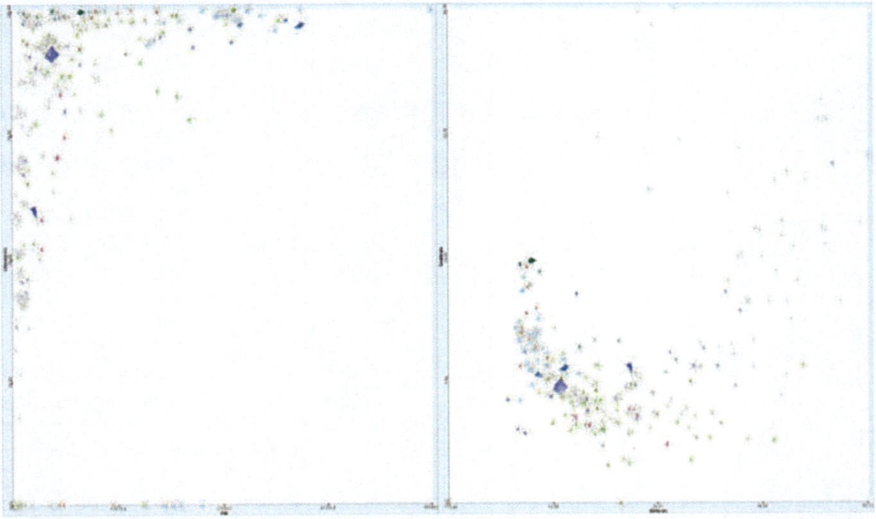

Fig. 16 We use Star plots to show several additional attributes in the visualization, including *area, net migration, phones (per 1000), agriculture* and *industry*

Fig. 17 The visualization at a navigational stage where we only show the top 22 populous countries for better comprehension and comparison among the items

alphabetical order with the same size (left panel) and the *birth-rate* and *death-rate* are still shown on the x-axis and y-axis (right panel).

In Figs. 16 and 17, for example, we wish to compare the two most populous countries, i.e. China and India. The scatterplot visualization shows that China has

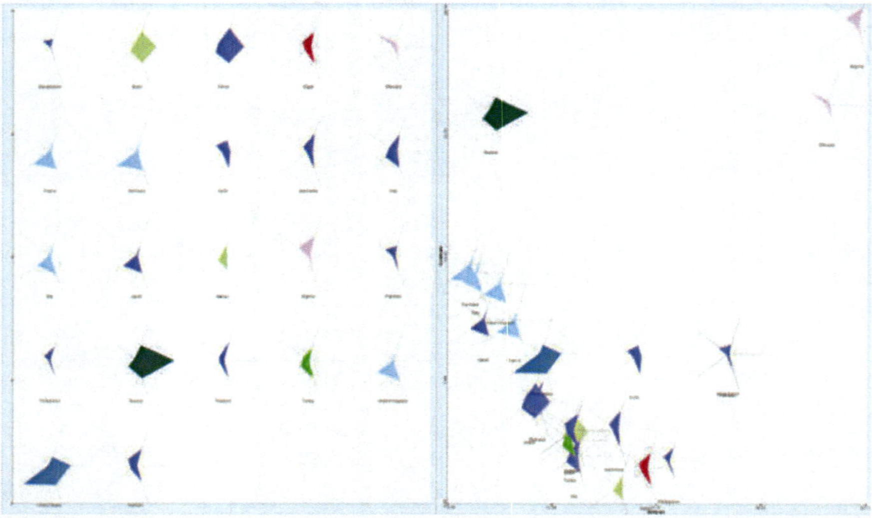

Fig. 18 An example Star plots showing additional attributes in alphabetical order (left) with larger and uniform size for better detail comparison among the countries

a marginally higher population, significantly higher literacy, and marginally higher gross domestic product per capita in comparison with India. Furthermore, the star plot visualization also indicates that China has a medium–low percentage of agriculture, a high percentage of the industry, large area, a moderate amount of net migration, and low birth-rate, while India has a medium–low percentage of both agriculture and industry, as well as moderate value for other attributes.

Besides, we can easily identify several insights from the single visualization, such as Russia has the largest area, Nigeria and Ethiopia have very high birth-rate, and the agriculture values are low in The United States, German, and Japan.

6 Conclusion and Future Work

We have presented a hybrid model and visualization technique that combines Scatterplots with Star plots to provide overview views of the correlation among data items, as well as additional information on each item as a star glyph respectively. Our experiments show the usefulness of the hybrid visualization, especially when we show additional information with a manageable number of selected items for detailed analysis and comparison. We are going to carry out a usability study to formally evaluate the effectiveness of this hybrid visualization, as well as identify the tasks and data that are suitable for this hybrid visualization. In addition, we will next implement the geographic visualization to utilize the ability to show multiple dimensions on the map using the star plot visualization.

References

1. Dzemyda, G., Kurasova, O., Zilinskas, J.: Multidimensional Data Visualization: Methods and Applications. Springer, Berlin (2012)
2. Ward, M.O.: A taxonomy of glyph placement strategies for multidimensional data visualization. Inf. Vis. **1**, 194–210 (2002)
3. Fisher, R.A.: The use of multiple measurements in taxonomic problems. Ann. Eugen. **7**, 179–188 (1936)
4. Konstorum, A., Jekel, N., Vidal, E., Laubenbacher, R.: Comparative analysis of linear and nonlinear dimension reduction techniques on mass cytometry data. bioRxiv, 273862 (2018). https://doi.org/10.1101/273862
5. Schölkopf, B., Smola, A., Müller, K.R.: Kernel principal component analysis. In: Artificial Neural Networks—ICANN'97, (1997)
6. Borg, I., Groenen, P.J.F.: Modern Multidimensional Scaling: Theory and Applications. Springer, New York (2005)
7. Belkin, M., Niyogi, P.: Laplacian eigenmaps for dimensionality reduction and data representation. Neural Comput. **15**, 373–1396 (2003)
8. Roweis, S.T., Saul, L.K.: Nonlinear dimensionality reduction by locally linear embedding. Science **290**, 2323–2326 (2000)
9. Van der Maaten L, Hinton G.: Visualizing data using t-SNE. J. Mach. Learn. Res. **9**, 2579–2605 (2008)
10. McInnes, L., Healy, J., Melville, J.: UMAP: Uniform Manifold Approximation and Projection for Dimension Reduction. arXiv:1802.03426 (2018). [stat.ML]
11. Sumithra, V., Surendran, S.: A review of various linear and non linear dimensionality reduction techniques. Int. J. Comput. Sci. Inf. Technol. **6**, 2354–2360 (2015)
12. Ho, T.K.: The random subspace method for constructing decision forests. IEEE Trans. Pattern Anal. Mach. Intell. **20**, 832–844 (1998)
13. Trutschl, M., Cvek, U., Grinstein, G.: Intelligently resolving Point Occlusion. In: IEEE Symposium On Infomation Visualization, pp. 131–136. Seattle, WA (2003)
14. Liu, S., Maljovec, D., Wang, W., Bremer, P.T., Pascucci, V.: Visualizing high-dimensional data: advances in the past decade. IEEE Trans. Vis. Comput. Graph. **23**, 1249–1268 (2017)
15. Ward, M.O.: XmdvTool: integrating multiple methods for visualizing multivariate data. In: Proceedings of the Conference on Visualization. Los Alamitos, CA (1994)
16. Becker, R.A., Cleveland, W.S., Shyu, M.J.: The design and control of trellis display. J. Comput. Stat. Graph. **5**, 123–155 (1996)
17. Steed, C.A., Ricciuto, D.M., Shipman, G., Smith, B., Thornton, P.E., Wang, D., Shi, X., Williams, D.N.: Big data visual analytics for exploratory earth system simulation analysis. Comput. Geosci. **61**, 71–82 (2013)
18. Chambers, J., Cleveland, W., Kleiner, B., Tukey, P.: Graphical Methods for Data Analysis. Wadsworth (1983)
19. Sangli, S., Kaur, G., Karki, B.B.: Star plot visualization of ultrahigh dimensional multivariate data. In: International Conference on Advances in Big Data Analytics, pp. 91–97 (2016)
20. Chernoff, H.: The use of faces to represent points in k-dimensional space graphically. J. Am. Stat. Assoc. **68**, 361–368 (1973)
21. Chambers, J.M.: Graphical Methods for Data Analysis (Statistics). Chapman & Hall, CRC (1983)
22. Nguyen, Q.V., Huang, M.L., Simoff, S.: Enhancing scatter plots with Start-plots for visualising multi-dimensional data. In: 24th International Conference on Information Visualisation, pp. 80–85 (2020)
23. Burch, M., Bott, F., Beck, F., Diehl, S.: Cartesian versus radial—a comparative evaluation of two visualization tools. In: International Symposium on Visual Computing, pp. 151–160 (2008)

24. Packham, I.S.J., Rafiq, M.Y., Borthwick, M.F., Denham, S.L.: Interactive visualisation for decision support and evaluation of robustness—in theory and in practice. Adv. Eng. Inform. **19**, 263–280 (2005)
25. Friendly, M., Denis, D.: The early origins and development of the scatter plot. J. Hist. Behav. Sci. **41**, 103–130 (2005)
26. Sedlmair, M., Munzner, T., Tory, M.: Empirical guidance on scatterplot and dimension reduction technique choices. IEEE Trans. Vis. Comput. Graph. **19**, 2634–2643 (2013)
27. Tory, M., Sprague, D., Wu, F., So, W.Y., Munzner, T.: Spatialization design: comparing points and landscape. IEEE Trans. Vis. Comput. Graph. **13**, 1262–1269 (2007)
28. Tory, M., Swindells, C., Dreezer, R.: Comparing dot and landscape spatialization for visual memory differences. IEEE Trans. Vis. Comput. Graph. **15**, 1033–1039 (2009)
29. Rensink, R.A., Baldridge, G.: The perception of correlation in scatter plot. Comput. Graph. Forum **29**, 1203–1210 (2010)
30. Cleveland, W.S., McGill, R.: The many faces of a scatterplot. J. Am. Stat. Assoc. **79**, 807–822 (1984)
31. Cui, Q., Ward, M.O., Rundensteiner, E.A.: Enhancing scatterplot matrices for data with ordering or spatial attributes. In: Visualization and Data Analysis (2006)
32. Nguyen, Q.V., Simoff, S., Qian, Y., Huang, M.L.: Deep exploration of multidimensional data with linkable scatterplots. In: 9th International Symposium on Visual Information Communication and Interaction, pp. 43–50. Dallas, Texas (2016)
33. Nguyen, Q.V., Qian, Y., Huang, M.L., Zhang, J.: TabuVis: a tool for visual analytics multidimensional datasets. Sci. China Inf. Sci. **052105**(12), (2013)
34. Nguyen, Q.V., Qian, Y., Huang, M.L., Zhang, J.: TabuVis: a light weight visual analytics system for multidimensional data. In: International Symposium on Visual Information Communication and Interaction, pp. 61–64 (2012). https://doi.org/10.1145/2397696.2397705
35. Nguyen, Q.V., Miller, N., Arness, D., Huang, W., Huang, M.L., Simoff, S.: Evaluation on interactive visualization data with scatterplots. Vis. Inf. (2020). https://doi.org/10.1016/j.visinf.2020.09.004
36. Huang, M.L., Nguyen, Q.V., Zhang, K. (eds.): Visual Information Communication. Springer, Berlin (2010)
37. Breiman, L.: Random forests. Mach. Learn. **45**, 5–32 (2001)
38. Bommert, A., Sun, X., Bischl, B., Rahnenführer, J., Lang, M.: Benchmark for filter methods for feature selection in high-dimensional classification data. Comput. Stat. Data Anal. **143**, 106839 (2020)
39. Schölkopf, B., Smola, A., Müller, K.-R.: Nonlinear component analysis as a kernel eigenvalue problem. Neural Comput. **10**, 1299–1319 (1998)
40. Yin, H.: Learning nonlinear principal manifolds by self-organising maps. In: Principal Manifolds for Data Visualization and Dimension Reduction. Lecture Notes in Computer Science and Engineering (LNCSE), vol. 58, pp. 68–95. Springer, Berlin (2007)
41. Belkin, M., Niyogi, P.: Laplacian eigenmaps and spectral techniques for embedding and clustering. Adv. Neural. Inf. Process. Syst. **14**, 586–691 (2001)
42. Van der Maaten L, Hinton G.: Visualizing data using t-SNE. J. Mach. Learn. Res. **9**(11), 2579–2605 (2008)
43. McInnes, L., Healy, J., Melville, J.: UMAP: Uniform Manifold Approximation and Projection for Dimension Reduction. arXiv:180203426 (2018). [stat.ML]
44. Shneiderman B.: The eyes have it: a task by data type taxonomy for information visualization. In: 1996 IEEE Symposium on Visual Languages, pp. 336–343 (1996)
45. Nguyen, Q.V., Nelmes, G., Huang, M.L., Simoff, S., Catchpoole, D.: Interactive visualization for patient-to-patient comparison. Genomics Inform. **12**, 263–276 (2014)
46. Nguyen, Q.V., Gleeson, A., Ho, N., Huang, M.L., Simoff, S., Catchpoole, D.: Visual analytics of clinical and genetic datasets of acute lymphoblastic leukaemia. In: 2011 International Conference on Neural Information Processing (ICONIP 2011), pp. 113–120. Shanghai, China (2011)

Optimization and Evaluation of Visualization

Extending a Genetic-Based Visualization: Going Beyond the Radial Layout?

Fatma Bouali, Barthélémy Serres, Christiane Guinot, and Gilles Venturini

Abstract We study in this work the properties of a new method called Gen-POIViz for data projection and visualization. It extends a radial visualization with a genetic-based optimization procedure so as to find the best possible projections. It uses as a basis a visualization called POIViz that uses Points of Interest (POIs) to display a large dataset. This visualization selected the POIs with a simple heuristic. In Gen-POIViz we have replaced this heuristic with a Genetic Algorithm (GA) which selects the best set of POIs so as to maximize an evaluation function based on the Kruskal stress. We continue in this chapter the study of Gen-POIViz by providing additional explanations and analysis of its properties. We study several possibilities in the use of a GA: we tested other layouts for the POIs (grid, any) as well as a different evaluation function. Finally, we consequently extend the experimental results to evaluate those possibilities. We found that alternative POI layout were not more efficient that the circle layout. As a conclusion, POIViz can deal with datasets that are often too large for other methods.

F. Bouali
University of Lille, IUT C, Dpt STID, and University of Tours, LIFAT (EA6300), Tours, France
e-mail: fatma.bouali@univ-lille.fr

B. Serres
University of Tours, ILIAD3 and LIFAT (EA6300), Tours, France
e-mail: serres@univ-tours.fr

C. Guinot
University of Tours, LIFAT (EA6300), Tours, France
e-mail: christiane.guinot@univ-tours.fr

G. Venturini (✉)
University of Tours, LIFAT (EA6300) and ILIAD3, Tours, France
e-mail: venturini@univ-tours.fr

B. Kovalerchuk et al. (eds.), *Integrating Artificial Intelligence and Visualization for Visual Knowledge Discovery*, Studies in Computational Intelligence 1014,
https://doi.org/10.1007/978-3-030-93119-3_21

1 Introduction

In data visualization techniques, methods that use dimension reduction or data projection to produce a 2D visualization are common and have been studied across many fields (Statistics, Information Visualization, Visual Data Mining, Visual Analytics, etc.). They can reduce the dimensionality of a dataset up to a 2D representation which can be visualized and presented to the user for analysis. Indeed, such methods try to minimize the loss of information of the dimension reduction and to give reliable insight about the initial multidimensional dataset, its distribution and its main characteristics. Some of these methods have a linear complexity, which is an important advantage over the others for dealing with large dataset, and in our work we concentrate on those methods. FastMap is a typical example of such approaches [14].

Radial approaches [8] use a circle-based representation for projecting data in 2D and are linear too. Among them, a representative method called Radviz has been extensively studied [10]. In the past, we have also proposed an alternative in radial approaches, called POIViz [12]. POIViz uses anchors laid around a circle to determine the location of the remaining data. Anchors in POIViz are data, unlike RadViz in which anchors are dimensions. Hence, POIViz uses the similarity (or distance) between the remaining data and the POIs to place them within the circle. The efficiency of POIViz depends on the choice of the anchors. In our previous work, this was done with a heuristic, and then the user was able to adjust the POIs using interactions (removing one POI, replacing it with another, etc.). It is this visualization that we have used in this work as a starting point. Our goal is to add an optimization procedure to optimize this visualization and to automatically select the anchors.

Hence, a third focus of this work is the use of optimization procedures to improve visualizations. Instead of letting the user search for relevant visualizations by interactively adjusting parameters, optimization procedures aim at automatically finding relevant representation thus avoiding this time-consuming step to the user. Usually, optimization of visualization requires to define a criterion that evaluates mathematically the efficiency of a visualization. The Kruskal stress can be used for instance, in the context of data projection. This criterion can guide the search procedure towards the best visualization (see some examples in Sect. 3.4). In our work, we study how a genetic-based method can be used to improve a radial visualization.

Recently, we have suggested how a method called POIViz could be extended with a Genetic Algorithm (GA). The new method called Gen-POIViz was described in a short paper [5]: a GA can optimize the POIs that determine the data projection. In that initial work, we fixed many options. In this chapter, we first provide many additional explanations and details about Gen-POIViz. We explain with more details its principles. We relax the initial constraints by testing important alternative options in Gen-POIViz: we study other layouts for the POIs, going beyond the initial circular representation, and we test another evaluation function with a smaller computational cost.

The remaining of the chapter is organized as follows: we present a domain overview and POIViz in Sect. 2. We give many details about the genetic-based optimization of Gen-POIViz in Sect. 3 and we propose alternative options (POIs layout, evaluation function). We evaluate those alternatives in the experimental Sect. 4. In Sect. 5, we conclude those tests by fixing the final setup of GenPOIViz.

2 State of the Art

2.1 *Linear and Radial Methods with Optimization*

The three points of focus of our work are: (1) linear methods for data projection, (2) radial approaches, (3) with the use of an optimization procedure. Linear methods have a linear complexity to produce a data projection. FastMap [14] determines mainly four points (two axes) to project the data in 2D. A heuristic is used to find those axes. Radial approaches use the circular geometry to project data. RadViz is a famous example [10] in which each anchor represents on dimension. Then each anchor "attracts" each data to be displayed. Another example of a radial and linear method is Star Coordinates [11] (see [15] for a comparison with RadViz). Star coordinates consists in using each dimension as a vector. All vectors start from the center of the circle. The layout of a data is determined by the sum of its vectors.

Optimization methods have a long history in the domain of data visualization. Bertin [3] suggested in his seminal work to improve a matrix display by reordering the rows and columns. Optimizing a visualization requires to define a mathematical criterion to be optimized. This criterion evaluates the quality of the visualization and represents an important issue to solve. As a consequence, the definition of such criteria is the subject of many research work like [1, 4]. These evaluation functions can even be defined with expert knowledge [13] or with user observation [2]. A very interesting example of the use of such criteria is the ScagExplorer system [7]. In this work, scatter plots are evaluated with many evaluation functions. Each function provides a numerical value. A scatter plot is thus evaluated with a vector of values. Then, scatter plots can be clustering according to their vectorial representation. In the final representation, ScagExplorer represents only the main representative of each found cluster. The user can thus observe a whole summary of the visualization search space.

The use of an optimization framework also implies to define a search procedure. This algorithm must sample the visualization search space efficiently (without performing an exhaustive search as done in ScagExplorer for instance). It must find a visualization that will maximize (or minimize) the evaluation criterion by sampling the search space. Many visualizations are using this principle, such as force-directed methods for graphs layout, or seriation methods for reordering row or columns of a matrix. They can use gradient search or other heuristics or even metaheuristics. Radial approaches have been optimized using such principles [1, 16], and their initial visual representation was greatly improved with optimization.

2.2 A Radial Visualization as a Starting Point

Let D denotes the dataset. We consider that a distance or dissimilarity value can be defined between every couple of data, either from a matrix or with a distance function computed over the data attributes. Our approaches can deal with numeric attributes but also with symbolic ones. For numeric attributes, we use the Euclidean distance, and for symbolic attributes, we use the Hamming distance. Then, the overall distance between two data d_i and d_j equals:

$$Dist_n(d_i, d_j) = Dist_{Euclidean}(d_i, d_j) + Dist_{Hamming}(d_i, d_j)$$

where $Dist_{Euclidean}(d_i, d_j)$ is computed using the numerical attributes available in the data representation (and respectively, $Dist_{Hamming}(d_i, d_j)$ is computed on the symbolic attributes). In the following, we will mainly use the Euclidean distance.

POIViz can be defined as follows (see Fig. 1): specific data are placed on a 2D circle. These data are called Points Of Interests (POIs) and will be denoted by $POI_1, ..., POI_k$. In general, in POIViz those POIs do not need to be real data, they can be any object provided that a distance to the data can be computed. For instance, a POI could be an hypothesis like "male with age over 50". However, in

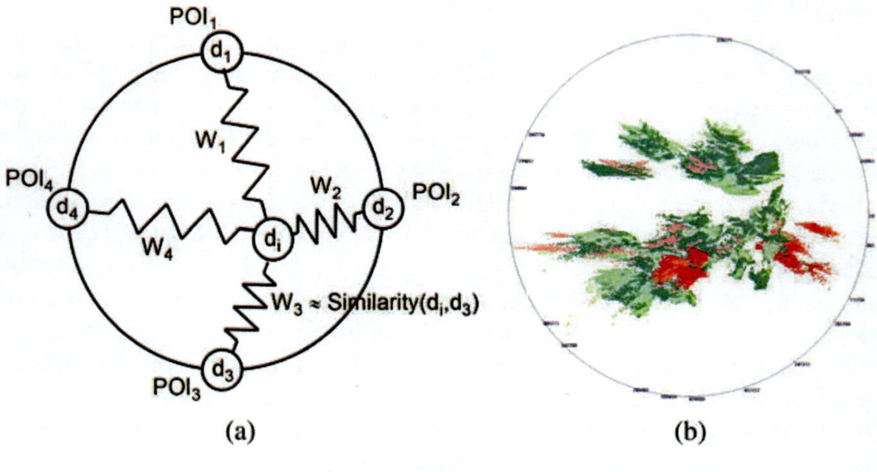

Fig. 1 In **a**, the main principles of POIViz/Gen-POIViz: selected data (here d_1 to d_4) serve as anchors and are called POIs. Those anchors are located around a circle. The remaining data (here d_i) are placed within the circle at a location that depends on their similarities with the POIs. Weights are proportional to this similarity, and if an anchor is close to d_i in the original space, then this anchor will attract d_i to its 2D location. Those weights can be seen as springs which strength depends on the similarity between the considered data and the POIs. In **b**, an example of a visual representation of the Forest Cover type dataset (582000 data with 54 dimensions) obtained with Gen-POIViz. A GA is used to search for an efficient set of POIs that accurately represent in 2D the original (and multidimensional) data

the following and in the context of data projection, we will consider that POIs are data (i.e., $\{POI_1, ..., POI_k\} \subset D$) and leave the other possibility for future work.

We denote by $(X(POI_j), Y(POI_j))$ the 2D display coordinates of POI_j. Each data d_i is displayed at a 2D location which is the weighted barycenter of the POIs (see the springs represented in Fig. 1(a)). The weight associated to each POI is proportional to the similarity between d_i and that POI. So the display coordinates $X(d_i)$ and $Y(d_i)$ are computed as follows:

$$X(d_i) = \sum_{j=1}^{k} W_{(i,j)} \times X(POI_j)$$

$$Y(d_i) = \sum_{j=1}^{k} W_{(i,j)} \times Y(POI_j)$$

where

$$W(i, j) = \frac{Simil(d_i, POI_j)}{\sum_{p=1}^{k} Simil(d_i, POI_p)}$$

$$Simil(d_i, POI_j) = 1 - \frac{Dist_n(d_i, POI_j)}{Max\ Distance}$$

These display coordinates can be post-processed using a central homothety (a proportional zoom) so as to enlarge the 2D graph as much as possible within the circle. In Fig. 1b, we show a visualization of the Forest Cover Type dataset (see Sect. 4.4 for more details) obtained with our approach.

In such a visualization, the user can interact with the data. A POI can be removed or added, which changes the data projection. The user can select data so as to obtain additional information. The overall complexity of this visualization is in $O(n \times k)$. So POIViz is linear in the number of data n. It does not require to compute a $n \times n$ distance matrix. In our previous work, we provided a fast and simple heuristic to select POIs: we initially selected with a random search procedure two POIs that are highly distant from each other. Then we used an insertion procedure to add other POIs one by one. For each insertion, we selected the most distant POI, again using a random search.

2.3 Proposed Extensions

We study in this chapter how to improve the selection of the POIs and also their layout, two points which greatly determine the quality of the data visualization. The number of POIs also influences the time complexity of the method: as far as running times are concerned, the smaller the number of POIs, the better, because the final computation of the data coordinates will be even faster. So we would like to propose algorithms to choose an efficient number of POIs: a small number of POIs that accurately represents in 2D the distances between data in the original space. In the previous version of our visualization, we were focused on user interactions and on the GPU implementation. The layout of POIs was on a circle only. The heuristic for POIs selection was not formally evaluated. Instead, we were relying on user interactions to improve the visualization (for instance, adding/removing POIs).

In this work, we would like to go further and to propose a new method that automatically selects a relevant and possible small set of POIs among the available data. We would like also to study alternative layout to the circle (e.g., grid or a layout determined by the optimization procedure). To achieve this, we introduce a search procedure to optimize a set of POIs. This procedure is stochastic and based on GAs. We use an evaluation criterion inspired by the Kruskal stress to guide the search towards better POIs. In addition, we propose a comparative experimental study to better highlight the strengths and weaknesses of our approach on several datasets.

3 The Genetic Optimization Approach

3.1 GA Main Principles

In our GA, an individual will be a set of POIs, with a variable number of POIs and possibly with a variable layout. The evaluation function will take into account the adjustment between the nD and the 2D distances. The overall algorithm is a steady state GA [18]:

1. Generate an initial population of Pop_{size} individuals using the random generation operator (see the next section), and evaluate each individual with the cost function,
2. Repeat
   a. Select two parents P_1 and P_2 with a specific binary tournament selection that favors good individuals with less POIs,
   b. With probability P_{cross}, apply a uniform crossover operator on the two parents to generate an offspring O_1,
   c. Apply a mutation operator to O_1,
   d. Evaluate $Cost(O_1)$ and Insert O_1 in the population if $Cost(O_1)$ is "better" that the worst individual of the population,
3. Until $Iteration_{max}$ individuals have been generated.

3.2 Genetic Representation and Selection Operator

One individual or solution represents a set of k POIs with $k \in [3, Kmax]$. Each individual is represented as a fixed set of $Kmax$ genes $\{g_1, ..., g_{Kmax}\}$ where each gene g_i is either a data present in the original dataset or a specific "null" value. This "null" value allows the GA to optimize a variable number of POIs. The layout of these POIs is by default a circle with $Kmax$ slots (see Fig. 2), but other layouts can be optimized too (see Sect. 3.6). Each slot can be either empty ("null" value) or filled with a POI. Slots are regularly placed on the circle.

We selected the binary tournament selection and we added a slight selective pressure to solutions with less POIs, with the aim of reducing their number. To select one parent, two individuals are randomly chosen in the population, and let us denote them i_1 and i_2 with $Cost(i_1) <= Cost(i_2)$. Individual i_1 (i.e., the best individual according to the evaluation function) is selected by default except if i_2 has a similar cost but with less POIs. So we exceptionally select i_2 if $Cost(i_2) < Tolerance \times Cost(i_1)$ and $\#POIs(i_2) < \#POIs(i_1)$, with $Tolerance \in [1, 1 + \epsilon]$. With this selection, we accept less POIs if the variation of cost is negligible. In our experiment we used $\epsilon = 0.005$.

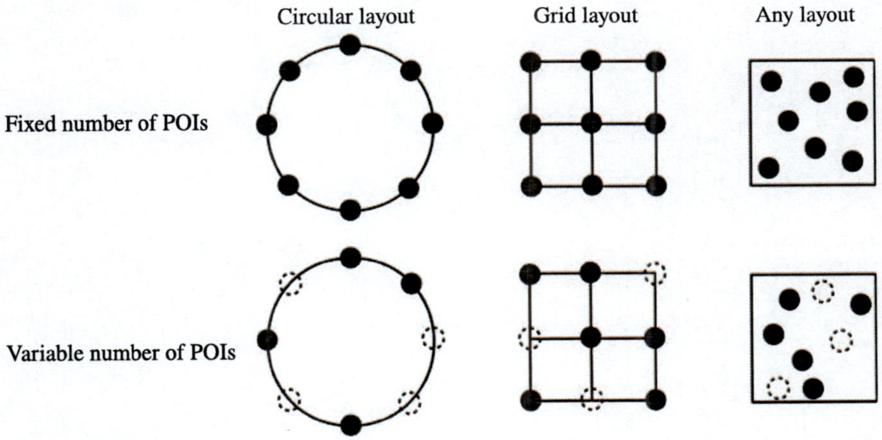

Fig. 2 The different layouts of POIs that can be optimized with our GA, possibly with a fixed or variable number of POIs. The first raw of layout figures concerns a fixed number of POIs. In the grid layout, the number of POIs is a power of 2. In the "Any layout", the POIs coordinates are optimized by the GA. In the second raw, we show in addition the effect of a variable number of POIs. POIs slots can be empty as determined by the GA. With both options (any layout, variable number of POIs), the GA can optimize the entire POIs set of the visualization

3.3 Genetic Operators

The uniform crossover (UX) [17] can exchange any genes of the parent individuals (see Fig. 3), which is the reason why we selected it. In order to generate O_1, the UX has a 0.5 probability of selecting a gene either from P_1 or P_2. Here genes represent POIs.

The mutation operator selects one gene g_i and modifies it (see Fig. 3). Initially, we were replacing the selected POI by a randomly selected data. We observed that this mutation was too disruptive and did not improve the individuals as often as we wanted. Hence we found a more efficient mutation: instead of selecting any other data, we selected a data that is close to g_i: we randomly choose 20 possible candidate data, and we select the data that is the closest to g_i.

Finally, the insertion procedure, that decides if a newly generated individual is inserted or not in the population, uses the same tolerance test as described earlier. In this test, the individual to be inserted is compared to the worst individual in the population. This method guarantees that the best individual in the population will always survive at each iteration.

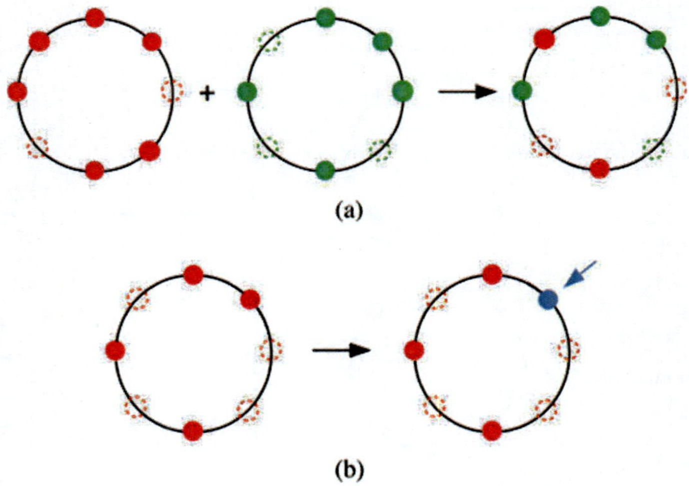

Fig. 3 Illustration of how the genetic operators can produce offspring. In **a**, the uniform crossover exchanges any genes (i.e., POIs) at corresponding positions. In **b**, the mutation considers one gene and replaces it with a new data from D (this data is selected using a stochastic heuristic, see text for explanations). In this example, we consider the circular layout but the same principle can be used for the grid layout. In addition, these operators can be extended to optimize any layout of the POIs (see text for explanations)

3.4 Evaluation Criterion: Adapting the Kruskal Stress

To define the evaluation criterion, we denote by $Dist_2(d_i, d_j)$ the 2D Euclidean distance in the final low dimensional space. The Kruskal stress is defined as follows:

$$Stress = \frac{\sum_{i,j\ j>i}(Dist_n(d_i, d_j) - Dist_2(d_i, d_j))^2}{\sum_{i,j\ j>i} Dist_n(d_i, d_j)^2}$$

in which value of 0 represents the optimum.

However, this stress has a major drawback for most dimension reduction techniques, including ou methods and those we use in Sect. 4. It is very sensitive to the scale factor between $Dist_n$ and $Dist_2$. If a perfect adjustment is found, but a proportional one such that $Dist_2 = \alpha Dist_n$ in which α is a positive constant, then replacing $Dist_2$ by $\alpha Dist_n$ in the stress formula results in a residual stress of $|(1-\alpha)|$, which is not 0 and thus not optimal. This forces the dimension reduction methods to scale properly the 2D coordinates. In POIViz, the coordinates of the POIs belong to [0, 1], and should thus be divided by a factor of α. To determine this scaling factor, we estimate $\tilde{\alpha} = \frac{\overline{Dist_2}}{\overline{Dist_n}}$. Hence the cost function is:

$$Cost = \frac{\sum_{i,j\ j>i}(\tilde{\alpha} \times Dist_n(d_i, d_j) - Dist_2(d_i, d_j))^2}{\sum_{i,j\ j>i}(\tilde{\alpha} \times Dist_n(d_i, d_j))^2}$$

So, to compute the final Kruskal stress, we adjust the scaling factor α with a simple and fast optimization procedure inspired by Newton's method. It is a simple gradient descent because the stress has a convex shape with respect to $\tilde{\alpha}$.

3.5 Optimizing the Number of POIs

The GA can optimize the number of POIs with a $[3, Kmax]$ interval (see Fig. 2) by considering that POIs can be added or deleted in an individual. In the genetic representation, the values of a POIs g_i are the data indexes and belong to $[1, m]$. To this set of values, we added a new one denoted by $null$ to obtain a new set $[null, 1, ..., m]$. In this way, by using this null value, the genetic representation can delete a POIs, and thus the GA can optimize the layout. In the genetic operators, wee have given to this special value a probability of 25% for appearing (creation or mutation operators).

3.6 From Radial and Grid to Any Layout

To study the influence of the 2D layout of the POIs, we have extended the initial genetic representation in which the display coordinates of each POI were constrained by a circle. As shown in Fig. 2, POIs can also be arranged in a grid. It seems that this grid layout was also suggested in the initial studies about RadViz [9], but it was not further developed.

Other layouts could be possible so we had then the idea to include the POIs display coordinates in the optimization procedure. We wanted to test whether other 2D configurations of the POIs could emerge and be efficient. So the extended genetic representation of an individual in Gen-POIViz with the "any layout" option is:

$$\{g_1, X(g_1), Y(g_1), g_2, X(g_2), Y_(g_2)..., g_k, X(g_k), Y(g_k)\}$$

where g_i is a data index and $X(g_i) \in [0, 1]$ and $Y(g_i) \in [0, 1]$ are the display coordinates of this POI in 2D.

The genetic operators are adapted to this new representation: the creation operator generates $X(g_i)$ and $Y(g_i)$ in the [0, 1] interval. The crossover operator can exchange the coordinates between two individuals. The mutation operator also apply to the coordinates:

$$Mutation(X(g_i)) = X(g_i) + Uniform\ Noise$$

$$Mutation(Y(g_i)) = Y(g_i) + Uniform\ Noise$$

where $Uniform\ Noise \in [0, 0.2]$ and such that the mutated values are kept in [0, 1].

If one combines the choice of the layout and a variable number of POIs, then the GA can select the best configuration that minimizes the cost function. Indeed, it greatly enlarges the search space and makes the optimization problem more difficult. Considering the user point of view and the computation time needed to obtain a result, we decided to keep the same number of generations as for the fixed layout option. In such a way, the user will wait a similar amount of time, whatever the selected option (fixed or variable layout).

3.7 Using a Data Sample

To speed-up our methods, we have considered the use of a data sample DS rather than the complete dataset D, with $|DS| << |D|$, for searching an optimal set of POIs. This will reduce the computational cost of the method (evaluation function) and it will also reduce the size of the search space for the GA. Once an optimized set of POIs is found by the GA, the remaining data not present in DS can be displayed: our method can easily cope with such a setup because it is a parametric method (the display coordinates of the data are determined by the choice of parameters, here the POIs). So once the POIs are selected in DS, the display of the complete data from D is done in linear time using those POIs. We just need to compute the data 2D coordinates according to the selected POIs, and not to re-optimize the whole visualization. For several other data projection methods, working on a sample, or dealing with continuously incoming or incremental data, is not possible.

3.8 1-Norm

From the optimization point of view, GAs are very versatile algorithms that do not make specific assumptions on the cost function, unlike other methods (especially those based on gradient descent). For instance, in a GA, the evaluation function does not have to be derivable. As a consequence, in a GA it is very easy to change the evaluation function of individuals. In our context, we have used the 2-norm for the error computation in the stress function (see Sect. 3.4). The 2-norm uses squares and square roots functions. We wanted to test the impact of using a 1-norm because this error function uses only absolute values, and is thus expected to have smaller computation times. This alternative cost function is thus:

$$Cost_{1-norm} = \frac{\sum_{i,j\ j>i} |\tilde{\alpha} \times Dist_n(d_i, d_j) - Dist_2(d_i, d_j)|}{\sum_{i,j\ j>i} (\tilde{\alpha} \times Dist_n(d_i, d_j))}$$

It is interesting to notice that both versions of the cost function, either 2-norm or 1-norm, have the same minimum. So the GA will look for the same optimal set of POIs. However, the evaluation function has an impact on the ability to reach this optimum. So we want to study if the 1-norm can be as efficient as the 2-norm, and to what extend it can reduce the computation times.

3.9 Parallelization

GAs have the advantages of being easily parallelized. Several models exist [6], such as the island model in which sub-populations of individuals evolve separately with genetic exchanges at some times. Those models are more complex than the sequential GA we use in Gen-POIViz. So alternatively, a simpler option consists in parallelizing the computation of the evaluation function. In our context, the evaluation function (i.e., Kruskal stress) is a fine-grained computation with a double loop on the data, either from the complete dataset or from the data sample (see the previous section). Therefore, we decided to keep the sequential structure of the GA but to parallelize the stress computation. This is done by using Parallel-for instructions on the CPU (for instance on the main loop on the data). This parallelization has the advantage of performing runs that are exactly similar to the sequential computation. It scales to

CPUs whatever the number of cores and without changing the model and the results of the algorithm. It can be applied to standard laptop computers as those used in the results section. In future work we could also consider the parallelization on a GPU.

4 Results

4.1 Experimental Setup

The datasets we used in our experiment were selected from the UCI Machine Learning repository.[1] They are represented in Tables 1, 2 and 3, with an increasing size. For the first group with small datasets (see Table 1), the complete distance matrix and the Kruskal stress can be computed. For the second group (see Table 2), the complete distance matrix is not necessary for the method, but it will be computed however to evaluate the exact Kruskal stress on D for comparison purposes. In this way, we can check precisely the effects of using a data sample rather than the complete dataset. The third group represents datasets for which computing the distance matrix or the exact Kruskal stress is intractable (see Table 3). Still, we evaluated a partial Kruskal stress (on 10000 data) so as to give an indication about the quality of the results. For some datasets (Shuttle Small, Forest-cover type and KKD Cup 99 datasets), one (or more) attribute had very large values compared to the others, which required to normalize each attribute in the [0, 1] interval.

The apparatus used in our tests is a PC equipped with an i7 processor (8 cores, 16GB of RAM). All results were averaged over 10 runs. In order to determine the GA parameters (population size, probabilities of operators), we performed many tests with Pop_{size} (from 1, 5, 10, 20, 40), P_{cross} (from 0, 0.25, 0.5, 0.75, 1 and the fact that mutation is performed systematically after the crossover or not). The best results were obtained with Pop_{size} between 20 or 40 (in the following we selected 30) and a systematic use of the crossover (in the following we used therefore $P_{cross} = 1$ and the fact that each crossover is followed by a mutation). Another parameter to set was $Kmax$ (the maximum number of POIs in individuals). We made exhaustive tests with values from 3 to 50 POIs. We found that above 40 POIs no improvements were observed. Therefore, in the following we selected $Kmax = 40$ in the genetic representation. Finally, we limited the number of generated individuals to 4000 so as to obtain running times in the order of 2 min.

We compared our approach to several other methods that are implemented in Python and integrated in the Orange software (see orange.biolab.si). In Orange, we used the three components (PCA, MDS, tSNE), respectively denoted by O-PCA, O-MDS and O-tSNE. We used the Manifold Learning component (Isomap, LLE,

[1] The interested reader may refer to archive.ics.uci.edu to obtain the references of each dataset and their donors.

Table 1 Results on small datasets. We summarize the dataset dimensions. Several different parameters are tested (different layouts, 1-norm). Values are the Kruskal stress averaged over 10 runs, with standard deviations in parenthesis, mean running times in seconds and mean number of POIs

Datasets	CNAE-9	MicroMass	Asian religion
Nb of data	1080	931	590
Nb of attributes	856	1300	8266
O-MDS	0.321	0.214	0.237
O-PCA	0.446	0.528	0.570
O-tSNE	0.457	0.817	0.931
O-ML-Isomap	0.490	0.535	0.582
O-ML-LLE	0.863	0.671	0.745
O-ML-Spectral	0.573	0.698	0.667
O-ML-tSNE	0.444	0.729	0.777
POIViz	0.572	0.522	0.694
$K = 40$ POIs	(0.005)	(0.008)	(0.009)
	<1s	<1s	<1s
Gen-POIViz	0.427	0.340	0.388
2-norm, Circle	(0.005)	(0.008)	(0.009)
$Kmax = 40$ POIs	138s	105s	47s
	29 POIs	29 POIs	29 POIs
Gen-POIViz	0.427	0.333	0.393
2-norm, Grid	(0.005)	(0.009)	(0.006)
$Kmax = 49$ POIs	137s	105s	47s
	34 POIs	35 POIs	31 POIs
Gen-POIViz	0.430	0.342	0.389
2-norm, Any layout	(0.004)	(0.012)	(0.006)
$Kmax = 40$ POIs	137s	105s	47s
	30 POIs	29 POIs	28 POIs
Gen-POIViz	0.430	0.351	0.430
1-norm, Circle	(0.005)	(0.020)	(0.022)
$Kmax = 40$ POIs	81s	63s	30s
	27 POIs	28 POIs	28 POIs

Spectral, tSNE and the same MDS version as O-MDS, denoted by O-ML-Isomap, O-ML-LLE, O-ML-Spectral and O-ML-tSNE). The RadViz component of Orange was not selected because the results were not consistent. This was probably due to the large number of dimensions in the datasets we used. RadViz can visualize a few dimensions only.

Table 2 Results with POIs determined from data samples ($|DS|$ denotes the size of the sample). Values represent the Kruskal stress on all $|D|$ data (mean values averaged over 10 runs). Values in parenthesis are the standard deviations. Other values are the average running times in seconds and the average number of POIs found. Several different parameters are tested (different layouts, 1-norm)

Datasets	Wave	Isolet1234	Shuttle small
Nb of data	5000	6238	14500
Nb of attributes	21	617	9
O-MDS	0.205	0.302	0.177
	120s	720s	3360s
O-PCA	0.225	0.349	0.221
O-tSNE	0.238	0.349	Error
O-ML-Isomap	0.241	0.377	0.898
O-ML-LLE	0.310	0.761	Error
O-ML-Spectral	0.306	0.485	0.421
O-ML-tSNE	0.239	0.533	0.447
POIViz	0.336	0.404	0.246
2-norm, Circle layout	(0.011)	(0.019)	(0.009)
$K = 40$ POIs	<1s	<1s	<1s
Gen-POIViz	0.225	0.338	0.219
2-norm, Circle layout	(0.001)	(0.003)	(0.01)
$Kmax = 40$ POIs	10s	10s	10s
$\|DS\| = 125$	22 POIs	25 POIs	25 POIs
"	0.224	0.334	0.226
	(0.001)	(0.002)	(0.017)
	16s	16s	16s
$\|DS\| = 250$	20 POIs	22 POIs	23 POIs
"	0.224	0.333	0.219
	(0.001)	(0.003)	(0.026)
	38s	38s	39s
$\|DS\| = 500$	20 POIs	25 POIs	26 POIs
"	0.224	0.332	0.188
	(0.003)	(0.002)	(0.014)
	119s	120s	120s
$\|DS\| = 1000$	20 POIs	25 POIs	21 POIs
"	0.224	0.333	0.185
	(0.001)	(0.002)	(0.035)
	436s	434s	447s
$\|DS\| = 2000$	21 POIs	19 POIs	21 POIs
Gen-POIViz	0.224	0.333	0.199
2-norm, Grid layout	(0.001)	(0.003)	(0.021)
$Kmax = 49$ POIs	117s	117s	117s
$\|DS\| = 1000$	23 POIs	27 POIs	28 POIs

(continued)

Table 2 (continued)

Datasets	Wave	Isolet1234	Shuttle small
Gen-POIViz	0.224	0.334	0.204
2-norm, Any layout	(0.001)	(0.002)	(0.023)
$Kmax = 40$ POIs	116s	117s	117s
$\|DS\| = 1000$	20 POIs	23 POIs	23 POIs
Gen-POIViz	0.225	0.336	0.189
1-norm, Circle layout	(0.001)	(0.003)	(0.019)
$Kmax = 40$ POIs	70s	70s	70s
$\|DS\| = 1000$	20 POIs	23 POIs	26 POIs

Table 3 Results of the comparative tests with large datasets (for which the computation of an exact Kruskal stress is intractable). Values represent the stress on 10000 data (the same sample for all methods), averaged over 10 trials

Datasets	Forest cov. typ.	Gas sensor	KDD cup 1999
Nb of data	582000	919438	4898431
Nb of attributes	54	11	42
O-PCA	0.398	0.223	Not appl
	8s	12 s	
POIViz	0.417	0.307	0.714
2-norm, circle layout	(0.005)	(0.011)	(0.038)
$K = 40$ POIs	<1s	<1s	<1s
Gen-POIViz	0.325	0.225	0.619
2-norm, circle layout	(0.004)	(0.004)	(0.005)
$Kmax = 40$ POIs	118s	118s	114s
$\|DS\| = 1000$	28 POIs	22 POIs	27 POIs
Gen-POIViz	0.327	0.229	0.623
2-norm, grid layout	(0.004)	(0.006)	(0.006)
$Kmax = 49$ POIs	117s	117s	112s
$\|DS\| = 1000$	27 POIs	25 POIs	31 POIs
Gen-POIViz	0.327	0.232	0.622
2-norm, any layout	(0.006)	(0.008)	(0.006)
$Kmax = 40$ POIs	117s	117s	118s
$\|DS\| = 1000$	29 POIs	14 POIs	22 POIs
Gen-POIViz	0.338	0.254	0.645
1-norm, circle layout	(0.009)	(0.019)	(0.013)
$Kmax = 40$ POIs	71s	71s	66s
$\|DS\| = 1000$	27 POIs	22 POIs	30 POIs

4.2 Results on Small Datasets

The results are presented in Table 1. Gen-POIViz are ranked second and are between those of O-MDS and O-PCA. To explain those results, one must consider that O-MDS is a force directed method that can precisely adjust the 2D layout of each data independently using attraction/repulsion forces that are exerted on each data. It is a non-parametric method. Gen-POIViz uses POIs to set the layout in a more global way: the coordinates of each data are not directly and individually optimized, but rather the POIs contribute globally to the layout. It is a parametric method. So for these datasets, Gen-POIViz does not reach the quality of O-MDS, and needs also higher computation times. But our approach is the second best method, and its results are also improved over the initial POIViz.

4.3 Results with a Data Sample

With those larger datasets, Gen-POIViz works with a data sample. At the end of a run, the exact Kruskal stress is computed on the whole dataset (and not on the data sample only) for comparison purposes. Results are presented in Table 2. We observed that our approach is again at the second rank amond the tested method: the final stress of Gen-POIViz is higher than O-MDS but smaller or equal to O-PCA. However, considering the running times, with the use of a data sample, our method becomes much faster compared to O-MDS. The difference increases greatly with the dataset size. O-MDS has a higher complexity (in the order of $|D|^2$), while our method is linear. Working with a small sample seems to be efficient even with a few hundred data, i.e. $|DS| \in [250, 1000]$. In the forthcoming tests on larger datasets, we used $|DS| = 1000$ as a comprise to keep the computation times within 2 min and to obtain relevant stress values. Some errors were observed in the Python components for the Shuttle-small dataset, which underlines the limit of the implementation of some methods when the data becomes too large.

4.4 Results on Larger Datasets

To further evaluate Gen-POIViz, we used even larger datasets (see Table 3) for which many other standard method failed. This was due to the size of the datasets which does not allow to compute a distance matrix. Gen-POIViz was able to process and to visualize those data in a reasonable time (2 min for the selection of POIs, and a few seconds for the display), thanks to the use of a data sample. Figure 1b shows a typical result obtained with the Forest cover type dataset. For the KDD Cup 1999 datasets, qualitative attributes are present, hence O-PCA cannot be applied. This highlights the flexibility of our approach. The results of Gen-POIViz are improved over the initial POIViz.

4.5 Other Layouts

Three different layouts (circular, grid, any) of the POIs have been tested with all datasets. Results are presented in the middle of Table 1 for small datasets, in Table 2 for medium datasets and in Table 3 for large datasets. In addition, we give in Table 4 more details on the results with different sizes of data samples for medium datasets. In Fig. 4, we show typical visualizations obtained with the Shuttle Small dataset.

From a quantitative point of view, the three studied layouts seem to have equivalent stress values. No significant differences can be observed: the GA seems to be able to adapt the set of POIs to each of the layout options. In Table 4, the sensitivity to the sample size is very similar to the one of Table 2. Overall, only a small difference can be noticed for the any layout option. A slight increase in the stress function is observed, so the quality is decreased a bit. In fact, we kept the number of genetic iterations to a constant value (4000) for all layouts, so as to keep the computation

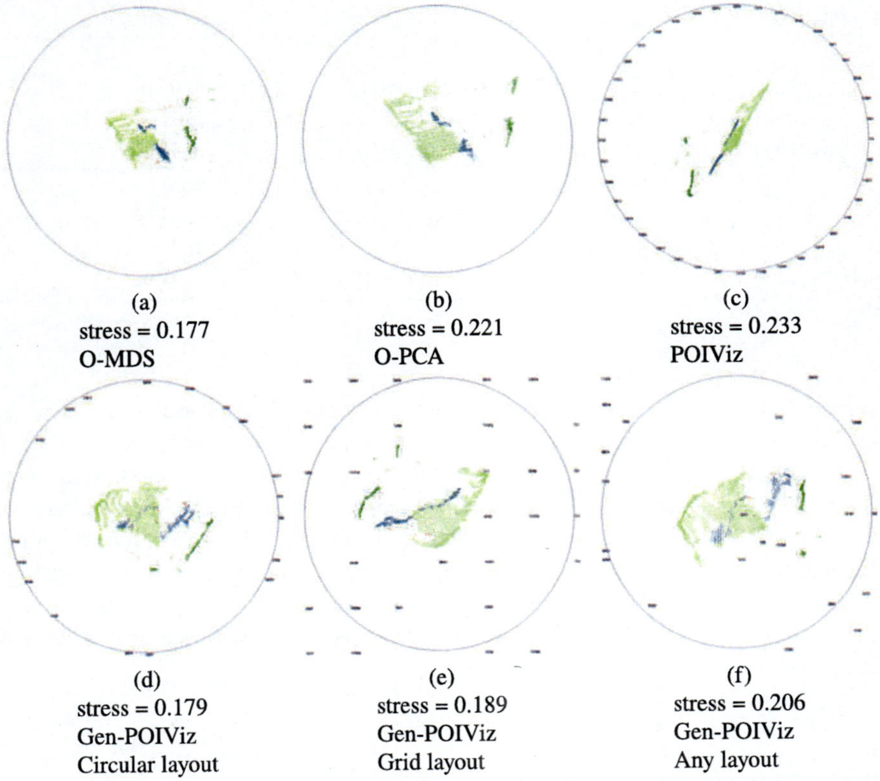

Fig. 4 A visual comparison of the different layouts in Gen-POIViz with the Shuttle Small dataset (14500 data). The first raw, from **a** to **c**, shows typical results obtained by other methods. The second raw, from **d** to **f**, shows different layouts obtained by our approach (circular, grid, any). We give the value of the Kruskal stress for each visualization

Table 4 Results with a grid layout and with POIs determined from data samples ($|DS|$ denotes the size of the sample)

Datasets	Wave	Isolet1234	Shuttle small
Nb of data	5000	6238	14500
Nb of attributes	21	617	8
Gen-POIViz	0.224	0.338	0.241
2-norm, Grid layout	(0.003)	(0.005)	(0.029)
$Kmax = 49$	9s	9s	9s
$\|DS\| = 125$	28 POIs	31 POIs	28 POIs
"	0.224	0.336	0.215
	(0.002)	(0.003)	(0.013)
	16s	17s	16s
$\|DS\| = 250$	21 POIs	27 POIs	30 POIs
"	0.224	0.336	0.201
	(0.002)	(0.002)	(0.017)
	37s	37s	37s
$\|DS\| = 500$	28 POIs	30 POIs	27 POIs
"	0.224	0.332	0.197
	(0.001)	(0.003)	(0.016)
	117s	118s	118s
$\|DS\| = 1000$	25 POIs	26 POIs	30 POIs
"	0.224	0.333	0.190
	(0.001)	(0.003)	(0.017)
	440s	440s	437s
$\|DS\| = 2000$	27 POIs	27 POIs	29 POIs

times in the order of 2 min. But with the any layout option, the search space is larger for the GA, which would probably require additional generations.

From a qualitative and user point of the view, and according to our past experience, the best layout is the circle. In our previous work, we showed that with a circle the user can easily understand interactions, such as promoting a data to a POI by dragging it out of the circle. POIs are clearly separated from the data and are thus more visible. Also, the visualization can be easily rotated so as to normalize the presentation of the data. So we argue that interactions would be less clear with the grid or the any layout options.

4.6 2-Norm Versus 1-Norm

We have tested the two different evaluation functions, the default one with 2-norm, and an alternative one with 1-norm (see Sect. 3.8). Results are presented at the bot-

Table 5 Results with the 1-norm stress function and with POIs determined from data samples ($|DS|$ denotes the size of the sample)

Datasets	Wave	Isolet1234	Shuttle small
Nb of data	5000	6238	14500
Nb of attributes	21	617	9
Gen-POIViz	0.226	0.337	0.221
1-norm, Circle layout	(0.002)	(0.004)	(0.020)
$Kmax = 40$	9s	9s	9s
$\|DS\| = 125$	23 POIs	25 POIs	25 POIs
"	0.226	0.335	0.226
	(0.003)	(0.004)	(0.024)
	12s	12s	12s
$\|DS\| = 250$	17 POIs	24 POIs	27 POIs
"	0.224	0.333	0.196
	(0.001)	(0.004)	(0.021)
	25s	25s	25s
$\|DS\| = 500$	21 POIs	24 POIs	24 POIs
"	0.225	0.336	0.189
	(0.001)	(0.003)	(0.019)
	70s	70s	70s
$\|DS\| = 1000$	20 POIs	23 POIs	26 POIs
"	0.226	0.336	0.189
	(0.001)	(0.003)	(0.024)
	246s	243s	245s
$\|DS\| = 2000$	22 POIs	23 POIs	23 POIs

tom of Table 1 for small datasets, Table 2 for medium datasets and Table 3 for large datasets. We also give in Table 5 more details on the results with different sizes of data samples.

For medium datasets, the difference of stress between the two functions is not noticeable. However, for small datasets, the difference increases with the number of dimensions in the datasets. For large datasets, the 2-norm evaluation function seems to have a slight advantage and smaller Kruskal stress values over 1-norm. Indeed, the computation time is almost divided by 2 with the 1-norm. So the choice between 1-norm or 2-norm depends on the compromise the user wants to achieve between computation times and quality of the results.

5 Conclusion

In this chapter, we have further study the use of an optimization procedure to improve a radial visualization. We have shown how to apply a GA to a visualization method. For this purpose, we have given additional details about how to define a genetic representation and genetic operators, and an evaluation function. We have proposed several extensions to the initial model: a variable number of POIs, a versatile layout of those anchors, the use of a data sample, and two different evaluation functions. In the comparative results, and for various datasets of increasing size, we have shown that the tested options improve the visualization used as a starting point (POIViz), and that Gen-POIViz was in general the second best method from the stress point of view. Also, we have shown that for larger datasets, or for a dataset with a symbolic and numeric representation, only our approach could be used. We tested also several alternatives. Despite our efforts in finding and testing different options for the POIs layout, the circular layout seems to be the most interesting in terms of performance, with a slight advantage only. From our previous work, it is also the most interesting layout for user interactions. So our future work, we believe that the best options to select with Gen-POIViz are to use the GA, with a variable number of POIs arranged on a circle, and with a data sample rather than the complete dataset. In addition, with this set-up our approach can deal with large datasets.

Several perspectives could be derived from this study. Working with a data sample seems to be an important option for dealing with large datasets. So other sampling strategies could be tested, as alternatives to the actual random selection. Specific data could be selected. Also, we would like to better understand which POIs are selected by the GA. We could then try to define a faster heuristic to search specifically for those data. Finally, as in our previous work, we would like to implement Gen-POIViz on the GPU, because this would significantly speed-up the optimization process.

References

1. Albuquerque, G., Eisemann, M., Lehmann, D.J., Theisel, H., Magnor, M.: Improving the visual analysis of high-dimensional datasets using quality measures. In: 2010 IEEE Symposium on Visual Analytics Science and Technology, pp. 19–26. IEEE (2010)
2. Albuquerque, G., Eisemann, M., Magnor, M.: Perception-based visual quality measures. In: IEEE Conference on Visual Analytics Science and Technology (VAST), pp. 13–20. IEEE (2011)
3. Bertin, J.: Graphics and Graphic Information Processing (translated by Wj Berg and P. Scott). Walter de Gruyter. Berlin, New York (1981)
4. Bertini, E., Tatu, A., Keim, D.: Quality metrics in high-dimensional data visualization: an overview and systematization. IEEE Trans. Visual Comput. Graphics **17**(12), 2203–2212 (2011)
5. Bouali, F., Serres, B., Guinot, C., Venturini, G.: Optimizing a radial visualization with a genetic algorithm. In: 2020 24th International Conference Information Visualisation (IV), pp. 409–414. IEEE (2020)

6. Cantu-Paz, E.: Efficient and Accurate Parallel Genetic Algorithms. Kluwer Academic Publishers, Norwell (2000)
7. Dang, T.N., Wilkinson, L.: Scagexplorer: exploring scatterplots by their scagnostics. In: Pacific Visualization Symposium (PacificVis), pp. 73–80. IEEE (2014)
8. Draper, G.M., Livnat, Y., Riesenfeld, R.F.: A survey of radial methods for information visualization. IEEE Trans. Visual Comput. Graph. **15**(5), 759–776 (2009)
9. Hoffman, P., Grinstein, G.: Visualizations for high dimensional data mining-table visualizations (1997)
10. Hoffman, P., Grinstein, G., Pinkney, D.: Dimensional anchors: a graphic primitive for multidimensional multivariate information visualizations. In: Proceedings of the 1999 Workshop on New Paradigms in Information Visualization and Manipulation in Conjunction with the Eighth ACM International Conference on Information and Knowledge Management, pp. 9–16 (1999)
11. Kandogan, E.: Star coordinates: A multi-dimensional visualization technique with uniform treatment of dimensions. In: IEEE Symposium on Information Visualization, vol. 2000, pp. 4–8 (2000)
12. Liu, T., Bouali, F., Venturini, G.: On visualizing large multidimensional datasets with a multithreaded radial approach. Distrib. Parallel Databases **34**(3), 321–345 (2016)
13. Pineo, D., Ware, C.: Data visualization optimization via computational modeling of perception. IEEE Trans. Visual Comput. Graph. **18**(2), 309–320 (2012)
14. Reina, G., Ertl, T.: Implementing fastmap on the gpu: considerations on general-purpose computation on graphics hardware. In: Theory and Practice of Computer Graphics, vol. 5, pp. 51–58 (2005)
15. Rubio-Sánchez, M., Raya, L., Diaz, F., Sanchez, A.: A comparative study between radviz and star coordinates. IEEE Trans. Visual Comput. Graph. **22**(1), 619–628 (2015)
16. Russell, A., Daniels, K., Grinstein, G.: Voronoi diagram based dimensional anchor assessment for radial visualizations. In: 2012 16th International Conference on Information Visualisation, pp. 229–233. IEEE (2012)
17. Syswerda, G.: Uniform crossover in genetic algorithms. In: Proceedings of the 3rd International Conference on Genetic Algorithms, pp. 2–9. Morgan Kaufmann Publishers Inc., San Francisco (1989)
18. Whitley, L.D.: The genitor algorithm and selection pressure: why rank-based allocation of reproductive trials is best. In: J.D. Schaffer (ed.) ICGA, pp. 116–123. Morgan Kaufmann (1989)

Dual Y Axes Charts Defended: Case Studies, Domain Analysis and a Method

Richard Brath, Craig Hagerman, and Eugene Sorenson

Abstract There is an assumption in the visualization community that dual axis charts should be avoided. This is backed up by research suggesting that use of dual axes cannot be justified and may even be deleterious. We refute these claims through examples showing dual axes charts to be useful and valid. We show dual axes charts are shown to be necessary for fine grained correlation analysis that is not made obvious within single axis charts or other means. We survey 25 chart-dense published reports from financial services and show dual axes are prevalent, occurring in 19% of 944 timeseries charts. In addition, we suggest an approach for automating the creation of dual axis charts.

Keywords Visualization · Timeseries · Correlation · Leading indicators · Normalization · Financial charts

1 Introduction

Line charts, bar charts and scatterplots typically have a single x-axis and single y-axis. In some cases, the chart author wishes to show multiple datasets in the same chart, however, as the datasets may have very different scales, the author may include a second y-axis for the second dataset. The presence of multiple axes can create confusion for the viewer as to which axis to use to interpret the data. Many researchers and authors recommend against the use of dual axis charts. We believe this is an over-generalization and there are, in fact, valid reasons to use dual axis. In this paper we

R. Brath (✉) · C. Hagerman
Uncharted Software Inc, Toronto, Canada
e-mail: rbrath@unchartedsoftware.com

C. Hagerman
e-mail: chagerman@unchartedsoftware.com

E. Sorenson
Cosaic Inc., New York, NY, USA
e-mail: eugene@cosaic.io

B. Kovalerchuk et al. (eds.), *Integrating Artificial Intelligence and Visualization for Visual Knowledge Discovery*, Studies in Computational Intelligence 1014,
https://doi.org/10.1007/978-3-030-93119-3_22

present the case in favor of dual axis charts through case studies, domain analysis and best practices. We analyzed over 1000 financial charts showing frequent use of dual-axis charts by domain experts, indicating acceptance and use of these charts in specific domains. We also show a technical approach using timeseries similarity mining to automate the display of multiple y-axes.

2 Background

Charts are commonly used to curate and summarize quantitative data. Often the chart designer and the viewer are different people, and the chart is a visual means of conveying information to that user. For example, a news service may use a bar chart or a timeseries to compare economic metrics. The viewer does not typically have a role in designing or manipulating such a chart. For a casual viewer such charts can be informative or even persuasive. They might also—as is a criticism of dual axis charts in particular—be misleading.

The situation is markedly different for another class of chart users. Professionals such financial analysts, quality control specialists, and data scientists both create and consume charts as part of their work. In this case a chart serves a different purpose. Charts are for helping such professionals surface insights, understand relationships and patterns, confirm hypothesis, assess correlations and communicate insights to their peers.

Prior studies critical of dual axis charts have not focused on such professional chart users and thus their recommendations may best be applied to casual chart viewers only. For the professional user, multi-axis charts are highly applicable for deeper analysis in specific areas as we will show.

2.1 *The Case Against Dual Axis Charts*

The literature suggests there is no use case for dual axes charts. In "Dual-Scaled Axes in Graphs: Are They Ever the Best Solution" [1] Stephen Few takes the position that the use of dual axis charts is never justified. He notes in a timeseries chart with dual axes, lines may cross but the crossover point is entirely arbitrary. Changing the scale of either axes would change where the lines cross. However, although such line crossings are not semantically meaningful, they may be perceived that way by a viewer since they are visually salient. Few suggests that this is a serious issue that can be mitigated through the use of a single common scale on which normalized percentages are plotted.

Isenberg et al. in *A Study on Dual-Scale Data Charts* [2] are similarly opposed to the use of dual axes. In this research they split a *single* timeseries into two ranges such that each range had different characteristics. They created five such variants to examine the effect on position, length and slope. They included one variant with dual x-axes which had poor performance compared to other options they tested. From this, they concluded the generalization that the use dual axes charts is not justified (not just dual x-axis charts).

Kaiser Fung provides data visualization criticism at his website *junkcharts*. Fung notes that dual axes charts can be easily manipulated, whether intentionally or otherwise, to give a convey a different message [3]. Since there are two axes, each can be altered independently of the other. This could, for example, be done to accentuate a trend or otherwise mislead.

The author of *ggplot2*, Hadley Wickham, has been asked about the use of dual axes charts on several occasions [4–5]. On being asked about adding support for dual axes charts Wickham responded that he did not see a need and was not inclined to. In public postings he has admitted to narrow use cases but “I don’t think they’re useful enough for me to spend hours to implement them” [4]. He has also mentioned a number of issues he has with dual axes charts including the inherent ambiguity of plotting values when the data space involves differing scales.

A number of influential visualization and statistics web bloggers have written to express doubts around the use of dual axes charts. Rost [6] and Qui [7] have written how dual axes may be confusing. Most of the humor derived from charts displayed on Tyler Vigen’s *Spurious Correlations* blog [8] comes from dual axes.

The criticism of dual axis charts can be summed up by the following points:

- They are not aesthetically pleasing due to being cluttered.
- Choices have to be made about what scale to use for the different axes which can affect the message conveyed to the user.
- Crossover points due to scaling choices may mislead users into thinking they signify a relationship in the data rather than in the design choice.
- It is possible to deliberately mislead users due to the above.

2.2 *Narrow Exceptions and Issues for Use of Dual-Axis Charts*

Hadley Wickham is decidedly against dual axis charts. However, in a public form he has admitted there may be (to him) narrow use cases: “I agree that they can be useful when the axes are simple linear transformations of each other” [4] Not all ggplot2 users agree with Wickham and some have posted workaround solutions in blogs [9, 10]. Similarly, in Python plotting libraries matplotlib and Seaborn dual axes are not natively available but can be created nonetheless [11, 12].

Although Steven Few is opposed to dual axes, he too has recognized there may be some (narrow) justifiable use cases. “A graph should only include a dual-scaled axis (assuming for the moment that one might be useful on occasions) when needed to compare data sets that have different units of measure"[14]. However, he is not aware of such useful uses of dual axes. “Today, I can’t think of a single case when there isn’t a better solution than a graph with a dual-scaled axis”[14].

Isenberg et al. recommend the use of cut-out chart rather than the superimposed (i.e., dual axis) chart type. However, they do note that dual axis charts have the unique advantage of maximizing display space, so may be preferable for that reason. It is notable that roughly half of the participants in the study were students, and a third were infrequent users of charts. It is not clear how much experience the remainder had with charts, but we surmise that they were likely not the kind of users we will describe shortly—that is, subject matter experts who spend significant time analyzing charts. We hypothesis that the chart recommendations in Isenberg's study are partly recommendations for the casual chart reader and not necessarily recommendations for charts to be used by professionals for deep analyses.

Isenberg et al.'s study also used superimposed charts with multiple x axes rather than multiple y axes. Moreover, they split a timeseries into two date ranges. This is an unconventional way of using multiple axes. We have found hundreds of examples of dual axis charts but none with such a configuration. For example, financial analysts would commonly use a chart plotting two timeseries against a *single* range of dates on the x-axis and with separate right and left y-axes. Thus, it is not surprising that Isenberg et al.'s superimposed charts were confusing, even if their users were familiar with dual-axis charts.

2.3 Motivation

A more general criticism of dual axis charts is simply that they are unnecessary and can be designed as non-dual axis instead. In our experience we believe that dual axes charts are prevalent because they offer several benefits. It is our belief that stipulations that dual axes should be avoided are an over-generalization. As mentioned above dual axes may be implemented in popular R and Python plotting libraries. Isenberg et al. note that dual axis charts are available (and every commonly used!) in Microsoft Excel. Many financial software offerings allow for many axis charts, some of which allow for up to eight axes (e.g., Eikon, Bloomberg, FactSet, Yahoo! Finance, ChartIQ). One such example is shown in Fig. 1. In fact, support for multiple axis charts is a common feature requirement for financial analysts. Why make such functionality possible if there is no justification for its use? It seems clear that the answer is that there are valid use cases for dual axis charts which prior research has overlooked. We will provide counterexamples to disprove the generalization.

3 Dual Axis Use Cases

Dual axis charts are often useful for superimposing data that reveals a relationship. Use case examples of such include pareto charts, histograms and timeseries charts.

Fig. 1 Example stock price timeseries chart with multiple y-axes from Cosaic

3.1 Pareto Charts

Pareto charts are always displayed with dual axes. The left axis typically shows the unit of measurement while the right axis shows the cumulative total, usually as a percentage (Fig. 2). They are an essential chart type within quality control since they allow the user to quickly assess magnitude per category on the left axis and

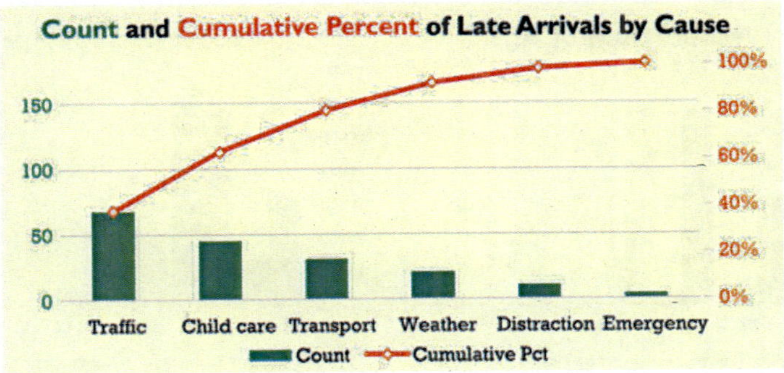

Fig. 2 Sample pareto chart. Left axis shows the nominal count, right axis shows percent of total

cumulative proportion on the right axis at the same time. When the two scales are aligned as in Fig. 2, counts or proportions for any bar or the cumulative total can be read, using the appropriate axis. The user does not need to do mental arithmetic to convert between these measures.

3.2 *Histograms*

One scenario described to the authors is web traffic analysis for web sites providing controlled goods, such as alcohol, tobacco, gambling, vehicles and so on. Some small subset of users are registered, while a much larger community is fairly anonymous, but may have partial data, such as age, which may be provided by a third-party service such as a survey.

By default, histograms may bin this data coarsely, as shown in Fig. 3. The left image shows the website population as a histogram with a single axis with the anonymous population in deep purple, and the registered population in teal. There is a big difference in the community sizes (only a small set of viewers are registered). Instead, the right chart shows a dual-axis histogram. Heights may be normalized to aid comparison. It can be seen that the histogram shapes are similar, except in the initial bucket, indicating ages 1–20.

With controlled goods, much finer binning is needed as the age of access to these goods is regulated and that age may occur inside a bin. Figure 4 shows the same histogram with a bin per each year of age. In this example, the two distributions are shown side-by-side (small multiples [13]), each with a different scale. The difference in the young ages is clearly visible: there is a sharp drop to zero for registered users (blue) at a specific age (i.e., the age of access to the good). That is, the site does not allow under-age people to register. The anonymous viewers however do show some under-age people accessing the website.

There is additional insight that can be gained by superimposing the two distributions as shown in Fig. 5. On the left is a single axis showing counts: heights are not

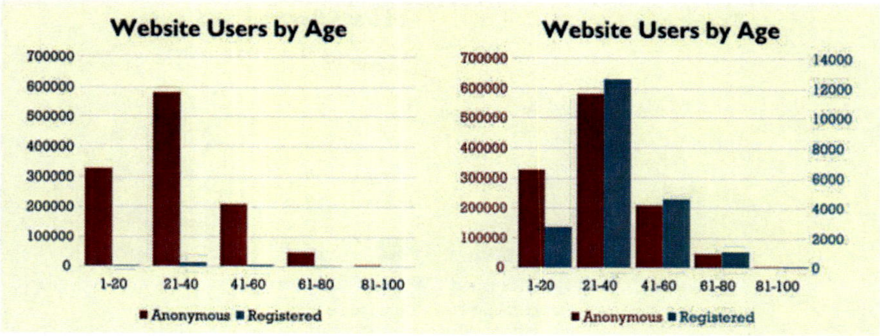

Fig. 3 Histograms, left single axis, right dual axis

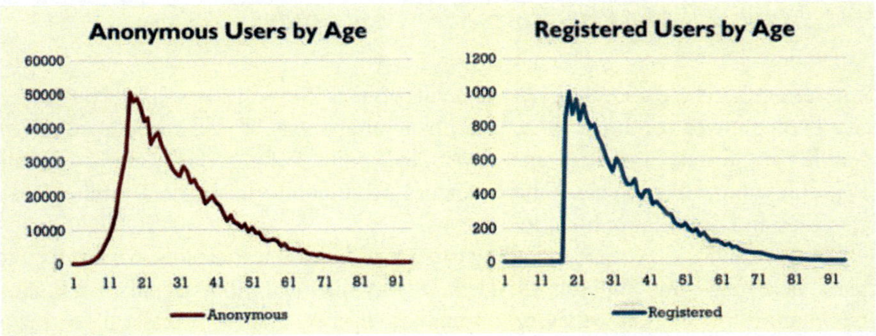

Fig. 4 Histograms, with small bins, split into small multiple charts

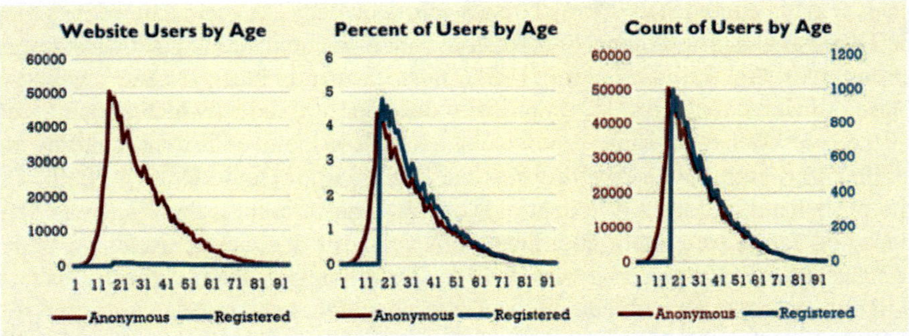

Fig. 5 Superimposed histograms, left single axis, middle normalized, right dual axis with counts

readable for the registered users. The middle image contains a single axis showing the given population percentage with the data normalized by height. This makes the difference between young users quickly apparent, and we can see that there are no registered users younger than 16 (the minimum registration age). Furthermore, it can be seen that the peak for anonymous viewers (purple) occurs slightly to the left of registered viewers (blue), i.e., the peak is under-age viewers. The superimposition of the two charts with separate scales enables these two points to occur close to each other, facilitating this observation which would be otherwise difficult to locate. This is potentially critical information, for example, there may be content on the site that is engaging under-age viewers.

The right chart shows counts on dual axes: while shape and patterns are similar to the normalized chart, there is additional information conveyed by the axes, i.e., the counts. At a glance, the viewer can recover the magnitudes for either distribution without reliance on interaction. In this example, it shows approximately 50,000 under-aged viewers below the legal age, whereas there are 1,000 registered viewers who are at the legal age.

3.3 *Financial Markets Timeseries Case Study*

Timeseries charts are ubiquitous in financial services. Several specialized chart types have evolved over centuries for use in financial analysis. These specialized charts include candlestick [14], kagi and renko [15], point & figure [16], market profile [17], equivolume [18]. These charts feature prominently in financial trading user interfaces and financial publications.

Charts and visualizations, including timeseries charts, scatterplots, bar charts, pie charts, heat maps, and 3D plots are used *daily* by financial professionals. Analyzing and interpreting market behavior through timeseries charts is integral to financial analysis. Professionals may even study to be certified as a Chartered Market Technician (*cmtassociation.org*) to recognize their analytic skills.

Financial analysis often involves the comparison of timeseries. For example, an analyst may be interested in broad trends across markets. As such, a timeseries may be better assessed in comparison, such as, another compared to another stock or an index. In Fig. 1, the price for NASDAQ stock market index (NDAQ), which is biased to tech companies, is compared to the S&P 500 index, which is composed of the top 500 largest capitalized stocks. Note how the overall trend patterns are similar, and both have sharp drops at the beginning of Covid (March 2020). On the other hand, in late 2017 (Sept-Nov), NDAQ has a modest drop whereas SPY does not. These patterns of coordinate movement, or not, can be used by a trader to make a relative wager as a *spread trade.* For example, if the trader believes tech will outperform the broad market, the trader can purchase NDAQ (go long) and sell SPY (go short). With this trade, if tech outperforms the broad market, the trader will make money regardless if the markets overall go up or down. A common analysis made by financial professionals is to identify and understand supply and demand imbalances across related markets, such as these two indices, a pair of commodities, a set of economic indicators and so on. For such analysis it is highly convenient and even necessary to be able to view multiple timeseries within the same plot.

4 Commodity Versus Currency Timeseries

We will now consider a financial markets case study where dual axis charts are necessary. The following example compares the price of crude oil to the exchange rate of the Canadian dollar between 2009 and 2016 as shown in Fig. 6. Both start low, reach a peak in 2011 and a low in early 2016. This comparison is highly relevant to commodities and foreign exchange traders, analysts, economists and other market participants as Canada is the fourth largest oil exporting nation and has the third largest proven oil reserves. Changes in the price of oil impact the Canadian economy and thus can impact the exchange rate of the Canadian dollar. While the side-by-side plots show the macro relationship, micro-comparisons such as localized time comparisons are difficult. Did oil hit bottom first or the Canadian dollar? Did one price

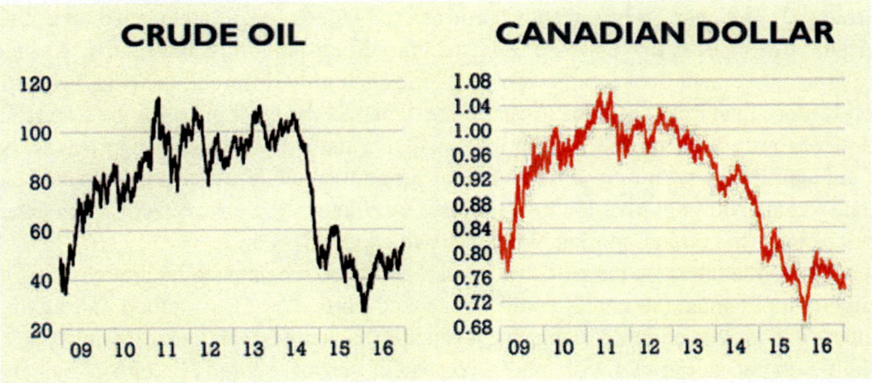

Fig. 6 Price of crude oil and exchange rate of Canadian dollar 2009–2016

continue in a trend for a short time while the other moved in a different direction? Local trends are important for market participants as these movements can impact billions of dollars of investments.

A naïve superimposition of the two charts with a single nominal axis is shown in Fig. 7. The price of oil ranges from $26.21 to $113.93 (per barrel) while the Canadian dollar ranges from $0.69 to $1.06 (value in US dollars). The chart is useless: A single nominal axis reduces the Canadian dollar to a flat line given the difference of magnitude of the nominal values.

A common approach to deal with the differences in magnitude is to normalize both timeseries to the start date and plot the percent difference relative to the start date. Normalizing timeseries with respect to date can be very effective, particularly when the movement is expected to be similar: both United Airlines and American Airlines are US passenger airlines, their stock prices should move similarly based on airline demand (unless one makes a significant business gaffe or has a brilliant

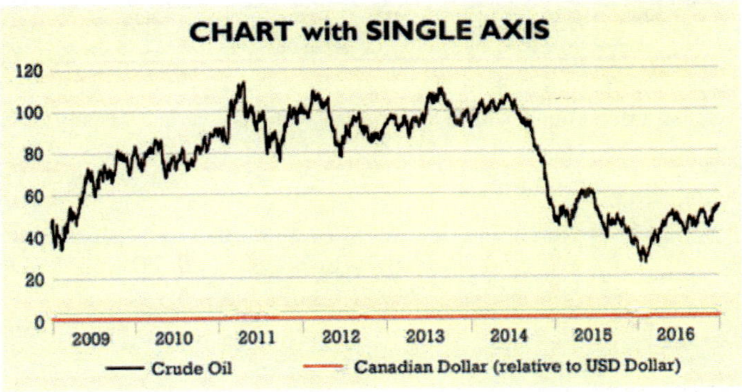

Fig. 7 Oil versus Canadian dollar with a single nominal axis. One line is flat, no comparisons are possible

strategy). The chart in Fig. 8 shows normalized values for oil and Canadian dollar. Again, this chart is useless with the Canadian dollar an almost flat line. The range in price movement between the two timeseries is highly different: the peak price for oil is more than four times the lowest price, whereas the peak price for the Canadian dollar is only 50% more than the lowest. The great difference in range leaves the Canadian dollar with only a tiny vertical range to draw. Furthermore, normalized data values add cognitive load on the user who needs to convert from normalized prices back to nominal market prices prior to executing a trade.

To provide the same range for each chart and align the dates to facilitate localized time comparisons, the charts could be stacked vertically. One method of stacking charts is a horizon chart. Horizon charts separate each timeseries vertically into tracks with a shared horizontal axis, and a consistent vertical scale [21]. Horizon charts uniquely overlay intervals for large movements, as shown in Fig. 9. For example, the price of crude, starts going down with a bit of red, then quickly climbs more than 50%, at which point it wraps to darker shade of blue starting at the baseline and again wrapping shortly thereafter as the price of oil climbs. The most salient visual cue are the darker bands which draw attention to magnitude, or the longer red time period on the Canadian dollar. Neither of these are relevant localized time comparisons.

Figure 10 provides two traditional timeseries charts stacked vertically with the same vertical space for each chart, and both aligned to a common date axis. Visual

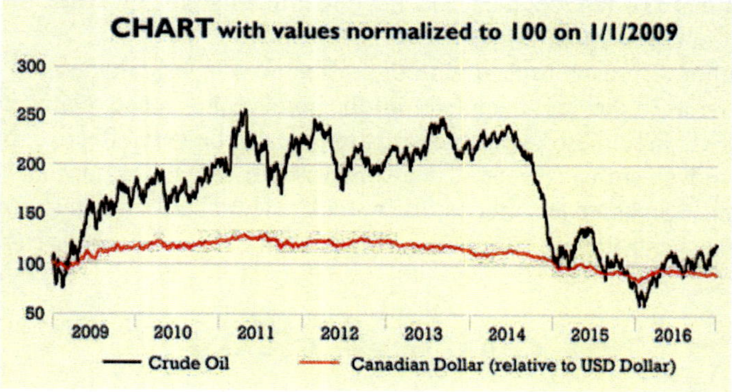

Fig. 8 Oil versus Canadian dollar with a single normalized axis. One line is nearly flat, comparisons are difficult

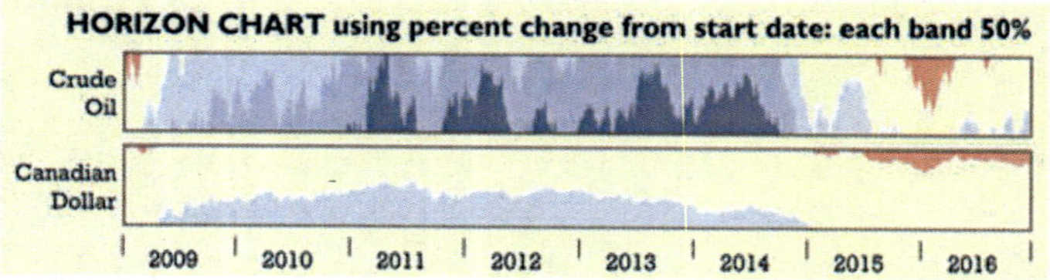

Fig. 9 Oil versus Canadian dollar as a horizon chart. Changes with darker steps are most salient

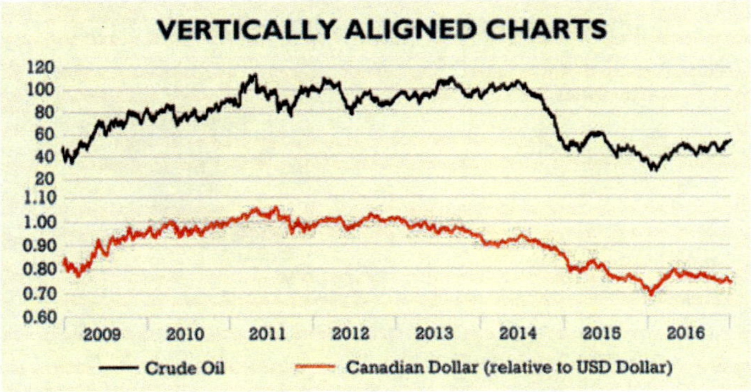

Fig. 10 Oil versus Canadian dollar as two vertically aligned line charts. Some temporal comparisons are feasible

comparisons by time can be made between the two charts, for example, in mid 2013 the price of oil is increasing while the price of the dollar is decreasing (and similarly at the end of 2016). These periods of contrary movement are *correlation breakdowns*, and trading strategies that expected coordinated movements are susceptible to financial losses during these periods. However, the viewer must continually scan up and down between the two lines to assess where the prices are moving together or diverging or whether once security is leading the other. Each series is more than 2000 points long—subtle comparisons across a few points is not feasible.

Much of the analysis discussed is with regards to the correlation of the two series. Correlations can be explored with scatterplots instead of line charts. Figure 11 shows a simple scatterplot for correlation analysis. Each point indicates the prices on a day, with the x-axis indicating the price of the Canadian dollar, and the y-axis indicating the price of crude oil each day. In this plot, the points are connected together by a thin line indicating the time sequence (i.e., a connected scatterplot). For a perfect correlation all the points would align to a single diagonal line. While the points lie somewhat along a diagonal, there are various outliers, and it can be seen the connections sometimes move horizontally or on the opposite slope. At a macro-level, the scatterplot can confirm that the prices are largely correlated across the overall eight-year period, as shown by general diagonal slope as well as the linear regression shown in the red line. However, local temporal patterns are difficult to perceive and would require tedious zoom and tooltip interactions to determine local activity.

One could attempt to better indicate time periods, for example, using color to encode sequence, as shown in Fig. 12. Unique colors are shown per year, as well as the linear regression. It can be immediately seen, for example that 2013 has a negative slope on the linear regression near the top-right indicating that 2013 had some inverse correlation, but from this plot it cannot be seen whether this occurred mostly early in the year, or late in the year, or throughout the year.

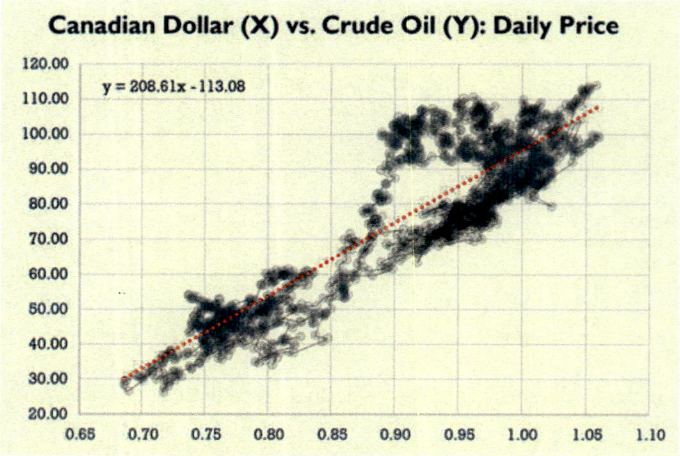

Fig. 11 Oil versus Canadian dollar plotted as a scatterplot

Correlations for a set of points can be quantified. A value of one indicates a perfect positive correlation, a value of negative one indicates perfect inverse correlation and zero indicates no correlation between the values. Using the oil and dollar dataset, correlations can be calculated on a daily basis using a set of prior days observations. This creates a timeseries of correlations, which can be plotted as a timeseries. Figure 13 shows a 20-day correlation and 50-day correlation for oil and Canadian dollar as timeseries. These timeseries are promising—the negative correlations in late 2013/early 2014 are clearly visible as is a sharp negative correlation in early 2011. Correlation calculations, however, are lagging indicators—the correlation for today is based on the prior days' observations. So, for example the low values in the chart in early 2014 are due to observations as well from 2013. Correlations can recover before they are indicated in the rolling correlation timeseries. For a financial market participant, this can confirm prior activity, but is not suitable for active trading.

Finally, we plot both price values timeseries on a line chart with two independent axes as shown in Fig. 14, with oil shown on the left axis and the Canadian dollar on the right axis. The proximity between the two series allows for close comparison of spatially proximate data points. Visually inspecting the time period for 2013 and early 2014 time period, it can be seen that there are a few occasions of divergence. For example, in mid 2013, the first yellow box highlights a sharp jump in the price of oil and a corresponding drop in the Canadian dollar. Note how this very quick sharp divergence in mid 2013 is not particularly noticeable in the 50-day correlation chart. Or, in late 2013 the second yellow box shows a period where the Canadian dollar is on a downward trend, while oil zigzags in an increasing price trend. Both lines are moving in the same direction in early 2014 while the correlation chart is still negative.

Other insights can be easily seen as well. For example, in early 2009, there is a local bottom in the prices, with oil bottoming prior to the Canadian dollar. Another

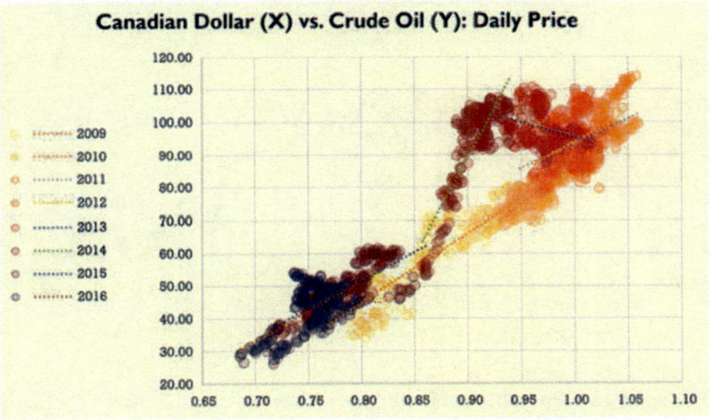

Fig. 12 Oil versus Canadian dollar plotted as a scatterplot, color-coded by year

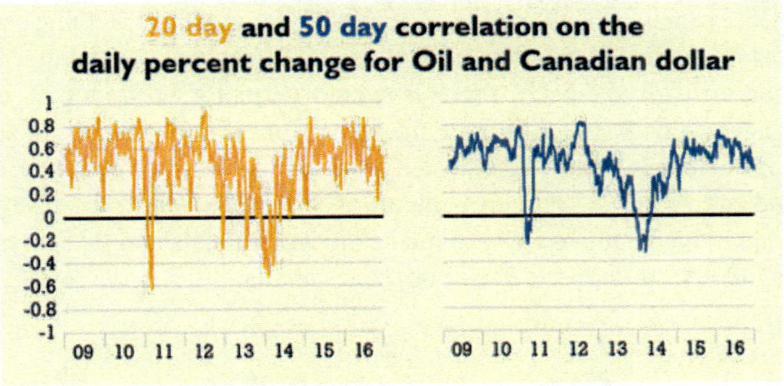

Fig. 13 Oil versus Canadian dollar rolling 20 day and 50-day correlations

bottom can be seen at the beginning of 2016, in this example, the Canadian dollar hits a low and starts bouncing back a few days prior to oil hitting a low.

Even though the correlation chart lags the price chart, both are useful, with the price chart providing early indication of divergence, and the correlation chart confirming divergence. Figure 15, shows a time aligned price chart and correlation chart. The highlights aid visually tracing between the price timeseries and the correlation timeseries. In practice, simple interactions on financial charts facilitate this vertical alignment through the use of cross-hairs which track mouse pointer location (or finger location on mobile devices).

The periods of correlation, non-correlation, and inverse correlation are transient, and can last for weeks, months or years. Trading strategies which assume correlation are at risk when correlations breakdown. A spread trade is susceptible to significant financial loss if the correlations significantly change. Figure 16 shows oil vs. dollar over a much longer timespan—note that the Canadian dollar hits a peak in mid-2007, while oil hits a peak nearly a year later. Similarly, for much of 2018 the two

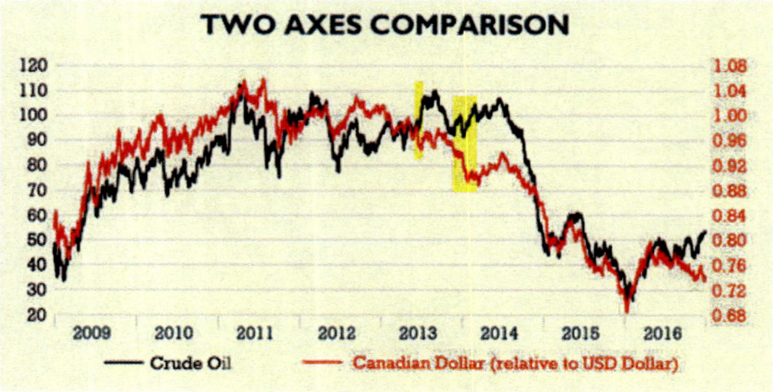

Fig. 14 Oil versus Canadian dollar two axis chart. Lines are close allowing for fine grain visual comparison between the series

were inversely correlated—oil rose while the Canadian dollar sank. Experienced financial professionals are attuned to various changes in the market place and adjust their trading strategies. In an interview with an oil trader [19], he explained: "On the trading floor, we'd all be closely tracking inventory data for a few months, and then it would change, we'd all switch to tracking storage, or refinery outages, or some other metric as a leading indicator driving our pricing." These changes in relationships to other indicators are not explicit—therefore early detection to the changes in the relative price movement is critical to financial professionals and this change can be visually facilitated through the use of dual axes charts.

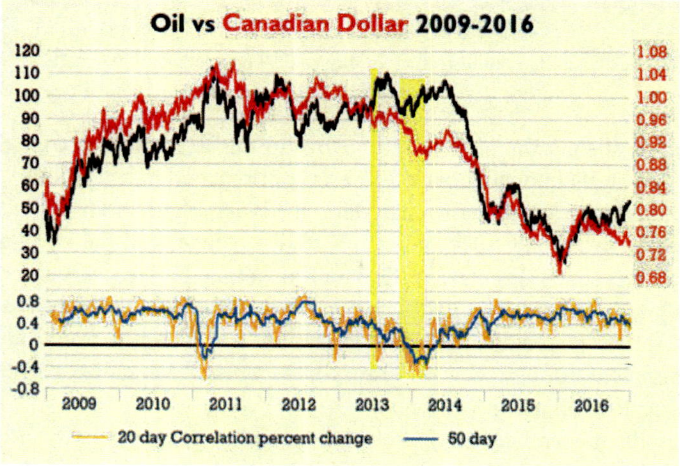

Fig. 15 Oil versus Canadian dollar, two axis price chart with time aligned correlation chart

Fig. 16 Oil versus Canadian dollar. Often they move in tandem, but sometimes move apart for a year

5 Inversely Correlated Timeseries

Some financial instruments typically have a negative correlation. Credit default swaps act like insurance: they will pay out if a company goes bankrupt. A company's stock price will sink if a company is perceived of at risk of bankruptcy, whereas credit default swaps will increase in price if a company is perceived at risk of bankruptcy (i.e., the cost of insurance is more if bankruptcy is more likely). Figure 17 shows the credit default swaps for Pemex (a Mexican oil producing company) and crude oil (which, if the price of crude goes down too low the risk of bankruptcy for an oil producing company increases). From the two timeseries, the inverse relationship is apparent, particularly with the big increase in the CDS price corresponding to the big drop in drop in crude oil price. But which one changes first? Was there advance notice of the impending price increase in the CDS?

A common approach in financial services is to invert one of the axes where there is an expected inverse correlation. Figure 18 show the two securities, with dual y axes, with the y-axis for crude oil inverted (and explicitly noted in the title). With the inverted axis for crude oil, the lines are spatially proximate to each other and it is easy to see when they move in tandem or apart. Visually it can be seen that the price for crude oil started to drop in early 2020, whereas the CDS price for Pemex also continued to drop for a month, then very quickly moved in the opposite direction. With the CDS price movement lagging the price movement in oil by a month, an astute financial market participant may have been able to profit from the divergence.

6 Discussion

The examples above show how dual axes facilitate analysis. Note that we are not deliberately manipulating axes for crossings, rather relying on extents of axes. We further aid the view with color coding of axes and labels to automatically associate each line with its corresponding axis. In the case of the timeseries analysis, the many

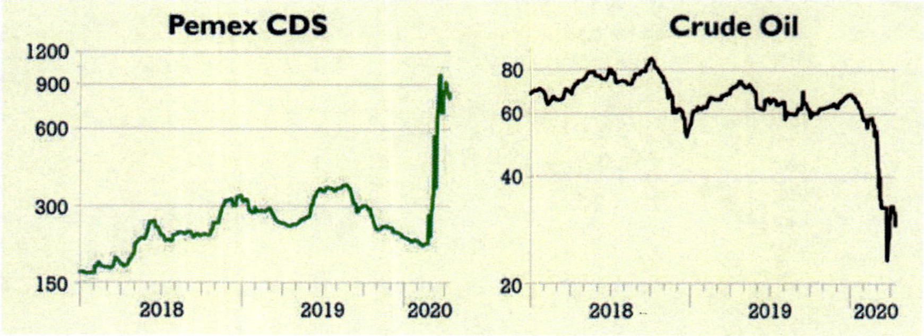

Fig. 17 Inversely correlated securities, when one goes down, the other is expected to go up

examples associated financial timeseries analysis shows how the fine grain local relative movement cannot be completed with other chart variations and requires the use of dual axes.

6.1 What Happens in the Real World

If dual axes charts are useful for professionals as we claim, then it should be feasible to find many examples of charts with dual axes in professional publications. We found 25 online chart-dense financial publications created by central banks, global investment banks and capital markets advisory firms [20–45]. We counted charts and focused specifically on timeseries charts with single axis or dual axes, as summarized in Table 1. Across these 25 publications, there were 1305 charts, of which 944 were timeseries charts. 179 of the 944 timeseries charts used dual axes, that is, 19% of timeseries charts were dual axis. Similarly, 92% of the publications use dual axes charts (23 of 25 publications). These publications show that timeseries charts with dual axes are prevalent in financial publications. This evidence implies that financial professionals are familiar with the charts and know how to interpret them.

Dual x-axes versus dual y-axes. 178 of the dual axis charts have dual y-axes (i.e., 19% of timeseries) and only one has a dual x-axis chart (0.1%) [29]. This confirms that dual-x-axes are rare—even in finance. Thus, the evaluation tests done by Isenberg et al. [1], which used a dual x-axis chart, may have been over-generalized to recommend against all dual axis charts and did not consider the common usage of dual y-axis charts among financial experts.

What techniques are used to make dual axis charts better? It is common practice to associate line color with axis labels as done in the examples here. Alternately axis label and color swatches can be positioned beside the corresponding axis. We recommend such design patterns.

Table 1 Use of dual axis charts from 25 fiancial pulications

Publication	Number pages	Number charts	Number timeseries charts	Number dual axis charts	Percent (%) dual axis charts
DBS	8	25	23	3	13.0
ANZ Research	8	32	30	10	33.3
BMO	10	12	4	4	100.0
Capital Economics	11	55	55	11	20.0
DWS	14	70	40	4	10.0
Deutsche Bank	15	7	6	2	33.3
Blackrock	16	12	5	2	40.0
Merk	18	13	13	11	84.6
Plumtree	19	27	3	0	0.0
Fathom Consulting Eikon	20	15	15	2	13.3
Advisor Perspectives	21	19	17	4	23.5
SGX	22	20	8	2	25.0
Morgan Stanley	22	14	6	2	33.3
QVM Invest	24	86	70	5	7.1
Baird Private Wealth	42	56	20	2	10.0
Fidelity	47	45	24	5	20.8
Unicredit	52	178	151	57	37.7
Credit Suisse	52	48	48	7	14.6
NatBank Fiji	55	65	57	9	15.8
NatBank Grc	56	49	42	9	21.4
Incrementum	60	36	31	5	16.1
JP Morgan	72	108	58	8	13.8
Ing Dutch	85	141	114	0	0.0
Ing German	85	78	43	4	9.3
Merrill	90	94	61	11	18.0
TOTAL	**924**	**1305**	**944**	**179**	**19.0**

6.2 *Automating Dual-Axis Charts*

The timeseries case study walks through conditions where the dual axis is needed—that is—data points which move together after one timeseries is linearly transformed to align with another (i.e. scale amplitude and translate a vertical offset). Timeseries distance measures use the same kinds of analysis to measure timeseries similarity. This is a well-researched area in statistics, data mining and machine learning [46–53]. Computing timeseries similarity can thus be used as part of a heuristic to automate when to display multiple axes.

Timeseries similarity can be computed by means of a distance measure. There are two primary distance metrics that may be used with timeseries: Euclidean distance and Dynamic Time Warping.

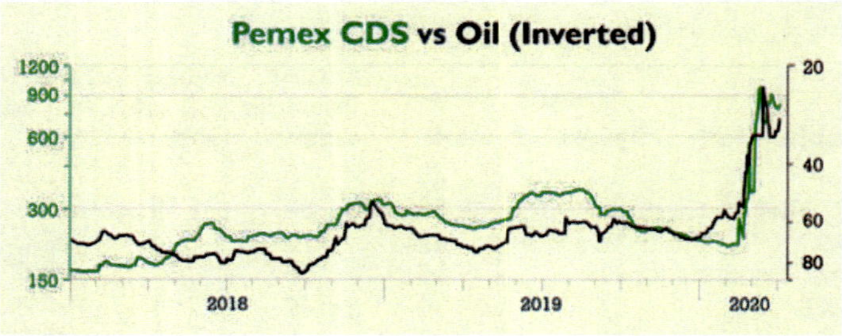

Fig. 18 Inversely correlated securities with dual axes: the second security has an inverted axis

Euclidean Distance is a linear mapping of one signal to another, or the distance between two points in each timeseries (Fig. 19 left). Euclidean space is sensitive to distortions in the data so the timeseries need to be preprocessed to remove noise, interpolate gaps, potentially smooth the trend (e.g., a moving average) and perform offset translation and amplitude scaling.

Dynamic Time Warping (DTW) is a technique used to measure similarity between timeseries that do not have fixed time axis alignment (Fig. 19 right). That is, a non-linear mapping between signals. For example, some sequence may be offset, vary in speed or be of a different length than the other. In financial timeseries this could be a lead/lag effect, for example, Japanese stock price movements may be closely correlated to the prior day's New York stock price movements.

Both Euclidean distance and DTW are used as distance metrics in timeseries similarity. While, DTW has a lower error rate than Euclidean, it is also very computationally expensive and can take a prohibitively long time to calculate on larger datasets. Fast approximations of DTW can be used to greatly speed up computation and make its use more tractable. Either Euclidean distance of DTW distance measures could be used to compare timeseries to determine the level of similarity and therefore the usefulness of plotting with two Y-axes.

Many caveats over the use of DTW in identifying timeseries similarity have to do with the excessive cost of performing DTW comparisons across multiple signals. This is not an issue in the envisioned use case which merely performs a single pairwise comparison of two signals in order to determine whether two Y-axes might be useful.

The proposed use case for automated decision making is in choosing whether to display a single or multiple charts in response to a user query involving two signals. Euclidean or approximate DTW distance measures below a given threshold would be defined as similar, and thus plotted together. We will illustrate this with a few examples shown in Fig. 20.

The top row contains three charts plotting one year of stock prices for two railroad companies. These are two highly similar timeseries whether using Euclidian or DTW measures. The leftmost chart has a singular nominal axis, which is not useful as both series have highly different range of values. The middle chart plots both series normalized to a starting value of 100: the similarity is visually evident. The rightmost chart shows the timeseries with two independent y-axes. The close plotting of the two

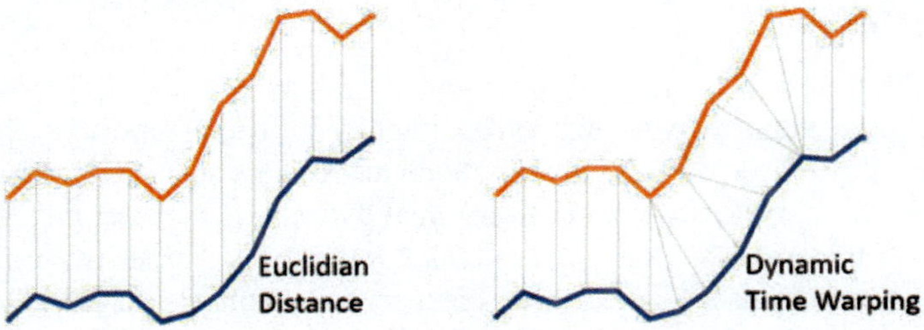

Fig. 19 Euclidean distance measures distance between two points in the same time interval. Dynamic time warping measures shortest distance between points in the two timeseries

series helps to highlight the small deviations that are difficult to see in the normalized chart. An additional benefit of the two axes is the price for both stocks can be read directly from the chart, that is, no reverse transformation is required.

The second row contains two financial services stocks, but in different industries (shopping malls and debt financing). Again the leftmost chart is not useful. The center chart aligns the starting point, but the overall trajectories diverge splitting the lines creating a lot of white space and reducing the ability to compare the series. The rightmost chart, with two y-axes, shows some periods of similarity: January–February, May–October; and periods of divergence, such as October–December. The two timeseries share similar patterns over the early part of the sequence and then diverge as the relationship between the securities deteriorates. The changing relationship of the securities serves as an important indicator to the financial analyst and will drive further research into the alignment and divergence. Note that an appropriate heuristic should consider timeseries similarity across subsets of timespans, then plot with two y-axes if there is a significant portion of time similarity.

The third row contains two completely unrelated stocks (a casino and a software company). The two timeseries would be deemed to be dissimilar by both Euclidean and DTW measures. The timeseries show no similarity across any of the plots, and the dual axis plot provides no benefit for a close reading between the two timeseries as there is no (or very low) similarity.

More examples could be shown for inverse relationships and leading/lagging indicators. However, it should also be noted that the appropriate axis type is related to the task—if the analyst is simply gauging which stock in each pair had the highest return over the last year, the normalized chart in the center is more appropriate than the dual axis chart.

7 Conclusions

Some literature within the visualization community has been skeptical of the benefits of multi axes displayed on a single chart. Such researchers have contended that there is either no valid use case or no need for dual axis charts and recommend against their use. In this paper we have presented a number of use cases for professionals such as quality control specialists, data scientists and financial analysts. The benefits for charts with dual axes, including:

- *Values can be read directly off axes.* Even if only a single measure is shown (e.g., Fig. 2), both nominal values and proportional values can be read directly from the chart without interaction. When two different datasets are compared, nominal values can be read using the axes (Fig. 5 right), rather than mentally converting the normalized values back to nominal values—which may not be feasible if the chart does not indicate the nominal values (e.g., Fig. 5 middle).
- *Local comparison.* The superimposition of two charts with two separate scales enables points in correlated datasets to be plotted close to each other, enabling ability to notice small meaningful differences such as local *minima and maxima* in distributions (Fig. 5 right), *divergence* and *lead/lag* effects (Fig. 14). As the image sequence in Figs. 6, 7, 8, 9, 10, 11, 12, 13 and14 show, the local analysis of fine grain patterns is impossible to otherwise see with other chart configurations.
- *Inverse relationships.* By inverting one of the two axes (Fig. 18), it is feasible to visually assess the degree of negative correlation exhibited by two timeseries, as well as patterns such as divergence.

We further show the broad use of dual axes charts in practice in the financial domain. These dual axes charts are used by professionals to surface insights, understand correlations and uncover patterns. This is a very different use case than, for example, a casual reader of a news story that might be deliberately or inadvertently manipulated by scaling effects of two y axes. As such we view the concerns of previous research to be an over-generalization which has been unaware of the ubiquitous use of dual axis charts within the financial analysis community. The evidence we have presented shows that such charts are appropriate when used by professionals for deeper analysis.

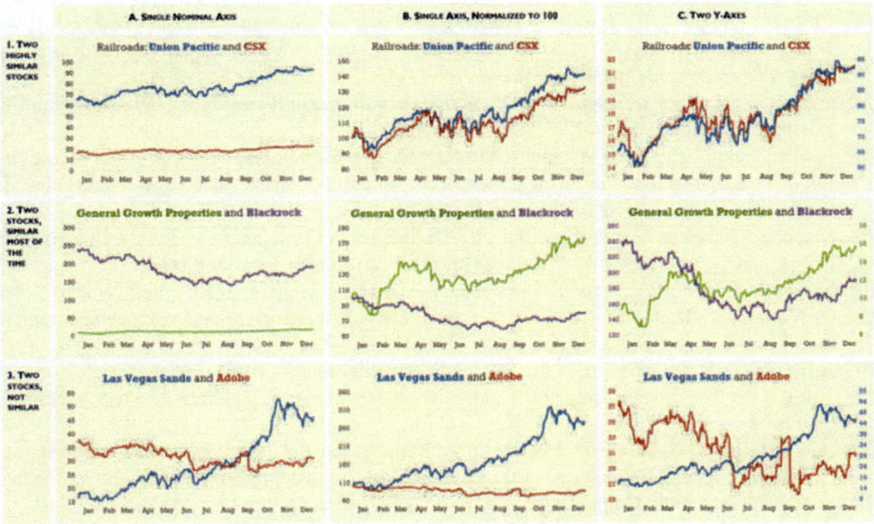

Fig. 20 Sample stock charts show with (A) single nominal axis, (B) single normalized axis, and (C) dual axes

References

1. Few, S.: Dual-scaled axes in graphs: are they ever the best solution. Perceptual Edge Newsletter, percep-tualedge.com/articles/visual_business_intelligence/dual-scaled_axes.pdf
2. Isenberg, P., Bezerianos, A., Dragicevic, P., Fekete, J.-D.: A study on dual-scale data charts. IEEE Trans. Visual Comput. Graph. **17**(12), 2469–2478 (2011)
3. Fung, K.: The hidden bad assumption behind most dual-axis time-series charts. Junk Charts: Recycling chart junk as junk art. junkcharts.typepad.com/junk_charts/axis/
4. Wickham, H.: I'm Hadley Wickham, Chief Scientist at RStudio and creator of lots of R packages (incl. ggplot2, dplyr, and devtools). I love R, data analysis/science, visualisation: ask me anything!". https://www.reddit.com/r/dataisbeautiful/comments/3mp9r7/im_hadley_wickham_chief_scientist_at_rstudio_and/
5. Wickham, H.: Nine visualisation challenges. Data Visualization New York at The New School, NYC, Nov. 17, 2016
6. ggplot with 2 y axes on each side and different scales. Stack Overflow, Nov 5, 2013. stackoverflow.com/a/3101876
7. Rost, L.C.: Why not to use two axes, and what to use instead. Datawrapper (2018). blog.datawrapper.de/dualaxis/
8. Qui, L.: Chart shown at Planned Parenthood hearing is misleading and 'ethically wrong. Politifact. Oct. 1, 2015. politi-fact.com/factchecks/2015/oct/01/jason-chaffetz/chart-shown-planned-parenthood-hearing-misleading-/
9. Vigen, T.: Spurious Correlations. Hachette Books, New York (2015)
10. Dual axis in ggplot2. RPubs. https://rpubs.com/kohske/dual_axis_in_ggplot2
11. Dual Y axis with R and ggplot2. The R Graph Gallery. https://www.r-graph-gallery.com/line-chart-dual-Y-axis-ggplot2.html
12. Chakerian, S.: How to plot two different scales on one plot in matplotlib (with legend). https://samchaaa.medium.com/how-to-plot-two-different-scales-on-one-plot-in-matplotlib-with-legend-46554ba5915a

13. Chicago Board of Trade. A Six-Part Study Guide to Market Profile. 1996. web.archive.org/web/20170829185705/incre. https://www.cmegroup.com/education/interactive/marketprofile/handbook.pdf
14. Second y-axis time series seaborn. https://stackoverflow.com/questions/47591650/second-y-axis-time-series-seaborn
15. Few, S.: Dual-Scaled Axes in Graphs. Are They Ever the Best Solution? March 2008. Perceptual Edge. Visual Business Intelligence Newsletter. https://www.perceptualedge.com/articles/visual_business_intelligence/dual-scaled_axes.pdf
16. Nison, S.: Japanese Candlestick Charting Techniques A Contemporary Guide to the Ancient Investment Techniques of the Far East. New York Institute of Finance (1991)
17. Nison, S.: Beyond Candlesticks: New Japanese Charting Techniques Revealed. Wiley (1994)
18. De Villiers, V.: The point and figure method of anticipating stock price movements: complete theory and practice. Stock market publications (1933)
19. Arms, R.W.: Volume Cycles in the Stock Market. Dow Jones—Irwin (1983)
20. Tufte, E.R.: The visual display of quantitative information, vol. 2. Graphics Press, Cheshire, CT (2001)
21. Heer, J., Kong, N., Agrawala, M.: Sizing the horizon: the effects of chart size and layering on the graphical perception of time series visualizations. Proceedings of the SIGCHI Conference on Human Factors in Computing Systems, pp. 1303–1312 (2009)
22. Sindel, B.: Navigating the Crypto Markets. TradersExpo. New York. March 11, 2019
23. Klok, M.: Stagnating Profitability for third year in a row: Dutch Economy Chart Book. ING Netherlands, Economic Department (2019)
24. Belski, B., Roccanova, N.: US Strategy Snapshot. BMO Financial Group, New York (2020)
25. Loufir, M., Romanos-Louizos, K., Troupi, L., Lampousis, A.: Emerging Markets Analysis Quarterly Chart Book, National Bank of Greece, Economic Analysis Division, Athens (2019)
26. CIO Chart Book. Bank of America, Merrill Lynch, New York (2018)
27. Wealth Management Perspectives. Morgan Stanley, New York (2020)
28. A selection of charts presented by Fathom Consulting at the UK Economic Outlook Seminar on 2nd August 2016. Thomson Reuters. London (2016)
29. Chartbook. Reserve Bank of Fiji, Suva, Fiji (2020)
30. Chartbook: SREITs & Property Trusts. SGX Research, Singapore Exchange, Singapore (2019)
31. Brzeski, C., Fechner, I.: Germany's economy chart book: on the brink of recession? Economic and Financial Analysis Division of ING DiBa (2019)
32. Emsbo-Mattingly, L., Weinstein, J., Hofschire, D., Carrigan, R.: Quarterly Market Update. Fidelity Investments, FMR LLC, Boston (2019)
33. Fixed Income Chartbook. DWS Investments UK Limited. London (2020)
34. Slok, T., Barnard, M.: March Economic Chart Book (Snapshot). Deutsche Bank Research, New York (2019)
35. Sneddon, D., Hine, C.: FX Chartbook: Core Themes & Favorite Trades. Credit Suisse, London (2017)
36. Reece, N.: U.S. Business Cycle Report, Merk Investments, LLC., New York (2020)
37. Stoeferle, R.-P., Valek, M.: Chartbook: In Gold we Trust 2017 Incrementum (2017)
38. Interest Rates Chart Book. CME Group, Chicago (2019)
39. Yeung, R., Liu, L.-G., Lam, L.: Hong Kong Monthly Chartbook. Australia and New Zealand Banking Group Limited (ANZ), Melbourne, Australia (2011)
40. Baig, T.: ASEAN-6 and Covid-19: DBS Chartbook. DBS Bank Ltd., Jakarta (2020)
41. Neilsen, E., Ingo Helmig, Valli, M. et al.: The UniCredit Economics Chartbook. UniCredit Research. Munich (2019)
42. QVM Chartbook: Conditions and Relative Performance. QVM Invest. Glastonbury, CT (2018)
43. Williams, M., Evans-Pritchard, J., Palmas, F.: China Chart Book. Capital Economics, Toronto (2019)
44. Azzarello, S. et al.: Guide to the Markets. J.P. Morgan, New York (2020)
45. Quarterly Market Review. Plum Street Advisors (2019)

46. Q2 2020 Market ChartBook, Baird Private Wealth Management, Robert W. Baird & Co., Charlotte, NC (2020)
47. Sonders, L.A.: Every Picture Tells a Story: Chartbook Look at Economy/Market. Advisor Perspectives, Charles Schwab & Co, San Francisco (2020)
48. Hildebrand, P., Bolvin, J., et al.: Midyear Outlook, p. 2020. Blackrock, New York (2020)
49. Mueen, A.:Similarity search on time series data: past, present and future. Tutorial slides. https://www.cs.unm.edu/~mueen/Tutorial/CIKM2016Tutorial.pdf
50. Gunopulos, D., Das, G.: Time series similarity measures and time series indexing. ACM SIGMOD Rec. **30**(2), 624 (2001)
51. Keogh, E.J.: Machine learning in time series databases (and everything is a time series!). Tutorial at the AAAI International Conference on Artificial Intelligence (2011)
52. Keogh, E.J., Kasetty, S.: On the need for time series data mining (2003)
53. Benchmarks: a survey and empirical demonstration. Data Mining and Knowledge Discovery 7, 349–371
54. Wang, X., Mueen, A., Ding, H., Trajcevski, G., Scheuermann, P., Keogh, E.J.: Experimental comparison of representation methods and distance measures for time series data. Data Mining and Knowledge Discovery In press. URL: (2012). https://doi.org/10.1007/s10618-012-0250-5
55. Serra, J., Arcos, J.L.: An empirical evaluation of similarity measures for time series classification. Knowl.-Based Syst. **1**(67), 305–314 (2014 Sep)
56. Abanda, A., Mori, U., Lozano, J.A.: A review on distance based time series classification. Data Min. Knowl. Disc. **33**(2), 378–412 (2019 Mar)

Hierarchical Visualization for Exploration of Large and Small Hierarchies

Alexander Macquisten, Adrian M. Smith, and Sara Johansson Fernstad

Abstract This paper presents a comparison of visualization methods for hierarchical data, and extends it by demonstrating an interactive approach that uses a combination of hierarchical visualization methods to support exploration and analysis of high-dimensional data. Hierarchical visualization can provide a useful abstraction for investigation of large and complex datasets, but the range of options for visualization can make it challenging for users to determine which is the most suitable for their datasets. Existing evaluations of hierarchical visualization methods typically compare the representation of hierarchical structures rather than the representation of underlying data, despite the underlying data commonly being a crucial part of the analysis. Five different visualization methods for hierarchical data are compared in this paper, with focus on understanding the underlying data features, represented by size and colour, for two different sizes of hierarchical structures. The study results indicate that Sunburst and Circular Treemap perform best for larger hierarchical structures, while Treemap, Icicle plot and Circular Treemap performed best for smaller hierarchies. The results from the study, which were first presented in Macquisten et al. [1], are in this paper used to inform the design of a visualization approach for high-dimensional data exploration that combines hierarchical visualization methods. The approach is demonstrated through a use case analysis of data from a study of the ocean microbiome.

A. Macquisten (✉) · S. J. Fernstad
Newcastle University, Newcastle upon Tyne NE1 7RU, UK
e-mail: a.macquisten2@newcastle.ac.uk

S. J. Fernstad
e-mail: Sara.Fernstad@newcastle.ac.uk

A. M. Smith
Unilever Research, Colworth, UK
e-mail: Adrian.M.Smith@unilever.com

B. Kovalerchuk et al. (eds.), *Integrating Artificial Intelligence and Visualization for Visual Knowledge Discovery*, Studies in Computational Intelligence 1014,
https://doi.org/10.1007/978-3-030-93119-3_23

1 Introduction

Ranging from the biological taxonomy of species to disk usage to business structure, hierarchically structured data is commonly occurring in a large variety of domains. Abstracting the data through a hierarchical structure can allow for an uncluttered visual representation of the full dataset. This can, for example, be helpful as an approach to organise and abstract high-dimensional and large data sets. Using methods such as hierarchical clustering or by using an inherent structure within the dataset, the challenges of high-dimensionality can to some extent be mitigated though a hierarchical structure that reveals meaningful relations between the dimensions.

For the exploration and analysis of large and complex datasets, hierarchical visualization is a powerful tool, both for the presentation of the data in a way that is understandable to the user and as a way for them to directly interact and manipulate the data. While purely computational methods, such as Principal Component Analysis [2] and Multidimensional Scaling [3], can efficiently process data and present solutions to well-defined questions, they are restricted by the data patterns that are pre-defined as interesting. For any analysis task where the question at hand is broad or where there are no set questions at all, the overview presented by hierarchical visualization can provide a good starting point for exploration and analysis, allowing the user to make use of their domain knowledge and findings made during analysis to determine which patterns and parts of the data that are particularly interesting.

It can be difficult for visualization designers and users to decide which hierarchical visualization method is appropriate to use for their data, as there exist a broad array of different visualization methods to help understand patterns and distributions in hierarchical data. In particular, the structure and size of the data may impact the appropriateness of visualization methods. Earlier comparisons [4–8] focus mainly on understanding the hierarchical structure. However, many hierarchical visualization methods are also able to represent data properties through visual features such as size and colour, and few of these evaluations compare the ability of hierarchical visualization methods to support understanding of properties of the underlying data. Nonetheless, the values and patterns in underlying data may often be a critical part of understanding the data as a whole.

The research presented here aims to address this by reporting and discussing the results of a usability evaluation that compares five visualization methods for representation of hierarchical data, initially presented by Macquisten et al. [1]. The five methods include Treemap [9], Icicle Plot [10], Sunburst Chart [11], Circular Treemap [12] and Bubble Treemap [13]. These are compared for different sized hierarchies and in context of their ability to clearly represent underlying data values. Building on the results from the study, this paper demonstrated how a combination of hierarchical visualization methods can be used for exploration of a high-dimensional data set from a study of the ocean microbiome [14]. To summarize, the main contributions of the research presented here are:

- A usability study that compares five hierarchical visualization methods that have not previously been compared in a joint study, in context of size of hierarchy and representation of data features;
- Recommendations based on study results and literature to support visualization designers and users in selecting appropriate hierarchical visualization;
- A visualization approach that support exploration of high-dimensional data using a combination of hierarchical visualization methods, demonstrated through a use-case.

The remainder of this paper is structured as follows: Sect. 2 provides background to hierarchical visualization methods and previous evaluations of such methods, and Sect. 3 describes a comparative evaluation of five hierarchical visualization methods. Section 4 covers the raw results of the evaluation, while 5 goes over the papers recommendations of which methods to use. This is followed by a use-case demonstrating the combination of hierarchical visualization in Sect. 6, and conclusions in Sect. 7.

2 Background

Many methods have been designed for hierarchical data visualization [15], and it is not always obvious which are the most useful to suit the needs of the user. One user may need a static presentation of the full hierarchical structure of a dataset, while another may use a hierarchical method as part of a larger exploration tool, for example with the aim to find interesting subsets of the data. This section will provide an overview of the hierarchical visualization methods used in the study presented in Sect. 3 and introduce previous studies that compare their usability.

2.1 Hierarchical Visualization Methods

Hierarchical visualization methods can be implicit or explicit [16]. Explicit methods display their parent-child structure through links between nodes. They are somewhat limited in terms of size of hierarchy, since explicit methods such as node-link diagrams or dendrograms take up more screen space as the structure and size increase. The use of interaction, such as shrinking and expanding of nodes, mitigate this issue to some extent [17–19], but it can still be challenging to display and explore complex hierarchical datasets using explicit methods as the space needed to show the full structure tends to grow exponentially.

Implicit hierarchical methods on the other hand are less affected by this issue [15] as their hierarchical structure is represented through node arrangement, allowing them to be more compact. They however still suffer from clutter issues as the data complexity increases, hence, filtering and interaction are also beneficial to make full use of implicit hierarchical visualization. A benefit of implicit methods is also their

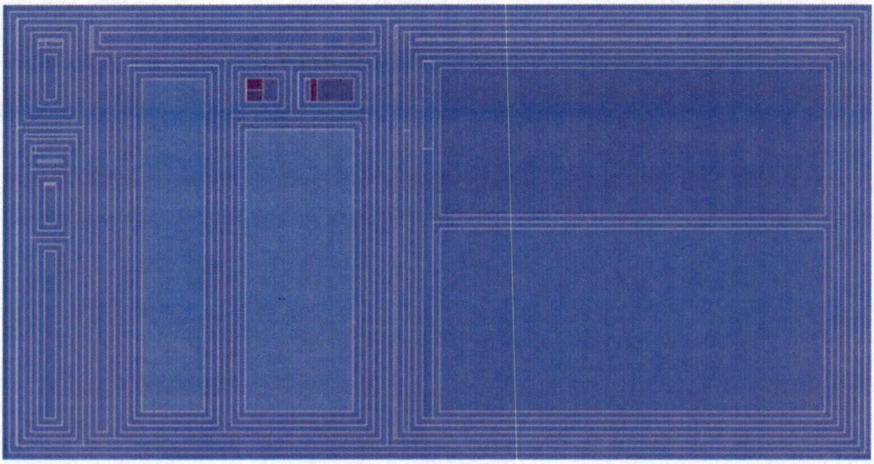

Fig. 1 An Treemap showing a dataset from a study of the gut microbiome of Preterm Infants [20]

ability to display additional data features, commonly through size and colour, which facilitates gaining of further insight from the underlying data. Due to this, the focus of the research presented here lies on implicit hierarchical visualization methods. While explicit methods in theory can represent features using both node size and colour, the displays would quickly become cluttered.

Treemap [9] is a space efficient visualization method that represents the nodes of a hierarchical structure as a series of nested rectangles (Fig. 1). The root node being the outermost rectangle, with child nodes displayed internally, and leaf nodes as the innermost rectangles. The area of each leaf node is commonly relative to a size or frequency value determined by a space filling algorithm, with the area of parent nodes being the total of their children. This provides visual information regarding both the structure and the size of each node without requiring any additional information, allowing for quick comparison of nodes to spot potential outliers or nodes of interest. While space efficient, a limitation of the Treemap layout is the risk of losing overview of the hierarchical structure itself, as the challenge of keeping track of multiple nested layers may become more difficult with an increasing hierarchical depth. Figures 1, 2, 3, 4 and 5 display data from a study of the gut microbiome of Preterm Infants [20]. The dataset consists of 516 biological entities across 867 samples, with the hierarchical structure consisting of 7 levels. In the Figures, the lowest 15% of the dataset is filtered out for the purpose of increasing perceivability of patterns. It is visible from the size of the innermost rectangles in the Treemap in Fig. 1 that four of the leaf nodes (corresponding to biological entities) are considerably more abundant than others.

Icicle Plot [10] displays hierarchical structure through the adjacency of nodes vertically downwards, with the root at the top, child nodes below and leaf nodes at the very bottom (Fig. 2). The width of each node is relative to a size or frequency value, resulting in the root node being the width of the entire plot. The hierarchical structure is presented clearly in Icicle, with an appearance similar to that of a stacked

Fig. 2 An Icicle Plot displaying the same dataset as shown in Fig. 1

bar chart. It is easy to perceive at what level a node is, as well as its respective children and parents' nodes. Though without interactivity, small nodes can be difficult to distinguish for large and complex data sets. Furthermore, given the size required for representing the root and lower levels in the hierarchy, Icicle plot is considerably less space efficient than, for example, Treemap. Four larger leaf nodes are easily spotted at the bottom level in the Icicle Plot in Fig. 2, which are also clearly visible in Fig. 1. However, the structure of the four nodes, of which two considerably larger, can be identified at the second level from the root in Fig. 2, is considerably harder to distinguish in the Treemap.

Sunburst Charts [11] are structured similarly to Icicle, using the adjacency of nodes to show parent-child relationships, but are displayed radially (Fig. 3). The root node is at the centre with child nodes expanding outwards towards the edge. Size or frequency is shown through the width of each arc relative to other arcs at the same level. Nodes further out in the structure may be larger than inner nodes, despite representing a smaller frequency. Through this, the Sunburst becomes more space efficient than Icicle for larger hierarchies with a higher number of leaf nodes, since less space is used for less detailed levels (i.e. the root) and relatively more space is used for more detailed levels (i.e. leaf nodes), although comparisons across multiple levels are not possible. The previously identified patterns of four large leaf nodes and four nodes at the second level from the root, can also be identified in the Sunburst in Fig. 3. Compared to the Icicle Plot in Fig. 2, the smaller leaf nodes are provided relatively more space in Fig. 3.

Circular Treemaps [12, 21] is a variation of Treemap that uses nested circles to represent nodes (Fig. 4). The outer circle represents the root, with child nodes as the inner nested circles. While circle areas are used to show size or frequency values, similarly to the other methods, the areas does not hold a direct one-to-one relationship to the frequency value at all levels, as the total area of a parent node is larger than the sum of the areas of its child nodes. A potential benefit of Circular

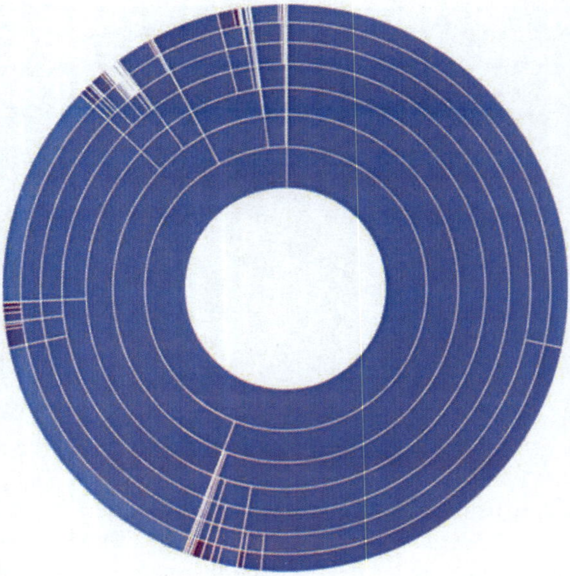

Fig. 3 A Sunburst Chart displaying the same dataset as shown in Fig. 1

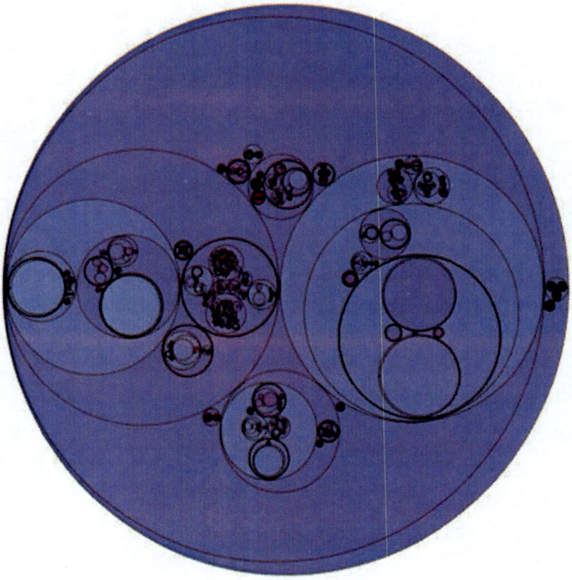

Fig. 4 A Circular Treemap displaying the same dataset as shown in Fig. 1

Fig. 5 A Bubble Treemap with the same dataset shown in Fig. 1

Treemap, in comparison to Treemap, is that the hierarchical structure may be more easily perceived due to the extra space emphasizing parent–child relationships.

Bubble Treemaps [13] display leaf nodes as circles, and the rest of the structure encloses the leaf nodes within a series of contours, to show the higher levels in the hierarchy (Fig. 5). Doing so allows it to consume less space than Circular Treemap but may still present a clearer hierarchical structure than Treemap, while still having the room to present additional information. Similarly to the Treemap in Fig. 1, the leaf node patterns can be fairly easily identified in the Circular Treemap (Fig. 4) and the Bubble Treemap (Fig. 5). Meanwhile, the diversities at other levels are harder to overview. Compared to Circular Treemap, Treemap and Bubble Treemap have more similar size representations and more clearly display the four considerably larger leaf nodes.

Some examples exist of 3D hierarchical visualization methods, such as Cone Tree [22], Cushion Treemaps [23], Botanical visualization [24] and Information Pyramids [25]. While acknowledging their potential usefulness, it was decided to focus the study presented in this paper on 2D visualization, to avoid the impact of additional visual variables on the results. A recent approach to visualization of multivariate hierarchical data is presented by Zheng and Sadlo [26]. Their method is highly interesting and promising, but has not been included in this work since it, due to its complexity, is not comparable to the fundamental representations evaluated here.

2.2 *Previous Evaluation of Hierarchical Visualization*

The study presented in this paper compare the performance of Treemap, Icicle, Sunburst, Circular Treemap and Bubble Treemap. Whilst they have not been compared together in a single study previously, there have been evaluations of some combinations of these methods. This paper is, however, unaware of any studies to date comparing Bubble Treemap to any other hierarchical visualization methods.

Stasko et al. [4] compared Sunburst against Treemap within the context of file search tasks. The evaluation was measured on the correctness of results and completion time, asking users questions such as comparing the size of nodes, and finding the path to nodes. The study found that Sunburst provided a greater understanding of the hierarchical structure than Treemap. Another study by Muramalla et al. [5] looked at Icicle and Sunburst, measuring user performance through a similar set of questions. They found that users performed better with Icicle, but preferred the layout of Sunburst.

More recently, Müller et al. [6] compared Icicle against Treemap, alongside Nodelink diagrams. Each method was shown at two different levels of hierarchical complexity (2 and 3 levels), and tasks consisted of counting and comparing nodes, with user performance measured in accuracy, time to completion and through eye tracking. The study found Icicle and Nodelink performed comparably while Treemap performed worse, but found through eye tracking that Treemaps draw visual attention better despite its performance. Long et al. [7] compared Treemaps with Circular Treemaps, alongside Cushion Treemaps and 3D Treemaps as a comparison of the effectiveness of different tree structures, finding Circular Treemaps to perform better than normal Treemaps, with 3D Treemaps performing best of all and Cushion Treemaps performing the worst. Woodburn et al. [8] compared Treemaps, Icicle and a their own method, a semicircular variant of Sunburst called a sundown chart. Sunburst was also compared but only within the pilot study. Their study evaluated performance for basic navigation tasks, node attribute (size) comparison, and understanding of hierarchy structure. Results found slight user preference in Icicle to their own method, and generally better performance for Icicle and sundown chart than for Treemap.

These previous evaluations of hierarchical visualization methods have mainly focused on tasks directly related to the hierarchical structure, while the evaluation presented here focus on the presentation of underlying data represented by node attributes (size and colour). The results from Woodburn et al. [8] indicate that Treemap and Icicle may perform better than Sundown chart for size comparison tasks, but their results were not statistically significant. Furthermore, the study presented in this paper compare the performance of the methods for different sized hierarchies. While Müller et al. [6] compared hierarchies of different complexity, the most complex still included only three levels, which is considerably less than in many hierarchical datasets.

3 Evaluation of Hierarchical Visualization Methods

The usability of hierarchical visualization methods and a users ability to extract meaningful information from them may often depend on the size of the hierarchy and on patterns in the underlying data. The study presented here aims to evaluate a set of comparable hierarchical visualization methods to determine their performance when it comes to identification of features of the underlying data, and to determine which may be most suitable for analysis of hierarchical data of different size. With the overarching goal of providing recommendation as to which hierarchical visualization methods are best suited for visual analysis of different scales of hierarchical datasets. The evaluation focused on five methods for comparison: Treemap, Icicle, Sunburst Chart (Sunburst), Circular Treemap (Circle) and Bubble Treemap (Bubble), in an online usability evaluation. The methods were considered comparable in terms of being implicit space filling methods that are able to handle relatively large hierarchies, and that are able to represent features of the data through node size and colour.

3.1 Data

In the study, the underlying data features were defined as a measure of abundance, which was represented by size, and a measure of significance (or interestingness), which was represented by colour. A light blue to dark purple sequential colour scheme was used, with darker colour representing a higher value of interest and with lighter values progressively showing lower values (Fig. 6). The Colour scheme used was the '3-class BuPu' scheme from Colorbrewer2 [27], which was tested for colour vision impairment using Color Oracle [28].

Each method was portrayed using two different hierarchical sizes, referred to as large and small. Large presents the full hierarchical structure the way each method traditionally does, while the small format focus on a single subsection of the hierarchical structure, only showing a parent node and its corresponding child nodes. 64 Synthetic datasets were generated for the experimental questions. With datasets for large hierarchies consisting of six hierarchical levels with 2–5 child nodes per

Fig. 6 Colour scale used to show how interesting a node is, with lighter nodes being comparably less interesting than darker nodes

level. While datasets generated for the small hierarchies consisted of two hierarchical levels, containing 10–20 child nodes.

3.2 Tasks

Multiple choice questions were used and participant were presented with four images of a visualization method with different datasets, and asked to compare them based on the question. The following questions were asked:

For the following views...

1. Which contains the leaf node with the highest abundance?
2. Which contains the node with the highest significance?
3. On the level after the root node, which contains the node with the highest abundance?
4. Which is the overall most significant?

While interaction may have a considerable impact on the usability of a visualization method, the focus of this study was on evaluating the most appropriate visual representation for different sized hierarchical structures. Hence, static images were used to represent the visualization methods to avoid the influence of interactive variations on performance.

3.3 Experimental Design and Procedure

The study was based on two hypotheses: **H1**: For large hierarchies, Sunburst would perform better due to it being space efficient and providing an easily overviewable and uncluttered representation of the hierarchy and data at the same time. **H2**: Treemap and Circle will perform best for small hierarchies, as the clutter issues they face in larger hierarchies will be mitigated, and their space efficiency can provide clear detail for a larger number of child nodes.

The experiment was designed as a within-subject study with the visualization method and display format (large/small) as the main factors. The study was conducted online, in order to standardise the range of screen sizes used, small devices, such as phones and tablets, were blocked from being used. JavaScript/D3 was used to implement the visualization methods, while the Gorilla experiment builder (www.gorilla.sc) was used to build and host the experiment [29].

Participants were provided with an initial background section covering the basic concepts of hierarchical visualization, the different hierarchical size formats that would be used and an explanation of the abundance and significance/interestingness data features. Then a training section for each method was presented, questions where in the same format as the main experiment questions, however users where given the results of their training answers and were allowed to repeat each training section if they wished. Following this was the experimental phase, where each method had 16 questions, with questions divided evenly between large and small hierarchies, and each of those between abundance and significance tasks. A Latin Square procedure [30] was used to counterbalance the order users received each methods, to ensure participants received a balanced variation in ordering of methods, and to reduce the potential learning impact on the results. For each question, the accuracy of the response and the response time was recorded. Additionally, after each question, participants were asked how confident they were with their answer, using a 5 point likert scale (1 = not confident, 5 = very confident).

4 Evaluation Results

In total, there were 17 participants who finished the study, 8 female and 9 male. The most abundant age group was 25–34 years (47.06%), followed by 35–44 (23.53%), 18–24 (17.65%) and 45–54 (11.76%). 41.18% of participants specialised in bio-science, 29.41% specialised in Computer Science and 29.41% were in other disciplines. Participants were asked to rank their experience of (1) Data Visualization and (2) Hierarchical data, using 5 point likert scales (1 = No prior experience, 5 = Expert). 41.18% of participants listed their experience with Data Visualization as high (4 or 5) while 35.3% listed it as low (1 or 2). For hierarchical visualization, 23.5% listed their experience as high, while 64.7% listed it as low.

Analysis of results was divided between large and small hierarchies, with additional analysis on size and colour results for each hierarchy. The data wasn't normally distributed, so Friedman test was used for significance testing. This was followed by post-hoc tests using Wilcoxon signed rank test, with a Bonferroni correction applied resulting in a significance level set at $p < 0.01$, and pairwise comparison to identify for which combinations of visualization methods there were significant difference in the accuracy, the response time for all answers and the response time for only correct answers. The separation between response time for all answers and response time for only correct answers was made to distinguish if quick but erroneous answers may impact response times. The results of perceived confidence is reported to review if performance results agree with participant confidence. Self reported confidence is, however, an unreliable measure of performance.

4.1 Large Hierarchies

Results for large hierarchies are reported in Figs. 7, 8 and 9, where medians, interquartile distances, minimum and maximum values are represented as box plots, and where red dots and white lines show the mean and standard deviation respectively. Significant differences were confirmed for overall accuracy ($\chi^2(16) = 36.078$, $p < 0.001$), accuracy for size ($\chi^2(16) = 42.559$, $p < 0.001$) and accuracy for colour ($\chi^2(16) = 22.676$, $p < 0.001$); as well as for overall response time of correct

Fig. 7 Accuracy: large hierarchies

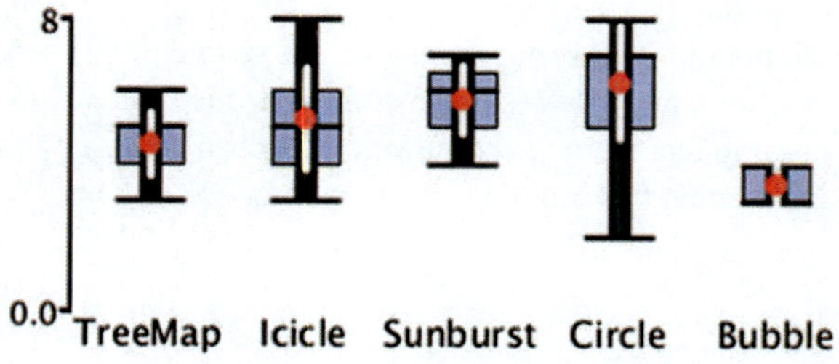

(a) Overall results.

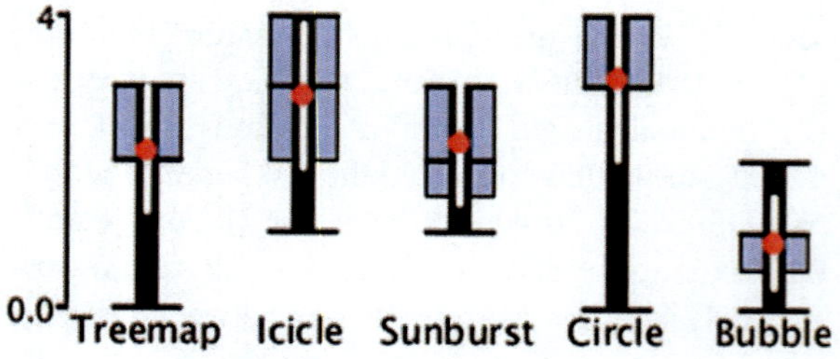

(b) Results for size related tasks.

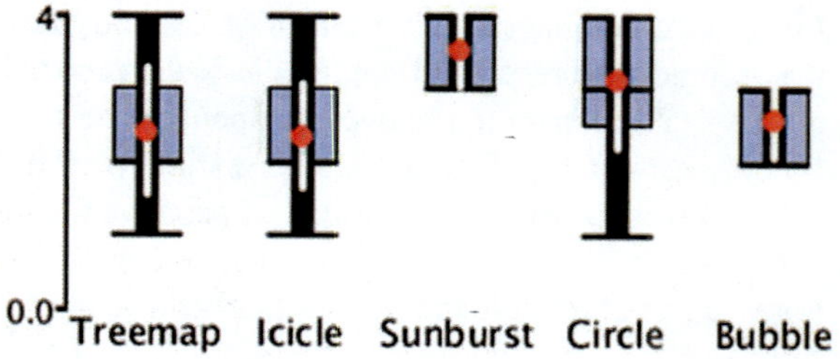

(c) Results for colour related tasks.

Fig. 8 Response time (ms): correct answers, large hierarchies

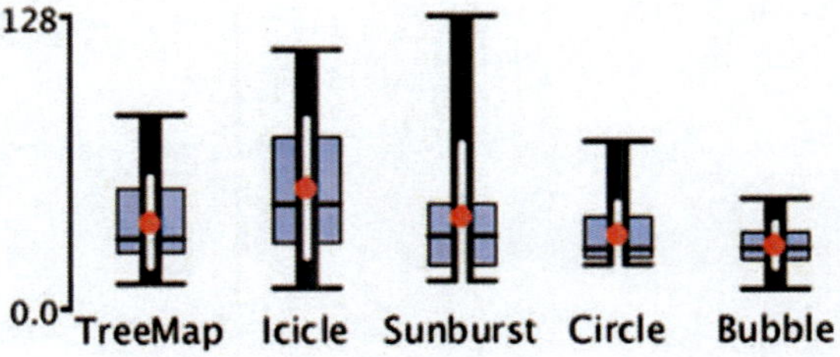

(a) Overall results.

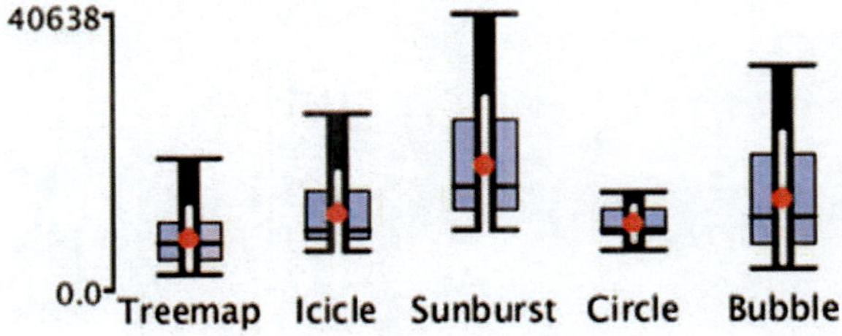

(b) Results for size related tasks.

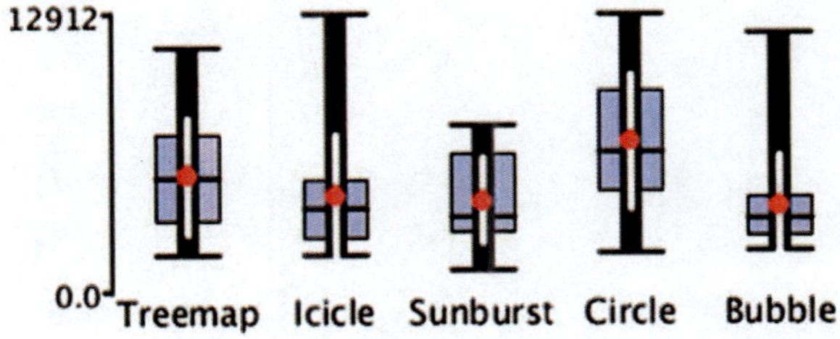

(c) Results for colour related tasks.

answers ($\chi^2(16) = 21.694$, $p < 0.001$) and response time of correct answers for size and colour separately ($\chi^2(16) = 18.462$, $p = 0.001$ and $\chi^2(16) = 26.071$, $p < 0.001$), though not for all response times($\chi^2(16) = 4.235$, $p = 375$). Post-hoc analysis, with results reported in Tables 1 and 2, revealed statistically significant differences for accuracy for some visualization methods. More specifically, the lower overall accuracy of Bubble was significant in relation to all other visualization methods. This was expected as the accuracy value for Bubble was 3, which was the lowest median accuracy value. Statistically significant differences for overall accuracy were also shown for Sunburst versus Treemap and Circle versus Treemap, where Treemap performed worse than both Sunburst and Circle. Whilst Circle performed

Fig. 9 Response time (ms): all answers, large hierarchies

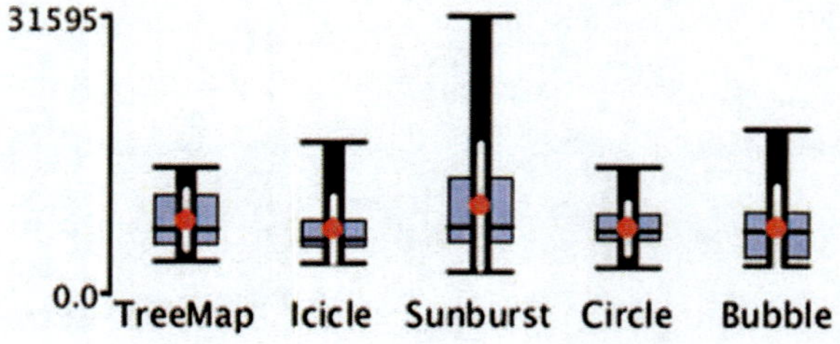

(a) Overall results.

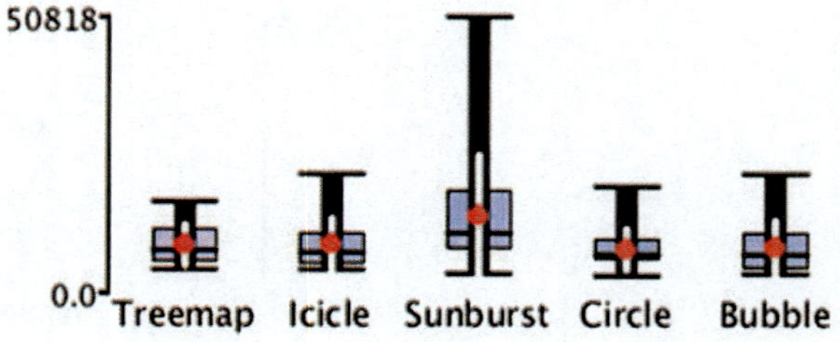

(b) Results for size related tasks.

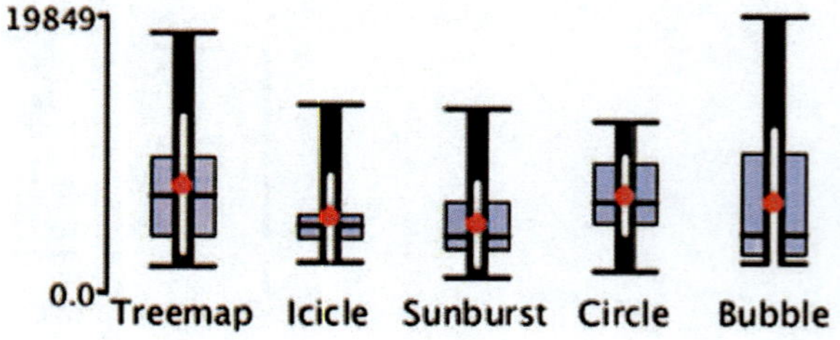

(c) Results for colour related tasks.

slightly better than Sunburst with median values of 7 and 6 respectively, there was no significant difference between them. The differences between Circle and Icicle, and Sunburst and Icicle were also not significant. When separating size and colour related tasks, accuracy results for size where at large the same, but with slightly worse performance for Sunburst, with Circle performing significantly better than Sunburst, while there were no significant difference with Treemap. For colour related tasks, on the other hand, the accuracy of Sunburst was higher with significantly better accuracy than Treemap, Icicle and Bubble. This in part supports **H1**, although Icicle and Circle appear to perform roughly as well as Sunburst, and better for size related questions. The post-hoc tests of the overall response time for correct answers

Table 1 Accuracy: large hierarchies

Methods	Overall		Size		Colour	
	Z	P	Z	P	Z	P
Icicle-Tree	–1.656	0.098	–2.345	0.019	–0.228	0.82
Sun-Tree	–3.384	**0.001**	–0.577	0.564	–3.272	**0.001**
Circle-Tree	–3.085	**0.002**	–2.951	**0.003**	–1.753	0.08
Bubble-Tree	–3.166	**0.002**	–3.247	**0.001**	–0.584	0.559
Sun-Icicle	–1.221	0.222	–2.145	0.032	–3.132	**0.002**
Circle-Icicle	–2.461	0.015	–0.794	0.427	–2.365	0.018
Bubble-Icicle	–3.086	**0.002**	–3.663	**<0.001**	–1.155	0.248
Circle-Sun	–1.383	0.167	–2.83	**0.005**	–1.645	0.1
Bubble-Sun	–3.689	**<0.001**	–3.493	**<0.001**	–3.358	**0.001**
Bubble-Circle	–3.484	**<0.001**	–3.472	**0.001**	–1.727	0.084

Table 2 Response time: correct answers, large hierarchies

Methods	Overall		Size		Colour	
	Z	P	Z	P	Z	P
Icicle-Tree	–2.438	0.150	–1.448	0.148	–1.444	0.149
Sun-Tree	–0.118	0.906	–3.051	**0.002**	–2.249	0.025
Circle-Tree	-1.160	0.246	–1.293	0.196	–2.249	0.025
Bubble-Tree	–2.296	0.022	–2.201	0.028	–2.201	0.028
Sun-Icicle	–2.154	0.031	–3.148	**0.002**	–0.26	0.795
Circle-Icicle	–2.959	**0.003**	–1.034	0.301	–3.006	**0.003**
Bubble-Icicle	–3.432	**0.001**	–0.384	0.701	–1.16	0.246
Circle-Sun	–1.018	0.309	–2.741	**0.006**	–3.574	**<0.001**
Bubble-Sun	–1.775	0.076	–1.293	0.196	–0.639	0.523
Bubble-Circle	–1.823	0.068	–1.083	0.279	–3.432	**<0.001**

revealed significance between Circle versus Icicle and Bubble versus Icicle, with worse performance from Icicle, but no other significant differences. For size related questions, Sunburst performed significantly worse than Treemap, Icicle and Circle. While Circle performed significantly worse than Icicle; Sunburst and Bubble for colour related questions. The confidence reported by participants was fairly similar for all visualization methods, with slightly higher confidence reported for Icicle.

4.2 Small Hierarchies

Results for small hierarchies are reported in Figs. 10, 11 and 12. Significant differences were confirmed for overall accuracy ($\chi^2(16) = 40.700$, $p < 0.001$) and accuracy for size questions ($\chi^2(16) = 48.121$, $p < 0.001$); for overall response time ($\chi^2(16) = 14.824$, $p = 0.005$), response time for size ($\chi^2(16) = 21.271$, $p < 0.001$) and for colour questions ($\chi^2(16) = 17.318$, $p = 0.002$); as well as for overall response time of correct answers ($\chi^2(16) = 36.329$, $p < 0.001$) and response time for correct answers for size and colour questions ($\chi^2(16) = 30.187$, $p < 0.001$ and

Fig. 10 Accuracy: small hierarchies

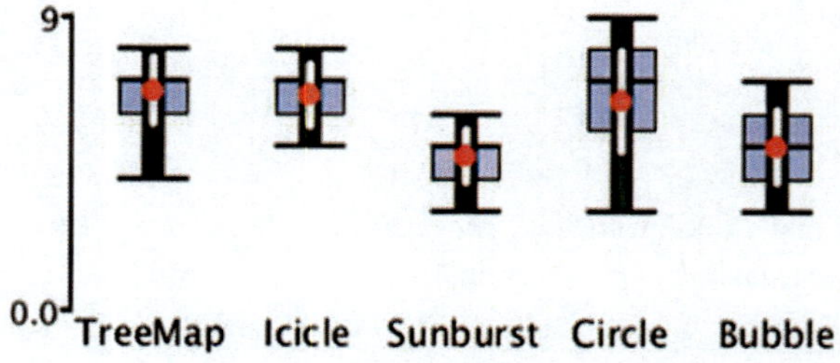

(a) Overall results.

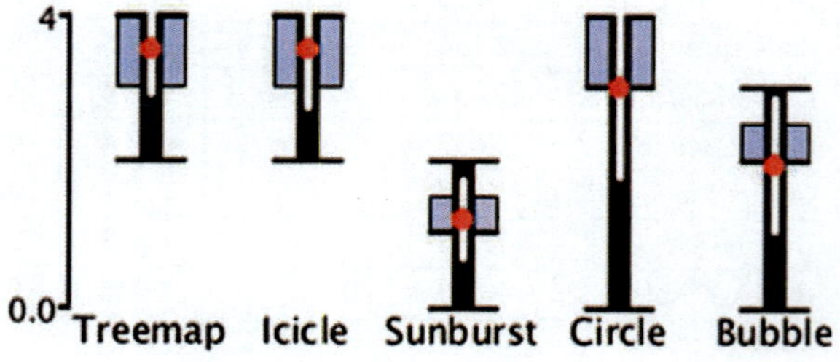

(b) Results for size related tasks.

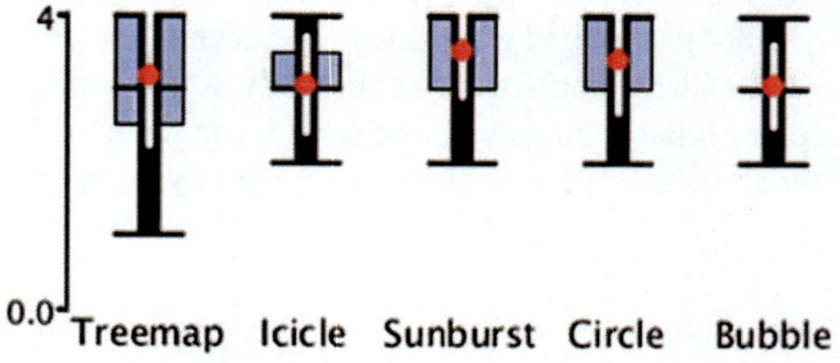

(c) Results for colour related tasks.

Fig. 11 Response time (ms): correct answers, small hierarchies

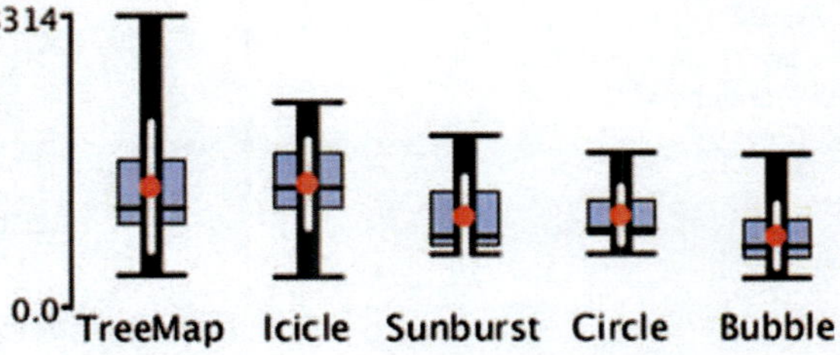

(a) Overall results.

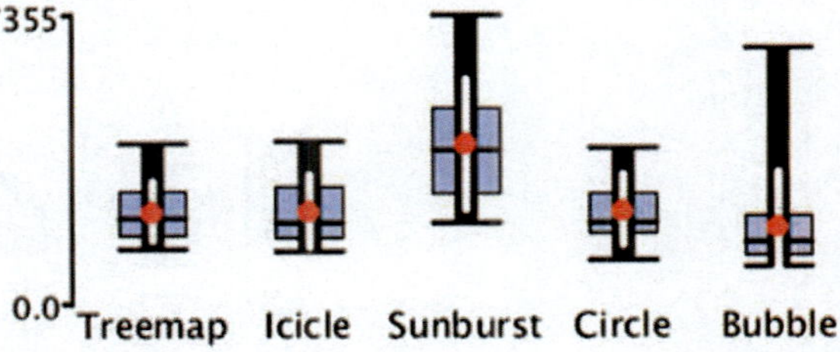

(b) Results for size related tasks.

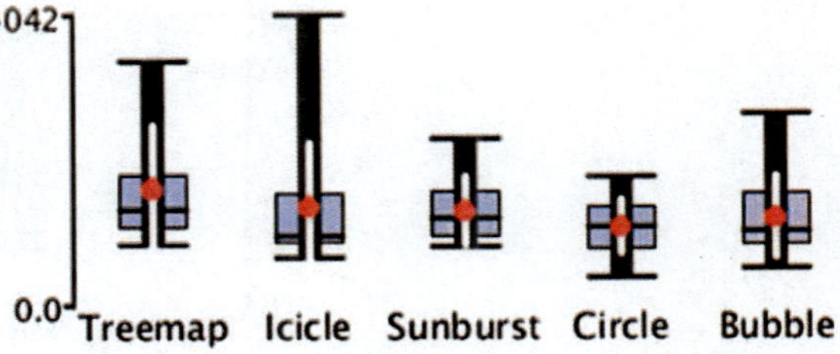

(c) Results for colour related tasks.

$\chi^2(16) = 14.588$, $p = 0.006$). Post-hoc analysis, with results reported in Tables 3, 4 and 5, revealed that for overall accuracy the Treemap, Icicle and Circle methods performed significantly better than Sunburst and Bubble, which mainly confirms **H2**. No other significant results were detected for overall accuracy. The results were comparable between overall accuracy and size related questions, but with significantly better results also for Bubble compared to Sunburst. Post-hoc analysis on the overall correct response time found significant differences for all comparisons except Icicle versus Treemap and Sunburst versus Circle. With shortest time for Bubble and longest time for Treemap and Icicle. For size related questions Sunburst performed significantly worse than all other methods, while it for colour related questions was only significantly slower than Circle. Treemap performed significantly worse than Circle and Bubble for colour questions. Conversely, post-hoc analysis for overall response time

Fig. 12 Response time (ms): all answers, small hierarchies

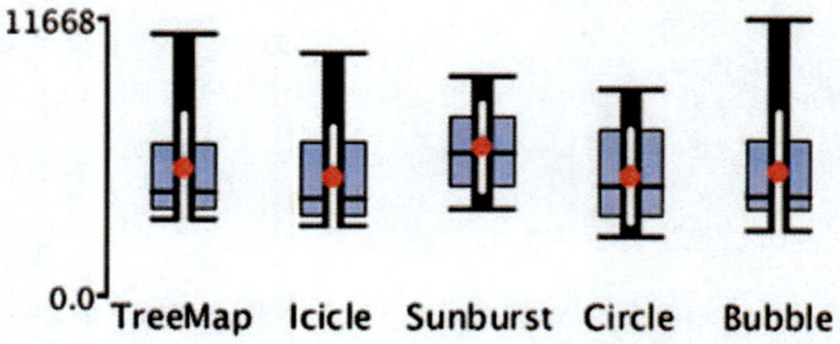

(a) Overall results.

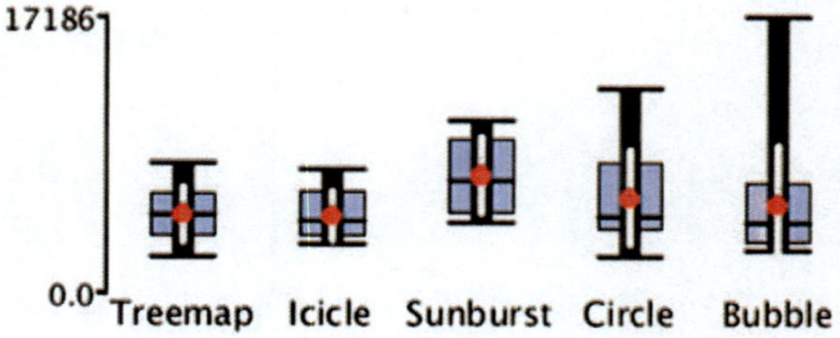

(b) Results for size related tasks.

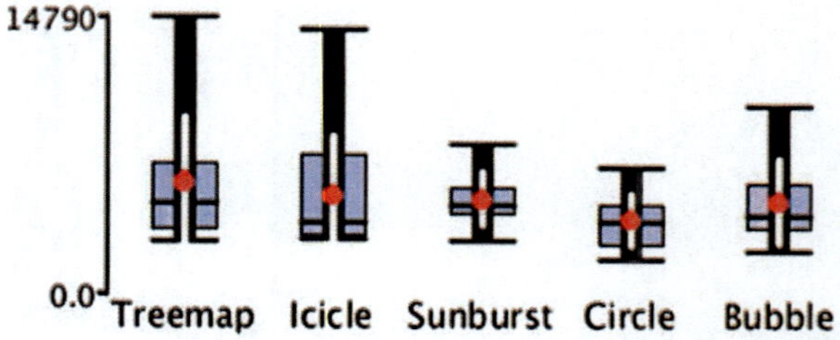

(c) Results for colour related tasks.

for all answers showed Sunburst as requiring significantly longer to use compared to Treemap, Icicle and Circle, with no other significant results detected. For size related questions, Sunburst performed significantly worse than Treemap and Icicle, while Circle performed significantly better than Treemap and Sunburst for colour related questions. Participants were confident using most methods but reported a slightly stronger confidence using Bubble.

Table 3 Accuracy: small hierarchies

Methods	Overall		Size	
	Z	P	Z	P
Icicle-Tree	–0.461	0.623	–0.061	0.951
Sun-Tree	–3.564	**<0.001**	–3.619	**<0.001**
Circle-Tree	–1.008	0.313	–1.511	0.131
Bubble-Tree	–3.568	**<0.001**	–3.588	**<0.001**
Sun-Icicle	–3.561	**<0.001**	–3.695	**<0.001**
Circle-Icicle	–0.690	0.490	–1.998	0.046
Bubble-Icicle	–3.248	**0.001**	–3.404	**0.001**
Circle-Sun	–3.170	**0.002**	–3.36	**0.001**
Bubble-Sun	–1.406	0.160	–2.804	**0.005**
Bubble-Circle	–2.729	**0.006**	–2.612	**0.009**

Table 4 Response time: correct answers, small hierarchies

Methods	Overall		Size		Colour	
	Z	P	Z	P	Z	P
Icicle-Tree	–0.450	0.653	–0.686	0.492	–2.249	0.025
Sun-Tree	–2.722	**0.006**	–3.413	**0.001**	–1.538	0.124
Circle-Tree	–2.580	**0.010**	–0.052	0.959	–3.195	**0.001**
Bubble-Tree	–3.006	**0.003**	–2.215	0.027	–2.959	**0.003**
Sun-Icicle	–2.817	**0.005**	–3.464	**0.001**	–0.45	0.653
Circle-Icicle	–2.959	**0.003**	–0.207	0.836	–1.254	0.21
Bubble-Icicle	–3.527	**<0.001**	–1.817	0.069	–0.781	0.435
Circle-Sun	–0.639	0.523	–3.516	**<0.001**	–2.675	**0.007**
Bubble-Sun	–2.722	**0.006**	–2.897	**0.004**	–1.065	0.287
Bubble-Circle	–2.675	**0.007**	–2.045	0.041	–0.828	0.407

5 Results and Recommendation

Based on these initial results and our observations from the literature, this paper provides the following recommendations: For large hierarchies, Sunburst and Circle performed best in the study when it came to accuracy results. With Treemap performing worse than Sunburst and Circle, which is consistent with performance results in other evaluations of Treemap for showing large hierarchical structures [4, 6–8]. While Bubble performed the worst out of all the large hierarchies. While differences between Icicle and Sunburst, and Icicle and Circle were not significant, Icicle performed worse than both in terms of the median accuracy, which was equal to the median accuracy of Treemap. This is contrary to the other evaluations where Icicle typically performed better than Sunburst [5, 8], though in the case of [8],

Table 5 Response time: all answers, small hierarchies

Methods	Overall		Size		Colour	
	Z	P	Z	P	Z	P
Icicle-Tree	–1.160	0.246	–0.639	0.523	–1.728	0.084
Sun-Tree	–1.917	**0.005**	–3.006	**0.003**	–0.923	0.356
Circle-Tree	–0.781	0.435	–1.065	0.287	–3.243	**0.001**
Bubble-Tree	–1.302	0.193	–0.26	0.795	–1.965	0.049
Sun-Icicle	–3.290	**0.001**	–3.621	**<0.001**	–0.355	0.723
Circle-Icicle	–0.166	0.868	–1.728	0.084	–2.391	0.017
Bubble-Icicle	–0.118	0.906	–0.544	0.586	–0.402	0.687
Circle-Sun	–2.817	**0.005**	–2.059	0.039	–3.574	**<0.001**
Bubble-Sun	–2.391	0.17	–2.391	0.017	–0.781	0.435
Bubble-Circle	–0.260	0.765	–0.828	0.407	–2.059	0.039

the two were only compared in a pilot study. One possible explanation to this may be that their studies mainly focused on tasks related to understanding of the hierarchical structures, rather than representation of data values which was the focus of this study. For correct response time, Icicle performed slightly better overall, with significant differences between Circle, Bubble and itself. Accuracy is however considered a more important performance measure than response time. When accuracy is divided into questions regarding size and colour, the only overlapping significant results between the two are between Sunburst and Bubble. When considering only size related tasks, Icicle and Circle perform the best with Circle also being significantly better than Sunburst, but when looking at colour tasks, Sunburst performs best. This also suggests that colour may be the best visual feature for representing important data features in Sunburst, while size may be a better choice for Icicle and Circular Treemap.

For the results of small hierarchies, Treemap, Icicle and Circle performed best in terms of accuracy. These results were significant in relation to Sunburst and Bubble. When dividing the results between size and colour tasks, the results held for size related tasks, while no significant accuracy results were found for colour related tasks, though it is worth noting that the performance of Sunburst was equal to that of Circle. For response time of accurate answers, Circle performed better than Icicle, Sunburst had significant differences between itself and both Icicle and Treemap, while Bubble performed significantly better than all the other methods. When comparing response time results for size and colour tasks separately, the most notable result again appear to be that Sunburst perform better for colour tasks than for size tasks, while Treemap and Icicle perform better for size related tasks. Overall, these results only in part agree with previous studies [4, 5] that focused on file search tasks, but not with

Müller et al. [6] who found Icicle to perform better than Treemap. The response times only for correct answers may, in the study presented here, however have been affected by the relatively low number of correct answers for Sunburst and Bubble in comparison to the other methods, hence results being based on a very small number of answers. This especially seems the case for Sunburst, as with all the response times, there was significant differences between Sunburst and Icicle (Icicle better) and Circle and Sunburst (Circle better).

Comparing the visual structure of Treemap to that of Icicle and Circle, the layout of Treemap is more space efficient. Circle uses additional space to represent higher levels in the hierarchy, and Icicle use a relatively large space to represent higher levels, while Treemap make use of all space available for the most detailed level displayed, suggesting that it may be the most appropriate visualization when leaf node information is of high importance or for hierarchies with a large number of leaf nodes. Icicle and Circle may however be considered to more clearly represent the hierarchical structure than the Treemap, which however may be less important in smaller hierarchical structures. Interestingly, there were no significant differences for accuracy for colour questions for any of the small hierarchy methods, with all visualization methods rendering good results.

6 Combined Hierarchical Visualization for Exploration of High-Dimensional Data

This section will demonstrate how a combination of visualization methods representing large and small hierarchical structures can be coordinated to support the exploration and analysis of high-dimensional hierarchical data. Specifically, this use-case visualize data from a study of the ocean microbiome, using data from the Tara Oceans microbial reference catalogue [14], which includes data from 139 samples taken from different oceanic regions, with a total of 35650 detected species (here treated as data dimensions). Based on the hierarchical taxonomy of the biological classification system, this data set has a hierarchical structure consisting of 8 levels (*Domain, Kingdom, Phylum, Class, Order, Family, Genus, Species*). To reduce the visual impact of biologically less abundant species, species (dimensions) with low abundance count (below 15% of the maximum abundance value) has been removed in the examples provided here, resulting in a data set with 150 dimensions.

High-dimensionality is a major challenge when visualizing this type of data. The aim of the visualization approach presented here is to utilise the hierarchical structure to overcome this challenge and enable overview and exploration of such data. In the spirit of the Visual Information Seeking Mantra [31], this paper suggests to combine a visual overview of the full hierarchy and its main features, with the ability to highlight and zoom into areas of potential interest through coordination with a more detailed hierarchical view.

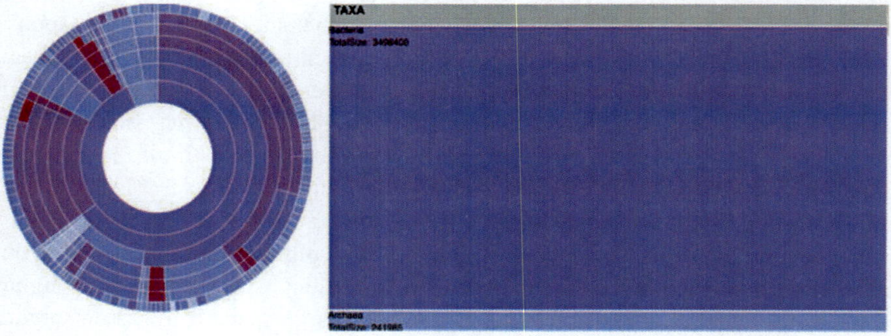

Fig. 13 Combination of Sunburst and Treemap showing the Tara Oceans data at the root of the structure

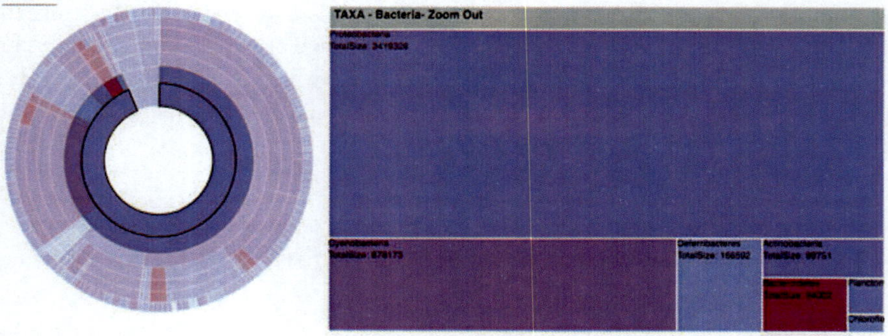

Fig. 14 Selection of the node Bacteria. In the Sunburst,the node is highlighted and nodes not part of its path are faded out. In the Treemap, the path is displayed at the top of the visual representation

Based on the results of the usability study and literature review, an overview of the full hierarchical structure is displayed using a Sunburst Chart. The Sunburst overview is complemented by a Treemap that shows a subset of the hierarchy, consisting of a selected taxon of interest and its local hierarchical structure in terms of its child nodes. Figures 13, 14 and 15 provide examples of the combination, using colour to represent a measure of interestingness. In this example use-case, interestingness is represented by a measure of abundance difference between sample groups, as described in Johansson Fernstad et al. [32].

The combination of Sunburst and Treemap was chosen based on their performance in the usability study, and as they complement each other well. Sunburst performed well for large hierarchies, and was selected over Icicle plot and Circular Treemap as overview visualization. While both Icicle and Circular Treemap performed comparably well in the study and performed better for size related tasks, Sunburst performed better for colour related tasks. In this particular multiple views setup, Sunburst is used for identification of interesting data subsets, with interestingness represented by colour. Thus, the performance for colour tasks was considered more important.

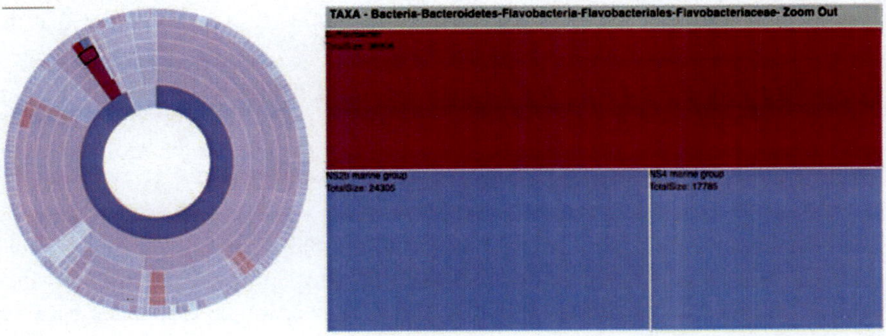

Fig. 15 Selection of the node Flavobacterlaceae

Furthermore, with the data set in question being so large, there are a large number of outer nodes competing for space. Telling the difference between two small and comparably sized nodes may be difficult without additional information. Meanwhile, colour is a highly salient visual feature and thus colour information may more clearly highlight an area of interest even if it is small.

Treemap, Icicle Plot and Circle Packing performed comparably well for small hierarchies in the evaluation. Of these, Treemap is the most space efficient, while Circle Packing and Icicle dedicate more space to showing the hierarchical structure. Displaying hierarchy is less relevant in the detailed view in this use-case, since only a small number of hierarchical levels are shown and with main focus being the individual nodes. The potential drawback of the hierarchical representation in Treemap not being as clear is also overcome by combining and linking it to Sunburst, which represents hierarchy clearly and allows the Treemap to be less cluttered.

In this combined hierarchical visualization, Sunburst provides an easily understandable overview of the hierarchical structure. It is also able to highlight and draw attention to patterns of potential interest in the hierarchy through colouring, as exemplified in the usability study. Through this it can provide guidance for exploration and allows for fast identification of taxon that may be particularly interesting to explore in more detail. A taxon of interest can then be selected in the Sunburst, resulting in its path in the hierarchical structure being highlighted. The Treemap is interactively linked to the Sunburst, and displays only the taxon of interest, which is selected in the Sunburst, and its child nodes. This provides a less cluttered view where the relationships and differences between child nodes in a particularly interesting subset of the hierarchy can be better understood and patterns easier compared, since more space is available for each individual child node in the subset.

The default overview of the hierarchical structure is shown in Fig. 13. Here, the selected node is the root of the hierarchical structure, thus the only child nodes shown in the Treemap are the two *domain* categories *Bacteria* and *Archaea*. By navigating through the Sunburst in Fig. 13 to the largest area of high interest (the largest dark purple node) at the *domain* level, the user is presented with the view shown in Fig. 14. The path in the hierarchy of the selected node is highlighted in the Sunburst. In the

Treemap, the user can see from the taxonomic path at the top of the plot that the taxon they selected are *Bacteria*. The Treemap also displays clearly that in this data set *Bacteria* has seven children: *Proteobacteria, Cyanobacteria, Deferribacteres, Actinobacteria, Bacteroidetes, Planctomycetes and Chloroflexi*. The *Proteobacteria* node has the by far largest rectangle, showing that it is clearly the most abundant taxon at this level, its relatively dark colouring suggests there is also some level of interestingness in terms of abundance difference between sample groups, although less than for some of the other child nodes. While the *Bacteroidetes* node is considerably less abundant, it may be the most interesting to investigate further, since the dark purple colour indicate a relatively clear separation in abundance between samples for the entities of this taxon.

Exploring this further, by selecting the *Bacteroidetes* node and navigating further down the hierarchical structure until reaching the *Flavobacterlaceae* node at the *Family* level in the Sunburst, its *Genus* level child nodes become visible in the Treemap (Fig. 15). While all the nodes at this level show a degree of abundance difference, the node *Sufflavibacter* at the top shows the strongest difference in abundance levels between sample groups, suggesting that this may be a genus of particular interest for the microbiologist, as it may be involved in driving abundance differences between the sample groups currently investigated. Similar interactive exploration of the hierarchical structure and subsets of interest can then continue to quickly identify other taxon of interest for further investigation.

7 Summary

This paper presented the first study that compares the five hierarchical visualization methods: Treemap, Icicle Plot, Sunburst Chart, Circular Treemap and Bubble Treemap, in their ability to represent underlying data features for large and small hierarchies. Previous evaluations have compared subsets of these methods and have mainly focused on tasks directly related to the hierarchical structure, while the study presented here investigated how well the methods represent underlying data through visual features.

The results presented in this paper determine that for large hierarchies, Sunburst Chart and Circular Treemap had the overall best results, with Sunburst Chart performing better for colour comparison tasks, while Icicle Plot and Circular Treemap performed better for size comparison tasks. While for small hierarchies, the results suggest that Treemap, Icicle plot and Circular Treemap are the best alternatives. The results put in context of previous research clearly confirms that tasks and data structure has to be taken into consideration when selecting appropriate hierarchical visualization method.

The potential of combining hierarchical visualization methods as a means to support exploration of interesting data features in high-dimensional hierarchical data, was demonstrated through a multiple coordinated views approach using Sunburst Chart and Treemap. Future work includes investigation of how representation of additional data features in hierarchical visualization can support exploration of such high-dimensional data, as well as more extensive user testing to further establish its potential.

References

1. Macquisten, A., Smith, A.M., Johansson Fernstad, S.: Evaluation of hierarchical visualization for large and small hierarchies. In: 24th International Conference Information Visualisation (IV), pp. 166–173 (2020). https://doi.org/10.1109/IV51561.2020.00036
2. Jolliffe, I.: Principal component analysis. In: Encyclopedia of Statistics in Behavioral Science (2005)
3. Cox, M.A., Cox, T.F.: Multidimensional scaling. In: Handbook of Data Visualization, pp. 315–347. Springer, Berlin, Heidelberg (2008)
4. Stasko, J., Catrambone, R., Guzdial, M., McDonald, K.: An evaluation of space-filling information visualizations for depicting hierarchical structures. Int. J. Human Comput. Stud. **53**, 663–694 (2000)
5. Muramalla, S., Altarawneh, R., Humayoun, S.R., Moses, R., Panis, S., Ebert, A.: Radial vs. Rectangular: evaluating visualization layout impact on user task performance of hierarchical data. IADIS Int. J. Comput. Sci. Inf. Syst. **12**, 17–31 (2017)
6. Müller, N.H. Liebold, B., Pietschmann, D., Ohler, P., Rosenthal, P.: Hierarchy visualization designs and their impact on perception and problem solving strategies. In: Proceedings Of The International Conference On Advances In Computer-Human Interactions, pp. 93–101 (2007)
7. Long, L.K., Hui, L.C., Fook, G.Y., Zainon, W.M.N.W.: A study on the effectiveness of tree-maps as tree visualization techniques. Procedia Comput. Sci. **124**, 108–115 (2017). https://doi.org/10.1016/j.procs.2017.12.136
8. Woodburn, L., Yang, Y., Marriott, K.: Interactive visualisation of hierarchical quantitative data: an evaluation. In: 2019 IEEE Visualization Conference, VIS 2019, pp. 96–100 (2019). https://doi.org/10.1109/VISUAL.2019.8933545
9. Shneiderman, B.: Tree visualization with tree-maps: a 2-d space-filling approach. ACM Trans. Gr. 92–99 (1992)
10. Kruskal, J.B., Landwehr, J.M.: Statistical computing: Icicle plots: better displays for hierarchical clustering. Am. Stat. **37**, 162–168 (1983). https://doi.org/10.1080/00031305.1983.10482733
11. Stasko, J., Zhang, E.: EFocus+ context display and navigation techniques for enhancing radial, space-filling hierarchy visualizations. In: IEEE Symposium on Information Visualization 2000. INFOVIS 2000. Proceedings, pp. 57–65 (2000)
12. Kai, W.: pebbles-using Circular Treemaps to visualize disk usage. http://lip.sourceforge.net/ctreemap.html
13. Görtler, J., Schulz, C., Weiskopf, D., Deussen, O.: Bubble treemaps for uncertainty visualization. IEEE Trans. Vis. Comput. Gr. **24**, 719–728 (2018)
14. Sunagawa, S., Coelho, L.P., Chaffron, S., Kultima, J.R., Labadie, K., Salazar, G., Djahanschiri, B., Zeller, G., Mende, D.R., Alberti, A., Cornejo-Castillo, F.M., Costea, P.I., Cruaud, C., d'Ovidio, F., Engelen, S., Ferrera, I., Gasol, J.M., Guidi, L., Hildebrand, F., Kokoszka, F., Lepoivre, C., Lima-Mendez, G., Poulain, J., Poulos, B.T., Royo-Llonch, M., Sarmento, H., Vieira-Silva, S., Dimier, C., Picheral, M., Searson, S., Kandels-Lewis, S., Tara Oceans Coordinators, Bowler, C., de Vargas, C., Gorsky, G., Grimsley, N., Hingamp, P., Iudicone, D., Jaillon, O., Not, F., Ogata, H., Pesant, S., Speich, S., Stemmann, L., Sullivan, M.B., Weissenbach, J., Wincker, P., Karsenti, E., Raes, J., Acinas, S.G., Bork, P., Ocean plankton: Structure and function of the global ocean microbiome. Science. **348**(6237), 1261359 (2015). https://doi.org/10.1126/science.1261359. PMID: 25999513

15. Schulz, H.J., Hadlak, S., Schumann, H.: The design space of implicit hierarchy visualization: a survey. IEEE Trans. Vis. Comput. Gr. **17**, 393–411 (2011). https://doi.org/10.1109/TVCG.2010.79
16. Schulz, H.J., Schumann, H.: Visualizing graphs-a generalized view. In: Proceedings of The International Conference on Information Visualisation, pp. 166–173 (2006). https://doi.org/10.1109/IV.2006.130
17. Dinkla, K., Westenberg, M.A., Timmerman, H.M., van Hijum, S.A.F.T., van Wijk, J.J.: Comparison of multiple weighted hierarchies: visual analytics for microbe community profiling. Comput. Gr. Forum **30**, 1141–1150 (2011). https://doi.org/10.1111/j.1467-8659.2011.01963.x
18. Huson, D.H., Auch, A.F., Qi, J., Schuster, S.C.: MEGAN analysis of metagenomic data. Genome Res. **17**, 377–386 (2007). https://doi.org/10.1101/gr.5969107
19. Huson, D.H., Richter, D.C., Mitra, S., Auch, A.F., Schuster, S.C.: Methods for comparative metagenomics. In: BMC Bioinf. **10** (2009). https://doi.org/10.1186/1471-2105-10-S1-S12
20. Stewart, C.J., Embleton, N.D., Clements, E., Luna, P.N., Smith, D.P., Fofanova, T.Y., Nelson, A., Taylor, G., Orr, C.H., Petrosino, J.F., Berrington, J.E., Cummings, S.P.: Cesarean or vaginal birth does not impact the longitudinal development of the gut microbiome in a cohort of exclusively preterm infants. Front Microbiol. **8**, 1008 (2017). https://doi.org/10.3389/fmicb.2017.01008. MID: 28634475; PMCID: PMC5459931
21. Zhao, H., Lu, L.: Variational circular treemaps for interactive visualization of hierarchical data. In: 2015 IEEE Pacific Visualization Symposium (PacificVis), pp. 81–85 (2015)
22. Robertson, G.G., Mackinlay, J.D., Card, S.K.: Cone trees: animated 3D visualizations of hierarchical information. In: Proceedings of The SIGCHI Conference on Human Factors in Computing Systems, pp. 189–194 (1991)
23. Van Wijk, J.J., Van de Wetering, H.: Cushion treemaps: visualization of hierarchical information. In: Proceedings 1999 IEEE Symposium on Information Visualization (InfoVis' 99), pp. 73–79 (1999)
24. Kleiberg, E., van De Wetering, H., Van Wijk, J.J.: Botanical visualization of huge hierarchies. In: IEEE Symposium on Information Visualization. INFOVIS 2001, pp. 87–94 (2001)
25. Andrews, K.: Visual exploration of large hierarchies with information pyramids. In: Proceedings of the International Conference on Information Visualisation, pp. 793–798 (2002)
26. Zheng, B., Sadlo F.: On the visualization of hierarchical multivariate data. In: Proceedings of on 14th IEEE Pacific Visualization Symposium (PacificVis) (2021). https://doi.org/10.1109/PacificVis52677.2021.00026
27. Brewer, C.A.: Color use guidelines for data representation. In: Proceedings of the Section on Statistical Graphics, American Statistical Association, pp. 55–60 (1999). http://www.colorbrewer.org
28. Jenny, B., Kelso, N.V.: Designing maps for the colour-vision impaired. Bull. Soc. Cartogr. **41**, 9–12 (2007)
29. Anwyl-Irvine, A.L., Massonnié, J., Flitton, A., Kirkham, N., Evershed, J.K.: Gorilla in our midst: an online behavioral experiment builder. Behavior Res. Methods **52**, 388–407 (2019). https://doi.org/10.3758/s13428-019-01237-x
30. Graziano, A.M., Raulin, M.L.: Research Methods: A Process Of Inquiry, 2nd ed., pp. xviii, 446–xviii, 446 (1993)
31. Shneiderman, B.: The eyes have it: a task by data type taxonomy for information visualizations. Craft Inf. Vis. **37**, 364–371 (2003)
32. Johansson Fernstad, S., Macquisten, A., Berrington, J., Embleton, N., Stewart, C.: Quality metrics to guide visual analysis of high dimensional genomics data. In: EuroVis Workshop On Visual Analytics (EuroVA) (2020). https://doi.org/10.2312/eurova.20201083

Geometric Analysis Leads to Adversarial Teaching of Cybersecurity

Christian Servin, Olga Kosheleva, and Vladik Kreinovich

Abstract As time goes, our civilization becomes more and more dependent on computers and therefore, more and more vulnerable to cyberattacks. Because of this threat, it is very important to make sure that computer science students—tomorrow's computer professionals—are sufficiently skilled in cybersecurity. In this paper, we analyze the need for teaching cybersecurity from the geometric viewpoint. We show that the corresponding geometric analysis leads to adversarial teaching—an empirically effective but not-well-theoretically-understood approach, when the class is divided into sparring mini-teams that try their best to attack each other and defend from each other. Thus, our analysis provides a theoretical explanation for the empirical efficiency of this approach.

1 Need for Teaching Cybersecurity: The Practical Problem and Its Natural Geometric Formalization

Cybersecurity is important. As time goes, our world relies on computers more and more. Computers are needed for us to communicate, to pay and get paid, to travel—in short, for all aspects of our lives. The ubiquity of computers makes all these aspects vulnerable to cyberattacks. This, in turn, makes cybersecurity—the ability to defeat such attacks—extremely important. It is therefore very important to teach cybersecurity skills to computer science students—and to other future (and present) computer professionals.

C. Servin
Computer Science and Information Technology Systems Department, El Paso Community College (EPCC), 919 Hunter Dr., El Paso, TX 79915-1908, USA
e-mail: cservin1@epcc.edu

O. Kosheleva · V. Kreinovich (✉)
University of Texas at El Paso, 500 W. University, El Paso, TX 79968, USA
e-mail: vladik@utep.edu

O. Kosheleva
e-mail: olgak@utep.edu

B. Kovalerchuk et al. (eds.), *Integrating Artificial Intelligence and Visualization for Visual Knowledge Discovery*, Studies in Computational Intelligence 1014,
https://doi.org/10.1007/978-3-030-93119-3_24

A natural geometric approach to cybersecurity. During the cybersecurity training, students are exposed to several possible situations. Let us denote the number of such situations by n, and the situations themselves by $s_1, \ldots, s_n$. Let us also denote by S the set of all possible situations.

In real life:

- when the correspondingly trained person encounters a situation s which is similar to one of the situations s_i with which he/she is familiar,
- then, with high probability, the person will be able to adequately react to the new situation s as well.

The closer the situations s and s_i, the higher the probability that this person's reaction will be adequate. Vice versa, the larger the difference between the two situation, the higher the probability that the cyberattack will be successful.

It is usually possible to describe the degree of difference between the two situations s and s' by a number $d(s, s')$. This number can be determined by experts, or it can be based on the empirical data. In this paper, we will assume that these numbers are determined in such a way that the following natural properties are satisfied:

- $d(s, s) = 0$ for all s,
- $d(s, s') > 0$ for all $s \neq s'$,
- $d(s, s') = d(s', s)$ for all s and s', and
- the triangle inequality $d(s, s'') \le d(s, s') + d(s', s'')$ for all s, s', and s''.

In mathematical terms, this means that the corresponding function $d(s, s')$ is a *metric*.

The closer a new situation s to one of the previous situations s_i, i.e., the smaller the value

$$d(s, \{s_1, \ldots, s_n\}) \stackrel{\text{def}}{=} \min_i d(s, s_i), \tag{1}$$

the more efficient the defence against this situation. Vice versa, the larger the value (1), the less efficient the defense, and the larger the probability that the attack will succeed.

The attackers are usually familiar with what potential defenders know. So, to attack, they create a situation s for which the probability of success is the largest, i.e., equivalently, for which the value (1) is the largest. Thus, the probability that an attack will be successful is determined by the value

$$\max_{s \in S} d(s, \{s_1, \ldots, s_n\}). \tag{2}$$

The larger this value, the larger the probability that the attack will be successful.

In mathematical terms, the maximum (2) can be naturally described in terms of the *Hausdorff distance* between the two sets A and B. This distance is defined as the smallest value $\varepsilon > 0$ for which:

- for each element $a \in A$, there exists an element $b \in B$ for which $d(a, b) \le \varepsilon$, and
- for each element $b \in B$, there exists an element $a \in A$ for which $d(a, b) \le \varepsilon$.

In this terms, the value (2) is equal to the *Hausdorff distance*

$$d_H(S, \{s_1, \ldots, s_n\}) \tag{3}$$

between:

- the set S of all possible situations, and
- the set $\{s_1, \ldots, s_n\}$ on which the defenders have been trained.

The larger the number n of situations that the student study, the better the students will be prepared to defend against cyberattacks. However, in practice, the number n of different situations is limited by the need to describe all of them within the given training time. From this viewpoint, it is reasonable to assume that the value n is fixed. In this case, our objective is:

- *given n*,
- to *minimize* the probability of the attack's success, i.e., equivalently, to find the situations $s_1, \ldots, s_n$ for which the expression (2) and (3) is the smallest.

We will denote the optimal value of the expression (2) and (3) by $\varepsilon_n(S)$:

$$\varepsilon_n(S) \stackrel{\text{def}}{=} \min_{s_1,\ldots,s_n} d_H(S, \{s_1, \ldots, s_n\}). \tag{4}$$

In this paper, we analyze how this minimization problem can be solved.

An alternative formulation. Alternatively, we can consider a different version of this problem: instead of fixing the training time (and thus, fixing n):

- we can fix the desired probability that the defense will be successful—i.e., equivalently, the desired value $\varepsilon > 0$ of the quantity (2)—and
- we want to find out how to train defenders to reach this efficiency within the smallest possible time—i.e., via the smallest possible number of training situations n.

In this alternative formulation, we want to find the set of situations $\{s_1, \ldots, s_n\}$ for which every other situation $s \in S$ is ε-close to one of them, i.e., for which, for each $s \in S$, there exists an i for which $d(s, s_i) \le \varepsilon$.

In mathematical terms, such a set is known as an ε-*net*; see, e.g., [7, 8]. The smallest possible number of elements in an ε-net for a set S is usually denoted by $N_\varepsilon(S)$. The binary logarithm $H_\varepsilon(S) \stackrel{\text{def}}{=} \log_2(N_\varepsilon(S))$ of this number is called ε-*entropy*; see, e.g., [7, 8].

The two formulations are equivalent. From the mathematical viewpoint, both formulation represent, in effect, the same optimization problem. Indeed, one can check that $H_\varepsilon(S)$ is an inverse function to $\varepsilon_n(S)$, in the following precise sense:

$$H_\varepsilon(S) \le n \Leftrightarrow \varepsilon \ge \varepsilon_n(S).$$

We cannot feasibly solve the corresponding optimization problem. It is known that problem of finding the smallest ε-net is, in general, NP-hard; see, e.g., [1].

This means, in particular, that unless P = NP (which most computer scientists believe to be false; see, e.g., [3]), it is not possible to have a feasible algorithm that:

- given a set S with a metric $d(s, s')$,
- always returns the optimal ε-net for this set.

So, we cannot rely on any general feasible algorithm to find the optimal set of training situations.

The best we can do is to find a solution which is as *close to optimal* as feasibly possible.

Comment. This paper is a revised and extended version of our shorter conference paper [10]; specifically, we made the following two extensions:

- first, in this paper, we explain the geometric model in more detail that in the conference version;
- second, we also explain how to best implement the corresponding teaching strategy.

2 A Natural Algorithm for Solving Our Problem

A natural algorithm. In the alternative formulation, we do not know how many training situations we will need to achieve the desired proficiency, i.e., the given value ε. So, let us add such situations one by one, until we reach the desired proficiency.

We start with selecting the first situation s_1. If this the only situation we learn, then the quality of this learning can be described by the value $d_H(S, \{s_1\}) = \max_s d(s, s_1)$. So, we need to find the situation $s_1 \in S$ for which this value is the smallest possible. In mathematics, such a point s_1 is known as a *center* (or a *Chebyshev center*) of the set S.

If for this center, we already have $d_H(S, \{s_1\}) \le \varepsilon$, then we are done. But what if this inequality has not been reached yet? In this case, as we have mentioned, the attacker can reach the highest probability of the attack's success if he/she selects a situation s_2 for which the distance $d(s_1, s_2)$ is the largest:

$$d(s_1, s_2) = \max_s d(s_1, s).$$

A natural idea is therefore to train the students on this most-probable attack situation. This way, the most potentially damaging attack will be prevented.

Similarly:

- if we have already trained the students on situations $s_1, \ldots, s_n$, and we still have not yet reached the desired level of proficiency ε, i.e., if $\max_s d(s, \{s_1, \ldots, s_n\}) > \varepsilon$,
- then a natural idea is to next train the students on the situation s_{n+1} which is potentially the most damaging, i.e., for which

$$d(s_{n+1}, \{s_1, \ldots, s_n\}) = \max_s d(s, \{s_1, \ldots, s_n\}). \quad (5)$$

We continue this procedure until we reach the desired proficiency level, i.e., until we reach the desired inequality

$$\max_s d(s, \{s_1, \ldots, s_n\}) \le \varepsilon. \quad (6)$$

Comment. Note that we only select the next situation s_{n+1} if

$$\max_s d(s, \{s_1, \ldots, s_n\}) > \varepsilon.$$

In this case, the situation s_{n+1} is selected so as to maximize this value. So, we have $d(s_{n+1}, \{s_1, \ldots, s_n\}) > \varepsilon$, i.e., $d(s_{n+1}, s_i) > \varepsilon$ for all $i \le n$.

In general, this means that if we select the training situations s_i by using the above algorithm, then we will have $d(s_i, s_j) > \varepsilon$ for all $i \ne j$. (So, all training situations generated by our algorithm are reasonably different.)

We stop when $d_H(S, \{s_1, \ldots, s_n\}) \le \varepsilon$. Thus, we have

$$d(s_i, s_j) > d_H(S, \{s_1, \ldots, s_n\})$$

for all $i \ne j$.

3 Natural Algorithm: Visualization

To make it easier to understand, let us give two simple geometric illustrations of the above algorithm. These examples are similar to examples provided in [6] for a different application of a similar idea—to selecting benchmarks for testing different numerical algorithms.

1D example. Let us start with the simplest example of a metric space S—namely, the interval [0, 1]:

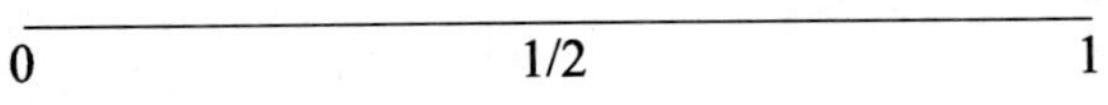

It is reasonable to select the midpoint 1/2 as s_1:

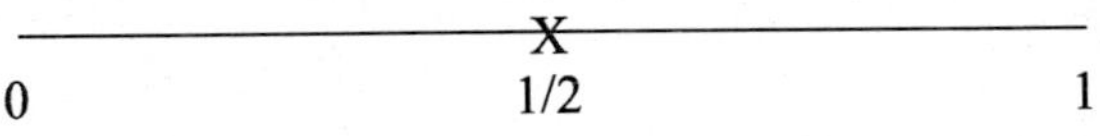

There are two points that are the farthest from s_1: the left endpoint 0 and the right endpoint 1. Without losing generality, let us select $s_2 = 0$:

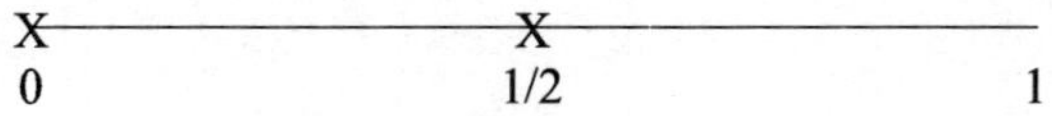

Now, $s_3 = 1$ is the point with the largest value of

$$d(s, \{s_1, s_2\}) = \min(d(s, s_1), d(s, s_2)) :$$

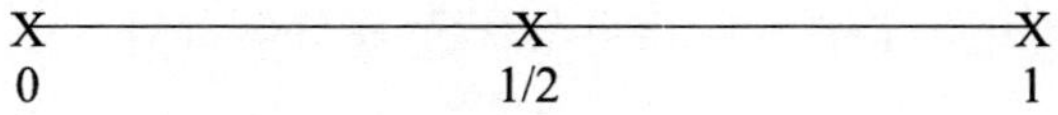

At this stage, the midpoints between 0 and 1/2 and between 1/2 and 1 are the farthest from the set $\{s_1, s_2, s_3\} = \{0, 1/2, 1\}$, so, after two stages, we add them both:

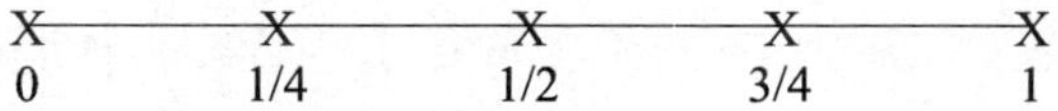

Now, the largest possible value of

$$d(s, \{s_1, s_2, s_3, s_4, s_5\}) = d(s, \{0, 1/4, 1/2, 3/4, 1\})$$

is 1/8. So, at the next stage, we add one of the points in between the existing ones, e.g., the first one (1/8):

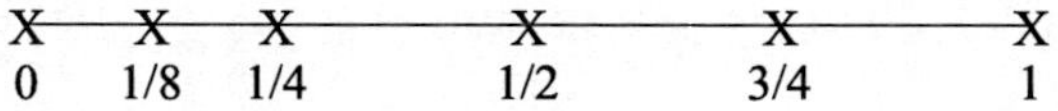

After three more stages, we add all midpoints, so we arrive at the following configuration:

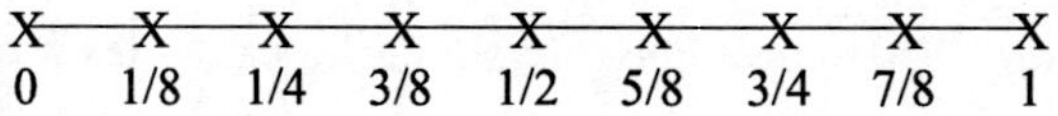

2D example: square. For a unit square, we get a similar situation. First, let us pick the midpoint as s_1:

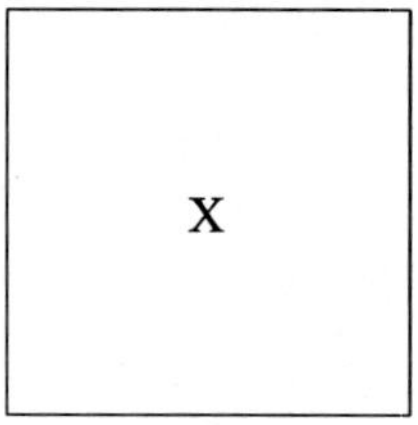

Then, the next four selections s_i are the vertices:

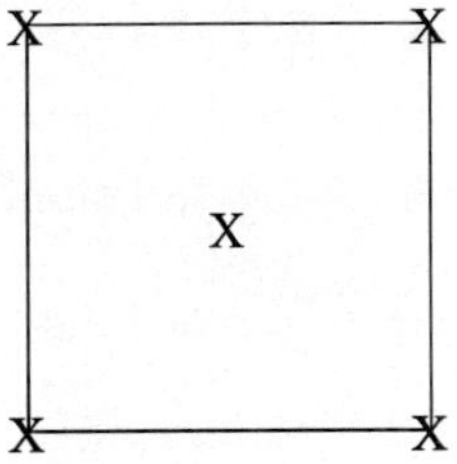

After this, the next four selected points s_i are he midpoints of the four edges:

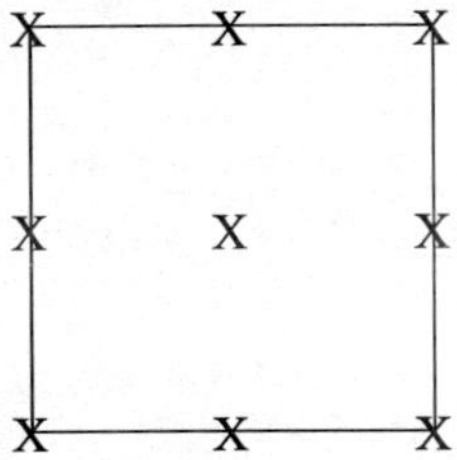

Here, we have, in effect, four sub-squares. On the next stage, the same procedure is repeated for each sub-square, etc.

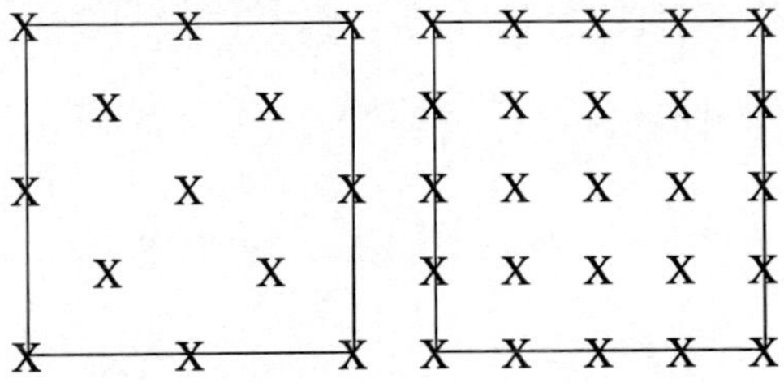

4 Our Algorithm is Efficient

Our algorithm eventually leads to success. Indeed, the set S of possible types of situations is finite. At each step, we add one more select to this set. So, the number of steps cannot exceed the number of elements in the set S—in other words, this algorithm will eventually stop.

Our algorithm is asymptotically optimal. Our algorithm is feasible, so it cannot be optimal—since our problem is NP-hard. Indeed, one can find examples of metric spaces in which this algorithm does not lead to the optimal arrangement.

It is possible to prove, however, that this algorithm is *asymptotically optimal* in the following precise sense. Recall that for each n, our objective is to minimize the value $d_H(S, \{s_1, \ldots, s_n\})$, i.e., make it as close as possible to the smallest possible value $\varepsilon_n(S)$ of this quantity.

- Due to NP-hardness, we cannot feasibly find the optimizing situations s_i for which $d_H(S, \{s_1, \ldots, s_n\})$ is exactly equal to $\varepsilon_n(S)$.
- It turns out that our algorithm increases the value ε, crudely speaking, by no more than a factor of 2.

To be more precise, for the situations s_i generated by our algorithm, the following inequality is satisfied:

$$d_H(S, \{s_1, \ldots, s_n\}) \le 2\varepsilon_{n-1}(S).$$

Let us prove this inequality by contradiction. Indeed, let us assume that, vice versa, $d_H(S, \{s_1, \ldots, s_n\}) > 2\varepsilon_{n-1}(S)$, i.e., that

$$\varepsilon_{n-1}(S) < \frac{1}{2} \cdot d_H(S, \{s_1, \ldots, s_n\}),$$

and let us show that this assumption leads to a contradiction.

By definition of $\varepsilon_{n-1}(S)$, this means that in the set S, there exists a $\varepsilon_{n-1}(S)$-net $s'_1, \ldots, s'_{n-1}$. By the definition of an ε-net, this means, in particular, that for every i, there exists an index i' for which

$$d(s_i, s'_{i'}) < \frac{1}{2} \cdot d_H(S, \{s_1, \ldots, s_n\}).$$

Let us denote one of such indices by $e(i)$; then, we have

$$d(s_i, s'_{e(i)}) < \frac{1}{2} \cdot d_H(S, \{s_1, \ldots, s_n\}).$$

For two different values i and j, we cannot have $e(i) = e(j)$—otherwise, we will have

$$d(s_i, s_j) \le d\left(s_i, s'_{e(i)}\right) + d\left(s'_{e(i)}, s_j\right) <$$

$$\frac{1}{2} \cdot d_H(S, \{s_1, \ldots, s_n\}) + \frac{1}{2} \cdot d_H(S, \{s_1, \ldots, s_n\}) = d_H(S, \{s_1, \ldots, s_n\}).$$

On the other hand, we have shown, in the comment after formula (6), that we must have

$$d(s_i, s_j) > d_H(S, \{s_1, \ldots, s_n\}).$$

for all $i \ne j$. Thus, for different value $i \ne j$, we must have $e(i) \ne e(j)$.

So:

- to each of n situations s_i ($i = 1, \ldots, n$), there corresponds a different situation $s'_{e(i)}$,

- however, this is impossible, since this would mean that we have at least n different situations s'_j, and we only have $n-1$ of them.

This contradiction shows that our assumption

$$\varepsilon_{n-1}(S) < \frac{1}{2} \cdot d_H(S, \{s_1, \ldots, s_n\})$$

is false. So, tthe desired inequality is proven.

5 How Can We Implement This Algorithm

How can we implement the above algorithm: idea. Selecting s_1 may be a problem, but once we have already selected $s_1, \ldots, s_n$, selecting the next situation s_{n+1} is straightforward. Indeed, as we have mentioned, this is exactly how the actual attacker naturally selects the way to attack: by selecting the situation s for which his/her probability of success is the largest, i.e., by selecting the situation s_{n+1} as described by the formula (5). Thus, all we have to do to implement this algorithm is to divide students into competing teams, and let the team which is currently attacking select the next situation.

This leads to the following way to implement the above algorithm in teaching cybersecurity.

Resulting teaching strategy. The instructor divides the class into one or more pairs of sparring mini-teams. In each pair, the teams interchangingly try to attack each other and to defend their team from a other team's attack.

This way, no matter what the original situation s_1 is, after each n situations $s_1, \ldots, s_n$, the attacking team will naturally select the situation s_{n+1} that satisfies the property (5).

This strategy has been tried—and it works. This teaching strategy has indeed been tried—under the name of *adversarial teaching* —and it works; see, e.g., [4, 5].

A similar strategy works in engineering design. Interestingly, a similar approach works in military engineering as well. For example, according to [9], new fighter planes are designed as follows—by using a program that simulates dogfights between different planes:

- The first stage is natural: we consider several possible designs, and for each of them, we simulate how this design will perform in a possible confrontation with the fighter planes used by the existing adversaries. We continue doing this until we find a design that can beat all the possible opponents.
- At first glance, this may seem to be sufficient, but, on second thought, it is not: it is not enough for a future plane to be better that what the opponent has now, we need to have a design that will be better than what the opponent will have in the future. To design such a plane, we perform the second stage of the design process:

namely, we design a plane that will be better than not only the current planes, but also better than our first-stage design.
- Then, we design a plane that will be better than the second-stage design, etc.

At the end, we get an almost perfect future plane—and this is what is then implemented and tested.

Our geometrical model explains why this strategy works. Adversarial teaching works, but why it works has, so far, remained largely a mystery.

In this paper, we explain why it works. Moreover, we explain that this strategy leads to asymptotically optimal results.

Comment. The above text confirms that a healthy and productive competition between students is good for education.

This is a known fact of education is general, not just in teaching cybersecurity; see, e.g., [2]. Moreover, competition is good in general, not just for learning.

Indeed, competition means that if one team is somewhat ahead, the other team tries to catch up, and vice versa. Without a competition, the folks may stagnate and not progress too much. Theoretically, this can also happen in a competitive environment, when two teams' performance is exactly the same. However, there are usually many random factors affecting each team's performance, these factors are independent, and the probability that two independent random variables are exactly equal is 0.

Thus, one of the teams will be always slightly ahead, thus providing an incentive for the other team to catch up—and this way, stagnation will be avoided.

6 How to Best Implement This Algorithm

How to best arrange the competition? In the previous text, we simply explained why the competition-related arrangement works well. Now that we know that it works well, a natural next question is: how to best arrange this competition so as to make the education most effective?

There are many aspects to this question: how many teams to select, how exactly to divide students into teams, etc. How better to assign students to teams depends on the affinity between the students. How many situations can be handled within the same time period depends of the students' motivations and stress level, etc. In this section, we deal with only one aspect of this question: into how many teams should we divide the students, and how many students should be in each team. Is it better to divide students into two competing teams—the case we analyzed in the previous text—or is it better to divide them into three or more teams?

For this aspect, it seems reasonable to conclude that, in the first approximation, e.g., the motivations do not change much whether we divide them into two or three or more teams, but what changes a lot is interaction between students, their feelings towards each other—collaboration or competition (and the resulting stress level). So, in this section, for simplicity, we assume that the students' feelings are the main factor that changes when we change the number of teams and the assignment of students to teams. Based on this assumption, we construct yet another (simple) mathematical model of the corresponding phenomenon. We will then use this model to make recommendations about the number of teams and the number of students in each team.

It is important to take student feelings into account. As all teachers know, effectiveness is not the only criterion that we use when selecting a teaching strategy. We also need to make sure that students feel good about this way of teaching. For example, if we force the students to study 14 h a day, they will probably learn more—but they will feel stressed and upset.

From the viewpoint of student feelings, which way of dividing the students into teams works better? Let us reformulate this question in precise mathematical terms.

Towards a mathematical model of student feelings. If we divide students into competing teams, then naturally students have:

- positive feeling towards their teammates, but
- somewhat negative feeling towards their competitors—i.e., students from competing teams.

Vice versa, a student gets:

- positive feedback from other students from his/her team and
- negative feedback feeling from students from competing teams.

Since a priori, we do not have any reason to believe that some of these feedbacks are more important than others, it is reasonable to view all these positive and negative feedbacks as equally important. From this viewpoint, for each student:

- each member of his/her team (and this student himself) brings to this student 1 positive feedback unit, and
- each member of each of the competing teams bring to this student 1 negative feedback unit.

If we add up all these units, we conclude that as a natural measure of the student's feelings we can take the difference between the number of other students in this student's team (who bring positive feedback) and the number of students in all competing teams (who bring negative feedback).

Let us describe this in precise terms.

Definitions and the first result

Definition 1 *Let a positive integer* N be given. This integer will be called the *number of students in the class*. The set $S = \{1, \ldots, N\}$ will be called the *set of students*; integers from this set will be called *students*.

- By a *division* into competing teams, we mean a tuple $D = (T_1, \ldots, T_k)$ of non-empty subsets of the set of all students such that the sets T_i are disjoint ($T_j \cap T_{j'} = \emptyset$ when $j \neq j'$), and their union is equal to the whole set S: $T_1 \cup \ldots \cup T_k = S$.

- The sets T_j are called *teams*.
- For each student i, let us denote, by $D(i)$, the number j of the team that contains this student: $i \in T_j = T_{D(i)}$.
- For each division D and for each student i, by the *feeling* $f_i(D)$ of the i-th student, we mean the value

$$f_i(D) = (|T_{D(i)}| - 1) - (|T_1| + \ldots + |T_{D(i)-1}| + |T_{D(i)+1}| + \ldots + |T_k|), \quad (7)$$

where $|T|$ denotes the number of elements in a set T.
- We say that a division D' is *better* than a division D if for all students i, we have $f_i(D) \le f_i(D')$, and for some students i, we have $f_i(D) < f_i(D')$.
- We say that the division D is optimal if no other division is better than D.

Proposition 1 *A division D is optimal if and only if it divides students into two teams $k = 2$.*

Comment. This result explains that, from the viewpoint of student feelings, the best way to organize competition is to divide students into two competing teams.

Proof
1°. Let us first prove that if any division $D = (T_1, T_2, T_3, \ldots, T_k)$ that divides students into $k > 2$ teams is not optimal.

To prove this, we will show that the division $D' \stackrel{\text{def}}{=} (T_1 \cup T_2, T_3, \ldots, T_k)$ is better than D. Indeed, for all students from the teams T_3 through T_k, the feeling does not change when we go from D to D', but for students from the teams T_1 and T_2 the feelings become better:

- students from team T_1 used to get negative feelings from students from the team T_2 (term $-|T_2|$ in $f_i(D)$), now this feeling is positive, while all other feelings remain unchanged;
- similarly, students from team T_2 used to get negative feelings from students from the team T_1 (term $-|T_1|$ in $f_i(D)$), now this feeling is positive, while all other feelings remain unchanged.

2°. In Part 1 of this proof, we showed that only divisions into two teams can be optimal. To complete the proof, let us show that every division into two teams is optimal.

To prove this, we need to show that no division into two teams can be better that another division into two teams.

Indeed, let us consider a division $D = (T_1, T_2)$ into two teams. Without losing generality, let us assume that the size of the first team is smaller than or equal to the size of the second team. So, this division divides students into a team $n \stackrel{\text{def}}{=} |T_1|$ of size $n \le N/2$ and the remaining team T_2 of size $N - n$. In this division:

- students from the first team get the feeling

$$f_i(D) = (n-1) - (N-n) = 2n - N - 1 \le -1,$$

while

- students from the second team get the feeling

$$f_i(D) = ((N-n)-1) - n = N - 2n - 1 \ge -1.$$

Let us consider three possible cases.

2.1°. If two divisions D and D' have the same first-team size n, then they have the same feeling for the same number of students, so the sum of all the feelings is the same—but if D' was better than D, we would have $\sum_{i=1} f_i(D) < \sum_{i=1} f_i(D')$. So, in this case, D cannot be better than D'.

2.2°. Let us now consider the case when the division $D' = (T_1', T_2')$ has more students in its first team than D, i.e., when $n' = |T_1'| > n = |T_1|$. In this case, $N - 2n' < N - 2n$, so $N - 2n' - 1 < N - 2n - 1$. So, every student who had a feeling $N - 2n \ge -1$ in the division D can only get smaller positive feeling in the division D'. So, such D' cannot be better than D.

2.3°. Finally, if D' has fewer students in its first team, i.e., if $n' < n$, then we have $2n' - N < 2n - N$, thus $2n' - N - 1 < 2n - N - 1$. So in the division D', some students will get lower negative feelings than in the division D.

2.4°. In all three cases, D' is not better than D.

The statement is proven, and so is the proposition.

Comment. This proposition makes sense: even if we divide the class into more than 2 teams, since the bigger team has an advantage, some team will naturally start cooperating—to increase their chances of succeeding, and we will end up with fewer and fewer competing teams—until it gets to two.

Visual representation of this result. Let us assume that originally, we had four competing teams: T_1, T_2, T_3, and T_4. Students have positive feelings towards their teammates and negative feeling towards students from other teams:

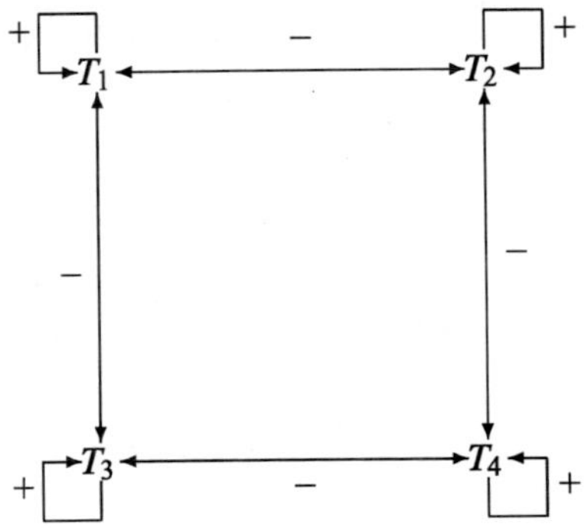

If we merge teams T_1 and T_2 into a single team $T_1 \cup T_2$, then:

- between students of the two original terms T_1 and T_2, the attitude changes from negative to positive, while
- all other feelings remain the same.

Thus, with this merger, the number of positive connections increases:

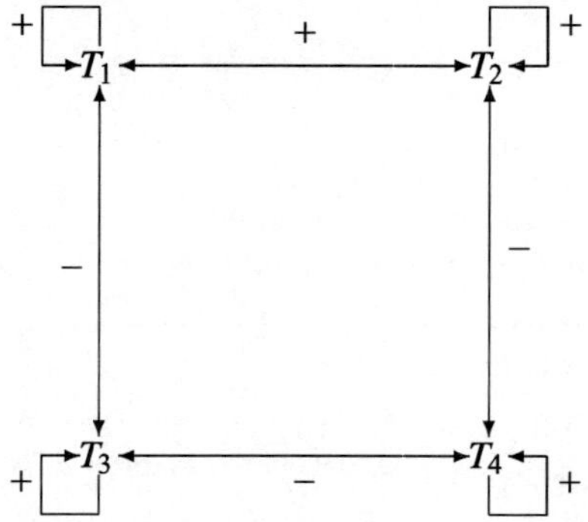

Finally, if we merge teams T_3 and T_4 into a single team $T_3 \cup T_4$—thus leaving only two competing teams: $T_1 \cup T_2$ versus $T_3 \cup T_4$—then:

- between students of the original terms T_3 and T_4, the attitude changes from negative to positive, while
- all other feelings remain the same.

Thus, with this merger, the number of positive connections increases even further:

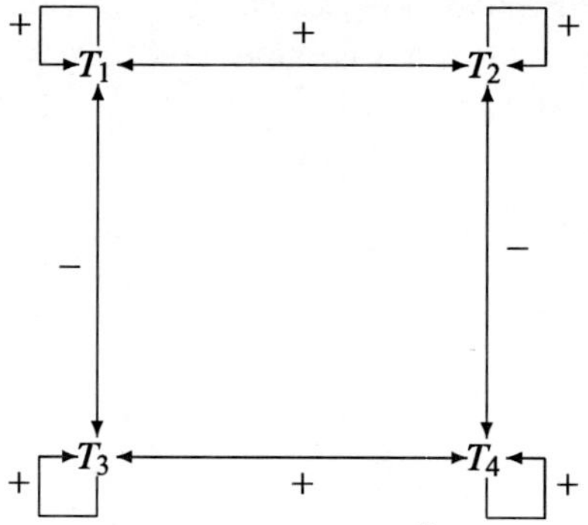

So which division into two teams should we select? A natural idea is to select a division in which the worst-case feeling

$$w(D) \stackrel{\text{def}}{=} \min_i f_i(D). \tag{10}$$

is the largest possible. This is described by the following result.

Proposition 2

- *For even N, the division with the largest possible value of $w(D)$ is the division into two equal size teams.*
- *For odd $N = 2m + 1$, the division with the largest possible value of $w(D)$ is the division into teams of size m and $m + 1$.*

Proof According to the formulas that we obtained when proving Proposition 1, for any division into teams of size n and $N - n$, the value $w(D)$ is equal to

$$w(D) = \min(2n - N - 1, N - 2n - 1).$$

Since $N - 2n = -(2n - N)$, this means that $w(D) = -|2n - N| - 1$. Thus, the largest possible $w(D)$ corresponds to the smallest possible value of $|2n - N| - 1$.

For even N, the smallest possible value is clearly 0, when $n = N/2$. For $N = 2m + 1$, we cannot have 0, but we can have 1, when $n = m$. The proposition is proven.

Visual illustration. Let us illustrate the above idea on a simple example. Suppose that initially, we divided 6 students into two unequal groups: a group of 2 and a group of 4:

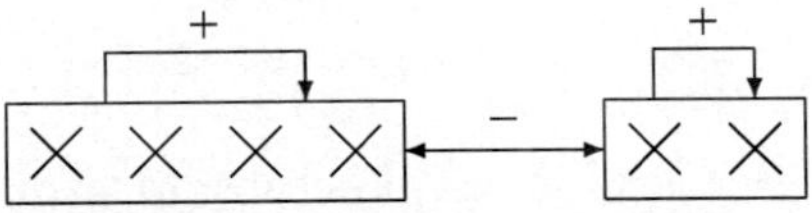

In this division, each student from the smaller group has:

- negative feelings from 4 students from the larger groups and
- positive feelings from his/her partner in the smaller group.

So here $w(D) = 1 - 3 = -2$.

On the other hand, if we instead divide 6 students into two equal groups, then we will get, for each student:

- 3 negative feelings and
- 2 positive feelings.

So here $w(D) = 2 - 3 = -1$m which is larger than -2:

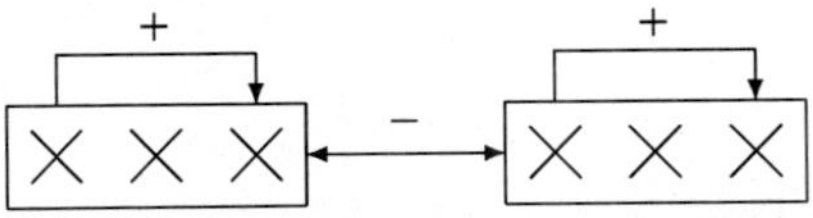

This example shows from the viewpoint of student feelings, the division into equal groups is indeed better.

7 What Next?

What we did. In this paper, we provided a *simplified* mathematical model that explains why adversarial teaching works—and show that, in some reasonable sense, adversarial teaching is indeed a close-to-optimal teaching strategy.

The existence of such an explanation made us (and will hopefully make others) more confident that this method is a right one.

Can we do better? Teaching with more confidence is good, but it would nice to have a model that helps us teach *better*.

For this, we need a more realistic model. Such model should take into account that some attacks are more difficult to defend against, while others are easier. Such models should take into account that in adversarial teaching, it is often not individual against individual, but rather a team against a team—so we need to take into account team dynamics, etc.

We hope that our simplified model will provide a starting point for developing such more realistic models.

How to motivate? This paper presents a mathematical demonstration of how adversarial learning *can be* beneficial for teaching cybersecurity topics. But how to make sure that adversarial learning *is* beneficial?

In this paper, we concentrated on the technical part, on *what* to teach—implicitly assuming that students have the needed motivation (and, of course, the needed background).

In reality, while some students are always eager to learn, but for other students, it is important to keep them motivated. In our experience, when properly organized, competitive environments like hackathons are great motivators—but, on the other hand, pedagogy teaches us that many students do not perform well in competitive environments.

How to best motivate students remains an important open problem.

Acknowledgements This work was supported in part by the US National Science Foundation grants 1623190 (A Model of Change for Preparing a New Generation for Professional Practice in Computer Science) and HRD-1242122 (Cyber-ShARE Center of Excellence). The authors are greatly thankful to Boris Kovalerchuk for his encouragement, and to the anonymous referees for valuable suggestions.

References

1. Cormen, T.H., Leiserson, C.E., Rivest, R.L., Stein, C.: Introduction to Algorithms, MIT Press. MA, and Mc-Graw Hill Co., New York, Cambridge (2009)
2. Foster, S.R., Esper, S., Griswold, W.G.: From competition to metacognition: designing diverse, sustainable educational games. In: Proceedings of the Proceedings of the ACM SIGCHI Conference on Human Factors in Computing Systems CHI'2013, Paris, France, April 27–May 2, pp. 99–108 (2013)
3. Gasarch, W.: The third P=?NP poll. ACM SIGACT News **50**(1), 40–59 (2019)

4. Hamman, S.T., Hopkinson, K.M.: Teaching adversarial thinking for cybersecurity. J. Colloq. Inf. Syst. Edu. (CISSE) **4**(1) (September 2016)
5. Katz, F.: Adversarial thinking: teaching students to think like a hacker. In: Proceedings of the 2019 Kennesaw State University Conference on Cybersecurity Education, Theory and Practice, Kennesaw, Georgia, October 11–12, 2019 (2019)
6. Kreinovich, V., Starks, S.A.: Why benchmarking is an (asymptotically) optimal approach to numerical methods: a geombinatoric proof. Geombinatorics **13**(3), 131–138 (2004)
7. Lorentz, G.G.: Approximation of Functions. Reinhart, and Winston, New York, Halt (1966)
8. Lorentz, G.G.: The 13-th problem of Hilbert. In: Browder, F.E. (ed.) Mathematical Developments Arising from Hilbert's Problems, American Mathematical Society, Providence, Rhode Island, Part 2, pp. 419–430 (1976)
9. Moiseev, N.: Elements of the Theory of Optimal Systems. Nauka, Moscow (1975).(in Russian)
10. Servin, C., Kosheleva, O., Kreinovich, V.: Adversarial teaching approach to cybersecurity: a mathematical model explains why it works well. In: Proceedings of the 24th International Conference on Information Visualisation IV'2020, Vienna and Melbourne, September 7–11, 2020, pp. 313–316 (2020)

Applications and Evaluations of Drawing Scatterplots as Polygons and Outlier Points

Asuka Nakabayashi and Takayuki Itoh

Abstract There have been a variety of techniques on multi-dimension data visualization techniques. Scatterplot matrix and parallel coordinate plots are typical and famous ones, but they have a problem that they may need a very large screen space when a dataset has a large number of dimensions. We suppose there are two types of solutions for this problem: one of them is to effectively select interesting sets of dimensions, and the other is to develop drawing techniques that realize comprehensive representations in the limited screen space. Based on this direction, we propose a method for selecting important scatterplots from all scatterplots generated from input datasets and for drawing the scatterplots as "outliers" and "regions enclosing non-outlier plots." The technique is useful for users to determine whether to delete outliers from the datasets and form mathematical models of non-outlier plots. This chapter extends the authors' conference paper by additionally introducing an example with a design optimization dataset and the user evaluation of this technique.

1 Introduction

Multi-dimensional data visualization is one of the most active research topics in the visualization community because there has been an enormous number of multi-dimensional datasets in our daily life and application domains. We often need to observe the features and regularity of multi-dimensional data to completely understand and effectively utilize the data. The development of effective visualization techniques is important to easily discover the features and trends of multi-dimensional data.

Scatterplot matrix (SPM) [2] and parallel coordinate plots (PCPs) [5] are well-known multi-dimensional data visualization techniques. Given a n-dimensional

A. Nakabayashi · T. Itoh (✉)
Ochanomizu University, 2-1-1 Otsuka, Bunkyo-ku, Tokyo 112-8610, Japan
e-mail: itot@is.ocha.ac.jp

A. Nakabayashi
e-mail: namabayashi.asuka@is.ocha.ac.jp

B. Kovalerchuk et al. (eds.), *Integrating Artificial Intelligence and Visualization for Visual Knowledge Discovery*, Studies in Computational Intelligence 1014,
https://doi.org/10.1007/978-3-030-93119-3_25

dataset, SPM generates scatterplots that assign all possible pairs of dimensions to their axes and arranges them as a $n \times n$ matrix. Meanwhile, PCP draws n parallel axes, plots the values on the axes, and connects them by polygonal lines. These techniques have a serious problem that they may need a large screen space when users want to visualize all the dimensions in case the input dataset has a large number of dimensions. Also, it is a bothering task if we need to visually observe all the dimensions of input datasets even though interesting features and trends are not necessarily discovered in all dimensions. To solve this problem, many visualization techniques that selectively display a small number of meaningful dimensions have been presented recently.

Mathematical modeling is an important process to utilize multi-dimensional data. We often discuss what types of models are appropriate to the multi-dimensional datasets, or what types of outlier removal methods we should apply. Machine learning is a typical task in which users may need to select appropriate mathematical models for multi-dimensional data. Visualization is effective to support the mathematical model construction for multi-dimensional datasets.

This chapter presents our multi-dimensional data visualization technique addressing the above problems. This technique consists of the following two processing steps.

- Selection and display of important scatterplots from all possible scatterplots.
- Drawing of scatterplots as "outliers" and "regions enclosing non-outlier plots."

This chapter is an extended version of the authors' conference paper [11]. We added an additional case study on the design optimization process of airplane wings, and the user experiment and evaluation. Also, we totally revised the conference paper.

The remainder of this chapter is as follows. Followed by the introduction of related work in Sects. 2, 3 describe the processing flow of the presented technique. Section 4 introduces two case studies applying datasets of the retail transactions and the design optimization process. Section 5 shows the user experiment and evaluation, and Sect. 6 concludes this chapter.

2 Related Work

2.1 *Multi-dimensional Data Visualization with Dimension Selection*

Every dimension in multidimensional datasets is not always interesting. Therefore, it is practical and efficient to select a set of interesting dimensions and visualize the lower-dimensional subspaces consisting of the selected dimensions. This idea can be realized by selectively displaying a set of meaningful low-dimensional PCPs or scatterplots. One of the typical techniques is presented by Suematsu et al. [23] that generated low-dimensional PCPs from meaningful groups of dimensions and

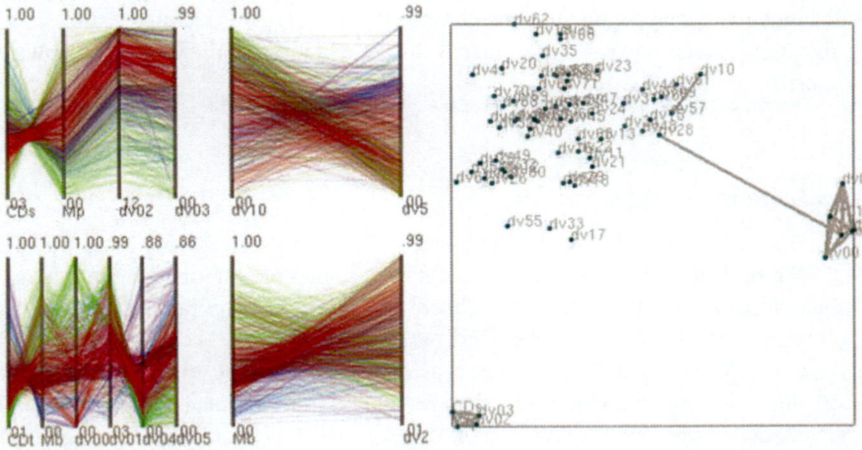

Fig. 1 A snapshot of visualization by Itoh et al. [6]

arranging them based on their dissimilarity. Zheng et al. [28] also presented a similar technique that selected scatterplots based on user-specified metrics and arranged based on the dissimilarity among the selected scatterplots.

One of the issues of the above approach is to interactively control the number of represented dimensions. Several studies have visualized low-dimensional subspaces based on the interactive dimension selection. Lee et al. [8] and Liu et al. [9] applied dimension reduction schemes to interactively select subsets of the high-dimensional data. Nohno et al. [13] presented a technique to interactively contract highly-correlated dimensions to adjust the number of axes displayed in PCPs. However, these technique did not generate arbitrary numbers of PCPs or scatterplots.

Itoh et al. [6] presented a technique that solves this issue. As shown in Fig. 1, this technique renders a set of low-dimensional PCPs on the left side of the window, while it displays the dimension graph that provides an interactive mechanism to select user-interested sets of dimensions on the right side of the window. Watanabe et al. [26] presented an extended technique that selects a combination of low-dimensional PCPs and scatterplots.

The technique presented in this chapter also provides an interactive mechanism to specify the number of displayed scatterplots; however, it does not generate PCPs while our previous techniques [6, 26] generate PCPs.

2.2 Evaluation and Selection of Scatterplots

Numeric evaluation of scatterplots has been an active research topic. Wilkinson et al. [27] proposed nine features of scagnostics based on the appearance of the scatterplots. Wang et al. [25] proposed the new scagnostics considering human perception. Several other studies focus on specific metrics of scatterplots, including correlation [4, 20] and class separation [19, 21].

Scagnostics can be applied to the technique presented in this chapter while selecting the scatterplots; however, our current implementation applies different criteria as described in Sect. 3.1.

2.3 Drawing of Scatterplots

Visual comprehensibility of scatterplots is an important problem for effective analysis of multi-dimensional data. There have been several studies on drawing scatterplots that improve their visual comprehensibility.

Bagplot [15] presented by Rousseeuw et al. extends boxplots to two-dimensional scatterplots. This method clearly divides outlier and other points by a simple visual representation; however, it often does not work well when the dense points form concave regions.

Continuous Scatterplots [1] presented by Bachthaler et al. paints the two-dimensional spaces based on the density of points. This is an epoch-making study because the technique applies a rendering technique well-applied to scientific visualization fields into information visualization fields. The technique may be time-consuming for some types of datasets.

Schreck et al. [17, 18] presented techniques on drawing the distribution of point in the scatterplots by simpler objects such as convex hulls. The techniques do not always display objects corresponding to all the points while our technique represents the distribution of all the points.

Sunspot plots [24] presented by Trauntner et al. combines the continuous rendering in dense regions and usual point drawing in sparse regions. This technique applies colors to represent the density of points, and therefore, it is difficult to represent the distribution of categorical values in addition to the distributions of points in the screen space.

3 Selection and Drawing of Scatterplots

This section describes the processing flow of our technique on the election and drawing of scatterplots. The technique firstly generates scatterplots from all the possible pairs of dimensions of the input multidimensional dataset. Then, the technique selects the user-specified number of important scatterplots applying one of the algorithms described in Sect. 3.1. Finally, the technique displays the selected scatterplots as a set of outlier points and regions enclosing non-outlier plots as described in Sect. 3.2. Figure 2 illustrates the processing flow of our technique.

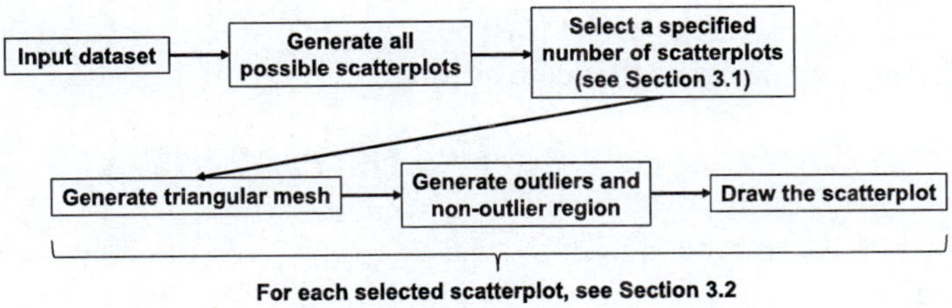

Fig. 2 Processing flow of the presented technique

3.1 Selection of Scatterplots

Our technique selects the specified number of important scatterplots generated from input multi-dimensional datasets. Our current implementation provides the following three types of algorithms for scatterplot selection. The first and second algorithms are also implemented by Itoh et al. [6].

The first algorithm calculates correlation coefficients between arbitrary pairs of dimensions and selects scatterplots formed from pairs of dimensions which have larger absolute values of correlation coefficients. Let d_1 and d_2 be dimensions, the absolute value of the correlation coefficient d_{d_1,d_2} between d_1 and d_2 is defined as

$$d_{d_1,d_2} = |1.0 - f_c(d_1, d_2)| \quad (1)$$

where $f_c(d_1, d_2)$ denotes Spearman's rank correlation coefficients.

The second algorithm calculates the entropy of positions of individuals on a scatterplot treating values of a categorical dimension as labels: this mechanism is useful to select scatterplots those specific labels of individuals are distantly positioned from others. In particular, we compute the entropy H for all pairs of dimensions

$$H(d_1, d_2) = -\frac{1}{N}\sum_{i=1}^{N}\sum_{c=1}^{C} p(y_i = c|x_i^{d_1,d_2}) \log p(y_i = c|x_i^{d_1,d_2}) \quad (2)$$

where x_i represents the i-th vector from an n-dimensional input dataset Ds defined as $Ds = \{x_1, \ldots, x_N\}$, y_i represents the label assigned to the i-th vector, N and C represent the number of vector and the number of labels, and $p(y_i = c|x_i^{d_1,d_2})$ represents the probability that the c-th class Y_c is assigned to the i-th vector x_i. Moreover, $x_i^{d_1,d_2}$ is a two-dimensional vector containing the dimensions d_1 and d_2 of x_i. This value represents the separation of labels in a scatterplot generated by d_1 and d_2.

The third algorithm measures the thinness of the region where the individuals place in the scatterplot as Wilkinson et al. [27] did. Our implementation generates a Delaunay triangular mesh T [22] connecting the individuals in a scatterplot and then removes all triangles which have at least one edge that is longer than a pre-defined threshold. Then. we calculate the score as follows:

$$T(d_1, d_2) = 1 - \sqrt{4\pi A_{rea}(T)}/P_{erimeter}(T) \quad (3)$$

where $A_{rea}(T)$ is the total area of T, and $P_{erimeter}(T)$ is the total length of the boundary of T.

3.2 *Drawing Scatterplots as Outliers and Regions Enclosing Non-outlier Plots*

Our technique draws scatterplots as "outliers" and "regions enclosing non-outlier plots." The current implementation applies Delaunay triangulation to connect adjacent dots in the scatterplots and detect outliers from the edges of the triangular mesh. Delaunay triangulation is a generic method which connects vertices scattered in a 2D space and generates a triangle mesh. We apply an incremental triangulation algorithm which firstly generates a large rectangle surrounding every vertex in a scatterplot as the initial mesh, then adds the vertices one-by-one, and connects the vertices to refine the triangular mesh.

After constructing the mesh, our implementation deletes edges which are longer than a user-specified threshold and also deletes triangles which contain such edges. Here, we extract vertices that are not connected to any vertices as outliers as shown in Fig. 3. Users can control the threshold to adjust the number of outliers. Then, the technique represents the regions enclosing non-outlier plots by drawing the region boundary of the triangles that connect the non-outlier plots by a dark color and filling the triangular mesh by a light color.

Also, users can delete arbitrary outliers from input datasets and redraw the scatterplots as shown in Fig. 4. When the outlier at the upper right in Fig. 4 (left) is deleted, the scale of the vertical axis is updated, and the non-outlier region is also updated as shown in Fig. 4 (right). The non-outlier regions can be enlarged when we delete the distant outliers. Therefore, users can discover new features, trends, and potential outliers that are previously undiscovered.

4 Examples

We implemented the presented technique with Java Development Kit (JDK) 1.8.0, reusing the implementation of Itoh et al. [6] for scatterplots selection and user interfaces. This section introduces two visualization examples applying our technique.

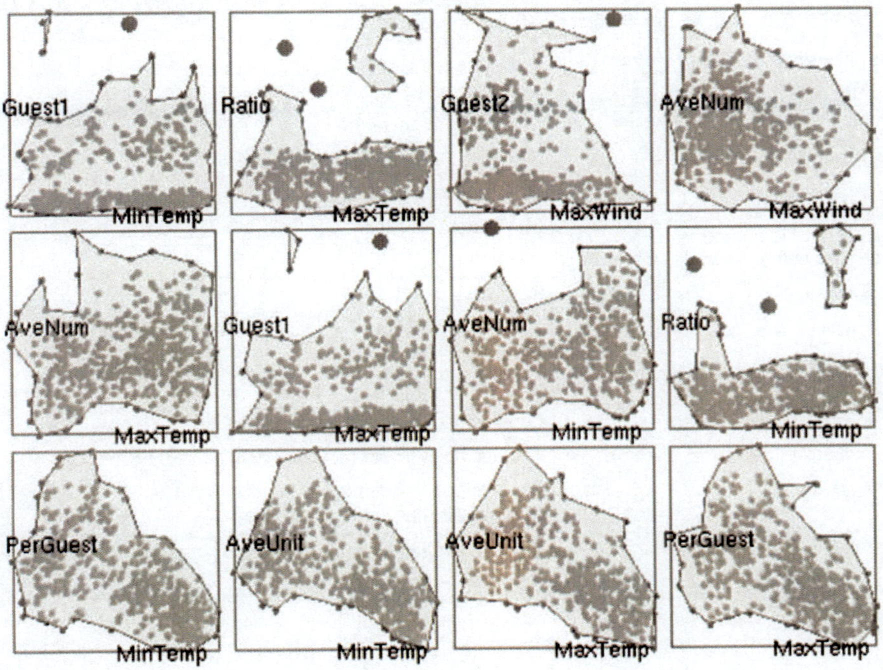

Fig. 3 An example of drawing the scatterplots as "outliers" and "regions enclosing non-outlier plots"

Fig. 4 An example of deleting an outlier. (left) before deleting an outlier. (right) after deleting an outlier and then rescaling the vertical axis

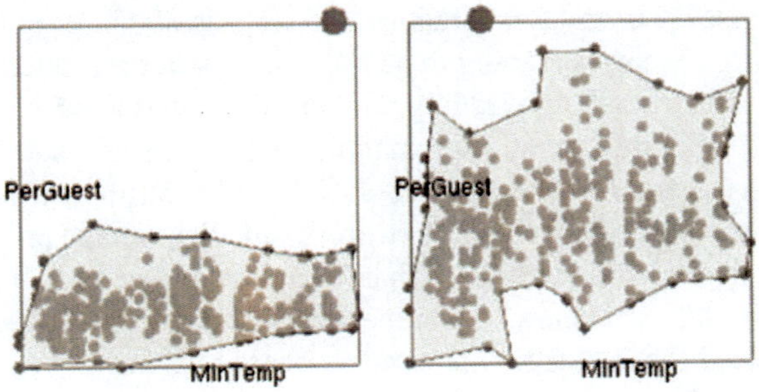

4.1 Example 1: Retail Transaction

This section introduces the first example applying a retail transaction dataset and climate values. Table 1 shows the explanatory variables (climate values) assigned to the horizontal axis and the objective functions (retail transaction values) assigned to the vertical axis in this dataset. These values are recorded in the dataset day-by-day.

The dataset owner recorded the revenue (described as Revenue in Table 1), the number of customers and visitors (described as Guest1 and Guest2 in Table 1), and the total number of purchased items (described as NumItem in Table 1) in their shop.

Table 1 The explanatory variables and the objective functions

Explanatory variables (Climate values)	
MinTemp	Minimum temperature
MaxTemp	Maximum temperature
SumRain	Precipitation
SumSunTime	Sunshine duration
MaxWind	Maximum wind speed
Objective functions (Retail transaction values)	
Revenue	Revenue
Guest1	Number of customers who bought something and had payments
Guest2	Number of visitors who got into the shop
Ratio	Conversion rate defined as Guest1/Guest2
PerGuest	Average revenue per customer defined as Revenue/Guest1
AveUnit	Average price of purchased items defined as Revenue/NumItem
AveNum	Average number of purchased items per customer defined as NumItem/Guest1

The dataset then calculated Ratio as Guest1/Guest2, PerGuest as Revenue/Guest1, AveUnit as Revenue/NumItem, and AveNum as NumItem/Guest1.

We generated 35 scatterplots consisting of five horizontal axes and seven vertical axes. The dataset contained the records of 457 days from May 1, 2016, to July 31, 2017, corresponding to 457 data points in the scatterplots. Remark that this dataset is perturbated by adding random real values to each column of the original dataset.

Figure 5 shows an example of a numerical distribution in which one attribute (weekday) is shaped to enclose the other attribute (holiday). This figure depicts that the average revenue per customer and the average price of purchased items are more dispersed on weekdays than on holiday.

Figure 6 shows the numerical distribution of the average price of purchased items and the average number of purchased items from September to November. As the winter approaches, the average price of purchased items tends to increase. This is presumably because the thicker clothes are, the higher the prices of items are. On the other hand, the average number of purchased items tends to decrease. This is also presumably because the prices of items are higher than those of clothes for spring or summer, and therefore customers get more careful while purchasing items.

Figure 7 shows the numerical distribution of the conversion rate in February, July, and November. This figure shows that there are some periods in which the conversion rate is prominently high only in early February and late July. One of the reasons for this phenomena is that a much larger number of customers than visitors for window shopping came there due to special events such as over-stock sales. The figure also depicts that there is only one day with an unusually high conversion rate in November.

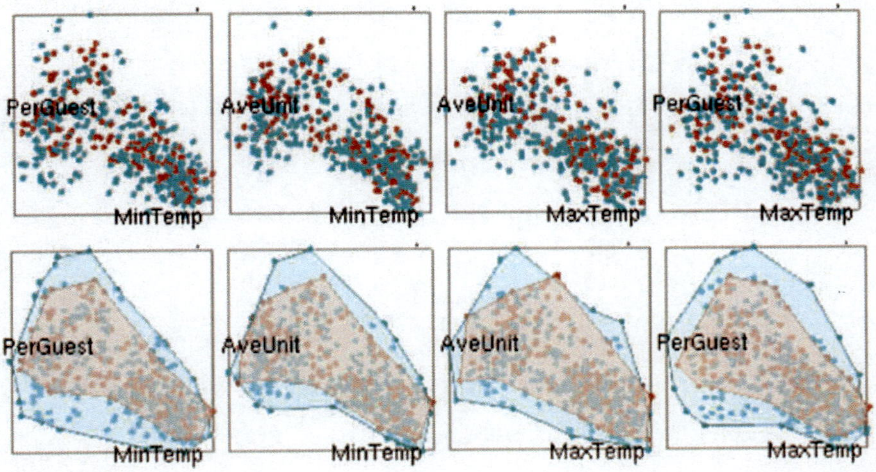

Fig. 5 An example of a numerical distribution in which a non-outlier region of a specific attribute shapes to enclose the non-outlier region of the other attribute. The light blue region represents the distribution on weekdays while the red region represents the distribution on holidays

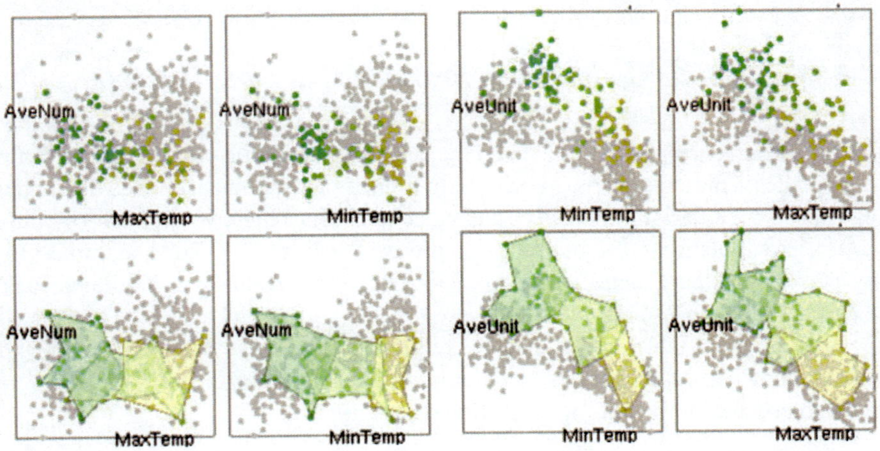

Fig. 6 The numerical distribution of the average price of purchased items and the average number of purchased items from September to November. The yellow region represents the distribution in September, the yellow green region represents in October and the green region represents in November. The gray dots represent the distribution in the other months. (upper) before enclosing. (lower) after enclosing

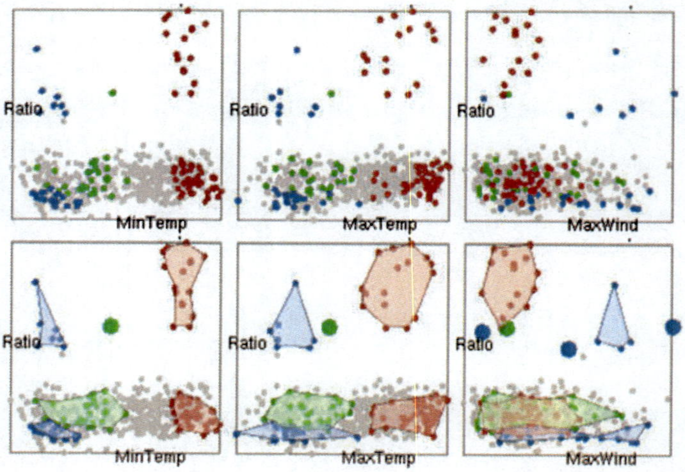

Fig. 7 The numerical distribution of the conversion rate in February, July, and November. The blue region represents the distribution in February, the red region represents in July and the green region represents in November. The gray dots represent the distribution in the other months. (left) before enclosing. (right) after enclosing

4.2 Example 2: Optimization Process of Airplane Design

As the second example, we visualized the design optimization process for the wings of airplanes. This experiment applied the optimization dataset that defined the shapes of wings by 72 design parameters and calculated four objective functions by fluid dynamics simulations. The optimization process obtained 776 Pareto solutions [16] by performing the genetic algorithm. We treated this result as a 76-dimensional dataset containing 776 samples.

The design variables in the dataset are given by the range from dv_{00} to dv_{71}. It is well-known in aerospace design community that the following six variables were most important for the optimum solution discovery:

- dv_{00}, dv_{01}: Span lengths of the inboard/outboard wing panels.
- dv_{02}, dv_{03}: Leading-edge sweep angles.
- dv_{04}, dv_{05}: Root-side chord lengths.

Other design variables included the following:

- dv_{06} to dv_{25}: Variables to define the inner surface connecting corresponding points on upper and lower surfaces of the wing, to design the warping of the wing.
- dv_{26} to dv_{32}: Variables to design the twist of the wing.
- dv_{33} to dv_{71}: Variables to design the thickness of the wing.

The objective functions are as follows:

- CD_t: Drag coefficient during the transonic cruise.
- CD_s: Drag coefficient during the supersonic cruise.
- M_b: Bending moment at the wing root during the supersonic cruise.
- M_p: Pitching moment during the supersonic cruise.

Figure 8 shows 24 scatterplots that depict the pairs of highly-correlated design parameters and objective functions. This result shows $dv02$,$dv03$ and Mp,CDs have especially high correlations. Also, $dv00$,$dv01$,$dv04$,$dv05$ and Mb,CDt have high correlations. These results have been well-known among the specialists on airplane design. Our previous visualization method [6] have also visualized such correlations; however, the previous method did not well represent the details of correlations and distributions of outliers. The presented technique well represents such detailed distribution since it applies scatterplots.

Figure 9 shows 20 scatterplots that depict the pairs of weakly correlated design parameters and objective functions. The design parameters $dv16$, $dv17$, $dv21$, $dv33$, $dv60$ appeared in this result has not been well discussed among the researchers working on this optimization process. The visualization result may bring new knowledge on fluid dynamics simulation and its optimization for wings of airplanes.

Since it is well-known that design parameters $dv00 \sim dv05$ highly correlate with objective functions, analysts may suppose that scatterplots those horizontal axes are assigned to such design parameters are not worth observing. Thereat, we added categorical variables that denote the range of one of the above design parameters and assigned colors according to the values of categorical variables. Figure 10 shows the scatterplots that represent the relationships between the objective function CDt and design parameters except $dv00 \sim dv05$, where samples are colored in skyblue if the value of $dv03$ is small, in yellow if the value is medium, or in purple if it is large. The result draws small skyblue triangles when one of the design parameters $dv33, dv43, dv18, dv51$, or $dv66$ is assigned to the horizontal axis. These scatterplots may suggest that the optimization process did not sufficiently explore the subspace where $dv03$ is small and CDt is large, and therefore more Pareto solutions may be discovered if the analysts adjust the optimization process.

Figure 11 shows the scatterplots that represent the relationships between Mp/CDs and design parameters except $dv00 \sim dv05$, where samples are colored in skyblue if the value of $dv00$ is small, in yellow if the value is medium, or in purple if it is large. Some of the scatterplots depict that skyblue samples have negative correlations while yellow samples have positive correlations. Some other scatterplots depict that skyblue samples have positive correlations while yellow samples have negative correlations. Such knowledge may lead analysts to improve the efficiency of the optimization process and assists in the discovery of a larger number of Pareto solutions.

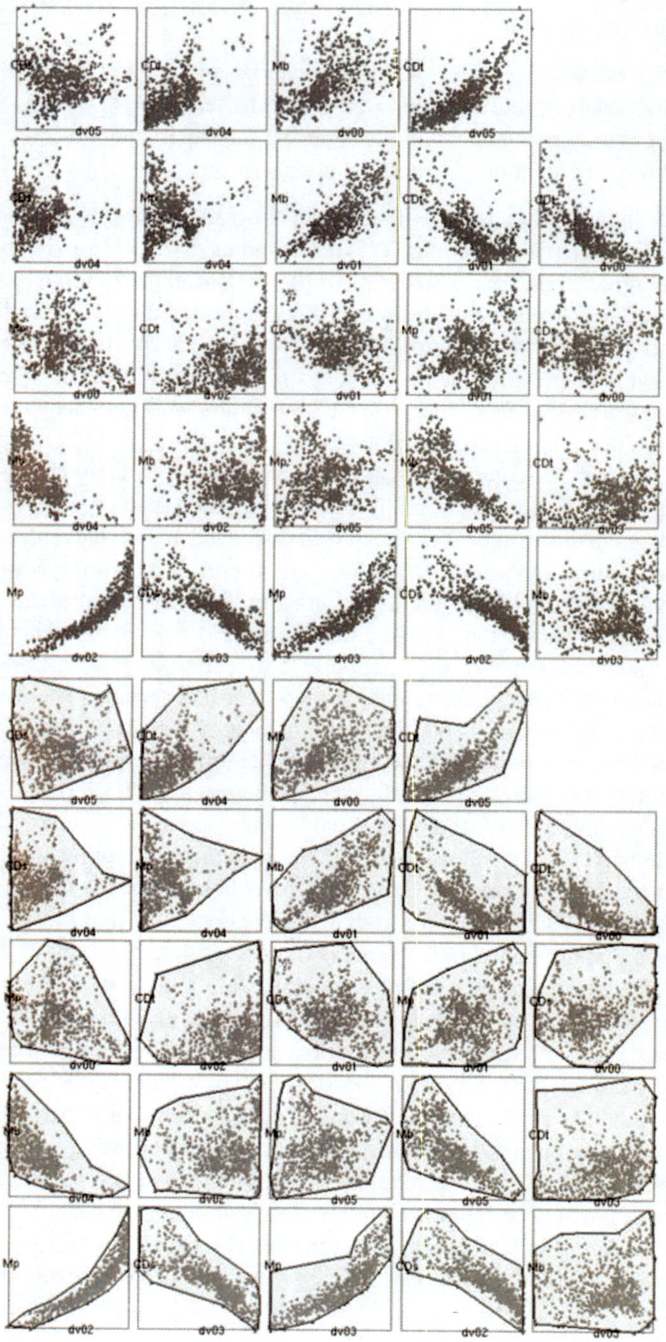

Fig. 8 Scatterplots that represent high correlations between design parameters and objective functions. The horizontal axes in some scatterplots are $dv02$ or $dv03$ while their vertical axes are Mp or CDs, which have especially high correlations. The horizontal axes in other scatterplots are $dv00$, $dv01$, $dv04$ or $dv05$ while their vertical axes are Mb or CDt, which also have high correlations

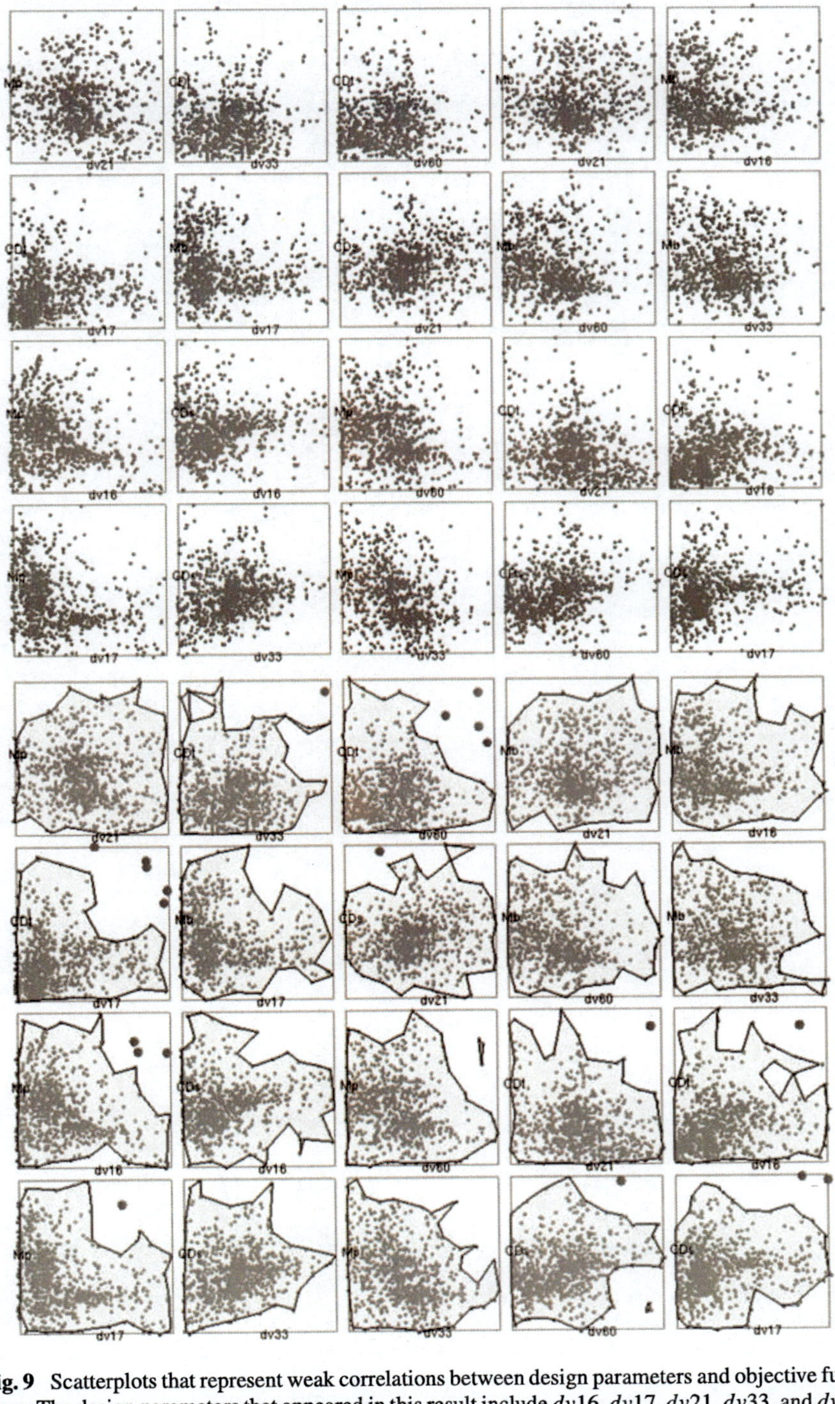

Fig. 9 Scatterplots that represent weak correlations between design parameters and objective functions. The design parameters that appeared in this result include $dv16$, $dv17$, $dv21$, $dv33$, and $dv60$, which have not been well discussed among the researchers working on this optimization process. This visualization may bring new knowledge of this research field

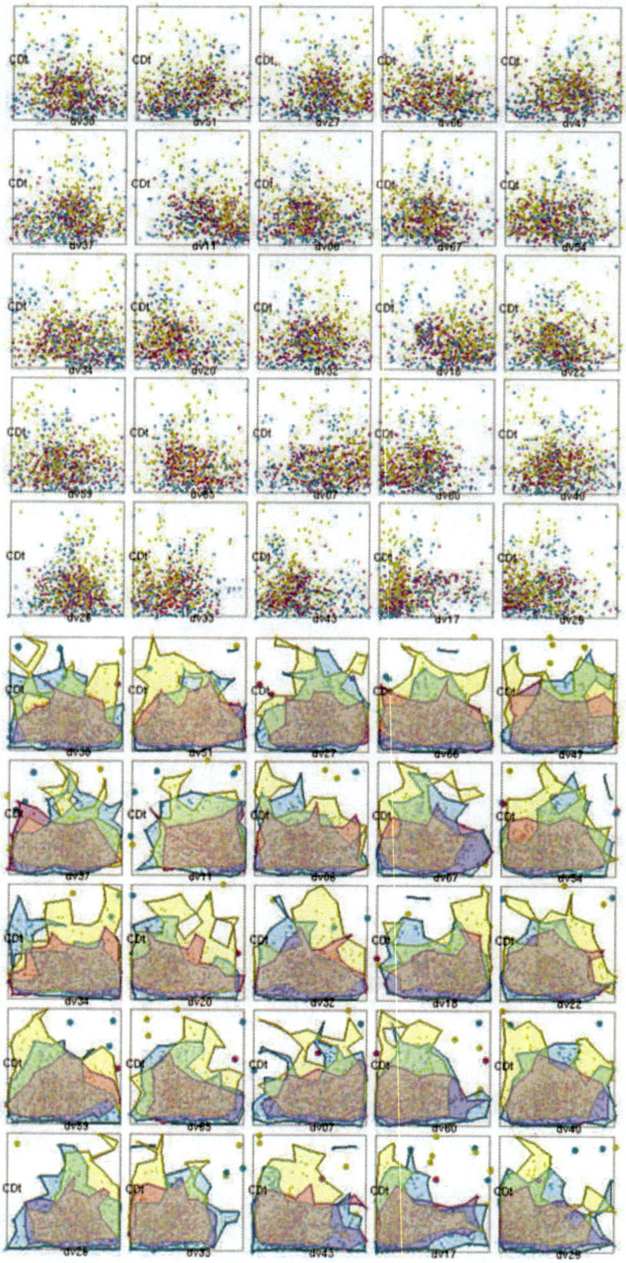

Fig. 10 Scatterplots that design parameters except $dv00 \sim dv05$ are assigned to the horizontal axes while CDt is assigned to the vertical axes. Samples are colored in skyblue if the value of $dv03$ is small, in yellow if the value is medium, or in purple if it is large. The result draws small skyblue triangles when one of the design parameters $dv33$, $dv43$, $dv18$, $dv51$, or $dv66$ is assigned to the horizontal axis. This result may suggest insufficient optimization processes

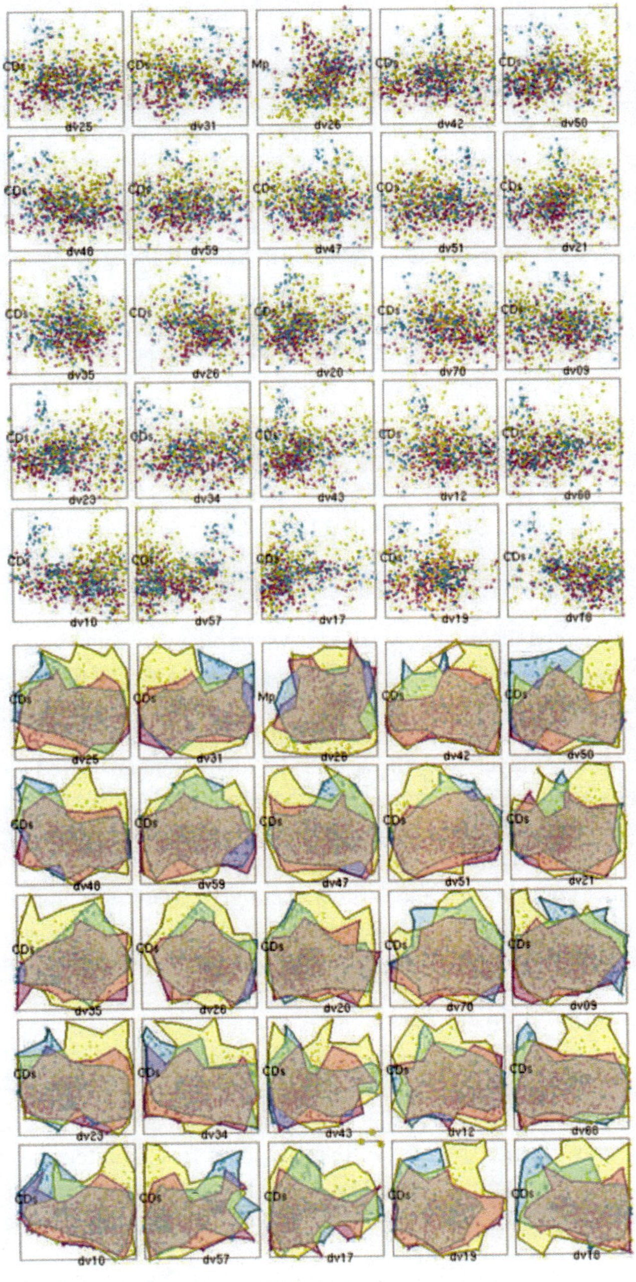

Fig. 11 Scatterplots that design parameters except $dv00 \sim dv05$ are assigned to the horizontal axes while Mp or CDs is assigned to the vertical axes. Samples are colored in skyblue if the value of $dv00$ is small, in yellow if the value is medium, or in purple if it is large. Some of the scatterplots depict that skyblue samples have negative correlations while yellow samples have positive correlations. Some other scatterplots depict that skyblue samples have positive correlations while yellow samples have negative correlations. This knowledge may contribute to improving the optimization processes

5 User Evaluation

This section introduces the user experiment and evaluation of the presented technique. We invited eight student participants in their twenties majoring in computer science. We provided and explained to the participants an instruction of our implementation of the technique, and asked them to install the program of our technique as well as the program of simple scatterplot matrix as the baseline method. Then, we asked the participants to observe the dataset introduced in Sect. 4.1 using the two programs and to answer the questions regarding the knowledge discovered from the visualization results. The participants observed all the scatterplots before answering the questions since the scatterplot matrix program displayed all of them. The following are the questions:

1. Which scatterplots can you discover negative correlations?
2. Which objective functions can you find the separation of samples between weekdays and weekends?
3. Which scatterplots can you see multiple clusters in February and July?
4. Where can you discover outliers?
5. Which pairs of scatterplots do you feel similar?

In addition, we asked the participants to answer which program was preferred from the viewpoints of questions 1–3. We also asked to answer which functions of the program they used while searching for outliers.

We divided the eight participants into two groups during the experiment. The former group firstly played with the scatterplot matrix and answered questions 1–3, and then played with the presented technique and answered questions 1–3. The latter group firstly played with the presented technique and answered questions 1–3, and then played with the scatterplot matrix and answered questions 1–3. We asked both groups to answer questions 4 and 5 after playing with both programs. We did not limit the time to answer all the questions.

Table 2 shows the statistics of the correctness of the answers and preferences of the participants. Here, we found that it was difficult to define correct answers for question 3 as described later and therefore we did not count the number of correct/incorrect answers for question 3.

For question 1, we defined the correct answers as the four scatterplots shown in Fig. 12 since they have larger d_{d_1,d_2} values. The correctness of answers between the two programs is quite similar, while many participants preferred our technique as shown in Table 2. Figure 13 shows examples of incorrect choices for question 1 where many participants selected especially while looking at polygons enclosing non-outlier points in our technique.

For question 2, Fig. 14 shows the definition of the correct answers as the three scatterplots where three objective functions, Revenue, Guest1, Guest2, are assigned to the vertical axis since the scatterplots have larger $H(d_1, d_2)$ values. Our technique got a larger number of correct answers, while many participants did not have strong preferences between two programs as shown in Table 2.

Table 2 Correctness of the answers and preferences of participants for questions 1–3

Question		1	2	3
Scatterplot matrix	Correct	6	4	–
	Incorrect	2	4	–
	Unanswered	0	0	–
Our Technique	Correct	5	6	–
	Incorrect	3	1	–
	Unanswered	0	1	–
Strongly prefer scatterplot matrix		1	0	1
Weakly prefer scatterplot matrix		1	1	0
No preference		0	3	2
Weakly prefer our technique		1	3	1
Strongly prefer our technique		5	1	4

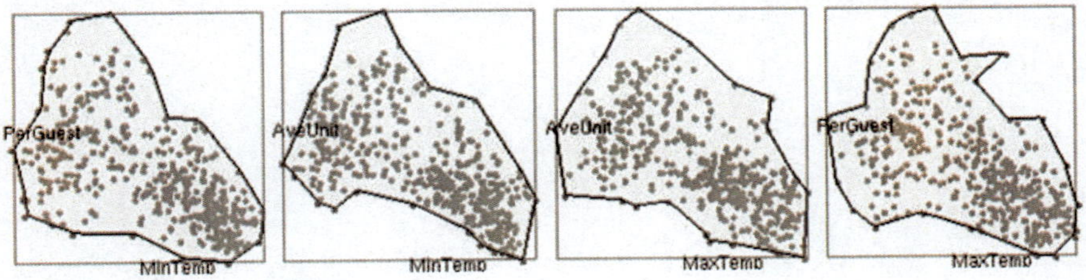

Fig. 12 Correct choices of scatterplots for question 1

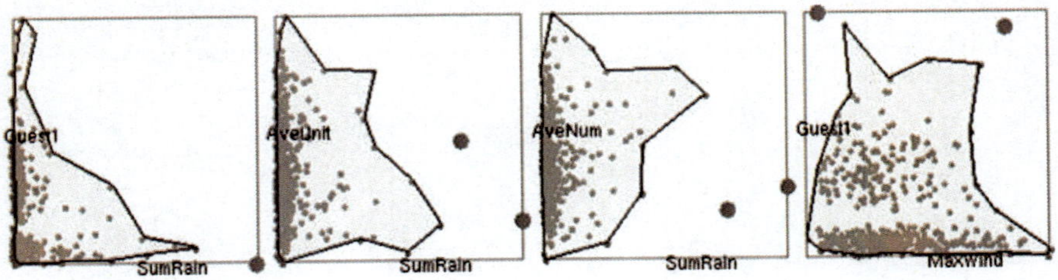

Fig. 13 Examples of incorrect choices for question 1

For question 3, the scatterplots shown in Fig. 15 are defined as the correct answers since the scatterplots have larger $T(d_1, d_2)$ values. On the other hand, four participants selected other scatterplots shown in Fig. 16. We suppose such choices of the participants are reasonable and therefore agree to make these choices as correct answers, but the current implementation did not select such scatterplots. This is the reason why we did not count the number of correct/incorrect answers for question 3 as shown in Table 2. We would like to improve the implementation so that such scatterplots shown in Fig. 16 can be also automatically selected.

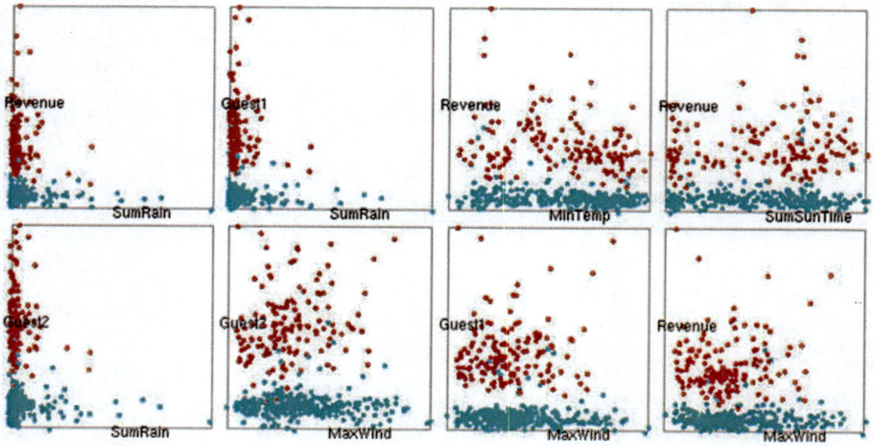

Fig. 14 Correct choices of scatterplots for question 2

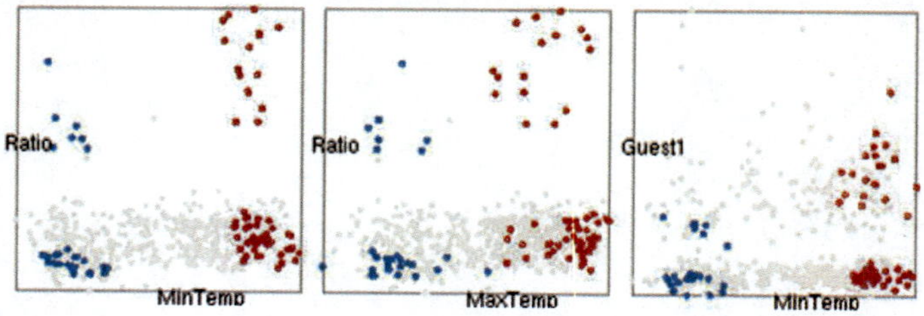

Fig. 15 Scatterplots that we firstly supposed as correct ones for question 3

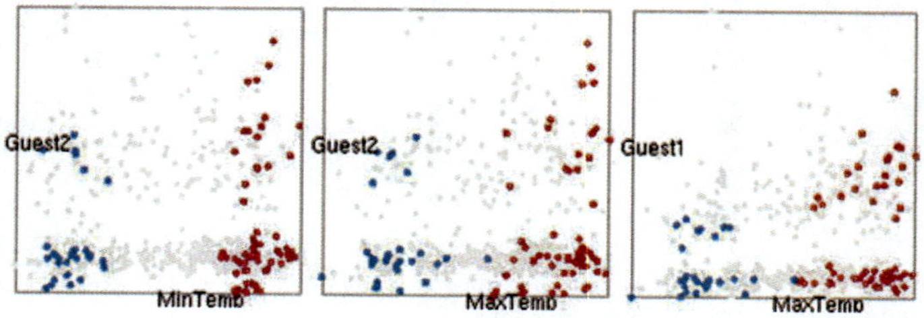

Fig. 16 Four participants thought these scatterplots satisfy question 3

Selections of outlier points among the participants for question 4 were consistent. Six of the participants preferred to use our technique to find outlier points. It demonstrates that our technique is effective in finding outlier points.

We also measured the similarity of scatterplots from three metrics and defined two sets of similar scatterplots as shown in Fig. 17. Many participants correctly selected these sets while several participants additionally selected other sets of scatterplots shown in Fig. 18.

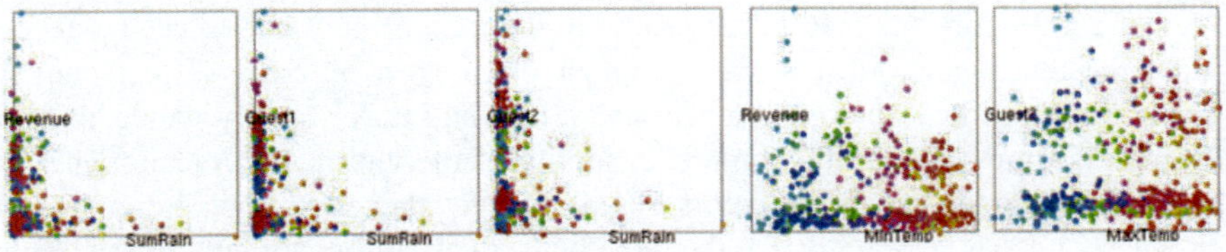

Fig. 17 We measured the similarity of scatterplots and defined two sets of scatterplots (left three and right two) are similar

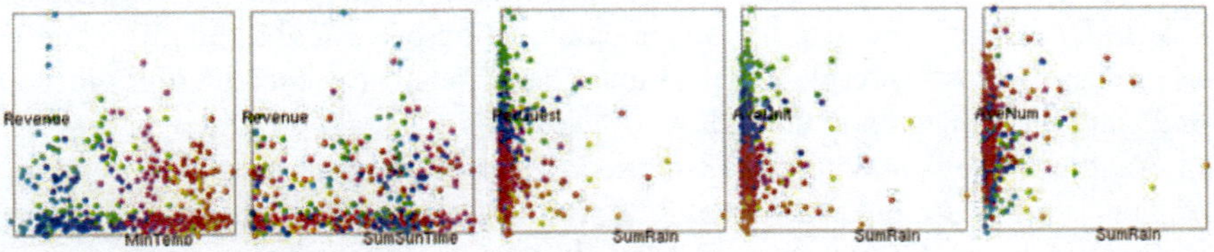

Fig. 18 Several participants additionally selected that the left two or right three scatterplots are similar

As a result of the above experiments, we found our technique is effective for the discovery of scatterplots that represents correlations and clusters. Also, we found our technique is helpful for the discovery of outlier points. Meanwhile, participants sometimes focused on non-important scatterplots while applying our scatterplot drawing technique.

6 Conclusion

This chapter proposed a visualization technique that effectively represents interesting dimensions by multiple scatterplots. The technique firstly selects important scatterplots from all possible scatterplots generated from input datasets. We implemented three metrics for the scatterplot selection: correlation, entropy, and thinness. The technique then draws the scatterplots as "outliers" and "regions enclosing non-outlier plots." This representation makes users easier to discover outliers and understand the trends of the densely distributed plots.

The chapter also introduced two examples. From the visualization of a retail transaction dataset, we found the difference of the transaction between weekdays and weekends, the correlation between temperature and transactions, and extremely high conversion rate in early February and late July. From the visualization of the dataset of the design optimization process for the wings of airplanes, we could find well-known correlations between the design parameters and objective functions.

Finally, the chapter introduced the user experiment and evaluation. As a result, we found our technique is effective for the discovery of scatterplots that represent correlations and clusters, and the discovery of outlier points.

Our current implementation has several limitations. First of all, we need to reconsider the representation of a larger number of categorical values. Our technique assigns independent colors to categorical values and may vaguely paint the overlapping portions. Therefore, comprehensibility of the visualization results may be worse when the three or more colors of polygons overlap each other.

We found experimentally that the three types of algorithms of our current implementation for scatterplot selection do not necessarily select all scatterplots that we subjectively determine important. Thus, we would like to improve the algorithms and also develop additional algorithms so that such scatterplots are automatically selected. Another approach will be the combination of such algorithms so that we can select various scatterplots at the same time. We already studied on this idea [7] and would like to apply to this study.

Another issue is how to extract outliers. Our current implementation to extract outliers is just based on distances on scatterplots. Data items are regarded as outliers if they are distant from all other data items on one of the scatterplots because we extract outliers using Delaunay triangular meshes. We would like to reconsider the implementation of outlier detection since there have been various techniques for extracting outliers directly from multi-dimensional data [3, 10, 14].

Participants sometimes focused on non-important scatterplots during the experiments as described in Sect. 5. We suppose one of the reasons of this problem is that the drawing technique does not finely represent the density of points since all the triangles are painted in the same color and opacity, and therefore we would like to adjust the implementation of the painting process. Also, we would like to analyze the recognition for the similarity of scatterplots and discuss the improvement of our implementation.

Scatterplot selection and outlier extraction are independently developed in the current implementation. The technique may bring better visualizations if these components tightly work together. For example, more interesting scatterplots may be discovered by extracting outliers in order. This discussion is one of our future directions.

Acknowledgements We appreciate ABEJA, Inc. for providing the retail transaction dataset. Also, we appreciate Prof. Shigeru Obayashi of Tohoku University, Japan, for providing the design optimization dataset.

References

1. Bachthaler, S., Weiskopf, D.: Continuous scatterplots. IEEE Trans. Visual Comput. Graphics **14**(6), 1428–1435 (2008)
2. Carr, D.B., Littlefield, R.J., Nicholson, W.L., Littlefield, J.S.: Scatterplot matrix techniques for large N. J. Am. Stat. Assoc. **82**, 424–436 (1987)
3. Cousineau, D., Chartier, S.: Outliers detection and treatment: a review. Int. J. Psychol. Res. **3**(1), 58–67 (2010)
4. Harrison, L., Yang, F., Franconeri, S., Chang, R.: Ranking visualizations of correlation using weber's law. IEEE Trans. Vis. Comput. Graph. **20**(12), 1943–1952 (2014)

5. Inselberg, A., Dimsdale, B.: Parallel coordinates: a tool for visualizing multi-dimensional geometry. In: IEEE Visualization (VIS1990), pp. 361–378 (1990)
6. Itoh, T., Kumar, A., Klein, K., Kim, J.: High-dimensional data visualization by interactive construction of low-dimensional parallel coordinate plots. J. Vis. Lang. Comput. **43**, 1–13 (2017)
7. Itoh, T., Nakabayashi, A., Hagita, M.: Scatterplot selection applying a graph coloring problem (2020). arXiv preprint, arXiv:2009.07342
8. Lee, J.H., McDonnellm, K.T., Zelenyuk, A., Imre, D., Mueller, K.: A structure-based distance metric for high-dimensional space exploration with multidimensional scaling. IEEE Trans. Comput. Graph. **20**(3), 351–364 (2013)
9. Liu, S., Wang, B., Bremer, P.-T., Pascucci, V.: Distortion-guided structure-driven interactive exploration of high-dimensional data. Comput. Graph. Forum **33**(3), 101–110 (2014)
10. Lu, C.T., Chen, D., Kou, Y.: Algorithms for spatial outlier detection. In: Proceedings of the Third IEEE International Conference on Data Mining (ICDM'03) (2003)
11. Nakabayashi, A., Itoh, T.: A technique for selection and drawing of scatterplots for multi-dimensional data visualization. In: 23rd International Conference on Information Visualisation (IV2019), pp. 62–67 (2019)
12. Nhon, D.T., Wilkinson, L.: ScagExplorer: exploring scatterplots by their scagnostics. In: IEEE Pacific Visualization Symposium (PacificVis 2014), pp. 73–80 (2014)
13. Nohno, K., Wu, H.-Y., Watanabe, K., Takahashi, S., Fujishiro, I.: Spectral-based contractible parallel coordinates. In: 18th International Conference on Information Visualisation, pp. 7–12 (2014)
14. Penny, K.I., Jolliffe, I.T.: A comparison of multivariate outlier detection methods for clinical laboratory safety data. J. R. Stat. Soc.: Ser. D (Stat.) **50**(3), 295–308 (2001)
15. Rousseeuw, P.J., Ruts, I., Tukey, J.W.: The bagplot: a bivariate boxplot. Am. Stat. **53**(4), 382–387 (1999)
16. Sasaki, D., Obayashi, S., Nakahashi, K.: Navier-Stokes optimization of supersonic wings with four objectives using evolutionary algorithm. J. Aircr. **39**(4), 621–629 (2002)
17. Schreck, T., Panse, C.: A new metaphor for projection-based visual analysis and data exploration. In: IS&T/SPIE Conference on Visualization and Data Analysis (2007)
18. Schreck, T., Schüßler, M., Worm, K., Zeilfelder, F.: Butterfly plots for visual analysis of large point cloud data. In: International Conference in Central Europe on Computer Graphics, Visualization and Computer Vision (WSCG'08), pp. 33–40 (2008)
19. Sedlmair, M., Tatu, A., Munzner, T., Tory, M.: A taxonomy of visual cluster separation factors. Comput. Graph. Forum **31**(3), 1335–1344 (2012)
20. Shao, L., Mahajan, A., Schreck, T., Lehmann, D.J.: Interactive regression lens for exploring scatter plots. Comput. Graph. Forum **36**(3), 157–166 (2017)
21. Sips, M., Neubert, B., Lewis, J.P., Hanrahan, P.M.: Selecting good views of high-dimensional data using class consistency. Comput. Graph. Forum **28**(3), 831–838 (2009)
22. Sloan, S.W.: A fast algorithm for constructing Delaunay triangulations in the plane. Adv. Eng. Softw. **9**(1), 34–55 (1987)
23. Suematsu, H., Zheng, Y., Itoh, T., Fujimaki, R., Morinaga, S., Kawahara, Y.: Arrangement of low-dimensional parallel coordinate plots for high-dimensional data visualization. In: 17th International Conference on Information Visualisation (IV2013), pp. 59–65 (2013)
24. Trautner, T., Bolte, F., Stoppel, S., Bruckner, S.: Sunspot plots: model-based structure enhancement for dense scatter plots. Comput. Graph. Forum **39**(3), 551–563 (2020)
25. Wang, Y., Wang, Z., Liu, T., Correll, M., Cheng, Z., Deussen, O., Sedlmair, M.: Improving the robustness of scagnostics. IEEE Trans. Vis. Comput. Graph. **26**(1), 759–769 (2020)
26. Watanabe, A., Itoh, T., Kanazaki, M., Chiba, K.: A scatterplots selection technique for multi-dimensional data visualization combining with parallel coordinate plots. In: 21st International Conference on Information Visualisation (IV2017), pp. 78–83 (2017)
27. Wilkinson, L., Anand, A., Grossman, R.: Graph-theoretic scagnostics. In: IEEE Symposium on Information Visualization, pp. 157–164 (2005)
28. Zheng, Y., Suematsu, H., Itoh, T., Fujimaki, R., Morinaga, S., Kawahara, Y.: Scatterplot layout for high-dimensional data visualization. J. Vis. **18**(1), 111–119 (2015)

Supply Chain and Decision Making: What is Next for Visualisation?

Catherine Xiaocui Lou, Alessio Bonti, Maria Prokofieva, and Mohamed Abdelrazek

Abstract Supply chain is a hot topic with strong links in the industry and business applications. The overflowing information generation, increasing complexity of businesses, digitalisation of the supply chain, and introduction of advanced analytics capabilities are all topical issues in the supply chain. Visualization of supply chain inform action in this regards is more than ever important and critical: it provides an easy way to understand and act upon solutions for decision makers, reduces the cognitive load and brings strategic benefits to the business. The development of data analytics and visualization techniques have been booming while little attention was given in the academic literature to structure the landscape and draft the road for further development. The present paper addresses this gap by providing a comprehensive review of the current literature in the use of visualisation in this growing area of supply chain and logistics. The paper employs the PRISMA methodology to identify the main theme, particular areas of development and suggests the future directions for research. Such a structural view developed on the basis of top academic and industry publications, leverages its contribution by provision of a brief structural view of available directional developments and links them to practical applications.

1 Introduction

Supply chain and logistics are essential elements in the current business world where distribution channels and the ability to deliver the product to the customer become more important than having a superior product on the market. Recent events with COVID-19 put the supply chain and logistics at the utmost forefront of the business. Escalating complexities in information channels, as well as the sheer volume of such

C. X. Lou (✉) · M. Prokofieva
Institute for Sustainable Industries and Liveable Cities, Victoria University, Melbourne, Australia
e-mail: Catherine.lou@vu.edu.au

A. Bonti · M. Abdelrazek
School of Information Technology, Faculty of Science Engineering and Built Environment, Deakin University, Burwood, Australia

B. Kovalerchuk et al. (eds.), *Integrating Artificial Intelligence and Visualization for Visual Knowledge Discovery*, Studies in Computational Intelligence 1014,
https://doi.org/10.1007/978-3-030-93119-3_26

information and underlying corporate and technological structures, put communication of such information and effectiveness of decision making based on information insights at stake. Effective communication relates to easy, quick and complete transfer of information, and visualisation of data plays a vital role in this regard via the provision of effective ways of portraying information through visual elements, such as diagrams, graphs, charts and maps.

Large companies usually have supply chain structures located across large geographical areas. However, small companies, less fortune with resources, also have non-stop complex information flows and require agile decision making to fight competition and industry demands. Such circumstances put pressures on companies and demand to look for solutions where decision makers can draw important insights at less cost and less time. With a hope to resolve ongoing operational and investment concerns companies more and more often tend to look at visualising their data, promote resilience and support their operations [1]. In this respect, visualisation offers dramatic, though yet fully explored, the potential for practical implementation as well as academic research. This is the motivation of the current paper which is to address the growing demand and current gap in research in the visualisation of the supply chain.

The contribution of the paper is in landscaping the current literature in the use of visualisation in the supply chain and decision making and identifying the main themes and areas of development as well as proposing future directions for research. Using a dataset of top academic and industry publications, the paper contributes to the literature by providing a clear structural view of available directional developments and links them to practical applications. The findings of the paper suggest one major application of visualisation in the current supply chain literature in supporting decision making. The directions include visualisation of supply chain data using different presentation graphics and tools, data flow presentations through visualisation of business processes in the supply chain, contextualised operational visualisation of daily activities, risk management and real time monitoring, together with interactive visualisation. The paper provides an overview of each direction as well as provide an exemplar study to guide future researchers. This paper extends [2] by providing a more solid methodological foundation and more comprehensive updated discussion of processes visualisation, analysis of tools for supply chain visualisation, and visualisation applications.

The rest of the paper is organised as follows. The next section presents the research design with methodology and detailed research process. Section 3 discusses the research findings of the visualisation in supply chain with different tools and supporting decision making. It is followed by the visualisation in application. The paper concludes by overviewing the direction for further research and limitations.

2 Research Design

To address the aims of the study, the paper used the PRISMA methodology which is one of the most popular and versatile approaches to conducting a comprehensive literature review. The two-stage design was implemented when during the first stage, a literature search was conducted and the second stage proceeded with analysis and identification of the themes and mapping of the literature. The workflow of the research design is presented in Fig. 1. Overall, 262 papers were located and included in the analysis.

The PRISMA (i.e. Preferred Reporting Items for Systematic Reviews and Meta-Analyses) presents an evidence-based minimum set of items for the development of systematic reviews and meta-analyses. The framework improves the clarity and transparency of systematic reviews by applying the QUOROM (QUality Of Reporting Of Meta-analysis) Statement (1999). The PRISMA methodology relates to a set of conditions that ensures the development of high-quality literature reviews and the generation of insights for further research [3]. The framework includes a 27-item checklist and is accompanied by a four-phase flow diagram. A copy of the checklist and its implementation are provided in Appendix A. It ensures that the research is guided through the review of a clearly formulated question (i.e. what are the uses of data visualisation in supply chain and decision-making literature) and uses systematic and explicit methods to identify, select, and critically evaluate relevant research, as well as to collect and analyse data from the collected studies.

As prescribed by the PRISMA framework, the study started with the aim to collect empirical evidence that accommodates the selected eligibility criteria and addresses a specific research question. Using systematic methods, the study attempted

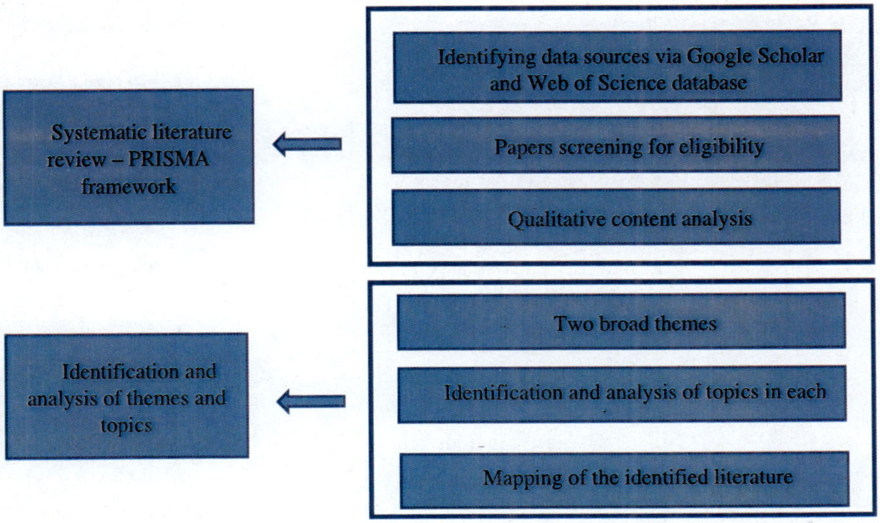

Fig. 1 Research design

to minimise bias and deliver reliable findings for conclusions and further research. Further components of the framework include the following:

- a clear set of objectives and a reproducible methodology,
- a systematic search guided by the eligibility criteria,
- assessment of the findings regarding their validity,
- systematic presentation of the findings.

Figure 2 presents the alignment of the PRISMA framework with the research design of the study.

During the first stage, the research team employed Google Scholar and Web of Science databases to search for academic publications in the area of study. The selection criteria were that the publication venue is peer-reviewed and/or top-ranked, relevant to the studied topic and the publication is in English. To identify the publications the following search terms were used: "visual", "visualisation", "supply chain", "supply chain decision making". The timeframe of the search was set between 2003 and 2021 and based on the availability of the data. The specified timeframe also considered developments in the IT and visualisation field as well as growth in the demand for supply chain research and applications. The reviewed studies covered the area of Science, Technology and Social Sciences (as identified by Web of Science). Overall, the search revealed 262 publications. The snapshot of the results is presented on Panel A, B and C in Fig. 3.

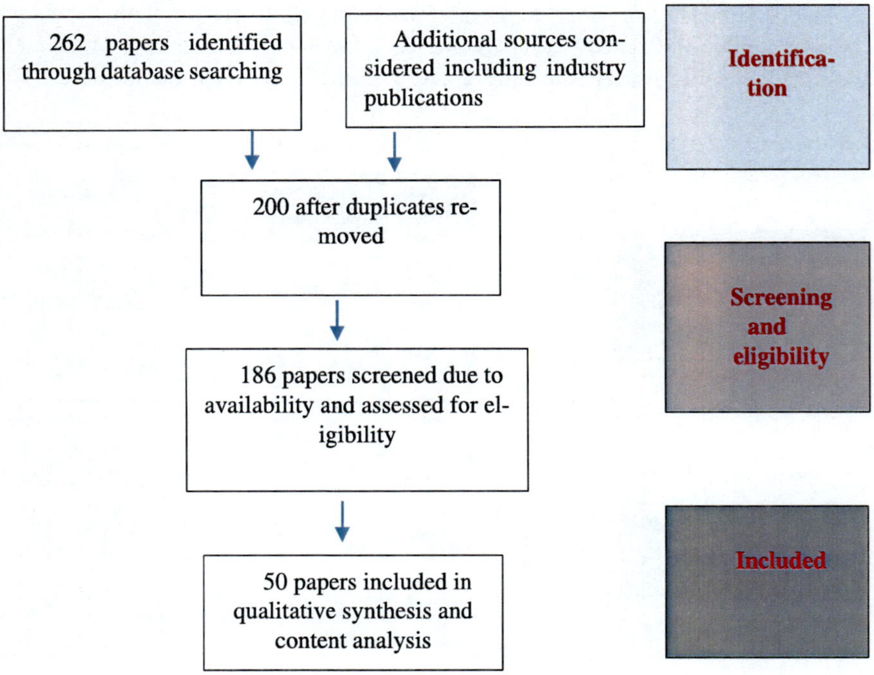

Fig. 2 Alignment of research design with the PRISMA framework

Panel A. Distribution of the original sample of papers across discipline areas

Panel B. Distribution of the original sample of papers across publication years

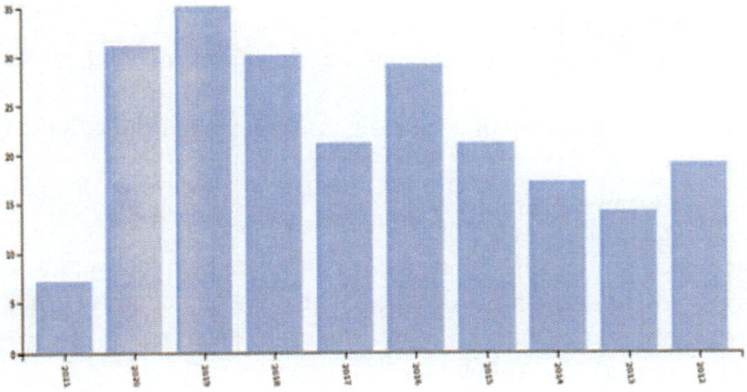

Panel C. Distribution of the original sample of papers across countries (based on authors' information)

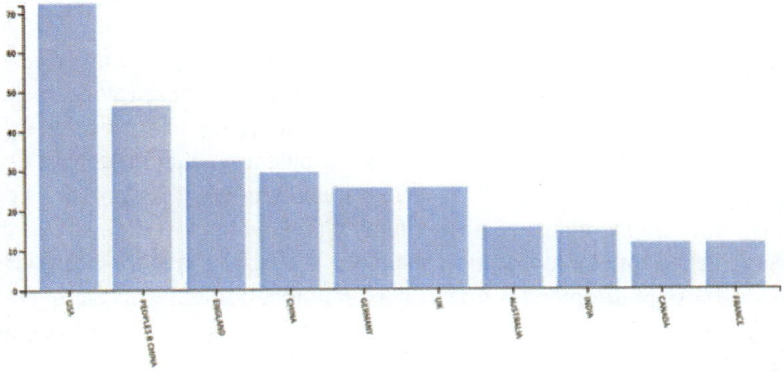

Fig. 3 Visual analysis of the results of the PRISMA framework application (Panel A, Panel B, and Panel C)

Overall, fifty papers were selected for qualitative synthesis and content analysis and fourteen publications were included in the final sample that satisfied the requirement of the quality publication and publication venue expectations. The identified papers were carefully documented and cross-referenced. The identified meta-data, including abstracts location, DOI, keywords and authors information, were recorded and examined independently by two researchers to identify its relevance to the study. Identified papers were analysed using the qualitative content analysis which is a research method for subjective analysis and evaluation of the text data [4]. Using this approach the identified sample papers were organised to elicit meaning and draw conclusions. Using the manifest analysis version of the content analysis (Merriam et al. 2016), the papers were decontextualised, re-contextualised, categorised and compiled into different topics presented in the following section.

The manifest content analysis aims to organise and derive meanings from the data to arrive at realistic conclusions. Conducting the manifest analysis results in surveying a broad surface structure to elicit deep meanings (latent analysis). The research passed through four distinct stages, including:

Stage 1—decontextualisation: identifying meaning units, including developing a coding system.

Stage 2—recontextualisation: understanding of the content through comparison of the original data.

Stage 3—categorisation: identifying homogeneous groups and triangulation of the results by independent investigators.

Stage 4—compilation: developing realising conclusions and deliverable findings.

The achieved results were presented to the group of academics and industry professionals to seek their advice and validation. The final outcomes of the systematic literature review are discussed below.

3 Visualisation of Supply Chain Decision Making

Supply chain plays a crucial role in the development and operations of any organisation. Irrespective of the size a well-organised supply chain provides benefits for the organisation and ensures the sustainability of its functions and further development. An inefficient supply chain weakens the business and can potentially create a high hidden cost detrimental for business and hindering its growth. For better understanding and maintenance of its supply chain, an organisation strives to create reliable and efficient communication channels. Visualisation in the supply chain not only provides opportunities for effective communication but also facilitates understanding and decision making and assists ineffective management of operations in supply.

3.1 Visualisation Tools to Discover Business Process

It is tradition in business decision making to separate management of business operations and business process management. The separation, however, creates an information gap and increases the value of communication between different parts of the business. Reporting and creating business insights from supply chain data are usually automated with user reports and dashboards. This approach puts pressure on the usability and efficiency of communicating data insights through respective data. Varying levels of requested granularity of information presents challenges to the final implementation of the systems. An example of a successful solution to the problem is proposed in [5], which provides an overview of the visualisation tool similar to Disco [6] and Celonis [7]. Making information available at various levels of granularity with different visibility layers is demonstrated to facilitate the ease of use and efficiency of the decision making. The tool displays information at the desired level of granularity and offers a critical advantage for the supply chain at large organisations to leverage high information volume and a greater number of information flows and levels. Figure 4 [5] demonstrates the implementation of the tool using such business entity and presents respective information flows and layers.

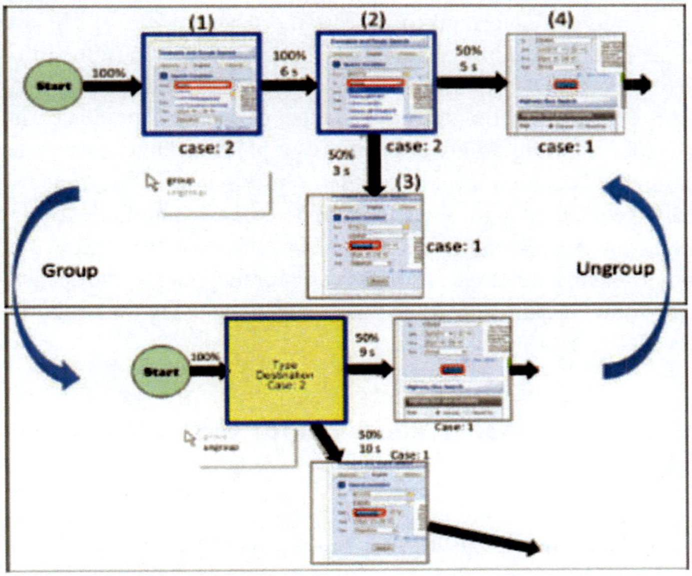

Fig. 4 Example of granularity modification [5]

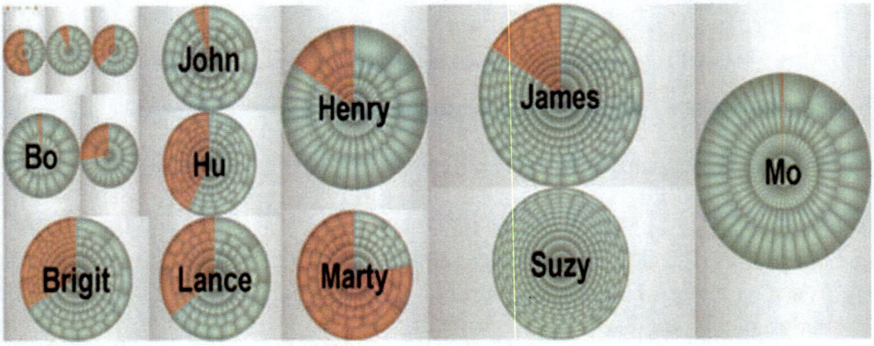

Fig. 5 Sample treemap generated through a generalised algorithm [8]

3.2 *Visualising Business Data with Generalised Tree Maps*

In businesses, data is collected to inform and facilitate decision making. The traditional way to present the data is in the visual forms of bar charts and pie charts. Treemaps are widely used to present the hierarchical data but are rarely used as they are less familiar and effective than business graphics [8]. Vliegen et al. [8] presented a generalised treemap algorithm to address the development of effective tools to work with treemaps in supply chain data. The proposed extensions to the treemap algorithm offer a way to visualise supply chain data that is intuitive and appealing to business users while inheriting the look and feel of business graphics. The proposed treemaps combine strong points of business graphics, such as ease of use, ease of understanding and detailed level of access to information. The proposed approach provides different layout algorithms that can be utilised at multiple levels to achieve maximum advantages. Such presentation allows identification, extraction of useful insights while leaving the scope of similar work for mapping on the geographic maps. Figure 5 [8] shows a sample treemap that is generated using the proposed method.

3.3 *The Event Tunnel Visualisation for Business Process Data*

In the contemporary world, business processes are becoming more event-driven and event-based loosely coupled distributed systems, which means with time the event state changes and backtracking mechanism is needed to detect any irregularities or error occurred [9]. Figure 6 [18] represents an event tunnel where information about business activities is stored. The event data is stored in a 3D cylinder which gives two visualisations for the user. The first one is the top view that has events along time axis and the second one is the side view which visualise the temporal order of the event. The 3D view for the top-view event tunnel is mapped to radial 2D rendering.

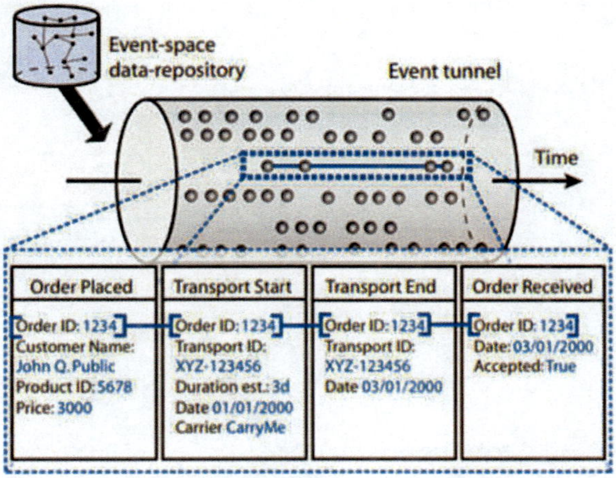

Fig. 6 Correlated events in business process [9]

Most recent events and business processes are at the top of the tunnel in top-view visualisation. These events are also more pertinent for the business analyst as they are recent events. It also helps to display correlated events close to one another.

Second visualisation is the side view which tells the order of the event and correlates event sequence. The visualisation bears a resemblance to process chart like GANTT diagram where schedule of every event is described. Figure 7 [9] is a dashboard where (a) represent the event tunnel side view (b) a text view and other (c,d,e,f,g,h,i,j) are various configuration console. This type of visualisation is suited for a large data in a business process with up to 40,000 events. The event tunnel visualisation can be used in real-time resource and logistics management. This has two advantages that an event group can subside into a single visual sequence pattern and correlation between different business processes can be procured.

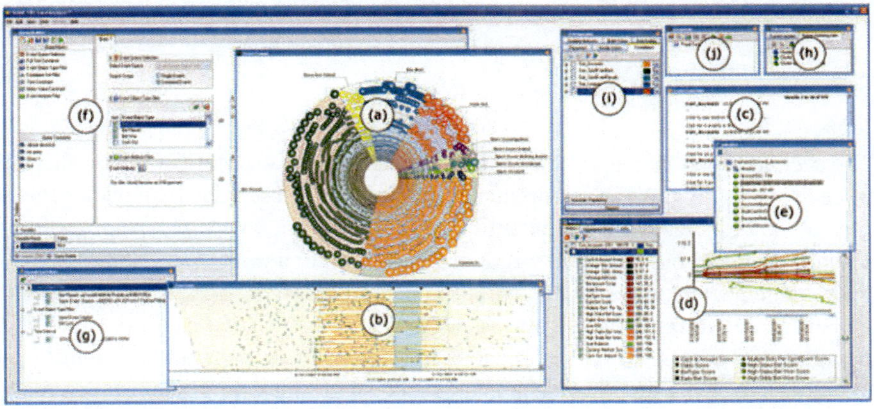

Fig. 7 Application interface of correlated events in business process [9]

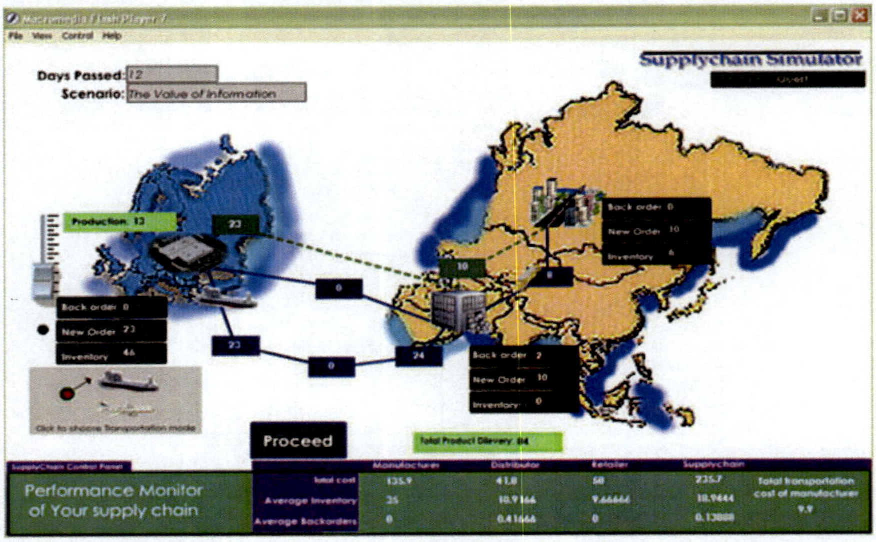

Fig. 8 Simulator Interface [10]

3.4 *Supply Chain Simulator*

Further application of visualisation approaches for the supply chain is using supply chain simulations. A good example of such papers is [10]. The proposed simulator presents a supply chain simulation model, which features the visualisation of the real behaviour of the dynamic system and detailed information for the user to enhance the process of decision making [10]. The model displays a simulated supply chain network of facilities and distribution systems that carries out the task of procurement and transformation of materials from manufacturer to customer. Figure 8 [10] shows the interface of the proposed model that gives access to the supply chain simulator.

3.5 *Visualisation of Decision Making*

The second broad theme identified through the content analysis is the visualisation of decision making. As an important step in logistics and supply chain management, this theme deserves particular attention as it draws from a wide perspective of data visualisation and information system management literature.

Management activities in the organisations rely mostly on the availability of data and the generation of data insights. Visualisation of organisational data presents wide opportunities across all areas of logistics and supply chain. Separation of control and operations presents significant challenges for decision making where those in charge need to rely on data insights, their validity, credibility and agility. The ability to react quickly to fast-changing conditions in the supply chain puts an organisation at high stakes. Visualisation in this context provides a way to present a large volume of data in a fast and multifaceted format so that information is conveyed to the user more effectively, which leads to optimal decision making.

3.6 Visualisation in Process Management for Daily Operational Activities

Supply chain offers close integration with the manufacturing process which allows cross-pollination of the recent trends in manufacturing into supply chain decision making. One such recent development is Just in Time Lean Manufacturing. Widely implemented in the manufacturing environment, Just in Time Lean Manufacturing meets the demands of businesses: allows cost reduction and increases the efficiency of operations. However, the above benefits come with an increased risk in these models. For example, the occurrence of any outliers in the production tends to line up and can result in heavy expenditures afterwards. Several papers have been proposed to address this problem, including [11]. They proposed a system that focuses on the process and display of manufacturing sequence data in such a way that the production status can be quickly communicated to the management, supervision and operational personnel.

In manufacturing process management, organising good logistics on daily activities is also important for the business to avoid any expensive investments, unnecessary costs, any outages and to save time. Though a detailed plan of production line does the job to have a production line up and running, any visualisation on the maintenance and status of the same would produce the best results in maintaining the best results of the manufacturing process. [12] came up with the idea of visualising the manufacturing data in the calendar view map (Fig. 9). This proposed method has the capability of exposing any possible outliers in the production line which helps the business to act quickly and bring up the optimal production line.

Geographic Information System (GIS) is applied into supply chain planning to get real-time data from the road [13, 14, 15] so that operators can obtain a very clear view of the application. Based on the GIS data information, the graphical model proposed by Lou et al. [16] can support classified domestics garbage' optimal transportation planning, and eventually could support on energy efficiency, which was partly presented in Figs. 10 and 11. Figure 10 [16]below shows a straightforward view of the quantities of waste associated with Waste Transport Centres while the graphic model presents the different transportation routines (Fig. 11).

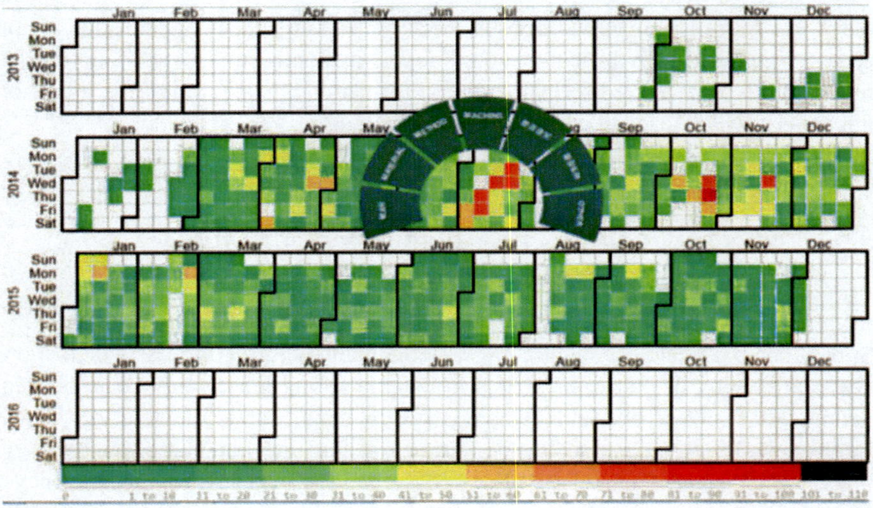

Fig. 9 Calendar view map [12]

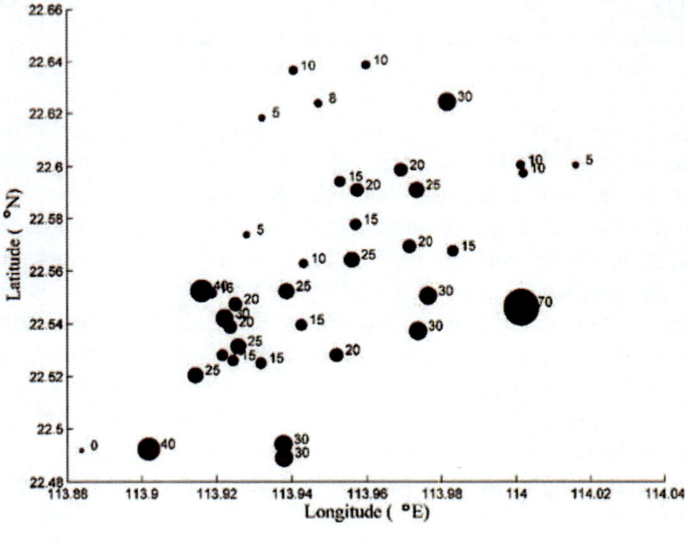

Fig. 10 The quantity (ton) of waste received at each Waste Transport Centre [16]

3.7 Interactive Visualisation

One of the topics identified under the broad category of visualisation of decision making is interactive visualisation. In high demand in the industry due to the inherent benefits of interactivity, one of the examples of papers in this topic is Singh and his

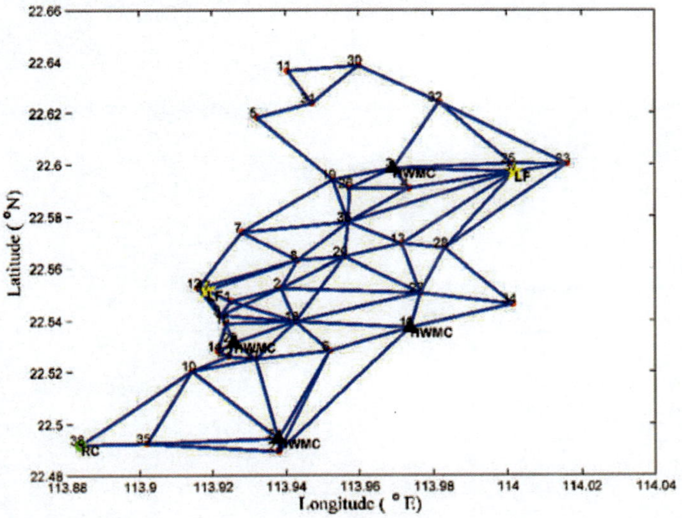

Fig. 11 A graphical model. The number by the vertex is the ID for the Waste Transport Centre [16]

team [17]. They proposed a web-based interactive visualisation prototype for visualising data on the supply chain. Developed in the Django framework, the prototype targets to facilitate the decision-making process for managers by providing different data views of the same supply chain. This prototype offers opportunities for visual analytics to the user by integrating data visualisation with analytical capabilities to generate meaningful insights. Each visual layout provides insights into specific aspects of the supply chain network. The prototype gives access to a single point to visual layouts and ensures no loss of essential information for decision making. Figure 12 [17] provides a flow diagram of the proposed prototype.

3.8 Visualisation with Risk Management and Real-Time Monitoring

With an increase in the complexity of the supply chain, the visibility of the supply chain network becomes an issue to tackle. For example, Goh et al. [18] proposed a supply chain visualisation platform, RiskVis, which is based on a multi-hierarchical modular designed to address complex supply chain network visibility issues. The primary aim of the tool is risk management and real-time monitoring for more efficient and effective supply chain risk management. The proposed supply chain visualiser platform provides a clear view and data insights of supply chain operation to the management in either local, regional and global settings. This platform identifies Spatio-temporal connectivity patterns of entities in the supply chain. Additionally, based on the hierarchical model, the platform allows identifying real-time risks

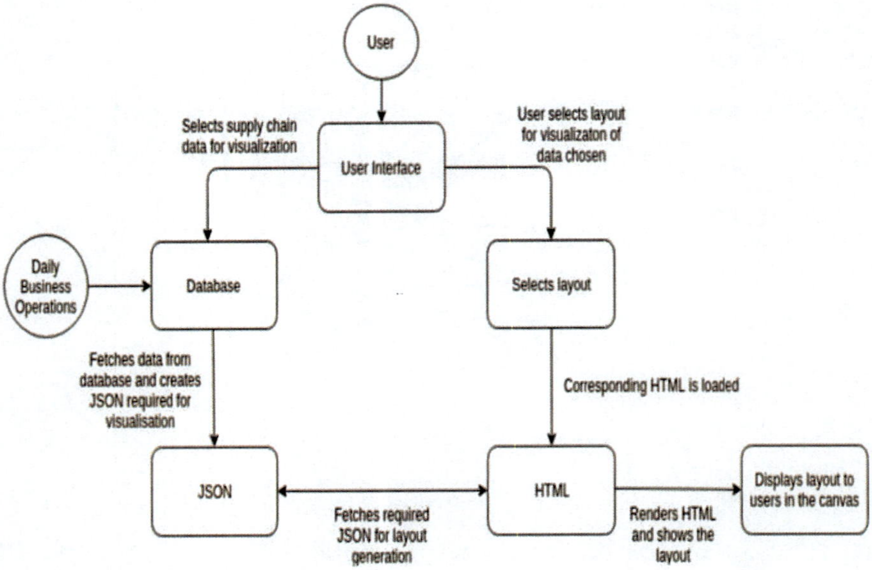

Fig. 12 Flow diagram of the proposed web-based interactive visualisation prototype [17]

related to data collection, risk monitoring and data insights generation. The proposed visualisation platform provides effective ways to explore large datasets, including information of a focal company and possible substitutions of suppliers. Another application of visualisation in the platform is real-time anomaly detection and the deployment of staff promptly to handle anomalies. Figures 13 and 14 [18] provide an example of the possible use of the platform for supply chain geared visualisations.

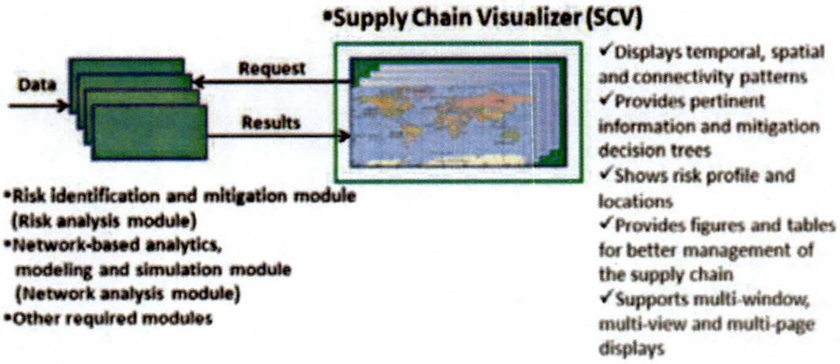

Fig. 13 Supply chain visualiser [18]

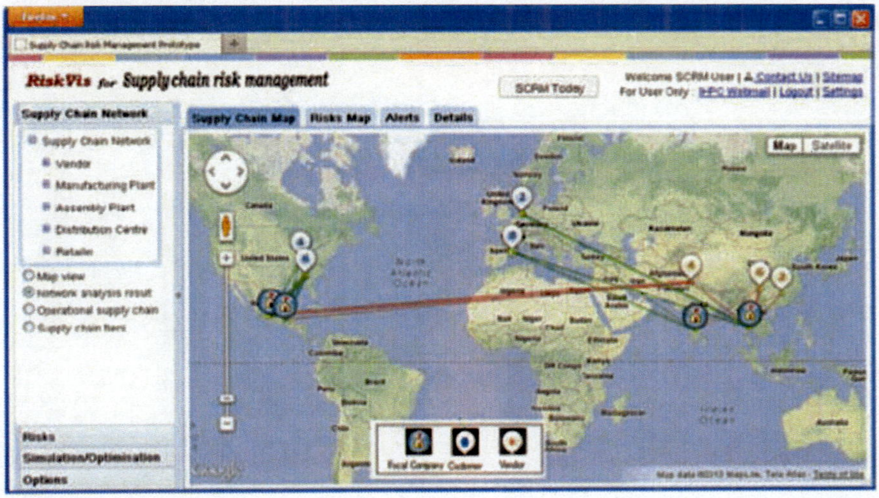

Fig. 14 RiskVis Interface [18]

4 Visualisation in Application

Visualisation can help a person to rapidly draw a straightforward conclusion when a large amount of data is analysed properly. Kriglstein and Rinderle-Ma [19] did a survey where different visualisation technique was compared against the familiarity of these techniques. Figure 15 [19] shows that people are more familiar with charts, animation, graphs, and tree visualisation, whereas pixel-oriented techniques and icon

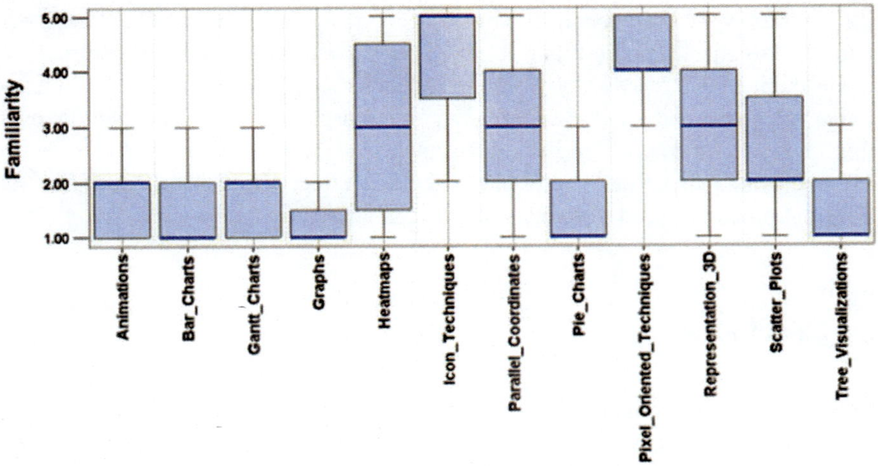

Fig. 15 Familiarity of the participants with different visualisation techniques (from Excellent (1) to Not Familiar (5)) [19]

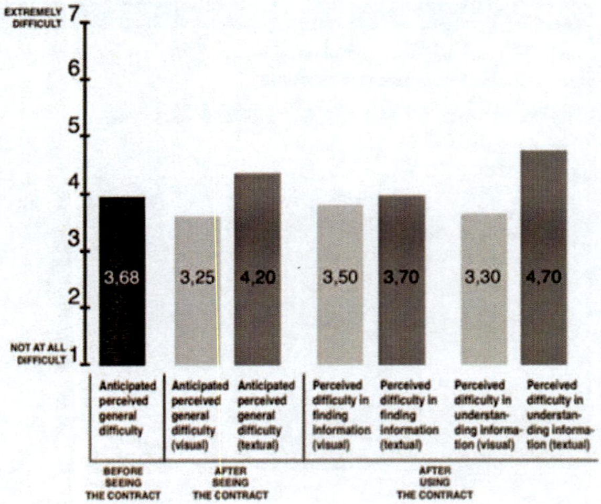

Fig. 16 Difficulty level before and after using the textual, visual contract [20]

techniques show less familiarity in the supply chain process. The study also shows that 86% of the participants use wide range of visualisation tools like ADONIS, ARIS Express, IBM Blueworks, Airista flow.

Companies face decision-making problems when there is an issue with design a problem; thus, to overcome this issue, visualisation plays a crucial role in identifying hot spots (region of interest) and non-competing criteria. Passera [20] has discussed the correlation between user experience, usability and visualisation through experiments. The participants were from different departments, ranging from legal, sales and supply chain. They were shown a traditional plain textual B-2-B contract document and a visualised contract document. Figure 16 [20] gives a graphical representation of the difficulty level in understanding the text and visual contract after using the contract. Visualised document improved company goals and increased their trust with the suppliers. People found it easy to understand the visualised contract and made them more trustworthy. Some participants said these contracts highlighted important points and made the company more transparent. 81% of the people found this approach more user friendly and thus increased their confidence in the company. In conclusion, visualisation presents a positive effect on business processes and different processes need different visualisation approaches.

5 Conclusion

Presentation of information plays a vital role in an administrative person taking decisions in a business. Particularly in today's time when the business world is dominated by large and complex supply chain networks, management must have timely access to business insights to make decisions quickly and efficiently. The current literature provides an overview of the landscape of the supply chain and decision-making literature by focusing on the visualisation of information to obtain and deliver meaningful insights. The paper identifies two broad themes that current literature delivers including visualisation of the supply chain and visualisation of the decision making. Within these broad themes, seven topics were identified to give researchers and industry a perspective of current developments. It is important to note that studies in these areas are still at the emerging stage as limited literature were identified. With the current strong demand for academic research, it is critical to accelerating research to address industry expectations. Importantly, the current studies exhibit a trend as the majority of the research is at the early prototyping stage. Future research may focus on identifying the business value of employing data visualisation in the supply chain as well as proposing validation techniques that can be used to fast the implementation and validation of the proposed prototypes. Development of core points to address the presentation and communication of data insights, projections, and possible outliers in the supply network is another promising area of further research. There are multiple tools for visualisation in the market which helps in enhancing the process of decision making in supply chain and can be tailored to fit individual business model as well as platform ecosystems.

The current study is subject to several limitations, including the scope of the academic research and validation of the findings in the identified literature. Developing future scope by design and implementation of more modular and interconnected systems for data visualisation and management of supply chain networks may consider the velocity of growth in data dimensions and interconnections between the entities in a supply chain network.

Appendix A

The adjusted PRISMA framework checklist and its implementation in the study

Section/topic	#	Checklist item
Title	1	Identify the report as a systematic review, meta-analysis, or both
Structured summary	2	Provide a structured summary including background; objectives; data sources; and other details of the study
Rationale	3	Describe the rationale for the review in the known context
Objectives	4	Provide an explicit statement of questions/aims being addressed
Protocol	5	Indicate if a review protocol exists and provides its details
Eligibility criteria	6	Specify study characteristics and report characteristics used as criteria for eligibility, giving rationale

(continued)

(continued)

Section/topic	#	Checklist item
Information sources	7	Describe all information sources (e.g., databases with dates of coverage, contact with study authors to identify additional studies) in the search
Search	8	Present full electronic search strategy, including limits used
Study selection	9	State the process for selecting studies, such as screening and eligibility
Data collection process	10	Describe method of data analysis
Data items	11	List and define all analytical items for which data were sought
Risk of bias	12	Describe methods used for assessing risk of bias and synthesis of information
Summary measures	13	Describe the principal summary measures
Synthesis of results	14	Describe the methods of handling data and combining results of studies
Risk of bias across studies	15	Specify any assessment of risk of bias that may affect the cumulative evidence
Additional analyses	16	Describe methods of additional analyses
Study selection	17	Give numbers of studies screened
Study characteristics	18	For each study, present characteristics for which data were extracted and provide the citations
Risk of bias within studies	19	Discuss the risk of bias for reviewed studies
Results of individual studies and synthesis of results	20 21	A brief of individual studies and summary
Risk of bias across studies	22	Discuss risk of bias across studies
Additional analysis	23	Give results of additional analyses
Summary of evidence	24	Summarize the main findings, including their relevance to key stakeholders
Limitations	25	Discuss limitations at study
Conclusions	26	Provide a general interpretation of the results in the context of other evidence, and implications for future research
Funding	27	Describe sources of funding for the systematic review and other support (e.g., supply of data); role of funders for the systematic review

References

1. Nath, A.: Reshaping supply chain and logistics for resilience. Supply Chain Brain (2020). https://www.supplychainbrain.com/blogs/1-think-tank/post/31444-reshaping-supply-chain-and-logistics-for-resilience.
2. Lou, C.X., Bonti, A., Prokofieva, M., Abdelrazek, M., Chowdary, K.: Literature review on visualization in supply chain decision making, in Proceedings of the International Conference on Information Visualisation 2020 (pp. 746–750) (2020)
3. Moher, D., Liberati A., Tetzlaff, J., Altman, DG.: The PRISMA group. 2009. Preferred reporting items for systematic reviews and meta-analyses: the PRISMA statement. PLoS Med **6**(7) (2009)
4. Bengtsson, M.: How to plan and perform a qualitative study using content analysis. NursingPlus Open **2**, 8–14 (2016)
5. Urabe, Y., Yagi, S., Tsuchikawa K., Masuda, T.: Visualizing user action data to discover business process. In: 20th Asia-Pacific Network Operations and Management Symposium, pp. 1–4. Matsue, Japan (2019)
6. Gunther, C.W., Rozinat, A.: Disco: discover your process. In: Demonstration Track of the 10th International Conference on Business Process Management, pp. 40–44 (2012)
7. Geyer-Klingeberg, J., Nakladal, J., Baldauf, F., Veit, F.: Process mining and robotic process automation: a perfect match. In: The 16th International Conference on Business Process Management (2018)
8. Vliegen, R., van Wijk, J.J., van der Linden, E.: Visualizing business data with generalized Treemaps. IEEE Trans. Vis. Comput. Graph. **12**(5), 789–796 (2006)
9. Suntinger, M., Obweger, H., Schiefer, J., Groller, M.E.: The event tunnel: interactive visualisation of complex event streams for business process pattern analysis. 2008 IEEE Pacific Visualisation Symposium. Kyoto, Japan **2008**, 111–118 (2008)
10. Siddiqui, A., Khan, M., Akhtar, S.: Supply chain simulator: a scenario-based educational tool to enhance student learning. Comput. Educ. **51**(1), 252–261 (2008)
11. Sackett, P., Williams, D.: Data visualisation in manufacturing decision making. J. Adv. Manuf. Syst. **02**(02), 163–185 (2003)
12. Chankhihort, D., Choi, S., Lee, G.J., Im, B.M., Ahn, D., Choi, E., Nasridinov, A., Kown, S., Lee, S., Kang, J., Park, K., Yoo, K.: Integrative manufacturing data visualisation using calendar view map, Eighth International Conference on Ubiquitous and Future Networks (ICUFN), Vienna, pp. 114–116 (2016)
13. Karadimas, N.V., Doukas, N., Kolokathi, M., Defteraiou, G.: Routing optimization heuristics algorithms for urban solid waste transportation management. WSEAS Trans. Comput. **7**(12), 2022–2031 (2008)
14. Nguyen-Trong, K., Nguyen-Thi-Ngoc, A., Nguyen-Ngoc, D., Dinh-Thi-Hai, V.: Optimization of municipal solid waste transportation by integrating GIS analysis, equation-based, and agent-based model. Waste Manage. **59**, 14–22 (2017)
15. Singh, A.: Remote sensing and GIS applications for municipal waste management. J. Environ. Manage. **243**, 22–29 (2019)
16. Lou, C.X., Shuai, J., Luo, L., Li, H.: Optimal transportation planning of classified domestic garbage based on map distance. J. Environ. Manag. **254** (2020)
17. Singh, S.K., Jenamani, M., Garg C., Alirajpurwala, H.: Multi-echelon supply network analysis with interactive visualisation. In: 2019 International Conference on Machine Learning, Big Data, Cloud and Parallel Computing (COMITCon), pp. 481–484. Faridabad, India (2019)

18. Goh, R.S.M., Wang, Z., Yin, X., Fu, X., Ponnambalam, L., Lu, S., Li, X.: RiskVis: Supply chain visualisation with risk management and real-time monitoring. In: 2013 IEEE International Conference on Automation Science and Engineering, Madison, pp. 207–212 (2013)
19. Kriglstein, S., Rinderle-Ma, S.: Change visualisation in business processes- requirements analysis. In: Proceedings of the International Conference on Computer Graphics Theory and Applications and International Conference on Information Visualisation Theory and Applications, vol 1: IVAPP, (VISIGRAPP 2012), pp. 584–593 (2012)
20. Passera, S.: Enhancing contract usability and user experience through visualization - an experimental evaluation. In: 16th International Conference on Information Visualisation, pp. 376–382. Montpellier, France (2012)

Author Index

B. Kovalerchuk et al. (eds.), *Integrating Artificial Intelligence and Visualization for Visual Knowledge Discovery*, Studies in Computational Intelligence 1014,
https://doi.org/10.1007/978-3-030-93119-3